Oscar Zariski: Collected Papers
Volume IV
Equisingularity on Algebraic Varieties

Oscar Zariski

Collected Papers

Volume IV
Equisingularity on Algebraic Varieties

Edited by
J. Lipman and B. Teissier

The MIT Press
Cambridge, Massachusetts, and London, England

Library of Congress Cataloging in Publication Data

Zariski, Oscar, 1899-
Collected papers.

(Mathematicians of our time, v.16)
Vol. 3 edited by M. Artin and B. Mazur; v. 4, by J. Lipman and B. Teissier.
Includes bibliographies.
CONTENTS: v. 1. Foundations of algebraic geometry and resolution of singularities.—v. 2. Holomorphic functions and linear systems.—v. 3. Topology of curves and surfaces, and special topics in the theory of algebraic varieties. [etc.]
1. Mathematics—Collected works. I. Series.
QA3.Z37 510'.8 73-171558
ISBN 978-0-262-24022-2 (hc.: alk. paper)—ISBN 978-0-262-51954-0 (pb. : alk. paper)

The MIT Press is pleased to keep this title available in print by manufacturing single copies, on demand, via digital printing technology.

Mathematicians of Our Time

Gian-Carlo Rota, series editor

Paul Erdös: The Art of Counting
edited by Joel Spencer [5]

Einar Hille: Classical Analysis and Functional Analysis
Selected Papers of Einar Hille
edited by Robert R. Kallman [11]

Mark Kac: Probability, Number Theory, and Statistical Physics
Selected Papers
edited by K. Baclawski and M. D. Donsker [14]

Charles Loewner: Theory of Continuous Groups
notes by Harley Flanders and Murray H. Protter [1]

Percy Alexander MacMahon: Collected Papers
Volume I
Combinatorics
edited by George E. Andrews [13]

George Pólya: Collected Papers
Volume I
Singularities of Analytic Functions
edited by R. P. Boas [7]

George Pólya: Collected Papers
Volume II
Location of Zeros
edited by R. P. Boas [8]

Collected Papers of Hans Rademacher
Volume I
edited by Emil Grosswald [3]

Collected Papers of Hans Rademacher
Volume II
edited by Emil Grosswald [4]

Note: Series number appears in brackets.

Stanislaw Ulam: Selected Works
Volume I
Sets, Numbers, and Universes
edited by W. A. Bayer, J. Mycielski, and G.-C. Rota [9]

Norbert Wiener: Collected Works
Volume I
Mathematical Philosophy and Foundations; Potential Theory; Brownian Movement, Wiener Integrals, Ergodic and Chaos Theories; Turbulence and Statistical Mechanics
edited by P. Masani [10]

Norbert Wiener: Collected Works
Volume II
Generalized Harmonic Analysis and Tauberian Theory; Classical Harmonic and Complex Analysis
edited by P. Masani [15]

Oscar Zariski: Collected Papers
Volume I
Foundations of Algebraic Geometry and Resolution of Singularities
edited by H. Hironaka and D. Mumford [2]

Oscar Zariski: Collected Papers
Volume II
Holomorphic Functions and Linear Systems
edited by M. Artin and D. Mumford [6]

Oscar Zariski: Collected Papers
Volume III
Topology of Curves and Surfaces, and Special Topics in the Theory of Algebraic Varieties
edited by M. Artin and B. Mazur [12]

Oscar Zariski: Collected Papers
Volume IV
Equisingularity on Algebraic Varieties
edited by J. Lipman and B. Teissier [16]

Oscar Zariski
(Photograph by Nan White)
© 1979

Contents

(Bracketed numbers are from the Bibliography)

Preface

The series "Mathematicians of Our Time" embraces, at least in principle, the works of living mathematicians; therefore, the term "collected works," as applied to this series stands of necessity for an open-ended entity, because the author—contrary to the old cliché that "mathematics is a young man's game"—may still be actively engaged in research and therefore continue to produce papers while the "collected" works is being printed. Thus, in my case, the bibliography of papers that appeared in the first volume did not include two papers, one of which was in course of publication and another of which was in preparation. (These two papers form, together with the last paper [89] of the list printed in the first volume, a sequence of three papers under the common title "General theory of saturation and of saturated local rings.") The paper that was then "in course of publication" appears in the second volume as the last published paper [90] in the bibliography, while the third paper of that series, which was in course of publication at the time of publication of the second volume, appears in the third volume as paper [93] of the bibliography. The bibliography in the present (final) volume includes three papers that have been written since the appearance of the third volume. These are [96], published in 1978, [97] and [97a], published in 1979. Paper [96] is the only one in this volume that does not deal with equisingularity. The complete set of four volumes includes all my published works, with the following exceptions (the numbers in brackets refer to the bibliography as printed in this volume):

1. Books [6,25,72,75].
2. Lecture notes [87,92].
3. Expository articles in fields to which I have made no original contribution myself [1,3,5,11]. All these articles deal with the foundations of set theory, and I wrote them in my early postgraduate years in Rome at the urging and with the encouragement of my teacher F. Enriques, whose primary interest at the time was in the philosophy and history of science and who was editor of a series of books entitled "Per la Storia e la Filosofia delle Matematiche." As the reader can see, the book listed under [6] was published in that series.

The editorial preparation and the writing of introductions to each volume is entrusted to the capable hands of younger men, who are experts in the field of algebraic geometry and who at one time or another have been either my students at Harvard or have been closely associated with me in some capacity at Harvard or elsewhere. Thus, the editors of the first volume, H. Hironaka and D. Mumford, as well as M. Artin, who joined D. Mumford as editor of the second volume, are truly leaders in the field of algebraic geometry and have studied at Harvard. B. Mazur, who joined M. Artin as editor of the third volume, is now my colleague at Harvard and as an expert topologist made an important contribution in his joint introduction with M. Artin. Of the two editors of the present fourth volume, J. Lipman studied at Harvard and is now professor at Purdue University, very active in the field of algebraic geometry, while B. Teissier, professor at the Ecole Polytechnique in Paris, has visited Harvard on various occasions. I had very close scientific contact with Teissier when I was a Visiting Professor at the Ecole Polytechnique in 1973.

While all the papers printed in these collected works belong, without exception, to algebraic geometry, the reader will undoubtedly notice that beginning with the year 1937 the nature of my work underwent a radical change. It became strongly *algebraic* in character, both as to methods used and as to the very formulation of the problem studied (these problems, nevertheless, always have had, and never ceased to have in my mind, their origin and motivation in algebraic *geometry*). A few words on how this change came about may be of some interest to the reader. When I was nearing the age of 40, the circumstances that led me to this radical change of direction in my research (a change that marked the beginning of what was destined to become my chief contribution to algebraic geometry) were in part personal in character, but chiefly they had to do with the objective situation that prevailed in algebraic geometry in the 1930s.

In my early studies as a student at the University of Kiev in the Ukraine, I was interested in algebra and also in number theory (by tradition, the latter subject is strongly cultivated in Russia). When I became a student of the University of Rome in 1921, algebraic geometry reigned supreme in that university. I had the great fortune of finding there on the faculty three great mathematicians, whose very names now symbolize and are identified with classical algebraic geometry: G. Castelnuovo, F. Enriques, and F. Severi. Since even within the classical framework of algebraic geometry the algebraic background was clearly in evidence, it was inevitable that I should be attracted to that field. For a long time, and in fact for almost ten years *after* I left Rome in 1927 for a position at the Johns Hopkins University in Baltimore, I felt quite happy with the kind of "synthetic" (an adjective dear to my Italian teachers) geometric proofs that constituted the very life stream of classical algebraic geometry (Italian style). However, even during my Roman period, my algebraic tendencies were showing and were clearly perceived by Castelnuovo, who once told me: "You are here with us but are not one of us." This was said not in reproach but good-naturedly, for Castelnuovo himself told me time and time again that the methods of the Italian geometric school had done all they could do, had reached a dead end, and were inadequate for further progress in the field of algebraic geometry. It was with this perception of my algebraic inclination that Castelnuovo suggested to me a problem for my doctoral dissertation, which was closely related to Galois theory (see [2,12] and section 2 of the introduction by M. Artin and B. Mazur to volume III).

Both Castelnuovo and Severi always spoke to me in the highest possible terms of S. Lefschetz's work in algebraic geometry, based on topology; they both were of the opinion that topological methods would play an increasingly important role in the development of algebraic geometry. Their views, very amply justified by future developments, have strongly influenced my own work for some time. This explains the topological trend in my work during the period 1929 to 1937 (see [15,16,17,20,22,27,28,29,31] and sections 5 and 7 of the introduction by M. Artin and B. Mazur to volume III). During that period I made frequent trips from Baltimore to Princeton to talk to and consult with Lefschetz, and I owe a great deal to him for his inspiring guidance and encouragement.

The breakdown (or the breakthrough, depending on how one looks at it) came when I wrote my Ergebnisse monograph *Algebraic Surfaces* [25]. At that time (1935) modern algebra had already come to life (through the work of Emmy Noether and the important treatise of B. L. van der Waerden), but while it was being applied to some aspects of the foundations of algebraic geometry by van der Waerden, in his series of papers *"Zur algebraischen Geometrie,"* the deeper aspects of *birational* algebraic geometry (such as the problem of reduction of singularities, the properties of fundamental loci and exceptional varieties of birational transformations, questions pertaining to complete linear systems and complete "continuous" systems of curves on surfaces, and so forth) were largely, or even entirely, virgin territory as far as algebraic exploration was concerned. In my Ergebnisse monograph I tried my best to present the underlying ideas of the ingenious geometric methods and proofs with which the Italian geometers were handling these deeper aspects of the whole theory of surfaces, and in all probability I succeeded, but at a price. The price was my own personal loss of the geometric paradise in which I so happily had been living. I began to feel distinctly unhappy about the rigor of the original proofs I was trying to sketch (without losing in the least my admiration for the imaginative geometric spirit that permeated these proofs); I became convinced that the whole structure must be done over again by purely algebraic methods. After spending a couple of years just studying modern algebra, I had to begin somewhere, and it was not by accident that I began with the problem of local uniformization and reduction of singularities. At that time there appeared the Ergebnisse monograph *Ideal-*

theorie of W. Krull, emphasizing valuation theory and the concept of integral dependence and integral closure. Krull said somewhere in his monograph that the *general* concept of valuation (including, therefore, nondiscrete valuations and valuations of rank > 1) was not likely to have applications in algebraic geometry. On the contrary, after some trial tests (such as the valuation-theoretic analysis of the notion of infinitely near base points; see title [35]), I felt that this concept could be extremely useful for the analysis of singularities and for the problem of reduction of singularities. At the same time I noticed some promising connections between integral closure and complete linear systems; a systematic study of these latter connections later led me to the notions of normal varieties and normalization. However, I also concluded that this program could be successful only provided that much of the preparatory work be done for ground fields that are not algebraically closed. I restricted myself to characteristic zero: for a short time, the quantum jump to $p \neq 0$ was beyond the range of either my intellectual curiosity or my newly acquired skills in algebra; but it did not take me too long to make that jump; see for instance [48,49,50] published in 1943 to 1947.

I carried out this initial program of work primarily in the four papers [37,39,40,41] published in 1939 and 1940. From then on, for more than 30 years, my work ranged over a wide variety of topics in algebraic geometry. It is not my intention here, nor is it the purpose of this preface, to brief the reader on the nature of these topics and the results obtained or the manner in which my papers can be grouped together in various categories, according to the principal topics treated. This is the task of the editors of the various volumes. I will say only a few words about the four volumes.

The papers collected in the first volume are divided in two groups: (1) *foundations,* meaning primarily properties of normal varieties, linear systems, birational transformations, and so on, and (2) *local uniformization and resolution of singularities.* These two subdivisions correspond precisely to the twofold aim I set to myself in my first concerted attack on algebraic geometry by purely algebraic methods—an undertaking and a state of mind about which I have already said a few words earlier. As a matter of fact, of the four main papers that I mentioned earlier as being the chief fruit of my first huddle with modern algebra and its applications to algebraic geometry, exactly two [37,40] belong to "foundations," while the other two [39,41] belong to the category "resolution of singularities and local uniformization."

The papers collected in the second volume are also divided in two groups: (1) *theory of formal holomorphic functions on algebraic varieties* (in any characteristic), meaning primarily analytic properties of an algebraic variety V, either in the neighborhood of a point (strictly *local* theory) or—and this is the deeper aspect of the theory—in the neighborhood of an algebraic subvariety of V (semi-global theory); (2) *linear systems, the Riemann-Roch theorem and applications* (again in any characteristic), the applications being primarily to algebraic surfaces (minimal models, characterization of rational or ruled surfaces, etc.).

My work on formal holomorphic functions was a natural outgrowth of my previous work on the local theory of singularities and their resolution. In the course of this previous work I developed an absorbing interest in the formal aspects of Krull's theory of local rings and their completions. In particular, I gave much thought to the possibility of extending to varieties V over arbitrary ground fields the classical notion of analytic continuation of a holomorphic function defined in the neighborhood of a point P of V. I sensed the probable existence of such an extension provided the analytic continuation were carried out along an algebraic subvariety W of V passing through P. It was wartime, and my heavy teaching load at Johns Hopkins University (18 hours a week) left me with little time for developing these ideas. Fortunately I was invited in January 1945 to spend at least one year at the University of São Paolo, Brazil, as exchange professor under the auspices of our Department of State. My light teaching schedule at São Paolo gave me the necessary leisure time to concentrate on an abstract theory of holomorphic functions. The year spent at São Paolo also presented me with a superlative audience consisting of one person—André Weil (who spent two or three years in São Paolo)—to whom I could speak about these ideas of mine during our frequent walks. The full theory of holomorphic functions—in the difficult case of complete (projective) varieties—was developed by me in my 1951 Transactions memoir [58]. However, the germ of this theory, in the easier case of affine varieties, appears already in my 1946 paper [49] written and published in Brazil. The key ingredient of the theory developed in this earlier paper is the concept of certain special rings, which later were named "Zariski rings," and properties of the completion of these rings. It is also this earlier Brazilian paper that led me to the discovery of a connection between the general theory of holomorphic functions and the connectedness theorem on algebraic varieties (and, in particular, the so-called principle of degeneration of Enriques). This connection was fully developed in my memoir [58] mentioned above. To a more strictly local frame of reference belong such papers as [52], [53], and [59] which deal with analytic properties of normal points of a variety.

As to the third volume, it includes *all* my papers that have not been included in the first two volumes (other than books, lecture notes, and certain expository articles mentioned at the beginning of the preface) *with one general exception*: the entire set of papers that deal either with the theory of equisingularity or with the theory of saturation is included in volume IV. The set of these papers (all published since 1964) consists of fifteen titles in the bibliography, namely the titles [80]–[97a], excluding the titles [83], [84], [87], [92], and [96] (paper [83] appears in the third volume, while paper [84] appears in the first volume since it deals with the reduction of singularities; the titles [87] and [92] are lecture notes; paper [96] deals with the resolution of singularities of an algebraic surface). In particular, all the papers I wrote before 1937 (except the paper [26], which belongs to the theory of algebraic surfaces and was therefore included in the second volume), whether in Rome (before 1927) or at the Johns Hopkins University (on or after 1928), appear in the third volume. The bulk of these papers, published during the period 1928–1937, is topological in nature, as I mentioned earlier in the preface. The reader will find in the introduction by M. Artin and B. Mazur an illuminating discussion of these papers and of their impact on later work by other mathematicians. Their discussion includes, in particular, my papers dealing with the following three topics: (1) solvability in radicals of equations of certain plane curves; (2) the fundamental group of the residual space of plane algebraic curves; (3) the topology of the singularities of plane algebraic curves. In one of my papers on the latter topic, namely, paper [22] ("On the topology of algebroid singularities"), I have found a number of misprints and (in section 5) a proof that is incomplete. These are dealt with in an addendum immediately following the paper, which contains a list of the errata and a complete proof of the result stated in the last five lines.

As I pointed out earlier, all the papers in the present volume IV (except title [96]) deal with the theory of equisingularity and also with a special case of equisingularity: the theory of saturation. The reader will find in the introduction of J. Lipman and B. Teissier an excellent exposition of (and I quote from that introduction), "the vision toward which the papers in this volume point." Suffice it to say that the theory of equisingularity, which I have initiated with two papers in 1965 and a third paper in 1968, under the common title "Studies in equisingularity" (titles [81], [82] and [85]) and on which I and other mathematicians continued to work up to the present time, is still largely an open field, and a definitive and convincing complete general theory of equisingularity is still not available, at least not in print. The last paper in this volume [97] develops a general theory of equisingularity. By a "general" theory, I mean one which is based on a "satisfactory" definition of equisingularity of a given r-dimensional variety V (of embedding dimension $\leq r + 1$, locally at each of its points; in particular, of a hypersurface V) along any irreducible subvariety W of V, of codimension > 1, i.e., such that $\dim W < r - 1$. (The case of codimension 1 has been completely settled in my paper [82].) By a "convincing" or "satisfactory" general theory of equisingularity, I mean one which, in the first place, is not contradicted, sooner or later, by counter-examples, and, in the second place, agrees with what one would expect from equisingularity when tested in examples against the behavior of V under a monoidal transformation centered in W.

In my student days in Rome algebraic geometry was almost synonymous with the theory of algebraic surfaces. This was the topic on which my Italian teachers lectured most frequently and in which arguments and controversy were also most frequent. Old proofs were questioned, corrections were offered, and these corrections were—rightly so—questioned in their turn. At any rate, the general theory of algebraic surfaces was very much on my mind in subsequent years, as witnessed—on a expository level—by my monograph [25] on algebraic surfaces, and—on a more significant research level—by the connection which I have found exists for varieties V of any dimension between normal (respect., arithmetically normal) varieties and the property that the hypersurfaces of a sufficiently high order (respect., of all orders) cut out on V complete linear systems. With this result as a starting point and with the conviction, indelibly impressed in my mind by my Italian teachers, that the theory of algebraic surfaces is the apex of algebraic geometry, it is no wonder that as soon as I realized that further progress in the problem of resolution of singularities would probably take years and years of further effort on my part, I decided that it was time for me to come to grips with the theory of algebraic surfaces. I felt that this would be the real testing ground for the algebraic methods which I had developed earlier. In his introduction to the second part of volume II Mumford says that he believes that my research on linear systems was to me "something like a dessert" (after the arduous efforts of the previous phase of my work). Objectively, Mumford may be right, but to me, subjectively, the proposed new work on linear systems felt more like the "main course." This work was also, in part, an answer to the following challenge sounded by Castelnuovo in his 1949 introduction to the treatise "Le superficie algebriche" of Enriques: "Verrà presto il continuatore dell'opera delle scuole italiana e francese il quale riesca a dare alla teoria delle superficie algebriche la perfezione che ha raggiunto la teoria delle curve algebriche?" (Note Castelnuovo's answer: "Lo spero ma ne dubito".) With this challenging question of Castelnuovo in mind, the reader will read with particular interest Mumford's analysis of my papers on linear systems and of later work done by others in the theory of algebraic surfaces. The reader will then realize that the theory of surfaces is still a very lively topic of research and that everything points to the likelihood that this theory will reach the degree of perfection dreamed of by Castelnuovo, except that this will not be the work of one "continuatore," but of many.

In 1950 I gave a lecture at the International Congress of Mathematicians at Harvard; the title of that lecture was "The fundamental ideas of abstract algebraic geometry" [60]. This is a good illustration of how relative in nature is what we call "abstract" at a given time. Certainly that lecture was very "abstract" for *that* time when compared with the reality of the Italian geometric school. Because it dealt only with projective varieties, that lecture, viewed at the present time, however, after the great generalization of the subject due to Grothendieck, appears to be a very, very concrete brand of mathematics. There is no doubt that the concept of "schemes" due to Grothendieck was a sound and inevitable generalization of the older concept of "variety" and that this generalization has introduced a new dimension into the conceptual content of algebraic geometry. What is more important is that this generalization has met what seems to me to be the true test of any generalization, that is, its effectiveness in solving, or throwing new light on, old problems by generalizing the terms of the problem (for example: the Riemann-Roch theorem for varieties of any dimension; the problem of the completeness of the characteristic linear series of a complete algebraic system of curves on a surface, both in characteristic zero and especially in characteristic $p \neq 0$; the computation of the fundamental group of an algebraic curve in characteristic $p \neq 0$).

But a mathematical theory cannot thrive indefinitely on greater and greater generality. A proper balance must ultimately be maintained between the generality and the concreteness of the structure studied, and usually this balance is restored after a period in which it was temporarily (and understandably) lost. There are signs at the present moment of the pendulum swinging back from "schemes," "motives," and so on toward concrete but difficult unsolved questions concerning the old pedestrian concept of a projective variety (and even of algebraic surfaces). There is no lack of such problems. It suffices to mention such questions as (1) criteria of rationality of higher varieties; (2) the study of cycles of codimension > 1 on any given variety; (3) even for divisors D on a variety there is the question of the behavior of the numerical function of n: dim $|nD|$; and finally (4) problems, such as reduction of singularities or the behavior of the zeta function, which are still unsolved when the ground field is of characteristic $p \neq 0$ (and is respectively algebraically closed or a finite field). These are new tasks that face the younger generation; I wholeheartedly wish that generation good speed and success.

Oscar Zariski

Cambridge, Massachusetts

Bibliography of Oscar Zariski

(The volume number in which a particular entry can be found is indicated by the number of asterisks.)

[1] *I fondamenti della teoria degli insiemi di Cantor,* Period. Mat., serie 4, vol. 4 (1924) pp. 408-437.

***[2] *Sulle equazioni algebriche contenenti linearmente un parametro e risolubili per radicali,* Atti Accad. Naz. Lincei Rend., Cl. Sci. Fis. Mat. Natur., serie V, vol. 33 (1924) pp. 80-82.

[3] *Gli sviluppi più recenti della teoria degli insiemi e il principio di Zermelo,* Period. Mat., serie 4, vol. 5 (1925) pp. 57-80.

***[4] *Sur le développement d'une fonction algébroide dans un domaine contenant plusieurs points critiques,* C. R. Acad. Sci., Paris, vol. 180 (1925) pp. 1153-1156.

[5] *Il principio de Zermelo e la funzione transfinita di Hilbert,* Rend. Sem. Mat. Roma, serie 2, vol. 2 (1925) pp. 24-26.

[6] *R. Dedekind, Essenza e Significato dei Numeri. Continuità e Numeri Irrazionali, Traduzione dal tedesco e note storico-critiche di Oscar Zariski* ("Per la Storia e la Filosofia delle Matematiche" series), Stock, Rome, 1926, 306 pp. The notes fill pp. 157-300.

***[7] *Sugli sviluppi in serie delle funzioni algebroidi in campi contenenti più punti critici,* Atti Accad. Naz. Lincei Mem., Cl. Sci. Fis. Mat. Natur., serie VI, vol. 1 (1926) pp. 481-495.

***[8] *Sull'impossibilità di risolvere parametricamente per radicali un'equazione algebrica* $f(x,y) = 0$ *di genere* $p > 6$ *a moduli generali,* Atti Accad. Naz. Lincei Rend., Cl. Sci. Fis. Mat. Natur., serie VI, vol. 3 (1926) pp. 660-666.

***[9] *Sulla rappresentazione conforme dell'area limitata da una lemniscata sopra un cerchio,* Atti Accad. Naz. Lincei Rend., Cl. Sci. Fis. Mat. Natur., serie VI, vol. 4 (1926) pp. 22-25.

***[10] *Sullo sviluppo di una funzione algebrica in un cerchio contenente più punti critici,* Atti Accad. Naz. Lincei Rend., Cl. Sci. Fis. Mat. Natur., serie VI, vol. 4 (1926) pp. 109-112.

[11] *El principio de la continuidad en su desarrolo histórico,* Rev. Mat. Hisp.-Amer., serie 2, vol. 1 (1926) pp. 161-166, 193-200, 233-240, 257-260.

***[12] *Sopra una classe di equazioni algebriche contenenti linearmente un parametro e risolubili per radicali,* Rend. Circolo Mat. Palermo, vol. 50 (1926) pp. 196-218.

***[13] *On a theorem of Severi,* Amer. J. Math., vol. 50 (1928) pp. 87-92.

***[14] *On hyperelliptic θ-functions with rational characteristics,* Amer. J. Math., vol. 50 (1928) pp. 315-344.

***[15] *Sopra il teorema d'esistenza per le funzioni algebriche di due variabili,* Atti Congr. Internaz. Mat. 2, Bologna, vol. 4 (1928) pp. 133-138.

***[16] *On the problem of existence of algebraic functions of two variables possessing a given branch curve,* Amer. J. Math., vol. 51 (1929) pp. 305-328.

***[17] *On the linear connection index of the algebraic surfaces* $z^n = f(x,y)$, Proc. Nat. Acad. Sci. U.S.A., vol. 15 (1929) pp. 494-501.

***[18] *On the moduli of algebraic functions possessing a given monodromie group,* Amer. J. Math., vol. 52 (1930) pp. 150-170.

***[19] *On the non-existence of curves of order 8 with 16 cusps,* Amer. J. Math., vol. 53 (1931) pp. 309-318.

***[20] *On the irregularity of cyclic multiple planes,* Ann. of Math., vol. 32 (1931) pp. 485-511.

***[21] *On quadrangular 3-webs of straight lines in space,* Abh. Math. Sem. Univ. Hamburg, vol. 9 (1932) pp. 79-83.

***[22] *On the topology of algebroid singularities,* Amer. J. Math., vol. 54 (1932) pp. 453-465.

***[23] *On a theorem of Eddington,* Amer. J. Math., vol. 54 (1932) pp. 466-470.

***[24] *Parametric representation of an algebraic variety,* Symposium on Algebraic Geometry, Princeton University, 1934-1935, mimeographed lectures, Princeton, 1935, pp. 1-10.

[25] *Algebraic Surfaces,* Ergebnisse der Mathematik, vol. 3, no. 5., Springer-Verlag, Berlin, 1935, 198 pp.; second supplemented edition, with

appendices by S. S. Abhyankar, J. Lipman, and D. Mumford, Ergebnisse der Mathematik, vol. 61, Springer-Verlag, Berlin-Heidelberg-New York, 1971, 270 pp.

**[26] (with S. F. Barber) *Reducible exceptional curves of the first kind,* Amer. J. Math., vol. 57 (1935) pp. 119-141.

***[27] *A topological proof of the Riemann-Roch theorem on an algebraic curve,* Amer. J. Math., vol. 58 (1936) pp. 1-14.

***[28] *On the Poincaré group of rational plane curves,* Amer. J. Math., vol. 58 (1936) pp. 607-619.

***[29] *A theorem on the Poincaré group of an algebraic hypersurface,* Ann. of Math., vol. 38 (1937) pp. 131-141.

***[30] *Generalized weight properties of the resultant of $n + 1$ polynomials in n indeterminates,* Trans. Amer. Math. Soc., vol. 41 (1937) pp. 249-265.

***[31] *The topological discriminant group of a Riemann surface of genus p,* Amer. J. Math., vol. 59 (1937) pp. 335-358.

***[32] *A remark concerning the parametric representation of an algebraic variety,* Amer. J. Math., vol. 59 (1937) pp. 363-364.

***[33] (In Russian) *Linear and continuous systems of curves on an algebraic surface,* Progress of Mathematical Sciences, Moscow, vol. 3 (1937).

*[34] *Some results in the arithmetic theory of algebraic functions of several variables,* Proc. Nat. Acad. Sci. U.S.A., vol. 23 (1937) pp. 410-414.

*[35] *Polynominal ideals defined by infinitely near base points,* Amer. J. Math., vol. 60 (1938) pp. 151-204.

*[36] (with O. F. G. Schilling) *On the linearity of pencils of curves on algebraic surfaces,* Amer. J. Math., vol. 60 (1938) pp. 320-324.

*[37] *Some results in the arithmetic theory of algebraic varieties,* Amer. J. Math., vol. 61 (1939) pp. 249-294.

*[38] (with H. T. Muhly) *The resolution of singularities of an algebraic curve,* Amer. J. Math., vol. 61 (1939) pp. 107-114.

*[39] *The reduction of the singularities of an algebraic surface,* Ann. of Math., vol. 40 (1939) pp. 639-689.

*[40] *Algebraic varieties over ground fields of characteristic zero,* Amer. J. Math., vol. 62 (1940) pp. 187-221.

*[41] *Local uniformization on algebraic varieties,* Ann. of Math., vol. 41 (1940) pp. 852-896.

*[42] *Pencils on an algebraic variety and a new proof of a theorem of Bertini,* Trans. Amer. Math. Soc., vol. 50 (1941) pp. 48-70.

*[43] *Normal varieties and birational correspondences,* Bull. Amer. Math. Soc., vol. 48 (1942) pp. 402-413.

*[44] *A simplified proof for the resolution of singularities of an algebraic surface,* Ann. of Math., vol. 43 (1942) pp. 583-593.

*[45] *Foundations of a general theory of birational correspondences,* Trans. Amer. Math. Soc., vol. 53 (1943) pp. 490-542.

*[46] *The compactness of the Riemann manifold of an abstract field of algebraic functions,* Bull. Amer. Math. Soc., vol. 45 (1944) pp. 683-691.

*[47] *Reduction of the singularities of algebraic three dimensional varieties,* Ann. of Math., vol. 45 (1944) pp. 472-542.

*[48] *The theorem of Bertini on the variable singular points of a linear system of varieties,* Trans. Amer. Math. Soc., vol. 56 (1944) pp. 130-140.

**[49] *Generalized semi-local rings,* Summa Brasiliensis Mathematicae, vol. 1, fasc. 8 (1946) pp. 169-195.

*[50] *The concept of a simple point of an abstract algebraic variety,* Trans. Amer. Math. Soc., vol. 62 (1947) pp. 1-52.

***[51] *A new proof of Hilbert's Nullstellensatz,* Bull. Amer. Math. Soc., vol. 53 (1947) pp. 362-368.

**[52] *Analytical irreducibility of normal varieties,* Ann. of Math., vol. 49 (1948) pp. 352-361.

**[53] *A simple analytical proof of a fundamental property of birational transformations,* Proc. Nat. Acad. Sci. U.S.A., vol. 35 (1949) pp. 62-66.

**[54] *A fundamental lemma from the theory of holomorphic functions on an algebraic variety,* Ann. Mat. Pura Appl., series 4, vol. 29 (1949) pp. 187-198.

**[55] *Quelques questions concernant la théorie des functions holomorphes sur une* *variété algébrique,* Colloque d'Algèbre et Théorie des Nombres, Paris, 1949, pp. 129-134.

**[56] *Postulation et genre arithmétique,* Colloque d'Algèbre et Théorie des Nombres, Paris, 1949, pp. 115-116.

**[57] (with H. T. Muhly) *Hilbert's characteristic function and the arithmetic genus of an algebraic variety,* Trans. Amer. Math. Soc., vol. 69 (1950) pp. 78-88.

**[58] *Theory and applications of holomorphic functions on algebraic varieties over arbitrary ground fields,* Mem. Amer. Math. Soc., no. 5 (1951) pp. 1-90.

**[59] *Sur la normalité analytique des variétés normales,* Ann. Inst. Fourier (Grenoble), vol. 2 (1950) pp. 161-164.

***[60] *The fundamental ideas of abstract algebraic geometry,* Proc. Internat. Cong. Math., Cambridge, Massachusetts, 1950, pp. 77-89.

**[61] *Complete linear systems on normal varieties and a generalization of a lemma of Enriques-Severi,* Ann. of Math., vol. 55 (1952) pp. 552-592.

*[62] *Le problème de la réduction des singularités d'une variété algébrique,* Bull. Sci. Mathématiques, vol. 78 (January-February 1954) pp. 1-10.

**[63] *Interprétations algébrico-géométriques du quatorzième problème de Hilbert,* Bull. Sci. Math., vol. 78 (July-August 1954) pp. 1-14.

***[64] *Applicazioni geometriche della teoria delle valutazioni,* Rend. Mat. e Appl., vol. 13, fasc. 1-2, Roma (1954) pp. 1-38.

***[65] (with S. Abhyankar) *Splitting of valuations in extensions of local domains,* Proc. Nat. Acad. Sci. U.S.A., vol. 41 (1955) pp. 84-90.

**[66] *The connectedness theorem for birational transformations,* Algebraic Geometry and Topology (Symposium in honor of S. Lefschetz), edited by R. H. Fox, D. C. Spencer, and A. W. Tucker, Princeton University Press, 1955, pp. 182-188.

***[67] *Algebraic sheaf theory* (Scientific report on the second Summer Institute), Bull. Amer. Math. Soc., vol. 62 (1956) pp. 117-141.

***[68] (with I. S. Cohen) *A fundamental inequality in the theory of extensions of valuations,* Illinois J. Math., vol. 1 (1957) pp. 1-8.

**[69] *Introduction to the problem of minimal models in the theory of algebraic surfaces,* Publ. Math. Soc. Japan, no. 4 (1958) pp. 1-89.

**[70] *The problem of minimal models in the theory of algebraic surfaces,* Amer. J. Math., vol. 80 (1958) pp. 146-184.

**[71] *On Castelnuovo's criterion of rationality* $p_a = P_2 = 0$ *of an algebraic surface,* Illinois J. Math., vol. 2 (1958) pp. 303-315.

[72] (with Pierre Samuel and cooperation of I. S. Cohen) *Commutative Algebra,* vol. I, D. Van Nostrand Company, Princeton, N.J., 1958.

***[73] *On the purity of the branch locus of algebraic functions,* Proc. Nat. Acad. Sci. U.S.A., vol. 44 (1958) pp. 791-796.

**[74] *Proof that any birational class of non-singular surfaces satisfies the descending chain condition,* Mem. Coll. Sci., Kyoto Univ., series A, vol. 32, Mathematics no. 1 (1959) pp. 21-31.

[75] (with Pierre Samuel) *Commutative Algebra,* vol. II, D. Van Nostrand Company, Princeton, N.J., 1960.

***[76] (with Peter Falb) *On differentials in function fields,* Amer. J. Math., vol. 83 (1961) pp. 542-556.

**[77] *On the superabundance of the complete linear systems* $|nD|$ *(n-large) for an arbitrary divisor D on an algebraic surface,* Atti del Convegno Internazionale di Geometria Algebrica tenuto a Torino, Maggio 1961, pp. 105-120.

*[78] *La risoluzione delle singolarità delle superficie algebriche immerse,* Nota I e II, Atti Accad. Naz. Lincei Rend., Cl. Sci. Fis. Mat. Natur., serie VIII, vol. 31, fasc. 3-4 (Settembre-Ottobre 1961) pp. 97-102; e fasc. 5 (Novembre 1961) pp. 177-180.

**[79] *The theorem of Riemann-Roch for high multiples of an effective divisor on an algebraic surface,* Ann. Math., vol. 76 (1962) pp. 560-615.

****[80] *Equisingular points on algebraic varieties,* Seminari dell'Istituto Nazionale di Alta Matematica, 1962-1963, Edizioni Cremonese, Roma, 1964, pp. 164-177.

****[81] *Studies in equisingularity I. Equivalent singularities of plane algebroid curves,* Amer. J. Math., vol. 87 (1965) pp. 507-536.

****[82] *Studies in equisingularity II. Equisingularity in co-dimension 1 (and characteristic zero),* Amer. J. Math., vol. 87 (1965) pp. 972-1006.

***[83] *Characterization of plane algebroid curves whose module of differentials has maximum torsion,* Proc. Nat. Acad. Sci. U.S.A., vol. 56 (1966) pp. 781-786.

*[84] *Exceptional singularities of an algebroid surface and their reduction,* Atti Accad. Naz. Lincei Rend., Cl. Sci. Fis. Mat. Natur., serie VIII, vol. 43, fasc. 3-4 (Settembre-Ottobre 1967) pp. 135-146.

****[85] *Studies in equisingularity III. Saturation of local rings and equisingularity,* Amer. J. Math., vol. 90 (1968) pp. 961-1023.

****[86] *Contributions to the problem of equisingularity,* Centro Internazionale Matematico Estivo (C.I.M.E.), Questions on Algebraic varieties. III ciclo, Varenna, 7-17 Settembre 1969, Edizioni Cremonese, Roma, 1970, pp. 261-343.

[87] *An Introduction to the Theory of Algebraic Surfaces,* Lecture Notes in Mathematics, No. 83, Springer-Verlag, Berlin, 1969.

****[88] *Some open questions in the theory of singularities,* Bull. Amer. Math. Soc., vol. 77 (1971) pp. 481-491.

****[89] *General theory of saturation and of saturated local rings, I. Saturation of complete local domains of dimension one having arbitrary coefficient fields (of characteristic zero),* Amer. J. Math., vol. 93 (1971) pp. 573-648.

****[90] *General theory of saturation and of saturated local rings, II. Saturated local rings of dimension 1,* Amer. J. Math., vol. 93 (1971) pp. 872-964.

****[91] *Quatre exposés sur la saturation,* Notes prises par J. J. Risler, Astérisque, 7 et 8 (1973) pp. 21-29.

[92] *Le problème des modules pour les branches planes,* Cours donné au Centre de Mathématiques de l'Ecole Polytechnique, octobre-novembre (1973). Rédigé par François Kmety et Michele Merle, pp. 1-144. Avec un Appendice de Bernard Teissier.

****[93] *General theory of saturation and of saturated local rings, III. Saturation in arbitrary dimension and, in particular, saturation of algebroid hypersurfaces,* Amer. J. Math., vol. 97 (1975) pp. 415-502.

****[94] *On equimultiple subvarieties of algebroid hypersurfaces,* Proc. Nat. Acad. Sci. U.S.A., vol. 72 (1975) pp. 1425-1426.

****[95] *The elusive concept of equisingularity and related questions,* Algebraic geometry; The Johns Hopkins Centennial Lectures (supplement to the American Journal of Mathematics) (1977) pp. 9-22.

****[96] *A new proof of the total embedded resolution theorem for algebraic surfaces (based on the theory of quasi-ordinary singularities),* Amer. J. Math., vol. 100 (1978) pp. 411-442.

****[97] *Foundations of a general theory of equisingularity on r-dimensional algebroid and algebraic varieties, of embedding dimension r + 1,* Amer. J. Math., vol. 101 (1979) pp. 453-514.

****[97a] *Abstract of the paper* "Foundations of a general theory of equisingularity on r-dimensional algebroid and algebraic varieties, of embedding dimension r + 1," *Symposia Matematica,* vol. XXIV (Instituto Nazionale di Alta Matematica Francesco Severi, Rome, 1979).

Introduction by J. Lipman and B. Teissier

The vision toward which the papers in this volume point is this: to seek a natural way of stratifying any algebraic or complex analytic variety X so that X is *equisingular along the strata,* i.e., the singularities which X has at the various points of each stratum are *equivalent* in some convincing sense.

An excellent introduction to the theories of equisingularity and saturation created in [81,82,85,89,90,93,97], is provided by Zariski himself in the expository papers [80,86,88,91,95]. In addition to reviewing salient features, we will indicate here some of the developments that have grown out of those theories. For more along these lines, see the report of Teissier.[31]

One of Zariski's basic ideas is that the equisingularity of a *hypersurface* $X \subset \mathbf{C}^N$ along a nonsingular subspace $Y \subset X$ around a point $0 \in Y$ should be defined (inductively, on the codimension of Y in X) by the equisingularity of the branch locus (i.e., reduced discriminant variety) $B_\pi \subset \mathbf{C}^{N-1}$ along $Y_\pi = \pi(Y)$ around $\pi(0)$, $\pi{:}X \to \mathbf{C}^{N-1}$ being a suitably general finite projection. The underlying feeling of a strong link between singularities and their "generic branch loci" can be traced back to Zariski's papers on fundamental groups [16–20,28,29]. (It is interesting to compare Zariski's ideas in these papers with recent proofs of the local and global versions of the Zariski-Lefschetz theorem given by Cheniot,[6] Hamm-Lê,[8] and Varchenko.[36]) The theory of saturation received some of its initial motivation from Whitney's work on topological triviality of analytic varieties along smooth subvarieties (cf. [85,§6]). Altogether, topology plays an important backstage role in Zariski's theory; and the conjunction of discriminant and topology provides the basis for many connections between his work on algebro-geometric equisingularity and recent work of others on monodromy, singularities of differentiable mappings, and analysis on singular varieties (cf. Teissier,[34] Varchenko,[40] and the end of section 3 below).

A really satisfactory theory of equisingularity exists only for the case when Y has codimension one in X. Here equisingularity of X along Y at 0 means that for some π, B_π is nonsingular. (It should be understood that we are thinking always of reduced *algebroid* varieties, or of *germs* of reduced complex analytic spaces). This case, studied in detail in [82], serves as a model for all further work in the area. There are many different criteria for equisingularity in codimension one, of algebro-geometric, differential-geometric, or topological nature. Each of these provides a theme for further development in higher codimension. But there some of the beautiful interconnections between the criteria vanish, and of others only a shadow is visible; the search for more substance remains a major challenge.

More specifically (cf. [82]), when B_π is nonsingular, Zariski showed that the singular locus $S = \mathrm{Sing}(X)$ is mapped isomorphically to B_π by π and that $\pi^{-1}(B_\pi) = S$ (this is the "non-splitting principle", cf. Teissier, p. 616[31]) and in particular S is nonsingular and of codimension one. Taking $Y = S$, we see that there exist retractions $X \to Y$ and that each of them displays X as a family of reduced plane curve germs parametrized by Y. The nonsingularity of B_π means that the irreducible components of any two curves in this family can be matched up in such a way that corresponding components have the same characteristic Puiseux exponents, and corresponding pairs of components have the same intersection multiplicity. (This germinal fact is implicit in Jung's work on local uniformization.[11]) This in turn means that the curves are *equivalent* from the viewpoint of resolution of singularities or, topologically, as embedded germs in $\mathbf{C}^2$. Conversely, any family of plane curve germs that has all its fibers reduced and equivalent in one of the above senses has singular locus (say Y) isomorphic to the parameter space (which is assumed nonsingular), and all projections $X \to Y \times \mathbf{C}$ "transversal" to X (i.e., in a direction not tangent to X at the origin) have nonsingular branch locus. Finally, given a hypersurface X whose singular locus Y is nonsingular and of codimension one, X is equisingular along Y if and only if the famous conditions (a) and (b) of Whitney[45] hold for the pair $(X - Y, Y)$. It can be shown in numerous ways that for any nonsingular Y of codimension one in X, the points of Y where X is equisingular form a dense Zariski-open subset of Y.

Complete as the theory is in codimension one, it does not exhaust all plausible notions of "non-variation of singularity type." For instance there are examples of Pham (cf. also Berthelot[4]) showing that the topology of a plane curve does not determine the topology of the versal unfolding of an equation of this curve and that very simple geometric features of the discriminant of this unfolding can change as the curve is deformed in an equisingular way.[21]

The codimension-one theory does not work as it stands for varieties over fields of characteristic >0, cf. Abhyankar.[1a,b] Perhaps in positive characteristic the concept of equivalence of plane curve singularities given in [81] (and further developed by Lejeune-Jalabert,[46] Moh,[47] and Fischer[48]) is not the definitive one.

We will now describe briefly some attempts to adapt the various equivalent ways of looking at codimension-one equisingularity to the case where the smooth subvariety Y has arbitrary codimension in the hypersurface X and also to the case where X is not a hypersurface. (Schickhoff even looks at some of these matters in the context of Banach spaces.[24])

1. Branch loci

The significance of the existence of equisingular projections π (those for which B_π is equisingular along Y_π) is not entirely clear. The theorem in [94] (cf. also Speder, p. 574[26]) is encouraging. Stronger positive evidence is provided first of all by the theorem of Varchenko, which states that if a family of hypersurface germs admits an equisingular projection (Y being the family of origins), then the family itself is topologically isomorphic to $X_0 \times Y$ for a suitable embedded germ X_0.[36,37,38] Secondly, Speder has shown that if X is "generically equisingular" along Y, then the pair (X-Sing X, Y) satisfies the Whitney conditions (i.e., X is "differentially equisingular" along Y [88], definition 2).[26] ("Generic" equisingularity is defined inductively by the condition that for "almost all" π, B_π is generically equisingular along Y_π).

For families X of isolated singularities of surfaces in $\mathbf{C}^3$, with smooth singular locus Y (of codimension two), Briançon[3] and Speder[27] show that the existence of *one* transversal equisingular projection already implies differential equisingularity. This also follows from a result of Lipman (unpublished) to the effect that the existence of such a projection implies the existence of a strong simultaneous resolution of the singularities of the family X, and Teissier's result:[33] "strong simultaneous resolution implies that the Whitney conditions hold" (the first result is proved only for families of surfaces, while the second holds quite generally). On the other hand, there are the following two examples of Briançon and Speder:

(1) Let $X \subseteq \mathbf{C}^4$ be given by

$$z^5 + ty^6z + y^7x + x^{15} = 0,$$

and let Y be the singular locus $x = y = z = 0$. For the projection $\pi(x,y,z,t) = (z,y,t)$, B_π is equisingular along Y_π. But X is *not* differentially equisingular along Y. Hence *no* transversal projection is equisingular; this answers negatively problem 1 on the second last page of [88].

(2) $X\colon z^3 + tx^4z + x^6 + y^6 = 0$

$$Y\colon x = y = z = 0.$$

Here X *is* differentially equisingular along Y; but again it can be shown that no transversal projection is equisingular.

The idea of using discriminants also appears from a completely different direction in the study of singularities, namely when it is realized that the number of vanishing cycles $\mu^{(N+1)}(X,0)$—or Milnor number[17]—of a hypersurface germ $(X,0) \subset (\mathbf{C}^{N+1},0)$ with isolated singularity is the order of vanishing of the discriminant of some map $(\mathbf{C}^{N+1},0) \rightarrow (\mathbf{C},0)$ having $(X,0)$ as fiber and that the discriminant of a general projection $f:(X,0) \rightarrow (\mathbf{C},0)$ vanishes to order $\mu^{(N+1)}(X,0) + \mu^{(N)}(f^{-1}(0),0)$ (cf. Teissier, section 5.5[34]). Actually $\mu^{(N)}(f^{-1}(0),0)$ does not depend on f, so we write $\mu^{(N)}(X,0)$ instead. Now it is easy to prove that $\mu^{(N+1)}(X,0)$ depends only on the topological type of the hypersurface $(X,0)$, but this is *not* so for $\mu^{(N)}(X,0)$. In fact, in the above example (1) of Briançon and Speder, considered as a family X_t of surfaces (with parameter t), the topological type of X_t does not depend on t (for t small), whereas $\mu^{(2)}(X_t,0)$ is different for $t \neq 0$ than for $t = 0$. In particular this shows that the topological type of a hypersurface germ does *not* determine the topological type of its general hyperplane section. To come back to $\mu^{(N+1)}$, Lê and Ramanujam proved that if it is constant in a family of hypersurfaces $(X_t,0)$ with isolated singularities and $N \neq 2$, then the fibers $(X_t,0)$ all have the same topological type.[12] Timourian showed that this implies that the family is locally topologically trivial.[35] However, as we have just seen, topological triviality does not imply that the family of discriminants of general projections $(X_t,0) \rightarrow (\mathbf{C},0)$ is trivial.

Apropos, there is a remarkable equivalence between *differential equisingularity* of a family of hypersurfaces with isolated singularities and *constancy of the sequence of Milnor numbers* of the members of the family together with their general linear sections of various dimensions. This area of investigation was opened up by Teissier[30] and further developed by Briançon and Speder. In fact, one of the ideas introduced in Teissier is the relationship between Zariski's discriminant conditions and the feeling that a "good" notion of equisingularity should have the following property:[31] if a hypersurface $X \subset \mathbf{C}^N$ is equisingular along Y, then for a sufficiently general nonsingular hypersurface $H \subset \mathbf{C}^N$ with $H \supset Y$, the intersection $X \cap H$ is equisingular along Y.

2. Saturation

There is nevertheless a fascinating theory when B_π is equisingular along Y_π in the most trivial sense, viz. B_π is *analytically a product* along Y_π. This is the theory of *equisaturation*. In order to capture algebraically the topological type in situations more general than that of plane curves, Zariski invented the notion of the *saturation* $\tilde{\mathfrak{o}}$ of a local ring $\mathfrak{o}$. For brevity, we deal here with "absolute saturation," in which case $\tilde{\mathfrak{o}}$ is defined by Zariski only when $\mathfrak{o}$ is either the local ring of a point on a hypersurface ([85, theorem 8.2],* [93, theorem 3.4]), or $\mathfrak{o}$ is one-dimensional ([93, appendix A]; cf. also Lipman[15] and Böger, Satz 5[5]). This $\tilde{\mathfrak{o}}$ is a local ring between $\mathfrak{o}$ and its normalization, and $\tilde{\mathfrak{o}}$ is radicial over $\mathfrak{o}$ [85,§4] (also Lipman[16]), so that in the analytic case the germs X and $\tilde{X}$ corresponding to $\mathfrak{o}$ and $\tilde{\mathfrak{o}}$ are locally homeomorphic, [85,§5] (also Seidenberg[25]). If $\mathfrak{o}_1$ and $\mathfrak{o}_2$ are the local rings of two hypersurface germs X_1 and X_2, and if the saturations $\tilde{\mathfrak{o}}_1$ and $\tilde{\mathfrak{o}}_2$ are isomorphic, then X_1 and X_2 are topologically equivalent as embedded germs [85,§6]; and the converse is true if X_1 and X_2 are plane curves [90,§7]. Given $X \supset Y$ as before, and a retraction $\rho{:}X \rightarrow Y$, (X,ρ) is *equisaturated* along Y if the fibers $\rho^{-1}(y)$ $(y \in Y)$ have isomorphic saturations at their origins (this is a loose translation of [85, definition 7.3]). A basic fact is that X is equisaturated along Y if and only if for some sufficiently general π, B_π is analytically a product along Y_π [85,§7]. In particular, when Y has codimension one in X then equisaturation of X along Y is equivalent with equisingularity of X along Y.

Equisaturation of X (with local ring $\mathfrak{o}$) along Y also means that the germ $\tilde{X}$ corresponding to the local ring $\tilde{\mathfrak{o}}$ is analytically a product along Y. Since X and $\tilde{X}$ are homeomorphic, this implies that X is topologically a product along Y. Zariski proves more, namely that the *pair* $X \subset \mathbf{C}^N$ is topologically trivial along Y [85,§7].

Here a rather curious thing happened. Pham and Teissier tried to interpret Zariski's work, starting from the idea that topological triviality should be proved by integrating Lipschitz vector fields on X since they have the property of being integrable, of course, but also of extending locally to the ambient $\mathbf{C}^N$, by a pretty result of Banach. They were encouraged by the fact that Zariski's computations in [85] looked like the use of Lipschitz conditions. Therefore Pham and Teissier introduced a purely algebraic description, using the concept of integral dependence on ideals, of the sheaf of *locally Lipschitz meromorphic functions* $\tilde{\mathcal{O}}_X$ on a reduced space $(X,\mathcal{O}_X)$ and defined the absolute Lipschitz saturation of $\mathcal{O}_{X,x}$ as $\tilde{\mathcal{O}}_{X,x}$.[23] They indicated that in the case of *hypersurfaces,* Zariski saturation and Lipschitz saturation *coincide* (a counter-example in the non-hypersurface case was given by Zariski [93, introduction]). The relation between analytic triviality of B_π along Y_π and topological triviality of X along Y had then the following simple analytical explanation: any vector field on Y_π extends to a vector field on $\mathbf{C}^{N-1}$ tangent to B_π, and this can be lifted to a vector field on X with coefficients

---

* Theorem 8.3 of [85] does not hold as stated, but it does hold for hypersurfaces (cf. first paragraph in introduction to [85], and Böger, p. 247[5]).

which are *meromorphic* but satisfy (locally) a Lipschitz inequality and therefore are bounded; this lifted vector field extends to the ambient space, and if we take a basis of constant vector fields on Y to start with, the integration of corresponding vector fields in $\mathbf{C}^N$ provides a topological (even Lipschitz) trivialization of $X \subset \mathbf{C}^N$ along Y.

Lipschitz and Zariski saturation also coincide in the one-dimensional case (Lipman, p. 808, remark ii[16]). In [89] and [90] Zariski investigates thoroughly the structure and automorphisms of one-dimensional saturated local rings. The simplest version of the "structure theorem" [85, theorem 1.12] may be interpreted as saying that complete equicharacteristic one-dimensional saturated (i.e., equal to their saturation) local domains over, say, **C**, are of the form $\mathbf{C}[[t^{a_1}, t^{a_2}, \ldots, t^{a_n}]]$ for suitable integers $a_1, \ldots, a_n$ (cf. *Math Reviews*, vol. 38, no. 5775). A class of one-dimensional rings which includes the saturated ones is studied by Lipman.[14] Notable among the algebraic features of one-dimensional saturated rings is the existence of "many automorphisms" ([93, appendix A]; also Böger, Satz 5[5]). This reflects the fact that saturation kills the *moduli* [92] of plane curve germs, in the previously mentioned sense that such germs have equivalent singularities at their origins if and only if they have isomorphic saturations. It is particularly enlightening, compared to the "Lipschitz" definition of saturation, to look at this result geometrically. Consider a nonplanar curve germ Γ. "Almost all" plane projections of Γ will have the same saturation as Γ. (This is the geometric meaning of the *existence of saturators* [85, proposition 1.6], a fact which also underlies the equality of one-dimensional Lipschitz and Zariski saturation.) Hence these plane projections have equivalent singularities at their origins. But conversely, *all* plane branches belonging to the same equivalence class can be obtained—up to isomorphism—as sufficiently general projections of a *single* Γ, namely the germ whose local ring is the saturation of the local ring of any one of the equivalent plane curves.

The algebraic theory of Lipschitz-saturation was taken up and improved on by Lipman[15,16] and Böger.[5,5a] Stutz provided new insight into the meaning of Lipschitz equisaturation when X is no longer a hypersurface, but Y is still of codimension one.[28,29] The joint theory of Zariski and Lipschitz saturation has been used by Nobile to prove an interesting theorem that implies in particular that any germ of reduced complex analytic surface is Lipschitz equivalent to an algebraic surface germ;[18] and by Teissier to give an algebraic proof of the fact that the constancy of Milnor's number in a family of plane curve germs implies that the family is equisingular (a result proved topologically by Lê-Ramanujam, see above).[32] To conclude on this topic, let us note that for a reduced complex analytic space X, the existence of a partition of X into nonsingular constructible strata X_α such that X is locally Lipschitz-trivial (and not just topologically trivial) along each stratum, seems to be completely open when X has dimension ≥ 3. However, Verdier has proved the existence of a stratification with "rugose" triviality, and rugosity is a Lipschitz-like condition, but relative to the stratification.[41]

3. Simultaneous resolution of singularities

Zariski proposes to geometers the following program for resolving singularities. Find for each complex analytic variety $X \subset \mathbf{C}^N$ a natural stratification with the following property: if $H \subset \mathbf{C}^N$ is a smooth hypersurface which avoids the "exceptional points" (i.e., the zero-dimensional strata), and cuts all the positive-dimensional strata transversally, then some suitable process of resolving the singularities of $X \cap H$ should propagate along the strata to resolve all the singularities of X outside the exceptional points. Thus, by induction on dimension, one would resolve almost all singularities of X. The remaining step would be to transform exceptional points into nonexceptional ones. For surfaces in $\mathbf{C}^3$, this idea is realized in [78] (cf. also [84, introduction]); the underlying fact is that the singularities of an equisingular family of plane curves can be simultaneously resolved by monoidal transformations [82, theorem 7.4]. A big problem in higher dimensions is that no *canonical* process for resolving singularities is known. Nevertheless, one would hope at least to be able to find natural stratifications which are "monoidally stable" in the sense that a certain class of permissible blow-ups $f:X' \to X$ would be *stratified* relative to the canonical stratifications on X' and X (i.e., f maps any stratum of X' smoothly onto some stratum of X).

There are several possible definitions of simultaneous resolution (cf. Teissier[33]), the strongest of which, as mentioned above, implies differential equisingularity. (Note: the family given in the above example (1) of Briançon-Speder admits a "weak" simultaneous resolution, though it is not differentially equisingular.) Several authors have approached equisingularity from the point of view of simultaneous resolution, and in some cases have succeeded in showing the nonsingularity of the local moduli space, a result which is not obvious even for plane curves (cf. Wahl,[42,43] Nobile,[19,20] and Teissier [92, appendix]).

The problem of finding criteria for simultaneous resolution in more general situations has its interest enhanced by the discovery by Arnol'd(§11)[2], Kushnirenko,[13] and Varchenko,[39,40] of very interesting "equiresolvable" families of functions having isolated critical points where the resolution can be explicitly constructed by a toroidal map $Z \to \mathbf{C}^{N+1}$ from the datum of the Newton polyhedron of the function, the families in question being made of "almost all" (in a precise sense) functions having a given Newton polyhedron. They also computed from the Newton polyhedron many invariants of the singularities, which are in fact invariants of the resolution, e.g., the zeta function of the monodromy, the "initial exponent" of the Fuchsian equation corresponding to the Gauss-Manin connection associated to the singularity, etc.

4. Differential equisingularity (X not necessarily a hypersurface)

For non-plane-curve germs, there is no satisfactory theory of equivalence of singularities. For example, any two irreducible curve germs in $\mathbf{C}^N$ ($N > 2$) are topologically equivalent. Nevertheless Stutz was able to generalize large portions of Zariski's equisingularity theory to the case where X is no longer a hypersurface, but $Y = \mathrm{Sing}(X)$ is still nonsingular and of codimension one;[28] and he brought out connections between equisingularity and the tangent cones C_4, C_5 of Whitney (§3).[44] We mentioned above the equivalence of equisingularity and differential equisingularity for families of plane curves. The following generalization is a distillation of results in Stutz[28] and Abhyankar.[1]

Theorem: Let X be a reduced d-dimensional complex analytic variety, let $Y = \mathrm{Sing}(X)$ be smooth and of dimension $d - 1$, and let $0 \in Y$. The following are equivalent:

(1) The *Whitney conditions* (a), (b) hold for the pair $(X - Y, Y)$ at every point in a neighborhood of 0 in Y.

(2) Every irreducible component X' of X at 0 contains Y (at least near 0), and the Zariski tangent cone T' ($\equiv$ Whitney's C_3) of X' at 0 is a d-plane; furthermore, if a sequence of points $x_i \in X' - Y$ approaches 0 and the tangent planes $T_{x_i}X'$ have a limit, then that limit is T'. (In other words $C_3 = C_4$ at 0; this is essentially a generalization of the *Jacobian criterion* [82,§5].)

(3) (*Simultaneous resolution*) If $\nu{:}\bar{X} \to X$ is the normalization of X, then (after replacing X by a suitable neighborhood of 0) we have:
(a) X is equimultiple along Y;
(b) $\bar{X}$ is nonsingular; and
(c) ν induces an etale covering $(\nu^{-1}(Y))_{\mathrm{reduced}} \to Y$.

Moreover, when these equivalent conditions hold, then for every projection $\pi{:}X \to \mathbf{C}^d$ transversal to X, the branch locus B_π is nonsingular, and $\pi^{-1}(B_\pi) = Y$ (near 0). Conversely, if there exists a projection π with B_π nonsingular of dimension $d - 1$, and if every component of X contains and is equimultiple along $\pi^{-1}(B_\pi)$ near 0 (which is automatically so if X is a hypersurface), then the above conditions hold.

Teissier studies a refinement of differential equisingularity, called "c-cosecance,"[34] that grew out of ideas of Hironaka.[9] This equisingularity condition *is* stable for generic hypersurface sections (cf. end of section 1 above).

We mention also, in closing, a still open question of Zariski [88]: if two hypersurface germs have the same (embedded) topological type, do they have the same multiplicity? A step toward an affirmative answer is taken by Ephraim.[7] In this vein there is an intriguing result of Hironaka's that in a Whitney stratification of an arbitrary X, the closure of any stratum has the same multiplicity (possibly zero) at all points of any other fixed stratum.[10]

5. Paper [97] (a general theory of equisingularity): Branch loci revisited

In this fundamental paper, Zariski introduces the concept of the dimensionality type $\text{d.t.}_k(V,Q)$ of an algebroid hypersurface V at a point Q of V, with respect to a fixed coefficient field k of the local ring $\mathfrak{o}$ of V at its closed point P. The definition is by induction on the dimension, as follows: Set $\mathfrak{o} = k[[x_1,\ldots,x_{r+1}]] = k[[X_1,\ldots,X_{r+1}]]/(f)$, where $f(X_1,\ldots,X_{r+1}) = 0, f \in k[[X_1,\ldots,X_{r+1}]]$ is the equation of an algebroid hypersurface. Zariski introduces a new algebraic concept of a "generic projection" by adding infinitely many new independent variables $u_{i,A} = (u)$ where $A \in \mathbf{Z}_0^{r+1}$, $1 \leq i \leq r$, and considering formal power series

$$x_i^* = \sum_{A \in \mathbf{Z}_0^{r+1},\ |A| \geq 1} u_{i,A} x^A \qquad (1 \leq i \leq r)$$

which in a precise sense define a generic projection into the affine space $A_{k^*}^r$ over the field k^* generated over k by the $u_{i,A}$. Zariski then defines the discriminant Δ_u^* of V with respect to this generic projection π_u; Δ_u^* is an algebroid hypersurface defined over k^*, and he defines the "image" Q^* of Q by π_u: He then defines inductively

$$\text{d.t.}_k(V,Q) = 1 + \text{d.t.}_{k^*}(\Delta_u^*, Q^*)$$

and $\text{d.t.}_k(V,Q) = 0$ if Q is a simple point of V. Intuitively $\text{d.t.}_k(V,P)$ is the codimension in V of the equisingularity stratum of P in V. The main theorem in [97] is that indeed the subsets $V(\sigma)$ of V consisting of the points of V where the dimensionality type of V is equal to a given integer σ form a stratification of V (by nonsingular subvarieties). Given now an algebraic hypersurface V over a field k, and defining the dimensionality type via the completions of the local rings, Zariski can then define an equisingular stratification for any algebraic hypersurface (an important fact, the semicontinuity of the dimensionality type for the Zariski topology along any algebraic subvariety W/k of V, is proved by Hironaka[49]). One of the beauties of Zariski's stratification is that, being defined by the constancy of a numerical invariant, it is uniquely defined (once k is fixed). Indeed the question of the independence of $\mathrm{d.t.}_k(V,P)$ on the field of representatives k is still open in general, although Zariski himself has important (unpublished) partial results. Another outstanding question is whether the dimensionality type can be computed by using only generic *linear* projections. In the complex-analytic framework this has been proved in the case where $\dim V = 3$ and V has a singular locus of dimension 1 by Briançon and Henry.[50]

References

1. S. S. Abhyankar, *A criterion of equisingularity,* Amer. J. Math., vol. 90(1968) pp. 342-345.

1a. S. S. Abhyankar, *Remarks on equisingularity,* Amer. J. Math., vol. 90(1968) pp. 108-144.

1b. S. S. Abhyankar, *Note on coefficient fields,* Amer. J. Math., vol. 90(1968) pp. 346-355.

2. V. I. Arnol'd, *Critical points of smooth functions,* Proc. Internat. Congress of Mathematicians, Vancouver 1974, vol. 1, pp. 19-40, Canadian Math. Congress, 1976.

3. J. Briançon, *Contribution à l'étude des déformations de germes de sous-espaces analytiques de* $\mathbf{C}^n$, Thèse, Université de Nice, 1976.

4. P. Berthelot, *Classification topologique universelle des singularités d'après F. Pham,* Astérisque, vol. 16(1974) pp. 174-213.

5. E. Böger, *Zur Theorie der Saturation bei analytischen Algebren,* Math. Ann., vol. 211 (1974) pp. 119-143.

5a. E. Böger, *Über die Gleichheit von absoluter und relativer Lipschitz-Saturation bei analytischen Algebren,* Manuscripta Math., vol. 16 (1975) pp. 229-249.

6. D. Cheniot, *Une démonstration du théorème de Zariski sur les sections hyperplanes d'une hypersurface projective . . . ,* Compositio Math., vol. 27 (1973) pp. 141-158.

7. R. Ephraim, *C^1 preservation of multiplicity,* Duke Math. J., vol. 43 (1976), pp. 797-803.

8. H. Hamm, Lê D. T., *Un théorème de Zariski du type de Lefschetz,* Ann. Sci. École Norm. Sup. 4e série, vol. 6 (1973) pp. 317-366.

9. H. Hironaka, *Equivalences and deformations of isolated singularities,* Woods Hole conference on Algebraic Geometry, Amer. Math. Soc., 1964.

10. H. Hironaka, *Normal cones in analytic Whitney stratifications* Inst. Hautes Études Sci. Publ. Math., vol. 36 (1969) pp. 127-138.

11. H. W. E. Jung, *Darstellung der Funktionen eines algebraischen Körpers zweier unabhängigen Veränderlichen . . . ,* J. Reine Angew. Math., vol. 133 (1908) pp. 289-314.

12. Lê D. T., C. P. Ramanujam, *The invariance of Milnor's number implies the invariance of the topological type,* Amer. J. Math., vol. 98 (1976) pp. 67-78.

13. A. G. Kushnirenko, *Polyèdres de Newton et nombres de Milnor,* Inventiones Math., vol. 32 (1976), pp. 1-32.

14. J. Lipman, *Stable ideals and Arf rings,* Amer. J. Math., vol. 93 (1971) pp. 649-685.

15. J. Lipman, *Absolute saturation of one-dimensional local rings,* Amer. J. Math., vol. 97 (1975) pp. 771-790.

16. J. Lipman, *Relative Lipschitz-saturation,* Amer. J. Math., vol. 97 (1975) pp. 791-813.

17. J. Milnor, *Singular points of complex hypersurfaces,* Annals of Math. Studies, vol. 61, Princeton University Press, 1968.

18. A. Nobile, *On saturations of embedded analytic rings,* Thesis, Massachusetts Institute of Technology, 1971 (to appear).

19. A. Nobile, *Equisingular deformations of Puiseux expansions,* Trans. Amer. Math. Soc., vol. 214 (1975) pp. 113-135.

20. A. Nobile, *On equisingular deformations of plane curve singularities,* preprint.

21. F. Pham, *Remarques sur l'équisingularité universelle,* preprint, Université de Nice, 1971.

22. F. Pham, *Fractions Lipschitziennes et saturation de Zariski des algèbres analytiques complexes,* Actes du Congrès International des Mathématiciens, Nice 1970, vol. 2, pp. 649-654, Gauthier-Villars, Paris, 1971.

23. F. Pham, B. Teissier, *Fractions Lipschitziennes d'une algèbre analytique complexe et saturation de Zariski,* preprint, Centre de Math., École Polytechnique, 1969.

24. W. Schickhoff, *Whitneysche Tangentenkegel, Multiplizitätsverhalten, Normal-pseudoflächheit und Aquisingularitätstheorie für Ramissche Räume,* Schriftenreihe Math. Instit. Universität Münster, 2. Serie, Heft 12, 1977.

25. A. Seidenberg, *Saturation of an analytic ring,* Amer. J. Math. vol. 94 (1972) pp. 424-430.

26. J.-P. Speder, *Équisingularité et conditions de Whitney,* Amer. J. Math. vol. 97 (1975) pp. 571-588.

27. J.-P. Speder, *Équisingularité et conditions de Whitney,* Thèse, Université de Nice, 1976.

28. J. Stutz, *Equisingularity and equisaturation in codimension* 1, Amer. J. Math. vol. 94 (1972) pp. 1245-1268.

29. J. Stutz, *Equisingularity and local analytic geometry,* Proc. Symposia in Pure Math. vol. 30, pp. 77-84, Amer. Math. Soc. 1977.

30. B. Teissier, *Cycles évanescents, sections planes, et conditions de Whitney,* Astérisque, vols. 7, 8 (1973) pp. 285-362.

31. B. Teissier, *Introduction to equisingularity problems,* Proc. Symposia in Pure Math. vol. 29, pp. 581-632, Amer. Math. Soc. 1975.

32. B. Teissier, *Sur diverses conditions numériques d'équisingularité des familles de courbes . . . ,* preprint, Centre de Math., École Polytechnique, 1975.

33. B. Teissier, *Sur la résolution simultanée comme condition d'équisingularité,* preprint, Centre de Math., École Polytechnique, 1975.

34. B. Teissier, *The hunting of invariants in the geometry of discriminants,* Proc. Nordic Summer School on Singularities (Oslo, 1976), Sijthoff and Noordhoff, Groningen, 1977.

35. J. G. Timourian, *The invariance of Milnor's number implies topological triviality,* Amer. J. Math. vol. 99 (1977) pp. 437-446.

36. A. N. Varchenko, *Theorems on the topological equisingularity of families of algebraic varieties and families of polynomial mappings,* Math. USSR Izvestija, vol. 6 (1972) pp. 949-1008.

37. A. N. Varchenko, *The relation between topological and algebro-geometric equisingularities according to Zariski,* Functional Anal. Appl., vol. 7 (1973) pp. 87-90.

38. A. N. Varchenko, *Algebro-geometrical equisingularity and local topological classification of smooth mappings,* Proc. Internat. Congress of Mathematicians, Vancouver 1974, vol. 1, pp. 427-431, Canadian Math. Congress, 1976.

39. A. N. Varchenko, *Zeta-function of monodromy and Newton's diagram,* Inventiones Math., vol. 37 (1976) pp. 253-262.

40. A. N. Varchenko, *Newton polyhedra and estimation of oscillating integrals,* Functional Anal. Appl., vol. 10 (1976) pp. 175-196.

41. J.-L. Verdier, *Stratifications de Whitney et théorème de Bertini-Sard,* Inventiones Math., vol. 36 (1976) pp. 295-312.

42. J. Wahl, *Equisingular deformations of plane algebroid curves,* Trans. Amer. Math. Soc., vol. 193 (1974), 143-170.

43. J. Wahl, *Equisingular deformations of normal surface singularities,* I, Annals of Math., vol. 104 (1976), 325-356.

44. H. Whitney, *Local properties of analytic varieties,* in "Differential and Combinatorial Topology" (Marston Morse symposium, edited by S. S. Cairns) pp. 205-244, Princeton University Press, 1965.

45. H. Whitney, *Tangents to an analytic variety,* Annals of Math., vol. 81 (1965) pp. 496-549.

46. M. Lejeune-Jalabert, *Sur l'équivalence des singularités des courbes algébroides planes,* Thèse, Université de Paris VII, 1973.

47. T. T. Moh, *On characteristic pairs of algebroid plane curves for characteristic p,* Bull. Inst. Math. Acad. Sinica, vol. 1 (1973) pp. 75-91.

48. K. G. Fischer, *The decomposition of the module of n-th order differentials in arbitrary characteristic,* Can. J. Math., vol. 30 (1978) pp. 512-517.

49. H. Hironaka, *On Zariski dimensionality type,* Amer. J. Math., vol. 101 (1979).

50. J. Briançon and J. P. G. Henry, *Equisingularité générique des familles de surfaces à singularités isolées,* Preprint, Centre de Math., École Polytechnique, No. M367.0778 (July 1978).

Reprints of Papers

Estratto dai "*Seminari dell'Istituto Nazionale di Alta Matematica 1962-63*"

EQUISINGULAR POINTS ON ALGEBRAIC VARIETIES (*)

by OSCAR ZARISKI (**)

§ 1. Introduction.

Now that the general problem of reduction of singularities of an algebraic variety has been settled by Hironaka (at least in the case of characteristic zero), the interest of algebraic geometers may very well shift to singularities themselves, to their classifications. My own interest in the classification of singularities goes back to 1954. At that time, *I* presented at the Princeton symposium in honor of Professor Lefschetz a tentative procedure for dividing the set of points of an algebraic variety into strata of singularities of increasing complexity, or — as *I* shall say — *strata of equisingularity*. This problem of stratification, in the general context of analytic varieties (real or complex), has attracted recently the attention of topologists and differential geometers, notably *R.* Thom and Hassler Whitney. In general terms, this problem can be described as follows:

Let V be a variety (algebraic or analytic), of dimension r, over an algebraically closed ground field. We shall assume that each irreducible component of V has dimension r (V is *unmixed*). The first (and biggest) stratum will be the set E of simple points of V. Let $S = V - E$ be the singular locus of V. Let M be an irreducible component of S, of *positive* dimension, and let Q be a

(*) Conferenze tenute al Seminario di Algebra, Geometria e Topologia nei giorni 11, 12, 18, 19, 25, 26 gennaio e 1, 2 febbraio 1963.

(**) This research was supported (in part) by the Air Force Office of Scientific Research.

This article is the text of a lecture given on October 15, 1963 at the second Science Symposium at Yeshiva University, in New York City. The contents represent essentially a summary of the course lectures given at the Istituto Nazionale di Alta Matematica during the month of January and part of February, 1963.

point of M. The problem is to give a precise meaning to a statement such as this: *V has at Q a singularity which is not worse than the singularity that V has at a generic point P of M.* The term which I shall actually use is that of « *equisingularity* ». I ask therefore what is to be meant by sayng that *V is equisingular at Q, along M.* We want a definition, preferably algebraic, which will meet satisfactorily a series of stringent tests, whether algebraic, or algebro-geometric or (in the complex domain) topological in nature. By this we mean that whatever property is used in the definition of equisingularity, that property should be proved to be equivalent to each of a series of other properties which we intuitively associate with the concept of equisingularity, and which, together, cover just about everything that one could possibly expect from a correct definition of that concept. In the definition of equisingularity we must not ask for too much nor for too little. For instance, it would be too much to require that the points Q and P have *analyticaly isomorphic* neighborhoods. It would be too little to ask only for *equimultiplicity*, i. e., to ask only that P and Q have the same multiplicity for V.

Assuming that we have already a definition of equisingularity of V at Q, along M, we would expect, at the very least, that the points of M at which V is not equisingular along M form a proper subvariety S' of M. The set S' should conceivably include all the singular points of M and the points where M meets the other irreducible components of the singular locus S, but *may* include other points of M. We then would set $E' = M - S'$, and E' would be a *second stratum of equisingularity on V* (there would be one such stratum for each irreducible component M of S).

We repeat this procedure with each irreducible component M' of S', i. e., we ask for a definition of equisingularity of V, along M', at any given point Q' of M', we denote by S'' the set of points of M' where V is not equisingular along M', and we thus obtain a third stratum of equisingularity $E'' = M' - S'$. This process stops when we reach an irreducible subvariety $M^{(i)}$ of V which is either a single point or is such that V is equisingular, along $M^{(i)}$, at each point of $M^{(i)}$.

We limit the scope of our investigation by imposing *a priori* the following two restrictions: (1) *we shall assume that the ground field k is of characteristic zero*; (2) *we assume that at each point Q of V the variety V is locally embeddable in an affine $(r+1)$ space.*

Always remaining within this limited scope of our problem, my main object in this lecture will be to introduce and discuss the concept of equisingularity in the following special case: cod $M = 1$ (i. e., dim $M = r - 1$). In this case our results are as complete as could be desired, and the theory that is developed could very well serve as a model, illustrating what the general theory should be like.

After having dealt with the above special case, *I* shall present a general definition of equisingularity, state a number of consequences and formulate still a larger number of unsolved questions.

§ 2. Local parameters and critical varieties.

Let o the *complete* local ring of V at a given point Q (if V is algebraic, o is the completion of the local ring of V at Q), and let m be the maximal ideal of o. The assumption that V is locally, at Q, biregularly embeddable in affine $(r+1)$-space A_{r+1} is equivalent with the following: *m has a basis of $r+1$ elements.* Note that if Q is not a simple point, all such bases are necessarily minimal.

If $\{x_1, x_2, \dots, x_{r+1}\}$ be a basis of m, then o can be identified with the residue class ring of the formal power series ring $k[[X_1, X_2, \dots, X_{r+1}]]$, modulo a principal ideal (f), where f is a power series without multiple factors and where $x_i = (f)$-residue of X_i. The analytic hypersurface

$$f(X_1, X_2, \dots, X_{r+1}) = 0 \tag{1}$$

is then a biregular embedding of V in A_{r+1}, locally at Q. This affine embedding depends on the choice of the basis $\{x_1, x_2, \dots, x_{r+1}\}$. The point Q is now the origin. The power series f is uniquely determined up to a unit factor in $k[[X_1, X_2, \dots, X_{r+1}]]$.

We shall say that r elements $x_1, x_2, \dots, x_r$ of m are *local parameters of V at Q* if (a) $x_1, x_2, \dots, x_r$ are parameters of the local ring o and (b) the set $\{x_1, x_2, \dots, x_r\}$ can be extended to a basis $\{x_1, x_2, \dots, x_r, x_{r+1}\}$ of m. If $\{x_1, x_2, \dots, x_{r+1}\}$ if a basis of m and (1) is the associated embedding of V in affine A_{r+1}, then $x_1, x_2, \dots, x_r$ are local parameters of V at Q, if and only if $f(0, 0, \dots, 0, X_{r+1}) \neq 0$.

Assume that this is so. Then by the Weierstrass preparation theorem we may assume that f is a polynomial in X_{r+1}, with leading coefficient 1, and that $f(0, 0, \ldots, 0, X_{r+1}) = X_{r+1}^n$, where n is the degree of f in X_{r+1}. By this condition the power series f is uniquely determined.

We denote by $\pi (= \pi_{\{x_1, x_2, \ldots, x_r\}} = \pi_x)$ the (regular) projection map $(x_1, x_2, \ldots, x_{r+1}) \to (x_1, x_2, \ldots, x_r)$ of V onto the affine r-space A_r of the local parameters $x_1, x_2, \ldots, x_r$. The inverse of π is n-valued, except at the points which annihilate the discriminant $D (= D(X_1, X_2, \ldots, X_r))$ of f with respect to X_{r+1}. We denote by $\Delta (= \Delta_{\{x_1, x_2, \ldots, x_r\}} = \Delta_x)$ the analytical hypersurface in A_r defined by the equation $D = 0$ and we call Δ *the local critical variety of* V *at* Q, *relative to the local parameters* $x_1, x_2, \ldots, x_r$. The critical variety Δ is uniquely determined by the local parameters $x_1, x_2, \ldots, x_r$, being independent of manner in which the set of these parameters is extended to a basis $\{x_1, x_2, \ldots, x_r, x_{r+1}\}$. We emphasize that Δ is defined here as a *variety*, i. e., each irreducible component of Δ occurs with multiplicity 1 (in other words, we disregard the *multiplicity* of the various irreducible factors of D).

§ 3, Equisingularity and equisingular local parameters.

Let M be an irreducible component of the singular locus of V and let cod $M = 1$. Let Q be a point of M and let $\{x_1, x_2, \ldots, x_r\}$ be local parameters of V at Q. Let $\pi (= \pi_x)$ be the projection map of V into the space of parameters $x_1, x_2, \ldots, x_r$ and let $\pi(M) = M'$. Since M is singular for V, we have $M' \subset \Delta (= \Delta_x)$. Furthermore, dim $M' = r - 1$, and hence M' is an irreducible component of the hypersurface Δ. Let $Q' = \pi(Q)$ (Q' is the origin $X_1 = X_2 = \ldots = X_r = 0$).

DEFINITION 3.1 *V is equisingular at Q, along M, if there exist local parameters $x_1, x_2, \ldots, x_r$ of V at Q such that the critical variety $\Delta_{\{x_1, x_2, \ldots, x_r\}}$ has a simple point at the origin Q'.*

The condition stated in the above definition certainly implies that Δ_x is locally irreducible at Q' (as an analytical variety) and that consequently $\Delta = \pi_x(M) (= M')$, locally at Q'.

We shall now state (without proof) a number of consequences of the above definition.

THEOREM 3.2. *Equisingularity of V at Q, along M, implies equimultiplicity of V at Q, along M. In other words, if we denote,* *for any point R of V, the multiplicity of R on V by $m_V(R)$, then we have the following: if V is equisingular at Q, aloug M, and if P is a generic point of M, then $m_V(Q) = m_V(P)$. Furthermore, if V is equisingular at Q, along M, then Q is necessarily a simple point of the singular locus of V.*

The last part of the above theorem signifies that equisingu·larity of V at Q, along M, implies that Q is a simple point of M and that M is the only irreducible component of the singular locus S of V which contains Q.

DEFINITION 3.3. *We say that a given set of local parameters $x_1, x_2, \ldots, x_r$ of V at Q (where $Q \in M$) is a set of* **equisingular** *local parameters, if the critical variety Δ_x associated with these parameters has a simple point at the point $Q' = \pi_x(Q)$.*

Our definition of equisingularity of V at Q, along M (Definition 3.1) can thus be stated as follows: *V is equisingular at Q, along M, if there exist equisingular local parameters of V, at Q.*

One may ask now the following question: if V *is* equisingular at Q, along M, *which sets of local parameters of V at Q are equisingular?*

A first result in that direction is as follows:

Let us say that local parameters $x_1, x_2, \ldots, x_r$ of V at Q are *transversal,* if in the corresponding embedding (1) of V in A_{r+1} we have $f(0, 0, \ldots, 0, X_{r+1}) = X_{r+1}^n$, with $n = s = m_V(Q)$. In intrinsic terms of local rings, this means that the multiplicity of the ideal $o \cdot (x_1, x_2, \ldots, x_r)$ is equal to the multiplicity of the local ring o. Geometrically speaking, this means that the direction $x_1 = x_2 = \ldots = x_r = 0$ is not tangent to V at Q. Then we have

THEOREM 3.4 *If V is equisingular at Q, along M, then every set of transversal local parameters of V at Q is equisingular.*

We shall now give a complete characterization of equisingular local parameters or V, at Q ($Q \in M$).

Given a set of local parameters $x_1, x_2, \ldots, x_r$ of V at Q, let $M' = \pi_x(M)$ and let

$$g(X_1, X_2, \ldots, X_r) = 0 \tag{2}$$

be the equation of M' in the affine r-space of the local parameters $x_1, x_2, \ldots, x_r$ (in other words: if p is the prime ideal of M in o, set $p' = p \cap k[]x_1, x_2, \ldots, x_r]]$; then p' is a principal ideal, and we set $p' = (g(x_1, x_2, \ldots, x_r))$.

In the affine space A_{r+1}, equation (2) defines a (cylindrical) hyperspace H, passing through M, and we can consider the intersection multiplicity

$$i(V, H; M)$$

[In terms of local rings; let $O = o_p$ (where p is as above, and O is therefore a local ring of dimension 1). Then $i(V, H; M)$ is the multiplicity of the ideal $O \cdot g$].

The line $X_1 = X_2 = \dots = X_r = 0$ is a generator L of this cylinder, passing through Q, and we can consider the intersection multiplicity

$$i(V, L; Q).$$

Then we have

THEOREM 3.5. *If V is equisingular at Q, along M, and $x_1 x_2, \dots, x_r$ are local parameters of V, at A, then $x_1, x_2, \dots, x_r$ are equisingular local parameters at Q, if and only if (in the above notations)*

$$i(V, H; M) = i(V, L; Q).$$

§ 4. Criteria of equisingularity.

We shall now state a number of conditions each of which is both necessary and sufficient for V to be equisingular at Q, along M.

A. *The Jacobian criterion of equisingularity.*

Consider any basis $x_1, x_2, \dots, x_{r+1}$ of m and let $f(X_1, X_2, \dots, X_{r+1}) = 0$ be the associated embedding of V in A_{r+1}. Let $f_i = \left(\frac{\partial f}{\partial X_i}\right)_{X=x}$ and let J be the ideal generated in o by $f_1, f_2, \dots, f_{r+1}$. It is easy to see that this ideal is independent of the choice of the basis $x_1, x_2, \dots, x_{r+1}$. Let $\bar{o}$ be the integral closure of o in its total quotient ring. Then we have.

THEOREM 4. 1. *V is equisingular at Q, along M, if and only if the extended ideal $\bar{o}J$ is principal and Q is a simple point of the singular locus of V.*

From an algebraic point of view, this criterion of equisingularity is the most satisfactory, because it is expressed directly in terms of an intrinsic property of the local ring o. This criterion plays an important role in the proofs.

We can express this criterion in another form. From the fact that $\bar{o}J$ is principal does not follow necessarily that one of the $r+1$ elements $f_1, f_2, \ldots, f_{r+1}$ generated $\bar{o}J$ ($\bar{o}$ may not be an integral domain, i. e., V may not be analytically irreducible at Q). However, we achieve that end by replacing $x_1, x_2, \ldots, x_{r+1}$ by suitable linear combinations with coefficients in k. We may therefore assume that $\bar{o}J = \bar{o} \cdot f_{r+1}$. This is equivalent to $\partial x_{r+1}/\partial x_i \in \bar{o}$ for $i = 1, 2, \ldots, r$. Therefore, we may state

THEOREM 4. 1′. *V is equisingular at Q, along M, if and only if for a suitable choice of the basis $\{x_1, x_2, \ldots, x_r, x_{r+1}\}$ of m the partial derivative $\partial x_{r+1}/\partial x_i$ $(i = 1, 2, \ldots, r)$ are meromorphic functions at Q which assume only a finite number of values at Q* (i. e., at $x_1 = x_2 = \ldots = x_{r+1} = 0$).

B. *The Thom-Whitney criterion.*

In his manuscript « Local properties of analytic varieties » (Proceedings of the Princeton Symposium in honor of Professor Marston Morse), Hassler Whitney introduces certain differential-geometric conditions to be impcosed on any « good » stratification of an analytic variety (I understand that these conditions appear also in Thom's manuscript, same Proceedings). In the special case in which M is an irreducible component of the singular locus of V, of codimension 1, and the stratum E is the set of simple points of V, Whitney requires that the points Q of M which belong to the next stratum (and which form an open subset or M) satisfy the following two conditions:

(1) *$m_V(Q) = m_V(M)$, and Q is a simple point of M.*

(2) *Let P be a generic (hence simple) point of V, let π be the tangent r-space of V at P, let σ be the r-space which is the join of P and the tangent $(r-1)$-space of M at Q. Then in any specialization*

$$(P, \pi, \sigma) \longrightarrow (Q, \pi_0, \sigma_0),$$

we must have $\pi_0 = \sigma_0$.

I was able to verify that *these two conditions together are equivalent to equisingularity of V at Q, along M.* (The necessity of condition (1) is nothing but Theorem 3.2 above).

C. *The criterion of transversal sections.*

Let Q be a point of M which is simple for both M and the singular locus of V. Embedd V as a hypersurface in A_{r+1}, consider any surface F in A_{r+1} which has at Q a simple point and is such that the tangent plane of F at Q and the tangent $(r-1)$-space of M at Q are independent. Consider (locally at Q) the intersection cycle $F \cdot V$ (this cycle is well defined, since it can beshown that each irreducible component of $F \cap V$ which contains Q is of the «right» dimension 1), which is of dimension 1. We call $F \cdot V$ a *transversal section of* V, *at* Q, *relative to* M, or a section of V, at Q, *transversal* to M.

(The above definition of transversal sections could also be given in intrinsic terms, independently of any particular embedding of V. Namely, let $z_1, z_2, \ldots, z_{r-1}$ be elements of o whose traces on M form a regular system of parameters of the *regular* local ring. $o_Q(M)$. Then it can be shown that the ideal generated in o by $z_1, z_2, \ldots, z_{r-1}$ is unmixed, of dimension 1, and this allows us to define a 1-dimensional cycle].

Now, assume that we are dealing with a transversal section Γ which has no multiple components, hence is a *curve*. The curve Γ is then locally, at Q, embeddable in the affine plane, since $\Gamma \subset F$ and Q is simple for F.

In the local theory of plane analytic curves (with singularities) we can introduce the concept of *equivalent singularities*, as follows:

Let Γ and Δ be two plane analytic curves, defined locally at points P, Q of the plane. For Γ and Δ to have equivalent singularities at P and Q respectively, we require first of all that

(1) *the number of irreducible branches of* Γ *be the same as the number of analytical branches of* Δ.

Assuming that (1) is satisfied, we furthermore require that the irreducible branches $\Gamma_1, \Gamma_2, \ldots, \Gamma_q$ of Γ and the irreducible branches $\Delta_1, \Delta_2, \ldots, \Delta_q$ of Δ can be paired in a manner to be described below.

Let, for a suitable arrangement of indices, the paired branches be (Γ_i, Δ_i), $i = 1, 2, \ldots, q$. We impose the following two requirements on the pairing (Γ_i, Δ_i):

(2) *For each* $i = 1, 2, \ldots, q$, Γ_i *and* Δ_i *have the same multiplicity at* P *and* Q *respectively (in symbols:* $m_{\Gamma_i}(P) = m_{\Delta_i}(Q)$).

(3) *For* $i \neq j$, Γ_i *and* Γ_j *have the same tangent line if and only if* Δ_i *and* Δ_j *have the same tangent line.*

In view of (3), Γ and Δ have the same number of distinct tangent lines, say t. We now apply a locally quadratic transformation to Γ, with center P. The transform of Γ will then split into t analytical curves $\Gamma'^{(1)}, \Gamma'^{(2)}, \Gamma'^{(t)}$, having distinct origins $P^{(1)}, P^{(2)}, \ldots, P^{(t)}$. Similarly, we apply a locally quadratic transformation to Δ, with center Q, getting a transform consisting of t analytical curves $\Delta^{(1)}, \Delta^{(2)}, \ldots, \Delta^{(t)}$, with distinct origins $Q^{(1)}$, $Q^{(2)}, \ldots, Q^{(t)}$.

We denote by φ the (1, 1) mapping of the set of branches $\Gamma_1, \Gamma_2, \ldots, \Gamma_q$ onto the set of branches $\Delta_1, \Delta_2, \ldots, \Delta_q$ defined by the pairing (Γ_i, Δ_i) (thus $\varphi(\Gamma_i) = \Delta_i$). By (3), φ induces a (1, 1) mapping of the set of t tangents of Γ onto the set of t tangents of Δ. Thus φ induces a (1, 1) mapping of the set $\{\Gamma^{(1)}, \Gamma^{(2)}, \ldots, \Gamma^{(t)}\}$ onto the set $\{\Delta^{(1)}, \Delta^{(2)}, \ldots, \Delta^{(t)}\}$. We choose the indices for the $\Gamma^{(\alpha)}$ and the $\Delta^{(\beta)}$ in such a way that $\Gamma^{(\alpha)}$ and $\Delta^{(\alpha)}$ are paired in this (1,1) correspondence ($\alpha = 1, 2, \ldots, t$). Then φ induces a (1, 1) mapping of the set of branches of $\Gamma^{(\alpha)}$ onto the set of branches of $\Delta^{(\alpha)}$, for each $\alpha = 1, 2, \ldots, t$. We denote by $\varphi^{(\alpha)}$ this induced pairing of branches of $\Gamma^{(\alpha)}$ and $\Delta^{(\alpha)}$.

Since the singularity P of Γ can be resolved by a finite number of successive locally quadratic transformations, the singularities of the $\Gamma^{(\alpha)}$ are «simpler» than that of Γ. Similarly for the $\Delta^{(\alpha)}$ and Δ. We can therefore define equivalence of Γ and Δ by assuming that equivalence has already been defined for the less complex singularities of the $\Gamma^{(\alpha)}$ and the $\Delta^{(\alpha)}$.

This being so, we give the following definition:

DEFINITION 4.2. *Assuming that condition* (1) *is satisfied, a pairing φ between the q irreducible branches of Γ and the q irreducible branches of Δ is called an equivalence between Γ and Δ, if φ satisfies conditions* (2) *and* (3) *and if, furthermore, for each $\alpha = 1, 2, \ldots, t$, the pairing $\varphi^{(\alpha)}$ between the irreducible branches of $\Gamma^{(\alpha)}$ and the irreducible branches of $\Delta^{(\alpha)}$, induces by φ, is an equivalence between $\Gamma^{(\alpha)}$ and $\Delta^{(\alpha)}$. The singularities which Γ and Δ possees at P and Q respectively, are said to be equivalent, if there exists an equivalence between Γ and Δ.*

We can now state a criterion of equisingularity in terms of equivalence of transversal sections, relative to M, as follows:

THEOREM 4.3. *Assume that V is equisingular at Q, along M. Let P be a general point of M and let Γ be some fixed section of V at P, transversal to M. (Γ is certainly a curve,* i.e., *a 1-cycle, free* *from multiple components). Let Γ_0 be any section of V at Q, transversal to M. Then*

(a) *Γ_0 is a curve.*

(b) *The singularities which Γ and Γ_0 possess at P and Q respectively are equivalent.*

Conversely, we have the following:

THEOREM 4.4. *Assume that Q is a simple point of M and that there exists at least one section Γ_0 of V at Q, transversal to M, such that Γ_0 is a curve and that the singularities which Γ and Γ_0 possess at the points P and Q respictively are equivalent. Then V is equisingular at Q, along M.*

The criterion of equisingularity, by means of equivalence of transversal sections, given by theorems 4.3, and 4.4, is the only one which could possibly be used in the case of characteristic $p \neq 0$. All the other criteria, and our very definition of equisingularity (Definition 3.1), are untenable in that case.

D. *A criterion based on monoidal transformations.*

Let V be equisingular at Q, along M, and $T: V \rightarrow V'$ be the monoidal transformation of V, with center M. We know then that if P is a generic point of M, then $T\{P\}$ is a finite set of points, say $P_1', P_2', \dots, P_h'$. Since $m_V(Q) = m_V(M)$, it is also known that $T\{Q\}$ is a finite set of points $Q_1', Q_2', \dots, Q_g'$, $g = h$. Let $M' = T[M] =$ proper transform of M. Then M' is an unmixed subvariety of V', of codimension 1.

THEOREM 4.5. *If V is equisingular at Q, along M, then $g = h$, and for each $i = 1, 2, \dots, h$ the variety V' is equisingular at Q_i', along M'.* [Note: Equisingularity at Q_i', along the (possibly) reducible subvariety M', should be intended to mean the following: *Q_i' belongs to only irreducible component M_j' of M', and V' is equisingular at Q_i', along M_j'*).

We know that the singularity which V' possesses at the generic point of any irreducible component of M' is « simpler » than the singularity of V at the generic point P of M, so that a finite number of successive monoidal transformations with centers of codimension 1 will ultimately resolve the singularity P. Therefore the preceding theorem contains implicitly an inductive criterion of equisingularity, and if this criterion is used for the purpose of a definition of equisingularity one arrives at the following definition:

V is equisingular at Q, along M, if there exists a sequence of birational transformations

$$T_i:\ V_i \rightarrow V_{i+1}\ (i = 0, 1, \dots, N), \qquad (V_0 = V)$$

with the following properties.

1) *T_i is a monoidal transformation, whose center M_i is an irreducible subvariety of V_i, of codimension 1, and $M_0 = M$.*

2) *$f_i(M_i) = M$, where $f_i = T_0^{-1} \dots T_{i-2}^{-1}\, T_{i-1}^{-1}$.*

3) *If Q_i is any point of M_i such that $f_i(Q_i) = Q$, then $m_{V_i}(Q_i) = m_{V_i}(M_i)$ and Q_i is a simple point of M_i (here $Q_0 = Q$).*

4) *If Q_i is as in 3) and if P_i is a generic point of M_i then the sets $T_i\{P_i\}$ and $T_i\{Q_i\}$ have the same number of points.*

5) *All points of V_{N+1} which correspond to Q are simple for V_{N+1}.*

§ 5. A definition of equisingularity in the general case.

The assumption that V is locally, at each of its points Q, a hypersurface in an affine A_{r+1} still being in force, let Q be a point of V, let $x = \{x_1, x_2, \dots, x_{r+1}\}$ be a basic of the maximal ideal m of the local ring of V at Q, and let

$$x_i^u = u_{i1}\, x_1 + \dots + u_{i,r+1}\, x_{r+1}, \qquad i = 1, 2, \dots, r,$$

where the u_{ij} are indeterminates. With the understanding that we take as new ground field the algebraic closure of the field $k(u)$, it follows easily that the x_i^u are local parameters of V at Q. We denote by Δ_x^u the critical variety $\Delta_{x_1^u, \dots, x_r^u}$ (see § 2) and call Δ_x^u the *generic critical variety of* V at Q, *relative to the basis* $x = \{x_1, x_2, \dots, x_{r+1}\}$ of m.

Let M be *any* irreducible subvariety of V, of codimension s, and let Q be a point of M. We shall define equisingularity of V at Q, along M, *by induction on s.*

If M is a simple subvariety of V, then equisingularity of V at Q, along M, shall mean of course, that Q is a simple point of V. This settles the case $s = 0$, in which case M is an irreducible component of V, hence is simple for V. If M is singular for V, of codimension $s > 0$, and assuming that equisingularity has already been defined along subvarieties of codimension $s-1$, we observe if x is a basis of m and if π^u_x is the projection map of V onto the affine r-space of the local parameters $x^u_1, x^u_2, \ldots, x^u_r$, the irreducible variety $\pi^u_x(M)$ is a subvariety of Δ^u_x, of codimension $s-1$. We then give the following definition:

DEFINITION 5.1. *V is equisingular at Q, along M, if there exists a basis $x = \{x_1, x_2, \ldots, x_{r+1}\}$ of $m_V(Q)$ such that Δ^u_x is equisingular at the point $\pi^u_x(Q)$, along $\pi^u_x(H)$.*

Note that in the case $s=1$ this definition is not identical with the previous definition of equisingularity (Definition 3.1). It can be proved, however, that the two definitions are equivalent. I found it necessary (at least temporarily, i. e., at the present stage of my investigation) to use indeterminates in the general definition of equisingularity, in order to be able to prove some theorems and to go some distance in the theory. Thus, it is possible to prove the following results:

THEOREM 5.2. *Let Σ/k be an irreducible algebraic system whose generic element is a triple (V, M, Q), where V is an unmixed variety, of dimension r, M is an irreducible subvariety of V, and Q is a point of M at which V is locally a hypersurface. If V is equisingular at Q, along M, then for almost all specializations (V_0, M_0, Q_0) of (V, M, Q), over k, V_0 is equisingular at Q_0, along M_0.*

With the aid of this theorem it is possible to prove, for instance, the following two results:

THEORÊM 5.3. *If M is an irreducible subvariety of V, then the points Q of M sueh that V is not equisingular at Q, along M, form a proper subvariety of M.*

If V is equisingular at Q, along M, any basis $\{x_1, x_2, \ldots, x_{r+1}\}$ of $m_V(Q)$ which satisfies the condition of Definition 5.1 will be called *an equisingular basis* of $m_V(Q)$, *relative to M.*

By *equisingular local parameters* of V at Q, relative to M, we mean a set of local parameters $x_1, x_2, \ldots, x_r$ of V at Q, such $\Delta_{x_1, x_2, \ldots, x_r}$ is equisingular at $\pi(M)$, where π is the projection map of V onto the r-space of the parameters $x_1, x_2, \ldots, x_r$.

THEOREM 5.4. *If V is equisingular at Q, along M, and if $\{x_1, x_2, \ldots, x_{r+1}\}$ is an equisingular basis of $m_V(Q)$, relative to*

M, then for almost all constants c_{ij} in k the r elements $x_i^c = \sum_{j=1}^{r+1} c_{ij} x_j$ $(i = 1, 2, \ldots, r)$ are equisingular local parameters of V at Q, along M.

6. Open questions.

From a purely technical point of view, further *immediate* progress in the theory depends entirely on the affirmative solution of the following question.

PROBLEM 6.1. *If V is equisingular at Q, along M, is every basis $\{x_1, x_2, \ldots, x_{r+1}\}$ of $m_V(Q)$ equisingular, relative to M.*

Our definition of equisingularity and Theorem 5.3 allow us to introduce a well-defined stratification of V, as indicated in the introduction (§ 1). Let $E_0, E_1, E_2, \ldots, E_N$ be the various strata thus obtained, where $E_0 =$ set of simple points of V. The strata E_i, are completely described by the following properties:

(1) If $\bar{E}_i$ $(i \geqq 1)$ is the least variety containing E_i then $\bar{E}_i$ is irreducible and E_i is the set of points Q of $\bar{E}_i$ such that V is equisingular at Q, along $\bar{E}_i$ (and hence, by Theorem 5.3, $\bar{E}_i - E_i$ is a proper subvariety of $\bar{E}_i$).

(2) For each $i \geqq 0$, $\bar{E}_i - E_i$ is the union of certain strata E_α, and for each $\alpha \geqq 1$ there exists an index i such that $\bar{E}_\alpha$ is an irreducible component of $\bar{E}_i - E_i$.

PROBLEM 6.2. *Prove that $E_\alpha \cap E_\beta = 0$ if $\alpha \neq \beta$.*
Included in Problem 2 is, for instance the following special question: if M is an irreducible component of the singular locus S of V and if Q is a point of M such that V is equisingular at Q, along M, then prove that Q belongs to no other irreducible component of S.

PROBLEM 6.3. *Prove that each point of E_i is a simple point of $\bar{E}_i$.*
Again, a special case of Problem 3 is the following: if M and Q are as above, prove that Q is a simple point of M.

DEFINITION 6.4. *A singular point Q of V is said to be of dimensionality type s if there exists an irreducible subvariety M of V, of codimension s, such that $Q \in M$ and V is equisingular at Q, along M, while there does not exist a subvariety of codimension $< s$ with that same property.*

Problem 6.5. *If* cod $\overline{E}_i = s$, *then is it true that all points of* E_i *have dimensionality type* s? (By (1), all point of E_i have dimensionality type $\leq s$).

A number of other questions arise in regard to the behavior of the strata E_i under monoidal transformations, and also in regard to sections of V transversal to $\overline{E}_i$ at points of E_i. We shall not insist on formulating now these questions. Rather, we wish to formulate one last general question which arises *in the complex domain.*

Assume that Q is *a simple point* of M and that V is equisingular at Q along M. Let W_Q be a transversal section of V at Q, relative to M (dim $W = s = \text{cod}_V M$).

Conjecture C_s. *There exists a neighborhood of* Q *on* V *which is homeomorphic to the direct product*

$$M \times W_Q,$$

with the additional condition that the two systems of fibres $W_X (X \in M)$ *and* $M_Y (Y \in W_Q)$ *associated with this direct product decomposition of* V, *have the following properties*:

(a) W_X *is a transversal section of* V *at* X, *relative to* M.

(b) M_Y *is an analytic variety, and* V *is equisingular at each point of* M_Y, *along* M_Y.

This has been proved for $s = 1$ by Whitney, and I can also prove it for $s = 2$.

When one tries to prove C_s by induction on s, one finds that it is necessary to prove something more. Assume V embedded in affine $(r+1)$-space A_{r+1}. Imagine W_Q to be the section of V by a non-singular variety H_Q, of dim $s+1$ (transversal to M). Then

Conjecture C_s'. *There exists a neighborhood of* Q *in* A_{r+1} *which is homeomorphic to the direct product* $M \times H_Q$, *in such a way that the two systems of fibrei* $\{H_X\} (X \in M)$, $M_Y (Y \in H_Q)$ *have the following properties:*

(a) H_X *is an analytic variety, transversal to* M *at* X.

(b) M_Y *is analytic.*

(c) *These fibrations induce on* V *a fibration having properties* a), b) in C_s (*where* $W_X = H_X \cap V$, *and* $M_Y \subset V$ *for* $Y \in W_Q = H_Q \cap V$).

Ouly C_1' *has been proved* (by Whitney).

STUDIES IN EQUISINGULARITY I
EQUIVALENT SINGULARITIES OF PLANE ALGEBROID CURVES.

By OSCAR ZARISKI.*

With this paper we initiate a series of investigations on the concept of equisingularity of an algebraic variety V, along an irreducible (singular) subvariety W of V, at a given point of W. The actual definition of this concept, at least in certain special situations (and, in particular, in the case $\operatorname{cod}_V W = 1$), was given by us in an earlier paper ("Equisingular points on algebraic varieties," seminari dell'Istituto Nazionale di Alta Matematics 1962-63, Roma, 1964, pp. 164-177), and will be presented and discussed in full (i. e., including proofs) in our next paper in this series. The present paper is preliminary in nature. It deals first of all with various ways of defining equivalent singularities of plane algebroid curves (§§ 1-4 bis). We define equivalence by induction on the number of locally quadratic transformations which are necessary to resolve the given singularity. No explicit (or implicit) use of Puiseux expansions is made, and, in fact, the case of non-zero characteristic (where Puiseux expansions are not available) is included in our treatment. We give 3 different definitions of equivalence, and we prove that they are all equivalent to each other. [It is our belief, however, that in the case of non-zero characteristic our treatment is not definitive, as it lumps together, in one equivalence class, singularities which, on the basis of some internal evidence, should be further subclassified; we propose to come back to the case $p \neq 0$ in a subsequent paper in this series.]

In § 5 we establish some technical results on derivations in complete integrally closed, domains (Theorems 3 and 4) and we obtain, as a consequence, a criterion for such a domain to be a power series ring over a subring.

In §§ 6-7, the geometric results of §§ 1-4 bis and the purely algebraic results of § 5 are applied toward the derivation of a discriminant criterion which enables one to decide whether or not, given an analytic family of algebroid curves C_t: $f(x, y; t) = 0$, the special member C_0 of the family has a singularity which is equivalent to the singularity of the generic member C_t.

Received October 26, 1964.

* This research was supported in part by the Air Force Office of Scientific Research and in part by the National Science Foundation.

This criterion will be used in our next paper dealing with equisingularity in codimension 1.

1. Algebroid plane curves. Let k be an algebraically closed field (of arbitrary characteristic) and let $k[[x,y]]$ be the ring of power series of two independent variables x, y, with coefficients in k. By an *algebroid plane curve* C we mean a local ring of the form $k[[x,y]]/(f)$, where $f=f(x,y)$ is a non-unit element of $k[[x,y]]$, free from multiple factors. We shall say that $f(x,y)=0$ is an *equation* of the curve C (the power series f is determined by C uniquely, to within an arbitrary unit factor in $k[[x,y]]$). The point $P\colon x=y=0$ will be referred to as the origin of C.

If $f(x,y)$ is an irreducible element of $k[[x,y]]$, C is said to be an irreducible algebroid curve. If f is reducible, $f=\phi_1\phi_2\cdots\phi_h$, where each ϕ_i is an irreducible element, then the h irreducible algebroid curves $\gamma_i\colon \phi_i=0$ are the irreducible *components* or *branches* of C.

Assume that the power series f begins with terms of degree s: $f=f_s(x,y)$ + terms of degree $>s$, where f_s is a homogeneous polynomial of degree s (f_s = leading form of f). Then one says that P is an *s-fold point of* C, and we write $s=m_P(C)$. Clearly $m_P(C)=\sum_{i=1}^{h} m_P(\gamma_i)$.

It is known that if f is irreducible then f_s is the power of a linear form: $f_s=(ax+by)^s$ (Hensel's lemma). The line (or direction) $ax+by=0$ is then called the tangent line of C. If C is reducible, then the tangent lines of the various irreducible components γ_i of C shall be, by definition, the tangent lines of C (the number of distinct tangent lines of C is therefore $\leq h$).

2. The quadratic transform of C. Let $l_i\colon a_ix+b_iy=0$ be the tangent line of γ_i $(i=1,2,\cdots,h)$. Without loss of generality we may assume that the b_i are all different from zero. We set $x'=x$, $y'=y/x$, and

$$f(x',x'y')=x'^sf'(x',y').$$

Then f' is a power series in x', y', and in fact, f' is a power series in x', whose coefficients are *polynomials* in y' (with coefficients in k):

$$(1)\qquad f'(x',y')=f_s(1,y')+x'f_{s+1}(1,y')+x'^2f_{s+2}(1,y')+\cdots$$

Let t be the number of *distinct* tangent lines p_ν of C, and let $y-\alpha_\nu x=0$ $(\nu=1,2,\cdots,t)$ be the equations of these lines (so that for each $i=1,2,\cdots,h$ the ratio $-a_i/b_i$ is equal to one of the α_ν). Let I_ν be the set of integers i $(1\leq i\leq h)$ such that $-a_i/b_i=\alpha_\nu$, and let C_ν be the union of the irreducible branches γ_i, $i\in I_\nu$. If we set $F_\nu(x,y)=\prod_{i\in I_\nu}\phi_i(x,y)$, then C_ν is the algebroid curve defined by the equation $F_\nu(x,y)=0$. The irreducible branches of C_ν are the γ_i, with $i\in I_\nu$, and they all have the same tangent line p_ν. We have $f(x,y)=\prod_{\nu=1}^{t}F_\nu(x,y)$, so that C is the union of the t algebroid curves C_ν. We call the C_ν the *tangential components* of C.

Let $r_i=m_P(\gamma_i), s_\nu=m_P(C_\nu)\,(=\sum_{i\in I_\nu} r_i)$,

$$\phi_i(x,y)=\phi_{ir_i}(x,y)+\phi_{i,r_i+1}(x,y)+\cdots,$$
$$F_\nu(x,y)=F_{\nu,s_\nu}(x,y)+F_{\nu,s_\nu+1}(x,y)+\cdots.$$

Since $\phi_{ir_i}(x,y)=(y-\alpha_\nu x)^{r_i}$ for $i\in I_\nu$ and $F_{\nu,s_\nu}(x,y)=(y-\alpha_\nu x)^{s_\nu}$, it follows that

$$\phi_i(x',x'y')=x'^{r_i}\phi_i'(x',y'),$$
$$F_\nu(x',x'y')=x'^{s_\nu}F_\nu'(x',y'),$$

where, for $i\in I_\nu$:

$$\phi_i'(x',y')=(y'-\alpha_\nu)^{r_i}+x'\phi_{i,r_i+1}(1,y')+\cdots,$$
$$F_\nu'(x',y')=\prod_{i\in I_\nu}\phi_i'(x',y')=(y'-\alpha_\nu)^{s_\nu}+x'F_{\nu,s_\nu+1}(1,y')+\cdots.$$

We regard ϕ_i' and F_ν' as power series in x' and $y'-\alpha_\nu$ and we denote by γ_i' and C_ν' the algebroid curves, with origin at the point $P_\nu'(0,\alpha_\nu)$, defined respectively by the equations $\phi_i'(x',y')=0$ and $F_\nu'(x',y')=0$. The curves γ_i' $(i\in I_\nu)$ are irreducible branches of C_ν'. We note that

$$f'(x',y')=\prod_{\nu=1}^{t}F_\nu'(x',y').$$

Hence the equation $f'(x',y')=0$ may be regarded as representing the set of (disjoint) algebroid curves $C_1',C_2',\cdots,C_t'$, having respectively as origins the points $P_1',P_2',\cdots,P_t'$. We denote by T the locally quadratic transformation (with center P) defined by $x'=x$, $y'=y/x$, we set $C'=\bigcup_{\nu=1}^{t}C_\nu'$, and we call C', C_ν' and γ_i' the *proper quadratic transforms* of C, C_ν and γ_i respectively:

$$C'=T(C),$$
$$C_\nu'=T(C_\nu),$$
$$\gamma_i'=T(\gamma_i).$$

We shall refer to $C_1',C_2',\cdots,C_t'$ are the *connected components* of C'.

Note. Although the above definitions are based on a specific (and explicit) choice of regular parameters x, y of the point P of the (x,y)-plane, they are easily seen to have an intrinsic invariantive meaning, as they could have been formulated in terms of local algebra (see Zariski-Samuel, Commutative Algebra, v. 2, Appendix 5). In such an intrinsic formulation one starts with a complete, noetherian, equicharacteristic local ring $\mathfrak{o}$, of dimension 1, whose residue field is algebraically closed and whose maximal ideal $\mathfrak{m}$ has a basis of two elements. If one, furthermore, assumes that $\mathfrak{o}$ has no proper nilpotent elements, then—for any given choice of a field k of representatives in $\mathfrak{o}$ and of a basis (ξ, η) of $\mathfrak{m}$—the local ring $\mathfrak{o}$ will be the local ring of a unique plane algebroid curve, defined over k. *The notion of the quadratic transform of $\mathfrak{o}$ is, however, independent of the choice of k, ξ and η.*

Similarly, given two local rings $\mathfrak{o}$, $\bar{\mathfrak{o}}$, satisfying the above conditions, and having the same (or isomorphic) residue fields, *then the definition of equivalence of $\mathfrak{o}$ and $\bar{\mathfrak{o}}$ given in* § 3 *below, remain meaningful and are independent of the choice of k, ξ and η.*

3. Three definitions of equivalence of algebroid curves. Let D be another plane algebroid curve, with some origin Q. We *assume that C and D have the same number h of irreducible branches,* and we denote by $\delta_1, \delta_2, \cdots, \delta_h$ the irreducible branches of D.

DEFINITION 1. *A $(1,1)$ mapping π of the set of branches $\gamma_1, \gamma_2, \cdots, \gamma_h$ of C onto the set of branches $\delta_1, \delta_2, \cdots, \delta_h$ of D is said to be a tangentially stable pairing $\pi\colon C \rightarrow D$ between the branches of C and those of D, if the following condition is satisfied: given any two branches γ_i and γ_j of C, the corresponding branches $\pi(\gamma_i)$ and $\pi(\gamma_j)$ of D have the same tangent if and only if γ_i and γ_j have the same tangent.*

Assume that there exists a tangentially stable pairing $\pi\colon C \rightarrow D$ between the branches of C and the branches of D. Then it is clear that C and D have the same number t of distinct tangent lines and that π induces a $(1,1)$ mapping of the set $\{p_1, p_2, \cdots, p_t\}$ of tangent lines of C onto the set $\{q_1, q_2, \cdots, p_t\}$ of tangent lines of D. We choose our indexing of these tangent lines in such a way that p_ν and q_ν are paired in this induced mapping, and we denote by C_ν (resp. D_ν) the tangential component of C (resp., D) associated with p_ν (resp., q_ν). Then it is clear that for each $\nu = 1, 2, \cdots, t$, π induces a $(1,1)$ mapping $\pi_\nu\colon C_\nu \rightarrow D_\nu$ of the set of branches of C_ν onto the set of branches of D_ν (the pairing π_ν is trivially tangentially stable, since both C_π and D_ν have only one tangent line).

Let π and π_ν be as above (π-tangentially stable), let T be a locally quadratic transformation with center at the origin P of C and let S be a locally quadratic transformation with center at the origin Q of D. Let $C' = T(C)$, $C_\nu' = T(C_\nu)$, $D' = S(D)$, $D_\nu' = S(D_\nu)$ be the proper transforms.

It is clear that π_ν induces a $(1,1)$ mapping π_ν' of the set of branches of C_ν' onto the set of branches of D_ν'. Namely, if we assume that the branches of C and D have been so indexed that $\pi(\gamma_i)=\delta_i$, for $i=1,2,\cdots,h$, then, in the notations of § 2, we set $\pi_\nu'(\gamma_i')=\delta_i'$ for all $i\in I_\nu$, where $\gamma_i'=T(\gamma_i)$ and $\delta_i'=S(\delta_i)$. The pairing $\pi_\nu'\colon C_\nu'\to D_\nu'$ between the branches of C_ν' and the branches of D_ν' is, however, not necessarily tangentially stable.

An algebroid curve C is *regular* if its origin P is a simple point of C, i.e., if $m_P(C)=1$. If P is a *singular* point (i.e., if $m_P(C)>1$), then we can resolve the singularity of C at P by a finite number of locally quadratic transformations. By a sequence of *successive quadratic transforms of C* we mean a sequence $\{C,C',C'',\cdots,C^{(i)},\cdots\}$ of algebroid curves $C^{(i)}$ such that for each i, $C^{(i+1)}$ is a connected component of the proper transform of $C^{(i)}$ under a locally quadratic transformation whose center is the origin of $C^{(i)}(C^{(0)}=C)$. The fact that the singularity of C can be resolved can then be stated as follows: there exists an integer N such that in *any* sequence of successive quadratic transforms of C, the curves $C^{(i)}$ are regular if $i\geqq N$. We denote $\sigma(C)$ the smallest integer N with the above property ($\sigma(C)=0$ if and only if C itself is a regular curve).

It is clear that if $C_1',C_2',\cdots,C_t'$ are the connected components of the proper quadratic transform $T(C)$ of C, and if $\sigma(C)>0$, then $\sigma(C_\nu')<\sigma(C)$ for $\nu=1,2,\cdots,t$. Our first definition of equivalence of algebroid curves (or—what is the same—of equivalence of algebroid singularities) proceeds by induction on $\sigma(C)$.

Let $\pi\colon C\to D$ be a pairing between the branches of C and the branches of D (it is already assumed that C and D have the same number h of branches). If C is regular (whence $\sigma(C)=0$), then C (and therefore also D) has only one branch, $\pi\colon C\to D$ is uniquely determined, and we say that π is an *(a)-equivalence* if also D is a regular curve. Assume that for all pairs of algebroid curves Γ, Δ with the same number of branches and such that $\sigma(\Gamma)<\sigma(C)$ it has already been defined what is to be meant by saying that a pairing $\Gamma\to\Delta$ between the branches of Γ and the branches of Δ is an (a)-equivalence. Then we define an (a)-equivalence between C and D as follows (we use the notations introduced earlier in this section):

DEFINITION 2. *An (a)-equivalence $\pi\colon C\to D$ is a pairing π between the branches of C and the branches of D having the following properties:*

1) *π is tangentially stable.*
2) *If $\delta_i=\pi(\gamma_i)$ $(i=1,2,\cdots,h)$, then $m_P(\gamma_i)=m_P(\delta_i)$.*
3) *The pairing $\pi_\nu'\colon C_\nu'\to D_\nu'$ $(\nu=1,2,\cdots,t)$ is an (a)-equivalence.*

Example. P is said to be an *ordinary* s-fold point of C if $s = m_P(C)$ and if the number t of distinct tangents of C is exactly equal to s. It follows that if P is an ordinary s-fold point of C, then C has exactly s branches $\gamma_1, \gamma_2, \cdots, \gamma_s$, these branches have distinct tangents $p_1, p_2, \cdots, p_s$, and each γ_i is a regular algebroid curve (since $s = \sum_{i=1}^{s} m_P(\gamma_i)$ and since $m_P(\gamma_i) \geqq 1$). The proper quadratic transform $C' = T(C)$ of C has then exactly s connected components $C_1', C_2', \cdots, C_s'$, where $C_i' = \gamma_i' = T(\gamma_i)$, and each C_i' is a regular curve (since it follows, quite generally, from (1), § 2, that $\sum_{\nu=1}^{t} m_{P_\nu'}(C_\nu') \leqq m_P(C)$, for any algebroid curve C). Now, suppose that P is an ordinary s-fold point of C and that a pairing $\pi\colon C \to D$ between the branches of C and D satisfies conditions 1) and 2) of Definition 2. Then it is clear that also Q is an ordinary s-fold point of D. Condition 3) is now automatically satisfied, since C_ν' and D_ν' are regular curves. Then π is an (a)-equivalence. Conversely, it is clear that if P is an ordinary s-fold point of C and if Q is an ordinary s-fold point of D, then any pairing $\pi\colon C \to D$ between the s branches of C and the s branches of D is an (a)-equivalence.

We now proceed to our second definition of equivalence between algebroid singularities. If T is our quadratic transformation, with center P (see § 2), then T blows up P into the line $x' = 0$ of the (x', y')-plane. We denote this line by $\mathcal{E}'$ and we refer to $\mathcal{E}'$ as the *exceptional curve* of T. If C_ν is a tangential component of C and $C_\nu' = T(C_\nu)$ is the proper T-transform of C_ν, then $\mathcal{E}'$ contains the origin P_ν' of C_ν', but $\mathcal{E}'$ is not a component of C_ν'. We denote by $C_\nu'^*$ the algebroid curve $C_\nu' \cup \mathcal{E}'$ and we call $C_\nu'^*$ the *total* T-transform of C_ν; in symbols: $C_\nu'^* = T\{C_\nu\}$. We set $C'^* = T(C) \cup \mathcal{E}'$ and we call C^* the *total* T-transform of C. Note that $m_{P_\nu'}(C_\nu'^*)$ is always $\geqq 2$.

It is known that after a finite number of successive quadratic transformations one can reach a stage where the total transform of C *has only ordinary double points.* More precisely: there exists an integer $N \geqq 0$ (depending on C) with the following property: if $\{C, C'^*, C''^*, \cdots, C^{(i)*}, \cdots\}$ is *any* sequence of algebroid curves such that for any i we have $C^{(i+1)*} = C^{(i+1)} \cup \mathcal{E}^{i+1}$, where $C^{(i+1)}$ is a connected component of the proper quadratic transform $T^{(i)}(C^{(i)*})$ of $C^{(i)*}$ ($T^{(i)}$ being a quadratic transformation with center at the origin $P^{(i)}$ of $C^{(i)*}$) and $\mathcal{E}^{(i+1)}$ is the exceptional curve of $T^{(i)}$, then for $i \geqq N$ the origin $P^{(i)}$ of $C^{(i)*}$ is an ordinary double point of $C^{(i)*}$. We denote by $\sigma^*(C)$ the smallest integer N having the above property.

It is clear that $\sigma^*(C) = 0$ if and only if the origin P of C is an ordinary double point of C. If C is a regular curve then a strict interpretation of our definition of $\sigma^*(C)$ would require to set $\sigma^*(C)=1$. However, we agree to set $\sigma^*(C)=0$ also if C is a regular curve (this could also have been achieved by a slight change in our general definition of $\sigma^*(C)$). It is easily seen that $\sigma^*(C)=1$ if and only if P is an ordinary s-fold point of C and $s>2$. If P is a tacnode of the first kind (i.e., if C consists of two regular branches having simple contact), then the *total* transform $C^*=T\{C\}$ of C has an ordinary triple point, and hence $\sigma^*(C)=2$ in this case. On the other hand, if P is an ordinary cusp of C, then $C'^*=T\{C\}$ has a tacnode of the first kind (while the *proper* transform $T(C)$ of C is regular), and thus $\sigma^*(C)=3$ in this case.

Let C and D have the same number of branches and let $\pi::C\rightarrow D$ be a pairing of the branches of C with the branches of D. If $\sigma^*(C)=0$, i.e., if P is either a simple point or an ordinary double point of C, then we shall say that π is a *(b)-equivalence between C and D* if and only if also $\sigma^*(D)=0$, i.e., if and only if the origin Q of D is a simple point or an ordinary double point of D according as P is a simple point or an ordinary double point of C. Assume that for all pairs Γ, Δ of algebroid curves, with the same number of branches, such that $\sigma^*(\Gamma)<\sigma^*(C)$, it has already been defined what is meant by saying that a pairing $\Gamma\rightarrow\Delta$ between the branches of Γ and the branches of Δ is a (b)-equivalence. Then we define a (b)-equivalence between C and D as follows:

DEFINITION 3. *A (b)-equivalence $\pi: C\rightarrow D$ is a pairing π between the branches of C and the branches of D having the following properties:*

1) *π is tangentially stable.*
2) *The pairings $\pi_\nu': C_\nu'\rightarrow D_\nu'$ $(\nu=1,2,\cdots,t)$ are (b)-equivalences.*
3) *If $\mathcal{E}'$ and E' are the exceptional curves of the quadratic transformations T and S respectively (having centers at P and Q), if $C_\nu'^*=C_\nu'\cup\mathcal{E}'$, $D_\nu'^*=D_\nu'\cup E'$, and if we extend the pairing π_ν' to a pairing $\pi_\nu'^*: C_\nu'^*\rightarrow D_\nu'^*$ by setting $\pi_\nu'^*(\mathcal{E}')=E'$, then $\pi_\nu'^*$ is a* (b)-equivalence.

Note that conditions 1) and 2) of this definition are identical with the condtions 1) and 3) of Definition 2; condition 2) of Definition 2 has been deleted and has been replaced in Definition 3 by condition 3). Thus the equality of the multiplicities of corresponding branches under π is not explicitly postulated in Definition 3.

We now give a third definition of equivalence of algebroid singularities, which we shall refer to as *formal equivalence.* Again we proceed by induction on $\sigma^*(C)$, where we agree that if $\sigma^*(C)=0$ formal equivalence coincides with (b)-equivalence.

17

DEFINITION 4. *Given two algebroid curves C, D, having the same number of branches, we say that C and D are formally equivalent if there exists a tangentially stable pairing $\pi: C \to D$ between the branches of C and the branches of D such that (in our previous notations):*

1) *C_ν' and D_ν' are formally equivalent* $(\nu = 1, 2, \cdots, t)$.
2) *$C_\nu'^*$ and $D_\nu'^*$ are formally equivalent* $(\nu = 1, 2, \cdots, t)$.

Note that this definition does not say anything about the nature of the pairings $\pi_\nu': C_\nu' \to D_\nu'$ and $\pi_\nu'^*: C_\nu'^* \to D_\nu'^*$ induced by π. Condition 1) merely requires that there exist, for each $\nu = 1, 2, \cdots, t$, some tangentially stable pairing $\rho_\nu': C_\nu' \to D_\nu'$ satisfying the conditions of the above inductive definition; and similarly, condition 2) requires that there exists a tangentially stable pairing $\rho_\nu'^*: C_\nu'^* \to D_\nu'^*$ satisfying similar conditions. It is not even required that $\rho_\nu'^*$ be an extension of ρ_ν'. For this reason, Definition 4 is the most subtle (and also the weakest) of our three definitions of equivalence. The fact (established below) that these three definitions are all equivalent to each other is therefore not devoid of interest.

W shall use the following notations:

If $\pi: C \to D$ is a pairing between the branches of C and the branches of D, then we write $\pi: C \overset{a}{\Longrightarrow} D$ or $\pi: C \overset{b}{\Longrightarrow} D$ according as π is an (a)-equivalence of a (b)-equivalence. We shall write $C \overset{f}{\equiv} D$ if C and D are formally equivalent.

4. Proof of equivalence of Definitions 2, 3 and 4. We recall first a few (well-known) elementary facts about the intersection number (C, D) of algebroid curves (having the same origin P) and about the behaviour of this number under a quadratic transformation T centered at P. We assume, of course, that C and D have no branches in common.

If $n = m_P(C)$ and $m = m_P(D)$ and if we assume that the line $x = 0$ is not tangent to either C or D, then, by the Weierstrass properation theorem, we may assume that C and D are defined by $f(x, y) = 0$ and $g(x, y) = 0$, where f and g *are monic polynomials in* y, of degree n and m respectively, with coefficients which are power series in x. If $y_1, y_2, \cdots, y_n$ are the roots of $f(x, y)$ and $\bar{y}_1, \bar{y}_2, \cdots, \bar{y}_m$ are the roots of $g(x, y)$ (in an algebraic closure of the field $k\{\{x\}\}$ of meromorphic functions of x), then (C, D), is defined as the order of the power series $\prod_{i=1}^{n} \prod_{j=1}^{m} (y_i - \bar{y}_j)$ in x (this power series is an element of $k[[x]]$ since the y_i and $\bar{y}_j$ are integral over $k[[x]]$). Since n and m are the degrees of the leading forms of f and g, it follows that

$$f = y^n + x a_1(x) y^{n-1} + \cdots + x^n a_n$$
$$g = y^m + x b_1(x) y^{m-1} + \cdots + x^m b_m(x),$$

with $a_i(x)$ and $b_j(x)$ in $k[[x]]$. If $f'(x',y') = 0$ and $g'(x',y') = 0$ are the equations of the proper transforms $C' = T(C)$ and $D' = T(D)$ of C and D (see § 2), then

$$f' = y'^n + a_1(x') y'^{n-1} + \cdots + a_n(x')$$
$$g' = y'^m + b_1(x') y'^{m-1} + \cdots + b_m(x'),$$

where $x' = x$. Thus, if $y_1', y_2', \cdots, y_n'$ and $\bar{y}_1', \bar{y}_2', \cdots, \bar{y}_m'$ are the roots of $f'(x',y')$ and $g'(x',y')$ (regarded as polynomials in y'), then the y_i' and $\bar{y}_j'$ are still integrally dependent on $k[[x]]$. Since $y_i = x y_i'$ and $\bar{y}_j = x \bar{y}_j'$, it follows that if $C_1', C_2', \cdots, C_t'$ are the connected components of C' and $D_1', D_2', \cdots, D_\tau'$ are the connected components of D', then

$$(2) \qquad (C, D) = mn + \sum_{\nu=1}^{t} \sum_{\mu=1}^{\tau} ((C_\nu', D_\mu').$$

Here $(C_\nu', D_\mu') > 0$ if and only if C_ν' and D_μ' have the same origin. By (2) if follows that $(C, D) \geqq mn$, with equality if and only if C and D have no common tangents. From the above expression of f' it follows also that if $\mathcal{E}'$: $x' = 0$ is the exceptional curve of T, then

$$(3) \qquad (C', \mathcal{E}') = \sum_{\nu=1}^{t} (C_\nu', \mathcal{E}') = n (= m_P(C)).$$

These are all the facts that we shall need in the sequel.

Let now C and D be algebroid curves, with origins P and Q, and with the same number of branches.

LEMMA 1. *Let* $\pi\colon C \overset{a}{\Longrightarrow} D$ *be an* (a)*-equivalence between* C *and* D *and let* Γ *be a regular curve through* P *and* Δ *a regular curve through* Q. *Assume that* Γ *is not a branch of* C, *that* Δ *is not a branch of* D, *and that for every pair* (γ_i, δ_i) *of corresponding branches under* π *we have*

$$(4) \qquad (\gamma_i, \Gamma) = (\delta_i, \Delta), \qquad i = 1, 2, \cdots, h.$$

Then the pairing $\rho\colon C \cup \Gamma \rightarrow D \cup \Delta$ *between the branches of* $C \cup \Gamma$ *and the branches of* $D \cup \Delta$ *which is the unique extension of* π *by* $\rho(\Gamma) = \Delta$, *is an* (a)*-equivalence.*

Proof. (by induction on $\sigma^*(C \cup \Gamma)$).

If $\sigma^*(C \cup \Gamma) = 0$, then C is a regular curve, and the regular curves C and Γ have distinct tangents. Hence, by (2), we have $(C, \Gamma) = 1$. Since

C and D are (a)-equivalent, also D is a regular curve, and it follows by (4) that $(D, \Delta) = 1$, whence $\sigma^*(D \cup \Delta) = 0$. Therefore ρ is an (a)-equivalence.

In the general case, we observe that $(\gamma_i, \Gamma) \geqq m_P(\gamma_i)$, with equality if and only if γ_i and Γ do not have the same tangent. Similarly, for (δ_i, Δ) and $m_Q(\delta_i)$. Since π is an (a)-equivalence, we have $m_P(\gamma_i) = m_Q(\delta_i)$. Hence it follows from (4) that γ_i and Γ have the same tangent if and only if δ_i and Δ have the same tangent. Therefore ρ *is a tangentially stable pairing*. Since $m_P(\Gamma) = m_Q(\Delta) = 1$, it is clear that any two corresponding branches under ρ have the same multiplicity. Thus ρ satisfies conditions 1) and 2) of the definition of an (a)-equivlance (Definition 2, § 3). We now check condition 3) of that definition.

Let $T(\Gamma) = \Gamma'$, $S(\Delta) = \Delta'$. We consider two cases.

First Case. Γ *is not tangent to any of the branches of* C (and hence Δ is not tangent to any of the branches of D). In this case, $C_1', C_2', \cdots, C_t'$ and Γ' are the connected components of the proper transform $T(C \cup \Gamma)$, and $D_1', D_2', \cdots, D_t'$ and Δ' are the connected components of $S(D \cup \Delta)$. If $\rho_\nu': C_\nu' \to D_\nu'$ $(\nu = 1, 2, \cdots, t)$, $\rho_{t+1}': \Gamma' \to \Delta'$ are the pairings induced by ρ, then $\rho_\nu' = \pi_\nu'$ is an (a)-equivalence (since π is an (a)-equivalence), and ρ_{t+1}' is trivially an (a)-equivalence (since Γ' and Δ' are regular curves). Thus condition 3) of Definition 2 is satisfied.

Second Case. The tangent of Γ *coincides with one of the* t *tangents* $p_1, p_2, \cdots, p_t$ of C, say p_1 is the tangent of Γ. In that case, the corresponding tangent line q_1 of D (under π) is the tangent of Δ. The tangential components of $C \cup \Gamma$ are now $C_1 \cup \Gamma$, $C_2, \cdots, C_t$, and the corresponding tangential components (under ρ) of $D \cup \Delta$ are $D_1 \cup \Delta, D_2, \cdots, D_t$. The connected components of $T(C \cup \Gamma)$ and $S(D \cup \Delta)$ are now $C_1' \cup \Gamma'$, $C_2', \cdots, C_t'$ and $D_1' \cup \Delta'$, $D_2', \cdots, D_t'$. Consider the induced pairing

$$\rho_1': C_1' \cup \Gamma' \to D_1' \cup \Delta', \quad \rho_\nu': C_\nu' \to D_\nu' \qquad (\nu = 2, 3, \cdots, t).$$

We have $\rho_\nu' = \pi_\nu'$ for $\nu \geqq 2$, and ρ_ν' is an (a)-equivalence for $\nu = 2, 3, \cdots, t$. Consider now ρ_1'. It is an extension of π_1', with $\rho_1'(\Gamma') = \Delta'$. Let γ_i' be any branch of C_1' $(i \in I_1)$ and let δ_i' be the corresponding branch of D_1' (under π_1'). Then $\gamma_i' = T(\gamma_i)$, $\delta_i' = S(\delta_i)$, with $\delta_i = \pi(\gamma_i)$. By (2), we have $(\gamma_i, \Gamma) = m_P(\gamma_i) + (\gamma_i', \Gamma')$, $(\delta_i, \Delta) = m_P(\delta_i) + (\delta_i', \Delta')$. Since $m_P(\gamma_i) = m_P(\delta_i)$ and $(\gamma_i, \Gamma) = (\delta_i, \Delta)$, it follows that $(\gamma_i', \Gamma') = (\delta_i', \Delta')$. Since $\sigma^*(C_1' \cup \Gamma') < \sigma(C \cup \Gamma)$ (if $\sigma^*(C \cup \Gamma) \neq 0$), it follows, by our induction hypothesis, that ρ_1' is an (a)-equivalence. This completes the proof of the lemma.

LEMMA 2. *If* $\pi\colon C \overset{a}{\Longrightarrow} D$ *and* $\pi(\gamma_i) = \delta_i$ $(i = 1, 2, \cdots, h)$, *then*

$$\gamma_i, \gamma_j) = (\delta_i, \delta_j) \text{ for all } i \neq j.$$

Proof. We have $(\gamma_i, \gamma_i) = m_P(\gamma_i) m_P(\gamma_j) + (\gamma_i', \gamma_j')$, where $\gamma_i' = T(\gamma_i)$, $(\delta_i, {}_j) = m_Q(\delta_i) m_Q(\delta_j) + (\delta_i', \delta_j')$. Here (γ_i', γ_j') is to be replaced by zero if γ_i' and γ_j' have distinct origins (and similarly for (δ_i', δ_j')). Since $m_P(\gamma_i) = m_Q(\delta_i)$ for all i, the lemma follows by induction on $\sigma(C)$.

THEOREM 1.

(a) *If* $\pi\colon C \overset{a}{\Longrightarrow} D$ *is an (a)-equivalence, then* π *is also a (b)-equivalence, and conversely.*

(b) *If there exists an (a)-equivalence (or a (b)-equivalence)* $\pi\colon C \Longrightarrow D$, *then* C *and* D *are formally equivalent.*

Proof. (a) The assertion is true if $\sigma^*(C) = 0$. We therefore use induction on $\sigma(C)$.

Assume that $\pi\colon C \Longrightarrow D$ is an (a)-equivalence. Using the notations of Definition 2, we have the (a)-equivalences $\pi_\nu'\colon C_\nu' \rightarrow D_\nu'$, $\nu = 1, 2, \cdots, t$. Therefore, by our induction hypothesis, the π_ν' are also (b)-equivalences. So we have only to show that (using the notations of Definition 3) the pairings $\pi_\nu'^*\colon C_\nu' \cup \mathcal{E}' \rightarrow D_\nu' \cup E'$ are (b)-equivalences. Since $\sigma^*(C_\nu' \cup \mathcal{E}') < \sigma^*(C)$ (if $\sigma^*(C) \neq 0$), it is sufficient to show (by our induction hypothesis) that $\pi_\nu'^*$ is an (a)-equivalence. Now $\pi_\nu'\colon C_\nu' \rightarrow D_\nu'$ is an (a)-equivalence, the curves $\mathcal{E}'$ and E' are regular, and $\pi_\nu'^*$ is an extension of π_ν', with $\pi_\nu'^*(\mathcal{E}') = E$. So, by Lemma 1, it is sufficient to show that if $\gamma_i' = \pi_\nu'(\gamma_i)$ are corresponding branches of C_ν' and D_ν', under π_ν', then $(\gamma_i', \mathcal{E}') = (\delta_i', E')$. Now, γ_i' and δ_i' are the proper transforms of the branches γ_i, δ_i of C and D respectively, and $\delta_i = \pi(\gamma_i)$. Since π is an (a)-equivalence we have $m_P(\gamma_i) = m_P(\delta_i)$, and since, by (3), we have $m_P(\gamma_i) = (\gamma_i', \mathcal{E}')$ and $m_Q(\delta_i) = (\delta_i', E')$, the equality $(\gamma_i', \mathcal{E}') = (\delta_i', E')$ is proved.

Conversely, assume that $\pi\colon C \rightarrow D$ is a (b)-equivalence. Then, by induction, the (b)-equivalences $\pi_\nu'\colon C_\nu' \rightarrow D_\nu'$ are also (a)-equivalences. There thus remains only to show that if $\delta_i = \pi(\gamma_i)$ then $m_P(\gamma_i) = m_Q(\delta_i)$. Now, consider the (b)-equivalences $\pi_\nu'^*\colon C_\nu' \cup \mathcal{E}' \rightarrow D_\nu' \cup E'$ (which are extensions of the π_ν'). By induction, π_ν^* is also an (a)-equivalence. Hence, by Lemma 2, we have $(\gamma_i', \mathcal{E}') = (\delta_i', E')$ for any $i = 1, 2, \cdots, h$. This implies, by (3), that $m_P(\gamma_i) = m_Q(\delta_i)$, as asserted.

From now on we may replace the terms (a)-equivalence and (b)-equivalence by the single term of *equivalence* (to be distinguished, for the moment, from *formal equivalence*).

(b) Assume that there exists an equivalence $\pi\colon C \underset{\equiv}{\Longrightarrow} D$. We have then the equivalences $\pi_\nu\colon C_\nu' \underset{\equiv}{\Longrightarrow} D_\nu'$ and $\pi_\nu'^*\colon C_\nu' \cup \mathcal{E}' \underset{\equiv}{\Longrightarrow} D_\nu' \cup E'$. Hence, by induction on $\sigma^*(C)$, it follows that C_ν' and D_ν' are formally equivalent and that also $C_\nu' + \mathcal{E}'$ and $D_\nu' + E'$ are formally equivalent. Thus C and D are formally equivalent.

4 bis. Continuation: proof that formal equivalence $C \overset{f}{\equiv} D$ implies the existence of a pairing equivalence $\pi\colon C \underset{\equiv}{\Longrightarrow} D$.

LEMMA 3. *Let C and D be algebroid curves having the same number of branches, let $\varepsilon_1, \varepsilon_2, \cdots, \varepsilon_m$ denote certain branches of C and assume that these branches ε_α are regular and have distinct tangents. Similarly, let $e_1, e_2, \cdots, e_m$ be regular branches of D, with distinct tangents. Let*

$$C = \Gamma \cup \varepsilon_1 \cup \varepsilon_2 \cup \cdots \cup \varepsilon_m, \qquad D = \Delta \cup e_1 \cup e_2 \cup \cdots \cup e_m.$$

Assume that Γ and Δ are equivalent and that for each $\alpha = 1, 2, \cdots, m$ also $\Gamma \cup \varepsilon_1 \cup \varepsilon_2 \cup \cdots \cup \varepsilon_\alpha$ and $\Delta \cup e_1 \cup e_2 \cup \cdots \cup e_\alpha$ are equivalent. Then there exists a pairing equivalence $\pi\colon C \underset{\equiv}{\Longrightarrow} D$ such that $\pi(\varepsilon_\alpha) = e_\alpha$, for $\alpha = 1, 2, \cdots, m$.

Proof. We consider two cases, according as $m = 1$ or $m > 1$.

Case I. $m = 1$. We divide this case into two subcases, according as *the tangent of ε* $(= \varepsilon_1)$ is not or is one of the tangents of C.

Case Ia. $C = \Gamma \cup \varepsilon$, $D = \Delta \cup e$, *and the tangent of ε is not a tangent of* Γ. Since $\Gamma \equiv \Delta$, Γ and Δ have the same number of distinct tangent lines. Similarly, $\Gamma \cup \varepsilon \equiv \Delta \cup e$ implies that $\Gamma \cup \varepsilon$ and $\Delta \cup e$ have the same number or distinct tangent lines. But since the number of distinct tangents of $\Gamma \cup \varepsilon$ is one greater than the number of distinct tangents of Γ, the same is true of $\Delta \cup e$ and Δ. *Hence the tangent of e is different from any tangent of* Δ. It follows at once from the definition of (a)-equivalence that given any equivalence $\sigma\colon \Gamma \underset{\equiv}{\Longrightarrow} \Delta$, the extended pairing $\pi\colon \Gamma \cup \varepsilon \underset{\equiv}{\Longrightarrow} \Delta \cup e$ defined by $\pi(\varepsilon) = e$ is also an equivalence.

Case Ib. The tangent of ε coincides with one of the tangents of Γ (and hence the tangent of e coincides with one of the tangents of Δ, by the reasoning of the Case Ia).

Let $\Gamma_1, \Gamma_2, \cdots, \Gamma_t$ and $\Delta_1, \Delta_2, \cdots, \Delta_t$ be the tangential components of Γ and Δ respectively, arranged in some arbitrary order, except that we assume that the tangent of ε coincides with the tangent of Γ_1 and the tangent of e coincides with the tangent of Δ_1. Then the tangential components of C are $\Gamma_1 \cup \varepsilon, \Gamma_2, \cdots, \Gamma_t$, and the tangential components of D are $\Delta_1 \cup e, \Delta_2, \cdots, \Delta_t$. We shall first show that

$$(5) \qquad \Gamma_1 \equiv \Delta_1 \text{ and } \Gamma_1 \cup \varepsilon \equiv \Delta_1 \cup e.$$

Let a be the number of curves in the set $\{\Gamma_1 \cup \varepsilon, \Gamma_2, \cdots, \Gamma_t\}$ which are equivalent to Γ_1 $(a \geqq 0)$. Since $\Gamma_1 \cup \varepsilon \not\equiv \Gamma_1$, it follows that $a+1$ is the number of curves in the set $\{\Gamma_1, \Gamma_2, \cdots, \Gamma_t\}$ which are equivalent to Γ_1. Since $\Gamma \equiv \Delta$, any Γ_i is equivalent to at least one Δ_j, and any Δ_i is equivalent to at least one Γ_j. Since equivalence of algebroid curves is obviously a transitive relation it follows that $a+1$ is also the number of curves in the set $\{\Delta_1, \Delta_2, \cdots, \Delta_t\}$ which are equivalent to Γ_1. Similarly, from $\Gamma \cup \varepsilon = \Delta + e$ follows that a is the number of curves in the set $\{\Delta_1 + e, \Delta_2, \cdots, \Delta_t\}$ which are equivalent to Γ_1. Therefore, we must have necessarily $\Gamma_1 \equiv \Delta_1$.

Similarly, if we denote by b the number of curves in this set $\{\Gamma_1, \Gamma_2, \cdots, \Gamma_t\}$ which are equivalent to $\Gamma_1 \cup \varepsilon$ $(b \geqq 0)$, then $b+1$ is the number of curves in the set $\{\Gamma_1 \cup \varepsilon, \Gamma_2, \cdots, \Gamma_t\}$ which are equivalent to $\Gamma_1 \cup \varepsilon$. Since $\Gamma \equiv \Delta$ and $\Gamma \cup \varepsilon \equiv \Delta + e$, it follows, by a similar argument as above, that $\Gamma_1 \cup \varepsilon \equiv \Delta_1 \cup e$.

We have thus shown that the assumptions of the lemma are satisfied when Γ and Δ are replaced by Γ_1 and Δ_1 respectively (and thus C and D are replaced by $\Gamma_1 \cup \varepsilon$ and $\Delta_1 \cup e$ respectively). With Γ and Δ replaced by Γ_1 and Δ_1 we haves a special case of the lemma, namely the case in which $m=1$ and *each of the two curves C and D has only one tangent.* Assume that the lemma has already been proved in this special case. Then we can assert that there exist an equivalence $\pi_1\colon \Gamma_1 \cup \varepsilon \rightarrow \Delta_1 \cup e$ such that $\pi_1(\varepsilon) = e$. We shall show now that π_1 can be extended to an equivalence $\pi\colon C \rightarrow D$. Consider the set $\{\Gamma_1 \cup \varepsilon, \Gamma_2, \cdots, \Gamma_t\}$ of tangential component of C and the set $\{\Delta_1 \cup e, \Delta_2, \cdots, \Delta_t\}$ of tangential components of D. Let M denote the set of those tangential components of C which are equivalent to $\Gamma_1 \cup \varepsilon$ (M contains at least one element, namely $\Gamma_1 \cup \varepsilon$). Similarly let N be the set of those tangeneial components of D which are equivalent to $\Delta_1 \cup e$. Since $\Gamma_1 \cup \varepsilon \equiv \Delta_1 \cup e$ and since equivalence is transitive, it follows that M and N have the same number of elements. Now, by assumption, we have $C \equiv D$, i.e., there exists some equivalence $\rho\colon C \underset{=}{\Rightarrow} D$. Fix such an equivalence ρ. For $\nu = 1, 2, \cdots, t$, ρ induces an equivalence $\rho_\nu\colon C_\nu \underset{=}{\Rightarrow} D_\nu$, where $C_1 = \Gamma_1 \cup \varepsilon$, $C_\nu = \Gamma_\nu$ for $\nu = 2, \cdots, t$, and $D_1, D_2, \cdots, D_t$ are the curves $\Delta_1 \cup e, \Delta_2, \cdots, \Delta_t$, in some order. We have already our equivalence $\pi_1\colon \Gamma_1 \cup \varepsilon \underset{=}{\Rightarrow} \Delta_1 \cup e$. We define π_ν, for $\nu = 2, \cdots, t$ as follows: a) If C_ν $(= \Gamma_\nu) \notin M$ (i.e., if

$\Gamma_\nu \not\equiv \Gamma_1 \cup \varepsilon$), then clearly also $D_\nu \notin N$ (since $\Gamma_\nu \equiv D_\nu$ and $\Gamma_1 \cup \varepsilon \equiv \Delta_1 \cup e$, whence $D_\nu \not\equiv \Delta_1 \cup e$); in this case we set $\pi_\nu = \rho_\nu$: $\Gamma_\nu \Longrightarrow D_\nu$). b) We set up an arbitrary (1,1) correspondence between the set of curves Γ_μ in M. *other than* $\Gamma_1 \cup \varepsilon$, and the set of curves Δ_μ in N, *other than* $\Delta_1 + e$. Assuming that corresponding curves are furnished with the same index μ, we have that $\Gamma_\mu \equiv \Delta_\mu$ for all μ such that $\Gamma_1 + \varepsilon \neq \Gamma_\mu \in M$ and $\Delta_1 + e \neq \Delta_\mu \in N$ (since $\Gamma_\mu \equiv \Gamma_1 \cup \varepsilon$, $\Delta_\mu \equiv \Delta_1 \cup e$ and $\Gamma_1 \cup \varepsilon \equiv \Delta_1 \cup e$). We then fix an *arbitrary equivalence* π_μ: $\Gamma_\mu \Longrightarrow \Delta_\mu$. This defines π_ν for all $\nu = 1, 2, \cdots, t$. Since $\Gamma_1 \cup \varepsilon, \Gamma_2, \cdots, \Gamma_t$ and $\Delta_1 \cup e, \Delta_2, \cdots, \Delta_t$ are the tangential components of C and D respectively, it follows that the ν equivalences π_ν patch up together to an equivalence π: $C \Longrightarrow D$ (use for instance the definition of (a)-equivalence).

Thus, in order to complete the proof of the lemma (in the case $m = 1$) we have only to show that (in the case Ib under consideration) there exists an equivalence π_1: $\Gamma_1 \cup \varepsilon \Longrightarrow \Delta_1 \cup e$ such that $\pi_1(\varepsilon) = e$. Let us apply our quadratic transformations T and S (with center P and Q respectively) to $\Gamma_1 \cup \varepsilon$ and $\Delta_1 \cup e$. If $\mathcal{E}'$ and E' are the exceptional curves of T and S, we will have the total transforms

$$T\{\Gamma_1 \cup \varepsilon\} = \Gamma_1' \cup \varepsilon' \cup \mathcal{E}',$$
$$S\{\Delta_1 \cup e\} = \Delta_1' \cup e' \cup E',$$

where $\Gamma_1' = T(\Gamma_1)$, $\varepsilon' = T(\varepsilon)$; $\Delta_1' = S(\Delta_1)$, $e' = S(e)$. Since $\Gamma_1 \equiv \Delta_1$, it follows that $\Gamma_1' \equiv \Delta_1'$ (by the definition of (a)-equivalence). Since $\Gamma_1 \cup \varepsilon \equiv \Delta_1 \cup e$, it follows that $\Gamma_1' \cup \varepsilon' \equiv \Delta_1' \cup e'$ and that

$$\Gamma_1' \cup \varepsilon' \cup \mathcal{E}' \equiv \Delta_1' \cup e' \cup E'$$

(by the definition of (b)-equivalence). Furthermore, ε' and $\mathcal{E}'$ are regular curves, with distinct tangents; similarly, e' and E' are regular curves, with distinct tangents. We have here therefore the case $m = 2$ of our lemma. Since $\sigma^*(\Gamma_1 \cup \varepsilon) < \sigma^*(C)$ (unless $e^*(C) = 0$, in which case the lemma is trivially true), we may assume, by induction, that there exists an equivalence

$$\pi_1^*: \Gamma_1' \cup \varepsilon' \cup \mathcal{E}' \Longrightarrow \Delta_1' \cup e' \cup E'$$

such that $\pi_1^*(\varepsilon') = e'$ and $\pi_1^*(\mathcal{E}') = E'$. Then π_1^* induces an equivalence π_1: $\Gamma_1 \cup \varepsilon \Longrightarrow \Delta_1 \cup e$ such that $\pi_1(\varepsilon) = e$.

Case II. $m > 1$. Let C_j be the tangential component of C determined by the tangent of ε_j (C_j may be ε_j itself). Similarly let D_j be the tangential component of D determined by the tangent of e_j. Then $C_1, C_2, \cdots, C_m$ are distinct tangential components of C, and similarly for $D_1, D_2, \cdots, D_m$ and D. Since $\Gamma \cup \varepsilon_1 \cup \cdots \cup \varepsilon_{j-1} \equiv \Delta \cup e_1 \cup \cdots \cup e_{j-1}$ and

$$\Gamma \cup \varepsilon_1 \cup \cdots \cup \varepsilon_{j-1} \cup \varepsilon_j \equiv \Delta \cup e_1 \cup e_2 \cup \cdots \cup e_{j-1} \cup e_j,$$

it follows, by the case $m = 1$ that there exists an equivalence

$$\sigma_j\colon \Gamma \cup \varepsilon_1 + \cdots \cup \varepsilon_{j-1} \cup \varepsilon_j \xrightarrow{\equiv} \Delta \cup e_1 \cup \cdots \cup e_{j-1} \cup e_j$$

such that $\sigma_j(\varepsilon_j) = e_j$. Then σ_j induces an equivalence $\pi_j\colon C_j \xrightarrow{\equiv} D_j$. In the course of the proof of the lemma in the case I we have shown that *if* $C \equiv D$ and if $\pi_1\colon C_1 \xrightarrow{\equiv} D_1$ is a *given* equivalence between two tangential components C_1 and D_1 of C and D respectively, then π_1 can be extended to an equivalence $\pi\colon C \xrightarrow{\equiv} D$. In a similar fashion it can be shown that the m equivalences $\pi_j\colon C_j \xrightarrow{\equiv} D_j$ can be extended to (i. e., are induced by) a single equivalence $\pi\colon C \xrightarrow{\equiv} D$. Since $\pi_j(\varepsilon_j) = e_j$, the proof of the lemma is complete.

COROLLARY. *If two algebroid curves C and D are formally equivalent, they are equivalent.*

In the notations of Definition 4 we have, by hypothesis, that C_ν' and D_ν' are formally equivalent, and also that $C_\nu' \cup \mathcal{E}'$ and $D_\nu' \cup E'$ are formally equivalent $(\nu = 1, 2, \cdots, t)$. By induction on $\sigma^*(C)$ we may therefore assume that $C_\nu' \equiv D_\nu'$ and $C_\nu' \cup \mathcal{E}' \equiv D_\nu' \cup E'$. Hence, by the preceding lemma, there exists an equivalence $\pi_\nu'\colon C_\nu' \cup \mathcal{E}' \longrightarrow D_\nu' \cup E'$ such that $\pi_\nu'(\mathcal{E}') = E'$. But that means that C and D are equivalent (namely (b)-equivalent).

Note. In a Lincei Note ("La risoluzione delle singolarità delle superficie algebriche immerse," Nota I. Accademia Nazionale dei Lincei, Rendiconti, vol. XXXI, fasc. 3-4, Settembre-Ottobre (1961)) we gave the following inductive definition of equivalence of algebroid singularities (we use the notatians of Definition 2, § 3):

$C \equiv D$ if the following three conditions are satisfied:

1) $m_P(C) = m_Q(D)$.

2) C and D have the same number t of distinct tangents.

3) For a suitable ordering of the connected components $C_1', C_2', \cdots, C_t'$ and $D_1', D_2', \cdots, D_t'$ of the proper quadratic transforms of C and D, it is true that $C_\nu' \equiv D_\nu'$ $(\nu = 1, 2, \cdots, t)$.

This definition is not "correct." In fact, even the following, more exacting. definition of equivalence would not be "correct":

$C \equiv D$ if there exist a tangentially stable pairing $\pi\colon C \to D$ such that

1') $m_P(\gamma_i) = m_Q(\delta_i)$, if $\delta_i = \pi(\gamma_i)$.

2') $C_\nu' \equiv D_\nu'$ for $\nu = 1, 2, \cdots, t$, where C_ν' and D_ν' are the proper transforms of tangential components C_ν and D_ν which are associated with each other under π.

Here is a counter-example.

Let C consist of two branches γ_1, γ_2, having the same tangent $(y = 0)$, where γ_1 and γ_2 are defined by the following power series:

$$\gamma_1\colon x = t^4, \qquad y = t^{10} + t^{11};$$
$$\gamma_2\colon x = t^6 + t^9, \qquad y = t^{10} + t^{13}.$$

Let also D consist of the following two branches δ_1, δ_2, with the same tangent $x = 0$:

$$\delta_1\colon x = t^{10} + t^{13}, \qquad y = t^4;$$
$$\delta_2\colon x = t^{10} + t^{11}, \qquad y = t^6 + t^7.$$

We have $m_P(\gamma_1) = m_P(\delta_1) = 4$ and $m_P(\gamma_2) = m_P(\delta_2) = 6$. Hence the pairing $\pi\colon C \to D$ defined by $\pi(\gamma_1) = \delta_1$ and $\pi(\gamma_2) = \delta_2$ satisfies condition (1') above. Applying to C the quadratic transformation $T\colon x' = x$, $y' = y/x$, we get that $C' = T(C) = \gamma_1' \cup \gamma_2'$, where $\gamma_1' = T(\gamma_1)$ and $\gamma_2' = T(\gamma_2)$ are defined by

$$\gamma_1'\colon x' = t^4, \qquad y' = t^6 + t^7;$$
$$\gamma_2'\colon x' = t^6 + t^9, \qquad y' = t^4.$$

Similarly, applying to D the quadratic transformation $S\colon x' = x/y$, $y' = y$, we get $D' = S(D) = \delta_1' \cup \delta_2'$, where $\delta_1' = S(\delta_1)$ and $\delta_2' = S(\delta_2)$ are defined by

$$\delta_1'\colon x' = t^6 + t^9, \qquad y' = t^4;$$
$$\delta_2'\colon x' = t^4, \qquad y' = t^6 + t^7.$$

Thus $\gamma_1' = \delta_2'$ and $\gamma_2' = \delta_1'$, whence $C' \equiv D'$. However, C and D are not equivalent in any reasonable algebro-geometric sense. This is so, because already the corresponding branches γ_1, δ_1 (both having a 4-fold point at P) are not equivalent. This can be seen by observing that the 4th quadratic transform $\gamma_1^{(4)}$ of γ_1 is a regular curve, while the 4th quadratic transform $\delta_1^{(4)}$ of δ_1 has a cusp. (Similarly it can be seen that γ_2, δ_2 are not equivalent).

5. Some auxiliary results on derivations. In this section we deal with an *integrally closed* (Noetherian) *local domain* R, which we assume to be *pseudo-geometric* (see Nagata, *Local Rings*, p. 131), and with the quotient field K of R.

We consider a finite separable algebraic extension K' of K, and we denote by R' the integral closure of R in K'. Thus R' is a finite R-module. (Later on, in this section, we shall consider the more general case of a ring R' which is a finite R-module, is integrally closed in its total ring of quotients and *is free from nilpotent elements.*)

THEOREM 2. *Let D be a derivation of K, with values in K, which is regular in R (i.e., such that $DR \subset R$), and let D' be the unique extension of D to K'. Assume the following:*

A. *If $\mathfrak{p}'$ is any minimal prime ideal of R' (i.e., $\mathfrak{p}'$ is a prime ideal of hight 1 in R'), then $\mathfrak{p}'$ is tamely ramified over R.*

B. *If $\mathfrak{p}'$ is any prime ideal of R', of hight 1, which is ramified over R and if $\mathfrak{p}' \cap R = \mathfrak{p}$, then there exists an element x in $\mathfrak{p}$ such that $x \notin \mathfrak{p}^{(2)}$ and $Dx = 0$.*

Under these assumptions we have $D'R' \subset R'$.

[*Note.* Following Abhyankar ("Tame coverings and fundamental groups of algebraic varieties, Part I," *American Journal of Mathematics*, vol. 81, no. 1 (1959), p. 53) we say that $\mathfrak{p}'$ is tamely ramified over R if the following is true:

Let $\mathfrak{p} = \mathfrak{p}' \cap R$, let $(S, \mathfrak{M})$ and $(S', \mathfrak{M}')$ be respectively the completions of the localizations $R_{\mathfrak{p}}$ and $R_{\mathfrak{p}'}'$, and let F, F' be the fields of quotients of S and S' respectively. (Note that since R is pseudogeometric, S and S' are integral domains.) Let F^* be the least Galois extensions of F which contains F' (note that since K' is separable algebraic over K, also S' is separable algebraic over S). Let $(S^*, \mathfrak{M}^*)$ be the integral closure of S in F^*. (Note that since S is complete, S^* is a local domain). Then

$$[F^*:F] \cdot [S^*/\mathfrak{M}^*: S/\mathfrak{M}]_s^{-1} \not\equiv 0 \pmod{p},$$

where p is the characteristic of S/M. Note that this condition is vacuous if $p = 0$ and that "$\mathfrak{p}'$ unramified" implies "$\mathfrak{p}'$ tamely ramified."]

Proof. Since $R' = \cap R_{\mathfrak{p}'}'$, where the intersection symbol is extended to all minimal prime ideals $\mathfrak{p}'$ of R', it is sufficient to show that $D'R'_{\mathfrak{p}'} \subset R'_{\mathfrak{p}'}$ for every minimal prime ideal $\mathfrak{p}'$ of R'. Now if $\mathfrak{p}' \cap R = \mathfrak{p}$, then $\mathfrak{p}$ is a minimal prime ideal in R; and if $\mathfrak{p}$ is any minimal prime ideal in R, then the integral closure of $R_{\mathfrak{p}}$ in K' is the intersection of all the $R'_{\mathfrak{p}'}$ such that $\mathfrak{p}' \cap R = \mathfrak{p}$. Since assumptions A and B remain valid if we replace R by $R_{\mathfrak{p}}$ ($\mathfrak{p}$-minimal in R) and R' by the integral closure of $R_{\mathfrak{p}}$ in K' we may assume that *R is a regular local ring, of dimension* 1. In this case, R' is a Dedekind domain having only a finite number of prime ideals $\neq (0)$, all lying over the maximal ideal $\mathfrak{p}$ of R.

Let $\mathfrak{p}'$ be any of the prime ideals of R', different from (0). We wish to show that $D'R'_{\mathfrak{p}'} \subset R'_{\mathfrak{p}'}$.

The case in which $\mathfrak{p}'$ is unramified over R is trivial. Namely, we have in this case: $[F^*: F] = [S'/\mathfrak{M}': S/\mathfrak{M}]_s = [S/\mathfrak{M}': S/\mathfrak{M}] =$ (say) n, and $S'/\mathfrak{M}' = R'/\mathfrak{p}'$. If ξ is any element of R' whose $\mathfrak{M}'$-residue is a primitive element of $S'/\mathfrak{M}'$ over $S/\mathfrak{M}$, then it is immediate that $S' = S[\xi]$ (since $S'\mathfrak{p}$ is the maximal ideal of S'). The derivation D has a unique extension to S, which we shall continue to denote by D, and D has a unique extension to the field of quotients F' of S' (since ξ is separable algebraic over F), which we shall continue to denote by D'. If $f(X)$ is the minimal polynomial of ξ over F, then $f(X) \in S[X]$ and $f'(\xi) \notin \mathfrak{M}'$. Since $f'(\xi)D'\xi - f^D(\xi) = 0$, where f^D is obtained by applying D to the coefficients of f, it follows that $D'\xi \in S'$. Thus D' is regular on S'. The assertion that $D'R'_{\mathfrak{p}'} \subset R'_{\mathfrak{p}'}$ now follows from the fact that $S' \cap K' = R'_{\mathfrak{p}'}$ and that $D'K' \subset K'$ (if $a, b \in R'_{\mathfrak{p}'}$ then $S'b \cap R'_{\mathfrak{p}'} = R'_{\mathfrak{p}'} \cdot b$, and hence b divides a in S' if and only if b divides a in $R'_{\mathfrak{p}'}$).

Assume now that $\mathfrak{p}'$ is ramified over R (hence tamely ramified). Again, $S^*/\mathfrak{M}^*$ is a finite separable algebraic extension of $S/\mathfrak{M}$. Upon replacing R by the integral closure of R in the inertia field of $\mathfrak{M}^*$, and using the unramified case already settled above, we may assume that $S^*/\mathfrak{M}^* = S/\mathfrak{M}$. In this case we have $S^*\mathfrak{M} = \mathfrak{M}^{*e}$, $e \not\equiv 0 \pmod p$, and $[F^*: F] = e$. Let x be as in assumption B. We have then $Sx = \mathfrak{M}$ and $S^*x = \mathfrak{M}^{*e}$. Since the Galois group of F^*/F is abelian, the proof can be reduced to the case of a cyclic extension F^*/F, of degree e. The adjunction to F and F^* of a primitive e-th root of unity does not destroy the validity of assumptions A and B. We may therefore assume that F contains a primitive e-th root of unity. Since S and S^* are complete rings, it is well-known then that S^* is of the form $S[t]$, where t is an element of S^* such that $t^e = ax$, a—a unit in S. We have $et^{e-1}D't = xDa = \frac{t^eDa}{a}$, i.e., $D't = \frac{Da}{ea} \cdot t \in S^*$. Hence $D'S^* \subset S^*$, and so also $D'R'_{\mathfrak{p}'} \subset R'_{\mathfrak{p}}$. This completes the proof.

We now generalize Theorem 2 as follows:

Let R be as above, and let R' be an overring of R which is integral over R and is a finite R-module (whence R' is a semi-local ring). We assume that

(a) *R' is integrally closed in its total quotient ring;*

(b) *R' has no nilpotent elements.*

As a consequence of (a) and (b), R' is a direct sum of local, integrally closed domains:

$$R' = R'_1 \oplus R'_2 \oplus \cdots \oplus R'_h,$$

where, if $1 = e_1 + e_2 + \cdots + e_h$ is the decomposition of 1 into idempotents e_i (with $e_ie_j = 0$, if $i \neq j$), then $R'_i = R'e_i$. Let $R_i = Re_i$, and let K'_i be the field of quotients of R'_i. Then R'_i is the integral closure of R_i in K'_i and is a finite R_i-module. Furthermore, K'_i is a finite algebraic extension of the field Ke_i. We assume furthermore that

(c) *K'_i is a separable extension of Ke_i $(i = 1, 2, \cdots, h)$.*

THEOREM 3. *Let D be a derivation of K, regular on R, and let D' be the unique extension of D to the total quotient ring K' $(= K'_1 \oplus K'_2 \oplus \cdots \oplus K'_h)$ of R' (the uniqueness of D' follows from* (c)*). Then under the same assumption A and B of Theorem* 1, *we have $D'R' \subset R'$.*

Proof. We have the decomposition of D' into the "direct" sum of derivations e_iD' of the K'_i:

$$D' = e_1D' \oplus e_2D' \oplus \cdots \oplus e_hD'.$$

If we set $D'_i = e_iD'$ and $D_i = e_iD$, where D_i is intended as a derivation of e_iK $(D_i(e_iu) = e_iD_iu)$, then D'_i is the unique extension of D_i to K'_i. Applying Theorem 2 to Re_i, $R'e_i$, we see at once that D'_i is regular on $R'e_i$, and this implies that D' is regular on R'.

The following theorem is a special case of Theorem 3:

THEOREM 4. *Let $R = k[[x_1, x_2, \cdots, x_r]]$ be a power series ring, over a field k, let K be the field of quotients of R and let.*

$$K' = K'_1 \oplus K'_2 \oplus \cdots \oplus K'_h \supset K$$

be a direct sum of fields $K'_i = K'e_i$, with K'_i a finite separable algebraic extension of Ke_i. Let R' be the integral closure of R in K'. Assume the following:

A. *Every prime ideal of R', of hight* 1, *is tamely ramified over R.*

B. *If Δ is a discriminant of a basis of K'/K consisting of elements of R', then Δ—up to a unit factor in R—is a power series which is independent of x_1.*

Under these assumptions, the derivation $\dfrac{\partial}{\partial x_1}$ of K' is regular on R'.

For the proof it is only necessary to observe that if an ideal $\mathfrak{p}'$ of R', of hight 1, is ramified over R, then the principal ideal $\mathfrak{p} = \mathfrak{p}' \cap R$ is generated by some irreducible factor ξ of Δ. We have then $\xi \notin \mathfrak{p}^{(2)}$ and $\frac{\partial \xi}{\partial x_1} = 0$, and thus assumption B of Theorem 3 is satisfied.

COROLLARY. *Let* $R = k[[x_1, x_2, \cdots, x_r]]$ *(as in Theorem 4), let* $f(Z)$ *be a monic separable polynomial in* $R[Z]$, *free from multiple factors, let* K' *be the total ring of quotients of the ring* $R[Z]/f(Z)$, *and let* R' *be the integral closure of* R *in* K'. *Assume that condition A of Theorem 4 is satisfied. Assume furthermore that the discriminant* Δ_0 *of* $f(Z)$ $(\Delta_0 \in R)$ *is of the form* ϵh, *where* ϵ *is a unit in* R *and* h *is a power series independent of* x_1. *Then the derivation* $\frac{\partial}{\partial x_1}$ *is regular on* R'.

Obvious consequence of Theorem 4, since the discriminant Δ of K'/K is a divisor of Δ_0 in R.

We have in mind a certain application of Theorem 4 to complete local rings. For that application we need a lemma:

LEMMA 4. *Let* $(\mathfrak{o}, \mathfrak{m})$ *be a complete semi-local ring of characteristic zero (with* $\mathfrak{m}$ *denoting the intersection of the maximal ideals of* $\mathfrak{o}$*) and let* D *be a derivation of* $\mathfrak{o}$ *with values in* $\mathfrak{o}$. *Assume that there exists an element* x *in* $\mathfrak{m}$ *of* $\mathfrak{o}$ *such that* Dx *is a unit in* $\mathfrak{o}$. *Then* $\mathfrak{o}$ *contains a ring* $\mathfrak{o}_1$ *of representatives of the (complete) local ring* $\mathfrak{o}/\mathfrak{o}x$, *having the following properties:* (a) D *is zero on* $\mathfrak{o}_1$; (b) x *is analytically independent on* $\mathfrak{o}_1$; (c) $\mathfrak{o}$ *is the power series ring* $\mathfrak{o}_1[[x]]$. [*It follows that* x *is not a zero advisor of* $\mathfrak{o}$, *and hence* $\dim \mathfrak{o}_1 = \dim \mathfrak{o} - 1$.]

Proof. Without loss of generality we may assume that $Dx = 1$ (replace D by $\frac{1}{Dx} \cdot D$). We consider the operator

$$e^{-xD} = I - xD + \frac{x^2}{2!} D^2 - \frac{x^3}{3!} D^3 + \cdots,$$

where I is the identity map of $\mathfrak{o}$ and $D^{(n)}$ denotes the n-th iterate of the derivation D. It is immediate that: (1) e^{-xD} is an endomorphism of $\mathfrak{o}$; (2) if $\mathfrak{o}_1 = \operatorname{Im} e^{-xD}$ then D is zero on $\mathfrak{o}$; (3) the kernel of e^{-xD} is the principal ideal $\mathfrak{o}x$ (it is obvious that kernel $\subset \mathfrak{o}x$; the opposite inclusion follows from $e^{-xD}(x) = 0$). From (2) follows that: (4) the restriction of e^{-xD} to $\mathfrak{o}_1$ is the identity map. From (1), (3) and (4) it follows that $\mathfrak{o}_1$ is a ring of representatives of $\mathfrak{o}/\mathfrak{o}x$. From $\mathfrak{o}_1 \cap \mathfrak{o}x = (0)$ all the remaining assertions of the lemma follow at once.

COROLLARY. *The notations and assumption being as in Theorem* 4 *(or* *as in its Corollary), assume furthermore that k is of characteristic zero. Then R' is a power series ring in x_1, with coefficients in a subring R'_1 of R' such that R'_1 is a ring of representatives of $R'/R'x_1$ and such that $\frac{\partial}{\partial x_1}$ is zero on R'_1.*

There is one essential complement to this corollary in the case dealt with in the corollary to Theorem 4. We have namely the following:

THEOREM 5. *The notations and the assumptions being as in the Corollary to Theorem 4, assume furthermore that k is of characteristic zero. Let $A' = R[z]$ be the subring $R[Z]/(f(Z))$ of R' (here z is the f-residue of Z), let $R_1 = k[[x_2, x_3, \cdots, x_r]]$, $\zeta = c^{-x_1 D}(z)$ (where $D = \frac{\partial}{\partial x_1}$) and $A'_1 = R_1[\zeta]$ $(= e^{-x_1 D}(A'))$. Then:* (a) *the ring R'_1 $(= e^{-x_1 D}R')$ is the integral closure of A'_1 (in the total quotient ring of A'_1)*; (b) *if $f_0(Z)$ is the polynomial obtained from $f(Z)$ by reduction module x_1, then $f_0(Z)$ is the minimal polynomial of ζ over $k\{\{x_2, x_3, \cdots, x_r\}\}$.*

Proof. (a) Since R' is integral over A' and e^{-xD_1} is an endomorphism of R', R'_1 is integral over A'_1. From the fact that x_1 is analytically independent over R'_1 and that $R' = R'_1[[x_1]]$, follows at once that the total quotient ring K'_1 of R'_1 is a subring of the total quotient ring K' of R' and that $R' \cap K'_1 = R'_1$. Hence, since R' is integrally closed in K', it follows that R'_1 is integrally closed in K'_1. To complete the proof of part (a) it remains only to show that K'_1 is also the total quotient ring of A'_1. This, however, will follow once (b) is proved. In fact, (b) implies that the total quotient ring of A'_1, as a vector space over the field $k\{\{x_2, x_3, \cdots, x_r\}\}$, has dimension n, where n is the degree of f. On the other hand, also K' has dimension n over the field $k\{\{x_1, x_2, \cdots, x_r\}\}$. This implies, in particular, that any $n+1$ element of R'_1 are linearly dependent over the ring $k[[x_1, x_2, \cdots, x_r]]$. Since x_1 is analytically independent over R'_1 (and since $k[[x_2, \cdots, x_r]] \subset R'_1$), it follows that any $n+1$ elements of R'_1 are linearly dependent over the ring $k[[x_2, x_3, \cdots, x_r]]$. Hence $\dim K'_1/k\{\{x_2, x_3, \cdots, x_r\}\} \leqq n$, and since we have just seen that the total quotient rings of A'_1 has exactly dimension n over $k\{\{x_2, x_3, \cdots, x_r\}\}$, it follows that this total quotient ring coincides with K'_1.

(b) Clearly, $f_0(\zeta) = 0$, and hence the minimal polynomial $\phi(Z)$ of ζ, over $k\{\{x_2, x_3, \cdots, x_r\}\}$, divides $f_0(Z)$. On the other hand, we have $0 = \phi(\zeta) = e^{-x_1 D}(\phi(z))$, whence $\phi(z) \in \operatorname{Ker} e^{-x_1 D}$, i.e., $\phi(z) \in R' \cdot x_1$. Now, if Δ_0 is the Z-discriminant of $f(Z)$, then $\Delta_0 R' \subset A'$. Hence $\Delta_0 \phi(z) = x_1 g(z)$, where $g(Z) \in k[[x_1, x_2, \cdots, x_r]][Z]$, and therefore we have an identity $\Delta_0 \phi(Z) - x_1 g(Z) = A(Z) f(Z)$, where again, $A(Z) \in k[[x_1, x_2, \cdots, x_r]][Z]$.

Setting $x_1 = 0$ in this identity, and observing $\Delta_0(0, x_2, \cdots, x_r) \neq 0$, we see that $\phi(Z)$ is divisible by $f_0(Z)$. Hence $\phi(Z) = f_0(Z)$, as asserted.

THEOREM 6. *The assumptions and notations being as in Theorem 5, let $f(Z) = \phi_1(Z)\phi_2(Z), \cdots, \phi_h(Z)$ be the factorization of f in irreducible factors in $R[Z]$, and let $\phi_{i0}(Z)$ $(\in R_1[Z])$ be the polynomial obtained from ϕ_j by reduction modulo x_1. Then:*

(a) *$f_0(Z) = \phi_{10}(Z)\phi_{20}(Z) \cdots \phi_{h0}(Z)$ is a factorization of $f_0(Z)$ in irreducible factors in $R_1[Z]$.*

(b) *If k is an algebraically closed field (always of characteristic zero) and if $\bar{F}_1$ is an algebraic closure of $k\{\{x_1\}\}$ then the $\phi_i(Z)$ are also irreducible in $\bar{F}_1\{\{x_2, x_3, \cdots, x_r\}\}[Z]$.*

Proof. (a) The integer h is the number of direct summands of the total quotient ring K' of $R[z]$. If $1 = e_1 + e_2 + \cdots + e_h$ is the decomposition of 1 into idempotents, then from $e_i^2 = e_i$ follows $2e_iDe_i = De_i$, whence $De_i = 0$ and $e^{-x_1De_i} = e_i$, showing that the e_i belong to the total quotient ring K'_1 of A'_1 (we use here the notations of the proof of Theorem 5). Hence also K'_1 is a direct sum of h fields, and this implies that $f_0(Z)$ is a product of h distinct factors in $R_1[Z]$. This proves (a).

(b) Let $F = k\{\{x_1, x_2, \cdots, x_r\}\}$. We have $K' = F[z]$ (where z is the f-residue Z) $=$ total quotient ring of the integral closure R' of R. Since $R' = R'_1[[x_1]]$ and $z \in R'$, it follows at once that $K' = F[\zeta]$ (note that the total quotient ring K'_1 of R'_1 is $k\{\{x_2, x_3, \cdots, x_r\}\}[\zeta]$). We now pass to any of the h direct summands $K'e_i$ of K'. We set $y = ze_i$, $\eta = \zeta e_i$ and (for simplicity) we identify $F \cdot e_i$ with F. Then $F[y] = F[Z]/(\phi_i(Z))$, $k\{\{x_2, x_3, \cdots, x_r\}\}[\eta] = k\{\{x_2, x_3, \cdots, x_r\}\}[Z]/(\phi_{i0}(Z))$, and we conclude at once that $F[y] = F[\eta]$. Let $\bar{F}$ denote the field $\bar{F}\{\{x_2, x_3, \cdots, x_r\}\}$. Our claim that the polynomial $\phi_i(Z)$ is irreducible over $\bar{F}$ is equivalent to the claim that the tensor product $F[y] \otimes_F \bar{F}$ is an integral domain. Since $F[y] = F(\eta)$, this claim is equivalent to asserting that the polynomial $\phi_{i0}(Z)$ is irreducible over $\bar{F}$. Now, the coefficients of ϕ_{i0} are in the field $k\{\{x_2, x_3, \cdots, x_r\}\}$, and ϕ_{i0} is irreducible over that field (by part (a) of the theorem). The irreducibility of ϕ_{i0} over $\bar{F}$ follows now from that fact that $k\{\{x_2, x_3, \cdots, x_r\}\}$ *is maximally algebraic in* $\bar{F}$.*

* *By a simple induction, it is sufficient to prove the following: if k and K are fields, $k \subset K$, and k is maximally algebraic in K, then the power series field* $k\{\{t\}\}$ *is maximally algebraic in $K\{\{t\}\}$. Proof.* The natural valuation V of $K\{\{t\}\}$ (with residue field K) is the extension of the natural valuation v of $k\{\{t\}\}$ (with residue field k).

6. Specializations of algebroid curves and equivalence. If $f(x,y)$ is a power series in x, y, with coefficients in a field k, and such that $f(0,0)=0$, and if this power series is free from multiple factors and is *regular in* y, then, by the Weierstrass preparation theorem, the total quotient ring K' of $k[[x,y]]/(f)$ is a direct sum of fields which are finite algebraic extensions of the field $k\{\{x\}\}$. If n is the dimension of K', regarded as a vector space over $k\{\{x\}\}$, then $1, y, y^2, \cdots, y^{n-1}$ (or—more precisely—their f-residues) form a basis of K'; we shall denote by $\Delta^y f$ the discriminant of that basis, and we shall refer to $\Delta^y f$ as the *y-discriminant of f*. The discriminant $\Delta^y f$ is a power series in x, hence up to a unit factor—is simply a power of x. If, by the Weierstrass preparation theorem, we write $f(x,y)=\epsilon(x,y)\phi(x,y)$, where $\epsilon(x,y)$ is a unit in $k[[x,y]]$ and $\phi(x,y)$ is a polynomial in y, of degree n, with coefficients in $k[[x]]$, then $\Delta^y f$ is simply the y-discriminant of the polynomial ϕ.

Let now $f(x,y;t)$ be a power series in x, y, t, with coefficients in an algebraically closed field k, *of characteristic zero.* We assume that $f(0,0;t)$ is identically zero and the the power series f is regular in y. We denote by $\bar{F}_t$ the algebraic closure of the quotient field $k\{\{t\}\}$ of $k[[t]]$, and we regard f as a power series in x, y, with coefficients in $\bar{F}_t$. We denote by $\Delta(x,t)$ the y-discriminant of $f(\Delta(x,t)\in k[[x,t]])$ and we assume that $\Delta(x,t)$ is not identically zero. In that case f has no multiple factors (in $\bar{F}_t[[x,y]]$), and we can interpret the equation $f=0$ as defining an algebroid curve C^t over the algebraically closed field $\bar{F}_t$, with origin at $P\colon x=y=0$.

Since $f(x,y;t)$ has been assumed to be regular in $y, f(x,y;0)$ is not identically zero. We set $f_0(x,y)=f(x,y;0)$, so that $f_0(x,y)$ is also regular in y. We shall also assume that the discriminant $\Delta(x,t)$ is not divisible by t. In that case, the y-discriminant $\Delta(x,0)$ of f_0 is not identically zero, f_0 has no multiple factors (in $k[[x,y]]$), and the equation $f_0(x,y)=0$ defines an algebroid curve C^0 over k (and hence, *a fortiori*, also over $\bar{F}_t$), with the same origin P as C^t. We shall say that C^0 *is a specialization of* C^t *over* $t\to 0$.

Our principal object in this section is the proof of the following theorem.

THEOREM 7. *Let k be of characteristic zero (and algebraically closed), and let C^0 be a specialization of C^t over $t\to 0$. Then:*

Let Σ be a finite algebraic extension of $k\{\{t\}\}$ in $K\{\{t\}\}$, and let w be the restriction of V to Σ. Then w is an extension of v, with the same value group as v, and the residue field of w is an algebraic extension of k. Since this residue field is contained in K (= residue field of V) and since k is maximally algebraic in K, it follows that w and v have the same residue field. Since $k\{\{t\}\}$ is a complete field (with respect to the valuation v), it follows that $\Sigma = k\{\{t\}\}$.

18

(a) *A sufficient condition that C^0 and C^t be equivalent (in the sense of § 3) is that the y-discriminant $\Delta(x,t)$ of $f(x,y;t)$ be of the form $\epsilon(x,t)x^N$ where $\epsilon(x,t)$ is a unit in $k[[x,t]]$.*

(b) *If the line $x=0$ is not a tangent of C^0, then the above condition on $\Delta(x,t)$ is also necessary for the equivalence of C^t and C^0.*

Proof. We shall need two lemmas which we shall now state.. In order not to interrupt the proof of the theorem, we shall assume for the moment these lemmas and will prove them in the next section.

LEMMA 5. *Let $C\colon f(x,y)=0$ and $D\colon g(x,y)=0$ be two algebroid curves, over an algebraically closed ground field k of characteristic zero (with common origin $P\colon x=y=0$). We assume that the power series f and g are regular in both x and y, that C and D have the same number h of irreducible branches, and that the line $x=0$ is neither a tangent of C nor a tangent of D. We assume that $\Delta^x f=\Delta^x g=y^M$ (apart from unit factors in $k[[y]]$). We furthermore assume that there exists a pairing $\pi\colon C\to D$ between the branches $\gamma_1,\gamma_2,\cdots,\gamma_h$ of C and the branches $\delta_1,\delta_2,\cdots,\delta_h$ of D having the following properties:*

a) *If $\pi(\gamma_i)=\delta_i$ then $m_P(\gamma_i)=m_P(\delta_i)$, and the intersection numbers (l,γ_i), $(l,\delta_i)$are equal $(i=1,2,\cdots,h)$; here l denotes the line $y=0$.*

b) *If $\phi_i(x,y)=0$ is an irreducible equation of γ_i and $\psi_i(x,y)$ is an irreducible equation of δ_i $(=\pi(\gamma_i))$, then $\Delta^x\phi_i=\Delta^x\psi_i=y^{M_i}$ (apart from unit factors in $k[[y]]$).*

Then $\Delta^y f=\Delta^y g$ (apart from unit factors in $k[[x]]$).

Note that the assumption $(l,\gamma_i)=(l,\delta_i)$ in a) can also be expressed as follows: if we assume—as we may—that the ϕ_i and ψ_i in b) are monic polynomial in x, then for each $i=1,2,\cdots,h$, the two polynomials ϕ_i and ψ_i are of the same degree.

LEMMA 6. *Let $C\colon f(x,y)=0$ and $D\colon g(x,y)=0$ be two algebroid curves, over an algebraically closed field k of characterictic zero (with common origin $P\colon x=y=0$). We assume that the power series f and g are regular in y and that there exists an equivalence $\pi\colon C \underset{\equiv}{\to} D$ having the following property: if $\pi(\gamma_i)=\delta_i$ $(i=1,2,\cdots,h)$ then $(m,\gamma_i)=(m,\delta_i)$; here m denotes the line $x=0$. Then $\Delta^y f=\Delta^y g$ (apart from unit factors in $k[[x]]$).*

Note the following immediate Corollary of Lemma 6:

COROLLARY. *If two algebroid curves* $C: f(x,y)=0$, $D: g(x,y)=0$ *(over an algebraically closed ground field* k, *of characteristic zero) are equivalent and if the line* $x=0$ *is neither a tangent of* C *nor a tangent of* D *(whence* f *and* g *are certainly regular in* y, *i.e., neither* f *nor* g *is divisible by* x*), then* $\Delta^y f=\Delta^y g$ *(apart from unit factors in* $k[[x]]$*).*

For if $\pi: C \Longrightarrow D$ is any equivalence between C and D and $\pi(\gamma_i)=\delta_i$, then $(m,\gamma_i)=m_P(\gamma_i)=m_P(\delta_i)=(m,\delta_i)$ (since the line $m: x=0$ is neither a tangent of γ_i nor a tangent of δ_i).

We now proceed with the proof of the theorem.

(a) Our basic assumption in this part of the theorem was that the y-discriminant $\Delta(x,t)$ of $f(x,y;t)$ is of the form $\epsilon(x,t)x^N$, with $\epsilon(x,t)$ a unit in $k[[x,t]]$. This was also the basic assumption made in the corollary to Theorem 4 (with $r=2$, $x_1=t$, $x_2=x$ and $h=x^N$). Thus Theorems 5 and 6 are applicable. Theorem 6 tells us that—upon assuming that $f(x,y;t)$ is a polynomial in y—the factorization $f=\phi_1\phi_2\cdots\phi_h$ of f into irreducible factors in $k[[t,x]][y]$ is also a factorization of f in irreducible factors in $\bar{F}_t[[x]][y]$. Thus the *irreducible branches* γ_i of C are given by $\gamma_i: \phi_i(x,y;t)=0$. Theorem 6 also tells us that if $\phi_{i0}(x,y)=\phi_i(x,y;0)$, then the irreducible branches δ_i of C^0 are given by $\delta_i: \phi_{i0}(x,y)=0$. This yields then a natural *specialization pairing* $\pi: C^t\rightarrow C^0$ between the branches of C^t and the branches of C^0: $\pi(\gamma_i)=\delta_i$. Upon replacing y by $y+cx$, where c is a suitable element of k (this substitution has no effect on the discriminant $\Delta^y f$), we may assume that the line $y=0$ is not tangent to C^0 (whence f_0, and hence also f, is regular in x). We shall now show that all the assumptions of Lemma 5 are satisfied if C and D are replaced by C^t and C^0 respectively and if the roles of x and y in Lemma 5 are interchanged.

Let $\gamma: \phi(x,y;t)=0$ be any of the irreducible branches of C^t and let $\delta=\pi(\gamma): \psi(x,y)=0$, where $\psi(x,y)=\phi(x,y;0)$. We have of course $\Delta^y\phi=\Delta^y\psi$ (apart from unit factors in $k[[x,t]]$), since $\Delta^y\phi$, as a factor of $\Delta^y f$, is still of the form $\epsilon_1(x,t)x^M$, with ϵ_1 a unit in $k[[x,t]]$, and so $\Delta^y\psi=\epsilon_1(x,0)x^M$. We pass to the local domain $k[[x,y]]/(\psi)$ of δ and we denote by η^0 the ψ residue of y and by $k(\delta)$ the field of quotients $k\{\{x\}\}[\eta^0]$ of that local ring. Similarly, we denote by η the ϕ-residue of y and by $\bar{F}^t(\gamma)$ the field of quotients $\bar{F}^t\{\{x\}\}[\eta]$ of the local ring $\bar{F}^t[[x,y]]/(\phi)$ of γ. From the proof of part (b) of Theorem 6 follows that $\bar{F}^t(\gamma)$ is obtained from $k(\delta)$ by the ground field extension $k\rightarrow\bar{F}^t$, i.e.,

$$\bar{F}^t\{\{x\}\}[\eta]=k\{\{x\}\}[\eta^0]\otimes_{k\{\{x\}\}}\bar{F}^t\{\{x\}\}.$$

If $[k(\delta):k\{\{x\}\}]=n$, then $k(\delta)=k\{\{x^{1/n}\}\}$ and $\bar{F}^t(\gamma)=\bar{F}^t\{\{x^{1/n}\}\}$. Let us denote by v_0 and v the natural valuations of the complete field $k(\delta)$ and $\bar{F}^t(\gamma)$ respectively. Setting $\xi=x^{1/n}$, we have $v_0(\xi)=v(\xi)=1$, and v is the extension of v_0.

We next show that

$$v(\eta)=v_0(\eta^0). \tag{6}$$

We know from Theorem 5 that η is given by a power series of this form

$$\eta=\eta^0+u_1t+u_2t^2+\cdots, \tag{7}$$

where the u_i are elements of $k(\delta)$ which are *integral* over $k[[x]]$, i.e., the u_i are elements of $k[[\xi]]$. Taking conjugates $\eta_1,\eta_2,\cdots,\eta_n$ of η (over $\bar{F}^t\{\{x\}\}$; this amounts to taking conjugates of $\eta^0,u_1,u_2,\cdots$ over $k\{\{x\}\}$), we observe that, for $\alpha\neq\beta$, $\eta_\alpha-\eta_\beta$ divides $\Delta^y f$ in $\bar{F}^t[[\xi]]$. So $\eta_\alpha-\eta_\beta$ is—apart from a unit in $\bar{F}^t[[\xi]]$—a power of ξ. Since $\eta_\alpha{}^0-\eta_\beta{}^0$ is a non-unit in $k[[\xi]]$, it follows at once that for each $i\geqq 1$, $u_{i\alpha}-u_{i\beta}$ must be divisible by $\eta_\alpha{}^0-\eta_\beta{}^0$ in $k[[\xi]]$ (here $u_{i1},u_{i2},\cdots,u_{in}$ are the conjugates of u_i over $k\{\{x\}\}$). This implies that $v_0(u_i)\geqq\min\{n,v_0(\eta^0)\}$. (Recall that we have assumed that $f(0,0;t)$ is identically zero; this implies that all the u_i are non-units in $k[[\xi]]$. Let $u_i=a_{i0}\xi^{\nu_i}+a_{i,1}\xi^{\nu_i+1}+\cdots$, $\nu_i>0$, $a_{i0}\neq 0$ $(a_{ij}\in k)$. If $\nu_i<n$, then there exists a conjugate u_{i2} such that $v_0(u_{i1}-u_{i2})=\nu_i$. So, in this case, $\nu_i\geqq v_0(\eta_1{}^0-\eta_2{}^0)\geqq v_0(\eta^0)$.). Since the line $y=0$ is not tangent to C, we have $v_0(\eta^0)\leqq v_0(x)=n$. Hence $v_0(u_i)\geqq v_0(\eta^0)$, and so $v(\eta)=v_0(\eta^0)$, as asserted.

On the other hand, we have $v_0(x)=v(x)=n$. Since

$$m_P(\gamma)=\min\cdot\{v(x),v(\eta)\}\text{ and }m_P(\delta)=\min\{v_0(x),v_0(\eta^0)\},$$

it follows from (6) that

$$m_P(\gamma)=m_P(\delta). \tag{8}$$

This holds for any two branches γ, δ such that $\delta=\pi(\gamma)$. This is part of condition a) of Lemma 5. Note also that equality (6) implies that the line $y=0$ is not tangent to any branch γ of C^t (since that line was assumed not to be a tangent of C^0). This is one of the conditions imposed in Lemma 5 (with x and y interchanged).

Note also that since $v(x)=v_0(x)=n$, we have $(l,\gamma)=(l,\delta)$, where now l is the line $x=0$. Thus both parts of condition (a) of Lemma 5 are satisfied.

It has already been pointed out above that $\Delta^y\phi=\Delta^y\psi$ (apart from unit factors in $k[[x,t]]$). Thus also condition b) of Lemma 5 is satisfied.

We can therefore conclude that $\Delta^x f = \Delta^x g$ (apart from unit factors in $k[[y,t]]$).

Since $y = 0$ is not tangent to either C^t or C^0, the above conclusion tells us that in order to prove that C^t and C^0 are equivalent, *we may add to our original assumption* $\Delta^y f = \Delta^y g$ *the assumption that the line* $x = 0$ *is not a tangent of* C^0 (and therefore also not a tangent of C^t, in view of the fact that we have $v(x) = v_0(x)$ for any two corresponding branches γ and $\delta = \pi(\gamma)$).

The rest of the proof is fairly simple and will consist in showing, by induction on the exponent M of x in the discriminant $\Delta^y f$, that the *natural specialization pairing* $\pi\colon C^t \to C^0$ *introduced above is an equivalence* (if $M = 0$, then both C^t and C^0 are regular curves).

First of all we show that π *is tangentially stable.* If we set $\Delta^y \phi_i = \Delta^y \phi_{i0} = x^{M_i}$ (apart from unit factors), then the expression of the discriminant as the square of the product of the differences of the roots of the polynomial and—on the other hand—the definition of intersection multiplicity given in § 4, show that $M = \sum_{i=1}^{h} M_i + \sum_{i<j} (\gamma_i, \gamma_j)^2$, where $\Delta^y f = x^M$. Similarly, $M = \sum_{i=1}^{h} M_i + \sum_{i<j} (\delta_i, \delta_j)^2$. It is also clear, since δ_i is the specialization of γ_i over $t \to 0$, that $(\gamma_i, \gamma_j) \leqq (\delta_i, \delta_j)$. Hence it follows that $(\gamma_i, \gamma_j) = (\delta_i, \delta_j)$, for all $i, j = 1, 2, \cdots, h, i \neq j$. This shows that π is tangentially stable, since γ_i and γ_j (respectively δ_i and δ_j) have the same tangent if and only if $(\gamma_i, \gamma_j) > m_P(\gamma_i) m_P(\gamma_j)$ (respectively, if and only if $(\delta_i, \delta_j) > m_P(\delta_i) m_P(\delta_j)$), and we have just shown that $m_P(\gamma_i) = m_P(\delta_i)$ for all $i = 1, 2, \cdots, h$.

We now apply a quadratic transformation T, with center P. Since the line $x = 0$ is not tangent of either C^t or C^0, we may take

$$x' = x, \qquad y' = y/x$$

as the equations of T. Let $C_\nu'^t$, $C_\nu'^0$ be the connected components of $T(C^t)$ and $T(C^0)$ respectively. Let $(0, \alpha_\nu)$ and $(0, \alpha_{\nu 0})$ be the origins of $C_\nu'^t$ and $C_\nu'^0$ respectively. The $\alpha_{\nu 0}$ are, of course, elements of the algebraically closed field k. As to the α_ν, they are—*a priori*-elements of $\bar{F}^t$. However—and this is an important point—*the* α_ν *are actually elements of the power series ring* $k[[t]]$. In fact, let $\gamma_i\colon \phi_i(x, y; t) = 0$ be any of the irreducible branches of the tangential component C_ν^t of C^t, let $s_i = m_P(\gamma_i)$ and let $\phi_{i,s_i}(x, y; t)$ be the leading form of $\phi_i(x, y; t)$. .This form has coefficients in $k[[t]]$, and the coefficient of y^{s_i} is a unit ϵ_i in $k[[t]]$. Since $\phi_{i,s_i} = \epsilon_i (y - \alpha_\nu x)^{s_i}$, it follows that $\alpha_\nu \in k[[t]]$, as asserted.

It follows that if we set $y_\nu' = y' - \alpha_\nu$, then the equation of $C_\nu'^t$ will be of the form

$$f_\nu'(x, y_\nu'; t) = 0,$$

where $f_\nu'(x, y_\nu'; t)$ is an element of $k[[x,t]][y_\nu']$, regular in y_ν'. The equation of $C_\nu'^0$ will be

$$f_{\nu 0}'(x, y_{\nu 0}') = 0. \qquad y_{\nu 0}' = y' - \alpha_\nu(0),$$

where $f_{\nu 0}'(x, Y) = f_\nu'(x, Y; 0)$, and $C_\nu'^0$ is a specialization of $C_\nu'^t$ over $t \to 0$. The product of the discriminants $\Delta^{y'\nu} f_\nu'$ is a divisor of the discriminant $\Delta^y f$, since, if we set

$$f(x, xy'; t) = x^s f'(x, y'; t),$$

where $s = m_P(C^t)$, then $\Delta^y f = x^{s(s-1)} \Delta^{y'} f'$. Thus, the discriminant $\Delta^{y'\nu} f_\nu'$ is a power of x (apart from a unit in $k[[x,t]]$. The pairing $\pi_\nu' : C_\nu'^t \to C_\nu'^0$ between the branches of $C_\nu'^t$ and the branches of $C_\nu'^0$, induced by our original specialization pairing π, is obviously still a pairing by specialization $t \to 0$. It follows, by our induction hypothesis, that each π_ν' is an equivalence. This proves that π is an equivalence.

(b) This part of Theorem 7 is an immediate consequence of the corollary to Lemma 6. For, let $\Delta^y f = A(x,t) x^N$, with $A(0,t) \neq 0$. Then $\Delta^y f_0 = A(x,0) x^N$. Since the line $x = 0$ is not tangent to C^0, it is certainly not tangent to C^t (the assumption that $C^t \equiv C^0$ implies that $m_P(C^t) = m_P(C^0)$, whence the leading form of f_0 is the specialization of the leading form of f, over $t \to 0$). It follows therefore, by the corollary to Lemma 6, that x^N is the highest power of x which divided $\Delta^y f$. Hence $A(0,0) \neq 0$, i.e., $A(x,t)$ is a unit in $k[[x,t]]$.

This completes the proof of Theorem 7.

7. Proofs of the Lemmas 5 and 6.

Proof of Lemma 5. We consider the equalities used already in the proof of Theorem 7:

$$M = \sum_{i=1}^{h} M_i + \sum_{i<j} (\gamma_i, \gamma_j)^2,$$

$$M = \sum M_i + \sum_{i<j} (\delta_i, \delta_j)^2.$$

They imply that

(9) $$\sum_{i<j} (\gamma_i, \gamma_j)^2 = \sum_{i<j} (\delta_i, \delta_j)^2.$$

Let now (apart from unit factors in $k[[x]]$):

$$\Delta^y \phi_i = x^{N_i}, \qquad \Delta^y f = x^N;$$

$$\Delta^y \psi_i = x^{N_i'}, \qquad \Delta^y g = x^{N'}.$$

From

$$N = \sum_{i=1}^{h} N_i + \sum_{i<j} (\gamma_i, \gamma_j)^2, \tag{10}$$

$$N' = \sum_{i=1}^{h} N_i' + \sum_{i<j} (\delta_i, \delta_j)^2 \tag{10'}$$

and from (9), it follows that in order to prove that $N = N'$ it will be sufficient to prove that $N_i = N_i'$, $i = 1, 2, \cdots, h$. Now, for fixed i, let v and v' denote the natural valuations of the fields of quotients of the local rings of γ_i and δ_i respectively, and let dx, dy and $d'x$ and $d'y$ denote the differentials of x and y on γ_i and δ_i respectively. We have

$$v(dx) = v'(d'x) \; (= m_P(\gamma_i) - 1 = m_P(\delta_i) - 1), \tag{11}$$

since the line $x = 0$ is not tangent to γ_i, nor to δ_i. We also have

$$v(dy) = v'(d'y) \; (= (l, \gamma_i) - 1 = (l, \delta_i) - 1). \tag{12}$$

Finally,

$$\begin{cases} M_i = v\left(\dfrac{\partial \phi_i}{\partial x}\right) = v'\left(\dfrac{\partial \psi_i}{\partial x}\right), \\ N_i = v\left(\dfrac{\partial \phi_i}{\partial y}\right),\; N_i' = v'\left(\dfrac{\partial \psi_i}{\partial y}\right). \end{cases} \tag{13}$$

The equality $N_i = N_i'$ follows now from (11), (12), (13) and from the relations

$$\frac{\partial \phi_i}{\partial x}\, dx + \frac{\partial \phi_i}{\partial y}\, dy = 0 \qquad (\text{on } \gamma_i),$$

$$\frac{\partial \psi_i}{\partial x}\, d'x + \frac{\partial \psi_i}{\partial y}\, d'y = 0 \qquad (\text{on } \delta_i).$$

Proof of Lemma 6. This time (9) is valid because π is an equivalence (see § 4, Lemma 2). Hence, by (10) and (10′), it is sufficient to show that $N_i = N_i'$ $(i = 1, 2, \cdots, h)$. Hence, we may assume that C and D are irreducible curves. We shall now proceed by induction on the numerical character $\sigma(C)$ $(= \sigma(D)$; see § 3).

First case. The line $m: x = 0$ is not tangent to C (and therefore also not tangent to D, since we have assumed that $(m, C) = (m, D)$). Without loss of generality, we may assume that $y = 0$ is tangent to both C and D.

Then the quadratic transformation T, with cented $P\colon x=y=0$, gives irreducible proper transforms

$$C'\colon f'(x,y')=0$$
$$D'\colon g'(x,y')=0,$$

having origins at $x=y'=0$ (here $y'=\frac{y}{x}$). We have $C'\equiv D'$, and if m' denotes the line $x=0$ in the (x,y')-plane (m' is the exceptional curve of T), then we know (see (3), § 4) that $(m',C')=(m',D')=m_P(C)$ $(=m_P(D))$. Hence, by our induction hypothesis, we have $\Delta^{y'}f'=\Delta^{y'}g'$. Since

$$\Delta^{y}f=x^{s(s-1)}\Delta^{y'}f',$$
$$\Delta^{y}g=x^{s(s-1)}\Delta^{y'}g'.$$

where $s=m_P(C)$ $(=m_P(D))$, the proof in this case is complete.

Second Case: the line $x=0$ is tangent to C (and hence also to D).

We may assume that the line $y=0$ is not tangent to C, nor tangent to D. Then by the first case we have $\Delta^{x}f=\Delta^{x}g$ (apart from unit factors in $k[[y]]$). Using the relations

$$\frac{\partial f}{\partial x}\,dx+\frac{\partial f}{\partial g}\,dy=0 \qquad \text{on } C$$
$$\frac{\partial g}{\partial x}\,dx+\frac{\partial g}{\partial y}\,dy=0 \qquad \text{on } D,$$

and relations analogous to (11), (12) and (13), with γ_i, δ_i replaced by C and D, and with x and y interchanged, we conclude that $\Delta^{y}f=\Delta^{y}g$.

HARVARD UNIVERSITY.

STUDIES IN EQUISINGULARITY II.*
EQUISINGULARITY IN CODIMENSION 1 (AND CHARACTERISTIC ZERO).

By Oscar Zariski.

Introduction. We deal with an r-dimensional algebroid hypersurface V which has a singular point at its origin O. We consider the singular locus W of V and we assume that

(A) *W has codimension* 1.

Under this assumption we introduce the concept of equisingularity of V at O, along W. For V to be equisingular at O, along W, we first of all require that

(B) *O be a simple point of W.*

When this condition is satisfied we can speak of *W-transversal sections of V*, at O (Definition 3.3). Any such section is an embedded (see Definition 2.1) algebroid scheme (in the sense of § 1), of dimension 1 or 2 (see Proposition 3.5; the case of dimension 2 may occur, but then only for "special" transversal sections). There is only one W-transversal section V_P of V at the *general* point P of W (see Note (b) at the end of § 3), and V_P is always an embedded algebroid *curve* (hence essentially a plane algebroid curve). We define equisingularity of V at O, along W, by the following condition:

(C) *There exists a W-transversal section V_0 of V, at O, such that V_0 is a curve and such that the two plane algebroid curves V_P and V_0 have equivalent singularities at P and O respectively (Definition* 4.1).

Here, equivalence of singularities of plane algebroid curves is intended in the sense defined in our paper [3].

When conditions (A), (B) and (C) are satisfied, we say that V has at its origin O a *singularity of dimensionality type* 1.

Pending further investigation of the concept of equivalence of singularities of plane algebroid curves in the case of characteristic $p \neq 0$ (see [3], Introduction), we restrict ourselves in this paper to the case of ground fields

Received February 5, 1965.
* This research was supported by grants from the National Science Foundation and the Air Force Office of Scientific Research.

which have characteristic zero (and are algebraically closed). It is proved (Corollary 5.3) that if *V has at O a singularity of dimensionality type 1, then all W-transversal section of V at O are curves and have equivalent singularities at O.*

The bulk of the paper consists of the derivation of a number of *criteria of equisingularity* in codimension 1 (or—what is the same thing—criteria for singularities of dimensionality type 1). We derive 4 such criteria, namely:

(a) Existence of equisingular local parameters (Definition 4.3 and Theorem 4.4).

(b) A Jacobian criterion (Theorem 5.1).

(c) An inductive criterion, based on the beshavior of V under a monoidal transformation centered at W (Theorem 7.4).

(d) A criterion due to Whitney and Thom (Theorem 8.1).

With these and other results proved in this paper, the theory of equisingularity in codimension 1 can now be regarded as complete (in characteristic zero).

In the complex domain, one can prove (as Whitney does; see [7], § 12; see also [5]) the existence of a "nice" fibration of V, in the neighborhood of W, a fibration induced by a fibration of the ambient affine $(r+1)$-space. In another paper of this series we shall prove a more general theorem, and for this reason we do not discuss, in this paper, the topological implication of equisingularity in the complex domain.

1. Algebroid schemes and varieties. For the purposes of this paper, by an *algebroid scheme* S we mean the spectrum of a complete noetherian equicharacteristic local ring $\mathfrak{o}$. If $k \subset \mathfrak{o}$ is any field of representatives of $\mathfrak{o}$, then we shall say that S is defined over k. The maximal ideal $\mathfrak{m}$ of $\mathfrak{o}$, as a point of the space S, will be called *the origin* of S. Since $\mathfrak{o}$ is always the homomorphic image of a formal power series ring over k, it follows that if S is an algebroid scheme then, for some integer n, we have

$$(1) \qquad S = \mathrm{Spec}(k[[X_1, X_2, \cdots, X_n]]/\mathfrak{A}),$$

where $\mathfrak{A}$ is an ideal in the power series ring $k[[X]]$ $(= k[[X_1, X_2, \cdots, X_n]])$ of n indeterminates. We denote by $\mathfrak{o}_S$ the fundamental structure sheaf of S:

$$\mathfrak{o}_{S,x} = \mathfrak{o}_x = \text{ring of quotients of } \mathfrak{o}, \text{ with respect to the prime ideal } x \text{ of } \mathfrak{o},$$

and by $\rho_y{}^x$ the canonical homomorphism of $\mathfrak{o}_x$ into $\mathfrak{o}_y$ $(x, y \in S; x \supset y$, i.e., x is a specialization of $y)$.

The *dimension* of S is defined as the Krull dimension of the local ring $\mathfrak{o}$ (= maximum of the dimensions of the prime ideals of $\mathfrak{A}$). If all the prime ideals of the zero ideal in $\mathfrak{o}$ have the same dimension r, then we say that S is unmixed, of dimension r.

An *algebroid variety* V is a reduced algebroid scheme. If $S = V$ is an alegbroid variety then the zero ideal in $\mathfrak{o}$ is an (irredundant) finite intersection if prime ideals:

$$(0) = \mathfrak{p}_1 \cap \mathfrak{p}_2 \cap \cdots \cap \mathfrak{p}_h. \tag{2}$$

Let V be an algebroid variety. The total ring of quotients K of $\mathfrak{o}$ is then a direct sum of fields:

$$K = \bigoplus_{i=1}^{h} K_i,$$

where h is the integer which occurs in (2), and we have

$$\mathfrak{p}_i = \mathfrak{o} \cap \mathfrak{P}_i,$$

where

$$\mathfrak{P}_i = \bigoplus_{j \neq i} K_j.$$

For any set (α) of distinct indices $\alpha_1, \alpha_2, \cdots, \alpha_q$ $(1 \leqq \alpha_\mu \leqq h)$ we set

$$K_{(\alpha)} = \bigoplus_{\mu=1}^{q} K_{\alpha_\mu} \quad (= \text{set of all elements } a_1 + a_2 + \cdots + a_h \text{ of } K,\ a_i \in K_i, \text{ such that } a_i = 0 \text{ if } i \notin (\alpha)),$$

and we denote by $\phi_{(\alpha)}$ the canonical surjection $K \to K_{(\alpha)}$. If (α) and (β) are two sets of indices and if $(\alpha) \supset (\beta)$, we denote by $\phi_{(\beta)}{}^{(\alpha)}$ the canonical surjection $K_{(\alpha)} \to K_{(\beta)}$. We have

$$\operatorname{Ker} \phi_{(\alpha)} = \bigoplus_{j \notin (\alpha)} K_j = \bigcap_{\mu=1}^{q} \mathfrak{P}_{\alpha_\mu};$$

$$\phi_{(\beta)} = \phi_{(\beta)}{}^{(\alpha)} \circ \phi_{(\alpha)}.$$

If $x \in V$ is a prime ideal of $\mathfrak{o}$, let $\mathfrak{p}_{\alpha_1}, \mathfrak{p}_{\alpha_2}, \cdots, \mathfrak{p}_{\alpha_q}$ be those prime ideals $\mathfrak{p}_i$ in (2) which are contained in x, let $(\alpha) = (\alpha_1, \alpha_2, \cdots, \alpha_q)$ and let

$$\mathfrak{N} = \bigcap_{\mu=1}^{q} \mathfrak{p}_{\alpha_\mu}.$$

Then $\mathfrak{N}$ is the kernel of the canonical homomorphism of $\mathfrak{o}$ into $\mathfrak{o}_x$. Since $\mathfrak{N}$ is also the kernel of $\phi_{(\alpha)} \mid \mathfrak{o}$, we have a canonical injection $\psi_x \colon \mathfrak{o}_x \to K_{(\alpha)}$.

If y is another point of V such that $x \supset y$, and if $(\beta) = (\beta_1, \beta_2, \cdots, \beta_s)$ is the set of indices such that $\mathfrak{p}_i \subset y$ if and only if $i \in (\beta)$ (whence $(\alpha) \supset (\beta)$), then one sees at once that

$$\phi_{(\beta)}{}^{(\alpha)} \circ \psi_x = \psi_y \circ \rho_y{}^x.$$

It follows that we can identify each $\mathfrak{o}_x$ with its ψ_x-image in $K_{(\alpha)}$, and when that is done then $\rho_y{}^x$ becomes identifed with $\phi_{(\beta)}{}^{(\alpha)} \mid \mathfrak{o}_x$. We assume that these identifications have been carried out. So now all the local rings $\mathfrak{o}_x$ $(x \in V)$ are subrings of K, and, for given x, the set (α) of indices, defined above, can be characterized as follows: $K_{(\alpha)}$ is the smallest subring of K which is a sum of direct summands K_i and which contains $\mathfrak{o}_x$.

If $h = 1$, i.e., if K is a field, then the algebroid variety V is *irreducible.* If $h > 1$, each of the h prime ideals $\mathfrak{p}_i$ in (2) defines an irreducible component V_i of V, and we have $V = \bigcup_{i=1}^{h} V_i$. More generally, if $(\alpha) = (\alpha_1, \alpha_2, \cdots, \alpha_q)$ is any set of distinct indices $(1 \leqq \alpha \leqq h)$ then we set $V_{(\alpha)} = \bigcup_{\mu=1}^{q} V_{\alpha_\mu}$. Then it follows at once that $\mathfrak{o}_x \subset K_{(\alpha)}$ if and only if no V_i, $i \notin (\alpha)$, contains x, and that $\mathfrak{o}_x$ can then be identified with $\mathfrak{o}_{V_{(\alpha)},x}$.

2. Embedded algebroid varieties ("hypersurfaces"). From now on we shall denote the elements (points) of an algebroid scheme S by capital Latin letters. In particular, the origin of S, i.e., the maximal ideal $\mathfrak{m}$ of $\mathfrak{o}$, will be denoted by O. We shall assume from now on that $\mathfrak{o}$ is *not* a regular ring. We denote by r the Krull dimension of $\mathfrak{o}$.

Definition 2.1. *An algebroid scheme $S = \mathrm{Spec}(\mathfrak{o})$, of dimension r, is embedded if the maximal ideal $\mathfrak{m}$ of $\mathfrak{o}$ has a basis of $r+1$ elements.*

If $\mathfrak{m}$ has a basis of $r+1$ elements, any such basis is minimal (since we have assumed that $\mathfrak{o}$ is not a regular ring). Hence S is an embedded scheme if and only if $\dim_k(\mathfrak{m}/\mathfrak{m}^2) = r+1$ ($k =$ any field of representatives of $\mathfrak{o}$). Any minimal basis of $\mathfrak{m}$ will be then called *a system of local coördinates of S at O.*

Let S be embedded, let $x_1, x_2, \cdots, x_{r+1}$ be local coördinates of S at O and let k be a field of representatives of $\mathfrak{o}$, which we fix once and for always. Then $\mathfrak{o}$ is a homomorphic image of a power series ring $k[[X_1, X_2, \cdots, X_{r+1}]]$:

$$\mathfrak{o} = k[[x_1, x_2, \cdots, x_{r+1}]] = k[[X_1, X_2, \cdots, X_{r+1}]]/\mathfrak{B},$$

where $\mathfrak{B}$ is an ideal in $k[[X]]$. Let us now assume that S is unmixed. Since $\dim S = r$, it follows at once that $\mathfrak{B}$ is a *principal ideal* $(f(X_1, X_2, \cdots, X_{r+1}))$, where $f(X)$ is a non-unit power series in $k[[X_1, X_2, \cdots, X_{r+1}]]$. We shall write symbolically

$$f(X_1, X_2, \cdots, X_{r+1}) = 0, \tag{3}$$

and we shall say that equation (3) represents an embedding of S into the affine $(r+1)$-space A_{r+1}. For a given set of local coördinates $x_1, x_2, \cdots, x_{r+1}$, the power series f is uniquely determined to within an arbitrary unit factor in $k[[X]]$. The transition to another set of local coördinates $y_1, y_2, \cdots, y_{r+1}$ will lead to another embedding of S, obtained from (3) by a formal analytic transformation of affine coördinates, biholomorphic at the origin.

The scheme S defined by (3) is a variety, if and only if f has no multiple factors.

From now on we assume that S is an unmixed embedded variety (denoted by V), of dimension r. We fix once and for always a field k of representatives in $\mathfrak{o}$.

Definition 2.2. *A set of r elements $x_1, x_2, \cdots, x_r$ of $\mathfrak{m}$ is called a system of local parameters of V at O if* (a) *$x_1, x_2, \cdots, x_r$ are parameters of $\mathfrak{o}$* (i.e., *if the ideal generated by $x_1, x_2, \cdots, x_r$ is primary for $\mathfrak{m}$*) *and* (b) *if the set $\{x_1, x_2, \cdots, x_r\}$ can be extended to a system $\{x_1, x_2, \cdots, x_r, x_{r+1}\}$ of local coördinates of V at O.*

If (3) represents an embedding of V in A_{r+1}, relative to a given system of local coördinates $x_1, x_2, \cdots, x_r, x_{r+1}$, then $\{x_1, x_2, \cdots, x_r\}$ is a system of local parameters if and only if $f(0, 0, \cdots, 0, 1) \neq 0$, i.e., if and only if the power series f is regular in X_{r+1}.

Assuming that f is regular in X_{r+1}, we write $f(0, 0, \cdots, 0, X_{r+1}) = cX_{r+1}{}^{\nu}$ + term of degree $> \nu$ $(c \in k, c \neq 0)$. Then ν is the intersection multiplicity of V with the linear space $X_1 = X_2 = \cdots = X_r = 0$; or—in intrinsic terms of local algebra—ν is the multiplicity of the (primary) ideal generated in $\mathfrak{o}$ by the local parameters $x_1, x_2, \cdots, x_r$. We have $\nu > 1$, since $\mathfrak{o}$ is not a regular ring. By the Weierstrass preparation theorem, we can choose the arbitrary unit factor in f in such a way that f becomes a monic polynomial in X_{r+1}, with coefficients in $k[[X_1, X_2, \cdots, X_r]]$:

$$f = X_{r+1}{}^{\nu} + \sum_{i=1}^{\nu} A_i(X_1, X_2, \cdots, X_r) X_{r+1}{}^{\nu-i}, \qquad A_i(0, 0, \cdots, 0) = 0. \tag{4}$$

Let s be the degree of the leading form of f:

$$f = f_s(X_1, X_2, \cdots, X_{r+1}) + f_{s+1}(X_1, X_2, \cdots, X_{r+1}) + \cdots,$$

where f_i is a homogeneous polynomial of degree i, with coefficients in k. The integer s is the multiplicity of the local ring $\mathfrak{o}$, or—in geometric terminology—O is an s-fold point of V.

Definition 2.3. *A system of local parameters* $x_1, x_2, \cdots, x_r$ *of* V *at* O *is called transversal if the multiplicity of the ideal* $\mathfrak{o}x_1 + \mathfrak{o}x_2 + \cdots + \mathfrak{o}x_r$ *is equal to the multiplicity* s *of* $\mathfrak{o}$.

Thus, in (4), we have $\nu = s$ if and only if the local parameters $x_1, x_2, \cdots, x_r$ are transversal. In that case (and in that case only) the leading form of each power series $A_i(X_1, X_2, \cdots, X_r)$ $(i = 1, 2, \cdots, \nu)$ is of degree $\geqq i$.

From the general theory of local rings it is known that there always exist transversal local parameters. We shall not elaborate this point, since in the set-up which interests us, the field k will always be infinite (even algebraically closed), and in that case it is obvious that if $\{x_1, x_2, \cdots, x_r, x_{r+1}\}$ is any system of local coördinates then there exists a matrix $\| c_{ij} \|$, with r rows and $r+1$ columns, with elements in k, such that the r elements $x_i' = \sum_{j=1}^{r+1} c_{ij} x_j$ form a system of transversal local parameters.

Let f be of the form (4), with $x_1, x_2, \cdots, x_r$ as local parameters and $x_1, x_2, \cdots, x_r, x_{r+1}$ as local coördinates. We say that $x_1, x_2, \cdots, x_r$ are *separating* local parameters if f is a separable polynomial in X_{r+1} (separating local parameters always exist if k is a perfect field). If $x_1, x_2, \cdots, x_r$ are separating local parameters then the discriminant $D(X_1, X_2, \cdots, X_r)$ of f with respect to X_{r+1} is a non-zero power series and is a non-unit (since $\nu > 1$, and $X_{r+1} = 0$ is a ν-fold root of $f(0, 0, \cdots, 0, X_{r+1})$. This power series depends only (up to a unit factor in $k[[X_1, X_2, \cdots, X_r]]$) on the choice of the local separating parameters $x_1, x_2, \cdots, x_r$ (and not on the choice of x_{r+1}), for if x'_{r+1} is any other element of $\mathfrak{m}$ such that $x_1, x_2, \cdots, x_r, x'_{r+1}$ is a system of local coördinates, then

$$\mathfrak{o} = \sum_{i=0}^{\nu-1} k[[x_1, x_2, \cdots, x_r]] x_{r+1}{}^i = \sum_{i=0}^{\nu-1} k[[x_1, x_2, \cdots, x_r]] x'_{r+1}{}^\nu,$$

and hence the discriminants of the two bases

$$\{1, x_{r+1}, \cdots, x_{r+1}{}^{\nu-1}\}, \quad \{1, x'_{r+1}, \cdots, x'_{r+1}{}^{\nu-1}\}$$

of $\mathfrak{o}$ over $k[[x_1, x_2, \cdots, x_r]]$ differ only by a unit factor in $k[[x_1, x_2, \cdots, x_r]]$.

We denote by D_0 the product of the *distinct* irreducible factors of D and by Δ the embedded variety in A_r defined by the equation $D_0 = 0$. We call Δ the *critical variety* relative to the local (separating) parameters $x_1, x_2, \cdots, x_r$.

3. Transversal sections. In this section V stands for an unmixed algebroid variety, of dimension r, not necessarily embedded. Let W be an irreducible subvariety of V, let P be the general point of W, and let A be an arbitrary *simple* point of W. Let $\rho = \mathrm{cod}_V P$ $(= \mathrm{cod}_V W = \dim \mathfrak{o}_{V,P})$, $\sigma = \mathrm{cod}_V A$ (whence $\sigma \geqq \rho$ and $\dim \mathfrak{o}_{W,A} = \sigma - \rho$). We set

$$\mathfrak{O} = \mathfrak{o}_{V,A},$$
$$\mathfrak{o}' = \mathfrak{o}/P,$$

whence

$$\mathfrak{o}_{W,A} = \mathfrak{O}/\mathfrak{O}P = \mathfrak{o}'_{A/P}. \tag{5}$$

Definition 3.1. *If* $y_1, y_2, \cdots, y_{\sigma-\rho}$ *are elements of* $\mathfrak{m}$ *and if* $y'_1, y'_2, \cdots, y'_{\sigma-\rho}$ *are their P-residues in* $\mathfrak{o}'$, *then the elements* $y_1, y_2, \cdots, y_{\sigma-\rho}$ *are called W-transversal parameters of V at A if the elements* $y'_1, y'_2, \cdots, y'_{\sigma-\rho}$ *form a system of regular parameters of the (regular) local ring* $\mathfrak{o}_{W,A}$.

The existence of W-transversal parameters at A follows from the fact that the element of A/P generate the maximal ideal of $\mathfrak{o}_{W,A}$. It is clear, on the other hand, that any set of W-transversal parameters of V at A consists of elements of A.

In the sequel, if L is any subset of $\mathfrak{o}$ we write $\mathfrak{O}L$ instead of $\mathfrak{O}\cdot\psi(\mathrm{L})$, where ψ is the canonical map of $\mathfrak{o}$ into $\mathfrak{O}$.

COROLLARY 3.2. *If* $y_1, y_2, \cdots, y_{\sigma-\rho} \in \mathfrak{m}$, *then the y's are W-tranversal parameters of V at A if and only if*

$$\mathfrak{O}\cdot(y_1, y_2, \cdots, y_{\sigma-\rho}) = \mathfrak{O}A. \tag{6}$$

This is an immediate consequence of (5).

Definition 3.3. *Let* $y_1, y_2, \cdots, y_{\sigma-\rho}$ *be W-transversal parameters of V at A, and let* $\mathfrak{O}^*$ *be the completion of the local ring* $\mathfrak{O}$ $(= \mathfrak{o}_{V,A})$. *We set*

$$\mathfrak{O}^*_{(y)} = \mathfrak{O}^*/(\mathfrak{O}^* y_1 + \mathfrak{O}^* y_2 + \cdots + \mathfrak{O}^*_{\sigma-\rho}), \tag{7}$$

$$V^*_{(y)} = \mathrm{Spec}(\mathfrak{O}^*_{(y)}), \tag{7'}$$

and we call the algebroid scheme $V^*_{(y)}$ *a W-transversal section of V at the point A (more precisely: the W-transversal section of V at A, relative to the W-transversal parameters* $y_1, y_3, \cdots, y_{\sigma-\rho}$).

Note that if we set

$$\mathfrak{O}_{(y)} = \mathfrak{O}/(\mathfrak{O}y_1 + \mathfrak{O}y_2 + \cdots + \mathfrak{O}y_{\sigma-\rho}), \tag{8}$$

$$V_{(y)} = \mathrm{Spec}(\mathfrak{O}_y), \tag{8'}$$

then $\mathfrak{O}^*_{(y)}$ is the completion of the local ring $\mathfrak{O}_{(y)}$, and $V^*_{(y)}$ is the completion of the scheme $V_{(y)}$.

The local ring $\mathfrak{O}^*_{(y)}$ may have nilpotent elements, and thus $V^*_{(y)}$ is not necessarily an algebroid *variety*. All we can say is that $V^*_{(y)}$ is an algebroid scheme of dimension $\geqq \rho$ (since $\dim \mathfrak{O} = \sigma$); it is defined over the $\boldsymbol{k}_A$, where $\boldsymbol{k}_A$ is any field of representatives of the local ring $\mathfrak{O}_{V,A}$. Since k is algebraically closed, there exist fields of representatives $\boldsymbol{k}_A$ which contain the field k, *and it is only such field* $\boldsymbol{k}_A$ *that will be allowed in the sequel.* Later on, in applications, we may want to consider $V^*_{(y)}$ over an algebraic closure $\bar{\boldsymbol{k}}_A$ of $\boldsymbol{k}_A$, by passing from $\mathfrak{O}^*_{(y)}$ to $\mathfrak{O}^*_{(y)} \otimes_{\boldsymbol{k}_A} \bar{\boldsymbol{k}}_A$.

The schemes W and $V_{(y)}$ are subschemes of V, and we can speak of the intersection of their supports. In this sense we have

$$W \cap V_{(y)} = A,$$

as follows at once from Corollary 3.2.

PROPOSITION 3.4. *Assume that* $\mathfrak{o}$ *is a Macaulay ring and that* $y_1, y_2, \cdots, y_{\sigma-\rho}$ *are W-transversal parameters of V at A. If* $V^*_{(y)}$ *is of dimension exactly* ρ, *then* $V^*_{(y)}$ *is unmixed. A necessary and sufficient condition that* $V^*_{(y)}$ *be of dimension* ρ *is that* $\{\psi(y_1), \psi(y_2), \cdots, \psi(y_{\sigma-\rho})\}$ *be a prime sequence in* $\mathfrak{O}$; (*here* ψ *is the canonical homomorphism of* $\mathfrak{o}$ *into* $\mathfrak{O}$), *and a sufficient condition is that* $\{y_1, y_2, \cdots, y_{\sigma-\rho}\}$ *be a prime sequence in* $\mathfrak{o}$.

Proof. Since $\mathfrak{o}$ is a Macaulay ring, also $\mathfrak{O}$ is a Macaulay ring ([4], Theorem 2, Corollary 4, p. 399), and so is $\mathfrak{O}^*$ ([4], Theorem 2, Corollary 6, p. 400). The first two assertions of the proposition are then well-known statements on Macaulay rings. The last statement is a consequence of the easily verified fact that if $\{a_1, a_2, \cdots, a_q\}$ is a prime sequence in $\mathfrak{o}$, consisting of elements of A, then $\{\psi(a_1), \psi(a_2), \cdots, \psi(a_q)\}$ is a prime sequence in $\mathfrak{o}_A$ $(= \mathfrak{O})$.

Example. Let V be the affine cone $X_1X_2 - X_3X_4 = 0$ in $\boldsymbol{A}_4$, let W be the plane $X_1 = X_3 = 0$ (whence $\rho = 1$) and let A be the origin. We have $\mathfrak{o}/P = k[[x_1, x_2, x_3, x_4]]/(x_1, x_3) = k[[x_2, x_4]]$. Hence x_2, x_4 are W-transversal parameters at O. However, the corresponding W-transversal section is the plane $X_2 = X_4 = 0$, hence has dimension > 1 $(= \rho)$. On the other hand, if we set $y_1 = x_2 - x_1$, $y_2 = x_4 - x_3$, then also y_1, y_2 are W-transversal parameters, and this time we get a W-transversal section which is exactly of dimension 1 (namely, the pair of lines $X_1 = X_2 = X_3 = X_4$ and $X_1 = X_2 = -X_3 = -X_4$). Possibly one should restrict the class of W-transversal sections by allowing only such sections which have the right dimension ρ.

PROPOSITION 3.5. *If V is an embedded variety then the maximal ideal of $\mathfrak{O}_{(y)}$ has a basis of $\rho+1$ elements. Hence $V^*_{(y)}$ is either of dimension ρ or $\rho+1$; in the latter case, $\mathfrak{O}^*_{(y)}$ is a regular ring, and in either case $V^*_{(y)}$ is an embedded unmixed scheme (since $\mathfrak{o}$ is a Macaulay ring).*

Proof. If V is embedded, then in the notations of § 2 we have $\mathfrak{o}=k[[X_1,X_2,\cdots,X_{r+1}]](f)$ (and thus $\mathfrak{o}$ is a Macaulay ring). Let $S=\text{Spec}(k[[X_1,X_2,\cdots,X_{r+1}]])$. Then $\mathfrak{o}_{S,A}$ is a *regular* ring of dimension $\sigma+1$. If P' is the prime ideal in $k[[X]]$ which contains f and is mapped onto P, then $\mathfrak{o}_{S,A}/\mathfrak{o}_{S,A}P'$ $(=\mathfrak{o}_{W,A})$ is a *regular* ring of dimension $\sigma-\rho$. It follows at once that $\mathfrak{o}_{S,A}P'$ has a basis of $\rho+1$ elements. Passing to $\mathfrak{o}_{V,A}=\mathfrak{O}=\mathfrak{o}_{S,A}/\mathfrak{o}_{S,A}f$, we conclude that $\mathfrak{O}P$ has a basis of $\rho+1$ elements. It follows therefore from (6) and from the definition (8) of $\mathfrak{O}_{(y)}$, that the maximal ideal of $\mathfrak{O}_{(y)}$ has a basis of $\rho+1$ elements. This completes the proof.

We note the following special cases:

(a) $A=O$ $(=\mathfrak{m})$. In this case $V_{(y)}$ is a scheme defined over k. We have $\sigma=r$, $(y)=(y_1,y_2,\cdots,y_{r-\rho})$, and $\mathfrak{O}=\mathfrak{O}^*=\mathfrak{o}$.

(b) $A=P$ (= general point of W). In this case the empty set is the only set of W-transversal parameters (since $\sigma=\rho$), there is only one W-transversal section of V at P, namely $V^*_P=\text{Spec}(\mathfrak{O}^*)$, where $\mathfrak{O}^*$ is the completion of $\mathfrak{o}_{V,P}$; it is an algebroid unmixed *variety*, of dimension ρ, defined over the field $\boldsymbol{k}_P$ $(\cong \mathfrak{o}_{V,P}/\mathfrak{m}_{V,P})$.

4. Definition and a basic criterion of equisingularity in codimension 1 (and characteristic zero). From now on we assume that the ground field k is of characteristic zero and algebraically closed. We assume that V is an embedded algebroid variety of dimension r and that the origin O of V is a singular point of V. We consider an irreducible algebroid subvariety W, of codimension 1 on V, having a simple point at O. There always exist W-transversal sections $V^*_{(y)}$ of V at O, which are of dimension 1 (and not 2), and, by Proposition 3.5, any such a section is an embedded algebroid scheme, hence as we may say, a 1-dimensional "algebroid cycle" in this affine plane, which may have multiple components. $V^*_{(y)}$ is defined over k and has origin O. On the other hand, *the* W-transversal section of V at the general point P of W is an algebroid embedded curve, having origin P and defined over any field of representatives $\boldsymbol{k}_P$ of $\mathfrak{o}^*_{V,P}$; we shall denote this curve of V^*_P. We note that since $k\subset\boldsymbol{k}_P$, $V^*_{(y)}$ is also defined over $\boldsymbol{k}_P$.

Definition 4.1. *V is said to be equisingular at O, along W, if there* *exists a W-transversal section* $V^*_{(y)}$ *of V at O, such that* $V^*_{(y)}$ *is a curve and such that* $V^*_{(y)}$ *and* V^*_P *have equivalent singularities at O and P respectively. (It is implicit in this definition that O is a simple point of W).*

[In regard to this definition, we recall that we have proved in ([3], Section 2, Note) that the equivalence of two plane algebroid curves C, D, defined over some common ground field, is an intrinsic relationship between the local rings of C and D (at their respective origins). Thus, the equivalence (or non-equivalence) of V^*_P and $V^*_{(y)}$ is independent of the choice of the field of representatives $\boldsymbol{k}_P$ of $\mathfrak{o}^*_{V,P}$].

Definition 4.2. *We shall say that V has at the point O a singularity of dimensionality type* 1, *if there exists an irreducible subvariety W of V, of codimension* 1, *such that V is equisingular at O, along W.*

Our aim in this section is to prove a basic criterion of equisingularity in codimension 1. We first give the following definition:

Definition 4.3. *A system of local parameters* $x_1, x_2, \cdots, x_r$ *of V at O (see Definition* 2.2) *is said to be equisingular, if the critical variety* $\Delta_{(x)}$ *associated with these parameters (see the end of* § 2) *is a regular algebroid variety (i.e., has a simple point at its origin).*

The criterion is as follows:

THEOREM 4.4. *The following conditions are equivalent:*

(a) *The origin O of V is a singularity of dimensionality type* 1.

(b) *There exist equisingular systems of local parameters of V at O.*

(c) *There exist equisingular systems of transversal local parameters of V at O.*

In the course of the proof of this theorem, several other results will be established, and we summarize these results in the following:

THEOREM 4.5.

(1) *If V is equisingular, at O, along an irreducible subvariety W of codimension* 1, *then:* (1a) *W is the entire singular locus of V (and thus W is uniquely determined), and O is a simple point of W;* (1b) *V is equimultiple at O along W, i.e., if P denotes the general point of W then the multiplicities* $m_V(P)$, $m_V(O)$ *of V, at P and at O respectively, are equal.*[1]

[1] *However, equimultiplicity of V at O, along the singular locus W of V* (always assuming that W is of codimension 1 and has a simple point at *O*) *does not imply*

(2) *Let* $\{x_1, x_2, \cdots, x_r\}$ *be a system of local parameters of* V *at* O, *let* π_x *denote the natural (surjective) morphism of* V *into* $\mathrm{Spec}(k[[x_1, x_2, \cdots, x_r]]$ *(determined by the injection of* $k[[x_1, x_2, \cdots, x_r]]$ *into* $\mathfrak{o}$*), and let* Δ_x *be the critical variety relative to the parameters* $x_1, x_2, \cdots, x_r$. *Assume that* $\{x_1, x_2, \cdots, x_r\}$ *is an equisingular system. Then*: (2a) $\pi_x^{-1}\{\Delta_x\}$ *is an irreducible subvariety* W *of* V, *of codimension* 1, V *is equisingular at* O *along* W, *and* $\pi_x \mid W : W \rightarrow \Delta_x$ *is an isomorphism*; (2b) *if* ξ $(=\xi(x_1, x_2, \cdots, x_r))$ *is a generator of the principal ideal in* $k[[x_1, x_2, \cdots, x_r]]$ *which defines* Δ_x *and if* $\mathfrak{O} = \mathfrak{o}_{V,P}$, *where* P *is the general point of* W, *then the* $\mathfrak{m}_{V,P}$*-primary ideal* (ξ) *in* $\mathfrak{O}$ *and the* $\mathfrak{m}$*-primary ideal* $\mathfrak{o}x_1 + \mathfrak{o}x_2 + \cdots + \mathfrak{o}x_r$ *in* $\mathfrak{o}$ *have the same multiplicity.*

[*Note.* Upon extending $\{x_1, x_2, \cdots, x_r\}$ to a system of local coördinates $x_1, x_2, \cdots, x_r, x_{r+1}$, we realize V as a hypersurface in the affine space A_{r+1}. Let π denote the natural (surjective) morphism (projection) of $\mathrm{Spec}(k[[X_1, X_2, \cdots, X_r, X_{r+1})$ into $\mathrm{Spec}(k[[X_1, X_2, \cdots, X_r]]$. Then $\pi^{-1}\{\Delta_x\}$ is a (cylindrical) algebroid hypersurface H in A_{r+1}, containing W, and defined by the equation $\xi(X_1, X_2, \cdots, X_r) = 0$, and $\pi^{-1}\{O\}$ is the line $L : X_1 = X_2 = \cdots = X_r = 0$. Part (2b) of the theorem can then be stated in terms of intersection multiplicities as follows:

(8) $$i(V \cdot H, W; A_{r+1}) = i(V \cdot L, O; A_{r+1}).]$$

Proof. A) Assume that the origin O of V is a singularity of dimensionality type 1, and let then W be an irreducible subvariety of V, of codimension 1, such that V is equisingular at O along W. Let $(y_1, y_2, \cdots, y_{r-1})$ be a system of W-transversal parameters of V at O such that the corresponding W-transversal section $V^*_{(y)}$ of V is an algebroid *curve* and such that the singularity of $V^*_{(y)}$ at its origin O is equivalent to the singularity of V^*_P at P (see Definition 4.1). We know, by the proof of Proposition 3.5, that $\mathfrak{o}_P$ has a basis of 2 elements. We fix such a basis and we denote its elements by y_r, y_{r+1}. Then by (6), Section 3, we have $\mathfrak{m} = \mathfrak{o} \cdot (y_1, y_2, \cdots, y_{r-1}, y_r, y_{r+1})$, i. e., the $r + 1$ elements y_i form a system of local coördinates of V at O. Let

(9) $$f(Y_1, Y_2, \cdots, Y_{r+1}) = 0$$

be an equation of the corresponding embedding of V in affine $(r+1)$-space. From now we shall identify V with the hypersurface (9).

---

equisingularity of V *at* O, *along* W. For example, the singular locus W of the surface $V: z^2 - xy^2 = 0$, is the line $y = z = 0$, and we have $m_V(O) = m_V(W) = 2$. The W-transversal section of V at the general point P of W is a curve with an ordinary double point, but no W-transversal sections of V at O can have an ordinary double point (compare with footnote 4).

The subvariety W is now represented by the linear $(r-1)$-space

$$W: Y_r = Y_{r+1} = 0,$$

and thus $f(Y_1, Y_2, \cdots, Y_{r-1}, 0, 0)$ must be identically zero, since $W \subset V$. As field of representatives $\boldsymbol{k}_P$ of $\mathfrak{o}^*_{V,P}$ we can take the field

$$k_P = k\{\{y_1, y_2, \cdots, y_{r-1}\}\}.$$

The W-transversal section V^*_P of V at P is the plane algebroid curve

$$\Gamma^y: F^y(Y_r, Y_{r+1}) = 0, \tag{10}$$

where

$$F^y(Y_r, Y_{r+1}) = f(y_1, y_2, \cdots, y_{r-1}, Y_r, Y_{r+1}). \tag{10'}$$

Its origin $\bar{O}$ is the point $Y_r = Y_{r+1} = 0$. The coefficients of $F^{(y)}$ belong to the power series ring $k[[y_1, y_2, \cdots, y_{r-1}]]$ (it is clear that $y_1, y_2, \cdots, y_{r-1}$ are analytically independent over k).

The W-transversal section $V^*_{(y)}$ of V at O is the plane algebroid curve

$$\Gamma^0: F^0(Y_r, Y_{r+1}) = 0, \tag{11}$$

where

$$F^0(Y_r, Y_{r+1}) = f(0, 0, \cdots, 0, Y_r, Y_{r+1}). \tag{11'}$$

In the terminology of our paper [3; § 6], Γ^0 is a specialization of Γ^y over $y \to 0$ (where $y = (y_1, y_2, \cdots, y_{r-1})$).

Let s be the multiplicity of the curve Γ^y at its origin $\bar{O}$ $(Y_r = Y_{r+1} = 0)$. Since Γ^y and Γ^0 have equivalent singularities at the origin $Y_r = Y_{r+1} = 0$, also Γ^0 must have exactly an s-fold point at the origin. Hence, if $F_s^y(Y_r, Y_{r+1})$ is the leading form of F^y, then $F_s^0(Y_r, Y_{r+1})$ has to be the leading form of F^0, i.e., we must have $F_s^0(Y_r, Y_{r+1}) \neq 0$; in other words: the coefficients of $F_s^y(Y_r, Y_{r+1})$ (which are power series in $y_1, y_2, \cdots, y_{r-1}$), do not all vanish at $y = (0)$. This shows *that the leading form f_s of $f(Y_1, Y_2, \cdots, Y_r, Y_{r+1})$ is also of degree s and is independent of $Y_1, Y_2, \cdots, Y_{r-1}$*:

$$f_s(Y_1, Y_2, \cdots, Y_r, Y_{r+1}) = g_s(Y_r, Y_{r+1}), \tag{12}$$

where g_s is a binary form, of degree s, with coefficients in k.

Thus $s = m_V(O)$. But, by the definition of s, it is clear that $s = m_V(P) = m_{\bar{O}}(\Gamma^y))$, where P is the general point of W. This proves part **(1b)** of Theorem 4.5 (equimultiplicity of V at O, along W).

B) Upon replacing y_r and y_{r+1} by non-special linear combinations $ay_r + by_{r+1}$, $cy_r + dy_{r+1}$, with coefficients in k, we may assume that the line

$Y_r = 0$ is not tangent to Γ^0 at the origin $\bar{O}$. This being assumed, two consequences will follow. In the first place, we will have, by (12),

$$f_s(0, 0, \cdots, 0, Y_{r+1}) = g_s(0, Y_{r+1}) = cY_{r+1}{}^s, c \neq 0,$$

and hence $y_1, y_2, \cdots, y_r$ are *transversal local parameters* of V, at O. In the second place, since Γ^y and Γ^0 have equivalent singularities, it follows from Theorem 7, part (b), of [3], that if $D(Y_1, Y_2, \cdots, Y_r)$ is the discriminant of $f(Y_1, Y_2, \cdots, Y_r, Y_{r+1})$ with respect to Y_{r+1}, then $D(y_1, y_2, \cdots, Y_{r-1}, Y_r)$ is of the form $\epsilon(y_1, y_2, \cdots, y_{r-1}, Y_r) \cdot Y_r{}^N$, where $\epsilon(Y_1, Y_2, \cdots, Y_r)$ is a unit in $k[[Y_1, Y_2, \cdots, Y_r]]$; in other words (see Definition 4.3), $y_1, y_2, \cdots, y_r$ are *equisingular* local parameters. *This proves that* 4.4(a) $\Rightarrow$ 4.4(c).

C) Clearly 4.4(c) implies 4.4(b). Thus, in order to complete the proof of Theorem 4.4 we have only to show that 4.4(b) implies 4.4(a). Now, the implication 4.4(b) $\Rightarrow$ 4.4(a) is certainly included in part 2a of Theorem 4.5. So we shall now prove this part of Theorem 4.5.

Assume then that $x = (x_1, x_2, \cdots, x_r)$ is an equisingular system of local parameters of V at O. Upon extending this system to a system $(x_1, x_2, \cdots, x_r, x_{r+1})$ of local coördinates of V at O we obtain a well defined embedding of V as a hypersurface in $\boldsymbol{A}_{r+1}$:

$$V: f(X_1, X_2, \cdots, X_r, X_{r+1}Q = 0,$$

where f is a monic polynomial in X_{r+1}, say of degree ν, with coefficients in $k[[X_1, X_2, \cdots, X_r]]$. Furthermore, $X_{r+1} = 0$ is a ν-fold root of $f(0, 0, \cdots, X_{r+1})$. We denote by R the subring $k[[x_1, x_2, \cdots, x_r]]$ of $\mathfrak{o}$, by K the total ring of quotients of $\mathfrak{o}$, and by $\bar{\mathfrak{o}}$ the integral closure of $\mathfrak{o}$ in K ($\bar{\mathfrak{o}}$ is also the integral closure of R in K).

By assumption, the discriminant D of f, with respect to X_{r+1}, is of the form $\epsilon(X_1, X_2, \cdots, X_r) \cdot [h(X_1, X_2, \cdots, X_r)]^N$, where ϵ is a unit in $k[[X_1, X_2, \cdots, X_r]]$ and h is a power series whose leading form is of degree 1. We can therefore assume, without loss of generality, that h is the power series X_r: thus

$$D = \epsilon(X_1, X_2, \cdots, X_r) X_r{}^N. \tag{13}$$

We now set

$$F^0(X_r, X_{r+1}) = f(0, 0, \cdots, 0, X_r, X_{r+1}).$$

In view of the expression (13) of D, the discriminant of F^0, with respect to X_{r+1}, is not identically zero, whence F^0 has no multiple factors. We set

$$\mathfrak{o}_0 = k[[X_r, X_{r+1}]]/(F^0),$$
$$K_0 = \text{total ring of quotients of } \mathfrak{o}_0,$$
$$\bar{\mathfrak{o}}_0 = \text{integral closure of } \mathfrak{o}_0 \text{ in } K_0.$$

We now apply Theorem 5 of ([3], § 5). By that theorem, we have that $\bar{\mathfrak{o}}$ contains a subring isomorphic with $\bar{\mathfrak{o}}_0$ (which we shall identify with $\bar{\mathfrak{o}}_0$), such that $x_1, x_2, \cdots, x_{r-1}$ are analytically independent over $\bar{\mathfrak{o}}_0$ and such that

$$\bar{\mathfrak{o}} = \bar{\mathfrak{o}}_0[[x_1, x_2, \cdots, x_{r-1}]]. \tag{14}$$

Furthermore, with the above identification, x_r is the F^0-residue of X_r, so that $\mathfrak{o}_0 = k[[x_r]][\zeta]$, where ζ is the F^0-residue of X_{r+1}.

Using these facts, we shall prove first that there exists a (non-unit) power series $\phi(x_1, x_2, \cdots, x_{r-1})$ such that

$$f(x_1, x_2, \cdots, x_r, 0, X_{r+1}) = [X_{r+1} - \phi(x_1, x_2, \cdots, x_{r-1})]^\nu. \tag{15}$$

We first consider the case in which the algebroid variety V is irreducible. In this case $\bar{\mathfrak{o}}$, and hence also $\bar{\mathfrak{o}}_0$, is an integral domain. Since $\bar{\mathfrak{o}}_0$ is the integral closure of the local domain $k[[x_r, \zeta]]$, of dimension 1, $\bar{\mathfrak{o}}_0$ is the power series ring $k[[t]]$, where $t = \sqrt[\nu]{x_r}$. Thus, by (14), we have now

$$\mathfrak{o} = k[[x_1, x_2, \cdots, x_{r-i}, t]], \tag{16}$$

i.e., $\bar{\mathfrak{o}}$ *is a regular ring of dimension* r [*this shows, incidentally, that the normalization of* V *is a non-singular algebroid variety*; compare with Proposition 4.6 below]. In particular,

$$x_{r+1} = \sum_{i=0}^{\infty} a_i(x_1, x_2, \cdots, x_{r-1}) t^i, \qquad (t = \sqrt[\nu]{x_r}). \tag{17}$$

The conjugates $x_{r+1}^{(j)}$ of x_{r+1} over $k\{\{x_1, x_2, \cdots, x_{r-1}, x_r\}\}$ are obtained by replacing t in (17) by $\omega^j t$, where ω is a primitive ν-th root of unity. Since $f(x_1, x_2, \cdots, x_r, X_{r+1}) = \prod_{j=0}^{\nu-1} (X_{r+1} - x_{r+1}^{(j)})$, we conclude at once that (15) holds, with $\phi(x_1, x_2, \cdots, x_{r-1}) = a_0(x_1, x_2, \cdots, x_{r-1})$.

Now, if V is not irreducible, we factor f into its irreducible factors:

$$f = f_1 f_2 \cdots f_h.$$

The discriminant of each factor f_μ (with respect to X_{r+1}) is still a power of X_r. Hence, by the irreducible case, we have

$$f_\mu(x_1, x_2, \cdots, x_{r-1}, 0, X_{r+1}) = [X_{r+1} - \phi_\mu(x_1, x_2, \cdots, x_{r-1})]^{\nu_\mu},$$

where f_μ is the degree of ν_μ in X_{r+1} $(\nu = \nu_1 + \nu_2 + \cdots + \nu_h)$, and ϕ_μ is a non-unit power series. Applying (16) to each irreducible f_μ, we see that the splitting field of the polynomial $f(x_1, x_2, \cdots, x_r, X_{r+1})$, over $k\{\{x_1, x_2, \cdots, x_r\}\}$ is a field of the type

$$k\{\{x_1, x_2, \cdots, x_{r-1}, t\}\},$$

15

where $t^q = x_r$, for some integer q, and that each root of $f(x_1, x_2, \cdots, x_r, X_{r+1})$ belongs to $k[[x_1, x_2, \cdots, x_{r-1}, t]]$. Let, for $\mu \neq \mu'$, $X_{r+1} = \xi_{r+1}$ and $X_{r+1} = \xi'_{r+1}$ be the roots of f_μ and $f_{\mu'}$ respectively. Then

$$\xi_{r+1} = \sum_{i=0}^{\infty} a_i(x_1, x_2, \cdots, x_{r-1}) t^i, \qquad a_0 = \phi_\mu$$

$$\xi'_{r+1} = \sum_{i=0}^{\infty} a'_i(x_1, x_2, \cdots, x_{r-1}) t^i, \qquad a' = \phi_{\mu'}.$$

Since $\xi_{r+1} - \xi'_{r-1}$ must divide the discriminant x_r^N $(= t^{qN})$ in $k[[x_1, x_2, \cdots, x_{r-1}, t]]$, it follows at once that $\phi_\mu = \phi_{\mu'}$. This establishes (15), and proves (in the notations of Theorem 4.5, part 2a) that $\pi_x^{-1}\{\Delta\}$ is the irreducible variety W, of codimension 1, defined by

$$X_r = X_{r+1} - \phi(X_1, X_2, \cdots, X_{r-1}) = 0,$$

and that $\pi_x \colon W \to \Delta_x$ is an isomorphism.

For simplicity, we replace x_{r+1} by $x_{r+1} - \phi(x_1, x_2, \cdots, x_{r-1})$. Then W is defined by

$$X_r = X_{r+1} = 0.$$

Let $\boldsymbol{k}_x = k\{\{x_1, x_2, \cdots, x_{r-1}\}\}$ and let us consider the two plane algebroid curves

$$\Gamma^x \colon F^x(X_r, X_{r+1}) = 0, \qquad (F^x = f(x_1, x_2, \cdots, x_{r-1}, X_r, X_{r+1}));$$

$$\Gamma^0 \colon F^0(X_r, X_{r+1}) = 0, \qquad (F^0 = f(0, 0, \cdots, 0, X_r, X_{r+1})),$$

both defined over $\boldsymbol{k}_x$. It is clear that Γ^x is the W-transversal section of W at the general point $(x_1, x_2, \cdots, x_{r-1}, 0, 0)$ of W, and that Γ^0 is the W-transversal section of V at O, relative to the W-transversal parameter $x_1, x_2, \cdots, x_{r-1}$ of V at O.

In view of the expression (13) of the discriminant D, it follows from Theorem 7, (a) of [3] that Γ^x and Γ_0 have equivalent singularities at the origin $X_r = X_{r+1} = 0$. This shows that V is equisingular at O, along W, and completes the proof of Theorem 4.5, part 2a, and also of Theorem 4.4.

Part 2b of Theorem 4.5 is now obvious, since both multiplicities in question are equal to the degree ν of f in X_{r+1} (the element ξ is now x_r).

COROLLARY 4.6. *Let V have at O a singularity of dimensionality type* 1 *and let W be the singular locus of V (whence W is irreducible, non-singular and of codimension* 1). *Then any W-transversal section of V at O has dimension* 1.

For let $y_1, y_2, \cdots, y_{r-1}$ be W-transversal parameters of V at O. Then we can complete $(y_1, y_2, \cdots, y_{r-1})$ to a system of local coördinates $y_1, y_2, \cdots, y_{r+1}$ of V at O in such a way that the prime ideal of W in $\mathfrak{o}$ is given by $\mathfrak{o} \cdot y_r + \mathfrak{o} \cdot y_{r+1}$. Let $f(Y_1, Y_2, \cdots, Y_{r+1}) = 0$ be the associated embedding of V. Then we know that the leading form f_s of f is a binary form in Y_r, Y_{r+1} (equimultiplicity of V at O, along W). Hence $f(0, 0, \cdots, 0, Y_r, Y_{r+1})$ is not identically zero.

Note. In the course of the above proof we have established the following result:

PROPOSITION 4.7. *If the origin O of V is a singular point of dimensionality type 1 and if V has h irreducible components, then the normalization $\bar{V}$ of V consists of h (disjoint) non-singular irreducible algebroid varieties.*[2]

5. A Jacobian criterion of equisingularity. Let $(x_1, x_2, \cdots x_{r+1})$ be a system of local coördinates of V at O, and let

$$f(X_1, X_2, \cdots, X_{r+1}) = 0$$

be the corresponding embedding of V in affine $(r+1)$ space. Consider the following ideal J in $\mathfrak{o}$:

$$J = \mathfrak{o}f'_{x_1} + \mathfrak{o}f'_{x_2} + \cdots + \mathfrak{o}f'_{x_{r+1}}.$$

It is immediately seen that the ideal J is independent of the choice of the local coordinates $x_1, x_2, \cdots, x_{r+1}$; in other words, J is independent of the choice of the embedding of V in $\boldsymbol{A}_{r+1}$. We call J *the Jacobian ideal of V at O.*

THEOREM 5.1. (*Jacobian criterion of equisingularity*). *Let $\bar{\mathfrak{o}}$ be the integral closure of $\mathfrak{o}$ in the total ring of quotients K of $\mathfrak{o}$. In order that O be a singular point of V of dimensionality type 1, it is necessary and sufficient that the following two conditions be satisfied:*

(a) *The ideal $\bar{\mathfrak{o}}J$ is principal.*[3]

(b) *O is a simple point of the singular locus of V.*

In the course of the proof, also the following will be established:

---

[2] *However, V may have a non-singular normalization without necessarily having at O a singularity of dimensionality type* 1. Thus, in the example of footnote 1, the normalization of V is non-singular.

[3] *Condition* (a) *alone does not imply* (b). For instance, consider the surface $z^n = x^a y^b$, with $a \geqq n$, $b \geqq n$. Then $\partial z/\partial x$ and $\partial z/\partial y$ are integral functions of x and y, and hence $\bar{\mathfrak{o}}J = \bar{\mathfrak{o}}f'_z$, where $f = z^n - x^a y^b$. If $n > 1$, then the singular locus of V consists of the two lines $x = z = 0$ and $y = z = 0$.

THEOREM 5.2. *Let $x_1, x_2, \cdots, x_{r+1}$ and f be as above, and assume that V has at O a singularity of dimensionality type 1 and that $x_1, x_2, \cdots, x_r$ are local parameters. Then the following conditions are equivalent:*

1) *$x_1, x_2, \cdots, x_r$ are transversal parameters of V at O.*

2) $\mathfrak{d}J = \mathfrak{d}f'_{x_{r+1}}$.

Furthermore, either condition implies that

3) *the local parameters $x_1, x_2, \cdots, x_r$ are equisingular.*[4]

Proof of Theorems 5.1 *and* 5.2. We agree once and for always that if $x_1, x_2, \cdots, x_{r+1}$ are local coördinates of V at O, then $f_{\{x_1, x_2, \cdots, x_{r+1}\}}$ stands for the power series in $k[[X_1, X_2, \cdots, X_{r+1}]]$ such that $f = 0$ represents that embedding of V in A_{r+1} which is determined by the local coördinates x_i (this power series is uniquely determined, to within an arbitrary unit factor).

We first prove the following:

A) *If $\mathfrak{d}J$ is principal, then there exist local coördinates $x_1, x_2, \cdots, x_r, x_{r+1}$ of V at O, such that:*

(18) $\quad x_1, x_2, \cdots, x_r$ *are local transversal parameters of V at O.*

(19) $$\mathfrak{d}J = \mathfrak{d}f'_{x_{r+1}},$$

where $f = f_{\{x_1, x_2, \cdots, x_{r+1}\}}$.

For, start with any system of local coördinates $x_1, x_2, \cdots, x_{r+1}$. If $\| c_{ij} \|$ is any non-singular $(r+1)$-rowed square matrix whose element c_{ij} are in k:

(20) $$| c_{ij} | \neq 0,$$

then the elements $y_1, y_2, \cdots, y_{r+1}$ of $\mathfrak{m}$ defined by

$$x_i = \sum_{j=1}^{r+1} c_{ij} y_j, \qquad i = 1, 2, \cdots, r+1,$$

also constitute a system of local coördinates. We shall impose other inequalities on the c_{ij}.

Let s be the multiplicity of the singular point O of V and let $f_s(X_1, X_2, \cdots, X_{r+1})$ be the leading form of f. We shall require that

(20′) $$f_s(c_{1,r+1}, c_{2,r+1}, \cdots, c_{r+1,r+1}) \neq 0.$$

This inequality insures that $y_1, y_2, \cdots, y_r$ *are local transversal parameters.*

---

[4] This also shows that the surface V of footnote 1 is not equisingular at O, along the double line $y = z = 0$. For x, y are transversal parameters, without being equisingular parameters (the critical curve $\Delta_{(x,y)}$ is the pair of lines $x = 0$ and $y = 0$.

Let $\bar{\mathfrak{o}}J = \bar{\mathfrak{o}}t$, where $0 \neq t \in \bar{\mathfrak{o}}$, and let

$$\left.\begin{array}{l} f'_{x_i} = \alpha_i t, \quad (i = 1, 2, \cdots, r+1) \\ t = \sum_{i=1}^{r+1} \beta_i f'_{x_i} \end{array}\right\} \alpha_i, \beta_i \in \bar{\mathfrak{o}}.$$

Then

$$(1 - \sum_{i=1}^{r+1} \alpha_i\beta_i) t = 0. \tag{21}$$

We consider in $\bar{\mathfrak{o}}$ the ideal $\mathfrak{A} = (0) : \bar{\mathfrak{o}}t$ and we set $\bar{\bar{\mathfrak{o}}} = \bar{\mathfrak{o}}/\mathfrak{A}$. If $\xi \in \bar{\mathfrak{o}}$, we denote by $\bar{\xi}$ the $\mathfrak{A}$-residue of ξ. We have then by (21):

$$\sum_{i=1}^{r+1} \bar{\alpha}_i \bar{\beta}_i = 1. \tag{21'}$$

Let $\bar{\mathfrak{m}}_1, \bar{\mathfrak{m}}_2, \cdots, \bar{\mathfrak{m}}_q$ be the maximal ideals of the semilocal ring $\bar{\bar{\mathfrak{o}}}$. By (21′), for any $\nu = 1, 2, \cdots, q$, the $\bar{\alpha}_i$ are not all in $\bar{\mathfrak{m}}_\nu$. We impose on the constants $c_{i,r+1}$ the additional conditions

$$\sum_{i=1}^{r+1} c_{i,r+1}\bar{\alpha}_i \notin \bar{\mathfrak{m}}_\nu, \qquad \nu = 1, 2, \cdots, q. \tag{20''}$$

With this condition satisfied, the element $\sum c_{i,r+1}\bar{\alpha}_i$ is a unit in $\bar{\bar{\mathfrak{o}}}$. If, then, we set $\epsilon = \sum_{i=1}^{r+1} c_{i,r+1}\alpha_i$, there exists an element ϵ' in $\bar{\mathfrak{o}}$ such that $\epsilon\epsilon' - 1 \in \mathfrak{A}$, i.e., $t = \epsilon\epsilon' t$. Now, let

$$g(Y_1, Y_2, \cdots, Y_{r+1}) = f(\sum_{j=1}^{r+1} c_{1j}Y_j, \sum_{j=1}^{r+1} c_{2j}Y_j, \cdots, \sum_{j=1}^{r+1} c_{r+1,j}Y_j),$$

so that $g(Y_1, Y_2, \cdots, Y_{r+1}) = 0$ is the defining equation of V, relative to the local coördinates $y_1, y_2, \cdots, y_{r+1}$. We have

$$g'_{y_{r+1}} = \sum_{i=1}^{r+1} c_{i,r+1} f'_{x_i} = (\sum_{i=1}^{r+1} c_{i,r+1}\alpha_i) t = \epsilon t.$$

whence $\epsilon' g'_{y_{r+1}} = t$. This shows that $\bar{\mathfrak{o}}J = \bar{\mathfrak{o}}g'_{y_{r+1}}$, and the assertion A) above is proved.

We now prove the following:

B) *Let $x_1, x_2, \cdots, x_{r+1}$ be local coördinates of V at O, satisfying conditions* (18) *and* (19). *Assume furthermore that O is a simple point of the singular locus of V. Then O is a singular point of dimensionality type* 1, *and the transversal local parameters $x_1, x_2, \cdots, x_r$ are equisingular.*

By (19), the ideal $\bar{\mathfrak{o}}J$ is principal. Since we are assuming always that O is not a simple point of V, $\bar{\mathfrak{o}}J$ is not the unit ideal; hence it is unmixed, of dimension $r-1$. Since every isolated prime ideal of J is the contraction of a prime ideal of $\bar{\mathfrak{o}}J$, it follows the isolated prime ideals of J are of dimension $r-1$. Thus the singular locus of V is of pure dimension $r-1$. Now, in B) we have assumed that this locus is non-singular. It follows that the singular locus of V is an irreducible non-singular variety W, of dimension $r-1$.

We assert that W is the variety of the principal ideal $\mathfrak{o}\cdot f'_{x_{r+1}}$. It is clear that $W\subset \mathfrak{V}(\mathfrak{o}\cdot f'_{x_{r+1}})$. On the other hand, if $\mathfrak{p}$ is any prime ideal of $\mathfrak{o}\cdot f'_{x_{r+1}}$, there exists a prime ideal $\bar{\mathfrak{p}}$ in $\bar{\mathfrak{o}}$ such that $\mathfrak{p}\cap\mathfrak{o}=\mathfrak{p}$. For such a prime ideal $\bar{\mathfrak{p}}$ we will have $\bar{\mathfrak{o}}\cdot J=\bar{\mathfrak{o}}\cdot f'_{x_{r+1}}\subset\bar{\mathfrak{p}}$, and hence $J\subset\mathfrak{p}$, showing that $\mathfrak{V}(\mathfrak{p})\subset W\ (=\mathfrak{V}(J))$, as asserted.

From $W=\mathfrak{V}(\mathfrak{o}\cdot f'_{x_{r+1}})$ follows that $\Delta_x=\pi_x(W)$, where $x=(x_1,x_2,\cdots,x_r)$ and where the notations are the same as in Theorem 4.5, part (2). Since $x_1,x_2,\cdots,x_r$ are transversal parameters at O, the line $X_1=X_2=\cdots=X_r=0$ is not tangent to V, and hence, *a fortiori*, not tangent to W. Since O is a simple point of W, we conclude that the origin $\bar{O}=\pi(O)$ of Δ_x is a simple point of Δ_x. This completes the proof B).

From A) and B), follows the sufficiency of conditions (a) and (b) of Theorem 5.1.

C) Conversely, let us assume that O is a singular point of V, of dimensionality type 1. Let $(x_1,x_2,\cdots,x_r)$ be an equisinguar system of *transversal* local parameters of V at O (see Theorem 4.4, part (c). We shall use the notation of part C) of the proof of Theorems 4.4 and 4.5. We may assume that $X_r=0$ is the critical variety. Then we have (14), and from this it follows that the $r-1$ derivations $\frac{\partial}{\partial x_i}$ $(i=1,2,\cdots,r-1)$ are regular on $\bar{\mathfrak{o}}$. In particular, we have

$$\frac{\partial x_{r+1}}{\partial x_i}=-f'_{x_i}/f'_{x_{r+1}}\in\bar{\mathfrak{o}},\qquad i=1,2,\cdots,r-1.$$

If we now prove that also

$$\frac{\partial x_{r+1}}{\partial x_r}\ (=-f'_{x_r}/f'_{x_{r+1}})\in\bar{\mathfrak{o}},$$

then it will follow that $\bar{\mathfrak{o}}J=\bar{\mathfrak{o}}f'_{x_{r+1}}$, and the proof of Theorem 5.1 will be complete.

Let us first consider the case in which V is irreducible. Then we have (17), where $t^\nu=x_r$. Since the parameters $x_1,x_2,\cdots,x_r$ are transversal, we have $\nu=s=$ multiplicity of the singular point O of V. We may also assume that the singular locus W of V is $X_r = X_{r+1} = 0$; then the term $a_0(x_1, x_2, \cdots, x_{r-1})$ in (17) is missing. Since W is also s-fold for V (Theorem 4.5, part (1b)), we have for every term $cX_1^{i_1}X_2^{i_2}\cdots X_r^{i_r}X_{r+1}^{i_{r+1}}$ ($c \in k$) in the defining equation of V, the inequality $i_r + i_{r+1} \geqq s$. Since the term X_{r+1}^s occurs in f, it follows at once that in (17) the power series representing x_{r+1} begins with terms of degree $\geqq s$ in t. Since

$$s \frac{\partial x_{r+1}}{\partial x_r} t^{s-1} = \frac{\partial x_{r+1}}{\partial t},$$

it follows that $\partial x_{r+1}/\partial x_r \in \bar{\mathfrak{o}}$, as asserted.

Now, in the general case, if V has h irreducible components $V_1, V_2, \cdots, V_h$, $\bar{\mathfrak{o}}$ is the direct sum of h local domains:

$$\bar{\mathfrak{o}} = \bar{\mathfrak{o}}_1 \oplus \bar{\mathfrak{o}}_2 \oplus \cdots \oplus \bar{\mathfrak{o}}_h,$$

where $\bar{\mathfrak{o}}_j$ is the integral closure of the local domain $\mathfrak{o}_j$ of V_j at O. Let $1 = e_1 + e_2 + \cdots + e_h$ be the decomposition of 1 into mutually orthogonal idempotents ($e_j \in \bar{\mathfrak{o}}$) and let $\phi_j\colon \bar{\mathfrak{o}} \to \bar{\mathfrak{o}}_j$ be the canonical surjection of $\bar{\mathfrak{o}}$ into $\bar{\mathfrak{o}}_j$, defined by $\phi_j(\xi) = e_j\xi$ ($\xi \in \bar{\mathfrak{o}}$). Let $\phi_j(x_i) = x_{ij}$ ($i = 1, 2, \cdots, r+1$; $j = 1, 2, \cdots, h$). It is clear that for each $j = 1, 2, \cdots, h$, the elements $x_{1j}, x_{2j}, \cdots, x_{rj}$ are *transversal* local parameters of V_j at O, and furthermore these parameters are equisingular [since the critical variety of V_j, relative to $x_{1j}, x_{2j}, \cdots, x_{rj}$, is either empty (if O is a simple point of V_j) or is $X_r = 0$]. Hence, by the irreducible case, we have that

$$\frac{\partial x_{r+1,j}}{\partial x_{i,j}} \in \bar{\mathfrak{o}}_j.$$

Since it is obvious that $\dfrac{\partial x_{r+1}}{\partial x_r} = \sum e_j \dfrac{\partial x_{r+1,j}}{\partial x_{r,j}}$, we conclude that $\dfrac{\partial x_{r+1}}{\partial x_r} \in \bar{\mathfrak{o}}$. This completes the proof of Theorem 5.1.

We note that if $(x_1, x_2, \cdots, x_r)$ is an equisingular system of local parameters of V at O and if these parameters are *not* transversal, then f is a monic polynomial in X_{r+1} of degree ν *greater* than s. If V is irreducible, then the power series in (17) which represents x_{r+1} would begin with a term of degree s in t (assuming, as we may, that the term $a_0(x_1, x_2, \cdots, x_{r-1})$ is missing). Thus $\partial x_{r+1}/\partial x_r$ *would definitely not belong to* $\bar{\mathfrak{o}}$. If V is reducible, then there would have to exist at least one irreducible component V_j of V such that $x_{1j}, x_{2j}, \cdots, x_{rj}$ are *not* transversal local parameters of V_j at O. Then, by the irreducible case we would have

$$\partial x_{r+1,j}/\partial x_{r,j} \notin \bar{\mathfrak{o}}_j,$$

and this would imply that

$$\partial x_{r+1}/\partial x_r \notin \bar{\mathfrak{o}},$$

whence $\bar{\mathfrak{o}}J \neq \bar{\mathfrak{o}}f'_{x_{r+1}}$. We have therefore shown that if $x_1, x_2, \cdots, x_r$ are equisingular local parameters, then $\bar{\mathfrak{o}}J = \bar{\mathfrak{o}}f'_{x_{r+1}}$ if and only if these parameters are transversal. In other words: *if condition* 3) *of Theorem* 5.2 *is satisfied then conditions* 1) *and* 2) *are equivalent.* Assertion B), proved above, says that 1) and 2) together imply 3). So, in order to complete the proof of Theorem 5.2, *we have only to show that conditions* 1) *and* 2) *of that theorem are equivalent.*

We first deal with the case in which V is irreducible. We fix a system $(y_1, y_2, \cdots, y_{r+1})$ of local coördinates of V at O such that the following conditions are satisfied [compare with part C) of the proof of Theorems 4.4 and 4.5]:

a) $y = (y_1, y_2, \cdots, y_r)$ is a system of transversal equisingular parameters.

b) The critical variety Δ_y relative to these parameters is $Y_r = 0$.

c) If

$$g(Y_1, Y_2, \cdots, Y_r, Y_{r+1}) = 0$$

is the embedding of V in an affine $(r+1)$-space, determined by the local coördinates, $y_1, y_2, \cdots, y_r, y_{r+1}$, then the singular locus W of V is defined by $Y_r = Y_{r+1} = 0$; here g is a monic polynomial in Y_{r+1}, of degree $s = m_V(O)$. We have (see (16) and (17)) $\bar{\mathfrak{o}} = k[[y_1, y_2, \cdots, y_{r-1}, t]]$ (where $t^s = y_r$) and

$$(22) \qquad y_{r+1} = a_s(y_1, y_2, \cdots, y_{r-1})t^s + a_{s+1}(y_1, y_2, \cdots, y_{r-1})t^{s+1} + \cdots,$$

since the leading form g_s depends only on Y_r and Y_{r+1}. If we replace y_{r+1} by

$$y_{r+1} - a_s(y_1, y_2, \cdots, y_{r-1})y_r,$$

then we will have $a_s = 0$ in (22), and the leading form g_s of g will be $Y_{r+1}{}^s$. Hence the r partial derivatives

$$\partial y_{r+1}/\partial y_i, \qquad i = 1, 2, \cdots, r,$$

are non-units in $\bar{\mathfrak{o}}$. In other words, the quotients

$$g'_{y_i}/g'_{y_{r+1}}, \qquad i = 1, 2, \cdots, r,$$

are non-units in $\bar{\mathfrak{o}}$. By part C) of the proof we know that $\bar{\mathfrak{o}}J = \bar{\mathfrak{o}}g'_{y_{r+1}}$. We can write

$$y_i = \sum_{j=1}^{r+1} c_{ij}x_j + \text{terms of higher degree}, \qquad (i = 1, 2, \cdots, r+1)$$

where the c_{ij} are in k (and $|c_{ij}| \neq 0$).
The direction $X_1 = X_2 = \cdots = X_r = 0$ is not tangent to V if and only if $c_{r+1,r+1} \neq 0$.

Let

$$f(X) = g(\sum c_{1j}X_j + \cdots, \sum c_{2j}X_j + \cdots, \sum c_{r+1,j}X_j + \cdots).$$

Then

$$f(X) = 0$$

is an equation of the embedding of V, relative to the local coördinates $x_1, x_2, \cdots, x_{r+1}$. We have:

$$f'_{x_{r+1}} = \sum_{i=1}^{r} (c_{i,r+1} + \cdots) g'_{y_i} + (c_{r+1,r+1} + \cdots) g'_{y_{r+1}},$$

where the dots stand for non-units in $\mathfrak{o}$. Since $g'_{y_i}/g'_{y_{r+1}}$ is a non-unit in $\bar{\mathfrak{o}}$ for $i = 1, 2, \cdots, r$, it follows that $\bar{\mathfrak{o}} f'_{x_{r+1}} = \mathfrak{o} g'_{y_{r+1}}$ $(= \bar{\mathfrak{o}} J)$, if and only if $c_{r+1,r+1} \neq 0$, i. e., if and only if $x_1, x_2, \cdots, x_r$ are transversal parameters. This completes the proof in the irreducible case.

Now consider the case in which V is reducible. Let $V_1, V_2, \cdots, V_h$ be the irreducible components of V. In the notations of part C) of the proof, the local parameters $x_1, x_2, \cdots, x_r$ are transversal if and only if, for each $j = 1, 2, \cdots, h$, the local parameters $x_{1j}, x_{2j}, \cdots, x_{rj}$ of V_j at O are transversal. In view of the equalities

$$\frac{\partial x_{r+1}}{\partial x_i} = \sum_{j=1}^{h} e_j \frac{\partial x_{r+1,j}}{\partial x_{i,j}}, \qquad (i = 1, 2, \cdots, r)$$

we have $\partial x_{r+1}/\partial x_i \in \bar{\mathfrak{o}}$ if and only if $\partial x_{r+1,j}/\partial x_{i,j} \in \bar{\mathfrak{o}}_j$ for all j. In other words, $\bar{\mathfrak{o}} J = \mathfrak{o} f'_{x_{r+1}}$, if and only if

$$\bar{\mathfrak{o}} J_j = \bar{\mathfrak{o}}_j f'_{j;x_{r+1,j}}, \qquad j = 1, 2, \cdots, h,$$

where J_j is the Jacobian ideal of V_j at O and where f_j is the irreducible factor of f such that $f_j = 0$ is the equation of V_j. In view of the irreducible case settled above, this completes the proof of Theorems 5.1 and 5.2.

The following is an important consequence of Theorem 5.2:

COROLLARY 5.3. *If V has at O a singularity of dimensionality type* 1 *and if W is the singular locus of V, then every W-transversal section $V^*_{(y)}$ of V at O is an embedded curve (i. e., an embedded reduced algebroid scheme, of dimension* 1). *Furthermore, all W-transversal sections $V^*_{(y)}$ of V at O have equivalent singularities at O.*

For, let $(y_1, y_2, \cdots, y_{r-1})$ be any system of W-transversal parameters of V at O. We know already that $V^*_{(y)}$ is embedded (Proposition 3.4) and has dimension 1 (Corollary 4.6). We complete the set $(y_1, y_2, \cdots, y_{r-1})$ to a system of local coördinates $(y_1, y_2, \cdots, y_{r+1})$ in such a manner that $\mathfrak{o}\cdot y_r + \mathfrak{o}\cdot y_{r+1}$ is the prime ideal of W in $\mathfrak{o}$. Let $f(Y_1, Y_2, \cdots, Y_{r+1}) = 0$ be the associated embedding of V in affine $(r+1)$-space. We know then that the leading form f_s of f is a binary form $f_s(Y_r, Y_{r+1})$ in Y_r, Y_{r+1} (this follows from the equimultiplicity of V along W, at O). Without loss of generality we may assume that $f_s(0,1) \neq 0$ (replace y_r, y_{r+1} by "non-special" linear combinations of y_r, y_{r+1}, with coefficients in k). Then $y_1, y_2, \cdots, y_{r-1}, y_r$ are transversal parameters, and therefore—by Theorem 5.2—equisingular. We have here precisely the situation which was reached at the very end of the proof of Theorems 4.4 and 4.5, and this allows us to conclude that $V^*_{(y)}$ is a curve and that this curve has at O a singularity which is equivalent to the singularity which the W-transversal section of V at the general point P of W has at P.

We shall conclude this section by giving a characterization of equisingular local parameters $x_1, x_2, \cdots, x_r$ (under the assumption that it is known already that V has at O a singularity of dimensionality type 1).

Let V have at O a singularity of dimensionality type 1, let W be the singular locus of V and let $x_1, x_2, \cdots, x_r$ be local parameters of V at O. *We shall use the notations of the Note which follows immediately the statement of Theorem* 4.5, except that if $W' = \pi\{W\}$ then we denote by H the cylindrical hypersurface $\pi^{-1}\{W'\}$.

PROPOSITION 5.4. *In order that the local parameters $x_1, x_2, \cdots, x_r$ be equisingular, it is necessary and sufficient that the following equality hold:*

$$i(V\cdot H, W; \mathbf{A}_{r+1}) = i(V\cdot L, O; \mathbf{A}_{r+1}) \tag{23}$$

Proof. Assume $x_1, x_2, \cdots, x_r$ are equisingular. Since W' is part of the (*non-singular*) critical variety Δ_x and has dimension $r-1$, it follows that $W' = \Delta_x$. So in this case, H has the same meaning in (23) as it does in the cited Note, and (23) is now merely the equality (8).

Conversely, assume (23). Let ν denote the common value of both sides of (23). Then the equation $f(X_1, X_2, \cdots, X_r, X_{r+1}) = 0$ of the hypersurface V is monic in X_{r+1}, of degree ν. Let P be the prime ideal which defines W and let $\mathfrak{o}/P = k[[\xi_1, \xi_2, \cdots, \xi_r, \xi_{r+1}]]$. The fact that $i(V\cdot H, W; \mathbf{A}_{r+1}) = \nu$ signifies that ξ_{r+1} is a ν-fold root of the polynomial $f(\xi_1, \xi_2, \cdots, \xi_r; X_{r+1})$. Hence $\xi_{r+1} \in k[[\xi_1, \xi_2, \cdots, \xi_r]]$, say $\xi_{r+1} = a(\xi_1, \xi_2, \cdots, \xi_r)$. Upon replacing x_{r+1} by $x_{r+1} - a(x_1, x_2, \cdots, x_r)$ we may therefore assume that $\xi_{r+1} = 0$. Since

$\dim W' = r - 1$, W' is defined by an irreducible equation $g(X_1, X_2, \cdots, X_r) = 0$, where $g \in k[[X_1, X_2, \cdots, X_r]]$, and W is defined by $g = 0$, $X_{r+1} = 0$. Since O is a simple point of W, g begins with terms of degree 1. So, without loss of generality, we may assume that $g = X_r$ and that W is the linear space $X_r = X_{r+1} = 0$. Upon replacing x_{r+1} by $cx_r + dx_{r+1}$, where c and d are non special constants in k, we may assume that $x_1, x_2, \cdots, x_{r-1}, x_{r+1}$ *are transversal parameters* of V at O (hence equisingular by Theorem 5.2).

To show that $x_1, x_2, \cdots, x_r$ are equisingular parameters we have to show that the associated critical variety Δ_x is the space $X_r = 0$. Now, clearly, it will be sufficient to show that for each irreducible component V_j of V, the critical variety $\Delta_x{}^j$ of V_j, associated with the local parameters $x_{1j}, x_{2j}, \cdots, x_{rj}$ (see part C of the proof of Theorem 5.1), is $X_r = 0$ (or is empty). So *we may assume that V is irreducible* (since the validity of (22) for V implies the validity of the similar equality for each V_j). In that case we have

$$\bar{\mathfrak{o}} = k[[x_1, x_2, \cdots, x_{r-1} t]],$$

where

$$t^s = x_{r+1},$$

and, by (22),

$$x_r = a_\nu(x_1, x_2, \cdots, x_{r-1}) t^\nu + a_{\nu+1}(x_1, x_2, \cdots, x_{r-1}) t^{\nu+1} + \cdots$$

where the a_i are power series. Since the hyperplane $X_r = 0$ meets V only in W ($X_{r+1} = 0$ a ν-fold proof of $f(X_1, X_2, \cdots, X_{r-1}, 0; X_{r+1})$), it follows that a_ν is a unit in $k[[x_1, x_2, \cdots, x_{r-1}]]$. We have

$$(24) \qquad f'_{x_{r+1}} = -f'_{x_r} \cdot \frac{\partial x_r}{\partial x_{r+1}} = -f'_{x_r} \cdot t^{\nu - s} \text{ times a unit in } \bar{\mathfrak{o}}.$$

Now, by Theorem 5.2, the local parameters $x_1, x_2, \cdots, x_{r-1}, x_{r+1}$ are equisingular. Hence W is the only subvariety of V along which f'_{x_r} is zero (Theorem 4.5, part 2a). Since also t vanishes only on W ($X_{r+1} = 0$ being the critical variety Δ of V, relative to the equisingular parameters $x_1, x_2, \cdots, x_{r-1}, x_{r+1}$), it follows from (24) that $f'_{x_{r+1}}$ vanishes only on W. This shows that Δ_x is just $X_r = 0$ and completes the proof.

6. Generalities on dilatations. Let V be an algebroid variety (in the sense of § 1), and let $\mathfrak{p}_1, \mathfrak{p}_2, \cdots, \mathfrak{p}_h$ be the prime ideals of the zero ideal in $\mathfrak{o}$. Let $\mathfrak{D}$ be an ideal in $\mathfrak{o}$ and let D_0 denote the set of elements of $\mathfrak{D}$ which are not zero divisors (D_0 may be empty; this happens if and only if $\mathfrak{D} \subset \mathfrak{p}_j$, for some $j = 1, 2, \cdots, h$).

Consider the reduced scheme

$$S = \bigcup_{x \in D_0} \operatorname{Spec}(\mathfrak{o}[x^{-1}\mathfrak{D}]).$$

Here, if K denotes the total ring of quotients of $\mathfrak{o}$, $x^{-1}\mathfrak{D}$ stands for the set of all elements y/x of K, where y ranges over $\mathfrak{D}$. The canonical morphism $T\colon S \rightarrow V$ is called the *$\mathfrak{D}$-dilatation* of V.

If D_0 is empty, then $\mathfrak{o}[x^{-1}\mathfrak{D}]$ is simply $\mathfrak{o}$, $S = V$ and T is the identity.

Assume D_0 is not empty. An elementary argument shows that there exist bases of $\mathfrak{D}$ which are contained in D_0. Let $(u_1, u_2, \cdots, u_q)$ be such a basis. Then it is easily seen that

$$S = \bigcup_{\alpha=1}^{q} \operatorname{Spec}(\mathfrak{o}'_\alpha),$$

where

$$\mathfrak{o}'_\alpha = \mathfrak{o}[\frac{u_1}{u_\alpha}, \frac{u_2}{u_\alpha}, \cdots, \frac{u_q}{u_\alpha}].$$

If P is any point of V, we denote by $T^{-1}\{P\}$ the set of points P' of S such that $T(P') = P$. It follows immediately from the definition of S that if $P \notin \mathfrak{V}(\mathfrak{D})$ then $T^{-1}\{P\}$ consists of a single point P' and that $\mathfrak{o}_{V,P} = \mathfrak{o}_{S,P'}$ (*biregularity of T on $V - \mathfrak{V}(\mathfrak{D})$*).

We denote by S_P the scheme

$$S_P = \bigcup_{P' \in T^{-1}\{P\}} \operatorname{Spec}(\mathfrak{o}_{S,P'}).$$

We set $\mathfrak{O} = \mathfrak{o}_{V,P}$, we denote by ρ the natural homomorphism of $\mathfrak{o}$ into $\mathfrak{O}$ and we set $\rho(u_\alpha) = \bar{u}_\alpha$. Since the kernel of ρ consists entirely of those elements ξ of $\mathfrak{o}$ for which there exists an element η in $\mathfrak{o}$, $\eta \notin P$, such that $\xi\eta = 0$, and since no u_α is a zero divisor, it follows that no $\bar{u}_\alpha$ is a zero divisor in $\mathfrak{O}$. We set

$$\mathfrak{O}'_\alpha = \mathfrak{O}[\frac{\bar{u}_1}{\bar{u}_\alpha}, \frac{\bar{u}_2}{\bar{u}_\alpha}, \cdots, \frac{\bar{u}_q}{\bar{u}_\alpha}], \qquad (\alpha = 1, 2, \cdots, q) \tag{25}$$

we adjoin a transcendental z_1 to the total ring of quotients of $\mathfrak{O}$, we set $z_\alpha = z_1 \cdot \dfrac{\bar{u}_\alpha}{\bar{u}_1}$ and

$$R' = \mathfrak{O}[z_1, z_2, \cdots, z_q], \tag{25'}$$

whence R' is a homogeneous ring over $\mathfrak{O}$. It is then immediately seen that

$$S_P = \bigcup_{\alpha=1}^{q} \operatorname{Spec}(\mathfrak{O}'_\alpha), \tag{26}$$

or—what is the same thing—

$$S_P = \operatorname{Proj}(R'). \tag{26'}$$

Denote by $\mathfrak{D}_P$ the extension of the ideal $\mathfrak{D}$ to $\mathfrak{O}$, i.e., let $\mathfrak{D}_P = \mathfrak{O}\rho(\mathfrak{D})$. Let $V_P = \operatorname{Spec}(\mathfrak{O})$. Then it follows at once from (26) that *S_P is the transform of V_P by the $\mathfrak{D}_P$-dilatation*. This dilatation will be denoted by T_P.

From the expression (26′) of S_P it follows that if $\mathfrak{M}$ is the maximal ideal of $\mathfrak{O}$ then the fibre $T^{-1}\{P\}$, as a subspace of S, can be identified with $\operatorname{Proj}(R'/R'\mathfrak{M})$. Thus, *$T^{-1}\{P\}$ is a projective model, defined over the field $\boldsymbol{k}(P)$.*

In particular, S_O is a projective variety defined over k.

It is clear that the closed points of $T^{-1}\{P\}$ (and also of S_P) are those and only those points P' of $T^{-1}\{P\}$ for which $\boldsymbol{k}(P') = \boldsymbol{k}(P)$. In particular, *the closed points O' of S_O are those points O' of S_O for which $\boldsymbol{k}(O') = k$. However, it is easily seen that $S_O = S$* and that consequently *the closed points of S are the points O' such that $\boldsymbol{k}(O') = k$, and that all these points are in $T^{-1}\{O\}$.*

Proof. If $P' \in S$, we may assume that $P' \in \operatorname{Spec}(\mathfrak{o}'_1)$, and that if $\mathfrak{p}'_1$ is the prime ideal of $\mathfrak{o}'_1$ which represents the point P' then

$$\frac{u_i}{u_1} \notin \mathfrak{p}'_1, \text{ for } i = 1, 2, \cdots, n;$$

$$\frac{u_\nu}{u_1} \in \mathfrak{p}'_1, \text{ for } \nu = n+1, n+2, \cdots, q.$$

Then it follows at once that

$$P' \in \operatorname{Spec}(\mathfrak{o}'_i), \text{ for } i = 1, 2, \cdots, n;$$

$$u_j/u_i \text{ is a unit in } \mathfrak{o}_{S,P'}, \text{ for } i, j = 1, 2, \cdots, n;$$

$$u_\nu/u_i \in \mathfrak{m}_{S,P'}, \text{ for } \nu = n+1, n+2, \cdots, q; i = 1, 2, \cdots, n.$$

Let t_i be the $\mathfrak{p}'_1$-residue of u_i/u_1 $(i = 1, 2, \cdots, n)$. If $\mathfrak{p}'_i$ is the prime ideal of $\mathfrak{o}_i'$ which represents the point P' $(i = 1, 2, \cdots, n)$, then it is seen at once that

$$\text{(27)} \qquad \mathfrak{o}'_i/\mathfrak{p}'_i = (\mathfrak{o}/\mathfrak{p})\left[\frac{t_1}{t_i}, \frac{t_2}{t_i}, \cdots, \frac{t_n}{t_i}\right], \qquad (i = 1, 2, \cdots, n),$$

where $\mathfrak{p}$ is the prime ideal in $\mathfrak{o}$ which represents the point $P = T(P')$. Since $P' \notin \operatorname{Spec}(\mathfrak{o}'_\nu)$, for $\nu = n+1, n+2, \cdots, q$, the (27) show at once that the closure $\overline{\{P'\}}$ of P' in S' is given by

$$\bigcup_{i=1}^{n} \operatorname{Spec}\left(\mathfrak{o}/\mathfrak{p}\left[\frac{t_1}{t_i}, \frac{t_2}{t_i}, \cdots, \frac{t_n}{t_i}\right]\right),$$

and is a projective model over the local domain $\mathfrak{o}/\mathfrak{p}$. This model is reduced to a point if and only if $P = O$ and $\mathfrak{o}'_1/\mathfrak{p}'_1 = \mathfrak{o}/\mathfrak{m} = k$. This completes the proof.

If O' is a closed point of S, the local ring $\mathfrak{o}_{S,O'}$ is noetherian, equicharacteristic, and has k as residue field. If $\mathfrak{o}'^*$ denotes the completion of this local ring, then $\mathrm{Spec}(\mathfrak{o}'^*)$ is an algebroid unmixed variety, of dimension r, defined over k. We shall denote this algebroid variety by $S^*_{O'}$ and we shall refer to it as *the completion* of S at O'. It is the completions of S at its various closed points that we will primarily be concerned with in the sequel.

If W is a subvariety (= closed subset) of V, we mean by the *total transform of* W on S (in symbols: $T^{-1}\{W\}$) the set of all $P' \in S'$ such that $T(P') \in W$. It is immediately seen that $T^{-1}\{W\}$ is the underlying space of a closed subscheme of S, namely of

$$\bigcup_{\alpha=1}^{q} \mathrm{Spec}(\mathfrak{o}'_\alpha/\mathfrak{o}'_\alpha\mathfrak{A}),$$

where $\mathfrak{A}$ is any ideal in $\mathfrak{o}$ such that $W = \mathfrak{V}(\mathfrak{A})$.

In particular, consider the subvariety $W = F = \mathfrak{V}(\mathfrak{D})$. Since $\mathfrak{o}'_\alpha\mathfrak{D} = \mathfrak{o}'_\alpha u_\alpha$, it follows that if P' is any point of $T^{-1}\{F\}$, then $T^{-1}\{F\}$ is defined, locally at P', by the principal ideal $\mathfrak{o}_{S,P'} \cdot u_\alpha$, for a suitable $\alpha = 1, 2, \cdots, q$. Since it is clear that the closed points of $T^{-1}\{F\}$ are also closed in S, it follows that $T^{-1}\{F\}$ *is unmixed, of dimension* $r-1$, *at each of its closed points.*

If W is an *irreducible* sub-variety of V, we define the *proper transform* $T^{-1}[W]$ by

$$T^{-1}[W] = \text{Closure of } T^{-1}\{P\},$$

where P is the general point of W. It is easily seen that $T^{-1}[W]$ is the union of those irreducible components of $T^{-1}\{W\}$ whose general point lies in $T^{-1}\{P\}$.

PROPOSITION 6.1. *Let P be a point of V, let $\mathfrak{O} = \mathfrak{o}_{V,P}$ and let ρ be the natural homomorphism of $\mathfrak{o}$ into $\mathfrak{O}$. If $T^{-1}\{P\}$ is a finite set then there exists a subring $\mathfrak{O}'$ of the total ring of quotients K of $\mathfrak{O}$ such that* (1) $\mathfrak{O} \subset \mathfrak{O}'$ *and $\mathfrak{O}'$ is a finite $\mathfrak{O}$-module;* (2) $S_P = \mathrm{Spec}\,(\mathfrak{O}')$. [In other words, S_P is dominated by the normalization of $V_P = \mathrm{Spec}(\mathfrak{O})$.]

Proof. This proposition is well-known and is, in fact, a special (and elementary) case of the "main theorem" ([1]). In the general context of schemes, its proof can be found in ([6], II, 6.2). For convenience of the reader, we outline here a proof which is merely an adaptation of an argument found in our paper ([1], pp. 506-508).

We start with an arbitrary basis $(u_1, u_2, \cdots, u_q)$ of $\mathfrak{D}$. We have $S_P = \mathrm{Proj}(R')$, where R' is the homogeneous ring, over $\mathfrak{O}$, defined in (25').

Sinve $T^{-1}\{P'\}$ is the underlying space of $\mathrm{Proj}(R'/R'\mathfrak{M})$ (where $\mathfrak{M}$ is the maximal ideal of $\mathfrak{O}$), the assumption that $T^{-1}\{P\}$ is a finite set is equivalent with the following: *there is only a finite number of prime (homogeneous) ideals in R' which contain $R'\mathfrak{M}$.* We can therefore find in R' a homogeneous element, of some positive degree n, which is not contained in any of the *non-irrelevant prime* (homogeneous) ideals of $R'\mathfrak{M}$. Let η be such an element, and let, say $\eta = f(z_1, z_2, \cdots, z_q)$, where f is a form, of degree n, with coefficients in $\mathfrak{O}$. In view of the above stated property of $R'\mathfrak{M}$ and in view of our choice of η, it follows that the ideal $R'\mathfrak{M} + R'\eta$ is irrelevant (i.e., it contains a power of the ideal $\mathfrak{Z} = R'z_1 + R'z_2 + \cdots + R'z_q$). Now, it is clear that if $\mathfrak{P}$ is any maximal element in the set of all *non-irrelevant* prime homogeneous ideals of R', then $\mathfrak{P} \cap \mathfrak{O} = \mathfrak{M}$, whence $\mathfrak{P} \supset R'\mathfrak{M}$ and thus $\eta \notin \mathfrak{P}$. *It follows that $R'\eta$ is itself an irrelevant ideal.*

Let, then, d be a positive integer such that $\mathfrak{Z}^d \subset R'\eta$. We denote by $\Omega_1(z), \Omega_2(z), \cdots, \Omega_N(z)$ (in some order) the monomials in $z_1, z_2, \cdots, z_q$ of degree d. Then, if $w(z)$ is *any* monomial in $z_1, z_2, \cdots, z_q$ of degree n, we have relations of the form

$$(28) \qquad \Omega_i(z) \cdot w(z) = \eta \sum_{j=1}^{N} a_{ij}\Omega_j(z), \quad (a_{ij} \in \mathfrak{O}) \qquad (i = 1, 2, \cdots, N).$$

Since none of the elements $\bar{u}_1, \bar{u}_2, \cdots, \bar{u}_q$ is a zero division in $\mathfrak{O}$, it follows that also none of the elements $z_1, z_2, \cdots, z_q$ is a zero divisor in R'. Therefore, it follows from (28) that

$$| w(z) - \delta_{ij}a_{ij}\eta | = 0,$$

and hence

$$(29) \qquad \frac{w(z)}{\eta} \left(= \frac{w(\bar{u})}{f(\bar{u})}\right) \text{ is integral over } \mathfrak{O}.$$

We now observe that if $R'^n = \sum_{j=1}^{\infty} R'_{jn}$, then we also have $S_P = \mathrm{Proj}(R'^n)$. Since $R'^n = \mathfrak{O}[w_1(z), w_2(z), \cdots, w_N(z)]$, where het $w_i(z)$ are the various monomials, of degree n, in $z_1, z_2, \cdots, z_q$, and since $f(z)$ is a linear combination of the $w_i(z)$, with coefficients in $\mathfrak{O}$, we deduce from (29) (which holds for any $w = w_i$) that the following ring $\mathfrak{O}'$ satisfies all the conditions of this proposition:

$$\mathfrak{O}' = \mathfrak{O}\left[\frac{w_1(\bar{u})}{f(\bar{u})}, \frac{w_2(\bar{u})}{f(\bar{u})}, \cdots, \frac{w_{N'}(\bar{u})}{f(\bar{u})}\right]. \qquad \text{Q.E.D.}$$

7. Equisingularity and monoidal dilatations. Let F be an irreducible subvariety of V, and let $\mathfrak{p}$ be the prime ideal of F in $\mathfrak{o}$. By *the monoidal* *transformation of V with center F (or centered at F)* we means the $\mathfrak{p}$-dilatation of V.

LEMMA 7.1. *Let F be an irreducible subvariety of V and let $T: S \rightarrow F$ be the monoidal transformation of V, centered at F. If V is embedded and if O is a simple point of F, then also $S^*_{O'}$ is embedded for any closed point O' of S.*

Proof. If $\rho = \mathrm{cod}_V F$, then it follows from Proposition 3.5 and from the proof of that proposition, that there exists a system of local coördinates $x_1, x_2, \cdots, x_{r+1}$ of V at O such that $(x_1, x_2, \cdots, x_{\rho+1})$ is a basis of $\mathfrak{p}$. (The elements $x_{\rho+2}, x_{\rho+3}, \cdots, x_{r+1}$ are then F-transversal parameters of V at O). Now, if O' is a closed point of S, and if, say,

$$O' \in \mathrm{Spec}(\mathfrak{o}[\frac{x_2}{x_1}, \frac{x_3}{x_1}, \cdots, \frac{x_{\rho+1}}{x_1}]),$$

then $\boldsymbol{k}(O') = k$ and we may assume that $x_i/x_1 \in \mathfrak{m}' = \mathfrak{m}_{S,O'}$, for $i = 2, 3, \cdots, \rho + 1$. It follows immediately that the elements

$$x_1, \frac{x_2}{x_1}, \cdots, \frac{x_{\rho+1}}{x_1}, x_{\rho+2}, \cdots, x_{r+1}$$

form a basis of $\mathfrak{m}'$. This completes the proof.

PROPOSITION 7.2. *With the same assumptions as those of Lemma 7.1, assume furthermore that* $\mathrm{cod}\, F = 1$. *Then the following conditions are equivalent:*

(1) *V is equimultiple along F, at O.*

(2) *$T^{-1}\{O\}$ is a finite set.*

Proof. Let $x_1, x_2, \cdots, x_r, x_{r+1}$ be local coördinates of V at O, such that (x_r, x_{r+1}) is a basis of the prime ideal of F and such that $x_1, x_2, \cdots, x_r$ are local transversal parameters of V at O. Let

$$f(X_1, X_2, \cdots, X_{r+1}) = 0$$

be the associated embedding of V in an affine $(r+1)$-space; here f is a monic polynomial in X_{r+1}, of degree $s = m_V(O)$. Assume (1), i.e., assume that $m_V(P) = m_V(O) = s$, where P is the general point of F. Then the leading forms of f is a form of degree s, which depends only on X_r, X_{r+1}:

$$f = f_s(X_r, X_{r+1}) + \text{terms of higher degree}.$$

We have then:

$$f = X_{r+1}{}^s + \sum_{i=1}^{s} A_i(X_1, X_2, \cdots, X_r) X_r{}^i X_{r+1}^{s-i},$$

where A_i is a power series in $X_1, X_2, \cdots, X_r$. Clearly x_r is not a zero divisor in $\mathfrak{o}$. Setting $x'_{r+1} = x_{r+1}/x_r$ we find from $f(x_1, x_2, \cdots, x_{r+1}) = 0$:

$$\frac{f(x)}{x_r{}^s} = x'_{r+1}{}^s + \sum_{i=1}^{s} A_i(x_1, x_2, \cdots, x_r) x'^{s-i}_{r+1} = 0.$$

Thus x'_{r+1} is integral over $\mathfrak{o}$. Hence $\mathfrak{o}[\frac{x_{r+1}}{x_r}]$ is a finite $\mathfrak{o}$-module and

$$S = \operatorname{Spec}(\mathfrak{o}[\frac{x_{r+1}}{x_r}]),$$

which proves (2).

Conversely, assume (2). We apply Proposition 6.1. Due to the fact that k is infinite, we can choose as an element η, in the proof of that proposition, an element of the form $az_1 + bz_2$, with a, b in k and $z_1/z_2 = x_r/x_{r+1}$. So we may assume that $\eta = z_1$, and the proof of Proposition 6.1 tells us that $\mathfrak{o}[\frac{x_{r+1}}{x_r}]$ is integral over $\mathfrak{o}$. We may also assume that $x_1, x_2, \cdots, x_r$ are local parameters of V at O, and that consequently $\mathfrak{o}$ is integral over the power series ring $k[[x_1, x_2, \cdots, x_r]]$. We have then a relation of the form

$$\left(\frac{x_{r+1}}{x_r}\right)^n + \sum_{j=1}^{n} B_j(x_1 \cdots x_r) \left(\frac{x_{r+1}}{x_r}\right)^{n-j} = 0,$$

or

$$g(x_1, x_2, \cdots, x_{r+1}) = 0, \text{ where}$$

$$g(X_1, X_2, \cdots, X_{r+1}) = X^n_{r+1} + \sum_{j=1}^{n} B_j(X_1, \cdots, X_r) X_r{}^j X^{s-j}_{r+1}.$$

The leading form of g is of degree n, and depends only on X_r, X_{r+1}. Since f is a factor of g, it follows that also the leading form of f depends only on X_r, X_{r+1}, and this proves (1) and completes the proof of the proposition.

COROLLARY 7.3. *The assumption being as in Proposition 7.2, let $x_1, x_2, \cdots, x_r, x_{r+1}$ be local coördinates of V at O such that:* (1) (x_r, x_{r+1}) *is the prime ideal of F;* (2) $x_1, x_2, \cdots, x_r$ *are local transversal parameters of V at O. Let $f_s(X_r, X_{r+1})$ be the leading form of f and let, say, m be the number of distinct linear factors of f_s:*

$$f_s(X_r, X_{r+1}) = \prod_{\alpha=1}^{m} (X_{r+1} - c_\alpha X_r)^{\lambda_\alpha}, \quad c_\alpha \in k$$

$$c_\alpha \neq c_\beta \text{ if } \alpha \neq \beta.$$

16

Let $X'_{r+1} = X_{r+1}/X_r$ *and let*

$$f(X_1, X_2, \cdots, X_r, X_r X'_{r+1}) = X_r^s f'(X_1, X_2, \cdots, X_r, X'_{r+1}).$$

Then f' *factors into* m *distinct factors in* $k[[X_1, X_2, \cdots, X_r]][X'_{r+1}]$:

$$f'(X_1, X_2, \cdots, X_r, X'_{r+1}) = \prod_{\alpha=1}^{m} f'_\alpha(X_1, X_2, \cdots, X_r, X'_{r+1}),$$

where f'_α *is a monic polynomial in* X_{r+1}, *of degree* λ_α, *and* $f'_\alpha(0, 0, \cdots, 0, X'_{r+1}) = (X'_{r+1} - c_\alpha)^{\lambda_\alpha}$. *If we denote by* V'_α *the algebroid variety in the affine space of the variables* $X_1, X_2, \cdots, X_r, X'_{r+1}$ *centered at the point* $O'_\alpha = (0, 0, \cdots, 0, c_\alpha)$ *and defined by* $f'_\alpha(X_1, X_2, \cdots, X_r, X'_{r+1}) = 0$, *then the* T*-transform* S *of* V *is the union of the* m *(disjoint) algebroid varieties* $V'_1, V'_2, \cdots, V'_m$. *Furthermore,* $T\{O\} = (O'_1, O'_2, \cdots, O'_m)$.

Obvious.

THEOREM 7.4. *Assume that the singular locus* W *of* V *is of codimension* 1 *and has at* O *a simple point. Let* $T\colon S \to V$ *be the monoidal dilatation of* V, *centered at* W, *and let* P *be the general point of* W. *Then* O *is a singularity of dimensionality type* 1 *if and only if the following conditions are satisfied:*

(1) $T^{-1}\{O\}$ *is a finite set, and the number of points in* $T^{-1}\{O\}$ *is the same as that in* $T^{-1}\{P\}$.

(2) *If* $T^{-1}\{O\} = (O'_1, O'_2, \cdots, O'_m)$, *then, in the notations of Corollary* 7.3, *each point* O'_α *is either a simple point of* V'_α, *or is a singularity of* V'_α *of dimensionality type* 1.

Proof. A) Assume that V has at O a singularity of dimensionality type 1. Then V is equimultiple along W, at O (Theorem 4.5, part 1b), and hence $T^{-1}\{O\}$ is a finite set (Proposition 7.2). Let V^*_P be the W-transversal section of V at P. Let $x_1, x_2, \cdots, x_{r-1}, x_r, x_{r+1}$ be as in Corollary 7.3 and let $V^*_{(x)}$ be the W-transversal section of V at O, relative to the W-transversal parameters $x_1, x_2, \cdots, x_{r-1}$. Then $V^*_{(x)}$ and $V^{(*)}{}_P$ are algebroid *curves* having equivalent singularities at O and P respectively (Corollary 5.3). Furthermore, $V^*_{(x)}$ and V^*_P are defined respectively by the equations

$$f(0, 0, \cdots, 0; X_r, X_{r+1}) = 0,$$
$$f(x_1, x_2, \cdots, x_{r-1}; X_r, X_{r+1}) = 0,$$

both P and O being represented now by the origin in (X_r, X_{r+1})-plane. By Corollary 7.3, the number m of points in $T^{-1}\{O\}$ is the number of distinct tangent lines of $V^*_{(x)}$. A similar argument shows that the number of points of $T^{-1}\{P\}$ is equal to the number of distinct tangent lines of V^*_P (it is sufficient to observe that—in the notation of § 6—we have $T^{-1}\{P\} = T_P^{-1}\{P\}$ and that T_P is the monoidal transformation of V^*_P, centered at P). Since $V^*_{(x)}$ and V^*_P have equivalent singularities, assertion (1) follows.

To prove assertion (2), we first observe that, by Theorem 5.2, the parameters $x_1, x_2, \cdots, x_r$ are equisingular, and that consequently the discriminant Δ of f with respect to X_{r+1} is a power of X_r (apart from a unit factor). This implies—in the notations of Corollary 7.3—that, for each $\alpha = 1, 2, \cdots, m$, the discriminant Δ'_α of f'_α, with respect to X'_{r+1}, is a power of X_r. Thus $x_1, x_2, \cdots, x_r$ are equisingular parameters of V'_α at O'_α, showing (Theorem 4.4) that O'_α is a singularity of V'_α of dimensionality type 1 (or is a simple point of V'_α).

B) Assume now that conditions (1) and (2) are satisfied. Again we choose local coördinates $x_1, x_2, \cdots, x_{r+1}$ as in Corollary 7.3 and we use the notations of that corollary. Furthermore, we assume that the W-transversal section $V^*_{(x)}$ of V, at O, relative to the W-transversal parameter $x_1, x_2, \cdots, x_{r-1}$, is a *curve*.

The points of $T^{-1}\{P\}$ can be identified with the points $(x_1, x_2, \cdots, x_{r-1}, 0, \xi')$, where ξ' is any of the roots of the polynomial $f'(x_1, x_2, \cdots, x_{r-1}, 0, X'_{r+1})$ (in X'_{r+1}). By condition (1), there exist exactly m such roots. Hence, for each $\alpha = 1, 2, \cdots, m$, the polynomial $f'_\alpha(x_1, x_2, \cdots, x_{r-1}, 0, X'_{r+1})$ has exactly one root $X'_{r+1} = \xi'_\alpha$, necessarily λ_α-fold. So $\xi'_\alpha \in k[[x_1, x_2, \cdots, x_{r-1}]]$. Now, the singular locus of V lies above $X_r = 0$. Hence, the singular locus of V'_α lies above a subvariety of the hyperplane $X_r = 0$, hence is either empty (if O'_α is simple for V'_α) or projects onto $X_r = 0$ (if O'_α is a singularity of dimensionality type one; use condition (2) of the theorem). Since

$$f'_\alpha(0, \cdots, 0, 0, X'_{r+1}) = (X'_{r+1} - c_\alpha)^{\lambda_\alpha},$$

while

$$f'_\alpha(x_1, x_2, \cdots, x_{r-1}, 0, X'_{r+1}) = (X'_{r+1} - \xi'_\alpha)^{\lambda_\alpha},$$

it follows from Proposition 5.4 that $x_1, x_2, \cdots, x_r$ are equisingular parameters of V'_α at O'_α; in other words: the X'_{r+1}-discriminant of f'_α is a power of X_r, apart from a unit factor (this conclusion holds also if O'_α is simple for V'_α). Therefore, also the X_{r+1}-discriminant of f is a power of X_r, showing that $x_1, x_2, \cdots, x_r$ are equisingular parameters of V at O. This completes the proof of the theorem, in view of Theorem 4.4.

We know (Proposition 4.7) that if V has at O a singularity of dimensionality type 1 then the normalization $\bar{V}$ of V is non-singular. We can add to this the following pertinent consequence of Theorem 7.4:

COROLLARY 7.5. *If V has at O a singularity of dimensionality type* 1 *then the normalization $\bar{V}$ of V can be obtained from V by a sequence of monoidal transformations.*

For, if V'_α is still singular (for some $\alpha = 1, 2, \cdots, m$), its singular locus W'_α is defined by $X_r = 0$, $X'_{r+1} - \phi_\alpha(X_1, X_2, \cdots, X_r) = 0$, where $\phi_\alpha(x_1, x_2, \cdots, x_r) = \xi'_\alpha$. It is easy to see that the W'_α-transveral section $V'^*_{\alpha,(x)}$ of V'_α, at O'_α, relative to the W'_α-transversal parameters $x_1, x_2, \cdots, x_{r-1}$ of V'_α, is the quadratic transform of the W-transversal section $V^*_{(x)}$ of V, at O. We know that all W-transversal section of V at O have equivalent singularities (Corollary 5.3). Thus, with every V which has at O a singularity of dimensionality type 1, there is associated an equivalence class, $C(V)$, of singularities of embedded algebroid curves (which is *not* the class of simple points, as long as O is actually a singular point of V). We have just seen that if $S = \bigcup_{\alpha=1}^{m} V'_\alpha$ is the monodial transform of V, centered at the singular locus W of V, then also V'_α has at its origin O'_α a singularity of dimensionality type 1 (or a simple point), and the set of m equivalence classes $C(V'_1)$, $C(V'_2), \cdots, C(V'_m)$ is the quadratic transform of $C(V)$. From this the corollary follows.

8. The Whitney-Thom criterion. Let V be an algebroid hypersurface in affine $\boldsymbol{A}_{r+1}$, with origin O. If $V_1, V_2, \cdots, V_h$ are the irreducible components of V, we denote by P_j the general point of V_j and by σ_j the tangent r-space of V_j at P_j. *We assume that the singular locus W of V is of codimension* 1 *on V and has a simple point at O.* Let L be the tangent $(r-1)$-space of W at O and let τ_j be the r-space determined by L and P_j.

THEOREM 8.1 (*Whitney-Thom*). *The point O is a singularity of V of dimensionality type* 1 *if and only if the following conditions are satisfied*:

(1) *V is equimultiple along W, at O.*

(2) *In any specialization $(P_j, \sigma_j, \tau_j) \to (O, \bar{\sigma}, \bar{\tau})$, we have necessarily $\bar{\sigma} = \bar{\tau}$.*

Proof. We choose the local coördinates $X_1, X_2, \cdots, X_{r+1}$ in $\boldsymbol{A}_{r+1}$ at O in such a way that if $x_1, x_2, \cdots, x_{r+1}$ are their traces on V then $x_1, x_2, \cdots, x_r$ are local transversal parameters of V at O, while W is defined by $X_r = X_{r+1}$

$=0$. From the proof of Proposition 7.2 it follows that condition (1) is equivalent to the following:

(1′) $$\frac{x_{r+1}}{x_r} \in \bar{\mathfrak{o}} = \text{integral closure of } \mathfrak{o}.$$

Let $f(X_1, X_2, \cdots, X_{r+1}) = 0$ be the equation of V, and let $f = f_1 f_2 \cdots f_h$, where $f_j = 0$ is the equation of V_j. Let $K = K_1 \oplus K_2 \oplus \cdots \oplus K_h$ be the direct sum decomposition of the total ring of quotients of $\mathfrak{o}$, where $K_j = \boldsymbol{k}(P_j)$, and let ρ_j be the natural surjection of K into K_j. If $x^{(j)}{}_{r+1} = \rho_j(x_{r+1})$, then P_j is the point $(x_1, x_2, \cdots, x_r, x^{(j)}{}_{r+1})$. (Since the restriction of ρ_j to the subfield $\boldsymbol{k}\{\{x_1, x_2, \cdots, x_r\}\}$ of K is an isomorphism, we identify x_i with $\rho_j(x_i)$, for $i = 1, 2, \cdots, r$). The hyperplane σ_j is defined by the equation:

$$\sigma_j\colon \sum_{i=1}^{r} \left(\frac{\partial f}{\partial X_i}\right)_{P_j} (X_i - x_i) + \left(\frac{\partial f}{\partial X_{r+1}}\right)_{P_j} (X_{r+1} - x^{(j)}{}_{r+1}) = 0.$$

On the other hand, the hyperplane τ_j is defined by

$$\tau_j\colon \frac{x^{(j)}{}_{r+1}}{x_r} X_r - X_{r+1} = 0.$$

Upon division by $\dfrac{\partial f}{\partial x^{(j)}{}_{r+1}}$, the equation of σ_j can also be written as follows:

$$\sigma_j\colon \sum_{i=1}^{r} \frac{\partial x^{(j)}{}_{r+1}}{\partial x_i} (X_i - x_i) - (X_{r+1} - x^{(j)}{}_{r+1}) = 0.$$

If we set $\rho_j(\bar{\mathfrak{o}}) = \bar{\mathfrak{o}}_j$, and denote by $\bar{\mathfrak{m}}_j$ the maximal ideal of the local domain $\bar{\mathfrak{o}}_j$, then condition (2) of the theorem is equivalent to the following:

$$\frac{\partial x^{(j)}{}_{r+1}}{\partial x_i} \in \bar{\mathfrak{m}}_j, \quad \text{for } i = 1, 2, \cdots, r-1;$$

$$\frac{\partial x^{(j)}{}_{r+1}}{\partial x_r} - \frac{x^{(j)}{}_{r+1}}{x_r} \in \bar{\mathfrak{m}}_j,$$

for $j = 1, 2, \cdots, h$; or—equivalently, if we denote by $\bar{\mathfrak{m}}$ the intersection of the maximal ideals of the semilocal ring $\bar{\mathfrak{o}}$:

(2′) $$\begin{cases} \dfrac{\partial x_{r+1}}{\partial x_i} \in \bar{\mathfrak{m}}, \quad \text{for } j = 1, 2, \cdots, r-1; \\[2ex] \dfrac{\partial x_{r+1}}{\partial x_r} - \dfrac{x_{r+1}}{x_r} \in \bar{\mathfrak{m}}. \end{cases}$$

Now, (1′) and (2′) together imply at any rate that $\dfrac{\partial x_{r+1}}{\partial x_i} \in \mathfrak{o}$ for $i = 1, 2, \cdots, r$, or—equivalently—that

$$\bar{\mathfrak{o}}J = \bar{\mathfrak{o}}f'_{x_{r+1}},$$

where J is the Jacobian ideal of V at O. By Theorem 5.1, this shows that conditions (1) and (2) imply that V has at O a singularity of dimensionality type 1 (in view of the assumptions made in regard to the singular locus W of V).

To prove the converse, we may assume that V is irreducible, since equisingularity of V at O, along W, implies equisingularity of each V_j at O, along W. In that case, $\bar{\mathfrak{o}}$ is a local domain. We know already that equisingularity of V at O, along W, implies (1). We may assume that $\frac{x_{r+1}}{x_r}$ is a non-unit in $\bar{\mathfrak{o}}$. Then we have only to show that the partial derivatives $\frac{\partial x_{r+1}}{\partial x_i}$ $(i = 1, 2, \cdots, r)$ are non-units in $\bar{\mathfrak{o}}$. But this is precisely what has been shown in the course of the proof of Theorem 5.2 (where g is to be replaced by f, and $y_1, y_2, \cdots, y_{r+1}$ are to be replaced by $x_1, x_2, \cdots, x_{r+1}$).

REFERENCES.

[1] O. Zariski, "Foundations of a general theory of birational transformations," *Transcations of the American Mathematical Society*, vol. 53 (1943), pp. 496-542.

[2] ———, "Equisingular points on algebraic varieties," *Seminarii dell' Instituto Nazionale di Alta Matematica*, 1962-63 (Rome), pp. 164-177.

[3] ———, "Studies in equisingularity I. Equivalent singularities of plane algebroid curves," *American Journal of Mathematics*, vol. 87 (1965), pp. 507-536.

[4] O. Zariski and P. Samuel, *Commutative Algebra*, vol. II.

[5] O. Zariski, "Seminar on singularities, II." Lecture notes of the Summer Institute on Algebraic Geometry in Woods Hole, Massachusetts (1964), American Mathematical Society.

[6] A. Grothendieck, *Éléments de Géometrie Algébrique.* II.

[7] H. Whitney, "Local properties of analytic varieties," *Differential and Combinatorial Topology*, Princeton University Press (1965).

STUDIES IN EQUISINGULARITY III.*

Saturation of Local Rings and Equisingularity.

By OSCAR ZARISKI.

Contents.

Introduction. In this paper we introduce a certain operation on rings which we call *saturation*: it consists in passing from a given commutative ring A with identity (satisfying some mild conditions; see conditions a—e below) to a class of rings $\tilde{A}$ lying between A and the integral closure of A (in its total ring of fractions). Our main object in this paper is to show, by means of geometric applications to plane algebroid curves (Part I) and to algebroid hypersurfaces (Part II), the usefullness of this new operation for the theory of singularities, and in particular for the concept of equisingularity. In order not to obscure this main object by excessive technical material, we do not include in this paper a general treatment of the theory of saturation; we only develop as much of that theory as is strictly necessary for our present purposes. The general theory of saturated rings will be developed by us else-

Received May 15, 1967.

* This work was supported by a grant from the National Science Foundation.

[Reprinted from *American Journal of Mathematics*, Vol. XC, No. 3, pp. 961-1023.]

where, in a separate work. It will, however, already appear clearly from the present paper that certain questions, which arise naturally in dimension > 1, cannot be answered without a more general treatment of the theory of saturation. We shall come back briefly to this observation at the end of the introduction and, with more detail, in he last section (§ 8) of this paper.

If A is a commutative ring with identity, if F is the total ring of fractions of A and if K is a subfield of F, we shall define the *saturation* $\tilde{A}$ *of* A *with respect to* K only if the following 5 conditions are satisfied:

a. *A sas no nilpotent elements* (other than zero).

b. *F is a noetherian ring, hence (in view of a) a finite direct sum of fields:*

$$F = F_1 \oplus F_2 \oplus \cdots \oplus F_h.$$

c. *No element of K, different from zero, is a zero divisor in F, or equivalently: K contains the element 1 of F.*

d. *If ϵ_i is the identity of F_i then F_i is a finite separable extension of the field $K\epsilon_i$* (note that, in view of c, K and $K\epsilon_i$ are isomorphic fields).

e. *If $R = A \cap K$, then A is integral over R.*

Note that if A is an integral domain then conditions a—c are trivially satisfied.

We fix an algebraic closure Ω of K, and we consider the various K-homomorphisms of F into Ω. If ψ is any such homomorphism then it is clear that $\psi(\epsilon_i) = 1$ for some i, while $\psi(\epsilon_j) = 0$ for $j \neq i$. Then $\psi(F_j) = 0$ for $j \neq i$, while the restriction of ψ to F_i is an isomorphism, and we have $\psi(a\epsilon_i) = a$ for all a in K. It follows from d that the number of distinct K-homomorphisms ψ of F into Ω is finite. It is also clear that for any given i, the compositum $\hat{F}_i^*$ of the set of fields $\psi(F_i)$ (as ψ varies) is a finite Galois extension of K (the least Galois extension of K containing a given $\psi(F_i)$ with $\psi(\epsilon_i) = 1$). Similarly, the compositum $\hat{F}^*$ of the set of fields $\psi(F)$ is a finite Galois extension of K.

For convenience, we shall use the following terminology: if η, ζ are elements of F, then we say that η *dominates* ζ if for any two K-homomorphisms ψ_1, ψ_2 of F into Ω, the following is true: *if* $\psi_1(\zeta) \neq \psi_2(\zeta)$ *then the quotient* $[\psi_1(\eta) - \psi_2(\eta)]/[\psi_1(\zeta) - \psi_2(\zeta)]$ *is integral over R, while if* $\psi_1(\zeta) = \psi_2(\zeta)$ *then also* $\psi_1(\eta) = \psi_2(\eta)$.

We note that if, for a given i, ψ_1 and ψ_1' are two K-homomorphisms of F into Ω such that $\psi_1(\epsilon_i) = \psi_1'(\epsilon_i) = 1$, then $\psi_1' = \phi_0\psi_1$, where ϕ_0 is a K-monomorphism of $\psi_1(F)$ into Ω. This monomorphism ϕ_0 can be extended to a K-automorphism ϕ of $\hat{F}^*$, and thus, for any element η of F, it is true that the set of elements $\psi_1'(\eta)-\psi_2(\eta)$ is the set of ϕ-images of the elements $\psi_1(\eta)-\psi_2(\eta)$ (ψ_1 and ψ_1' being fixed, as above). It follows that if we *fix* for each $i=1,2,\cdots,h$ a K-homomorphism $\psi_1^{(i)}$ of F into Ω such that $\psi_1^{(i)}(\epsilon_i)=1$, then in order to check whether η dominates ζ it is sufficient to verify the above conditions of the definition of domination only for pairs (ψ_1,ψ_2) such that ψ_1 ranges over the set $\{\psi_1^{(1)},\psi_1^{(2)},\cdots,\psi_1^{(h)}\}$, while ψ_2 ranges over the set of all K-homomorphisms of F into Ω.

In particular, if A is an integral domain (whence $h=1$), we may assume $F\subset\Omega$, whence $\hat{F}^*$ is the least Galois extension F^* of K containing F. We can take for $\psi_1^{(1)}$ he identity map. When that is done, then the definition of domination to the following: *η dominates ζ if for any element τ of the Galois group of F^*/K the following is true: if $\zeta^\tau\neq\zeta$ then the quotient $(\eta^\tau-\eta)/(\zeta^\tau-\zeta)$ is integral over R, while if $\zeta^\tau=\zeta$ then also $\eta^\tau=\eta$.*

Let $\bar{A}$ be the integral closure of A in F.

Definition I. *The ring A is said to be saturated with respect to the field K if it contains every element of $\bar{A}$ which dominates an element of A ("A is stable, within $\bar{A}$, with respect to the relation of domination.").*[1]

We note that since A satisfies conditions *a*—*e*, every ring between A and $\bar{A}$ also satisfies these conditions. We note also that $\bar{A}$ is saturated with respect to K, and that consequently the set of rings between A and $\bar{A}$ which are saturated with respect to K is non-empty. It is obvious that the intersection of these rings is itself saturated with respect to K (and thus belongs to the set).

We therefore set down the following

PROPOSITION—*Definition* II. *There exists a smallest ring between A and $\bar{A}$ which is saturated with respect to K. This ring is called the saturation of A with respec to K and will be denoted by $\tilde{A}_K$.*

We give a construction of the ring $\tilde{A}_K$ which will be useful in the sequel. We define by induction on i an infinite ascending chain of rings A_i between

[1] Had we imposed the more stringent requirement that A be stable with respect to domination, *within the quotient field F of A*, then we would necessarily be led to the trivial case $A=F$. For in that case we would have in the first place $K\subset A$, since $\psi_1(\eta)=\psi_2(\eta)=\eta$ for all η in K and for all K-homomorphisms ψ_1, ψ_2 of F into Ω. Since $R=K\cap A$, it follows then that $R=K$, and thus A is integral *over the field K*. If, then, a is any element of A which is not a zero divisor and $a^n+a_1a^{n-1}+\cdots+a_n=0$ ($a_i\in K$) is a relation of integral dependence for a over K, *of least degree n*, then $a_n\neq0$ and $1/a=-1/a_n\cdot(a^{n-1}+a_1a^{n-2}+\cdots+a_{n-1})\in K[a]\subset A$.

A and $\bar{A}$, as follows: $A_0 = A$; if $A_0, A_1, \cdots, A_i$ have already been defined, then we denote by L_i the set of elements η of $\bar{A}$ such that η dominates some elements ζ of A_i (ζ may depend on η), and we set $A_{i+1} = A_i[L_i]$. It is then easily seen (the proof is straightforward) that

$$(0) \qquad \tilde{A}_K = \bigcup_{i=0}^{\infty} A_i.$$

We shall be primarily interested in the case in which A is a *complete (noetherian) local ring* $\mathfrak{o}$, free from nilpotent elements (other than zero). Thus, conditions a and b are satisfied. However, we shall also assume that $\mathfrak{o}$ is *equicharacteristic and equidimensional* [equidimensionality means that all the prime ideals of the zero ideal (or, in the Bourbaki terminology: all the associated prime ideals of $\mathfrak{o}$) have the same dimension]. The field K will be chosen as follows:

We fix a field k of representatives of $\mathfrak{o}$ and a set of parameters $x_1, x_2, \cdots, x_r$ of $\mathfrak{o}$ (where r is the Krull dimension of $\mathfrak{o}$). It is then well known that $x_1, x_2, \cdots, x_r$ are analytically independent over k *and that* $\mathfrak{o}$ *is a finite module over the power series ring* $R = k[[x_1, x_2, \cdots, x_r]]$. Now, our assumption of equidimensionality of $\mathfrak{o}$ implies that *no element* of R, *different from zero, is a zero divisor in* $\mathfrak{o}$. Hence the total ring of fractions F of $\mathfrak{o}$ contains the field of fractions of R. It is this field of fractions of R that we take as our field K. It is then immediate that conditions c and e are satisfied. We now must assume also that condition d is satisfied; we express this additional condition on the parameters $x_1, x_2, \cdots, x_r$ by saying that *the* r *elements* x_i *form a separating system of parameters of* $\mathfrak{o}$.

We can now speak of the saturation $\tilde{\mathfrak{o}}_K$ of $\mathfrak{o}$ with respect to the field $K = k((x_1, x_2, \cdots, x_r))$ of formal meromorphic function of $x_1, x, \cdots, x_r$. We shall also refer to the ring $\tilde{\mathfrak{o}}_K$ as the *k-saturation of* $\mathfrak{o}$ *with respect to the parameters* $x_1, x_2, \cdots, x_r$, and we shall denote it by $\tilde{\mathfrak{o}}^{(k)}{}_{(x_1, x_2, \cdots, x_r)}$, or $\tilde{\mathfrak{o}}^{(k)}{}_{(x)}$. The local ring $\mathfrak{o}$ is said to be *k-saturated with respect to the parameters* $x_1, x_2, \cdots, x_r$ if $\mathfrak{o} = \tilde{\mathfrak{o}}^{(k)}{}_{(x)}$. We shall say that $\mathfrak{o}$ *is k-saturated if* $\mathfrak{o}$ *is k-saturated with respect to some separating system of parameters of* $\mathfrak{o}$. If the field of representatives k is fixed once and for always and there is no danger of confusion, we shall write $\tilde{\mathfrak{o}}_{(x)}$ instead of $\tilde{\mathfrak{o}}^{(k)}{}_{(x)}$ and shall speak of saturation and saturated rings when we mean k-saturation and k-saturated rings. *A priori*, the ring $\tilde{\mathfrak{o}}_{(x)}$ depends not only on the choice of the parameters x_i but also on the choice of the field k of representatives of $\mathfrak{o}$. We shall have occasion to come back to this aspect of saturation later on in this introduction.

In Part I we deal with *irreducible* algebroid curves, defined over an *algebraically closed* ground field k, *of characteristic zero*. Therefore in Part I we are dealing with a local ring $\mathfrak{o}$ which satisfies the following additional conditions: $r=1$, $\mathfrak{o}$ is a local *domain, of characteristic zero*, and the field of representatives k is *algebraically closed* (and necessarily of characteristic zero, since $\mathfrak{o}$ is assumed to be equicharacteristic). In Section 1, the basic algebraic theorems on saturation are proved for this special case, and the results in this case are very complete. Thus, to mention some of the principal results, we prove the following:

1) If $\mathfrak{o}$ is saturated, it is saturated with respect to any transversal parameter of $\mathfrak{o}$ (Theorem 1.8; the definition of transversal parameters is given immediately before the statement of Theorem 1.8). Furthermore, if $\mathfrak{o}$ is saturated and x, x' are transversal parameters of $\mathfrak{o}$, then there exists a k-automorphism of $\mathfrak{o}$ which sends x into x' (Theorem 1.11, part (A)).

2) If x, x' are transversal parameters of $\mathfrak{o}$ then $\tilde{\mathfrak{o}}_x = \tilde{\mathfrak{o}}_{x'}$ (Corollary-Definition 1.9).

3) If k and k' are two fields of representatives of $\mathfrak{o}$ and x is a parameter of $\mathfrak{o}$, then $\tilde{\mathfrak{o}}_x^{(k)} = \tilde{\mathfrak{o}}_x^{(k')}$ (Corollary 2.16), and any isomorphism between k and k' can be extended to an automorphism of $\tilde{\mathfrak{o}}_x^{(k)}$ (Thsorem 1.16).

The *structure theorem* 1.12, together with its partial inverse, Theorem 1.15, give a complete characterization of saturated local domains of dimension 1 (satisfying all the conditions stated above) in terms of ideal theory, and exhibit these local domains very explicitly.

All the theorems proved in Section 1 *continue to hold true* (sometimes with very slight modifications) *if k is not algebraically closed*. This will be shown by us in another paper, where we shall also show that many of the above results continue to hold true *if $\mathfrak{o}$ is not a domain*. In particular, Corollary-Definition 1.9 and Theorem 1.8 are valid without exception. However, in all these theorems the assumption that the characteristic of $\mathfrak{o}$ is zero is essential.

The proofs of the various results established in Section 1 run very smoothly, and this is due, to a large extent, to he systematic use of the very simple but helpful Lemma 1.7.

The main result established in Section 2 is Theorem 2.1; it constitutes the first geometric application and the real motivation of the operation of saturation. This theorem says that a necessary and sufficient condition that two irreducible algebroid curves C and C_1 be equivalent (i.e., have equivalent singularities in the classical sense) is that the saturations $\tilde{\mathfrak{o}}$ and $\tilde{\mathfrak{o}}_1$ of their local rings $\mathfrak{o}$ and $\mathfrak{o}_1$ be isomorphic. The proof is by induction on the length of the $\mathfrak{o}$-module $\tilde{\mathfrak{o}}/\mathfrak{o}$, where $\tilde{\mathfrak{o}}$ is the integral closure of $\mathfrak{o}$, and requires a careful examination of the relationship between $\tilde{\mathfrak{o}}$ and $\widetilde{\mathfrak{o}'}$, where $\mathfrak{o}'$ is the quadratic transform of $\mathfrak{o}$. If the quadratic transformation does *not* lower the multiplicity of C, then the two operations $\tilde{}$ and $'$ commute ($\widetilde{\mathfrak{o}'} = \tilde{\mathfrak{o}}'$), and there is no difficulty. If the multiplicity of C *is* lowered by the quadratic transformation (this is Case II of the proof of Theorem 2.1) then the above operations do not commute, and it is this difficulty that requires for its elimination a number of preliminary technical propositions (2.2, 2.3, 2.4).

Theorem 2.1 remains valid also in the case of reducible plane algebroid curves, as we shall show elsewhere, but the proof in that case is much more difficult.

In Section 3, we are interested primarily in interpreting the numerical characters, which occur in the structure Theorem 1.12, by means of the *characteristic exponents* of the Puiseux expansion $y = y(x)$ of the curve C. We point out two incidental (more or less known) results which this interpretation leads to:

1) the characteristic exponents are independent of the choice of x, as long as x is a transversal parameter;

2) the *inversion formula* (see end of Section 3) which expresses the characteristic exponents of $x = x(y)$ in terms of the characteristic exponents of $y = y(x)$. This formula is almost a direct consequence of Proposition 2.2.

In Part II, we deal with algebroid varieties. In Section 4, we prove a general ideal-theoretic result on saturation, to the effect that the morphism $\operatorname{Spec} \tilde{A}_K \rightarrow \operatorname{Spec} A$ is *radicial* (Theorem 4.1). In the case of an algebroid equidimensional variety V, over a given ground field k, we have an operation of *saturation of V with respect to separating parameters* $x_1, x_2, \cdots, x_r$; it leads from V to a variety $\tilde{V}_{(x)}$ whose local ring is the saturation $\tilde{\mathfrak{o}}_{(x)}$ of the local ring $\mathfrak{o}$ of V, with respect to the parameters $x_1, x_2, \cdots, x_r$. The variety $\tilde{V}_{(x)}$ is dominated by the normalization $\bar{V}$ of V, and the natural morphism $\tilde{f}_{(x)} : \tilde{V}_{(x)} \rightarrow V$ is radicial. In the complex-analytic domain this leads to the conclusion that the *holomorphic map determined by $\tilde{f}_{(x)}$ is a local homeomorphism* between $\tilde{V}_{(x)}$ and V (Corollary 5.2). Thus, if two complex-analytic varieties V and V', defined locally at points O and O' (and having local rings $\mathfrak{o}$ and $\mathfrak{o}'$), have, at these points, isomorphic saturations (with respect to suitable parameters (x) and (x')), then V and V' are locally homeomorphic (at O and O'), the local homeomorphism being determined by the isomorphism between $\tilde{V}_{(x)}$ and $\tilde{V}_{(x')}$.

In Section 6, we consider the special case in which V and V' are hypersurfaces, embedded, say, in affine spaces $\mathcal{A}_{r+1}$ and $\mathcal{A}'_{r+1}$, and in which the parameters x_i (resp. x'_i) form part of a minimal basis of the maximal ideal $\mathfrak{m}$ (resp. $\mathfrak{m}'$) of $\mathfrak{o}$ (resp. $\mathfrak{o}'$). In this case we prove (Theorem 5.1) that if there exists an isomorphism ϕ between $\tilde{\mathfrak{o}}_{(x)}$ and $\tilde{\mathfrak{o}}'_{(x')}$ which transforms the parameters (x) into the parameters (x'), then the local homeomorphism between V and V' which is associated with ϕ can be extended to a local homeomorphism between the ambient affine space $\mathcal{A}_{r+1}$, $\mathcal{A}'_{r+1}$. Thus, in this case, the hypersurfaces V and V' are not only locally homeomorphic but also topologically equivalent, as embedded varieties. The proof has two parts. The first part is a purely algebraic preparation, in which, from the existence of the isomorphism ϕ, one draws the conclusions expressed by (4) and (4′) (Section 6). After this algebraic preparation is completed, the rest of the proof follows closely an argument of real function theory which was given by Whitney in a similar set-up (and essentially in the special case $r=1$; see [4], §§ 11-12).

In Section 7, where the ground field k is algebraically closed and of characteristic zero, we define a notion of *equisaturation of an algebroid variety V along a subvariety W, at a point O of W* (Definition 7.3), we establish a discriminant criterion of equisaturation for hypersurfaces (Theorem 7.4), and we show (by repeating the argument of the proof of Theorem 6.1) that if a hypersurface V is equisaturated along a subvariety W, at a point O of W, then V, as an embedded variety in the affine $\mathcal{A}_{r+1}$, is topologically equivalent, locally at O, to the direct product $W \times V_0$, where V_0 is a suitable section of V at O, transversal to W. Thus, equisaturation has, in some sense, all the earmarks of equisingularity. If $\mathrm{cod}_V W = 1$, then equisaturation and equisingularity are identical concepts; see Corollary 7.5.

In the last section (§ 8) we discuss a generalization of Corollary-Definition 1.9 to algebroid varieties of higher dimension (in characteristic zero). The proofs will be given elsewhere, but we say enough about the underlying idea of the proof to bring out clearly the necessity of proving first Corollary-Definition 1.9 for one-dimensional local rings having a residue field which is not algebraically closed. This will be done in a subsequent paper which we shall devote to a general treatment of the theory of saturation and saturated rings.

While the bulk of this work has been done at Harvard, many of its ideas and proofs have been made more precise and concise in our mind in the course of a seminar conducted on this subject during the Spring term of 1965-66 at Purdue University. We are also glad to say that the final version of this work has been prepared and written in the Spring of 1967 in Bures-sur-Yvette, where the author was at that time visiting member of the Institut des Hautes Etudes Scientifiques, and was presented in a series of lectures at the University of Rome in June, 1967.

Part I. Saturated Local Domains of Dimension 1.

1. Basic properties. Throughout Part I of the paper, $\mathfrak{o}$ denotes a complete, equicharacteristic local domain, of Krull dimension 1 and characteristic zero, having an algebraically closed residue field; the maximal ideal of $\mathfrak{o}$ will be denoted by $\mathfrak{m}$. We fix once and for always a field k of representatives of $\mathfrak{o}$. Thus, when we speak of the *saturation of* $\mathfrak{o}$ *with respect to a parameter* x *of* $\mathfrak{o}$, we mean the saturation of $\mathfrak{o}$ with respect to the field $k((x))$ of formal meromorphic functions of x, over k. Similarly, the exprssion "$\mathfrak{o}$ is saturated" means "$\mathfrak{o}$ is k-saturated" (see Introduction). Only at the end of this section (see Theorem 1.16 and Corollary 1.17) will we deal simultaneously with several fields of representatives of $\mathfrak{o}$, and in that case we shall use the more explicit expressions "k-saturation of $\mathfrak{o}$ wih respect to x," "k-saturated ring," etc.

The field of fractions of $\mathfrak{o}$ will be denoted by F, and $\bar{\mathfrak{o}}$ will denote the integral closure of $\mathfrak{o}$ in F. The ring $\bar{\mathfrak{o}}$ is a power series ring $k[[t]]$, and we have $F=k((t))$. The natural valuation of F, whose valuation ring is $\bar{\mathfrak{o}}$, will be denoted by v. The saturation of $\mathfrak{o}$ with respect to a parameter x will be denoted by $\tilde{\mathfrak{o}}_x$.

PROPOSITION 1.1. *The ring* $\tilde{\mathfrak{o}}_x$ *is invariant under all automorphisms of* $F/k((x))$.

Proof. Let η be any element of $\tilde{\mathfrak{o}}_x$ and let σ be any element of the Galois group $G=\mathcal{G}(F/k((x)))$. We have to show that $\eta^\sigma\in\tilde{\mathfrak{o}}_x$. Now, in the first place, $\eta^\sigma\in\bar{\mathfrak{o}}$, since $\bar{\mathfrak{o}}$ is invariant under σ. Let, now, τ be any element of G. Since G is commutative (cyclic), we have $(\eta^\sigma)^\tau-\eta^\sigma=(\eta^\tau)^\sigma-\eta^\sigma=(\eta^\tau-\eta)^\sigma$, and therefore $v((\eta^\sigma)^\tau-\eta^\sigma)=v(\eta^\tau-\eta)$. This completes the proof, since $\eta\in\tilde{\mathfrak{o}}_x$ and since $\tilde{\mathfrak{o}}_x$ is saturated with respect to x.

PROPOSITION 1.2. *The multiplicity of* $\tilde{\mathfrak{o}}_x$ *is the same as the multiplicity of* $\mathfrak{o}$.

Proof. We refer to the exprssion (0) of $\tilde{A}_K$ which was given in the Introduction and we denote this time by $\mathfrak{o}_1,\mathfrak{o}_2,\cdots,\mathfrak{o}_i\cdots$ the rings which were denoted by $A_1,A_2,\cdots,A_i\cdots$ in that expression. Since $\bar{\mathfrak{o}}$ is now a finite module over $\mathfrak{o}$, we are dealing this time with a finite ascending chain of rings, say

$$\mathfrak{o} \subset \mathfrak{o}_1 \subset \mathfrak{o}_2 \subset \cdots \subset \mathfrak{o}_h = \bar{\mathfrak{o}}_x.$$

We recall that $\mathfrak{o}_{i+1} = \mathfrak{o}_i[L_i]$, where L_i is the set of elements η of $\bar{\mathfrak{o}}$ such that η dominates some elements ζ of $\mathfrak{o}_i$. It will be sufficient to show that $\mathfrak{o}_i$ and $\mathfrak{o}_{i+1}$ have the same multiplicity. We shall show, for example, that $\mathfrak{o}$ and $\mathfrak{o}_1$ $(=\mathfrak{o}[L_0])$ have the same multiplicity. For every element η of L_0, let

$$\eta_0 = \frac{1}{[F: k((x))]} \cdot Tr_{F/k((x))}(\eta),$$

and let $\eta = \eta_0 + \eta'$. Since $\eta \in \bar{\mathfrak{o}}$, we have $\eta_0 \in k[[x]] \subset \mathfrak{o}$. Therefore, if we denote by L'_0 the set of elements η' obtained as above, by letting η range over L_0, then $\mathfrak{o}_1 = \mathfrak{o}[L'_0]$. Observing, that $Tr_{F/k((x))}(\eta') = 0$, we see that we may assume

$$(1) \qquad Tr_{F/k((x))}(\eta) = 0,$$

for every element η of L_0. *Observe that* (1) *implies that* $v(\eta) > 0$, *for every* η *in* L_0.

Since k is algebraically closed, the multiplicity of any local domain $\mathfrak{O}$ between $\mathfrak{o}$ and $\bar{\mathfrak{o}}$ is equal to min.$\{v(\xi)\}$ as ξ ranges over the maximal ideal of $\mathfrak{O}$. Let e = multiplicity of $\mathfrak{o}$. To prove that e is also the multiplicity of $\mathfrak{o}_1$, it is therefore sufficient to show that

$$(2) \qquad v(\eta) \geqq e, \text{ for every element } \eta \text{ in } L_0.$$

Recall that we have $v(\eta) > 0$ for all η in L_0. Let G be the Galois group of $F/k((x))$. We shall prove in a moment (see Lemma 1.3 below) that (1) implies that

$$(3) \qquad v(\eta^\tau - \eta) = v(\eta), \text{ for } \textit{some } \tau \text{ in } G.$$

On the other hand, since $\eta \in L$, there exists an element ζ in $\mathfrak{o}$ such that $v(\eta^\tau - \eta) \geqq v(\zeta^\tau - \zeta)$ for *all* τ in G. Upon subtracting from ζ a suitable element of k (this does not affect $\zeta^\tau - \zeta$) we may assume that $\zeta \in \mathfrak{m}$ (= maximal ideal of $\mathfrak{o}$). Since $v(\zeta^\tau) = v(\zeta)$, it follows that $v(\zeta^\tau - \zeta) \geqq v(\zeta)$, and since $\zeta \in \mathfrak{m}$, we have $v(\zeta) \geqq e$. Hence

$$(4) \qquad v(\eta^\tau - \eta) \geqq e, \textit{ for all } \tau \textit{ in } G.$$

Combining (3) and (4), we find (2). This completes the proof.

The lemma alluded to in the course of the above proof is the following:

Lemma 1.3. *Let η be a non-zero element of $\bar{\mathfrak{o}}$, let $n = [F: k((x))]$ $(= v(x))$ and let*

$$\eta_0 = \frac{1}{n} Tr_{F/k((x))}(\eta),$$

$$\eta' = \eta - \eta_0.$$

If $v(\eta^\tau - \eta) > v(\eta)$ for all τ in G $(= \mathfrak{G}(F/k((x))))$, then $v(\eta) < v(\eta')$ (and hence $\eta_0 \neq 0$ and $v(\eta) = v(\eta_0)$). Conversely, if there exists an element η'_0 in K such that $v(\eta) < v(\eta - \eta'_0)$, then $v(\eta^\tau - \eta) > v(\eta)$ for all τ in G.

Proof. The converse is obvious since $\eta^\tau - \eta = (\eta - \eta'_0)^\tau - (\eta - \eta'_0)$ and $v[(\eta - \eta'_0)^\tau - (\eta - \eta'_0)] \geqq v(\eta - \eta'_0)$ for all τ in G. For the proof of the direct assertion we may assume that $\eta \notin k((x))$ and that consequently $\eta' \neq 0$. We have

$$\eta_0 = \frac{1}{n} \sum_{\tau \in G} \eta^\tau,$$

and therefore

$$\eta' = \frac{1}{n} \sum_{\tau \in G} (\eta - \eta^\tau).$$

Since $v(\eta^\tau - \eta) > v(\eta)$ for all τ in G, it follows that $v(\eta') > v(\eta)$.

We shall need now a general lemma on saturation of a ring A.

Lemma 1.4. *The notations being the same as in the Introduction, assume that A is a simple ring adjunction of R:*

$$A = R[y].$$

Then $\tilde{A}_K$ coincides with the set of all elements η of $\bar{A}$ which dominate y (and hence, with reference to the expression (0) of $\tilde{A}_K$ given in the Introduction, we have $\tilde{A}_K = A_1 = L_0$).

Proof. Denote by S the set of all elements η of $\bar{A}$ which dominate y. Clearly, $A \subset S \subset \tilde{A}_K$. It is immediate that S is a ring [this follows from he identity

$$\psi_1(\xi\eta) - \psi_2(\xi\eta) = (\psi_1(\xi) - \psi_2(\xi))\psi_1(\eta) + (\psi_1(\eta) - \psi_2(\eta))\psi_2(\xi)].$$

To complete the proof we have only to show that S is saturated with respect to K. But this is obvious, since if $\zeta \in S$ and $\eta \in \bar{A}$ and if η dominates ζ, then multiplication of the quotient $[\psi_1(\eta) - \psi_2(\eta)]/[\psi_1(\zeta) - \psi_2(\zeta)]$ by $[\psi_1(\zeta) - \psi_2(\zeta)]/[\psi_1(y) - \psi_2(y)]$ shows that η dominates y, and that consequently $\eta \in S$.

Definition 1.5. *If x and y are elements of $\mathfrak{m}$, then y is called an x-saturator of $\mathfrak{o}$, if $\mathfrak{o}$ and its subring $k[[x,y]]$ have the same saturation with respect to x.*

PROPOSITION 1.6. *For any element x of $\mathfrak{m}$, there exists x-saturators y in $\mathfrak{m}$.*

Proof. We fix a finite basis $\{\xi_1, \xi_2, \cdots, \xi_h\}$ of $\mathfrak{o}$ over $k[[x]]$ and we denote by V the vector space $\sum k\xi_i$ over k. Let G be the Galois group of $F/k((x))$. For each τ in G, $\tau \neq 1$, we set

$$V_\tau = \{\eta \in V \mid v(\eta^\tau - \eta))\} > \min_{1 \leq i \leq h}. \{v(\xi_i{}^\tau - \xi_i)\}.$$

Then V_τ is a *proper* subspace of V. Since k is an infinite field, we can find an element y in V such that $y \notin \bigcup_{1 \neq \tau} V_\tau$. Then $y \in \mathfrak{o}$ and

$$(5) \qquad v(y^\tau - y) = \min_{1 \leq i \leq h}. \{v(\xi_i{}^\tau - \xi_i)\}, \text{ all } \tau \in G, \tau \neq 1.$$

Let $\mathfrak{o}' = k[[x,y]] = [[x]][y]$. Then we have, in view of (5): $\xi_i \in \tilde{\mathfrak{o}}'_x$, $1 \leq i \leq h$. Hence $\mathfrak{o} \subset \tilde{\mathfrak{o}}'_x$ (since $k[[x]] \subset \mathfrak{o}'$), and therefore $\tilde{\mathfrak{o}}_x \subset \tilde{\mathfrak{o}}'_x$. On the other hand, since $\mathfrak{o}' \subset \mathfrak{o}$, we have also $\tilde{\mathfrak{o}}'_x \subset \tilde{\mathfrak{o}}_x$. Thus $\tilde{\mathfrak{o}}'_x = \tilde{\mathfrak{o}}'_x$, and this completes the proof (if the element y is not in $\mathfrak{m}$, we replace it by $y - c$, were c is a suitable element of k).

Note. The proof of Proposition 1.2 could have been shortened if we had postponed its proof until after Proposition 1.6.

We shall now prove a simple lemma, which will, however, be of prime importance for the proofs of a number of theorems in this and later sections.

LEMMA 1.7. *Let f_1, f_2 be two automorphisms of F having the same restriction to k and let t be an element of F such that $v(t) = 1$ (whence $F = k((t))$). Assume that $v(f_2(t) - f_1(t)) = 1 + \lambda$, > 0. Then*

$$v(f_2(\eta) - f_1(\eta)) = v(\eta) + \lambda,$$

for all elements η of F such that $\eta \neq 0$ and $v(\eta) \neq 0$.

Proof. If we set $f = f_1{}^{-1}f_2$, then f is a k-automorphism of F. The assumption of the lemma is equivalent to $v(f(t) - t) = 1 + \lambda$, $\lambda > 0$ (or equivalently: $v(\frac{f(t)}{t} - 1) = \lambda > 0$). The conclusion of the lemma is equivalent to $v(f(\eta) - \eta) = v(\eta) + \lambda$, for all $\eta \in F$ such that $\eta \neq 0$ and

$v(\eta)\neq 0$. We may therefore assume that f_1 is the identity. We assume this, and we write f for f_2 (whence f is a k-automorphism).

By assumption, we have $f(t)=t\xi$, where $\xi=1+at^{\lambda}+\cdots, 0\neq a\in k$. If n is any positaive integer, then

$$f(t^n)-t^n=t^n(\xi^n-1)=t^n(\xi-1)\prod_{i=1}^{n-1}(\xi-\omega^i),$$

where ω is a primiive n-th root of unity. Since the characteristic of k is 0, it follows that $f(t^n)-t^n=\epsilon_n t^{n+\lambda}$, where ϵ_n is a unit in $\mathfrak{o}$. Now, let η be any element of F such that $\eta\neq 0$ and $v(\eta)=n>0$: $\eta=\sum_{i=0}^{\infty} c_i t^{n+i}$, $c_0\neq 0$. Then $f(\eta)-\eta=\sum_{i=0}^{\infty} c_i\epsilon_{n+i}t^{n+i+\lambda}$, and hence

$$v(f(\eta)-\eta)=n+\lambda=v(\eta)+\lambda.$$

If now η is an element of F such that $v(\eta)<0$, then we set $\zeta=\dfrac{1}{\eta}$, and we have

$$\begin{aligned}v(f(\eta)-\eta)-v(\eta)&=v\Big[\frac{1}{f(\zeta)}-\frac{1}{\zeta}\Big]+v(\zeta)\\&=v(\zeta-f(\zeta))-v(f(\zeta))=v(f(\zeta)-\zeta)-v(\zeta)=\lambda.\end{aligned}$$

This completes the proof of the lemma.

A parameter x of $\mathfrak{o}$ is called *transversal* if $v(x)=$ multiplicity of $\mathfrak{o}$. As a first application of the preceding lemma we prove the following theorem:

THEOREM 1.8. *If $\mathfrak{o}$ is saturated, then $\mathfrak{o}$ is saturated with respect to any transversal parameter of $\mathfrak{o}$.*

Proof. By assumption, $\mathfrak{o}$ is saturated with respect to some parameter x'. We set $n=$ multiplicity of $\mathfrak{o}$, $m=v(x')$ (whence $n\leqq m$), and we fix an x'-saturator y of $\mathfrak{o}$ (Proposition 1.6). Since $\mathfrak{o}$ is the saturation of $k[[x',y]]$ with respect to x', it follows that n is also the multiplicity of $k[[x',y]]$ (Proposition 1.2), *i.e.*,

$$n=\min(m,v(y)). \qquad (6)$$

Let x be any transversal parameter of $\mathfrak{o}$. *We have to prove that $\mathfrak{o}$ is saturated with respect to x.*

Let G' and G be the Galois groups of F over $k((x'))$ and $k((x))$ respectively; these are cyclic groups of order m and n respectively.

We fix in F elements t' and t such that $x'=t'^m$, $x=t^n$. For any σ in G we have

$$(7) \qquad t^{\sigma} = c_{\sigma} t, \quad c_{\sigma}^{n} = 1, \qquad (\sigma \in G).$$

Similarly, for any τ in G' we have

$$(7') \qquad t'^{\tau} = c'_{\tau} t', \quad c'_{\tau}{}^{m} = 1,, \qquad (\tau \in G').$$

Here the m-th root of unity c'_{τ} is uniquely determined by τ, and similarly for the n-th root of unity c_{σ} and σ. Since t' is a power series in t, of the form $at +$ terms of higher degree $(a \neq 0)$, it follows from (7) that

$$(8) \qquad v(t'^{\sigma} - c_{\sigma} t') \geqq 2, \text{ for all } \sigma \text{ in } G.$$

Let e be the highest common divisor of m and n, and let G_1 (resp., G'_1) be the subgroup of order e of G (resp., of G'). As τ ranges over G'_1, the coefficient c'_{τ} in (7′) ranges over the set of e-th roots of unity. Similarly, as σ ranges over G_1, the coefficitnt c_{σ} in (7) ranges over the same set of e-th roots of unity. We denote by ϕ the isomorphism $G'_1 \xrightarrow{\sim} G_1$ defined by stipulating hat $\sigma = \phi(\tau)$ if and only if $c_{\sigma} = c'_{\tau}$. By (7′) and (8) it follows that if $\sigma = \phi(\tau)$ then $v(t'^{\tau} - t'^{\sigma}) = 1 + \lambda$, $\lambda > 0$. Hence by Lemma 1.7 we have

$$(9) \qquad v(\eta^{\tau} - \eta^{\sigma}) = v(x^{\tau} - x) - v(x) + v(\eta), \qquad (\tau \in G'_1, \sigma = \phi(\tau) \in G_1)$$

for all elements η of F such that $\eta \neq 0$ and $v(\eta) \neq 0$.

We note now that G'_1 is a proper subgroup of G' if and only if $e < m$, i.e., if and only if $m > n$. Hence, by (6), we have that

$$(10) \qquad \text{“}G'_1 \neq G'\text{”} \Rightarrow \text{“}v(y) = n.\text{”}$$

In the case $G'_1 \neq G'$ we obesrve that if τ is any element of G' which does not belong to G'_1, then the order of τ does *not* divide n, whence $c'_{\tau}{}^{n} \neq 1$. Since, by (10), we have in this case $y = bt'^{n} + \cdots, b \neq 0$, and thus $y^{\tau} = bc'_{\tau}{}^{n} t'^{n} + \cdots$, it follows that

$$(11) \qquad \text{“}\tau \notin G'_1\text{”} \Rightarrow \text{“}v(y^{\tau} - y) = n.\text{”}$$

We now proceed to the proof that $\mathfrak{o}$ is saturated with respect to x. By definition of saturation, we have to show the following: if ζ is an element of $\mathfrak{o}$ and η is an element of $\bar{\mathfrak{o}}$ such that

$$(12) \qquad v(\eta^{\sigma} - \eta) \geqq v(\zeta^{\sigma} - \zeta), \text{ for all } \sigma \text{ in } G,$$

then $\eta \in \mathfrak{o}$. Since y is an x'-saturator of $\mathfrak{o}$, to say that $\eta \in \mathfrak{o}$ is equivalent to saying that (see Lemma 1.4)

$$(13) \qquad v(\eta^{\tau} - \eta) \geqq v(y^{\tau} - y), \text{ for all } \tau \text{ in } G'.$$

So we have to prove (13). In proving (13) we may assume (as we have done repeatedly on various previous occasions) that $v(\zeta) > 0$ and $v(\eta) > 0$. We may even assume (by subtracting from η a suitable element of $k[[x]]$) that $Tr_{F/k((x))}\eta = 0$ and that consequently (see Lemma 1.3)

$$(14) \qquad v(\eta^\sigma - \eta) = v(\eta), \text{ for some } \sigma \text{ in } G.$$

Since $\zeta \in \mathfrak{m}$, we have

$$(15) \qquad v(\zeta) \geqq n.$$

Since $v(\zeta^\sigma - \zeta) \geqq v(\zeta)$, it follows from (12) and (14) that

$$(15') \qquad v(\eta) \geqq n.$$

We now proceed to the proof of (13).

Let us first assume that $\tau \notin G'_1$. Then, by (11), $v(y^\tau - y) = n$. In this case, the inequality $v(\eta^\tau - \eta) \geqq v(y^\tau - y)$ follows therefore directly from $v(\eta^\tau - \eta) \geqq v(\eta)$ and from (15′).

Let now $\tau \in G'_1$, and let $\sigma = \phi(\tau)$. Then we use (9) and the following relation similar to (9), obtained by replacing η by ζ:

$$(16) \qquad v(\zeta^\tau - \zeta^\sigma) = v(x^\tau - x) - v(x) + v(\zeta).$$

Since $v(x) = n$, it follows from (9), (15′), (16) and (15) that $v(\eta^\tau - \eta^\sigma) \geqq v(x^\tau - x)$ and $v(\zeta^\tau - \zeta^\sigma) \geqq v(x^\tau - x)$. Since $x \in \mathfrak{o}$ and since y is an x'-saturator of $\mathfrak{o}$, we have by Lemma 1.4 that $v(x^\tau - x) \geqq v(y^\tau - y)$. Thus

$$(17) \qquad v(\eta^\tau - \eta^\sigma) \geqq v(y^\tau - y),$$

$$(17') \qquad v(\zeta^\tau - \zeta^\sigma) \geqq v(y^\tau - y).$$

Since $\zeta \in \mathfrak{o}$, we have $v(\zeta^\tau - \zeta) \geqq v(y^\tau - y)$. Therefore, by (17′), we find that $v(\zeta^\sigma - \zeta) \geqq v(y^\tau - y)$, and this inequality, in conjunction with (12), yields $v(\eta^\sigma - \eta) \geqq v(y^\tau - y)$. Combining this last inequality with (17), we obtain (13). This completes the proof.

Corollary—*Definition* 1.9. *The saturation of* $\mathfrak{o}$ *wih respect to a transversal parameter is a ring which is independent of the choice of that parameter. This ring is called the saturation of* $\mathfrak{o}$ *and will be denoted by* $\tilde{\mathfrak{o}}$.

For, let x and x' be two transversal parameters of $\mathfrak{o}$ and let $\tilde{\mathfrak{o}} = \tilde{\mathfrak{o}}_x$, $\tilde{\mathfrak{o}}' = \tilde{\mathfrak{o}}_{x'}$. Since $\mathfrak{o}$ and $\tilde{\mathfrak{o}}'$ have the same multiplicity (Proposition 1.2), x is also a transversal parameter of $\tilde{\mathfrak{o}}'$. Therefore, by the above theorem, $\tilde{\mathfrak{o}}'$ is saturated wih respct to x. By the minimality property of $\bar{\mathfrak{o}}_x$, it follows therefore that $\bar{\mathfrak{o}}' \supset \bar{\mathfrak{o}}$ (since $\bar{\mathfrak{o}}' \supset \mathfrak{o}$). Similarly, $\bar{\mathfrak{o}} \supset \bar{\mathfrak{o}}'$, and hence $\bar{\mathfrak{o}} = \bar{\mathfrak{o}}'$.

COROLLARY 1.10. *The ring $\tilde{\mathfrak{o}}$ is the smallest saturated ring (between $\mathfrak{o}$ and $\bar{\mathfrak{o}}$) which contains $\mathfrak{o}$.*

First of all, if z is any parameter of $\mathfrak{o}$ and x is a transversal parameter of $\mathfrak{o}$, then x is also a transversal parameter of $\tilde{\mathfrak{o}}_z$, and $\tilde{\mathfrak{o}}_z$ is herefore saturated with respect to x. Thus $\tilde{\mathfrak{o}}_z \supset \tilde{\mathfrak{o}}_x$ $(= \tilde{\mathfrak{o}})$; in other words, $\tilde{\mathfrak{o}}$ *is the smallest saturation of* $\mathfrak{o}$. Now, let $\mathfrak{o}'$ be any saturated local ring between $\mathfrak{o}$ and $\bar{\mathfrak{o}}$ (not necessarily a saturation of $\mathfrak{o}$ with respect to a parameter of $\mathfrak{o}$). Let x' be a transversal parameter of $\mathfrak{o}'$. Then $\mathfrak{o}'$ is saturated with respect to x'. For any integer $n \geqq 1$, the Galois group of $F/k((x'))$ is a subgroup of the Galois group of $F/k((x'^n))$. From this and from the very definition of saturated rings, it follows that $\mathfrak{o}'$ is also saturated with respect to x'^n. Now, we take n sufficiently large, so as to have $x'^n \in \mathfrak{o}$. Then $\mathfrak{o}' \supset \tilde{\mathfrak{o}}_{x'^n}$ (since $\mathfrak{o}' \supset \mathfrak{o}$), and on the other hand we have just established the fact that $\tilde{\mathfrak{o}}_{x'^n} \supset \tilde{\mathfrak{o}}$. Hence $\mathfrak{o}' \supset \tilde{\mathfrak{o}}$.

If x and x' are two elements of $\bar{\mathfrak{o}}$ and if $v(x) = v(x') = n > 0$, then there exist exactly n k-automorphisms of $\bar{\mathfrak{o}}$ which send x into x'. The next theorem expresses a remarkable property of our saturated local domains.

THEOREM 1.11. (A) *If $\mathfrak{o}$ is saturated and if x, x' are transversal parameters of $\mathfrak{o}$, then the k-automorphisms of $\bar{\mathfrak{o}}$ which send x into x' induce automorphisms of $\mathfrak{o}$.*

(B) *More generally, if $\mathfrak{o}$ is saturated with respect to a parameter x and if x' is an element of $\mathfrak{m}$, such that $v(x) = v(x')$, then a necessary and sufficient condition in order that there exist a k-automorphism of $\mathfrak{o}$ which send x into x' is that the following be true:*

$$(18) \qquad v(x'^{\tau} - x') \geqq v(y^{\tau} - y) + v(x) - n, \quad \textit{for all } \tau \textit{ in } \mathfrak{G}(F/k((x)));$$

here y denotes an x-saturator of $\mathfrak{o}$, and n is the multiplicity of $\mathfrak{o}$.

(C) *If* (18) *holds, then every k-automorphism of $\bar{\mathfrak{o}}$ which sends x into x' induces an automorphism of $\mathfrak{o}$; and, furthermore, $\mathfrak{o}$ is also saturated with respect to x'.*

Proof. If there exists a k-automorphism of $\mathfrak{o}$ which sends x into x', then it is obvious that $\mathfrak{o}$ is also saturated with respect to x'. Furthermore, Proposition 1.1 tells us that if there exists one k-automorphism of $\mathfrak{o}$ which send x into x', then all k-automorphisms of $\bar{\mathfrak{o}}$ which send x into x' induce automorphisms of $\mathfrak{o}$. Thus part (C) of the theorem is a simple consequence of Part (B). Also Part (A) of the theorem is an immediate consequence of Part (B) (and of Theorem 2.7), since if x is transversal then $v(x)=n$, and the inequality (18) is automatically satisfied since $x'\in\mathfrak{o}$ (see Lemma 1.4). Thus we have only to prove (B).

We fix an element t in F such that $x=t^m$, where $m=v(x)$. Let $x'=ct^m+$ terms of higher degree $(0\neq c\in k)$. We fix a k-automorphism f_1 of $\bar{\mathfrak{o}}$ such that $f_1(x)=x'$. Then

(19) $$f_1(t)=\gamma t+\text{terms of higher degree};\ \gamma^m=c.$$

Let τ be any element of the Galois group G of $F/k((x))$. We set $f_2=\tau f_1\tau^{-1}$. Then f_2 is again a k-automorphism of $\bar{\mathfrak{o}}$, and we deduce from (19) that

(19′) $$f_2(t)=\gamma t+\text{terms of higher degree.}$$

Thus $v(f_2(t)-f_1(t))=1+\lambda$, $\lambda>0$, and we can apply Lemma 1.7. We have therefore

$$v(f_2(\eta)-f_1(\eta))=v(f_2(x)-f_1(x))+v(\eta)-v(x),$$

for every element η of F such that $\eta\neq 0$ and $v(\eta)\neq 0$. Now, $f_1(x)=x'$ and $f_2(x)=x'^{\tau}$. Hence, we can re-write the above equality as follows:

(20) $$v(f_2(\eta)-f_1(\eta))=v(x'^{\tau}-x')+v(\eta)-v(x).$$

In order to show that f_1 induces an automorphism of $\mathfrak{o}$ it is sufficient to show that $f_1(\mathfrak{o})$ is contained in $\mathfrak{o}$, for if that is so then the equality $f_1(\mathfrak{o})=\mathfrak{o}$ follows by observing that $f_1(\mathfrak{o})\subset\mathfrak{o}\subset\bar{\mathfrak{o}}$ and that the $\mathfrak{o}$-module $\bar{\mathfrak{o}}/\mathfrak{o}$ and the $f_1(\mathfrak{o})$-module $\bar{\mathfrak{o}}/f_1(\bar{\mathfrak{o}})$ have the same length. Let then η be any element of $\mathfrak{m}$, and let $\eta'=f_1(\eta)$. Then $\eta'\in\mathfrak{o}$ if and only if

(21) $$v(\eta'^{\tau}-\eta')\geqq v(y^{\tau}-y),\text{ for all }\tau\text{ in }G(=\mathfrak{G}(F/k((x)))).$$

Now, $\eta'^{\tau}=f_2(\eta^{\tau})$, and $\eta'^{\tau}-\eta'=f_2(\eta^{\tau})-f_1(\eta)=f_2(\eta^{\tau}-\eta)+f_2(\eta)-f_1(\eta)$. Since $\eta\in\mathfrak{o}$, we have $v(\eta^{\tau}-\eta)\geqq v(y^{\tau}-y)$, and therefore also $v[f_2(\eta^{\tau}-\eta)]\geqq v(y^{\tau}-y)$. Hence (21) is equivalent to

(22) $$v(f_2(\eta)-f_1(\eta))\geqq v(y^{\tau}-y),$$

which, by (20), is in its turn equivalent to

(23) $$v(x'^{\tau}-x')\geqq v(y^{\tau}-y)+v(x)-v(\eta).$$

It must be shown then that the inequality (23) holds for all elements η of $\mathfrak{m}$. This, however, is obvious, since $\min\{v(\tau_i)\} = n$, and since therefore (23) is equivalent to the condition (18) stated in the heorem. This completes the proof.

We next give a fundamental structure theorem for our saturated local domains.

THEOREM 1.12. *Let $\mathfrak{o}$ be saturated and let x be a parameter such that $\mathfrak{o}$ is saturated with respect to x. Let $m = v(x)$, let y be an x-saturator of $\mathfrak{o}$, and let*

$$\beta_1 < \beta_2 < \cdots < \beta_g$$

be the set of integers $v(y^\tau - y)$ obtained by letting τ range over

$$G = \mathfrak{G}(F/k((x))), \quad \tau \neq 1.$$

Then

(A) *The integers β_ν depend only on the set $\mathfrak{m}v$ (= set of integers $v(\xi), \xi \in \mathfrak{m}$) and the integer m. Namely:*

β_1 is the smallest integer in $\mathfrak{m}v$ which is not divisible by m.

If $e_1 = (\beta_1, m)$, then β_2 is the smallest integer in $\mathfrak{m}v$ which is not divisible by e_1.

If $e_2 = (\beta_2, e_1)$, then β_3 is the smallest integer in $\mathfrak{m}v$ which is not divisibe by e_2.

$e_{g-1} > 1$, β_g is he smallest integer in $\mathfrak{m}v$ which is not divisible by e_{g-1}, and $(\beta_g, e_{g-1}) = 1$ $(= e_g)$.

[*The set of integers $m, \beta_1, \beta_2, \cdots, \beta_g$ will be called the x-characteristic of $\mathfrak{o}$.*]

(B) *For each $\nu = 0, 1, 2, \cdots, g$, let K_ν be the (uniquely determined) subfield of F such that $K_\nu \supset k((x))$ and $[K_\nu : k((x))] = \dfrac{m}{e_\nu}$ (here $e_0 = m$), whence*

$$k((x)) = K_0 < K_1 < K_2 < \cdots < K_g = F.$$

Let $m_\nu = \beta_\nu / e_\nu$ $(\nu = 1, 2, \cdots, g)$ and let $\mathfrak{M}_\nu = \bar{\mathfrak{m}} \cap K_\nu$ $(\nu = 0, 1, \cdots, g)$. Then

$$\mathfrak{m} = \mathfrak{M}_0 + \mathfrak{M}_1^{m_1} + \mathfrak{M}_2^{m_2} + \cdots + \mathfrak{M}_g^{m_g}, \tag{24}$$

no term in (24) being redundant, except $\mathfrak{M}_0$, which is redundant if and only if x is not a transversal parameter.

[*The decomposition (24) will be called the standard decomposition of $\mathfrak{m}$, relative to x.*]

21

Proof. We first wish to point out that if y' is another x-saturator of $\mathfrak{o}$, then $v(y^{\tau}-y)=v(y'^{\tau}-y')$ for all τ in G. For, by Lemma 1.4, we have $v(y'^{\tau}-y')\geqq v(y^{\tau}-y)$ and $v(y^{\tau}-y)\geqq v(y'^{\tau}-y')$. Thus, that the β's depend only on $\mathfrak{o}$ and x, is obvious. What Part (A) implies is that if $\mathfrak{o}$ is also saturated with respect to another parameter x' suct that $v(x)=v(x')$, and if y' is an x'-saturator of $\mathfrak{o}$, then the set of integers

$$\{v(y'^{\tau'}-y')\mid 1\neq\tau'\in G'=\mathfrak{G}(F/k((x')))\}$$

coincides with the set of integers $\beta_1,\beta_2,\cdots,\beta_g$.

Let

$$(25)\qquad G_\nu=\{\tau\in G\mid v(y^{\tau}-y)\geqq\beta_{\nu+1},\quad \nu=0,1,\cdots,g-1.$$

Clearly the G_ν form a strictly descending chain:

$$(26)\qquad G=G_0>G_1>\cdots>G_g=(1).$$

Furthermore, *each* G_ν *is a subgroup of* G; for if $\tau_1,\tau_2\in G_\nu$, then $y^{\tau_1\tau_2}-y=(y^{\tau_2}-y)^{\tau_1}+(y^{\tau_1}-y)$, and from this it follows that $\tau_1\tau_2\in G_\nu$; similarly, if $\tau\in G_\nu$, then from $y^{\tau^{-1}}-y=(y-y^{\tau})^{\tau^{-1}}$ follows that $\tau^{-1}\in G_\nu$.

Let K_ν be the fixed field of G_ν. Thus

$$(27)\qquad k((x))=K_0<K_1<\cdots<K_g=F.$$

We set $e_\nu=[F:K_\nu]$. For any element η of $\bar{\mathfrak{o}}$ we set

$$\eta'_\nu=\frac{1}{e_\nu}Tr_{F/K_\nu}(\eta),\quad \nu=0,1,\cdots,g;$$

$$\eta_\nu=\eta'_\nu-\eta'_{\nu-1},\quad \nu=1,2,\cdots,g;\eta_0=\eta'_0.$$

Since $\eta'_g=\eta$, it follows that

$$(28)\qquad \eta=\eta_0+\eta_1+\cdots+\eta_g,\quad \eta_\nu\in\bar{\mathfrak{o}}\cap K_\nu.$$

we have

$$\eta'_{\nu-1}=\frac{1}{[K_\nu:K_{\nu-1}]}Tr_{K_\nu/K_{\nu-1}}(\eta'_\nu),\quad \nu=1,2,\cdots,g,$$

and $\eta_\nu=\eta'_\nu-\eta'_{\nu-1}$. Hence

$$Tr_{K_\nu/K_{\nu-1}}(\eta_\nu)=0,$$

and therefore, by Lemma 1.3, we have that (assuming $\eta_\nu\neq0$)

$$(29)\qquad \exists\,\tau_{\nu-1}\in G_{\nu-1}\mid v(\eta_\nu^{\tau_{\nu-1}}-\eta_\nu)=v(\eta_\nu),\quad \nu=1,2,\cdots,g.$$

Now, suppose that $\eta \in \mathfrak{o}$. In that case, we have, by the definition of an x-saturator and by (25):

$$(30) \qquad v(\eta^\tau - \eta) \geqq \beta_\nu, \text{ for all } \tau \in G_{\nu-1}.$$

If $\tau \in G_{g-1}$, we have clearly $\eta^\tau - \eta = \eta_g{}^\tau - \eta_g$, and hence, by (30), we have $v(\eta_g{}^\tau - \eta_g) \geqq \beta_g$, for all $\tau \in G_{g-1}$. However, if, in particular, τ is the element τ_{g-1} which occurs in (29), for $\nu = g$, then $v(\eta_g{}^\tau - \eta_g) = v(\eta_g)$. Therefore, we conclude that

$$v(\eta_g) \geqq \beta_g, \text{ if } \eta \in \mathfrak{o}.$$

Assume that it has already been shown that if $\eta \in \mathfrak{o}$ then

$$(31) \qquad v(\eta_\mu) \geqq \beta_\mu, \text{ for } \mu = \nu + 1, \nu + 2, \cdots, g,$$

where ν is an integer, $1 \leqq \nu \leqq g$. Then we shall show that if $\eta \in \mathfrak{o}$ then also

$$(32) \qquad v(\eta_\nu) \geqq \beta_\nu.$$

Namely, we have clearly, for any $\tau \in G_{\nu-1}$:

$$(33) \qquad \eta^\tau - \eta = (\eta_\nu{}^\tau - \eta_\nu) + (\eta_{\nu+1}{}^\tau - \eta_{\nu+1}) + \cdots + (\eta_g{}^\tau - \eta_g).$$

We have $v(\eta_\mu{}^\tau - \eta_\mu) \geqq v(\eta_\mu)$, whence by (31): $v(\eta_\mu{}^\tau - \eta_\mu) \geqq \beta_\mu > \beta_\nu$ for $\mu = \nu + 1, \nu + 2, \cdots, g$. Hence, using (30), we find that

$$(34) \qquad v(\eta_\nu{}^\tau - \eta_\nu) \geqq \beta_\nu, \text{ for all } \tau \in G_{\nu-1}.$$

Now, by (29), there exists some τ in $G_{\nu-1}$ (namely $\tau_{\nu-1}$) such that $v(\eta_\nu{}^\tau - \eta_\nu) = v(\eta_\nu)$. From this and (34) follows (32).

We have therefore shown that if $\eta \in \mathfrak{o}$, *then in the decomposition* (28) *of* η *we have the inequalities* (32) (for $\nu = 1, 2, \cdots, g$). As to η_0, we have $\eta_0 \in \bar{\mathfrak{o}} \cap k((x)) = k[[x]]$; thus

$$(35) \qquad \eta_0 \in \mathfrak{M}_0.$$

Applying the decomposition (28) to the element y, we have

$$(36) \qquad y = y_0 + y_1 + \cdots + y_g, \quad v(y_\nu) \geqq \beta_\nu, \ \nu = 1, 2, \cdots, g.$$

If $\tau \in G_{\nu-1} - G_\nu$, then, by (25), $v(y^\tau - y) = \beta_\nu$; on the other hand, $y^\tau - y = (y_\nu{}^\tau - y_\nu) + (y_{\nu+1}{}^\tau - y_{\nu+1}) + \cdots + (y_g{}^\tau - y_g)$, and if $\mu > \nu$ then $v(y_\mu{}^\tau - y_\mu) \geqq v(y_\mu) \geqq \beta_\mu > \beta_\nu$. Hence $v(y_\nu{}^\tau - y_\nu) = \beta_\nu$. We have therefore

$$\beta_\nu \leqq v(y_\nu) \leqq v(y_\nu{}^\tau - y_\nu) = \beta_\nu,$$

and consequently

(37) $$v(y_\nu)=\beta_\nu, \qquad \nu=1,2,\cdots,g.$$

Since $y_\nu \in K_\nu$ and $[F: K_\nu]=e_\nu$, it follows from (37) that

(38) $$\beta_\nu \equiv 0 \ (\mathrm{mod}\, e_\nu),$$

and that if we set $m_\nu=\beta_\nu/e_\nu$, then from (35), (32) and (28) follows that

(39) $$\mathfrak{o} \subset k+\mathfrak{M}_0+\mathfrak{M}_1{}^{m_1}+\mathfrak{M}_2{}^{m_2}+\cdots+\mathfrak{M}_g{}^{m_g}.$$

On the other hand, if η is any element of $\mathfrak{M}_\nu{}^{m_\nu}$ and τ is any element of G, say $\tau \in G_{\mu-1}-G_\mu$, then $\eta^\tau=\eta$ if $\mu>\nu$; and if $\mu \leqq \nu$ then $v(\eta^\tau-\eta) \geqq v(\eta) \geqq \beta_\nu \geqq \beta_\mu = v(y^\tau-y)$. Therefore $\eta \in \mathfrak{o}$. Thus we ave also shown that

(40) $$\mathfrak{o} \supset k+\mathfrak{M}_0+\mathfrak{M}_1{}^{m_1}+\cdots+\mathfrak{M}_g{}^{m_g},$$

since $k+\mathfrak{M}_0=k[[x]] \subset \mathfrak{o}$. From (39) and (40) follows (24).

From $v(x)=m$ follows that $[F: k((x))]=m$, and hence $[K_\nu: k((x))] = \dfrac{m}{e_\nu}$.

Assume that x is a transversal parameter. Since $k[[x,y]]$ and $\mathfrak{o}$ have the same multiplicity (Proposition 1.2), it follows that $v(y) \geqq v(x)$. Hence there exists an element c in k such that if we set $y'=y-cx$, then $v(y')>v(x)$. Now $v(y^\tau-y)=v(y'^\tau-y') \geqq v(y')$. It follows that in this case all the *β_ν are greater than $v(x)$*. Since $\mathfrak{m}=\mathfrak{M}_0+\mathfrak{M}_1{}^{m_1}+\cdots+\mathfrak{M}_g{}^{m_g}$ and since $v(\mathfrak{m})=v(x)=v(\mathfrak{M}_0)$, while $v(\mathfrak{M}_\nu{}^{m_\nu})=\beta_\nu$, we conclude *that $\mathfrak{M}_0$ cannot be deleted in* (24) [in regard to the notation $v(\mathfrak{m})$, $v(\mathfrak{M}_0)$, etc., if E is any subset of $\bar{\mathfrak{o}}$ we mean by $v(E)$ the minimum of $v(\xi)$ as ξ ranges over E]. As a matter of fact, the inequalities $\beta_1<\beta_2<\cdots<\beta_g$ imply that neither is any term $\mathfrak{M}_\nu{}^{m_\nu}$ $(\nu=1,2,\cdots,g)$ superfluous in (24).

On the other hand, if x is not transversal, then $v(\mathfrak{M}_0)>v(\mathfrak{m})$, and then it follows from (24) and from $\beta_1<\beta_2<\cdots<\beta_g$ that $v(\mathfrak{m})=\beta_1$. In this case, we have then $v(\mathfrak{M}_0)>v(\mathfrak{M}_1{}^{m_1})$, and since $K_0 \subset K_1$ it follows that $\mathfrak{M}_0 \subset \mathfrak{M}_1{}^{m_1}$. Thus $\mathfrak{M}_0$ is superfluous in (24).

We shall now prove that

(41) $$(\beta_\nu, e_{\nu-1})=e_\nu, \qquad \nu=1,2,\cdots,g \ (e_0=m).$$

To see this, we set $e_{\nu-1}=e_\nu n_\nu$. Since $\beta_\nu=e_\nu m_\nu$, (41) is equivalent to

(42) $$(m_\nu, n_\nu)=1.$$

Let $K_{\nu-1}=k((\xi))$, where ξ is therefore any element of $K_{\nu-1}$ such that $\nu(\xi)=e_{\nu-1}$. Since $[K_\nu: K_{\nu-1}]=n_\nu$, we can set $\xi=u^{n_\nu}$, and then $K_\nu=k((u))$, with $v(u)=e_\nu$. Since $y_\nu \in K_\nu$ and $v(y_\nu)=\beta_\nu=e_\nu m_\nu$, it follows that

$y_\nu = cu^{m_\nu} +$ terms of degree $> m_\nu$ in u, with $c \neq 0$ $(c \in k)$. From (25), (36) and (37) follows that if $\tau \in G_{\nu-1} - G_\nu$, then $v(y_\nu{}^\tau - y_\nu) = \beta_\nu$. In other words: if $\sigma \in \mathfrak{G}(K_\nu/K_{\nu-1})$ and $\sigma \neq 1$, then $v(y_\nu{}^\sigma - y) = v(y_\nu)$ $(= \beta_\nu)$. Now for any such σ, we have $u^\sigma = \epsilon u$, where ϵ ranges over the set of *all* n-th roots of unity, different from 1, as σ ranges over $\mathfrak{G}(K_\nu/K_{\nu-1})$, $\sigma \neq 1$. The equality $v(y_\nu{}^\sigma - y_\nu) = v(y_\nu)$ implies that for *each* such ϵ, we must have $c(\epsilon^{m_\nu} - 1) \neq 0$, i.e., $\epsilon^{m_\nu} \neq 1$. This implies (42).

Let ξ be any element of $\mathfrak{m}$. Then by (24) we can write ξ in the form

$$\xi = \xi_{\nu_1} + \xi_{\nu_2} + \cdots + \xi_{\nu_s},$$

where $0 \leqq \nu_1 < \cdots < \nu_s$ and $\xi_{\nu_i} \in \mathfrak{M}_{\nu_i}{}^{m_{\nu_i}}$ $(m_0 = 1)$. If for some ν_i, ν_j we have $\nu_i < \nu_j$ and $v(\xi_{\nu_i}) \geqq v(\xi_{\nu_j})$, then certainly $\xi_{\nu_i} \in \mathfrak{M}_{\nu_j}{}^{m_{\nu_j}}$, and thus the sum $\xi_{\nu_i} + \xi_{\nu_j}$ can be written as a single term which is an element of $\mathfrak{M}_{\nu_j}{}^{m_{\nu_j}}$. We may therefore assume that

$$v(\xi_{\nu_1}) < v(\xi_{\nu_2}) < \cdots < (\xi_{\nu_s}),$$

and hence that $u(\xi) = v(\xi_{\nu_1})$. Therefore $v(\xi)$ is an integer of the form

$$\beta_{\nu_1} + he_{\nu_1}, \qquad h \geqq 0.$$

We have therefore shown that any element of $\mathfrak{m}v$ is an integer of the form

(43) $\quad \beta_\nu + he_\nu$, where $h \geqq 0$ and $\quad \nu = 0, 1, \cdots, g$. $(\beta_0 = m, e_0 = m)$.

Conversely, it is obvious that any integer of the form (43) is the value of an element in $\mathfrak{M}_\nu{}^{m_\nu}$, and hence belongs to $\mathfrak{m}v$. *Thus $\mathfrak{m}v$ is the set of all integers of the form* (43). This, in view of (41) and in view of the fact that $m > e_1 > e_2 > \cdots > e_g = 1$, implies that the integers $\beta_1, \beta_2, \cdots, \beta_g$ have the properties stated in Part (A) of the theorem. Thus all the assertions made in the theorem have been proved, and the proof is complete.

The following is an important consequence of Theorem 1.12 which we shall have occasion to use in the next section:

COROLLARY 1.13. *Let $\mathfrak{o}$ and $\mathfrak{o}'$ be two complete, equicharacteristic local domains, of Krull dimension* 1 *and characteristic zero, having in common an algebraically closed field of representatives k. If $\mathfrak{o}$ and $\mathfrak{o}'$ are saturated rings, then a necessary and sufficient condition that $\mathfrak{o}$ and $\mathfrak{o}'$ be k-isomorphic is that the two sets of integers $\mathfrak{m}v$ and $\mathfrak{m}'v'$ coincide (here $\mathfrak{m}$ and $\mathfrak{m}'$ denote the maximal ideals of $\mathfrak{o}$ and $\mathfrak{o}'$ respectively, v is the natural valuation of the fied F of fractions of $\mathfrak{o}$, while v' is the natural valuation of the field F' of fractions of $\mathfrak{o}'$).*

The necessity of the condition is obvious. Assume now that the condition $\mathfrak{m}v = \mathfrak{m}'v'$ is satisfied. Let x be a transversal parameter of $\mathfrak{o}$, and x'—a transversal parameter of $\mathfrak{o}'$. Since $v(x)$ is the smallest integer in $\mathfrak{m}v$ and $v'(x')$ is the smallest integer in $\mathfrak{m}'v'$, it follows that $v(x) = v'(x') =$ (say)n. Write the standard decomposition (24) of $\mathfrak{m}$, relative to x. The relative degrees $\dfrac{n}{e_\nu} = [K_\nu: k((x))]$ and the exponents $m_\nu = \beta_\nu/e_\nu$ depend only on n and on the β_ν's, hence depend only on the set $\mathfrak{m}v$ (by Theorem 1.12). Since $\mathfrak{m}v = \mathfrak{m}'v'$, it follows that the standard decompositions of $\mathfrak{m}'$, relative to x', will be similar in form to (24), i.e., we will have

$$\mathfrak{m}' = \mathfrak{M}_0' + \mathfrak{M}_1'^{m_1} + \mathfrak{M}_2'^{m_2} + \mathfrak{M}_g'^{m_g}, \tag{44}$$

with $K_0' = k((x'))$ and $[K_\nu': k((x'))] = [K_\nu: k((x))]$. If, then, we fix a k-isomorphism $\phi: F \xrightarrow{\sim} F'$ such that $\phi(x) = x'$, (such isomorphisms exist since $v(x) = v'(x')$ and since k is algebraically closed and of characteristic zero), then we will have necessarily $\phi(K_\nu) = K_\nu'$, $\phi(\mathfrak{M}_\nu) = \mathfrak{M}_\nu'$, and hence, by (24) and (44), $\phi(\mathfrak{m}) = \mathfrak{m}'$ and $\phi(\mathfrak{o}) = \mathfrak{o}'$.

The following is another Corollary of Theorem 1.12 which, in view of Theorem 1.8, is a generalization of Part (A) of Theorem 1.11 (and whose proof is similar to that of the preceding corollary):

COROLLARY 1.14. *If x and x' are two parameters of $\mathfrak{o}$ such that $v(x) = v(x')$, and if $\mathfrak{o}$ is saturated both with respect to x and with respect to x', then the k-automorphisms of $\bar{\mathfrak{o}}$ which send x into x' induce automorphisms of $\mathfrak{o}$.*

For, by Theorem 1.12, $\mathfrak{o}$ has the same x and x'-characteristic, and hence the standard decompositions of $\mathfrak{m}$ relative to x and x' will be respectively

$$\mathfrak{m} = \mathfrak{M}_0 + \mathfrak{M}_1^{m_1} + \mathfrak{M}_2^{m_2} + \cdots + \mathfrak{M}_g^{m_g}, \tag{45}$$

$$\mathfrak{m} = \mathfrak{M}_0' + \mathfrak{M}_1'^{m_1} + \mathfrak{M}_2'^{m_2} + \cdots + \mathfrak{M}_g'^{m_g}, \tag{45'}$$

and will be associated with the following ascending chains of fields:

$$k((x)) = K_0 < K_1 < \cdots < K_g = F,$$

$$k((x')) = K_0' < K_1' < \cdots < K_g' = F.$$

Since the relative degrees $[F: K_\nu]$ and $[F: K_\nu']$ are equal (namely, are equal to e_ν) it follows that any automorphism of F which sends K_0 into K_0', sends K_ν into K_ν', $\mathfrak{M}_\nu$ into $\mathfrak{M}_\nu'$, and hence, in view of (45), (45'), induces an automorphism of $\mathfrak{o}$.

For the sake of completeness, we add to the structure Theorem 1.12 the following strong converse, in order to obtain a structural *characterization* of our saturated local domains $\mathfrak{o}$:

THEOREM 1.15. *Given an strictly ascending sequence of fields*

$$K_1 < K_2 < \cdots < K_g = F$$

such that $k \subset K_1$ *and* $[F: K_\nu] = e_\nu < \infty$, *given any sequence of* g *positive integers* $m_1, m_2, \cdots, m_g$, *and setting* $\mathfrak{M}_\nu = \bar{\mathfrak{m}} \cap K_\nu$, *the set*

$$\mathfrak{o} = k + \mathfrak{M}_1^{m_1} + \mathfrak{M}_2^{m_2} + \cdots + \mathfrak{M}_g^{m_g} \tag{46}$$

is a saturated local domain.

Proof. We set $\beta_\nu = e_\nu m_\nu$. If for some pair of distinct indices ν, μ we have $\nu < \mu$ and $\beta_\nu \geqq \beta_\mu$, then $\mathfrak{M}_\nu^{m_\nu} \subset \mathfrak{M}_\mu^{m_\mu}$ and $\mathfrak{M}_\nu^{m_\nu}$ is superfluous in (46). We may therefore assume that

$$\beta_1 < \beta_2 < \cdots < \beta_g.$$

It is clear that the set (46) is a ring; this follows from the fact that if $\nu \leqq \mu$ then $\mathfrak{M}_\nu^{m_\nu}\mathfrak{M}_\mu^{m_\mu} \subset \mathfrak{M}_\mu^{m_\mu}$. We have $\mathfrak{M}_g = \bar{\mathfrak{m}}$, whence $\mathfrak{o} \supset \bar{\mathfrak{m}}^{m_g}$, showing that $\mathfrak{o}$ is a complete local domain, with F as field of fractions and $\bar{\mathfrak{o}}$ as integral closure.

Let x be any non-zero element of $\mathfrak{M}_1^{m_1}$. *We shall show that* $\mathfrak{o}$ *is saturated with respect to* x.

We set $K_0 = k((x))$ and $G = \mathfrak{G}(F/k((x)))$. Let ζ be an element of $\mathfrak{o}$, let η be an element of $\bar{\mathfrak{o}}$, and assume that

$$v(\eta^\tau - \eta) \geqq v(\zeta^\tau - \zeta), \text{ for all } \tau \text{ in } G. \tag{47}$$

We have to prove that $\eta \in \mathfrak{o}$.

Since $\zeta \in \mathfrak{o}$, we can write

$$\zeta = \zeta_1 + \zeta_2 + \cdots + \zeta_g, \ \zeta_\nu \in \mathfrak{M}_\nu^{m_\nu}.$$

We proceed as in the beginning of the proof of Theorem 1.12. We have also now a sequence of fields similar to the sequence (27). For the given element η of $\bar{\mathfrak{o}}$ we write the decomposition (28) obtained by means of the traces η_ν'; here we set $e_0 = [F: K_0]$. Then, upon setting $G_\nu = \mathfrak{G}(F/K_\nu)$, we obtain (29) by using Lemma 1.3. We then show, by induction from $\nu + 1$ to ν, that $\eta_\nu \in \mathfrak{M}_\nu^{m_\nu}$ $(\nu = 1, 2, \cdots, g)$. That part of the proof of Theorem 1.12 was based entirely on the inequalities (30). In the present case, these inequalities still hold true, but for a different reason: they follow from (47), from the fact that $v(\zeta_\nu) \geqq \beta_\nu$ (since $\zeta_\nu \in \mathfrak{M}_\nu^{m_\nu}$) and from the equality

$$\zeta^\tau - \zeta = (\zeta_\nu{}^\tau - \zeta_\nu) + (\zeta_{\nu+1}{}^\tau - \zeta_{\nu+1}) + \cdots + (\zeta_g{}^\tau - \zeta_g),$$

which is valid if $\tau \in G_{\nu-1}$.

As to η_0, we observe that since $\eta \in \bar{\mathfrak{o}}$, we have $\eta_0 \in k[[x]] \subset k + \mathfrak{M}_1{}^{m_1}$ (since $x \in \mathfrak{M}_1{}^{m_1}$, by assumption). This completes the proof.

Note that the above proof, in conjunction with Theorem 1.12, shows again that if $\mathfrak{o}$ is saturated then $\mathfrak{o}$ is necessarily also saturated with respect to some transversal parameter (a result which is, however, weaker than Theorem 1.8). For, assuming that in (46) we have $\beta_1 < \beta_2 < \cdots < \beta_g$ and taking x in $\mathfrak{M}_1{}^{m_1}$ but *not* in $\mathfrak{M}_1{}^{m_1+1}$, we have then that x is a transversal parameter of $\mathfrak{o}$.

We conclude this section with one final application of the structure theorems 1.12 and 1.15. Up to now we have always dealt with a fixed field k of representatives of $\mathfrak{o}$. We now allow a change of field of representatives and revert to the more precise terminology "k-saturation" etc. The application that we have in mind is the following theorem, which, in addition to showing that saturation is an intrinsic operation, independent of the choice of the field of representatives, puts also into evidence another remarkable property of our saturated rings $\mathfrak{o}$.

THEOREM 1.16. *Let k and k' be two fields of representatives of $\mathfrak{o}$ and let θ be an isomorphism $k \xrightarrow{\sim} k'$. Assume that $\mathfrak{o}$ is k-saturated with respect to some parameter x. Then θ can be extended to an automorphism ϕ of $\mathfrak{o}$ such that $\phi(x) = x$ (and consequently $\mathfrak{o}$ is also k'-saturated with respect to x).*

Proof. Let $v(x) = m$, set $x = t^m$ (whence $F = k((t)) = k'((t))$), and consider the standard decomposition (24) of m. We have the chain of fields $k((x)) = K_0 < K_1 < \cdots < K_g$, where

$$K_0 = k((t^{e_0})), \quad K_\nu = k((t^{e_\nu})). \quad (e_0 = m).$$

Consider the automorphism ϕ of $\bar{\mathfrak{o}}$ such that $\phi(t) = t$ and $\phi \mid k = \theta$. Let $\phi(K_\nu) = K_\nu'$, $\phi(\mathfrak{M}_\nu) = \mathfrak{M}_\nu'$. Then

$$\phi(\mathfrak{o}) = k' + \mathfrak{M}_0' + \mathfrak{M}_1'^{m_1} + \cdots + \mathfrak{M}_g'^{m_g}.$$

Now, $\mathfrak{M}_\nu'^{m_\nu}$ consists of all power series in t^{e_ν}, with coefficients in k', which begin with a term of degree $e_\nu m_\nu$ in t or higher. Now, all the monomials $t^{e_\nu i}$, $i \geqq m_\nu$, belong to $\mathfrak{o}$, and $\mathfrak{o}$ is complete. Hence $\mathfrak{M}_\nu'^{m_\nu} \subset \mathfrak{o}$, *whence* $\phi(\mathfrak{o}) \subset \mathfrak{o}$. On the other hand, $\phi(\bar{\mathfrak{o}}) = \bar{\mathfrak{o}}$ and the $\mathfrak{o}$-module $\bar{\mathfrak{o}}/\mathfrak{o}$ has the same length as the $\phi(\mathfrak{o})$-module $\bar{\mathfrak{o}}/\phi(\mathfrak{o})$. Hence $\phi(\mathfrak{o}) = \mathfrak{o}$. This completes the proof.

COROLLARY 1.17. *If k, k' are two fields of representatives of $\mathfrak{o}$, and x is a parameter of $\mathfrak{o}$, then the k-saturation of $\mathfrak{o}$ with respect to x coincides with the k'-saturation of $\mathfrak{o}$ with respect to x. In particular, the k-saturation of $\mathfrak{o}$ coincides with the k'-saturation of $\mathfrak{o}$.*

For, let $\tilde{\mathfrak{o}}_x$ and $\tilde{\mathfrak{o}}_x'$ denote respectively the k-saturation and the k'-saturation $\mathfrak{o}$ with respect to x. By the preceding theorem, $\tilde{\mathfrak{o}}_x'$ is also k-saturated with respect to x. Since $\tilde{\mathfrak{o}}_x' \supset \mathfrak{o}$, it follows that $\tilde{\mathfrak{o}}_x' \supset \tilde{\mathfrak{o}}_x$. Similarly, it follows that $\tilde{\mathfrak{o}}_x \supset \tilde{\mathfrak{o}}_x'$, and hence $\tilde{\mathfrak{o}}_x = \tilde{\mathfrak{o}}_x'$.

2. A saturation-theoretic criterion of equivalence of plane algebroid branches. In [6] we have given an inductive definition of equivalence between two plane algebroid curves C, C_1. This definition was essentially a streamlined version of the classical concept of equivalence, based on the composition of a singularity as defined by successive locally quadratic transformations. The chief object of this section is to establish the following saturation-theoretic criterion of equivalence between *irreducible* plane algebroid curves.[2]

THEOREM 2.1. *Let C, C_1 be irreducible plane algebroid curves, defined over an algebraically closed field of characteristic zero, and let $\mathfrak{o}$, $\mathfrak{o}_1$ be the local domains of C, C_1 respectively. A necessary and sufficient condition that C and C_1 be equivalent (or have equivalent singularities) is that he saturations $\tilde{\mathfrak{o}}$ and $\tilde{\mathfrak{o}}_1$ be isomorphic.*

Note. In the formulation of the above criterion we have deliberately omitted any reference to any particular field k of representatives of $\mathfrak{o}$ and $\mathfrak{o}_1$, for Corollary 1.17 tells us that any such reference is irrelevant. We note also that our definition of equivalence $C \equiv C_1$ given in [6] was actually a definition of equivalence between $\mathfrak{o}$ and $\mathfrak{o}_1$, independent of any particular choice of a field k of representatives as a field of reference. Equivalence $C \equiv C_1$ has been defined only under the assumption that $\mathfrak{o}$ and $\mathfrak{o}_1$ have isomorphic and algebraically closed fields of representatives. This is in good agreement with the above criterion, for if $\tilde{\mathfrak{o}}$ and $\tilde{\mathfrak{o}}_1$ are isomorphic then $\mathfrak{o}$ and $\mathfrak{o}_1$ must have isomorphic fields of representatives.

Before proceding to the proof of Theorem 2.1, we shall need three auxiliary propositions.

PROPOSITION 2.2. *Let $\mathfrak{o}$ be saturated, let x be a transversal parameter*

[2] This criterion is also valid for *reducible* curves, but as we are dealing in this paper only with local *domains* (of dimension 1), we are stating this criterion only in the irreducible case. The more general case will be treated elsewhere.

of $\mathfrak{o}$ *and let* y *be an* x*-saturator of* $\mathfrak{o}$. *Let* $\mathfrak{m}$ *be the maximal ideal of* $\mathfrak{o}$ *and let* $\tilde{\mathfrak{m}}_y$ *be the maximal ideal of* $\tilde{\mathfrak{o}}_y$. *Then*

$$(1)\quad \tilde{\mathfrak{m}}_y = \mathfrak{m}.\frac{x}{y} \cap \bar{\mathfrak{o}}x,$$

$$(2)\quad \mathfrak{m} \cap \bar{\mathfrak{o}}y = \tilde{\mathfrak{m}}_y.\frac{x}{y}.$$

Proof. Let $v(x) = n$, $v(y) = m$ and let $e = (n, m)$. We set

$$G = \mathfrak{G}(F/k((x))), \qquad G' = \mathfrak{G}(F/k((y))),$$

and we use the notations introduced in the course of the proof of Theorem 1.8, *with the understanding that the role of the two elements* x' *and* x *of that proof is now played by the elements* y *and* x *respectively.*

We note that the element y of the proof of Theorem 1.8 was any x'-saturator of $\mathfrak{o}$. Now in our present set up, we have that y is an x-saturator of $\mathfrak{o}$. It is easily seen that this, together with the transversality of x, *implies that* x *is a* y*-saturator of* $\mathfrak{o}$. In fact, we have $\mathfrak{o} \subset \tilde{\mathfrak{o}}_x = \widetilde{k[[x, y]]}_x \subset \widetilde{k[[x, y]]}_y$ (this last inclusion follows from Corollary 1.10). Hence $\tilde{\mathfrak{o}}_y \subset \widetilde{k[[x, y]]}_y$, and since the opposite inclusion is obvious, the assertion $\tilde{\mathfrak{o}}_y = \widetilde{k[[x, y]]}_y$ follows. Since our present element y plays the role of the element x' of the proof of Theorem 1.8, *we see that we can identify both elements* x *and* y *of that proof with our present element* x.

By analogy with relations (9) and (11) of the proof of Theorem 1.8, we now have respectively the relations

$$(3)\quad \begin{aligned} v(\eta^\tau - \eta^\sigma) - v(\eta) &= v(x^\tau - x) - v(x) \\ &= v(y - y^\sigma) - v(y), \quad (\tau \in G_1', \sigma = \phi(\tau) \in G_1), \end{aligned}$$

and

$$(4)\quad \text{“}\tau \in G', \tau \notin G_1'\text{”} \Rightarrow \text{“}v(x^\tau - x) = v(x)\text{”} \quad (= n).$$

Let η be any element of the right hand side of (1). Then $\eta = \xi x/y$, where $\xi \in \mathfrak{m}$ and $v(\xi) \geqq v(y)$ (since $\eta \in \bar{\mathfrak{o}}x$ and therefore $v(\eta) \geqq v(x)$). We have for all σ in G: $\eta^\sigma y^\sigma - \eta y = x(\xi^\sigma - \xi)$, and also $v(\xi^\sigma - \xi) \geqq v(y^\sigma - y)$ since $\xi \in \mathfrak{o} = \tilde{\mathfrak{o}}_x$ and y is an x-saturator. Hence we have the following inequality

$$(5)\quad v[\eta^\sigma(y^\sigma - y) + y(\eta^\sigma - \eta)] \geqq v(x) + v(y^\sigma - y), \quad (\sigma \in G).$$

Now since $v(\eta) \geqq v(x)$, we have also $v(\eta^\sigma) \geqq v(x)$, and hence

$$(6)\quad v[\eta^\sigma(y^\sigma - y)] \geqq v(x) + v(y^\sigma - y), \quad (\sigma \in G).$$

Now, assume that $\sigma = \phi(\tau)$, $\tau \in G_1'$. Then from (5) and (6) we deduce (using the second equality in (3)) that

$$(7) \qquad v(\eta^\sigma - \eta) \geqq v(x^\tau - x), \quad (\tau \in G_1', \sigma = \phi(\tau)).$$

Since $v(\eta) \geqq v(x)$, it follows also from (3) that

$$(8) \qquad v(\eta^\tau - \eta^\sigma) \geqq v(x^\tau - x), \quad (\tau \in G_1', \sigma = \phi(\tau)).$$

From (7) and (8) we deduce that

$$(9) \qquad v(\eta^\tau - \eta) \geqq v(x^\tau - x), \quad (\tau \in G_1').$$

On the other hand, if $\tau \in G'$ and $\tau \notin G_1'$ then we write $v(\eta^\tau - \eta) \geqq v(\eta) \geqq v(x)$, and hence, by (4), (9) holds also if $\tau \in G'$ and $\tau \notin G_1'$. Thus (9) holds for all τ in G' and this implies that $\eta \in \tilde{\mathfrak{o}}_y$ (since x belongs to $\mathfrak{o}$). We have thus shown that that hight-hand side of (1) is contained in $\mathfrak{m}_y$.

Conversely, let η be any element of $\tilde{\mathfrak{m}}_y$. We set $\xi = \eta y/x$. We have $v(\eta) \geqq v(\tilde{\mathfrak{m}}_y) = v(x)$ (by the transversality of x in $\mathfrak{o}$ and by Proposition 1.2). Therefore $\eta \in \bar{\mathfrak{o}}x$ and $v(\xi) \geqq v(y)$. Since $\eta \in \tilde{\mathfrak{o}}_y$ and since x is a y-saturator of $\mathfrak{o}$, the inequality (9) holds *for all* τ in G'. The inequality (8) is still valid since it is a consequence of (3) and of the inequality $v(\eta) \geqq v(x)$. From (8) and (9) we deduce (7). Again, also (6) is valid since it is a simple consequence of the inequality $v(\eta) \geqq v(x)$. From (1) and (6) we deduce (5), *for all* σ in G_1, i.e., we have $v(\eta^\sigma y^\sigma - \eta y) \geqq x(x) + v(y^\sigma - y)$, if $\sigma \in G_1$. Now $\eta y = x\xi$, and $x^\sigma = x$. Hence this last inequality can be written as follows:

$$(10) \qquad v(\xi^\sigma - \xi) \geqq v(y^\sigma - y), \text{ for all } \sigma \in G_1.$$

On the other hand, if $\sigma \in G$ and $\sigma \notin G_1$, we have, by analogy with (4): $v(y^\sigma - y) = v(y)$. Since $v(\xi) \geqq v(y)$, it follows that (10) holds for all $\sigma \in G$, and this implies that $\xi \in \mathfrak{m}$. Therefore $\eta \in \mathfrak{m}x/y$, and this completes the proof of (1). As to (2), it is a direct consequence of (1). This completes the proof of Proposition 2.2.

PROPOSITION 2.3. *Let $\mathfrak{o}$, $\mathfrak{o}_1$ be two complete, equicharacteristic, saturated local domains, of dimension 1, having in common a field of representatives k, which is algebraically closed and of characteristic zero. Let F be the field of fractions of $\mathfrak{o}$, let v be the natural valuation of F, let x be a transversal parameter of $\mathfrak{o}$ and let y be an x-saturator of $\mathfrak{o}$. Let F_1, v_1, x_1 and y_1 have a similar meaning for $\mathfrak{o}_1$. Assume that $v(y) = v_1(y_1)$ and that the rings $\tilde{\mathfrak{o}}_y$ and $\tilde{\mathfrak{o}}_{1,y_1}$ are k-isomorphic. Then also $\mathfrak{o}$ and $\mathfrak{o}_1$ are k-isomorphic.*

Proof. By Corollary 1.13 it will be sufficient to show that the two sets of integers $\mathfrak{m}v$ and $\mathfrak{m}_1 v_1$ are identical. By assumption, the sets $\mathfrak{m}_y v$ and $\mathfrak{m}_{1,y_1} v_1$ are identical. Let α be any element of $\mathfrak{m}v$.

We consider first the case in which $\alpha \geqq v(y)$. We fix an element ξ in $\mathfrak{m}$ such that $v(\xi) = \alpha$. Since $v(\xi) \geqq v(y)$, it follows from Proposition 2.2 that the element $\eta = \xi x/y$ belongs to $\tilde{\mathfrak{m}}_y$. Thus $\alpha + v(x) - v(y) \in \tilde{\mathfrak{m}}_y v$, or $\alpha + v(x) - v(y) \in \tilde{\mathfrak{m}}_{1,y_1} v_1$. Now, we have by assumption: $v(y) = v_1(y_1)$. We also know that $\tilde{\mathfrak{o}}_y$ *and* $\mathfrak{o}$ have the same multiplicity (Proposition 1.2), and hence x is a transversal parameter of $\tilde{\mathfrak{o}}_y$. Similarly, x_1 is a transversal parameter of $\tilde{\mathfrak{o}}_{1,y_1}$. Since $\tilde{\mathfrak{o}}_y$ and $\tilde{\mathfrak{o}}_{1,y_1}$ are isomorphic, it follows that $v(x) = v_1(x_1)$. Thus, $\alpha + v_1(x_1) - v_1(y_1) \in \mathfrak{m}_{1,y_1} v_1$. This implies, by Proposition 2.2, that $\alpha \in \mathfrak{m}_1 v_1$.

Assume now that $\alpha < v(y)$. Let $n = v(x) = v_1(x_1)$. By Theorem 1.12, part (A), it follows that α is divisible by n. In fact, in the notations of that theorem, we have $v(y^\tau - y) = \beta_1$ for some τ, and $v(y) \geqq \beta_1$, and consequently $\alpha < \beta_1$. It follows now that also α belongs to $\mathfrak{m}_1 v_1$.

Thus, we have shown that $\mathfrak{m}v \subset \mathfrak{m}_1 v_1$. Similarly, $\mathfrak{m}_1 v_1 \subset \mathfrak{m}v$, whence $\mathfrak{m}v = \mathfrak{m}_1 v_1$. This completes he proof.

PROPOSITION 2.4. *Let $\mathfrak{o}$ be saturated, let x be a transversal parameter of $\mathfrak{o}$ and let y, y_1 be two x-saturators of $\mathfrak{o}$ such that $v(y) = v(y_1)$. Then there exists a k-automorphism of $\tilde{\mathfrak{o}}_y$ which sends y into y_1 (and hence we have $\tilde{\mathfrak{o}}_y = \tilde{\mathfrak{o}}_{y_1}$).*

Proof. As was pointed out in the beginning of the proof of Proposition 2.2, the fact that y is an x-saturator of $\mathfrak{o}$ and that x is transversal implies that x is a y-saturator of $\mathfrak{o}$. Therefore we can apply Theorem 1.11, part (B), with $\mathfrak{o}$, x, x' and y replaced respectively by $\tilde{\mathfrak{o}}_y$, y, y_1 and x. Thus, the proposition will be proved if we prove that (we use the notations G, G_1, G' and G_1' of the proof of Proposition 2.2):

$$(11) \qquad v(y_1^\tau - y_1) \geqq v(x^\tau - x) + v(y) - v(x), \text{ for all } \tau \text{ in } G'.$$

If $\tau \notin G_1'$, then (11) follows from (4) and from $v(y_1^\tau - y_1) \geqq v(y_1) = v(y)$. Assume now that $\tau \in G_1'$ and let $\sigma = \phi(\tau)$ $(\sigma \in G_1)$. Then we have the analog of (3), with η being replaced by y_1:

$$(12) \qquad v(y_1^\tau - y_1^\sigma) = v(x^\tau - x) - v(x) + v(y_1), \ (\tau \in G_1', \sigma = \phi(\tau)).$$

Since both y and y_1 are x-saturators of $\mathfrak{o}$, we have (by the very definition of an x-saturator):

(13) $$v(y^\sigma - y) = v(y_1{}^\sigma - y_1), \text{ for all } \sigma \text{ in } G_1.$$

If we now apply (3) to the case $\eta = y$ and observe that $y^\tau = y$, we find

(14) $$v(y^\sigma - y) = v(x^\tau - x) - v(x) + v(y),$$

and now (11) follows from (12), (13) and (14). This completes the proof.

We now proceed to the proof of Theorem 2.1.

The proof will be by induction on the length of the $\mathfrak{o}$-module $\tilde{\mathfrak{o}}/\mathfrak{o}$. Let C', C_1' be the quadratic transforms of C and C_1 and let $\mathfrak{o}'$, $\mathfrak{o}_1'$ be the local domains of C' and C_1'. We may, of course, assume that C is *not* a regular branch, i. e., that $\mathfrak{o} \neq \tilde{\mathfrak{o}}$. Then $\mathfrak{o}$ is a proper subring of $\mathfrak{o}'$, the length of the $\mathfrak{o}'$-module $\tilde{\mathfrak{o}}/\mathfrak{o}'$ is less than the length of $\mathfrak{o}$-module $\tilde{\mathfrak{o}}/\mathfrak{o}$, and hence we may assume that the theorem holds for C' and C_1'. We choose a basis (x, y) of the maximal ideal $\mathfrak{m}$ of $\mathfrak{o}$ in such a way that x is a transversal parameter and that $v(y)$ is not an integral multiple of $v(x)$. In a similar way we choose a basis x_1, y_1 of the maximal ideal $\mathfrak{m}_1$ of $\mathfrak{o}_1$. We set

$$y' = y/x, \quad y_1' = y_1/x_1.$$

Then

$$\mathfrak{o} = k[[x, y]], \quad \mathfrak{o}_1 = k[[x_1, y_1]];$$

$$\mathfrak{o}' = k[[x, y']], \quad \mathfrak{o}_1' = k[[x_1, y_1']].$$

Here (x, y') is a basis of the maximal ideal $\mathfrak{m}'$ of $\mathfrak{o}'$; similarly, (x_1, y_1') is a basis of the maximal ideal $\mathfrak{m}_1'$ of $\mathfrak{o}_1'$. As usual, we denote by $\tilde{\mathfrak{o}}$ (resp., $\tilde{\mathfrak{o}}_1$) the saturation $\tilde{\mathfrak{o}}_x$ (resp., $\tilde{\mathfrak{o}}_{1,x_1}$) with respect to a transversal parameter. We denote by $\tilde{\mathfrak{o}}'$ (resp., $\tilde{\mathfrak{o}}_1'$) the quadratic transform of $\tilde{\mathfrak{o}}$ (resp., of $\tilde{\mathfrak{o}}_1$). We observe first of all that we have the following relations:

(15) $$\widetilde{(\mathfrak{o}')}_x = \tilde{\mathfrak{o}} : (x),$$

(15$_1$) $$\widetilde{(\mathfrak{o}_1')}_{x_1} = \tilde{\mathfrak{o}}_1 : (x_1).$$

Here $\tilde{\mathfrak{o}} : (x)$ stands for the set of element η in $\tilde{\mathfrak{o}}$ such that $x\eta \in \tilde{\mathfrak{o}}$; and similarly for $\tilde{\mathfrak{o}}_1 : (x_1)$ and $\tilde{\mathfrak{o}}_1$. Let us prove, for instance, (15). If $\eta \in \tilde{\mathfrak{o}}$, then $\eta \in \widetilde{(\mathfrak{o}')}_x$ if and only if $v(\eta^\tau - \eta) \geqq v(y'^\tau - y')$ for all τ in $G = \mathfrak{G}(F/k((x)))$. Since $y'^\tau - y' = (y^\tau - y)/x$, we see that $\eta \in \widetilde{(\mathfrak{o}')}_x$ if and only if $\eta \in \tilde{\mathfrak{o}}$ and $v[(x\eta)^\tau - x\eta] \geqq v(y^\tau - y)$, for all $\tau \in G$, i. e., if and only if $\eta \in \tilde{\mathfrak{o}}$ and $x\eta \in \tilde{\mathfrak{o}}$, and this proves (15).

We begin with the proof of the necessity of the condition stated in the theorem. Assume then that $C \equiv C_1$. We have then $e(C) = e(C_1)$ (where $e(C)$ stands for the multiplicity df the singular point O of C, i.e., $e(C)$ = multiplicity of the local ring $\mathfrak{o}$; similarly for $e(C_1)$ and $\mathfrak{o}_1$), and we also have (by our inductive definition of equivalence) $C' \equiv C_1'$. We shall deal separately with two cases:

Case I. $e(C') = e(C)$.

Case II. $e(C') < e(C)$.

Case I. In this case we also have $e(C_1') = e(C_1)$, since $e(C) = e(C_1)$ and since $e(C') = e(C_1')$ (in view of $C' \equiv C_1'$). Thus x is also a transversal parameter of $\mathfrak{o}'$, and x_1 is a transversal parameter of $\mathfrak{o}_1'$. We have therefore $\widetilde{\mathfrak{o}'} = (\widetilde{\mathfrak{o}'})_x$ and $\widetilde{\mathfrak{o}_1'} = (\widetilde{\mathfrak{o}_1'})_{x_1}$. Therefore, by (15) and (15_1), we have

$$(16) \qquad \widetilde{\mathfrak{o}'} = \bar{\mathfrak{o}} : (x),$$

$$(16_1) \qquad \widetilde{\mathfrak{o}_1'} = \bar{\mathfrak{o}}_1 : (x_1).$$

By or induction hypothesis, $\widetilde{\mathfrak{o}'}$ and $\widetilde{\mathfrak{o}_1'}$ are k-isomorphic. A k-isomorphism between $\widetilde{\mathfrak{o}'}$ and $\widetilde{\mathfrak{o}_1'}$ will send x into some transversal parameter of $\widetilde{\mathfrak{o}_1'}$. It follows therefore from Theorem 1.11, part (A), that there also exists a k-isomorphism $\phi: \widetilde{\mathfrak{o}'} \xrightarrow{\approx} \widetilde{\mathfrak{o}_1'}$ such that $\phi(x) = x_1$. By (16), the elements of $\bar{\mathfrak{o}}$ are the elements of the form $c + \xi' x$, with $\xi' \in \widetilde{\mathfrak{o}'}$ and $c \in k$. In fact, it is obvious from (16) that any element of that form belongs to $\bar{\mathfrak{o}}$. Conversely, if η is any elments of $\mathfrak{m}$ and if we set $\xi' = \eta/x$, then $\xi' \in \bar{\mathfrak{o}}$ (since x is a transversal parameter of $\mathfrak{o}$) and $\xi' x \in \bar{\mathfrak{o}}$. Hence, by (16), $\xi' \in \bar{\mathfrak{o}}'$, and thus η is of the indicated form. Similarly, the elements of $\bar{\mathfrak{o}}_1$ are the elements of the form $c + \xi_1' x_1$, with $\xi_1' \in \bar{\mathfrak{o}}_1'$ and $c \in k$. Since $\phi(x) = x_1$, it follows that the isomorphism $\phi: \widetilde{\mathfrak{o}'} \xrightarrow{\approx} \widetilde{\mathfrak{o}_1'}$ induces an isomorphism between the rings $\bar{\mathfrak{o}}$ and $\widetilde{\mathfrak{o}_1}$.

Case II. In this case we have also $e(C_1') < e(C_1)$. Hence $v(x) > v(y')$ and $v_1(x_1) > v_1(y_1')$, so that this time it is y' (and not x) that is a transversal parameter of the quadratic transform $\mathfrak{o}'$ of $\mathfrak{o}$; similarly, y_1' is a transversal parameter of $\mathfrak{o}_1'$. Thus, this time we have

$$\widetilde{\mathfrak{o}'} = (\widetilde{\mathfrak{o}'})_{y'} \text{ and } \widetilde{\mathfrak{o}_1'} = (\widetilde{\mathfrak{o}_1'})_{y_1'}.$$

As in Case I, our induction hypothesis implies that the rings $\widetilde{\mathfrak{o}'}$ and $\widetilde{\mathfrak{o}_1'}$ are k-isomorphic, and again, by Theorem 1.11, part (A), we may assume that there exists a k-isomorphism $\psi\colon \tilde{\mathfrak{o}}' \xrightarrow{\sim} \tilde{\mathfrak{o}}_1'$ such that $\psi(y') = y_1'$. Let $\psi(x) = x_1'$. Then

$$(17) \qquad \tilde{\mathfrak{o}}_1' = \overbrace{k[[x_1, y_1']]}_{y_1'} = \overbrace{k[[x_1', y_1']]}_{y_1'}.$$

[*Note.* The equality $\mathfrak{o}_1' = \overbrace{k[[x_1', y_1']]}_{y_1'}$ is obtained by applying ψ to the equality $\mathfrak{o}' = \overbrace{k[[x, y']]}_{y'}$].

Here we have a situation similar to the one which was present in Proposition 2.4, namely: we have a saturated ring $\tilde{\mathfrak{o}}_1'$, a transversal parameter y_1' of that ring, and two y_1'-saturators x_1, x_1'. Furthermore, we have $v_1(x_1) = v_1(x_1')$ (since $v_1(x_1') = v(x)$ and $v(x) = v_1(x_1) = e(C) = e(C_1)$). Hence, by Proposition 2.4, there exists a k-automorphism of $\tilde{\mathfrak{o}}_1'$ which sends x_1' into x_1. Composing the isomorphism ψ with this automorphism we obtain a k-isomorphism $\phi\colon \tilde{\mathfrak{o}}' \xrightarrow{\sim} \tilde{\mathfrak{o}}_1'$ such that $\phi(x) = x_1$. This isomorphism can be extended to an isomorphism (which we shall continue to denote by ϕ) between the saturation of $\tilde{\mathfrak{o}}'$ with respect to x and the saturation of $\tilde{\mathfrak{o}}_1'$ with respect to x_1. *The saturation of $\tilde{\mathfrak{o}}'$ with respect to x contains, of course, the saturation $\widetilde{(\mathfrak{o}')}_x$ of $\mathfrak{o}'$ with respect to x* (since $\tilde{\mathfrak{o}}' \supset \mathfrak{o}'$). On the other hand, by Corollary 1.10, we have that $\widetilde{(\mathfrak{o}')}_x \supset \tilde{\mathfrak{o}}'$ $(= \widetilde{(\mathfrak{o}')}_{y'})$, and therefore, by the minimality property of the saturation of $\tilde{\mathfrak{o}}'$ with respect to x, we have that $\widetilde{(\mathfrak{o}')}_x$ *contains the saturation of $\tilde{\mathfrak{o}}'$ with respect to x.* Thus, $\widetilde{(\mathfrak{o}')}_x$ is *equal* to the saturation of $\tilde{\mathfrak{o}}'$ with respect to x. Similarly, $\widetilde{(\mathfrak{o}_1')}_{x_1}$, is *equal* to the saturation of $\tilde{\mathfrak{o}}_1'$ with respect to x_1. The extended isomorphism ϕ is therefore a k-isomorphism $\widetilde{(\mathfrak{o}')}_x \xrightarrow{\sim} \widetilde{(\mathfrak{o}_1')}_{x_1}$, i.e., by (15) and (15$_1$), ϕ is a k-isomorphism $\tilde{\mathfrak{o}}\colon (x) \xrightarrow{\sim} \mathfrak{o}_1\colon (x_1)$. Since $\phi(x) = x_1$, we conclude, as in case I, that ϕ induces an isomorphism between $\tilde{\mathfrak{o}}$ and $\tilde{\mathfrak{o}}_1$.

We now proceed to the proof of the sufficiency of the condition stated in the theorem. Assume then that there exists a k-isomorphism $\phi\colon \tilde{\mathfrak{o}} \xrightarrow{\sim} \mathfrak{o}_1$. By Theorem 1.11, part (A), we may also assume that $\phi(x) = x_1$. We have then, in the first place, that $e(\mathfrak{o}) = e(\mathfrak{o}_1)$, i.e., $e(C) = e(C_1)$. So we have only to prove that $C' \equiv C_1'$. By our induction hypothesis, this amounts to proving that $\tilde{\mathfrak{o}}'$ and $\tilde{\mathfrak{o}}_1'$ are k-isomorphic. Again, we shall deal separately with the two cases I and II indicated above, but we first observe that in both cases, the isomorphism $\phi : \bar{\mathfrak{o}} \xrightarrow{\sim} \tilde{\mathfrak{o}}_1$ can be extended to an isomorphism $\bar{\mathfrak{o}} : (x) \xrightarrow{\sim} \tilde{\mathfrak{o}}_1 : (x_1)$ (which we shall continue to denote by ϕ), since $\phi(x) = x_1$ and since ϕ has an extension to an isomorphism between $\bar{\mathfrak{o}}$ and $\bar{\mathfrak{o}}_1$.

Case I. In this case, it follows from (16) and (16_1) that ϕ is the desired isomorphism between $\widetilde{\mathfrak{o}'}$ and $\widetilde{\mathfrak{o}_1'}$.

Case II. In this case, we have by (15) and (15_1):

$$\phi : \overbrace{k[[x, y']]}_x \xrightarrow{\sim} \overbrace{k[[x_1, y_1']]}_{x_1},$$

and $\phi(x) = x_1$. Proposition 2.3 is now applicable, since y' and y_1' are transversal parameters and $v(x) = v(x_1)$ $(= e(C) = e(C_1))$, and we conclude that $\widetilde{\mathfrak{o}'}$ $(= \overbrace{k[[x, y']]}_{y'})$ and $\widetilde{\mathfrak{o}_1'}$ $(= \overbrace{k[[x_1, y_1']]}_{y_1'})$ are k-isomorphic. This concludes the proof of the theorem.

We shall conclude this section by deriving some results concerning the behaviour of our saturated local domains under locally quadratic transformations and related also to the question of permutability of the two operations: saturation and locally quadratic transformation.

Proposition 2.5. *If $\mathfrak{o}$ is saturated, if x is a transversal parameter of $\mathfrak{o}$ and if $\mathfrak{o}'$ is the locally quadratic transform of $\mathfrak{o}$, then*

$$\mathfrak{o}' = \mathfrak{o} : (x), \tag{18}$$

where $\mathfrak{o} : (x)$ stands for the set of all elements η in $\bar{\mathfrak{o}}$ such that $x\eta \in \mathfrak{o}$. Furthermore, $\mathfrak{o}'$ is saturated with respect to x.

Proof. Since $v(x) \leqq v(\xi)$ for all ξ in $\mathfrak{m}$, it follows that $\mathfrak{o}'$ is generated over $\mathfrak{o}$ (as a ring) by the set of quotients ξ/x, $\xi \in \mathfrak{m}$. Since $\mathfrak{o} : (x)$ is precisely the set of these quotients ξ/x, it follows that $\mathfrak{o} : (x) \subset \mathfrak{o}'$. On the other hand, if we choose an x-saturator y of $\mathfrak{o}$, so that $\mathfrak{o} = \overbrace{k[[x, y]]}_x$, and assume furthermore (as we may) that $v(y) > v(x)$, then from (15) (where $\mathfrak{o}$ should now be replaced by $\overbrace{k[[x, y]]}$) *we deduce that* $\mathfrak{o} : (x)$ *is a ring* (namely the ring $\overbrace{k[[x, y']]}_x$, where $y' = y/x$). This shows that $\mathfrak{o} : (x)$ is saturated with respect to x. Since $\mathfrak{o} : (x)$ contains $\mathfrak{o}$ and also all the generators ξ/x of $\mathfrak{o}'$ over $\mathfrak{o}$, it follows that $\mathfrak{o}' \subset \mathfrak{o} : (x)$. This establishes (18) and completes the proof.

Corollary 2.6. *If x is a transversal parameter of $\mathfrak{o}$ (where now $\mathfrak{o}$ is not necessarily saturated) then*

$$(\tilde{\mathfrak{o}}_x)' = \widetilde{(\mathfrak{o}')}_x, \tag{19}$$

where ′ stands for the operation of taking the locally quadratic transform.

For, if y denotes an x-saturator of $\mathfrak{o}$ such that $v(y) > v(x)$ and if we set $y' = y/x$, then by (15) we have $\tilde{\mathfrak{o}}_x : (x) = \widetilde{k[[x, y']]}_x$. Since $y' \in \mathfrak{o}'$ it follows from this equality that $\tilde{\mathfrak{o}}_x : (x) \subset \widetilde{(\mathfrak{o}')}_x$. Now by (18), we have $\tilde{\mathfrak{o}}_x : (x) = (\tilde{\mathfrak{o}}_x)'$. Hence $(\tilde{\mathfrak{o}}_x)' \subset \widetilde{(\mathfrak{o}')}_x$. Since $\mathfrak{o} \subset \tilde{\mathfrak{o}}_x$ and since x is also a transversal parameted of $\tilde{\mathfrak{o}}_x$, it follows that $\tilde{\mathfrak{o}'} \subset (\tilde{\mathfrak{o}}_x)'$. Thus, we see that $(\tilde{\mathfrak{o}}_x)'$ is a local domain containing $\mathfrak{o}'$ contained in $\widetilde{(\mathfrak{o}')}_x$ and saturated with respect to x (by Proposition 2.5). The equality (19) now follows from the minimality property of $\widetilde{(\mathfrak{o}')}_x$.

Note that since x is a transversal parameter of $\mathfrak{o}$, we can write, in (19), $\tilde{\mathfrak{o}}$ instead of $\tilde{\mathfrak{o}}_x$. *If there is no drop in multiplicity from* $\mathfrak{o}$ *to* $\mathfrak{o}'$, then x is also a transversal parameter of $\mathfrak{o}'$, and (19) can then be written as follows:

(20) $$(\tilde{\mathfrak{o}})' = (\tilde{\mathfrak{o}'}) \quad (\text{if } e(\mathfrak{o}) = e(\mathfrak{o}')).$$

However, if $e(\mathfrak{o}') < e(\mathfrak{o})$ then we can only assert (see Corollary 1.10) that

(21) $$(\tilde{\mathfrak{o}})' \supset (\tilde{\mathfrak{o}'}).$$

We note that (20) *is always true if* $\mathfrak{o}$ *is saturated,* for in that case (20) becomes $\mathfrak{o}' = \tilde{\mathfrak{o}}'$, which is a true relation, since by Proposition 2.5, $\mathfrak{o}'$ is saturated. In the example $\mathfrak{o} = k[[t^4, t + t^7]]$, we have $(\tilde{\mathfrak{o}})' = k[[t^2, t^3]]$, $(\tilde{\mathfrak{o}'}) = k[[t^2 + t^3, t^4, t^5]]$, and (20) is false.

3. The characteristic of a Puiseux expansion and saturation. Let $\mathfrak{o}$ be the local domain of an irreducible plane algebroid curve C, over k, let F be the field of fractions of $\mathfrak{o}$ and let $\bar{\mathfrak{o}}$ be the integral closure of $\mathfrak{o}$ in F. We denote by v the natural valuation of the local field F. If (x, y) is a basis of the maximal ideal $\mathfrak{m}$ of $\mathfrak{o}$ and if, say, $v(x) = n$, then $\bar{\mathfrak{o}}$ contains an element t such that $t^n = x$, and we have $\bar{\mathfrak{o}} = k[[t]]$, $F = k((t))$. We have then for y a Puiseux expansion in $x^{1/n}$:

(1) $$y = \sum_{i=1}^{\infty} a_i x^{i/n}.$$

We shall assume that C is *not* a regular branch; or equivalently: $n > 1$, $v(y) > 1$. Let

(2) $$\frac{m_1}{n_1} < \frac{m_2}{n_1 n_2} < \cdots < \frac{m_g}{n_1 n_2 \cdots n_g} = \frac{m_g}{n},$$

where

(2′) $$(m_i, n_i) = 1;\ i = 1, 2, \cdots, g;\ g \geqq 1,$$

22

be the *characteristic exponents* of the series (1). We recall their definition.

$\frac{m_1}{n_1}$ is the first exponent in (1) which is not an integer.

If $n_1 < n$, $\frac{m_2}{n_1 n_2}$ is the first exponent in (1) which is not an integral multiple of $\frac{1}{n_1}$.

If $n_1 n_2 < n$, $\frac{m_3}{n_1 n_2 n_3}$ is the first exponent in (1) which is not an integral multiple of $\frac{1}{n_1 n_2}$; etc.

It follows that $n_i > 1$, $i = 1, 2, \cdots, g$, and that $n_1 n_2 \cdots n_g = n$. We note that the characteristic exponents determine uniquely the pairs (m_i, n_i), in view of (2) and (2′).

We re-write the expansion (1) by putting into evidence the characteristic terms:

$$y = \cdots + b_1 x^{\frac{m_1}{n_1}} + \cdots + b_2 x^{\frac{m_2}{n_1 n_2}} + \cdots + b_g x^{\frac{m_g}{n_1 n_2 \cdots n_g}} + \cdots, \tag{3}$$
$$b_\nu \neq 0, \nu = 1, 2, \cdots, g.$$

We shall now proceed to finding the saturation $\tilde{\mathfrak{o}}_x$ of $\mathfrak{o}$ with respect to the parameter x; more precisely with respect to the field $K = k((x))$. The field F is a cyclic extension of K, of degree n. Let G be the Galois group of F/K. More generally, let

$$\left.\begin{aligned} K_\nu &= k((x^{\frac{1}{n_1 n_2 \cdots n_\nu}})), \\ G_\nu &= \mathfrak{G}(F/K_\nu) \end{aligned}\right\} \quad \nu = 1, 2, \cdots, g.$$

Thus

$$K = K_0 < K_1 < \cdots < K_g = F; \tag{4}$$

$$G = G_0 > G_1 > \cdots > G_g = (1). \tag{4′}$$

If τ is any element of $G_{\nu-1}$, *not* in G_ν, then it follows from (3) that $y^\tau - y = b_\nu(\epsilon_\nu - 1) x^{\frac{m_\nu}{n_1 n_2 \cdots n_\nu}} +$ terms of higher degree. Here $\epsilon_\nu{}^n = 1$ and $\epsilon_\nu \neq 1$. Therefore

$$v(y^\tau - y) = \beta_\nu, \text{ for all } \tau \in G_{\nu-1} - G_\nu, \tag{5}$$

where we have set

$$\beta_\nu = \frac{m_\nu}{n_1 n_2 \cdots n_\nu} \cdot n = m_\nu n_{\nu+1} \cdots n_g, \quad (\nu = 1, 2, \cdots, g). \tag{6}$$

The set of integers $\{n; \beta_1, \beta_2, \cdots, \beta_g\}$ is called the *characteristic* of the Puiseux expansion (3); here n is $v(x)$, and $\frac{\beta_1}{n}, \frac{\beta_2}{n}, \cdots, \frac{\beta_g}{n}$ are the characteristic exponents of the expansion. From (5) it follows that *the set of integers* $\beta_1, \beta_2, \cdots, \beta_g$ *is the set of integers* $v(y^\tau - y)$ *obtained by letting* τ *range over* G (with $\tau \neq 1$).

We note that by (5) we have

(7) $$G_{\nu-1} = \{\tau \in G \mid v(y^\tau - y) \geqq \beta_\nu\}, \quad \nu = 1, 2, \cdots, g.$$

Again by (5), and applying Lemma 1.4, we find that

(8) $$\tilde{\mathfrak{o}}_x = \{\eta \in k[[t]] \mid v(\eta^\tau - \eta) \geqq \beta_\nu, \text{ for all } \tau \in G_{\nu-1} \text{ and for } \nu = 1, 2, \cdots, g\}.$$

In the terminology of Theorem 1.12 we can say that the *characteristic* $\{n, \beta_1, \beta_2, \cdots, \beta_g\}$ *of the Puiseux expansion* $y = y(x)$ *is the x-characteristic of the saturated local domain* $\tilde{\mathfrak{o}}_x$, *where* $\mathfrak{o} = k[[x, y]]$.

Corollary-*Definition* 1.9 tells us that *the characteristic* $\{n, \beta_1, \beta_2, \cdots, \beta_g\}$ (and therefore also the characteristic exponents) *of the Puiseux expansion* $y = y(x)$ of a given irreducible plane algebroid branch C is *independent of the choice of the basis* (x, y) of the maximal ideal $\mathfrak{m}$ of $\mathfrak{o}$, *as long as we stipulate that* $v(x) \leqq v(y)$, i.e., *as long as* x *is a transversal parameter of* $\mathfrak{o}$. The characteristic thus obtained may be called the *characteristic of the branch* C.

The characteristic of C uniquely determines the set $\tilde{\mathfrak{m}}v$, where $\tilde{\mathfrak{m}}$ is the maximal ideal of $\tilde{\mathfrak{o}}$ (see Theorem 1.12, Part (A). Therefore, by Theorem 2.1 and Corollary 1.13, *two irreducible plane algebroid branches* C, C_1 *have equivalent singularities if and only if they have the same characteristic.*

By means of the standard decomposition of $\mathfrak{m}$ relative to a given transversal parameter x of $\tilde{\mathfrak{o}}$ (see Theorem 1.12, Part (B)), we see that *the saturation* $\tilde{\mathfrak{o}}$ *of* $\mathfrak{o}$ *can be realized as the local domain of an irreducible algebroid curve* $\Gamma(C)$, *immersed in an affine space of dimension*

$$\sum_{\nu=2}^{g}\left[\frac{m_\nu}{n_\nu}\right] - \sum_{\nu=1}^{g-1} m_\nu + g + n - 1$$

(of dimension $\leqq n$ if $g = 1$), *and that* $\Gamma(C)$ *depends only on the equivalence class of* C.

For, $\tilde{\mathfrak{o}}$ admits the following ring generators over its subring $k[[x]]$:

$$t^{s_\nu n_{\nu+1} \cdots n_g}, \quad m_\nu \leqq s_\nu \leqq \left[\frac{m_{\nu+1}}{n_{\nu+1}}\right]; \quad \nu = 1, 2, \cdots, g-1;$$

$$t^{m_g + i}, \quad 0 \leqq i \leqq n - 1,$$

where one of the last n generators, namely the element t^{m_g+i} such that $m_g + i \equiv 0 \pmod{n}$, can be omitted (since it is a power of x).

All plane algebroid branches which are equivalent to C are projections of $\Gamma(C)$, and it is clear that also the generic projection of $\Gamma(C)$ into the plane is itself equivalent to C.

An example.

The local ring $\mathfrak{o}$ of a (singular) plane algebroid branch C is itself saturated if and only if C has a double point.

For, if C has a double point, then its parametric equations are of the form $x = t^2$, $y = t^m +$ terms of higher degree, m-odd. The characteristic of C is $(2, m)$, and $\tilde{\mathfrak{o}}$ is the ring $k + \mathfrak{M} + \bar{\mathfrak{m}}^m$, i. e., $\mathfrak{m}$ consists of all power series in t of the form $\xi(t^2) + a_1 t^m + a_2 t^{m+1} + \cdots$, where $\xi(t^2)$ is any power series in t^2, while $a_1, a_2, \cdots$ are arbitrary elements of k. It is immediate, however, that if $m' \geqq m$ then $t^{m'} \in k[[x, y]] = \mathfrak{o}$, and hence $\tilde{\mathfrak{o}} = \mathfrak{o}$.

If the multiplicity of C is $n > 2$, w may assume that the parametric equations of C *are*: $x = t^n$, $y = t^{m_1 n_2 \cdots n_g} + \cdots$ $(m_1 > n_1 > 1;\ (m_1, n_1) = 1)$. One finds at once that the element $t^{(m_1+1)n_2\cdots n_g}$ belongs to $\tilde{\mathfrak{o}}$ but not to $\mathfrak{o}$ provided $m_1 + 1 \not\equiv 0 \pmod{n_1}$. If, however, $m_1 + 1$ is divisible by n_1 then $t^{(m_1+2)n_2\cdots n_g} \notin \mathfrak{o}$, provided $n_1 \neq 2$. In the special case $n_1 = 2$ we have necessarily $n_1 < n$, and from this it follows easily that $t^{m_1 n_1 \cdots n_g} \notin \mathfrak{o}$.

One final remark, and that will concern a quantitative consequence of Proposition 2. 2. Namely, we have the following result (*inversion formula*):

Let $\{n, \beta_1, \beta_2, \cdots, \beta_g\}$ be the characteristic of a Puiseux expansion $y = y(x)$ and let $\{n', \beta_1', \beta_2', \cdots, \beta_{g'}'\}$ be the characteristic of the Puiseux expansion $x = x(y)$ (whence $n = v(x)$, $n' = v(y)$). Assuming, without loss of generality, that $n \leqq n'$, the following is true:

(a) *if $n < n' < \beta_1$ (whence n' is an integral multiple of n) then* $g' = g + 1$, $\beta_1' = n$, $\beta'_{\nu+1} + n' = \beta_\nu + n\, (\nu = 1, 2, \cdots, g)$.

(b) *In all other cases (i. e., if $n' = n$ or if $n' = \beta_1$), we have $g' = g$ and $\beta_\nu' + n' = \beta_\nu + n$ $(\nu = 1, 2, \cdots, g)$.*

Proof. We recall that, in our present notations we have: $\beta_1 =$ least integer of $\bar{\mathfrak{m}}_x v$ which is not a multiple of n, and $e_1 = (n, \beta_1)$; more generally, if $\beta_1, \beta_2, \cdots, \beta_i$ and $e_1, e_2, \cdots, e_i$ have already been defined and if $e_i > 1$, then β_{i+1} is the least integer of $\bar{\mathfrak{m}}_x v$ which is not a multiple of e_i, and $e_{i+1} = (e_i, \beta_{i+1})$. Similarly, we have: β_1' is the least integer of $\bar{\mathfrak{m}}_y v$ which is not a multiple of n', and $e_1' = (n', \beta_1')$; more generally, if $\beta_1', \beta_2', \cdots, \beta_i'$ and $e_1', e_2', \cdots, e_i'$ have already been defined and if $e_i' > 1$, then β'_{i+1} is the least integer of $\bar{\mathfrak{m}}_y v$ which is not a multiple of e_i', and $e'_{i+1} = (e_i', \beta'_{i+1})$. Here $\bar{\mathfrak{m}}_x$

is the maximal ideal of $\overbrace{k[[x,y]]}_x$ and $\tilde{\mathfrak{m}}_y$ is the maximal ideal of $\overbrace{k[[x,y]]}_y$.

We identify in Proposition 2.2 the saturated ring $\mathfrak{o}$ with the ring $\overbrace{k[[x,y]]}_x$ (whence y is an x-saturator of $\mathfrak{o}$), As was pointed out in the proof of that proposition, this implies that x is a y-saturator of $\mathfrak{o}$, and hence $\overbrace{k[[x,y]]}_y = \tilde{\mathfrak{o}}_y$. Hence, by Proposition 2.2, the set $\tilde{\mathfrak{m}}_y v$ consists of the set of integers

$$(9) \qquad \alpha + n - n' \mid \alpha \in \tilde{\mathfrak{m}}_x v, \quad \alpha \geqq n'.$$

Now assume first that $n < n' < \beta_1$. The smallest integer of $\tilde{\mathfrak{m}}_y v$ is n (since $n = e(k[[x,y]]) = e(\mathfrak{o}) = e(\tilde{\mathfrak{o}}_y)$), and $n < n'$. Hence $\beta_1' = n$. Moreover, since $n' < \beta_1$, we have $n' \equiv 0 \pmod{n}$, whence $e_1' = (n', \beta_1') = (n', n) = n$. Thus, β_2' is the smallest integer of $\tilde{\mathfrak{m}}_y v$ which is not divisible by n, and therefore, by (9), $\beta_2' = \beta_1 + n - n'$ (always taking into account that n' is divisible by n and that $n' < \beta_1$) and $e_2' = (\beta_2', n) = (\beta_1, n) = e_1$. In a similar way it follows now that $\beta_3' = \beta_2 + n - n'$ and $e_3' = e_2$, etc.

The case $n' = n$ is trivial, because in that case we have by Proposition 2.2, $\tilde{\mathfrak{m}}_x v = \mathfrak{m}_y v$ (and, actually, $\mathfrak{m}_x = \mathfrak{m}_y$, since now also y is a transversal parameter).

There remains the case $n' = \beta_1$ (the case $n' > \beta_1$ does not occur, because $n' = v(y) \leqq v(y^\tau - y)$ for all $\tau \in \mathcal{G}(F/k((x)))$ and $v(y^\tau - y) = \beta_1$ for some τ). Clearly, since $\beta_1 > n$, we have now again $\beta_1' = n$, thus $\beta_1' + n' = n + n' = \beta_1 + n$. We also have $e_1' = (n', \beta_1') = (\beta_1, n) = e_1$, and from this follows, in view of (9), that $\beta_2' = \beta_2 + n - n'$ (taking into account that both n and n' $(=\beta_1)$ are divisible by e_1), and thus $(\beta_2', e_1') = (\beta_2, e_1) = e_2$. Similarly one finds that $\beta_3' = \beta_3 + n - n'$ and $e_3' = e_3$, etc.

Part II. Saturation of Algebroid Varieties.

4. An ideal-theoretic property of saturation. Let A, F, K and $\tilde{A}$ $(=\tilde{A}_K)$ have the same meaning as in the Introduction. We shall prove the following general result:

THEOREM 4.1. *Assume that the ring $R = A \cap K$ is integrally closed in K. Then for every prime ideal $\mathfrak{p}$ in A there is one and only one prime ideal $\tilde{\mathfrak{p}}$ in $\tilde{A}$ such that $\tilde{\mathfrak{p}} \cap A = \mathfrak{p}$; furthermore, the field of fractions of $\tilde{A}/\tilde{\mathfrak{p}}$ is a purely inseparable extension of the field of fractions of $A/\mathfrak{p}$.*

When two rings A, $\tilde{A}$, where A is a subring of $\tilde{A}$, are such that the morphism f: Spec $\tilde{A} \to$ Spec A defined by $f(\tilde{\mathfrak{p}}) = \tilde{\mathfrak{p}} \cap A$ satisfies the conditions stated in the above theorem, (with one modification, however: the words "one and only one" are to be replaced by the words "at most one"), then f is called a *radicial* morphism ([3], Ch. I, Def. 3.5.4). As $\bar{A}$ is, in our case, integrally dependent on A, the existence of *at least one* $\bar{\mathfrak{p}}$ is assured.

Proof. We shall use the notations adopted in the Introduction. Since $\bar{A} = \bigcup_{i=0}^{\infty} A_i$ and $A_i \subset A_{i+1}$, it is sufficient to prove that each of the morphisms $\operatorname{Spec} A_{i+1} \to \operatorname{Spec} A_i$ is radicial, and it will be sufficient to prove this for $i = 0$, since A_{i+1} is obtained from A_i by the same process by which A_1 is obtained from A. It will be more convenient now to denote the ring A_1 by A'. We have therefore $A' = A[L]$, where L is the set of all elements η of $\bar{A}$ such that η dominates some element ζ of A (ζ may depend on η). Since A' is integral over A, there is at least one prime ideal in A which contracts to $\mathfrak{p}$. The first part of the theorem asserts that if $\mathfrak{p}_1'$, $\mathfrak{p}_2'$ are two prime ideals in A' such that $\mathfrak{p}_1' \cap A = \mathfrak{p}_2' \cap A = \mathfrak{p}$, then $\mathfrak{p}_1' = \mathfrak{p}_2'$.

By assumption (condition b, Introduction), we have a direct sum decomposition of F into fields F_i:

$$F = F_1 \oplus F_2 \oplus \cdots \oplus F_h.$$

We consider (in some algebraic closure of F_i) the least Galois extension F_i^* of $K\epsilon_i$ which contains F_i, and we set

$$F^* = F_1^* \oplus F_2^* \oplus \cdots \oplus F_h^*.$$

Let A^* be the integral closure of A in F^*. To prove the first part of our theorem (i.e., the above equality $\mathfrak{p}_1' = \mathfrak{p}_2'$) we have to show the following: if $\mathfrak{p}_1^*$, $\mathfrak{p}_2^*$ are prime ideals in A^* such that

$$\mathfrak{p}_1^* \cap A = \mathfrak{p}_2^* \cap A \; (= \mathfrak{p}), \tag{1}$$

then

$$\mathfrak{p}_1^* \cap A' = \mathfrak{p}_2^* \cap A'. \tag{2}$$

We set $\mathfrak{p}_i^* \cap A' = \mathfrak{p}_i'$ $(i = 1, 2)$. We shall prove, for instance, that

$$\mathfrak{p}_1' \subset \mathfrak{p}_2'. \tag{3}$$

We write the decomposition of the zero ideal in A and A':

$$(0) = \mathfrak{P}_1 \cap \mathfrak{P}_2 \cap \cdots \cap \mathfrak{P}_h, \text{in } A;$$

$$(0) = \mathfrak{P}_1' \cap \mathfrak{P}_2' \cap \cdots \cap \mathfrak{P}_h', \text{ in } A'.$$

Here we have $\mathfrak{P}_i = \{\eta \in A \mid \eta\epsilon_i = 0\} = A \cap \sum_{j \neq i} F_j$; similarly,

$$\mathfrak{P}_i' = \{\eta' \in A' \mid \eta'\epsilon_i = 0\} = A' \cap \sum_{j \neq i} F_j;$$

and $\mathfrak{P}_i = \mathfrak{P}_i' \cap A$. We fix two prime ideals $\mathfrak{P}_s', \mathfrak{P}_t'$ in the set $\{\mathfrak{P}_1', \mathfrak{P}_2', \cdots, \mathfrak{P}_h'\}$ such that

(4) $$\mathfrak{p}_1' \supset \mathfrak{P}_s', \quad \mathfrak{p}_2' \supset \mathfrak{P}_t'.$$

The ideals $\mathfrak{P}_s'$ and $\mathfrak{P}_t'$ need not be distinct. We have then

(5) $$\mathfrak{p} \supset \mathfrak{P}_s, \quad \mathfrak{p} \supset \mathfrak{P}_t.$$

We also set

(6) $$\mathfrak{p} \cap R = \mathfrak{p}_0.$$

We use the notations $\hat{F}_i^*$ $(i = 1, 2, \cdots, h)$, $\hat{F}^*$ of the Introduction. If ψ is any K-homomorphism of F into Ω, then $\psi(F) \subset \hat{F}^*$, and, in fact, $\psi(F) \subset \hat{F}_i^*$, if we choose i so that $\psi(\epsilon_i) = 1$ (in which case, we have necessarily $\psi(\epsilon_j) = 0$ if $j \neq i$). [We recall that each $\hat{F}_i^*$ is a Galois extension of K, that $\hat{F}^*$ is the compositum of the fields $\hat{F}_i^*$, and we note that the h fields $\hat{F}_i^*$ need not be distinct]. Any ψ, as above, can be extended to a homomorphism of F^* into $\hat{F}^*$. We shall therefore deal directly with K-homomorphisms ψ of F^* into $\hat{F}^*$. We note that if $\psi(\epsilon_i) = 1$ then $\psi(F^*) = \hat{F}_i^*$, and the restriction $\psi \mid F_i^*$ is an isomorphism $F_i^* \xrightarrow{\sim} \hat{F}_i^*$ which takes $a\epsilon_i$ $(a \in K)$ into a.

We now fix two K-homomorphisms ψ_1, ψ_2 of F^* into $\hat{F}^*$ such that

(7) $$\psi_1 \mid F_s^* \neq 0, \qquad \psi_2 \mid F_t^* \neq 0,$$

or equivalently: $\psi_1(\epsilon_s) = \psi_2(\epsilon_t) = 1$. We set, for $i = 1, 2$:

(8) $$\hat{A}_i = \psi_i(A), \qquad \hat{A}_i' = \psi_i(A'); \\ \hat{\mathfrak{p}}_i = \psi_i(\mathfrak{p}), \qquad \hat{\mathfrak{p}}_i' = \psi_i(\mathfrak{p}_i').$$

By (4) and (5), $\mathfrak{p}_i$ and $\mathfrak{p}_i'$ are prime ideal in $\hat{A}_i$ and $\hat{A}_i'$ respectively (since, by (7), $\mathfrak{P}_s'$ is the kernel of $\psi_1 \mid A'$, and $\mathfrak{P}_t'$ is the kernel of $\psi_2 \mid A'$).

We note that multiplication by ϵ_s and ϵ_t define surjective homomorphisms $A' \to A'\epsilon_s$ and $A' \to A'\epsilon_t$, with kernels $\mathfrak{P}_s'$ and $\mathfrak{P}_t'$ respectively; similarly for the homomorphisms $A \to A\epsilon_s$, $A \to A\epsilon_t$ and $\mathfrak{P}_s$, $\mathfrak{P}_t$. It follows at once from (4), (5) and (6) that

(9_1) $$\mathfrak{p}_1'\epsilon_s \cap A\epsilon_s = \mathfrak{p}\epsilon_s,$$

(9_2) $$\mathfrak{p}_2'\epsilon_t \cap A\epsilon_t = \mathfrak{p}\epsilon_t,$$

(9_3) $$\mathfrak{p}\epsilon_s \cap R\epsilon_s = \mathfrak{p}_0\epsilon_s, \quad \mathfrak{p}\epsilon_t \cap R\epsilon_t = \mathfrak{p}_0\epsilon_t.$$

Since $\psi_1 \mid A'\epsilon_s$ and $\psi_2 \mid A'\epsilon_t$ are isomorphisms $A'\epsilon_s \xrightarrow{\simeq} \hat{A}_1'$ and $A'\epsilon_t \xrightarrow{\simeq} \hat{A}_2'$ respectively, it follows from (9_1) and $9_2)$ that

$$(10) \qquad \hat{\mathfrak{p}}_i' \cap \hat{A}_i = \hat{\mathfrak{p}}_i, \quad i = 1, 2,$$

and since $\psi_i \mid R$ is the identity it follows from (9_3) that

$$(11) \qquad \hat{\mathfrak{p}}_1 \cap R = \mathfrak{p}_2 \cap R = \mathfrak{p}_0.$$

Let $\hat{A}^*$ be the integral closure of R in $\hat{F}^*$. We have, of course, $R \subset \hat{A}_i \subset \hat{A}_i' \subset \hat{A}^*$ $(i = 1, 2)$. We fix in $\hat{A}^*$ a prime ideal $\hat{\mathfrak{p}}_i^*$ $(i = 1, 2)$ such that

$$(12) \qquad \hat{\mathfrak{p}}_i^* \cap \hat{A}_i' = \hat{\mathfrak{p}}_i', \quad i = 1, 2.$$

Then we have by (10) and (11) that $\hat{\mathfrak{p}}_i^* \cap R = \mathfrak{p}_0$ $(i = 1, 2)$. Therefore $\hat{\mathfrak{p}}_1^*$ and $\hat{\mathfrak{p}}_2^*$ are conjugate prime ideals[3] over K, i.e., there exists a K-automorphism τ of $\hat{F}^*$ such that

$$(13) \qquad \hat{\mathfrak{p}}_2^* = (\hat{\mathfrak{p}}_1^*)^{\tau}.$$

For any such automorphism τ we set

$$(14) \qquad \bar{\psi}_1 = \tau\psi_1.$$

At this stage we shall need the following lemma:

LEMMA 4.2. *There exists a K-automorphism τ of $\hat{F}^*$ such that* (13) *holds and, furthermore, such that if $\bar{\psi}_1$ is defined as in* (14) *then*

$$(15) \qquad \psi_2(\zeta) - \bar{\psi}_1(\zeta) \in \hat{\mathfrak{p}}_2^*, \textit{ for all } \zeta \textit{ in } A.$$

In order not to interrupt the proof of the theorem, we postpone the proof of the lemma and admit it for the moment without proof. Our object now is to prove the inclusion (3). Let then η' be any element of $\mathfrak{p}_1'$. By (8), $\psi_1(\eta') \in \hat{\mathfrak{p}}_1 \subset \hat{\mathfrak{p}}_1^*$, and hence by (13) and (14):

$$(16) \qquad \bar{\psi}_1(\eta') \in \hat{\mathfrak{p}}_2^*.$$

To prove (3), i.e., that $\eta' \in \mathfrak{p}_2'$, is equivalent to proving that $\psi_2(\eta') \in \hat{\mathfrak{p}}_2^*$ [see (8) and (12)]. In view of (16), *everything is now reduced to proving that*

$$(17) \qquad \psi_2(\eta') - \bar{\psi}_1(\eta') \in \hat{\mathfrak{p}}_2^*.$$

Since $\eta' \in A' = A[L]$, we can write

[3] Here, and through the rest of the proof, we use known facts from Galois theory of (finite or infinite) normal algebraic extensions of the field of fractions K of an arbitrary integrally closed domain R. See [2], § 2, Proposition 6. See also [1].

$$\eta' = f(\eta_1, \eta_2, \cdots, \eta_N), \quad \eta_\nu \in L, \quad \nu = 1, 2, \cdots, N,$$

where f is a polynomial with coefficients in A. We denote by f^1 (resp. f^2) the polynomial obtained from f by applying $\bar{\psi}_1$ (resp., ψ_2) to the coefficients of f. Then we have:

$$\begin{aligned} &\psi_2(\eta') - \bar{\psi}_1(\eta') = [f^2(\psi_2(\eta_1), \psi_2(\eta_2), \cdots, \psi_2(\eta_N)) \\ (18) \qquad &- f^1(\psi_2(\eta_1), \psi_2(\eta_2), \cdots, \psi_2(\eta_N))] \\ &+ [f^1(\psi_2(\eta_1), \psi_2(\eta_2), \cdots, \psi_2(\eta_N)) - f^1(\psi_1(\eta_1), \psi_1(\eta_2), \cdots, \psi_1(\eta_N))]. \end{aligned}$$

By (15), the expression inside the first square brackets is an element of $\hat{\mathfrak{p}}_2^*$ (since the coefficients of f are in A). Now, since the elements η_ν are in L, there exist elements $\zeta_1, \zeta_2, \cdots, \zeta_N$ in A such that η_ν dominates ζ_ν $(\nu = 1, 2, \cdots, N)$. In particular, each quotient

$$[\psi_2(\eta_\nu) - \psi_1(\eta_\nu)]/[\psi_2(\zeta_\nu) - \psi_1(\zeta_\nu)]$$

belongs to $\hat{A}^*$, unless $\psi_2(\zeta_\nu) = \psi_1(\zeta_\nu)$, in which case we must have also $\psi_2(\eta_\nu) = \psi_1(\eta_\nu)$. Now, again by (15), each of the N elements $\psi_2(\zeta_\nu) - \psi_1(\zeta_\nu)$ belongs to $\hat{\mathfrak{p}}_2^*$. It follows that also each of the N elements $\psi_2(\eta_\nu) - \psi_1(\eta_\nu)$ belongs to $\hat{\mathfrak{p}}_2^*$. Consequently, also the expression sinside the second square bracket of (18) belongs to $\hat{\mathfrak{p}}_2^*$. This proves (17) and completes the proof of the first part of the theorem (modulo the proof of Lemma 4.2, still to be given).

We now proceed to the proof of the second part of the theorem. Let, as before, A^* be the integral closure of A in F^*, let $\mathfrak{p}^*$ be a prime ideal in A^* and $\mathfrak{p}' = \mathfrak{p}^* \cap A'$, $\mathfrak{p} = \mathfrak{p}^* \cap A$. Let Δ (resp., Δ') be the field of fractions of $A/\mathfrak{p}$ (resp., of $A'/\mathfrak{p}'$), whence Δ' is an algebraic extension of Δ. *We have to show that Δ' is purely inseparable over Δ.*

Among the h prime ideals $\mathfrak{P}_i^*$ of the zero in A^* we fix one, say $\mathfrak{P}_1^*$, which is contained in $\mathfrak{p}^*$. Then $\mathfrak{P}_1' \subset \mathfrak{p}'$ and $\mathfrak{P}_1 \subset \mathfrak{p}$. We fix a K-homomorphism ψ of F^* into $\hat{F}^*$ such that $\psi \mid F_1^* \neq 0$. We set $\psi(F) = \hat{F}_1$, $\psi(F^*) = \hat{F}_1^*$ $(= \psi(F_1^*))$, so that F_1^* is the least Galois extension of K which contains the field $\hat{F}_1$. If we set furthermore

$$(19) \qquad \hat{A} = \psi(A), \quad \hat{A}' = \psi(A'), \quad \hat{A}_1^* = \psi(A^*),$$

$$(20) \qquad \hat{\mathfrak{p}} = \psi(\mathfrak{p}), \quad \hat{\mathfrak{p}}' = \psi(\mathfrak{p}'), \quad \hat{\mathfrak{p}}_1^* = \psi(\mathfrak{p}^*),$$

then $\ddot{F}_1$ is the field of fractions of both $\hat{A}$ and $\hat{A}'$, $\hat{A}'$ is integal over $\hat{A}$, and $\hat{A}_1^*$ is the integral closure of $\hat{A}$ (hence also of R) in $\hat{F}_1^*$. Furthermore, $\hat{\mathfrak{p}}$, $\hat{\mathfrak{p}}'$ and $\hat{\mathfrak{p}}_1^*$ are prime ideal in their respective rings $\hat{A}$, $\hat{A}'$ and $\hat{A}_1^*$.

Let $\hat{L}$ denote the set of elements $\psi(\eta)$, $\eta \in L$. Then

$$\hat{A}' = \hat{A}[\hat{L}]. \tag{21}$$

If ψ_1 is any other K-homomorphism of F^* into $\hat{F}^*$ such that $\psi_1 \mid F_1^* \neq 0$, then ψ_1 and ψ have the same kernel (namely $\mathfrak{P}_1^* = F_2^* \oplus F_3^* \oplus \cdots \oplus F_h^*$), and thus $\psi_1\psi^{-1}$ is a K-automorphism τ of $\hat{F}_1^*$ $(= \psi(F^*) = \psi_1(F^*))$, i.e., τ is an element of the Galois group of $\hat{F}_1^*/K$. Conversely, if τ is any element of that group and if we set $\psi_1 = \tau\psi$, then ψ_1 is a K-homomorphism of F^* into $\hat{F}^*$ such that $\psi_1 \mid F_1^* \neq 0$. Now, by definition of the set L, we have that for η in L there is an element ζ in A such that for each K-homomorphism ψ_1 of F^* into $\hat{F}^*$ the quotient $[\psi_1(\eta) - \psi(\eta)]/[\psi_1(\zeta) - \psi(\zeta)]$ is integral over R, unless the denominator is zero, in which case also the numerator is zero. Applying this condition only to those K-homomorphisms ψ_1 which satisfy the condition $\psi_1 \mid F_1^* \neq 0$, we conclude that the elements $\hat{\eta}$ of $\hat{L}$ have the following property: *each element $\hat{\eta}$ of $\hat{L}$ dominates some element ζ of $\hat{A}$* (and, of course, $\hat{\eta}$ belongs to integral closure of $\hat{A}$ into its field of fractions $\hat{F}_1$). From (19) and (20) follows that the surjective homomorphism $\psi: A' - \hat{A}'$ induces an isomorphism $A'/\mathfrak{p}' \longrightarrow \hat{A}'/\hat{\mathfrak{p}}'$, and that this isomorphism maps $A/\mathfrak{p}$ onto $\hat{A}/\hat{\mathfrak{p}}$. Thus, if we denote by $\hat{\Delta}'$ and $\hat{\Delta}$ the field of fractions of $\hat{A}'/\hat{\mathfrak{p}}'$ and $\hat{A}/\hat{\mathfrak{p}}$ respectively, we have a natural isomorphism $\hat{\Delta}' \xrightarrow{\simeq} \Delta'$ which maps $\hat{\Delta}$ onto Δ. Thus it will be sufficient to show that $\hat{\Delta}'$ is purely inseparable over $\hat{\Delta}$. In conclusion, we can now replace A and A' $(= A[L])$ by the *integral domains* $\hat{A}$ and $\hat{A}'$ $(= \hat{A}[\hat{L}]$; see (21). We therefore may asume that our original rings A and A' are integral domains. Thus, F is now a field, and F^* is the least Galois extension of K containing F. We denote by G the Galois group of F^*/K and by $G_{\mathfrak{p}^*}$ the decomposition group of $\mathfrak{p}^*$.

Let Δ^* be the field of fractions of $A^*/\mathfrak{p}^*$. It is known that Δ^* is a normal extension of Δ. Let σ^* be any Δ-automorphism of Δ^*. We shall prove that $\sigma^* \mid \Delta'$ is the identity, and this will show that Δ' is a purely inseparable extension of Δ.

Let ϕ be the canonical homomorphism $A^* \to \Delta^*$. We set $\Delta_0 =$ field of fractions of $R/\mathfrak{p}_0$ (where $\mathfrak{p}_0 = \mathfrak{p} \cap R$). It is known that given any element τ of $G_{\mathfrak{p}^*}$, there is a well defined Δ_0-automorphism τ^* of Δ^* such that

$$\phi\tau = \tau^*\phi,$$

and that the mapping $G_{\mathfrak{p}^*} \to \mathfrak{G}(\Delta^*/\Delta_0)$, defined in this fashion, is surjective. So, in particular, we can find an element σ in $G_{\mathfrak{p}^*}$ such that

(22) $$\phi\sigma = \sigma^*\phi.$$

Since $\sigma^* \mid \Delta$ is the identity and since $A' = A[L]$, it follows that in order to show that $\sigma^* \mid \Delta'$ is also the identity it is sufficient to show that if $\eta \in L$ then $(\sigma^*\phi)(\eta) = \phi(\eta)$; or—equivalently (in view of (22))—that $\phi(\eta^\sigma) = \phi(\eta)$, i.e., that

$$\eta^\sigma - \eta \in \mathfrak{p}^*.$$

But this inclusion follows from the fact that η dominates some elements ζ of A, and that for each element ζ of A we have $\zeta^\sigma - \zeta \in \mathfrak{p}^*$ (this last assertion is a consequence of (22) and of the assumption that $\sigma^* \mid \Delta$ is the identity).

This completes the proof of the theorem. We now proceed to the proof of Lemma 4.2 which was used in the proof of the theorem.

Proof of Lemma 4.2. Let ϕ_2 denote the canonical homomorphism $\hat{A}^* \to \hat{A}^* | \hat{\mathfrak{p}}_2^*$, let Δ^* be the field of fractions of $\hat{A}^*/\mathfrak{p}_2^*$, and let τ_1 be some R-automorphism of $\hat{A}^*$ such that $\hat{\mathfrak{p}}_2^* = \hat{\mathfrak{p}}_1^{*\tau_1}$. We set $\phi_1 = \phi_2\tau_1$; thus ϕ_1 is also a homomorphism of $\hat{A}^*$ into Δ^*. We look at the effect of ϕ_1 and ϕ_2 on the subrings $\hat{A}_1$ and $\hat{A}_2$ respectively [see (8)]. We denote by Δ_i the field of fractions of $\phi_i(\hat{A}_i)$ and we set $\theta_i = (\phi_i \mid \hat{A}_i)\cdot\psi_i$ $(i = 1, 2)$. We have $\operatorname{Ker}(\phi_2 \mid \hat{A}_2) = \hat{\mathfrak{p}}_2^* \cap \hat{A}_2 = \hat{\mathfrak{p}}_2$, and $\psi_2^{-1}(\hat{\mathfrak{p}}_2) = \mathfrak{p}$. Hence $\operatorname{Ker}\theta_2 = \mathfrak{p}$, and similarly $\operatorname{Ker}\theta_1 = \mathfrak{p}$. Hence $\theta_1 = \rho_0^*\theta_2$, where $\rho_0^*\colon \Delta_2 \xrightarrow{\sim} \Delta_1$ is an isomorphism. We summarize all this in the following commutative diagram:

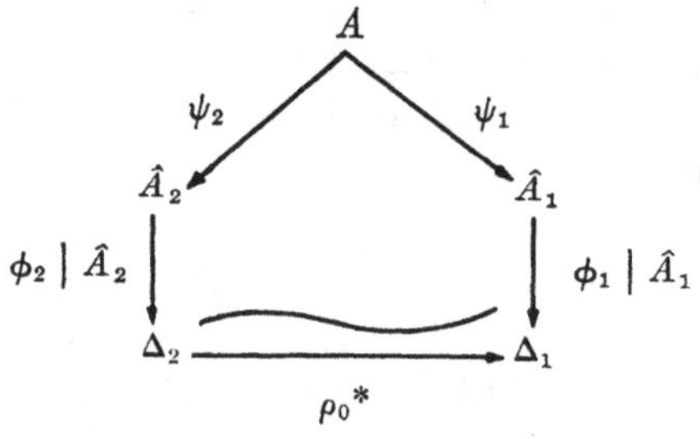

Since $\phi_1 = \phi_2\tau_1$ and since $\tau_1 \mid R$ is the identity, it follows that $\phi_1 \mid R = \phi_2 \mid R$. Since also $\psi_1 \mid R = \psi_2 \mid R$ (= the identity), it follows that $\theta_1 \mid R = \theta_2 \mid R$. Therefore, if Δ_0 denotes the field of fractions of $\phi_2(R)$ $(= R/\mathfrak{p}_0)$, then $\rho_0^* | \Delta_0$ is the identity. Thus ρ_0^* is a Δ_0-isomorphism $\Delta_2 \longrightarrow \Delta_1$. Since Δ^* is a normal extension of Δ_0, we can extend ρ_0^* to a Δ_0-automorphism ρ^* of Δ^*. Again, by known facts of Galois theory, there exists an element ρ in the decomposition group $G_{\hat{\mathfrak{p}}_2^*}$ such that $\phi_2\rho = \rho^*\phi$. Applying this equality to any element of the form $\psi_2(\zeta)$, $\zeta \in A$, and using the fact that we have on A:

$$\phi_2\rho\psi_2 = \rho^*\phi_2\psi_2 = \rho_0{}^*\theta_2 = \theta_1 = \phi_1\psi_1 = \phi_2\tau_1\psi_1,$$

we find that

$$(\rho\psi_2)(\zeta) - (\tau_1\psi_1)(\zeta) \in \hat{\mathfrak{p}}_2{}^*$$

for all ζ in A; or—equivalently:

$$\psi_2(\zeta) - \bar{\psi}_1(\zeta) \in \hat{\mathfrak{p}}_2{}^*, \text{ all } \zeta \text{ in } A,$$

where

$$\bar{\psi}_1 = \tau\psi_1 \qquad \tau = \rho^{-1}\tau_1.$$

Since $\hat{\mathfrak{p}}_2{}^* = \hat{\mathfrak{p}}_1{}^{*\tau_1}$ and $\rho \in G_{\hat{\mathfrak{p}}_2{}^*}$, we have $\hat{\mathfrak{p}}_2{}^* = \hat{\mathfrak{p}}_1{}^{*\tau}$, and the lemma is proved.

5. Application to algebroid or complex-analytic varieties. Let V be either an equidimensional formal algebroid variety, defined locally, at its center O, over an algebraically closed ground field k [whence k is a particular field of representatives of the (complete) local ring $\mathfrak{o}$ of V] or an equidimensional complex-analytic variety, defined locally at O (in which case $k = C =$ field of complex numbers). In the latter case we mean by the local ring $\mathfrak{o}$ of V the ring of germs of analytic functions on V at the origin O. In other words, in this latter case $\mathfrak{o}$ is a homomorphic image of the ring of convergent power series in a certain number n of independent complex variables $Z_1, Z_2, \cdots, Z_n$ (for instance, $n =$ embedding dimension of V). Let r be the Krull dimension of $\mathfrak{o}$ ($r =$ dimension of each irreducible component of V) and let $(x_1, x_2, \cdots, x_r)$ be a system of separating local parameters of $\mathfrak{o}$ (see Introduction). By Theorem 4.1, the saturation $\tilde{\mathfrak{o}}_{(x)}$ ($= \tilde{\mathfrak{o}}^{(k)}{}_{(x)}$) has only one maximal ideal, and is therefore a local ring between $\mathfrak{o}$ and the integral closure $\bar{\mathfrak{o}}$ of $\mathfrak{o}$. [It is understood that, in the complex-analytic case, the field $K = k((x_1, x_2, \cdots, x_r))$ with respect to which the saturation $\tilde{\mathfrak{o}}_{(x)}$ is defined, is the field of fractions of the ring of convergent power series in the r independent complex variables $x_1, x_2, \cdots, x_r$.] Thus $\tilde{\mathfrak{o}}_{(x)}$ is the local ring of an algebroid (or complex-analytic) variety $\tilde{V}_{(x)}$ which dominates V (since $\mathfrak{o} \subset \tilde{\mathfrak{o}}_{(x)}$), and the two varieties V and $\tilde{V}_{(x)}$ have the same normalization $\bar{V}$. We call $\tilde{V}_{(x)}$ the *saturation* of V *with respect to the parameters* $x_1, x_2, \cdots, x_r$. In our present formal set-up, V and $\tilde{V}_{(x)}$ are simply Spec $\mathfrak{o}$ and Spec $\tilde{\mathfrak{o}}_{(x)}$ respectively. The morphism $\tilde{f}_{(x)}: \tilde{V}_{(x)} \to V$ is proper (since $\tilde{\mathfrak{o}}_{(x)}$ is integral over $\mathfrak{o}$) and finite (since $\tilde{\mathfrak{o}}_{(x)}$ is a finite $\mathfrak{o}$-module). By Theorem 4.1, $\tilde{f}_{(x)}$ *is a radicial morphism.*

Assume now that the equidimensional algebroid variety V is complex-analytic. Then also $\tilde{V}_{(x)}$ is complex-analytic, and the morphism $\tilde{f}_{(x)}: \tilde{V}_{(x)}$

$\to V$ defines locally (or: is induced by) a continuous (holomorphic) map of some neighborhood of O on V onto some neighborhood of the center O of $\tilde{V}_{(x)}$, on $\tilde{V}_{(x)}$. We continue to denote by $\tilde{f}_{(x)}$ this continuous map.

PROPOSITION 5.1. *Let V and V' be two complex analytic varieties (not necessarily equidimensional), defined locally in the neighborhood of their centers O and O'. We make the following assumptions:* 1) *the local ring $\mathfrak{o}'$ of V' contains the local ring $\mathfrak{o}$ of V as a subring;* 2) *$\mathfrak{o}'$ is a finite $\mathfrak{o}$-module. Assume that the morphism f:* Spec $\mathfrak{o}'$—Spec $\mathfrak{o}$ *is radicial. Then the continuous map $V' \to V$ defined by f (and which we shall continue to denote by f) is a local homeomorphism.*

Proof. Let $r = \dim V$ ($= \dim V'$). (i.e., $r =$ maximum of the dimensions of the irreducible components of V). The proposition is trivial in the case $r = 0$ (in which case $V = O$ and $V' = O'$). We now use induction on r. We first show that $\mathfrak{o}$ and $\mathfrak{o}'$ have the same total ring of fractions. Let $\mathfrak{p}_1' \cap \mathfrak{p}_2' \cap \cdots \cap \mathfrak{p}_h'$ be the irredundant decomposition of the zero ideal in $\mathfrak{o}'$; here the $\mathfrak{p}_i'$ are prime ideals, since $\mathfrak{o}'$ has no nilpotent elements (other than zero). Let $\mathfrak{p}_i = \mathfrak{p}_i' \cap \mathfrak{o}$. Then $\mathfrak{p}_1 \cap \mathfrak{p}_2 \cap \cdots \cap \mathfrak{p}_h$ is the zero ideal in $\mathfrak{o}$, and this is an irredundant decomposition, since the morphism f is radicial. For the same reason, we have that if F_i is the field of fractions of $\mathfrak{o}/\mathfrak{p}_i$ and F_i' is the field of fractions of $\mathfrak{o}'/\mathfrak{p}_i'$, then $F_i = F_i'$. So $\mathfrak{o}$ and $\mathfrak{o}'$ have the same total ring of fractions, namely $F_1 \oplus F_2 \oplus \cdots \oplus F_h$.

Since $\mathfrak{o}'$ is a finite $\mathfrak{o}$-module, $\mathfrak{o}$ has a non-zero conductor $\mathfrak{C}$ in $\mathfrak{o}'$. Let W' (resp. W) be the subvariety of V' (resp., of V) defined by $\mathfrak{C}$. Then W and W' are proper analytic subvarieties of V and V', and have dimension $< r$. It is well known that the restriction of f to $V' - W'$ is a local homeomorphism $V' - W' \to V - W$.[4] Therefore it is sufficient to show that $f \mid W'$: $W' \to W$ is a local homeomorphism. Now, the local rings of W' and W are respectively $\mathfrak{o}'/\mathrm{Rad}'\,\mathfrak{C}$ and $\mathfrak{o}/\mathrm{Rad}\,\mathfrak{C}$, where the two radicals are taken in $\mathfrak{o}'$ and $\mathfrak{o}$ respectively. It follows at once that $\mathfrak{o}'/\mathrm{Rad}'\,\mathfrak{C}$ is a finite module over its subring $\mathfrak{o}/\mathrm{Rad}\,\mathfrak{C}$ and that the morphism $\mathrm{Spec}(\mathfrak{o}'/\mathrm{Rad}'\,\mathfrak{C}) \to \mathrm{Spec}(\mathfrak{o}/\mathrm{Rad}\,\mathfrak{C})$ is radicial.

The proposition now follows by our induction hypothesis.

COROLLARY 5.2. *If V, $\tilde{V}_{(x)}$ and $\tilde{f}_{(x)}: \tilde{V}_{(x)} \to V$ are as above, and V, $\tilde{V}_{(x)}$ are complex-analytic, then $\tilde{f}_{(x)}$ is a local homeomorphism.*

[4] It is a well-known fact [see S. Abhyankar, *Local Analytic Geometry* (Academic Press, 1964), § 46] that the local conductors $\mathfrak{C}_P$ ($P \in V$) form a coherent sheaf of ideals on V, and from this it follows that 1) the non-normal points of V form an analytic subvariety W of V and that 2) W is defined, locally at O, by the local conductor $\mathfrak{C}$ ($= \mathfrak{C}_O$).

COROLLARY 5.3. *If two equidimensional, complex-analytic varieties V and V' have isomorphic saturations (with respect to suitable parameters (x) and (x')), then V and V' are locally homeomorphic varieties.*

6. Special case of complex-analytic hypersurfaces. Let $\mathfrak{o}$ (resp., $\mathfrak{o}'$) be the local ring of a complex-analytic variety V (resp., V') at a point O (resp., O'). We assume the following: 1) both V and V' are equidimensional at O and O', of the same dimension r; 2) the maximal ideal $\mathfrak{m}$ of $\mathfrak{o}$ and the maximal ideal $\mathfrak{m}'$ of $\mathfrak{o}'$ have bases of $r+1$ elements. We fix a basis $(x_1, x_2, \cdots, x_r, x_{r+1})$ of $\mathfrak{m}$ and a basis $(x'_1, x'_2, \cdots, x'_{r+1})$ of $\mathfrak{m}'$. By means of these bases we regard V (resp., V') as embedded in the affine space $\mathcal{A}_{r+1}$ (resp., $\mathcal{A}'_{r+1}$), in the neighborhood of the origin O (resp., O'). We assume furthermore that $x_1, x_2, \cdots, x_r$ are parameters of $\mathfrak{o}$ and that $x'_1, x'_2, \cdots, x'_r$ are parameters of $\mathfrak{o}'$. By Corollary 5.2 we have the local homeomorphisms $\tilde{f}_{(x)}: \tilde{V}_{(x)} \to V$ and $\tilde{f}'_{(x')}: \tilde{V}'_{(x')} \to V'$. The object of this section is to prove the following result:

THEOREM 6.1. *If there exists a C-isomorphism $\phi: \tilde{\mathfrak{o}}_{(x)} \to \tilde{\mathfrak{o}}'_{(x')}$ such that $\phi(x_i) = x'_i$ $(i = 1, 2, \cdots, r)$ and we continue to denote by ϕ the associated local homeomorphism $V_{(x)} \to V'_{(x')}$, then the local homeomorphism*

$$\psi = \tilde{f}'_{(x')} \phi \tilde{f}_{(x)}^{-1}: V \to V' \tag{1}$$

can be extended to a local homeomorphism between the ambient affine spaces $\mathcal{A}_{r+1}$, $\mathcal{A}'_{r+1}$ (at their respective origins O, O').

Proof. The isomorphism ϕ can be extended (uniquely) to an isomorphism between the total rings of fractions F, F' of $\mathfrak{o}$, $\mathfrak{o}'$. Since $\phi(x_i) = x'_i$ $(i = 1, 2, \cdots, r)$, we shall identify F with F', x_i with x'_i $(i = 1, 2, \cdots, r)$ and $\tilde{\mathfrak{o}}_{(x)}$ with $\tilde{\mathfrak{o}}'_{(x')}$. We write y instead of x_{r+1} and y' instead of x'_{r+1}. Thus V and V' are now defined by equations of the form:

$$V: g(x_1, x_2, \cdots, x_r; y) = 0,$$
$$V': g'(x_1, x_2, \cdots, x_r; y') = 0,$$

where $g((x); Y)$ and $g'((x); Y)$ are monic polynomials in Y, free from multiple factors, with coefficients which are power series in $x_1, x_2, \cdots, x_r$, convergent in some neighborhood

$$U_\epsilon: |x_i| < \epsilon \quad (i = 1, 2, \cdots, r; \ \epsilon\text{-real positive}) \tag{2}$$

of the origin O_0: $x_1 = x_2 = \cdots = x_r = 0$, and zero at O_0 (except, of course, for the leading coefficients 1). By our identifications we have

$$(3) \qquad F = F' = k((x_1, x_2, \cdots, x_r))[y] = k((x_1, x_2, \cdots, x_r))[y'],$$

where $K = k((x_1, x_2, \cdots, x_r))$ is the field of germs of meromorphic functions of $x_1, x_2, \cdots, x_r$ at O_0. From (3) it follows that g and g' have the same degree in Y, say n, since this degree is the dimension of F, regarded as a vector space over K. If h is the number of irreducible components of V, then F is a direct sum of h fields:

$$F = F_1 \oplus F_2 \oplus \cdots \oplus F_h,$$

and conversely. Thus V and V' have the same number h of irreducible components, and h is also the number of irreducible factors of each of the power series $g((x);Y)$, $g'((x);Y)$.

We shall use the notations of the Introduction. The ring F, being a simple ring extension of $R = k[[x_1, x_2, \cdots, x_r]]$, has precisely n distinct K-homomorphisms into Ω, say $\psi_1, \psi_2, \cdots, \psi_n$. If we set

$$\psi_i(y) = y_i, \qquad \psi_i(y') = y'_i,$$

then $y_1, y_2, \cdots, y_n$ are the roots of $g((x);Y)$ in Ω, and similarly, $y'_1, y'_2, \cdots, y'_n$ are the roots of $g'((x);Y)$ in Ω. We have $\psi_i(F) = K[y_i] = K[y'_i]$, and the compositum F^* of the n fields $\psi_i(F)$ is both the splitting field

$$(4) \qquad \hat{F}^* = K(y_1, y_2, \cdots, y_n)$$

of the polynomial $g((x);Y)$ over K and the splitting field

$$(4') \qquad \hat{F}^* = K(y'_1, y'_2, \cdots, y'_n)$$

of the polynomial $g'((x);Y)$ over K.

The n roots $y_1, y_2, \cdots, y_n$ can be identified with the branches $y_i(x)$ of the n-valued algebroid function $y(x)$ defined in U_ϵ by the equation $g((x);y) = 0$. Similarly, $y'_1, y'_2, \cdots, y'_n$ can be identified with the branches $y'_i(x)$ of the n-valued algebroid function $y'(x)$ defined in U_ϵ by the equation $g((x);y') = 0$. Note that we have here a well defined and natural pairing between the branches of $y(x)$ and the branches of $y'(x)$: it is defined by associating $y_i = \psi_i(y)$ with $y'_i = \psi_i(y')$ $(i = 1, 2, \cdots, n)$. In view of (4) and (4′), and also in view of the equalities $K(y_i) = K(y'_i)$ $(= \psi_i(F))$ $(i = 1, 2, \cdots, n)$, it follows that the monodromy group of the function $y(x)$ (which is a group of substitutions on $y_1(x), y_2(x), \cdots, y_n(x)$) is isomorphic with the monodromy group of the function $y'(x)$, the isomorphism between the two groups of substitutions being obtained upon replacing each $y_i(x)$ by the corresponding $y'_i(x)$. Both groups are isomorphic with the Galois group of $\hat{F}^*/K$.

The two polynomials $g((x);Y)$ and $g'((x);Y)$ have thus the same splitting field, and the algebroid functions $y(x)$, $y'(x)$ defined by the equations $g=0$, $g'=0$ have the same monodromy group. But there is more than that in common between g and g'. Let $\hat{R}^*$ be the integral closure of R in $\hat{F}^*$. The fact that $\bar{\mathfrak{o}}_x$ and $\bar{\mathfrak{o}}'_x$ are identical implies, by Lemma 1.4, and is in fact equivalent to, the following conditions: for all $i, j = 1, 2, \cdots, n$, $i \neq j$,

$$(5) \qquad (y'_i - y'_j)/(y_i - y_j) \in \hat{R}^*,$$

$$(5') \qquad (y_i - y_j)/(y'_i - y'_j) \in \hat{R}^*.$$

In fact, by Lemma 1.4, (5) is equivalent with the inclusion $\mathfrak{o}' \subset \bar{\mathfrak{o}}_{(x)}$, and (5′) is equivalent with the inclusion $\mathfrak{o} \subset \bar{\mathfrak{o}}'_{(x)}$. *Thus the quotients* $(y'_i - y'_j)/(y_i - y_j)$ *are units in* $\hat{R}^*$. The conditions (5) and (5′) therefore imply that the discriminants $D(x)$ and $D'(x)$ of g and g' (regarded as polynomials in Y) *differ by a factor* $a(x_1, x_2, \cdots, x_r)$ *which is a unit in* R. We agree to take ϵ so small as to have $a(x) \neq 0$ in U_ϵ (see (2)). *In that case, the two algebroid functions* $y(x)$ *and* $y'(x)$ *have the same critical variety in* U_ϵ. We shall denote this critical variety by Δ.

Let π: $V \to \mathcal{A}_r$ be the projection defined generically by $(x_1, x_2, \cdots, x_r, y) \to (x_1, x_2, \cdots, x_r)$. Thus, if $(a_1, a_2, \cdots, a_r)$ is any point of U_ϵ and b is any of the roots of $g(a_1, a_2, \cdots, a_r; Y)$, then $(a_1, a_2, \cdots, a_r, b)$ is a point of V and $\pi(a_1, a_2, \cdots, a_r, b) = (a_1, a_2, \cdots, a_r)$. The inverse image $\pi^{-1}(U_\epsilon)$ is an open neighborhood of O on V, and its image under π is U_ϵ. Similarly, let π': $V' \to \mathcal{A}_r$ be the projection defined generically by $(x_1, x_2, \cdots, x_r, y') \to (x_1, x_2, \cdots, x_r)$. By definition of the mappings $f_{(x)}: V_{(x)} \to V$ and $\bar{f}_{(x)}: \bar{V}_{(x)} \to V'$, we have $\pi' \bar{f}_{(x)} \phi = \pi f_{(x)}$, and consequently, by (1): $\pi' \psi = \pi$. In other words, for each point $(a_1, a_2, \cdots, a_r, b)$ of V, where ψ is defined, we have $\psi(a_1, a_2, \cdots, a_r, b) = (a_1, a_2, \cdots, a_r, b')$, where b' is some root of $g'(a_1, a_2, \cdots, a_r, Y)$. The very existence of the local homeomorphism ψ guarantees that for all points $(a_1, a_2, \cdots, a_r)$, sufficiently near the origin, there is a way of associating with *each* root b of $f(a_1, a_2, \cdots, a_r; Y)$ a *definite* root b' of $g'(a_1, a_2, \cdots, a_r; Y)$ in such a way that the mapping $(a_1, a_2, \cdots, a_r, b) \to (a_1, a_2, \cdots, a_r, b')$ is a local homeomorphism between V and V'. With the aid of what we have seen above in regard to the monodromy group of the two functions $y(x)$, $y'(x)$ and their common critical variety, we can throw some light on the nature of ψ, from a function-theoretic point of view. In the first place, if the point (a) $[= (a_1, a_2, \cdots, a_r)]$ is in $U_\epsilon - \Delta$, then both polynomials $g((a); Y)$ and $g'((a); Y)$ have n distinct roots. Thus both π and π' define n-fold coverings of $U_\epsilon - \Delta$. This, together with the equality of the monodromy groups, leads us at once to a homeomorphism between $\pi^{-1}(U_\epsilon - \Delta)$ and $\pi'^{-1}(U_\epsilon - \Delta)$, and it is easily seen that this coincides with the restriction ψ_0 of ψ to $\pi^{-1}(U_\epsilon - \Delta)$. Now, in general, the fact that the two n-valued algebroid functions $y(x)$ and $y'(x)$ have, in the neighborhood of the origin $x_1 = x_2 = \cdots = x_r = 0$, the same critical variety Δ and the same monodromy group, does not in itself guarantee that ψ_0 can be extended to a local homeomorphism $\pi^{-1}(U_\epsilon) \to \pi'^{-1}(U_\epsilon)$, if ϵ is sufficiently small. For the existence of such an extension it is obviously necessary and sufficient that the following condition be satisfied in U_ϵ, for sufficiently small ϵ: if $(\bar{a}) = (\bar{a}_1, \bar{a}_2, \cdots, \bar{a}_n)$ *is any point of* Δ *in* U_ϵ *and if for some* $i, j = 1, 2, \cdots, n$, $i \neq j$, *we have that* $y_i(x) - y_j(x)$ *approaches zero as* (x) *approaches* $(\bar{a})$ *along a given arc* γ *(all of which lies in* $U_\epsilon - \Delta$, *except for the end point* $(\bar{a})$*), then also* $y'_i(x) - y'_j(x)$ *approaches zero as* x *approaches* $(\bar{a})$ *along that same arc* γ. This condition must therefore be satisfied *a priori* in our present case, since we know that ψ_0 does have the required extension ψ. The real and the only reason why this condition must be satisfied in our present case can therefore be due only to the fact that the saturations $\tilde{\mathfrak{o}}_x$ and $\tilde{\mathfrak{o}}'_x$ are equal; or equivalently —to the fact that *all the quotients* $[y_i(x) - y_j(x)]/[y'_i(x) - y'_j(x)]$ are units in $\hat{R}^*$ (see (5) and (5′)). Indeed, one could now verify directly that the above condition is satisfied if we take ϵ so small as to have the following: *at each point* a *of* U_ϵ *each of the values of each the algebroid functions* $[y_i(x) - y_j(x)]/[y'_i(x) - y'_j(x)]$ *is different from zero* $(i, j = 1, 2, \cdots, n;\ i \neq j)$.

The proof of the possibility of extending ψ to a local homeomorphism between the ambient space $\mathcal{A}_{r+1}$ and $\mathcal{A}'_{r+1}$ will follow substantially the lines of a proof by Whitney in a similar situation (Whitney [4], §§ 11-12.[5] We shall only outline the various steps of the proof, refering the to Whitney's paper for the necessary details.

We set E_ϵ = set all points $(a_1, a_2, \cdots, a_r, b)$ of $\mathcal{A}_{r+1}$ such that the point $(a) = (a_1, a_2, \cdots, a_r)$ belongs to U_ϵ. Thus E_ϵ is an infinite cylinder which has as basis U_ϵ (see (2)) and as axis the line $X_1 = X_2 = \cdots = X_r = 0$. We shall take ϵ as small as needed (and as will be specified) at the various steps

[5] Whitney's proof concerns essentially the case $r = 1$, and instead of dealing with *two* varieties V, V', of dimension r, he has to deal with an *analytic family* of *curves* $V(t)$, depending on parameters $t_1, t_2, \cdots, t_s$, and his object is to show the existence, under suitable conditions, of a continuous isotopic deformation of $V(0)$ into $V(t)$. The fact that the $V(t)$ are curves allows him to use *Puiseux expansions* (with coefficients which are power series in the parameters t). In the next section we shall generalize Whitney's result to certain analytic families of varieties of dimension r.

of the proof. We have already assumed that ϵ has been chosen so small as to have the following two conditions satisfied: 1) all the coefficients of the polynomials g, g' in Y are power series in $x_1, x_2, \cdots, x_r$ which converge in U_ϵ; 2) the quotient $D(x_1, x_2, \cdots, x_r)/D'(x_1, x_2, \cdots, x_r)$ of the discriminants of g and g' is holomorphic and $\neq 0$ at each point of U_ϵ. Now we outline the various steps of the proof.

A). If $z, z_1, z_2, \cdots, z_n$ are complex numbers and $z \neq z_i$ $(i = 1, 2, \cdots, n)$, we define: $\mu_i = 1/|z - z_i|$ and

$$\nu_i(z_1, z_2, \cdots, z_n, z) = \mu_i / \sum_{j=1}^{n} \mu_j.$$

The ν_i are real-valued, positive, real analytic functions of $z_1, z_2, \cdots, z_n, z$, defined outside the n hyperplanes $z = z_j$. If for a *given* i and j, $i \neq j$, we have $z_j \neq z_i$, then the function ν_i approaches the unique limit zero as the variable point $(Z_1, Z_2, \cdots, Z_n, Z)$ approaches $(z_1, z_2, \cdots, z_n, z_j)$. Thus, if $z_i \neq z_j$ then the function ν_i is also defined and continuous on the hyperplane $z = z_j$, and similarly ν_j is defined and continuous on the hyperplane $z = z_i$. Since, from the definition of the ν_i it follows that

$$\sum_{j=1}^{n} \nu_j = 1, \tag{6}$$

we see that if, for a given i, we have $z_i \neq z_j$ for *all* $j \neq i$, then ν_i is also defined and continuous on the hyperplane $z = z_i$, and its value at $(z_1, z_2, \cdots, z_n, z_i)$ is 1. We have therefore:

$$\text{(7)} \quad \nu_i(Z_1, Z_2, \cdots, Z_n, Z) \to 0 \text{ as } (Z_1, Z_2, \cdots, Z_n, Z) \to (z_1, z_2, \cdots, z_n, z_j), \text{ if } i \neq j \text{ and } z_i \neq z_j.$$

$$\text{(8)} \quad \nu_i(Z_1, Z_2, \cdots, Z_n, Z) \to 1 \text{ as } (Z_1, Z_2, \cdots, Z_n, Z) \to (z_1, z_2, \cdots, z_n, z_i), \text{ if } z_i \neq z_j \text{ for all } j \neq i.$$

B). Let $Z_1, Z_2, \cdots, Z_n, Z, U_1, U_2, \cdots, U_n$ be complex variables. Consider the function

$$\Psi = \sum_{i=1}^{n} \nu_i(Z_1, Z_2, \cdots, Z_n, Z) U_i.$$

Let $z_1, z_2, \cdots, z_n, z, u_1, u_2, \cdots, u_n$ be complex numbers such that

$$z_j \neq z, \text{ if } j = q+1, q+2, \cdots, n;$$

$$z_1 = z_2 = \cdots = z_q = z;$$

$$u_1 = u_2 = \cdots = u_q \ (\text{say } = u).$$

Then $\lim \Psi$ *exists as* $(Z_1, Z_2, \cdots, Z_n, Z, U_1, U_2, \cdots, U_n)$ *approaches*

$$(z, z, \cdots, z, z_{q+1}, z_{q+2}, \cdots, z_n, z, u, u, \cdots, u, u_{q+1}, u_{q+2}, \cdots, u_n),$$

and this limit is equal to u. For the proof, one writes Ψ in the form

$$\Psi = \sum_{i=1}^{q} \nu_i(Z_1, Z_2, \cdots, Z_n, Z)(U_i - u)$$

$$+ u \sum_{i=1}^{q} \nu_i(Z_1, Z_2, \cdots, Z_n, Z) + \sum_{j=q+1}^{n} \nu_j(Z_1, Z_2, \cdots, Z_n, Z) U_j,$$

and one observes that by (7) we have $\lim \nu_j = 0$, for $j \geqq q+1$. Hence, by (6), $\lim u \sum_{\nu=1}^{q} \nu_i = u$, and since $|\nu_i| \leqq 1$ and $U_i \to u$ for $i = 1, 2, \cdots, q$, the statement follows.

C). Let $z_1, z_2, \cdots, z_n, z, u_1, u_2, \cdots, u_n$ be complex numbers such that $z_i \neq z_j$ if $i \neq j$ $(i, j = 1, 2, \cdots, n)$. Let v be a complex number such that $|v| = 1$ and let λ be a real variable. Consider the function

$$\theta(\lambda) = \sum_{i=1}^{n} \nu_i(z_1, z_2, \cdots, z_n, z + \lambda v) u_i. \tag{9}$$

Assume that

$$|u_i - u_j| / |z_i - z_j| \leqq \gamma, \quad (i, j = 1, 2, \cdots, n; i \neq j) \tag{10}$$

where γ is some positive real number. *Then*

$$|\theta'(\lambda)| \leqq 4\gamma(n-1), \tag{11}$$

for all values of λ *at which the derivative* $\theta'(\lambda)$ is defined. (We note that $\theta'(\lambda)$ is certainly defined for all values of λ such that $z + \lambda v \neq z_i$, $i = 1, 2, \cdots, n$).

D). The following is an easy consequence of C): Let $z_1, z_2, \cdots, z_n, z, z'$, $u_1, u_2, \cdots, u_n$ be complex numbers such that $z_i \neq z_j$ if $i \neq j$ $(i, j = 1, 2, \cdots, n)$ and $z \neq z'$, and let γ be a positive real number. *If* (10) *is satisfied then*

$$\left| \sum_{i=1}^{n} [\nu_i(z_1, z_2, \cdots, z_n, z') - \nu_i(z_1, z_2, \cdots, z_n, z)] u_i \right| \leqq 4\gamma(n-1) |z' - z|. \tag{12}$$

For the proof, one sets $z' = z + av$, a-real $\neq 0$, $|v| = 1$ and one observes that, in the notations of step C, the left-hand side of (12) is equal to $|\theta(a) - \theta(0)|$, i. e., to $\left| \int_{\lambda=0}^{a} \theta'(\lambda) d\lambda \right|$, and now (12) follows from (11), in view of $|a| = |z' - z|$.

E). We define a mapping ψ of the cylinder E_ϵ into $\mathcal{A}'_{r+1}$ as follows: if $(\xi_1, \xi_2, \cdots, \xi_r, \eta)$ is any point of E_ϵ then

$$\psi(\xi_1, \xi_2, \cdots, \xi_r, \eta) = (\xi_1, \xi_2, \cdots, \xi_r, \phi(\xi_1, \xi_2, \cdots, \xi_r, \eta)),$$

where

$$\phi(\xi_1, \xi_2, \cdots, \xi_r, \eta) = \eta + \sum_{i=1}^{n} \sigma_i(\xi_1, \xi_2, \cdots, \xi_r, \eta)[y_i'(\xi) - y_i(\xi)]$$

and

$$\sigma_i(\xi, \eta) = \nu_i(y_1(\xi), y_2(\xi), \cdots, y_n(\xi), \eta), \quad i = 1, 2, \cdots, n.$$

By (7) and (8) it follows at once that the restriction of ψ to V is our original local homeomorphism ψ of V to V'. Using (12), it can be proved that the map ψ is a homeomorphism of E_ϵ if the following condition is satisfied:

$$(13) \qquad \left| \frac{y_i'(\xi) - y_j'(\xi)}{y_i(\xi) - y_j(\xi)} - 1 \right| < \frac{1}{4(n-1)}, \text{ for all point } (\xi_1, \xi_2, \cdots, \xi_r) \text{ in } U_\epsilon \text{ and for all } i \neq j \ (i, j = 1, 2, \cdots, n).$$

Now, condtion (13) is certainly satisfied, for ϵ sufficiently small, if each of the algebroid functions

$$[y_i'(x) - y_j'(x)]/[y_i(x) - y_j(x)] \quad (i \neq j; i, j = 1, 2, \cdots, n)$$

has a unique value at the origin $(x) = (0)$ and that value is 1 (Note that in view of (5) and (5′) each of these functions has only a finite number of limiting vaues at $(x) = (0)$ and that all these limiting values are different from zero). If, however, this is not the case, we proceed as follows:

We set

$$(14) \quad \epsilon_{ij}(x) = [y_i'(x) - y_j'(x)]/[y_i(x) - y_j(x)], \quad (i \neq j; i, j = 1, 2, \cdots, n)$$

and *we fix* an arc γ lying entirely in $U_\epsilon - \Delta$, except for its end point O_0 $(x_1 = x_2 = \cdots = x_r = 0)$. We set

$$(15) \qquad c_{ij} = \lim_{\substack{(x)\to(0)\\(\text{on } \gamma)}} \epsilon_{ij}(x).$$

By (5) and (5′) we have

$$(16) \qquad c_{ij} \neq 0, \infty \quad (i \neq j; i, j = 1, 2, \cdots, n).$$

It is clear that if we replace γ by another similar arc γ' (terminating at O_0) and if we denote by c'_{ij} the limit of $\epsilon_{ij}(x)$ as (x) approaches the origin (O_0) along the arc γ', then the set $\{c'_{ij}\}$ is merely a permutation of the set $\{c_{ij}\}$ (it is assumed that the set of pairs (i, j) is ordered in some definite way, say lexicographically).

Let t be a complex parameter and let

$$y^t(x) = (1 - t)y(x) + ty'(x).$$

For each value $\bar{t}$ of t, $y^{\bar{t}}(x)$ is an algebroid, n-valued function of $x_1, x_2, \cdots, x_r$, and it defines a hypersurfaces $V^{\bar{t}}$ in affine $(r+1)$-space. We thus have an analytic family of hypersurfaces V^t, such that $V^0 = V$ and $V^1 = V'$. The n branches of $y^t(x)$ are

$$y_i^t(x) = (1-t)y_i(x) + ty_i'(x),$$

and we have, by (14):

$$(17) \qquad y_i^t(x) - y_j^t) = [y_i(x) - y_j(x)][1 - t + \epsilon_{ij}(x)t].$$

We consider the set of constants c_{ij} in (15) which are different from 1 and we denote the corresponding constants $1/(1-c_{ij})$, in some order, by $\bar{t}_1, \bar{t}_2, \cdots, \bar{t}_h$. We denote by T the plane of the complex variable t, *punctured at the h points $\bar{t}_\nu$*. By (16), the points $t=0$ and $t=1$ belong to T. From (17) it follows that *if t, t' are any two points of T then the quotients*

$$(y_i^{t'} - y_j^{t'})/(y_i^t - y_j^t) \quad (i \neq j; i, j = 1, 2, \cdots, n)$$

are units in $\hat{R}^*$. Thus, if $\mathfrak{o}^t$ and $\mathfrak{o}^{t'}$ denote the local rings of V^t and $V^{t'}$, then the saturations $\tilde{\mathfrak{o}}^t{}_{(x)}$ and $\tilde{\mathfrak{o}}^{t'}{}_{(x)}$ are identical, and we therefore have a well defined local homeomorphism $\psi^{t,t'}$ of V^t onto $V^{t'}$. Now, it is clear, that for each $t \in T$ there exists a *positive* real number $\epsilon(t)$ such that

$$\left| \frac{y_i^{t'}(\xi) - y_j^{t'}(\xi)}{y_i^t(\xi) - y_j^t(\xi)} - 1 \right| < \frac{1}{4(n-1)}$$

for all $i \neq j$ $(i, j = 1, 2, \cdots, n)$ and all $\xi_1, \xi_2 \cdots, \xi_r$, t' such that $|\xi_\nu| < \epsilon(t)$ $(i = 1, 2, \cdots, r)$ and $|t' - t| < \epsilon(t)$. Therefore, by what has been proved above, $\psi^{t,t'}$ can be extended to a homeomorphism of $E_{\epsilon(t)}$ provided $|t'-t| < \epsilon(t)$. This being so, we have only to join $t=0$ to $t=1$ by an arc Γ in the punctured plane T, set $\epsilon = \min_{t \in \Gamma} \{\epsilon(t)\}$ (whence $\epsilon > 0$, since Γ is a closed arc) and insert in the arc Γ a sequence of intermediate points $t_1', t_2', \cdots, t_q'$ such that $|t_{i+1}' - t_i'| < \epsilon$ for $i = 0, 1, \cdots, q$ (here $t_0' = 0$, $t_{q+1}' = 1$). Since we know that for each $i = 0, 1, \cdots, q$, the local homeomorphism $\psi^{t_i', t_{i+1}'}: V^{t_i'} \rightarrow V^{t_{i+1}'}$ can be extended to a homeomorphism of the cylinder E_ϵ, we conclude that also the local homeomorphism $\psi: V \rightarrow V'$ (which is abviously equal to the product of the $q+1$ local homeomorphism $\psi^{t_i', t_{i+1}'}$) can be extended to a homeomorphism of the cylinder E_ϵ. This completes the proof of the theorem.

7. A special form of equisingularity: equisaturation. We keep the notation of § 6. In Theorem 6.1 we have assumed the existence of C-isomorphism $\phi\colon \tilde{\mathfrak{o}}_{(x)} \xrightarrow{\sim} \tilde{\mathfrak{o}}'_{(x')}$ such that $\phi(x_i) = x_i'$ $(i = 1, 2, \cdots, r)$. For our present purposes, we need first of all a simple lemma which gives an equivalent form of this assumption. We drop the assumption that the ground field k (of characteristic zero and algebraically closed) is the field C of complex numbers. We denote by F and F' the total ring of fractions of $\mathfrak{o}$ and $\mathfrak{o}'$ respectively. Since $x_1, x_2, \cdots, x_r$ as well as $x_1', x_2', \cdots, x_r'$ are analytically independent over k, we shall identify x_i with x_i' $(i = 1, 2, \cdots, r)$, and we shall assume therefore that the power series ring $R = k[[x_1, x_2, \cdots, x_r]]$ is a subring of both $\mathfrak{o}$ and $\mathfrak{o}'$, and that the field $K = k((x_1, x_2, \cdots, x_r))$ is a subfield of both F and F'. We recall that $\mathfrak{o} = R[y]$, $\mathfrak{o}' = R[y']$.

Assume that there exists a K-isomorphism $\phi\colon F \longrightarrow F'$. Then the polynomials $g((x);Y)$ and $g'((x');Y)$ *have* the same degree n, where $n = \dim_K F = \dim_K F'$. We have exactly n K-homomorphisms $\psi_1, \psi_2, \cdots, \psi_n$ of F into Ω and exactly n K-homomorphisms $\psi_1', \psi_2', \cdots, \psi_n'$ of F' into Ω, and ϕ establishes a natural pairing (ψ_i, ψ_i') between the ψ's and the ψ''s, namely: $\psi_i = \psi_i'\phi$ $(i = 1, 2, \cdots, n)$. If we set $y_i = \psi_i(y)$ and $y_i' = \psi_i'(y')$, then we have also a natural pairing (y_i, y_i') between the y_i and the y_i'. Note that for each $i = 1, 2, \cdots, n$ we have

$$K(y_i) = \psi_i(F) = \psi_i'(\phi(F)) = \psi_i'(F') = K(y_i').$$

LEMMA 7.1. *The following conditions are equivalent:*

1) *There exists an R-isomorphism $\phi\colon \tilde{\mathfrak{o}}_{(x)} \xrightarrow{\sim} \tilde{\mathfrak{o}}'_{(x)}$.*

2) *There exists a K-isomorphism $\phi\colon F \xrightarrow{\sim} F'$ such that if y_i, y_i' are paired as above, then the quotients $(y_i' - y_j')/(y_i - y_i)$ are units in the integral closure $\hat{R}^*$ of R in Ω, for all $i \neq j$ $(i, j = 1, 2, \cdots, n)$.*

Proof. That 1) implies 2) has already been proved in the preceding section, for 2) is equivalent with (5) and (5′) of the preceding section. Assume now condition 2). The fact that $(y_i - y_i)/(y_i' - y_j')$ is integral over R reads now as follows: $[\psi_i'(\eta') - \psi_j'(\eta')]/[\psi_i'(y') - \psi_j'(y')]$ is integral over R, where $\eta' = \phi(y)$. *In other words, η' dominates y'.* Therefore, $\eta' \in \tilde{\mathfrak{o}}'_{(x)}$, or equivalently: $\phi(\mathfrak{o}) \subset \tilde{\mathfrak{o}}'_{(x)}$. This implies that $\phi(\tilde{\mathfrak{o}}_{(x)}) \subset \tilde{\mathfrak{o}}'_{(x)}$. In a similar way it is shown that $\phi^{-1}(\tilde{\mathfrak{o}}'_{(x)}) \subset \tilde{\mathfrak{o}}_{(x)}$, and thus the restriction of ϕ to $\tilde{\mathfrak{o}}_{(x)}$ is an R-isomorphism $\tilde{\mathfrak{o}}_{(x)} \longrightarrow \tilde{\mathfrak{o}}'_{(x)}$.

We will need another lemma, namely the following:

LEMMA 7.2. *Let R be an integrally closed noetherian domain with field of fractions F, let $F[[t]] = F[[t_1, t_2, \cdots, t_s]]$ be a power series ring over F, and let $\xi_1, \xi_2, \cdots, \xi_h$ be elements of $F[[t]]$. Assume the following:*

(1) *the constant term of each powers series* $\xi_i(t)$ *is* 1;

(2) *there exists an element* λ *of* R *such that* $\lambda\xi_i \in R[[t]]$, $i = 1, 2, \cdots, h$;

(3) $\prod_{i=1}^{h} \xi_i \in R[[t]]$.

Then each ξ_i *belongs to* $R[[t]]$.

Proof. Since R is an intersection of discrete, rank 1, valuation rings, it is sufficient to prove the lemma under the assumption that R itself is a discrete, rank 1, valuation ring. For each $i = 1, 2, \cdots, h$ we denote by L_i the set of coefficients of the terms which actually occur in the power series $\xi_i(t)$, and we set $m_i = \min_{a \in L_i} \{v(a)\}$. By conditions (1) and (2) of the lemma we have $-\infty < m_i \leqq 0$. We order the totality of monomials $t_1^{n_1} t_2^{n_2} \cdots t_s^{n_s}$ ($n_i \geqq 0$, $i = 1, 2, \cdots, s$) lexicographically (this yields a well-ordered set), and for each $i = 1, 2, \cdots, s$, we denote by Ω_i the *first* monomial which actually occurs in $\xi_i(t)$ and whose coefficient in that power series has value m_i. We denote that coefficient by a_i.

We now consider in the power series $\xi_1\xi_2 \cdots \xi_h$ the coefficient of the monomial $\Omega_1\Omega_2 \cdots \Omega_h$. This coefficient is of the form $a_1a_2 \cdots a_h +$ a sum of terms $a_1'a_2' \cdots a_h'$, where each factor a_i' belongs to L_i (and hence $v(a_i') \geqq m_i$) and where, if $a_i'\Omega_i'$ is the corresponding term of the power series $\xi_i(t)$, then $\Omega_1'\Omega_2' \cdots \Omega_h' = \Omega_1\Omega_2 \cdots \Omega_h$. In view of this last equality, and given the nature of a lexicographical ordering, it follows that we cannot have $\Omega_i' \geqq \Omega_i$ for $i = 1, 2, \cdots, h$, without having $\Omega_i' = \Omega_i$ for all i, which is of course excluded (we have $a_i' \neq a_i$ for some value of i). Thus $\Omega_i' < \Omega_i$, and therefore $v(a_i') > m_i$, for at least one value of i. Since $v(a_i') \geqq m_i$ for *all* i, it follows that $v(a_1'a_2' \cdots a_h') > m_1 + m_2 + \cdots + m_h$. Since $v(a_1a_2 \cdots a_h) = m_1 + m_2 + \cdots + m_h$, we conclude that the if a is the coefficient of $\Omega_1\Omega_2 \cdots \Omega_h$ in $\xi_1\xi_2 \cdots \xi_h$, then $v(a) = m_1 + m_2 + \cdots + m_h$. By condition (3) of the theorem, it follows therefore that $m_1 + m_2 + \cdots + m_h \geqq 0$. Since $m_i \leqq 0$ for all i, it follows that $m_i = 0$ for all i. By the definition of m_i, this implies that all the coefficients of $\xi_i(t)$ belongs to R, i.e., $\xi_i \in R[[t]]$. This completes the proof.

After these preliminaries we now pass to what is the real object of this section. Let k be an algebraically closed ground field of characteristic zero and let V be an algebroid hypersurface in an affine $\mathcal{A}_{\rho+s+1}$, with center at the origin O. Let W be an irreducible subvariety of V having at O a simple point and of dimension ρ (of codimension s on V). We can then find a basis $(x_1, x_2, \cdots, x_s, y, t_1, t_2, \cdots, t_\rho)$ of the maximal ideal $\mathfrak{m}$ of the local ring $\mathfrak{o}$ of V such that:

1) $x_1 x_2, \cdots, x_s, t_1, t_2, \cdots, t_\rho$ are parameters of $\mathfrak{o}$;

2) W is defined by the equations $x_1 = x_2 = \cdots = x_s = y = 0$.

We fix such a basis and write the equation of V:

$$V: f(x;t;y) = 0,$$

where $f(x;t;Y)$ may be assumed to be a monic polynomial in Y; here x stands for $\{x_1, x_2, \cdots, x_s\}$ and t stands for $\{t_1, t_2, \cdots, t_\rho\}$. We may also assume that if f is of degree n in Y then $f(0;0;Y) = Y^n$. Furthermore, f is free from multiple factors. We set

$$f_0(X;Y) = f_0(X_1, X_2, \cdots, X_s; Y) = f(X;0;Y).$$

We assume that f_0 has no multiple factors, and we denote by V_0 the hypersurface in $\mathcal{A}_{s+1}$ defined by the equation $f_0(X;Y) = 0$. Thus V_0 is a section of V, transversal to W at the point O. We denote by k^* the field $k((t))$, $(= k((t_1, t_2, \cdots, t_\rho))$ of formal meromorphic functions of $t_1, t_2, \cdots, t_\rho$ and we call V_t the hypersurface in $\mathcal{A}_{s+1}$, *over* k^*, defined by the equation $f(x;t;y) = 0$. Thus V_t is the section of V, transversal to W at the generic point of W (see [7], § 3, p. 980).

We also regard V_0 as a hypersurface over k^* and we denote by $\mathfrak{O}_t$ and $\mathfrak{O}_0$ the local rings of V_t and V_0, *over* k^*. If $\mathfrak{p}$ denotes the prime ideal of W in $\mathfrak{o}$ then $\mathfrak{O}_t$ is the completion of the local ring $\mathfrak{o}_\mathfrak{p}$. Let $h_\mathfrak{p}$ be the canonical homomorphism $\mathfrak{o} \to \mathfrak{o}_\mathfrak{p}$ and let $h_\mathfrak{p}(x_i) = \bar{x}_i$, $h_\mathfrak{p}(y) = \bar{y}$. Since no element of $k[[t_1, t_2, \cdots, t_\rho]]$, other than zero, is a zero divisor in $\mathfrak{o}$, the restriction of $h_\mathfrak{p}$ to $k[[t_1, t_2, \cdots, t_\rho]]$ is an injection. We may therefore identify $h_\mathfrak{p}(t_j)$ with t_j. We have then:

$$\mathfrak{o} = k[[x;t;y]] = k[[x;t]][y]$$
$$\mathfrak{O}_t = k^*[[\bar{x};\bar{y}]] = k^*[[\bar{x}]][\bar{y}];$$
$$\mathfrak{O}_0 = k^*[[\xi;y^0]] = k^*[[\xi]][y^0],$$

where $\xi_1, \xi_2, \cdots, \xi_s, y^0$ are the residues of $x_1, x_2, \cdots, x_s, y$ modulo the ideal generated in $\mathfrak{o}$ by $t_1, t_2, \cdots, t_\rho$ (or—equivalently: they are the residues of $X_1, X_2, \cdots, X_s, Y$ modulo the principal ideal generated in

$$k^*[[X_1, X_2, \cdots, X_s, Y]]$$

by $f_0(X;Y)$).

DEFINITION 7.3. *V is equisaturated at O, along W, if there exists a k^*-isomorphism of saturated rings:*

$$\phi: \widehat{(\mathfrak{O}_t)_{k^*((\bar{x}))}} \xrightarrow{\sim} \widehat{(\mathfrak{O}_0)_{k^*((\xi))}},$$

such that $\phi(\bar{x}_i) = \xi_i$ $(i = 1, 2, \cdots, s)$.

A more precise way of phrasing the above definition would have been to say that V is equisaturated at O, along W, *if there exists a basis* $(x_1, x_2, \cdots, x_s, y, t_1, t_2, \cdots, t_\rho)$ *of* $\mathfrak{m}$, satisfying condition 1) and 2) stated earlier, and such that there exists a k^*-isomorphism ϕ as stated in the definition. The existence of the required isomorphism ϕ, for a given basis $(x; t; y)$ of $\mathfrak{m}$, does not at all imply the existence of a similar isomorphism for any other basis $(x'; t'; y')$. Geometrically speaking, the existence of ϕ is with reference to *a specific choice of the transversal section* V_0 *and to a specific choice of local parameters* $x_1, x_2, \cdots, x_s$ *of the generic transversal section* V_t.

THEOREM 7.4. *V is equisaturated at O, along W (more precisely: the isomorphism ϕ of Definition 7.3 exists) if and only if the discriminant $D(x; t)$ of $f(x; t; Y)$, with respect to Y, is of the following form:*

$$D(x; t) = D_0(x) \cdot \epsilon(x, t), \tag{1}$$

where $D_0(x)$ is a power series in $x_1, x_2, \cdots, x_s$ (with coefficients in k) and $\epsilon(x, t)$ is a unit in $k[[x; t]]$.

Proof. Assume that an isomorphism ϕ, of the type indicated in Definition 7.3, exists. We assert that $h_{\mathfrak{p}}$ is an injection (and that consequently $\bar{x}_i$ and $\bar{y}$ can be identified with x_i and y). For let $\bar{F}_t$ and F_0 be the total rings of quotients of $\mathfrak{O}_t$ and $\mathfrak{O}_0$ respectively. These rings are isomorphic (under an extension of ϕ). Since $F(0, 0, Y) = Y^n$, n is the dimension of F_0 as a vector over $k^*((\xi_1, \xi_2, \cdots, \xi_s))$. Hence also the dimension of $\bar{F}_t$ over $k^*((\bar{x}_1, \bar{x}_2, \cdots, \bar{x}_s))$ is n. This implies that $f(0; t_1, t_2, \cdots, t_\rho; Y) = Y^n$; in other words: every irreducible factor of $f(X; T; Y)$ must vanish for $X_1 = X_2 = \cdots = X_s = 0$ (or—geometrically speaking: every irreducible component of V must contain W). Thus every zero divisor in $\mathfrak{o}$ belongs to the prime ideal $\mathfrak{p}$, showing that $h_{\mathfrak{p}}$ is an injection.

We shall identify each $\bar{x}_i$ with the corresponding x_i and ξ_i $(i = 1, 2, \cdots, s)$. Let F_t and F_0 be the total ring of fractions of $\mathfrak{O}_t$ and $\mathfrak{O}_0$ respectively. Then ϕ has a unique extension to an isomorphism between F_t and F_0, an isomorphism which we shall continue to denote by ϕ. Let Ω be an algebraic closure of the field $k^*((x))$ and let $y_1, y_2, \cdots, y_n$ (resp., $y_1^0, y_2^0, \cdots, y_n^0$) be the roots of $f(x; t; Y)$ (resp., of $f_0(x; Y)$ in Ω, the roots y_i and y_i^0 being paired by the isomorphism ϕ (as explained in the beginning of this section). By Lemma 7.1, the quotients $(y_i - y_j)/(y_i^0 - y_j^0)$

$(i \neq j)$ are units in the integral closure of $k^*[[x]]$ in Ω. If then we denote by $D_0(x)$ the discriminant of $f_0(x;Y)$, then we will have

$$D(x,t) = D_0(x) \cdot H^*(x), \tag{2}$$

where $H^*(x)$ is a unit in $k^*[[x_1, x_2, \cdots, x_s]]$. Now, since $D(x,t) \in k[[x,t]]$ and $D_0(x) \in k[[x]]$, it follows easily from (2) that $H^*(x) \in k[[x,t]]$.[6] Since $D_0(x) = D(x;0)$, it follows that if we write $H^*(x) = \epsilon(x;t)$, then $\epsilon(x;0) = 1$, showing that $\epsilon(x;t)$ is a unit in $k[[x;t]]$.

Conversely, assume that (1) holds, with ϵ being unit in $k[[x;t]]$. By Theorem 5 of our paper [6] it follows that if we set

$$\mathfrak{o}_0 = k[[\xi_1, \xi_2, \cdots, \xi_s; y^0]]$$

and denote by $\bar{\mathfrak{o}}$ and $\bar{\mathfrak{o}}_0$ the integral closure of $\mathfrak{o}$ and $\mathfrak{o}_0$ respectively (in their total ring of fractions), then there exists a $k[[t]]$-isomorphism $\phi\colon \bar{\mathfrak{o}} \xrightarrow{\sim} \bar{\mathfrak{o}}_0[[t]]$ such that $\phi(x_i) = \xi_i$ $(i = 1, 2, \cdots, s)$. We shall therefore identify x_i with ξ_i and $\bar{\mathfrak{o}}$ with $\bar{\mathfrak{o}}_0[[t]]$. We have therefore:

$$\mathfrak{o}_0 = k[[x_1, x_2, \cdots, x_s; y]] \tag{3}$$

and

$$\bar{\mathfrak{o}} = \bar{\mathfrak{o}}_0[[t]]. \tag{4}$$

If then $y_1, y_2, \cdots, y_n$ are the roots of $f(x;t;Y)$ in Ω and if $y_1^0, y_2^0, \cdots, y^0_n$ are the *corresponding* roots of $f(x;0;Y)$ [the pairing y_i, y_i^0 of the roots being determined, in the usual way (see the beginning of this section), by the isomorphism ϕ, hence—essentially—by our identification (4)], we can write (by applying the $k[[t]]$-homomorphisms $\psi_1, \psi_2, \cdots, \psi_n$ of $\bar{\mathfrak{o}}$ into Ω):

$$y_i = y_i^0 + u_{i1}(t) + u_{i2}(t) + \cdots, \tag{5}$$

---

[6] *Proof.* Let $H^*(x) = \sum_{i=0}^{\infty} H_i^*(x)$, where $H_i^*(x)$ is a form in $x_1, x_2, \cdots, x_s$, of degree i, with coefficients in $k((t))$. Let, similarly, $D(x; t) = \sum_{i=q}^{\infty} D_i(x; t)$, where $D_i(x; t)$ is a form in $x_1, x_2, \cdots, x_s$, of degree i, with coefficients in $k[[t_1, t_2, \cdots, t_\rho]]$. We have $D_0(x) = D(x; 0) = \sum_{i=q}^{\infty} D_{i,0}(x)$, where $D_{i,0}(x) = D_i(x; 0)$. By (2) we have $D_q(x; t) = D_{q,0}(x)$, with $H_0^*(x) = A(t)/B(t)$, where $A(t)$, $B(t)$ belong to $k[[t]]$ and may be assumed to be relatively prime. We have therefore $B(t) D_q(x; t) = A(t) D_{q,0}(x)$. Since $B(t)$ and $D_{q,0}(x)$ are obviously relatively prime in $k[[x, t]]$, it follows from the above equality that $B(t)$ must divide $A(t)$ in $k[[t]]$. Hence $B(t)$ is a unit in $k[[t]]$, and thus $H_0^*(x) \in k[[t]]$. In the same way, using induction on i, one can show that all the coefficients of the form $H_i^*(x)$ belong to $k[[t]]$, and hence $H^*(x) \in k[[x; t]]$, as asserted.

where $u_{ij}(t)$ is a form of degree j in $t_1, t_2, \cdots, t_\rho$, with coefficients in the integral closure of $k[[x]][y_1^0, y_2^0, \cdots, y_n^0]$. Denote this integral closure by R. By (5) we have that the quotients $\eta_{ij} = (y_i - y_j)/(y_i^0 - y_j^0)$ $(i \neq j)$ are power series in $t_1, t_2, \cdots, t_\rho$, with coefficients in the field of fractions of R, and that the constant term in each of these power series is 1. Furthermore, it also follows from (5) that if we set $D_0(x) = \prod_{i<j} (y_i^0 - y_j^0)^2$, then $D_0 \cdot \eta_{ij} \in R[[t]]$, with $D_0 \in R$. Finally, by condition (1) of the theorem, we have that the product of all the η_{ij} belongs to $R[[t]]$. Hence, by Lemma 7.2, the quotients η_{ij} belong $R[[t]]$. Since $f(0,0;Y) = Y^n$, each y_i^0 belong to the radical of the ideal $R((x_1, x_2, \cdots, x_s)$. Since $f(0,t;0) = 0$, at least one y_i belongs to the radical of the ideal $R[[t]](x_1, x_2, \cdots, x_s)$. Since all η_{ij} arein $R[[t]]$, we conclude that *each* y_i belongs to the radical of the ideal $R[[t]](x_1, x_2, \cdots, x_s)$. In other words, we must have $f(0;t;Y) = Y^n$. This shows that $h_\mathfrak{p}$ is an injection. We can thus identify the $\bar{x}_i$ with the x_i. We note that the η_{ij} are integral over $k((t))[[x_1, x_2, \cdots, x_s]]$. Since the constant term in each of these quotients is 1, also the reciprocal $1/\eta_{ij}$ is integral over $k((t))[[x_1, x_2, \cdots, x_s]]$. It follows now from Lemma 7.1 that, with our identification (4), we have $\widetilde{(\mathfrak{O}_t)}_{k^*((x))} = \widetilde{(\mathfrak{O}_0)}_{k^*((x))}$. This completes the proof of the theorem.

COROLLARY 7.5. *If* $\operatorname{cod}_V W = 1$, *then* V *is equisaturated at* O, *along* W, *if and only if* V *is equisingular at* O, *along* W *(in the sense of our paper* [7]).

For if $s = 1$, then condition (1) of Theorem 7.4 now signifies that D, apart from a unit factor, is a power of x_1.

We shall now sketch the derivation of a topological property of equisaturation in the case in which V and W are complex-analytic varieties. The details are partly given in Section 6 (where the set-up was similar to the one we are dealing with now) and partly are to be found in the cited paper of Whitney (where the case $\operatorname{cod}_V W = 1$ was dealt with).

The roots y_i (see (5)) are now the n branches $y_i(x;t)$ of an algebroid function of the $s + \rho$ complex variables $x_1, x_2, \cdots, x_s, t_1, t_2, \cdots, t_\rho$, while $y_i^0 = y_i(x;0)$. If ϵ is positive real number we set

$$U_\epsilon = \{(\xi_1, \xi_2, \cdots, \xi_s) \mid |\xi_i| < \epsilon\},$$
$$E_\epsilon = \{(\xi_1, \xi_2, \cdots, \xi_s, a) \mid (\xi) \in U_\epsilon\}$$
$$T_\epsilon = \{(\tau_1, \tau_2, \cdots, \tau_\rho) \mid |\tau_j| < \epsilon\}.$$

We take ϵ sufficiently small so as to satisfy the following conditions:

(1) The coefficients of the polynomial $f(x;t;Y)$ are convergent at each point of $U_\epsilon \times T_\epsilon$.

(2) The unit power series $\epsilon(x;t)$ which occurs in (1) is different from zero at each point of $U_\epsilon \times T_\epsilon$.

(3) $|[y_i(x,t) - y_j(x,t)]/[y_i(x,0) - y_j(x,0)] - 1| < \dfrac{1}{4(n-1)}$ at all points of $U_\epsilon \times T_\epsilon$.

Under this condition, the considerations of § 6 show that for each $(\bar{t})$ in T_ϵ there is a well defined homeomorphism $\psi_{(\bar{t})}$ of E_ϵ into a neighborhood of the origin in $\mathcal{A}_{s+1} \times (\bar{t})$ such that $\psi_{(t)}$ transforms the hypersurface $V_{\bar{t}}$: $f(x;\bar{t};Y) = 0$ into $V_0 \times (\bar{t})$, where V_0 is the hypersurface $f(x;0;Y) = 0$. The homeomorphism $\psi_{(\bar{t})}$ varies continuously with $(\bar{t})$, and what we thus obtain, as $(\bar{t})$ varies in T_ϵ, is the homeomorphism ψ of the infinite cylinder $E_\epsilon \times T_\epsilon$ in $\mathcal{A}_{r+s+1}$ onto a neighborhood of the origin in $\mathcal{A}_{r+s+1}$, such that ψ, restricted to V, is a local homeomorphism of V with $V_0 \times T_\epsilon$. Thus, if V is equisaturated at O, along W, then V, as an *embedded hypersurface*, is topologically equivalent to the direct product of the transversal section V_0 at O and of the neighborhood T_ϵ of the origin in the space of variables $t_1, t_2, \cdots, t_\rho$.

8. A preview of a general treatment of saturation. We have repeatedly pointed out in the introduction that:

1) most of the theorems of Part I generalize to the case in which the residue field of $\mathfrak{o}$ is not algebraically closed (but still of characteristic zero) and $\mathfrak{o}$ is not an integral domain; and that:

2) such a generalization is necessary if one wishes to deal, in the case of local rings of dimension > 1, with some questions which have been answered in Part I for the case of dimension 1.

We should like to illustrate the second contention by discussing (without proofs, of course; these will be given by us elsewhere) a generalization to dimension > 1, of Corollary—Definition 1.9. We recall that that Corollary asserts that the saturation $\tilde{\mathfrak{o}}_x$ *with* respect to a *transversal* parameter x is independent of the choice of x.

The cited corollary remains valid, without any change, if $\mathfrak{o}$ is a complete equidimensional, equicharacteristic local ring of characteristic zero (*not necessarily a domain*; but still free from nilpotent elements), of dimension 1, and if the residue field of $\mathfrak{o}$ *is not algebraically closed.* We may as well explain here what we mean, in that case, by a transversal parameter x of $\mathfrak{o}$. The defintion is the following:

If $\mathfrak{o}$ is an integral domain, then the integral closure $\bar{\mathfrak{o}}$ of $\mathfrak{o}$ is a valuation ring, and if v denotes the corresponding valuation then x is a transversal parameter if and only if $v(x) = \min_{\xi \in \mathfrak{m}} \{v(\xi)\}$.

If $\mathfrak{o}$ is not an integral domain, we denote by $\mathfrak{p}_1, \mathfrak{p}_2, \cdots, \mathfrak{p}_h$ the prime ideals of the ideal (0) in $\mathfrak{o}$. By the equidimensionality of $\mathfrak{o}$, each of the rings $\mathfrak{o}/\mathfrak{p}_i$ is of dimension 1 $(i = 1, 2, \cdots, h)$. Then x is transversal if and only if for each $i = 1, 2, \cdots, h$, the $\mathfrak{p}_i$-residue of x is a transversal parameter of $\mathfrak{o}/\mathfrak{p}_i$.

Now, let $\mathfrak{o}$ be the local ring of an r-dimensional, equidimensional, algebroid variety V, defined over an algebraically closed ground field k of characteristic zero. Let $(x_1, x_2, \cdots, x_r)$ be a system of parameters of $\mathfrak{o}$, let $R = k[[x_1, x_2, \cdots, x_r]]$ and let $\mathfrak{p}$ be a prime ideal of height 1 in $\mathfrak{o}$. The intersection $\mathfrak{P} = R \cap \mathfrak{p}$ is then a principal ideal in R, say $\mathfrak{P} = R\xi$, $\xi \neq 0$. Since $\xi \in R$, it follows from the equidimensionality of $\mathfrak{o}$ that ξ is not a zero divisor in $\mathfrak{o}$, and, of course, ξ is also not a unit, since $\xi \in \mathfrak{p}$. Hence the *isolated* prime ideals of the principal ideal $\mathfrak{o}\xi$ are all of height 1, and, in particular, $\mathfrak{p}$ must be one of the isolated prime ideals of $\mathfrak{o}\xi$. We now pass to the ring of quotients $\mathfrak{o}_\mathfrak{p}$ of $\mathfrak{o}$ with respect to $\mathfrak{p}$. This is a local ring of dimension 1 and equidimensional. We have a canonical homomorphism $h_\mathfrak{p}: \mathfrak{o} \to \mathfrak{o}_\mathfrak{p}$, and the restriction of $h_\mathfrak{p}$ to R is an injection. Hence we may assume that $R \subset \mathfrak{o}_\mathfrak{p}$. By known facts on quotient rings ([5], v. 1, Ch. IV), the principal ideal $\mathfrak{o}_\mathfrak{p}\xi$ is primary for the maximal ideal $\mathfrak{o}_\mathfrak{p} \cdot h(\mathfrak{p})$ of $\mathfrak{o}_\mathfrak{p}$. Thus *$\xi$ is a parameter of $\mathfrak{o}_\mathfrak{p}$*.

Definition 8.1. *The parameters $x_1, x_2, \cdots, x_r$ are said to be transversal with respect to $\mathfrak{p}$ if ξ is a transversal parameter of $\mathfrak{o}_\mathfrak{p}$.*[7]

Let M denote the set of all prime ideals of $\mathfrak{o}$ which have height 1. Let F be the total ring of fractions of $\mathfrak{o}$, and let $F_\mathfrak{p}$ $(\mathfrak{p} \in M)$ denote the total ring of fraction of $\mathfrak{o}_\mathfrak{p}$. The homomorphism $h_\mathfrak{p}: \mathfrak{o} \to \mathfrak{o}_\mathfrak{p}$ has a unique extension to a homomorphism $F \to F_\mathfrak{p}$. We shall continue to denote by $h_\mathfrak{p}$ this extended homomorphism.

We now make an additional assumption concerning the local ring $\mathfrak{o}$:

(S) *If $\xi \in F$ and $h_\mathfrak{p}(\xi) \in \mathfrak{o}_\mathfrak{p}$ for all $\mathfrak{p}$ in M, then $\xi \in \mathfrak{o}$.*[8]

---

[7] This definition seems to depend (and probably does depend) on the choice of the field k of representatives of $\mathfrak{o}$.

[8] It is not difficult to see that condition S is equivalent to the following condition: *if an element b of $\mathfrak{o}$ is not a zero-divisor in $\mathfrak{o}$ then all the prime ideals of the principal ideal $\mathfrak{o}b$ are isolated (and hence of height* 1, if b is not a unit). This condition is

The theorem which can be proved and which is a generalization of Corollary-Definition 1.9 is the following:

THEOREM 8.2. *Let $W_1, W_2, \cdots, W_s$ be those irreducible components of the singular locus of V which have codimension 1, let $\mathfrak{p}_1, \mathfrak{p}_2, \cdots, \mathfrak{p}_s$ be the corresponding prime ideals (in M) and let $(x_1, x_2, \cdots, x_r)$, $(x_1', x_2', \cdots, x_r')$, be two sets of parameters of $\mathfrak{o}$. If both sets of parameters are transversal with respect to each of the s prime ideal $\mathfrak{p}_1, \mathfrak{p}_2, \cdots, \mathfrak{p}_s$, then the two saturations $\tilde{\mathfrak{o}}_{(x)}^{(k)}$ and $\tilde{\mathfrak{o}}_{(x')}^{(k)}$ are identical.*

We shall now give an idea of the proof, enough of it, at any rate, to justify the contention made in the beginning of this section.

Let $K = k((x_1, x_2, \cdots, x_r))$ = field of fractions of $R = k[[x_1, x_2, \cdots, x_r]]$. For any prime ideal $\mathfrak{p}$ in M we denote by $\xi_\mathfrak{p}$ a generator of the principal ideal $\mathfrak{P} = \mathfrak{p} \cap R$, and we consider the local ring $R_\mathfrak{P}$. This is a regular local ring of dimension 1, with $\xi_\mathfrak{p}$ as regular parameter. We denote by $\hat{R}_\mathfrak{P}$ the completion of $R_\mathfrak{P}$, by $\hat{\mathfrak{o}}_\mathfrak{p}$ the completion of $\mathfrak{o}_\mathfrak{p}$, and by $\hat{K}_\mathfrak{P}$ and $\hat{F}_\mathfrak{p}$ the total ring of fractions of $\hat{R}_\mathfrak{P}$ and $\hat{\mathfrak{o}}_\mathfrak{p}$ respectively. By known facts on completions of local rings we have that $\hat{K}_\mathfrak{P} \subset \hat{F}_\mathfrak{p}$, that no element $\hat{K}_\mathfrak{P}$ (different from zero) is a zero-divisor in $\hat{F}_\mathfrak{p}$ (this follows from the equidimensionality of $\mathfrak{o}_\mathfrak{p}$), that $\hat{\mathfrak{o}}_\mathfrak{p} \cap \hat{K}_\mathfrak{P} = \hat{R}_\mathfrak{P}$ and finally, that $\hat{\mathfrak{o}}_\mathfrak{p}$ is integral over $\hat{R}_\mathfrak{P}$. Hence the saturation

$$\widetilde{(\hat{\mathfrak{o}}_\mathfrak{p})}_{\hat{K}_\mathfrak{P}}$$

is defined (seee Introduction). We fix a field Δ_0 of representatives of $\hat{R}_\mathfrak{P}$. Thus

$$\hat{R}_\mathfrak{P} = \Delta_0[[\xi_\mathfrak{p}]],$$
$$\hat{K}_\mathfrak{P} = \Delta_0((\xi_\mathfrak{p})).$$

We now consider the algebraic closure Δ of Δ_0 in $\hat{\mathfrak{o}}_\mathfrak{p}$. Then Δ is a field of representatives of $\hat{\mathfrak{o}}_\mathfrak{p}$, and Δ is a finite algebraic extension of Δ_0. It is a theorem, which we can prove (in characteristic zero) for complete local rings of dimension 1, that as a consequence of the finiteness of Δ/Δ_0, the saturation $\widetilde{(\hat{\mathfrak{o}}_\mathfrak{p})}_{\Delta_0((\xi_\mathfrak{p}))}$ is meaningful and is equal to $\widetilde{(\hat{\mathfrak{o}}_\mathfrak{p})}_{\Delta((\xi_\mathfrak{p}))}$. As the latter ring is,

satisfied if, for instance, $\mathfrak{o}$ is a Macaulay ring; in particular, if V is an hypersurface. It seems likely that condition (S) is not really essential for the validity of our final conclusion, which is Theorem 8.2 below. We note also that cndition (S) *together with the equidimensionality of* $\mathfrak{o}$ implies the condition of Serre, designated as (S_2) by Grothendieck in [3], Ch. IV, Def. 5.7.2, and, in particular, Cor. 5.7.7.

by definition, the saturation of $\mathfrak{o}_\mathfrak{p}$ with respect to the parameter $\xi_\mathfrak{p}$, and as $\Delta_0((\xi_\mathfrak{p})) = \hat{K}_\mathfrak{P}$, we have

$$(1) \qquad \widetilde{(\hat{\mathfrak{o}}_\mathfrak{p})}_{(\xi_\mathfrak{p})} = \widetilde{(\hat{\mathfrak{o}}_\mathfrak{p})}_{\hat{K}_\mathfrak{P}}.$$

Now, we can also prove the following theorem:

THEOREM 8.3. *Under the assumption that condition* (S) *is satisfied, a necessary and sufficient condition that an element* ξ *of* F *belongs to* $\tilde{\mathfrak{o}}^{(k)}{}_{(x_1,x_2,\cdots,x_r)}$ *is that for any* $\mathfrak{p}$ *in* M *we have*

$$h_\mathfrak{p}(\xi) \in \widetilde{(\hat{\mathfrak{o}}_\mathfrak{p})}_{\hat{K}_\mathfrak{P}}, \text{ with } \mathfrak{P} = \mathfrak{p} \cap R;$$

or, symbolically (*if not very precisely*), *and using* (1):

$$(2) \qquad \tilde{\mathfrak{o}}^{(k)}{}_{(x_1,x_2,\cdots,x_r)} = \bigcap_{\mathfrak{p} \in M} (F \cap \widetilde{(\hat{\mathfrak{o}}_\mathfrak{p})}_{(\xi_\mathfrak{p})}.$$

Now, if $\mathfrak{p} \neq \mathfrak{p}_1, \mathfrak{p}_2, \cdots, \mathfrak{p}_s$, then $\hat{\mathfrak{o}}_\mathfrak{p}$ is a regular ring, whence $(\tilde{\mathfrak{o}}_\mathfrak{p})_{(\xi_\mathfrak{p})} = \hat{\mathfrak{o}}_\mathfrak{p}$. If we now assume that $x_1, x_2, \cdots, x_r$ are transversal with respect to $\mathfrak{p}_1, \mathfrak{p}_2, \cdots, \mathfrak{p}_s$, then $(\hat{\mathfrak{o}}_{\mathfrak{p}_i})_{(\xi_{\mathfrak{p}_i})}$ is *the saturation of* $\hat{\mathfrak{o}}_{\mathfrak{p}_i}$ *and depends only on* $\hat{\mathfrak{o}}_{\mathfrak{p}_i}$ (it is here that we use the assumption that Corollary-Definition 1.9 holds for local rings, of dimension 1, under the more general conditions, as stated earlier). This establishes Theorem 8.2.

HARVARD UNIVERSITY.

REFERENCES.

[1] S. Abhyankar, "Ramification theoretic methods in algebraic geometry," *Annals of Mathematics Studies* (Princeton), no. 43.

[2] N. Bourbaki, *Algèbre Commutative*, ch. 5.

[3] A. Grothendieck, *Eléments de Géométrie Algébrique*.

[4] H. Whitney, "Local properties of analytic varieties," *Differential and Combinatorial Topology*. A symposium in honor of Marston Morse (Princeton, 1965), pp. 205-244.

[5] O. Zariski and P. Samuel, *Commutative Algebra*.

[6] O. Zariski, "Studies in equisingularity I. Equivalent Singularities of plane algebroid curves," *American Journal of Mathematics*, vol. 87 (1965), pp. 507-536.

[7] ———, "Studies in equisinguladity II. Equisingularity in codimension 1 (and characteristic zero)," *American Journal of Mathematics*, vol. 87 (1965), pp. 972-1006.

CONTRIBUTIONS TO THE PROBLEM OF EQUISINGULARITY

by

Oscar Zariski

Harvard University, U.S.A.

INTRODUCTION. The general problem which we propose in these lectures is the following: given an irreducible subvariety W of the singular locus of an algebraic (or algebroid) variety V and given a simple point Q of W, give a precise meaning to the following intuitive statement: "the singularity which V has at the point Q is 'not worse' than (or is 'of the same type' as) the singularity which V has at the general point of W". We briefly phrase this statement as follows: "V is equisingular along W, at Q." It is understood that we require the solution to consist not merely of some plausible definition and some reasonable consequences, but above all of a body of criteria of various nature (algebro-geometric, topological and differentio-geometric) and of the proofs of equivalence of these various criteria.

In this lecture we give, in the first place, a complete solution of this problem in the special case of $\mathrm{cod}_V W = 1$. We also treat a special type of equisingularity which we call equisaturation; we are led to this concept by our algebraic theory of saturation and saturated local rings. For both of these topics we need a thorough analysis of some old and new aspects of the concept of equivalent singularities of plane algebroid curves. This analysis is developed in Sections 1-6. In the last section we discuss connections with the differentiogeometric conditions A and B of Whitney-Thom.

1. Plane Algebroid Curves

We shall begin with the theory of singularities of (formal) plane algebroid curves (more precisely: algebroid curves which admit an embedding in the plane). Our object will be to present some aspects - both old and new - of the classical concept of equivalence between such singularities. We shall restrict ourselves to ground fields k which are algebraically closed and of characteristic zero.

We shall be guided by the following intrinsic definition of a plane algebroid curve C (over k): it is a neotherian complete local ring $\mathfrak{o}$ having the following properties :

a) $\mathfrak{o}$ has Krull dimension 1.

b) $\mathfrak{o}$ has no nilpotent elements (other than zero).

c) The maximal ideal $\mathfrak{m}$ of $\mathfrak{o}$ can be generated by two elements.

d) k is a coefficient field of $\mathfrak{o}$.

We say then that $\mathfrak{o}$ is the local ring of C.

Condition (a) signifies that 1) there exist ideals $\mathfrak{q}$ primary for $\mathfrak{m}$ (i.e., which are primary and have $\mathfrak{m}$ as associated prime ideal) which are generated by a single element and that 2) the zero ideal is not primary for $\mathfrak{m}$. Any element x of $\mathfrak{o}$ such that the principal ideal $\mathfrak{o}x$ is primary for $\mathfrak{m}$ is called a parameter of $\mathfrak{o}$ (or of C). It is obvious, of course, that every parameter belongs to $\mathfrak{m}$ (hence is a non-unit). It also follows easily from 2) that a parameter of $\mathfrak{o}$ is never a zero-divisor.

Condition (b) signifies that the zero ideal is a finite (irredundant) intersection of prime ideals :

$$(1) \qquad (0) = \mathfrak{p}_1 \cap \mathfrak{p}_2 \cap \ldots \cap \mathfrak{p}_h \, ,$$

By (a) we must have $\mathfrak{p}_i < \mathfrak{m}$ $(i=1,2,\ldots,h)$, and it thus follows at once from (a) and (b) that <u>an element x of $\mathfrak{m}$ is a parameter if and only if x is not a zero-divisor (equivalently: if and only if</u> $x \notin \mathfrak{p}_i$, $i = 1,2,\ldots,h$). The h+1 prime ideals $\mathfrak{p}_1, \mathfrak{p}_2, \ldots, \mathfrak{p}_h, \mathfrak{m}$ are all the prime ideals of $\mathfrak{O}$. Thus the scheme Spec ($\mathfrak{O}$) has exactly h+1 points; of these, the point represented by $\mathfrak{m}$ is the only closed point.

Condition (d) signifies that k is a field contained in $\mathfrak{O}$ and that the injection $k \to \mathfrak{O}/\mathfrak{m}$ induced by the canonical homomorphism $\mathfrak{O} \to \mathfrak{O}/\mathfrak{m}$ is surjective, hence is an isomorphism $k \xrightarrow{\sim} \mathfrak{O}/\mathfrak{m}$.

The three conditions (a), (b) and (d) define the general concept of an algebroid curve C, over an arbitrary, algebraically closed ground field k (not necessarily of characteristic zero). We now look at condition (c).

Let $\{x_1, x_2, \ldots, x_n\}$ be any basis of $\mathfrak{m}$. Using (d) and the completeness of $\mathfrak{O}$ we find a natural surjective k-Homomorphism

$$\Phi : k[[X_1, X_2, \ldots, X_n]] \to \mathfrak{O} = k[[x_1, x_2, \ldots x_n]] ,$$

such that $\Phi(X_i) = x_i (i=1, =, \ldots, n)$; here $k[[X_1, X_2, \ldots, X_n]]$ denotes the ring of formal power series in the indeterminates X_i, with coefficients in k. Thus

$$\mathfrak{O} = k[[x_1, x_2, \ldots, x_n]] = k[[X_1, x_2, \ldots, X_n]]/\mathfrak{U} , \tag{2}$$

where $\mathfrak{U}$ is an ideal in $k[[X_1, X_2, \ldots, X_n]]$. Conditions (a) and (b) signify that $\mathfrak{U}$ is a finite intersection of h one-dimensional prime ideals in $k[[X_1, X_2, \ldots, X_n]]$, where h is the integer which occurs in (1) . We thus have an embedding of C as an algebroid curve in the affine n-space (over k) . Of particular interest are those embeddings of C which are associated with <u>minimal</u> bases of $\mathfrak{m}$ (all minimal bases of $\mathfrak{m}$ have the same number of elements; this number is the dimension of the k-vector space $\mathfrak{m}/\mathfrak{m}^2$, and is sometimes denoted by Emb $\mathfrak{o}$ and is called the <u>embedding dimension</u> of $\mathfrak{o}$, or of C) .

Condition (c) signifies that the embedding dimension of C is 1 or 2. If n=1, i.e., if $\mathfrak{m}$ is a principal ideal, the local ring $\mathfrak{o}$ is regular, C is a <u>regular branch</u>; equivalently: $\mathfrak{o}$ is a formal power series ring in one indeterminate, and - correspondingly - C, for a suitable embedding, is the affine line. We exclude this case, and we assume therefore that the embedding dimension of C is 2. If we then take a minimal basis $\{x_1, x_2\}$ of $\mathfrak{m}$, the power series ideal $\mathfrak{A}$ in (2) will be a principal ideal: $\mathfrak{A} = (f(X_1, X_2))$, and f a product of h distinct (non-associated) irreducible power series :

(3) $$f = f_1 f_2 \dots f_h \ .$$

We then write symbolically

(4) $$C : f(X_1, X_2) = 0 \ ,$$

and we say that (4) <u>is an equation of</u> C. The power series f is uniquely determined by the basis $\{x_1, x_2\}$ of $\mathfrak{m}$, up to an arbitrary unit factor in $k[[X_1, X_2]]$. The transition from one minimal basis $\{x_1, x_2\}$ to another minimal basis $\{x'_1, x'_2\}$ is effected by an analytic transformation of the form

$$x_i' = a_{i2}x_1 + a_{i2}x_2 + \text{terms of degree} > 1 \text{ in } x_1, x_2, \quad i = 1, 2,$$

where $a_{ij} \varepsilon k$ and $a_{11}a_{22} - a_{12}a_{21} \neq 0$.

We denote by F the total ring of quotients of $\mathfrak{O}$, and we write $F = k((C))$. If M denotes the complement of $\bigcup_{i=1}^{h} \mathfrak{p}_i$ in $\mathfrak{O}$ (i.e., if M is the set of elements of $\mathfrak{O}$ which are not zero-divisors), then F is the quotient ring $\mathfrak{O}_M$ of $\mathfrak{O}$ with respect to the multiplicative system M. We shall use freely properties of quotient rings with respect to multiplicative systems, and we refer for that to Zariski-Samuel, Commutative Algebra, Vol. I, pp. 221-231. Thus, for example, using an appropriate property of quotient rings (namely, Theorem 16, p. 224, loc. cit.), we see that $\mathfrak{O}$ is a subring of F, that the ideals $\mathfrak{f}_i = F.\mathfrak{p}_i$ are the only prime ideals of F, that these ideals are therefore maximal, and that $(0) = \mathfrak{f}_1 \cap \mathfrak{f}_2 \cap \ldots \cap \mathfrak{f}_h$, with $\mathfrak{f}_i \cap \mathfrak{O} = \mathfrak{p}_i$. If we set $F_i = \mathfrak{f}_1 \cap \mathfrak{f}_2 \cap \ldots \cap \mathfrak{f}_{i-1} \cap \mathfrak{f}_{i+1} \cap \ldots \cap \mathfrak{f}_h$, then each F_i is a field:

$$F_i \doteq F/\mathfrak{f}_i,$$

and F is the direct sum of the F_i:

$$F = F_1 \oplus F_2 \oplus \ldots \oplus F_h .$$

For a given equation (4) of C we have, using the factorization (3) of f, that F is the set of all quotients $A(x_1, x_2)/B(x_1, x_2)$ such that $B(X_1, X_2)$ is not divisible by $f(X_1, X_2)$ (here $A(X_1, X_2)$, $B(X_1, X_2)$ are formal power series); $\mathfrak{p}_i = \mathfrak{O} f_i(x_1, x_2)$, $\mathfrak{f}_i =$

$= F. f_i(x_i, x_2)$, F_i = set of all above quotients $A(x_1, x_2)/B(x_1, x_2)$ such that $A(X_1, X_2)$ is divisible by $f(X_1, X_2)/f_i(X_1, X_2)$. Finally $F_i \cong$ field of fractions of the integral domain $k[[X_1, X_2]]/(f_i(X_1, X_2))$.

If h = 1, i.e., if F is a field, then we say that C is irreducible, or that C is an algebroid branch. If $h > 1$, we denote by γ_i the algebroid branch whose local ring is $\mathfrak{O}/\mathfrak{p}_i (i=1, 2, \ldots, h)$, and we refer to $\gamma_1, \gamma_2, \ldots, \gamma_h$ as the branches of C. If $\mathfrak{m} = (x_1, x_2)$, then the maximal ideal of $\mathfrak{O}/\mathfrak{p}_i$ is generated by the $\mathfrak{p}_i$-residues of x_1, x_2, and with respect to the plane embedding of γ_i thus obtained, $f_i(X_1, X_2) = 0$ is an equation of γ_i.

It is to be observed that while the embedding dimension of any γ_i is $\leqslant 2$, it may very well happen that any given γ_i may be regular.

The multiplicity $m(C)$ of C (at its origin 0, i.e., at the point $X_1 = X_2 = 0$) can be defined intrinsically as the multiplicity $e(\mathfrak{O})$ of the local ring $\mathfrak{O}$ (see [10], volume 2, p. 294). If $s = m(C)$, then it can be seen that s is the degree of the initial form of f: $f = f_s + f_{s+1} + \ldots$, where f_i is a form of degree i, and $f_s \neq 0$. From (3) it follows that $m(C) = \sum_{i=1}^{h} m(\gamma_i)$. We have $m(C) = 1$ if and only if C is a regular branch.

We have the well-known concept of the graded ring $G(\mathfrak{O})$ of $\mathfrak{O}$ (see [10], volume 2, p. 249):

$$G(\mathfrak{O}) = \sum_{n=0}^{\infty} \mathfrak{m}^n/\mathfrak{m}^{n+1}$$

It follows easily that

$$G(\mathfrak{O}) \cong k[X_1, X_2]/(f_s(X_1, X_2)),$$

i.e., geometrically speaking $G(\mathcal{O})$ is a set of $\underline{s}$ (not necessarily distinct) lines in the plane passing through the origin 0 of C . These are called the tangent lines of C.

It C is irreducible, then it has only one tangent line; equivalently: if the power series f is irreducible then the initial form f_s is the power $(a_1X_1 + a_2X_2)^s$ of a linear form $(a_1, a_2, \in k)$.
We sketch the proof (which is based on the Weierstrass preparation theorem and on Hensel's lemma). We may assume $s > 1$. Since f is irreducible, f is not divisible by X_1, hence f is regular in X_2 . By the Weierstrass preparation theorem we may therefore assume that f is a monic polynomial in X_2 with coefficients in $k[[X_1]]$. The quotient $f(X_1, X_2'X_1)/X_1^s$ is a monic polynomial $g(X_1, X_2')$ in X_2', with coefficients in $k[[X_1]]$, and is irreducible in $k[[X_1]][X_2']$. We have $g(0, X_2' = f_s(1, X_2')$, and hence, by Hensel's lemma, we must have $f_s(1, X_2') = (X_2' - \alpha)^s$, $\alpha \in k$.

q. e. d.

Since the leading form f_s of f is the product of the leading forms of the irreducible factors $f_1, f_2, \ldots, f_h$ of f, the set of tangents of C coincides with the set of tangents of the branches of C .

The ring $\mathfrak{m}/\mathfrak{m}^2$ is a vector space over k, of dimension 2. For each element x of $\mathfrak{m}$ we denote by $\bar{x}$ the associated vector in $\mathfrak{m}/\mathfrak{m}^2$ ($\bar{x} = \mathfrak{m}^2$ - residue of x) . If $\mathfrak{m} = \mathcal{O}x_1 + \mathcal{O}x_2$, then $\bar{x}_1, \bar{x}_2$ form a basis of $\mathfrak{m}/\mathfrak{m}^2$, and thus $\bar{x}$ is a linear form $a_1\bar{x}_1 + a_2\bar{x}_2$, where $a_1 = a_2 = 0$ if and only if $x \in \mathfrak{m}^2$. We say that x is a transversal parameter of $\mathcal{O}$ if $x \in \mathfrak{m}$, $x \notin \mathfrak{m}^2$ and if the image of x in the graded algebra $G(\mathcal{O})$ (i.e., if the element $\bar{x}$, regarded as a homogeneous element of degree 1 of $G(\mathcal{O})$) is not a zero-divisor in $G(\mathcal{O})$. It is seen at once that x is a transversal parameter if and only if the linear form $a_1X_1 + a_2X_2$ (where $a_1\bar{x}_1 + a_2\bar{x}_2 = x$) is not a factor of the initial form f_s. In other words: x is a transversal parameter if and only if the vector $\bar{x}$ is not a tangent of C (equivalently: the line $a_1X_1 + a_2X_2 = 0$ is not a tangent of C). It is also immediate that a transversal parameter is indeed a parameter of $\mathfrak{O}$.

If x is expressed as a power series $g(x_1, x_2)$ of the basis elements x_1, x_2 of $\mathfrak{m}$ (whence $g(0,0) = 0$), and if we denote by D the analytic cycle defined by $g(X_1, X_2) = 0$, then x is transversal if and only if the intersection multiplicity (D, C) is equal to $\mathfrak{m}(C)$. It is also easily seen that x is transversal if and only if x is a superficial element of order 1 for $\mathfrak{m}$ (see [10], volume 2, p. 285). Finally, x is transversal if and only if the multiplicity of the principal ideal $\mathfrak{O}x$ (see [10], volume 2, p. 294) is equal to the multiplicity of C.

2. The Quadratic Transform of C

We first summarize some of the properties of quotient rings which we shall use most frequently in the sequel. For details see ([10], volume 1, pp. 221-231).

Let R be a commutative ring with identity, and let M be a multiplicative system (m. s.) in R, i.e., a non-empty subset of R which is closed under multiplication and does not contain the zero. There is then a natural homomorphism $\phi: R \to R_M$, and in this homomorphism the elements of $\Phi(M)$ are units in R_M. The quotient ring R_M is uniquely determined, up to an isomorphism, by the existence of Φ and by a suitable "universal property". The properties of the operation of quotient ring formation which we wish to summarize are the following :

(a) Φ is an injection if and only if no element of M is a zero divisor in R. In particular, no element of $\Phi(M)$ is a zero divisor in $\Phi(R)$. and thus R_M can be identified with $\Phi(R)_{\Phi(M)}$.

(b) If $\mathfrak{p}$ is a prime ideal in R, we set $\mathfrak{p}^e = R_M \Phi(\mathfrak{p})$; and if $\mathfrak{p}'$ is a prime ideal in R_M, we set $\mathfrak{p}'^c = f^{-1}\{\mathfrak{p}'\}$. The correspondence $\mathfrak{p}' \to \mathfrak{p}'^c$ sets up a (1, 1) mapping of the set of all prime ideals $\mathfrak{p}'$ in R_M onto the set of those prime ideals $\mathfrak{p}$ in R which are disjoint from M, and we have

$$(\mathfrak{p}'^c)^e = \mathfrak{p}' .$$

(c) If M' is a m. s. in R such that $M' \supset M$, then $\Phi(M')$ is a m. s. in R_M, and there is a canonical isomorphism $(R_M)_{\Phi(M')} \xrightarrow{\sim} R_{M'}$; this homomorphism is compatible with the canonical homomorphisms

$$R \to R_M, \; R \to R_{M'} \quad \text{and} \quad R_M \to (R_M)_{\Phi(M')} .$$

(d) If $M' \supset M$, as in (c), and if every element of M' is of the form em, where e is a unit and $m \in M$, then the canonical homomorphism $R_M \to (R_M)_{\Phi(M')}$ is an isomorphism. Therefore, we shall in this case identify R_M with $R_{M'}$.

(e) A special case of (d) leads to the following conclusion: if $\mathfrak{p}'$ is a prime ideal in R_M and if $\mathfrak{p} = \mathfrak{p}'^c$, then the two quotient rings $R_{\mathfrak{p}}$ ($=R_{R-\mathfrak{p}}$) and $(R_M)_{\mathfrak{p}'}$ are canonically isomorphic. This is seen by observing that the two multiplicative systems $R_M - \mathfrak{p}'$ and $\varphi(R-\mathfrak{p})$ in R_M are in the same relation as M' and M are in (d) (with R being replaced by R_M) and $R - \mathfrak{p} \supset M$.

Here are two preliminary applications of the above properties.

A) Let M be a m.s. in F and let $\mathfrak{U}^*$ be the kernel of the canonical homomorphism $\phi : F \to F_{M^*}$. If $\xi \varepsilon \sqrt{\mathfrak{U}^*}$ then $\xi^n m=0$ for some positive integer n and for some $m \varepsilon M^*$. Hence $(\xi m)^n = 0$, and therefore $\xi m = 0$, since F has no nilpotent elements (different from zero). This shows that $\xi \varepsilon \mathfrak{U}^*$, $\mathfrak{U}^* \sqrt{\mathfrak{U}^*}$, whence $\mathfrak{U}^* = \bigcap_{i \varepsilon S} \mathfrak{f}_i$, where S is a (non-empty) subset of the set of indices $\{1, 2, \ldots, h\}$. Therefore $\phi(F) \cong F/\mathfrak{U}^* \cong$ $\cong \bigoplus_{i \varepsilon S} F_i$, and we can identify $\phi(F)$ with $\bigoplus_{i \varepsilon S}$ and F_{M^*} with $(\bigoplus_{i \varepsilon S} F_i)_{\overline{M}^*}$, where $\overline{M}^*$ ($= (M^*)$) is a m.s. in $\bigoplus_{i \varepsilon S} F_i$, no elements of which is a zero=divisor. But then all elements of $\overline{M}^*$ are units, showing that $F_{M^*} = \bigoplus_{i \varepsilon S} F_i \subset F$. We have shown that every quotient ring of F can be canonically identified with a subring of F, in fact with a partial direct sum of some of the fields F_i.

B) Let $\mathfrak{O}'$ be any sub-ring of F and let M' be a m.s. in $\mathfrak{O}'$. Then $\mathfrak{O}'_{M'}$ can be canonically identified with a sub-ring of $F_{M'}$, i.e., with a sub-ring of F (by A). From now on all quotient rings $\mathfrak{O}'_{M'}$ ($\mathfrak{O} \subset F$) will be regarded (without ambiguity) as sub-rings of F. In particular, if $\mathfrak{O}' = \mathfrak{O}$ and $M' = \mathfrak{O} - \bigcup_{i \in S} \mathfrak{p}_i$, then one sees at once that $\mathfrak{O}'_M = \bigoplus_{i \in S} F_i$.

We define the quadratic transform T(C) of C intrinsically as the following (reduced) scheme :

(1) $$T(C) = \bigcup_x \operatorname{Spec}(\mathfrak{O}[x^{-1}\mathfrak{m}]),$$

where x ranges over the set of non-zero divisors in $\mathfrak{m}$ and where $x^{-1}\mathfrak{m}$ denotes the set of quotients z/x, $z \in \mathfrak{m}$. It is understood that the localizations of $\mathfrak{O}[x^{-1}\mathfrak{m}]$ are identified with well defined sub-rings of F, and that we identify points of $\operatorname{Spec}(\mathfrak{O}[x^{-1}\mathfrak{m}])$ and $\operatorname{Spec}(\mathfrak{O}[x'^{-1}\mathfrak{m}])$ if their local rings coincide.

Let us fix a basis $\{x_1, x_2\}$ of $\mathfrak{m}$ such that neither x_1 nor x_2 is a zero-divisor and such that x_1 is a transversal parameter of . We shall prove now that

(2) $$T(C) = \operatorname{Spec}(\mathfrak{O}[\frac{x_2}{x_1}]).$$

For any given x in $\mathfrak{m}$ (x - not a zero divisor), we set $x'_1 = x_1/x$, $x'_2 = x_2/x$ and $\mathfrak{O}' = \mathfrak{O}[x^{-1}\mathfrak{m}]$. We have then $\mathfrak{O}' = \mathfrak{O}[x'_1, x'_2]$. We also set

$$\mathfrak{O}'_1 = \mathfrak{O}[x_1{}^{-1}\mathfrak{m}] = \mathfrak{O}[x_2/x_1],$$
$$\mathfrak{O}'_2 = \mathfrak{O}[x_2{}^{-1}\mathfrak{m}] = \mathfrak{O}[x_1/x_2]$$

Let $\mathfrak{p}'$ be any prime ideal in $\mathfrak{O}'$, and let $x = u_1x_1 + u_2x_2 (u_i \in \mathfrak{O})$. We have then $1 = u_1x'_1 + u_2x'_2$, and hence either $x'_1 \notin \mathfrak{p}'$ or $x'_2 \notin \mathfrak{p}$. We shall prove the following :

C) If $x'_i \notin \mathfrak{p}'$ (where i is either 1 or 2) then

<u>there exists a prime ideal</u> $\mathfrak{p}_i'$ <u>in</u> $\mathfrak{O}_i'$ <u>such that</u> $(\mathfrak{O}_i')_{\mathfrak{p}_i'} = \mathfrak{O}'_{\mathfrak{p}'}$.

From C) it will follow already that

(3) $$T(C) = \mathrm{Spec}(\mathfrak{O}_1') \cup \mathrm{Spec}(\mathfrak{O}_2') .$$

We shall also show that

D) <u>If</u> x_1 <u>is a transversal parameter then</u> x_1/x_2 <u>is a unit in</u> o_2'. In view of D), we see that in the special case in which $x = x_2$ (and therefore $x_1' = x_1/x_2$) the assumption $x_i' \notin \mathfrak{p}'$ in C) is satisfied for $i = 1$ and for any prime ideal $\mathfrak{p}'$ in $\mathfrak{O}'(= \mathfrak{O}_2')$. Hence by C), we have $\mathrm{Spec}(\mathfrak{O}_2') \subset \mathrm{Spec}(\mathfrak{O}_1')$, and thus (2) follows from (3).

To prove C), let, say, $x_1' \notin \mathfrak{p}'$. We consider in $\mathfrak{O}'$ the multiplicative system $N' = \{x_1'^i, i \geq 0\}$, and we set $\mathfrak{O}'' = \mathfrak{O}_{N'}$. Since x_1' is not a zero-divisor, we have $\mathfrak{O}' \subset \mathfrak{O}''$, and since N' and $\mathfrak{p}'$ are disjoint, it follows (by property (b)) that if we set $\mathfrak{p}'' = \mathfrak{p}'^e (= \mathfrak{O}''\mathfrak{p}')$, then $\mathfrak{p}' = \mathfrak{p}''^c (= \mathfrak{p}'' \cap \mathfrak{O}')$. Hence, if we apply property (e) to $R = \mathfrak{O}'$, $M = N'$, with $\mathfrak{p}'$ replaced by $\mathfrak{p}''$, we find that

(4) $$\mathfrak{O}'_{\mathfrak{p}'} = \mathfrak{O}''_{\mathfrak{p}''} .$$

We now observe that $1/x_1' \in \mathfrak{O}_1'$, we consider in $\mathfrak{O}_1'$ the multiplicative system $N_1' = \{1/x_1'^i, i \geq 0\}$, and we set $\mathfrak{O}_1'' = (\mathfrak{O}_1')_{N_1'}$. We see at once that $\mathfrak{O}'' = \mathfrak{O}_1''$, since $\mathfrak{O}'' = \mathfrak{O}'[1/x_1'] = \mathfrak{O}[x_1/x, x_2/x, x/x_1] = \mathfrak{O}[x_1/x, x_2/x_1, x/x_1] = \mathfrak{O}[x_1/x, x_2/x_1] = \mathfrak{O}_1'[x_1/x] = \mathfrak{O}_1''$. It follows therefore from property (e) that $\mathfrak{O}''_{\mathfrak{p}''} = (\mathfrak{O}_1')_{\mathfrak{p}_1'}$

where $\mathfrak{p}'_1$ is the prime ideal in $\mathfrak{O}'_1$ which corresponds to the ideal $\mathfrak{p}''$ in the quotient ring $(\mathfrak{O}'_1)_{N_{1'}}$ $(= \mathfrak{O}1'' = \mathfrak{O}'')$.

It remains to prove D). We may assume that f is a monic polynomial in X_1. Since x_1 is transversal, the term X_2^s is present in the initial form $f_s(X_1, X_2)$. We have $x_1 = x_2 x'_1$, and thus $f(x_2 x_1', x_2) = 0$. This equation is of the form $x_2^s g(x'_1, x_2) = 0$, where g is a polynomial in x'_1, with coefficients in $k[[x_2]]$, and where $g(0,0) \neq 0$. Since x_2 is not a zero-divisor, it follows that $g(x'_1, x_2) = 0$, showing that $\mathfrak{O}'_2 x'_1 + \mathfrak{O}'_2 x_2 = (1)$. Now, let $\mathfrak{p}'$ be any prime ideal in $\mathfrak{O}'_2$. If $\mathfrak{p}' \cap \mathfrak{O} = \mathfrak{m}$, then $x_2 \in \mathfrak{p}'$, whence $x'_1 \notin \mathfrak{p}'$. If $\mathfrak{p}' \cap \mathfrak{O} = \mathfrak{p}_i$ $(1 \leq i \leq h)$, then $x'_1 \notin \mathfrak{p}'$, since x'_1 is not a zero-divisor. This proves D).

We may assume that f is a monic polynomial in X_2. We denote by $\mathfrak{O}'$ the ring $\mathfrak{O}[x'_2]$, $x'_2 = x_2/x_1$. Since x_1 is a transversal parameter, f is exactly of degree s in X_2. Thus the relation $f(x_1, x_2) = 0$ has the form

$$x_2^s + A_1(x_1)x_1 x_2^{s-1} + A_2(x_1)x_1^2 x_2^{s-2} + \dots + A_s(x_1)x_1^s = 0,$$

where the $A_i(x_1)$ are power series in x_1. This equation implies that $x_2/x_1 (= x'_2)$ <u>is integral over</u> $k[[x_1]]$. Since $\mathfrak{O} = k[[x_1]][x_2]$, it follows that $\mathfrak{O}'(= \mathfrak{O}[x_2/x_1]) = k[[x_1]][x_2/x_1]$ and that consequently $\mathfrak{O}'$ <u>is a complete semi-local ring.</u> Let $\mathfrak{m}'_1, \mathfrak{m}'_2, \dots, \mathfrak{m}'_d$ be the distinct maximal ideals of $\mathfrak{O}$. In view of the integral dependence of x'_2 over $k[[x_1]]$, k is a coefficient field of each of the d <u>local</u> complete rings $\mathfrak{O}'_{\mathfrak{m}'_j}$.

We denote these $\underline{d}$ rings by $\mathfrak{O}_1', \mathfrak{O}_2', \ldots, \mathfrak{O}_d'$. If c_j is the $\mathfrak{m}_j'$ - residue of x_2' in k, then $\mathfrak{m}_j' = \mathfrak{O}' x_1 + \mathfrak{O}'(x_2'-c_j)$. Thus, the d constants $c_1, c_2, \ldots, c_d$ are distinct, and each local ring $\mathfrak{O}_j'$ has embedding dimension $\leqslant 2$. Let C_j' be the plane algebroid curve defined by the local ring $\mathfrak{O}_j'$. The origin $0_j'$ of C_j' in the (X_1, X_2') - plane is the point $X_1 = 0, X_2' = c_j$. The constants $c_1, c_2, \ldots, c_d$ are the distinct roots of $f_s(1, X)$. Thus the integer d is the number of distinct tangent lines of C.

We call the union $\bigcup_{j=1}^d C_j'$ of the disjoint algebroid curves C_j' the <u>proper quadratic transform</u> T(C) of C.

The quadratic transformation T operates also on the (X_1, X_2)-plane: it blows up the origin 0 into the line $X_1 = 0$ of the plane (X_1, X_2'). We call this line the <u>exceptional</u> curve of T and we denote it by E'; this curve contains the $\underline{d}$ origins $0_1', 0_2', \ldots, 0_d'$. We set $C_j'^{*} = C_j' \cup E'$, and we call the union $\bigcup_{j=1}^d C_j'^{*}$ of the d algebroid curves $C_j'^{*}$ the <u>total quadratic transform</u> $T\{C\}$ of (

Let $p_1, p_2, \ldots, p_d$ be the distinct tangent lines of C, where we assume that p_j is associated with the root c_j of $f_s(1, X)$. We denote by C_j the union of those branches of C which have p_j as tangent line, and we call the algebroid curves $C_1, C_2, \ldots, C_d$ the <u>tangential components</u> of C. We have thus $C = \bigcup_{j=1}^d C_j$, and one sees easily that $C_j' = T(C)$, $C_j'^{*} = T\{C_j\}$, $j = 1, 2, \ldots, d$.

3. Three (equivalent) Definitions of Equivalence of Plane Algebroid Curves.

We denote by $\gamma_1, \gamma_2, \ldots, \gamma_h$ the h branches of C. Let D be another algebroid curve, which has the same number of branches as C. Let $\delta_1, \delta_2, \ldots, \delta_h$ be the branches of D.

A (1;1) mapping π of the set of branches of C onto the set of branches of D will be said to be a tangentially stable pairing between the branches of C and D if the following is true: for any pair of branches γ_α, γ_β of C the corresponding branches $\pi(\gamma_\alpha)$, $\pi(\gamma_\beta)$ have the same tangent if and only if $\gamma_\alpha, \gamma_\beta$ have the same tangent.

We shall use the notation: $\pi : (C) \to (D)$ to indicate a tangentially stable pairing between the branches of C and D.

It is clear that if there exists a tangentially stable pairing $\pi : (C) \to (D)$, then the number of distinct tangents of C is the same as the number of distinct tangents of D. If $p_1, p_2, \ldots, p_d$ are the tangents of C and $q_1, q_2, \ldots, q_d$ are the tangents of D, then, given $\pi : (C) \to (D)$, we have an induced (1, 1) mapping of the set $\{p_1, p_2, \ldots, p_d\}$ onto the set $\{q_1, q_2, \ldots, q_d\}$. We shall assume that p_j and q_j correspond in this mapping.

It is well known that by a finite number $\sigma(C)$ (respectively $\sigma^*(C)$) of successive quadratic transformations it is possible to obtain a proper transform of C which has no singular points (respectively, a total transform of C which has only ordinary, double points). This integer $\sigma(C)$, or the integer $\sigma^*(C)$, can be regarded as a measure of the complexity of the singularity of C. This fact is the basis of the various inductive definitions of equivalence $C \equiv D$ given below. Namely, in each of these definitions it is assumed that the meaning of the equivalence statements $C'_j \equiv D'_j$ and $C'^*_j \equiv D'^*_j$ is already known for $j = 1, 2, \ldots, d$ (as $\sigma(C'_j) < \sigma(C)$ and $\sigma^*(C'^*_j) < \sigma^*(C)$, etc.) and it is also stipulated that if C is a regular branch (respectively, if C has an ordinary double point) then $C \equiv D$ if and only also D is a regular branch (respectively, if also D has an ordinary double point).

Definition 3.1. : A tangentially stable pairing $\pi : (C) \to (D)$ is an (a) - equivalence if the following conditions are satisfied:

1) If $\delta_i = \pi(\gamma_i)$ $(i=1, 2, \ldots, h)$, then $m(\gamma_i) = m(\delta_i)$.

2) The pairings $\pi'_j : (C'_j) \to (D'_j)$ induced by π are (a) - equivalences.

It is understood that if $h = 1$ and C is a regular branch, in which case π is uniquely determined, then π is an (a) -equivalence if and only if condition 1) is satisfied, i.e., if and only if also D is a regular branch. This is also our understanding in Definition 3;2; given below.

In the above definition we have made use of the obvious fact that any tangentially stable pairing $\pi : (C) \to (D)$ induces a (1, 1) mapping π_j of the set of branches of any of the d tangential components C_j of C onto the set of branches of the corresponding tangential component D_j of D (the common tangent of the branches of C_j and the common tangent of the branches of D_j being corresponding tangents p_j and q_j). The mapping π_j induces a pairing π'_j between the branches of

C'_j and those of D'_j .

It should also be pointed out that according to the above definition, if a pairing $\pi:(C) \to (D)$ is an (a)-equivalence then it must be at any rate tangentially stable. Thus, condition 2) includes the tacit requirement that the π'_j be tangentially stable.

Let $\pi:(C) \to (D)$ be a tangentially stable pairing and let π'_j be defined as above. We extend π'_j to a (1, 1) mapping π'_j between the branches of C'^*_j $(=C'_j \cup E')$ and the branches of D'^*_j $(=D_j \cup E'_1)$, by setting $\pi'^*_j(E') = E'_1$; here E'_1 is the exceptional curve which corresponds to the origin of D .

<u>Definition 3.2: A tangentially stable pairing</u> $\pi:(C) \to (D)$ <u>is a (b)-equivalence if the following conditions are satisfied:</u>

1) <u>The pairings</u> $\pi'_j : (C'_j) \to (D'_j)$ <u>are</u> (b) <u>-equivalences.</u>
2) <u>The pairings</u> $\pi'^*_j:(C'^*_j) \to (D'^*_j)$ <u>are</u> (b-<u>equivalences</u>).
   (j=1, 2, . . . , d) .

It shall be understood that if h = 2 and C has an ordinary double point, then π is a (b)-equivalence if and only if also D has an ordinary double point. Note that in this case there are two (1, 1) mappings of the pair of branches (γ_1, γ_2) of C onto the pair of branches (δ_1, δ_2) of D and that both are tangentially stable pairings, hence are (b)-equivalences according to our convention.

<u>Definition 3.3; C and D are formally equivalent if there exists a tangentially stable pairing</u> $\pi:(C) \to (D)$ <u>such that:</u>

1) C'_j and D'_j are formally equivalent $(j = 1, 2, \ldots, d)$.

2) C'^{*}_j and D'^{*}_j are formally equivalent $(j = 1, 2, \ldots, d)$.

It is understood here that if C is a regular branch (respectively, if C has an ordinary double point), then C and D are formally equivalent if and only if also D is a regular branch (respectively, if also D has an ordinary double point).

It can be shown - and this is not entirely a trivial matter - that these three definitions are all equivalent; Definition 3.1 is the only one in which the multiplicities of the two curves C, D (and hence also, inductively, the multiplicities of the connected components of their successive quadratic transforms) are explicitly required to be the same [in view of condition 1) of that definition]. This definition is also the one which adheres most faithfully to the classical concept of equivalence which is due to Max Noether and which is based on the notion of multiple points of C which are "finitely near" the "proper" point 0 (the origin) and lie in the successive neighborhoods of order 1, 2 etc. of 0 (a pattern which is often referred to as the composition of the singular point 0 of C).

In Definitions 3.2. and 3.3 the multiplicites of the branches of C or of D (and hence also the multiplicities of C and D themselves-) are not mentioned at all; these two definitions are therefore more subtle than Definition 3.1 (and, for that very reason, more useful for some applications, as we shall see later on). On the other hand, in both Definitions 3.2. and 3.3, the inductive hypothesis includes not only the equivalence ((b)-equivalence in

Definition 3.2; formal equivalence in Definition 3.3) of the connected components C'_j, D'_j of the proper transforms of C and D, but also the equivalence of the connected components C'^{*}_j, D'^{*}_j of the total transforms of C and D. The difference between Definition 3.2 and 3.3 is this: a (b)-equivalence is by definition, a certain tangentially stable pairing $\pi:(C) \to (D)$, while formal equivalence is a certain condition on the pair $\{C.D.\}$ but is not a tangentially stable pairing $(C) \to (D)$. More precisely, while in Definition 3.3 we do assume the existence of a tangentially stable pairing $\pi:(C)\to(D)$, nothing is said about the nature of the pairings $\pi'_j : (C'_j) \to (D'_j)$ and $\pi'^{*}_j : (C'^{*}_j) \to (D'^{*}_j)$ induced by π. Definition 3.3 is the most subtle of the three definitions.

4. Outline of Proof of Equivalence

For any two algebroid curves C.D in the plane, free from common branches, we consider the intersection number (C, D). This number is zero if and only if the two curves have distinct origins.

Assuming that C and D have the same origin 0 and applying a quadratic transformation T with center 0, we have - in the notations of § 3 - the following well-known formula :

$$(1) \qquad (C, D) = m(C) \; m(D) + \sum_{\alpha=1}^{d} \sum_{\beta=1}^{\bar{d}} (C'_\alpha , D'_\beta) ,$$

where $C'_1, C'_2, \ldots, C'_d$ are the connected components of $T(C)$ and $D'_1, D'_2, \ldots, D'_{\bar{d}}$ are the connected components of $T(D)$. In particular, then, $(C, D) = m(C)m(D)$ if and only if C and D have no common tangents. If, furthermore, E' is the exceptional curve into which 0 is blown up by T, then we also have

$$(2) \qquad \sum_{\alpha=1}^{d} (C'_\alpha , E') = m(C)$$

The two formulas (1) and (2) are all that is technically needed for the proof, which consists of several steps.

(A) One first proves the following lemma :

Lemma 4.1. Let $\pi : (C) \rightarrow (D)$ be an (a)-equivalence, let Γ be a branch through the origin of C, let Δ be a branch through the origin of D and assume that Γ and Δ are regular branches, that Γ is not a branch of C and that Δ is not a branch of D. Assume furthermore that

$$(3) \qquad (\gamma_i, \Gamma) = (\delta_i, \Delta), \quad i = 1, 2, \ldots, h ,$$

where $\gamma_1, \gamma_2, \ldots, \gamma_h$ are the branches of C; $\delta_1, \delta_2, \ldots, \delta_h$ are the branches of D, and where $\delta_i = \pi(\gamma_i)$. Extend the mapping π to a mapping ρ of the set $\{\gamma_1, \gamma_2, \ldots, \gamma_h, \Gamma\}$ onto the set $\{\delta_1, \delta_2, \ldots, \delta_h, \Delta\}$ by setting $\rho(\Gamma) =$. Then $\rho : (C \cup \Gamma) \rightarrow (D \cup \Delta)$ is an (a)-equivalence.

The proof of this lemma is by induction on the character

$\sigma^*(C \cup \Gamma)$. If $\sigma^*(C \cup \Gamma) = 0$, then C is a regular branch, C and Γ have distinct tangents, $(C, \Gamma) = 1$, whence also $(D, \Delta) = 1$, by (3). Thus also $\sigma^*(D \cup \Delta) = 1$, and the assertion of the lemma is then trivial.

It follows directly from (3) that ρ is a tangentially stable pairing which satisfies condition 1) of (a)-equivalence. In the proof that ρ also satisfies condition 2) of Definition 3.1 one applies a quadratic transformation to C and D and one considers separately two cases accordingly as the tangent of Γ is or is not a tangent of C.

(B) The next step is the following

Lemma 4.2. If $\pi:(C) \to (D)$ is an (a)-equivalence and $\pi(\gamma_i) = \delta_i$ $(i = 1, 2, \ldots, h)$, then

$$(\gamma_i, \gamma_j) = (\delta_i, \delta_j), \text{ for all } i \neq j.$$

The proof is by induction on the numerical character $\sigma(C)$ and is a straightforward consequence of formula (1).

(C) Using Lemmas 4.1 and 4.2, one proves

Theorem 4.3. If $\pi:(C) \to (D)$ is an (a)-equivalence, then it is also a (b)-equivalence; and conversely.

The proof is by induction on $\sigma^*(C)$ and is rather straightforward. At this stage we drop the terms (a)-equivalence and (b)-equivalence and speak simply of equivalence pairings $\pi:(C) \to (D)$, and we say that C and D are equivalent if there exists an equivalence pairing $\pi:(C) \to (D)$. The proof that equivalence of C and D implies the formal equivalence of C and D is immediate, and is by induction on $\sigma^*(C)$. For if $\pi:(C) \to D$ is an equivalence pairing, then π being a (b)-equivalence pairings $\pi'_j:(C'_j) \to (D'_j)$ and $\pi'^*_j:(C'^*_j) \to (D'^*_j)$ $(j = 1, 2, \ldots, h)$. Therefore, by induction, C'_j and D'_j are formally equivalent and so are C'^*_j and D'^*_j. Hence C and D are formally equivalent.

More difficult is the proof of the converse, i.e., that formal equivalence implies equivalence. The key to the proof is the following lemma:

Lemma 4.4. Let C and D have the same number h of branches. Let $\gamma_1, \gamma_2, \ldots \gamma_m$ be certain m of the h branches of $C(m \leqq h)$ and assume that these m branches are regular and have distinct tangents. Similarly, let $\delta_1, \delta_2, \ldots, \delta_m$ be m of the m of the h branches of D, also regular and having distinct tangents, Let

$$C = \Gamma \cup \gamma_1 \cup \gamma_2 \cup \ldots \cup \gamma_m,$$
$$D = \Delta \cup \delta_1 \cup \delta_2 \cup \ldots \cup \delta_m,$$

where Γ, therefore, is the union of the remaining branches of C, and similarly for Δ and D. Assume that Γ and Δ are equivalent and that for each $\alpha = 1, 2, \ldots, m$ also $\Gamma \cup \gamma_1 \cup \gamma_2 \cup \ldots \cup \gamma_\alpha$ and $\Delta \cup \delta_1 \cup \delta_2 \cup \ldots \cup \delta_\alpha$ are equivalent. Then there exists an equivalence pairing $\pi:(C) \to (D)$, such that $\pi(\gamma_\alpha) = \delta_\alpha$, for $\alpha = 1, 2, \ldots, m$.

The proof of this lemma is divided into several cases:

Case 1. m = 1, and the tangent of γ_1 is not a tangent of

Γ. This case is straightforward: the equivalence of Γ and Δ implies that Γ and Δ have the same number of tangent lines; similarly, $\Gamma \cup \gamma_1$ and $\Delta \cup \delta_1$ have the same number of tangent lines. Since the number of tangent lines of $\Gamma \cup \gamma_1$ is one greater than the number of tangent lines of Γ, it follows that the tangent of δ_1 is not a tangent of Δ. From the definition of (a)-equivalence it follows at once that given an equivalence pairing $\rho : (\Gamma) \to (\Delta)$, the extended pairing $\pi : (C) \to (D)$ defined by $\pi(\gamma_1) = \delta_1$, is an (a)-equivalence.

Case 2. $m = 1$ and the tangent of γ_1 is a tangent of Γ (whence also the tangent of δ_1 is a tangent of Δ).

In this case, one first achieves a reduction to the case in which Γ (and therefore also C) has only one tangent line. Passing to the total quadratic transforms of C and D (a transition which adds a new regular branch to both total transforms, namely the exceptional curve created by the quadratic transformation), one finds oneself in the case in which $m = 2$, while $\sigma^*(C)$ has decreased. The result now follows by induction on $\sigma^*(C)$.

Case 3. $m > 1$. In this case the result follows easily by applying the case $m = 1$ to each of the tangential components $C_1, C_2, \ldots, C_m$ of C which are determined by the m tangents of $\gamma_1, \gamma_2, \ldots, \gamma_m$.

The proof that formal equivalence implies equivalence is now as follows :

In the notation of Definition 3. 3 (§ 3), we have, by hypothesis, that C_j' and D_j' are formally equivalent and that also $C_j' \cup E'$ and $D_j' \cup E'$ are formally equivalent (we assume here, without loss of generality, that C and D have the same origin 0). By induction on σ (C), we can conclude that C_j' and

D'_j are equivalent and that also $C'_j \cup E'$ and $D'_j \cup E'$ are equivalent. Therefore, by the case $m = 1$ of Lemma 4.4, there exists an equivalence $\pi'_j: (C'_j \cup E') \to (D'_j \cup E')$ such that $\pi'_j(E') = E'$. But that implies that C and D are equivalent (namely, (b)-equivalent).

Note: We have used here only the case $m=1$ of Lemma 4.4. However, the proof of the lemma for $m=1$ required that the lemma be proved also in the case $m = 2$ (see Case 2 above). So Lemma 4.4. is needed only for $m=1, 2$.

For later applications we ahall also need the following equivalence criterion:

Proposition 4.5. *Assume that C and D have the same number h of branches and that a pairing π between the branches γ_i of C and the branches δ_i of D has the following two properties:*

1) *Corresponding branches γ_i δ_i are equivalent* $(i=1, 2, \ldots, h)$.

2) *For all pairs* (i, j), $i = j$, *we have* $(\gamma_i, \gamma_j) = (\delta_i, \delta_j)$. *Then π is an equivalence.*

Proof. By 1) we have $m(\gamma_i) = m(\delta_i)$, $i = 1, 2, \ldots, h$. Then 2) implies that π is tangentially stable. Let $C_1, C_2, \ldots, C_d$ be the tangential components of C and let $D_1, D_2, \ldots, D_d$ be the corresponding tangential components of D (relative to π). Apply a quadratic transformation to both C and D, and let C'_j and D'_j be the proper transforms of C_j and D_j. Let π'_j be the pairing between the branches of C'_j and D'_j induced by π. If γ' is any branch of C'_j and $\delta' = \pi'_j(\gamma')$ is the corresponding branch of D'_j, then we have, by 1), that γ' and δ' are equivalent. If γ_1', γ_2' are any two branches of C_j' and δ_1', δ_2' are the corresponding branches of D_j', we have that γ_1', γ_2' are the quadratic transform of two branches γ_1, γ_2 of C_j, while δ_1', δ_2'

are the quadratic transforms of the corresponding branches δ_1, δ_2 of D_j.

We have $(\gamma_1', \gamma_2') = (\gamma_1, \gamma_2) - m(\gamma_1)m(\gamma_2)$, $(\delta_1', \delta_2') = (\delta_1, \delta_2) - m(\delta_1)m(\delta_2)$. Hence $(\gamma_1', \gamma_2') = (\delta_1', \delta_2')$. The proposition now follows by induction on the numerical character $\sigma(C)$.

In connection with Proposition 4.5, it is well to recall at this time the well known criterion of equivalence of algebroid branches γ, δ, in terms of the characteristic exponents of the Puiseux expansions of γ and δ.

Let (x, y) be a minimal basis of the local domain $\mathfrak{O}$ of γ and assume that x is a transversal parameter. Then, if n denotes the multiplicity of γ $(n > 1)$, we can expand y into a power series in $x^{i/n}$;

$$(4) \qquad y = \sum_{i}^{\infty} = n\, a_i x^{i/n} ,$$

where the highest common divisor of all the numerators i such that $a_i \neq 0$ is prime to n. (In the present case we actually have $a_i = 0$ for $i < n$.) The characteristic exponents.

$$(5) \qquad \beta_1/n < \beta_2/n < \ldots < \beta_g/n, g \geqq 1$$

of the Puiseux expansion (4) are defined by the following conditions:

a) $n > (n, \beta_1) > (n, \beta_1, \beta_2) > \ldots > (n, \beta_1, \beta_2, \ldots, \beta_g) = 1$;

b) $a_{\beta_j} \neq 0$, $j = 1, 2, \ldots, g$;

c) Each β_j is minimum (more precisely: if $\beta_1, \beta_2, \ldots, \beta_{j-1}$ have already been defined $(j \geqq 1)$ and if $(n, \beta_1, \beta_2, \ldots, \beta_{j-1}) > 1$, then β_j is the smallest of all the integers q such that $a_q \neq 0$ and $(n, \beta_1, \beta_2, \ldots, \beta_{j-1}) >$

$$> (n, \beta_1, \beta_2, \ldots, \beta_{j-1}, q)) .$$

Since n is the least common denominator of the rational numbers β_j/n, the characteristic exponents determine the integer n, hence the multiplicity of γ. Let

$$(6) \qquad \bar{y} = \sum_{i=n}^{\infty} \bar{a}_i \bar{x}^{i/\bar{n}} ,$$

be the Puiseux expansion of the branch δ, where $\bar{x}$ is also a transversal parameter of δ (whence $\bar{n} = m(\delta)$), and let

$$(7) \qquad \bar{\beta}_1/\bar{n} < \bar{\beta}_2/\bar{n} < \ldots \bar{\beta}_{\bar{g}}/\bar{n}$$

be the characteristic exponents of (6). We have the following well-known result :

Proposition 4.6. The branches γ and δ are equivalent if and only if the Puiseux expansion (4) and (6) have the same characteristic exponents.

[It is understood that it is implicitly assumed in this proposition that x and $\bar{x}$ are transversal parameters; or -equivalently - that $\beta_1 > n$ and $\bar{\beta}_1 > \bar{n}$].

A fast proof of this proposition, avoiding the complications arising from relating the compositinn of the singularity of γ to the expansions of the characteristic exponents into continued franctions, can be obtained by using the following inversion formula (see Abhyankar [2] ; Zariski [9] , p.996) :

Let $a_m x^{m/n}$ be the leading term of (4) ($m \geqq n$), and let

$$\beta'_1/m < \beta'_2/m < \ldots < \beta'_{g'}/m$$

be the characteristic exponents of the Puiseux expansion $x = \dot{x}(y^{1/m})$.
Then :

1) If $n < m < \beta_1$ (i.e., if $m>n$ and is an integral multiple of n), then $g' = g + 1$, $\beta'_1 = n$, $\beta'_{\nu+1} = \beta_\nu + n - m$ ($\nu = 1, 2, \ldots, g$).

2) In all other cases (i.e. , if . , $m = n$ or $m = \beta_1$), we have $g' = g$ and $\beta' = \beta_\nu + n - m$ ($\nu = 1, 2, \ldots, g$).

We now apply to γ and δ the quadratic transformations $y' = y/x$ and $\bar{y}' = \bar{y}/\bar{x}$ respectively. Without loss of generality, we may assume that the leading terms of (4) and (6) are $a_{\beta_1} x^{\beta_1/n}$ and $a_{\beta_1} \bar{x}^{\bar{\beta}_1/\bar{n}}$ The transformed branches γ', δ' will be centered at the origin, and will be given by the expansions:

(8) $$\gamma' : \quad y' = \sum_{i=\beta_1}^{\infty} a_i x^{(i-n)/n}$$

(9) $$\delta' : \quad \bar{y}' = \sum_{i=\bar{\beta}_i}^{\infty} \bar{a}_i \bar{x}^{(i-\bar{n})/\bar{n}} .$$

I. Assume $\gamma \equiv \delta$. Then $n = \bar{n}$ and $\gamma' \equiv \delta'$.

(a) If $\beta_1 > 2n$, then $m(\gamma') = m(\gamma)$ (=n) , whence also $m(\delta') = m(\delta)$, and thus $\bar{\beta}_1 > 2n$. In this case, x and $\bar{x}$ are still transversal parameters for γ' and δ', and using induction on $\sigma(\gamma)$, we conclude that the characteristic exponents of (8) and (9) are the same , i.e., $(\beta_i - n)/n = (\bar{\beta}_i - n)/n$, all i, and the proposition is proved.

(b) If $\beta_1 < 2n$, we have $m(\gamma') = \beta_1 - n < n$. Therefore also $m(\delta') < m(\delta)$, showing that $\bar{\beta}_1 = \beta_1$. This time, y' and $\bar{y}'$ are transversal parameters. The induction hypothesis implies now that the characteristic exponents of the two Puiseux expansions $x = x(y'^{1/(\beta_1 - n)})$, $\bar{x} = \bar{x}(\bar{y}'^{1/(\beta_1 - n)})$ are the same. The proposition now follows easily from the inversion formula.

II. Assume that (4) and (6) have the same characteristic exponents.

We have now again $n = \bar{n}$, whence $m(\gamma) = m(\delta)$. As in Case I, there are again the two cases (a) and (b) to be considered. In case (a), the equality of the characteristic exponents of (8) and (9) implies without further ado (and by induction hypothesis) that $\gamma' \equiv \delta'$, and hence $\gamma \equiv \delta$. In case (b), the equality of the characteristic exponents of (8) and (9) implies, by the inversion formula, the equality of the characteristic exponents of the expansions $x = x(y'^{1/(\beta_1 - n)})$, $\bar{x} = \bar{x}(\bar{y}'^{1/(\beta_1 - n)})$, and hence we find again that $\gamma' \equiv \delta'$.

Corollary. The characteristic exponents of (4) (with x-transversal) are independent of the choice of the transversal parameter x.

Apply Proposition 4.6 to the case $\gamma = \delta$. We can therefore speak of the characteristic exponents of the branch γ.

5. On Some Numerical Characters of C'

Given a plane algebroid curve C and given a parameter x such that x belongs to some minimal basis $\{x, y\}$ of the maximal ideal $\mathfrak{m}$ of the local ring $\mathfrak{O}$ of C, we shall associate with the pair $\{C, x\}$ an integer $M_x(C)$ defined as follows:

Let y be any element of $\mathfrak{m}$ such that $\{x, y\}$ is a basis of $\mathfrak{m}$ and let $f(X, Y) = 0$ be the equation of C relative to this basis. Then $f(X, Y)$ is regular in Y, and we may assume then that f is a monic polynomial in Y and that $f(0, Y)$ is a power Y^n of Y (these conditions determine f uniquely). In this sense we can speak of the Y-discriminant $\Delta^Y f$ of f; it will be a power series in X. Up to a unit factor in $k[[X]]$, this power series is independent of the choice of y, and hence depends only on the pair $\{C, x\}$. The integer $M_x(C)$ is the highest power of X which divides $\Delta^Y f$.

Assume C irreducible. Then the quotient field of the local domain $\mathfrak{O}$ is a field $k((C))$ of formal power series in one indeterminate $\{u\}$. We denote by v the natural valuation of $k((C))$. Assume that both elements x and y of the basis $\{x, y\}$ of $\mathfrak{m}$ are parameters; equivalently: $f(X, Y)$ is also regular in X. Then also the integer $M_y(C)$ is defined. It is easy to see that

$$(1) \qquad M_x(C) = v\left(\frac{\partial f}{\partial y}\right), \quad M_y(C) = v\left(\frac{\partial f}{\partial x}\right).$$

We have the relations $\frac{\partial f}{\partial x} dx + \frac{\partial f}{\partial y} dy = 0$, $v\left(\frac{dx}{du}\right) = v(x) - 1$, $v\left(\frac{dy}{du}\right) = v(y) - 1$. Hence, by (1), we find that

$$(2) \qquad M_x(C) - v(x) = M_y(C) - v(y).$$

Thus the integer $M_x(C) - v(x)$ is therefore a numerical character of the irreducible curve C. Using the well-known relation in the integral closure $\bar{\mathfrak{o}}$ of the local domain $\mathfrak{o}$:

$$(f_y) = \mathfrak{C}\,\mathfrak{D}_x ,$$

where $\mathfrak{C}$ is the conductor of $\mathfrak{o}$ in $\bar{\mathfrak{o}}$ and $\mathfrak{D}_x$ is the different of $\bar{\mathfrak{o}}$ in $\mathfrak{o}$, and recalling that (since k is of characteristic zero) $\mathfrak{D}_x = \bar{\mathfrak{m}}^{e-1}$, where $e = v(x)$ and $\mathfrak{m}$ is the maximal ideal of $\bar{\mathfrak{o}}$, it follows that

$$\mathfrak{C} = \bar{\mathfrak{m}}^{M_x(C)-v(x)+1}$$

Whether C is or is not irreducible, it is easy to see that any transversal parameter x belongs to some minimal basis. Hence the integer $M_x(C)$ is defined for all transversal parameters x of o . In the case of an irreducible C , what we have just said abov shows that (a) the integer $M_x(C)$ is the same for all transversal parameters, for if x is transversal then $v(x) = m(C)$; it is also obvious that (b) if $M_{x'}$ is defined and x' is not transversal then $M_x(C) < M_{x'}(C)$. We can show easily that the last two assertions are also valid if C is not irreducible. For, let $\gamma_1, \gamma_2, \ldots, \gamma_h$ be the branches of C. The local ring $\mathfrak{o}_i$ of γ_i is a homomorphic image of $\mathfrak{o}$, and if x_i, y_i are the images of x, y in $\mathfrak{o}_i$, then (x_i, y_i) is a basis of the maximal ideal of $\mathfrak{o}_i$, and x_i is a parameter of $\mathfrak{o}_i$. From the expression of $\Delta^Y f$ in terms of the roots $Y_1, Y_2, \ldots, Y_n$ of $f(X,Y)$ follows at once that

$$(3) \qquad M_x(C) = \sum_{i=1}^{h} M_{x_i}(\gamma_i) + 2 \sum_{i<j} (\gamma_i, \gamma_j) .$$

From this relation, the above assertions (a), (b) follow also if C is reducible. We denote by Δ(C) the integer $M_x(C)$ x - a transversal parameter of o. It follows from (3) that

$$(4) \qquad \Delta(C) = \sum_{i=1}^{h} \Delta(\gamma_i) + 2 \sum_{i<j} (\gamma_i, \gamma_j) .$$

Proposition 5.1. If $C \equiv D$, then $\Delta(C) = \Delta(D)$.

Proof. I view of (4) and of Lemma 4.2, it is sufficient to consider the case in which C and D are irreducible. We shall assume that C and D have been embedded in the same (X, Y)-plane, that they have the same origin $X = Y = 0$, and that the line $X = 0$ is neither a tangent of C nor a tangent of D. Let $\mathfrak{o}$, $\mathfrak{o}_1$ be the local domains of C and D, let $\{x, y\}$ be a minimal basis of the maximal ideal of $\mathfrak{o}$, and let $\{x_1, y_1\}$ be a minimal basis of the maximal ideal of $\mathfrak{o}_1$. We assume that both x and x_1 are transversal parameters. We may assume that neither y nor y_1 is transversal. If we then set $y' = y/x$, $y'_1 = y_1/x_1$, and call C', D' the quadratic transforms of C and D, we see that $\{x, y'\}$ is a basis of the maximal ideal of the local ring of C', and that similarly $\{x_1, y_1'\}$ is a basis of the maximal ideal of the local ring of D'. We set $s = m(C) = m(D)$. One finds at once that

$$M_x(C) = s(s-1) + M_x(C'),$$

$$M_{x_1}(C) = s(s-1) + M_{x_1}(D'),$$

So we have only to show that $M_x(C') = M_{x_1}(D')$.

If $m(C') = m(C)$, then also $m(D') = m(D)$ (since $C' \equiv D'$), x, x_1 are still transversal parameters for C' and D', $M_x(C') = \Delta(C')$, $M_{x_1}(D') = \Delta(D')$, and the proposition follows by induction on the integer $\Delta(C)$. Assume, however, that $m(C') < m(C)$, and that consequently $\underline{x}$ is <u>not</u> a transversal parameter for C' (nor, then, is x_1 transversal for D'). The local equation of the exceptional curve E', in the (X, Y')-plane, is $X = 0$, and this line is tangent to C', since $(E', C') = m(C)$ (formula (2), § 4) and $m(C) > m(C')$. Similarly, E' is tangent to D'. Therefore the line $Y' = 0$ is neither a tangent of C' nor a tangent of D'. By induction on $\Delta(C)$, we may therefore assume that $M_{y'}(C') = M_{y_1'}(D')$, since these two integers are equal respectively to $\Delta(C')$ and $\Delta(D')$. We have relations similar to (2):

$$M_x(C') - v(x) = M_{y'}(C') - v(y');$$

$$M_{x_1}(D') - v_1(x_1) = M_{y_1'}(D') - v_1(y_1'),$$

where v_1 is the natural valuation of the field $k((D))$. Now, $v(y') = m(C')$, $v_1(y_1') = m(D')$, thus $v(y') = v_1(y_1')$. Similarly $v(x) = v_1(x_1)$ $(= m(C) = m(D))$. The proposition now follows from $M_{y'}(C') = M_{y_1'}(D')$.

Proposition 5.1 can be strengthened as follows:

<u>Proposition 5.2. Let</u> (x, y) <u>be a minimal basis of the maximal ideal of the local ring of</u> C, <u>let</u> (x_1, y_1) <u>be a minimal basis of the maximal ideal of the local ring of</u> D. <u>Consider the associated embeddings of</u> C <u>and</u> D <u>in the</u> (X, Y)-<u>plane and denote by</u> ℓ <u>the line</u> $X = 0$. <u>Assume that</u> $C \equiv D$ <u>and</u>

that there exists an equivalence pairing $\pi : (C) \to (D)$, such that $(\ell, \gamma) = (\ell, \delta)$ for any pair of corresponding branches γ and $\delta = \pi(\gamma)$. Then $M_x(C) = M_{x_1}(D)$.

Proof. By (3) and by Lemma 4.2 , it is sufficient to consider the case in which C and D are irreducible. If ℓ is not tangent to C, then our assertion coincides with Proposition 5.1. Assume that ℓ is tangent to C, and hence also to D (since $m(C) = m(D)$) . By using relations similar to (2) and the notations of the proof of Proposition 5.1, we find that

$$M_x(C) = \Delta(C) + (\ell, C) - m(C),$$

$$M_{x_1}(D) = \Delta(D) + (\ell, D) - m(D),$$

and the proposition is proved.

6. Equisingular Analytic Families of Algebroid Curves: A Discriminant Criterion

We consider an analytic family

(1) $$\mathfrak{F}: f(X, Y; t) = 0$$

of plane algebroid curves, depending on a parameter t and having the same origin $X = Y = 0$ (whence $f(0, 0; t)$ is identically zero; without any change in what follows, except for more cumbersome notations, we could have assumed that the family $\mathfrak{F}$ depends on a finite number of parameters $t_1, t_2, \ldots, t_\rho$). Here f is a formal power series in X, Y, t, with coefficients in our algebraically closed ground field k (of characteristic zero). We make the following assumptions: f has no multiple factors in $k[[X, Y; t]]$, and if we set

$$f_o(X, Y) = f(X, Y; 0), \tag{2}$$

then f_o is not identically zero and has no multiple factors in $k[[X, Y]]$. We denote by K the algebraic closure of the field $k((t))$, and we regard (1) as an equation of an algebroid curve C_t, defined over K, and with origin $X = Y = 0$; C_t is then the general member of the family $\mathfrak{F}$. We denote by C_o the special member of $\mathfrak{F}$, defined by

$$C_o : f_o(X, Y) = 0, \tag{3}$$

where f_o is defined in (2). Thus C_o is an algebroid curve defined over k; but we shall regard it also as defined over K. We say that $\mathfrak{F}$ is <u>an equisingular family, if the two algebroid curves C_t and C_o</u> (both having their origin at $X = Y = 0$ and both being defined over K) <u>are equivalent</u> (i.e., have equivalent singularities). Our object in this section is to give a discriminant criterion for $\mathfrak{F}$ to be equisingular and to outline the proof of that criterion. This criterion is of basic importance for our theory of equisingularity in codimension 1 (see § 7).

We shall first introduce some notations and make some preliminary remarks. We denote by $\mathfrak{o}_t$ the (complete) local ring of C_t (over K), by $\mathfrak{o}$ the (complete) local ring of C_o (over k), and by $\mathfrak{D}$ the local ring $k[[X, Y; t]]/(f) = k[[x, y; t]]$, where x, y are the f-residues of X, Y (we identify t with its f-residue). Then (x, y) is a basis of the maximal ideal $\mathfrak{m}_t$ of

$\mathfrak{O}_t$. We have $\mathfrak{O} = \mathfrak{D}/\mathfrak{D}t = k[[\xi, \eta]]$, where ξ, η are the $\mathfrak{D}t$-residues of x, y. Then (ξ, η) is a basis of the maximal ideal $\mathfrak{m}$ of $\mathfrak{O}$. We assume that X does not divide $f_o(X, Y)$, or-equivalently - that ξ is a parameter of $\mathfrak{O}$. Then, a fortiori, X does not divide $f(X, Y;t)$, and x is a parameter of $\mathfrak{O}_t$. We denote by $\mathfrak{M}$ the maximal ideal (x, y, t) of $\mathfrak{D}$.

We shall not allow the minimal basis (x, y) of $\mathfrak{m}_t$ to range over the entire set of minimal bases of $\mathfrak{m}_t$. Rather, only such other bases (x', y') of $\mathfrak{m}_t$ will be allowed in which the element x', y' belong to $\mathfrak{D}$; or-equivalently - we must have $x' = Ax + By$, $y' = Cx + Dy$, where $A, B, C, D \in \mathfrak{D}$ and $AD - BC$ is a unit in $\mathfrak{D}$. If ξ', η' denote the $\mathfrak{D}t$-residues of x', y', then (ξ', η') is a basis of $\mathfrak{m}$. The further conditinn that ξ' be a parameter of $\mathfrak{O}$ (and hence that x' be a parameter, of $\mathfrak{O}_t$) remains in force.

Any basis (x, y) of $\mathfrak{m}_t$, fixed according to the above specifications, will determine an associated embedding (1) of C_t in the (X, Y)-plane, with $f(X, Y;t) \in k[[X, Y;t]]$, and an associated embedding (3) of C_o, where f_o is defined in (2) and where X does not divide f_o.

We consider the decomposition of f into its irreducible factors in $k[[X, Y;t]]$:

$$f = \prod_{i=1}^{h} f_i(X, Y;t) . \tag{4}$$

Since we have assumed that $f(0, 0;t) = 0$, we must have $f_i(0, 0;t) = 0$ for at least one of the factors f_i. Let s_i denote the degree of the initial form of $f_i(X, Y;t)$, where we now regard f_i as a power series in X, Y (over K). A priori, we may have

$s_i = 0$ for <u>some</u> i. If, say, $s_1 = 0$, then $f_1(X, Y;t)$ is a unit in $K[[X, Y]]$ and the factor f_1 could be deleted from the equation $f = 0$ without affecting C_t. We have

$$(5) \qquad f_o = \prod_{i=1}^{h} f_i(X, Y; 0) = \prod_{i=1}^{h} f_{i,o}(X, Y).$$

The degree $s_{i,o}$ of the initial form of $f_i(X, Y;0)$ is $\geq s_i$. Since $m(c_t) = \sum s_i$, $m(C_o) = \sum s_{i,o}$, it follows that <u>if</u> $C_t \equiv C_o$ <u>then necessarily</u> $s_i = s_{io}$ <u>for</u> $i = 1, 2, \ldots, h$. Now, each $f_i(X, Y;t)$ is irreducible, hence a non - unit in $k[[X, Y]];t$. Therefore $f_i(0, 0; 0) = 0$, showing that $s_{io} > 0$. Hence, <u>a necessary condition for the family</u> $\mathfrak{F}$ <u>to be equisingular is that</u> s_i <u>be positive for all</u> $i = 1, 2, \ldots, h$, i.e., <u>that</u> $f_i(0, 0;t)$ <u>be identically zero for each i.</u> However, we shall not impose this condition explicitly in our criterion; it will follow automatically from that criterion.

A final preliminary remark is the following :

If ℓ denotes the line $X = 0$ then $(\ell, C_t) \leq (\ell, C_o)$. If, then, $m(C_t) = m(C_o)$ (in particular, if $C_t \equiv C_o$) , and if we choose the embedding (1) of C_t in such a way that ℓ is <u>not</u> tangent to C_o (equivalently: ξ is a transversal parameter of $\mathfrak{O}$) , then necessarily ℓ will also not be a tangent of C_t . (and hence $(\ell, C_t) = (\ell, C_o)$) . Thus, <u>if</u> $C_t \equiv C_o$, <u>there always exist embeddings</u> (1) of C_t. <u>such that the intersection multiplicities</u> (ℓ, C_t) <u>and</u> (ℓ, C_o) <u>are equal</u> (where ℓ is the line $X = 0$); in particular, this will be so if ℓ is not tangent to C_o (always under the assumption that $C_t \equiv C_o$) .

Since f is regular in Y, we may assume that f is a monic polynomial in Y (with coefficients in $k[[X;t]]$), such that $f(0, Y; 0)$ is a power of Y, say $f(0, Y;0) = Y^n$.

(these conditions determine f uniquely). We denote by $D(X;t)$ the Y-discriminant of f; this will be a power series in X and t. We now state our criterion :

Theorem 6.1 The following two conditions are equivalent:

(A) C_t C_o and $(\ell, C_t) = (\ell, C_o)$ (where ℓ is the line $X = 0$).

(B) $D(X,t)$ is of the form $X^M \varepsilon(X,t)$, where $\varepsilon(X,t)$ is a is a unit in $k[[X,t]]$ (i.e., $\varepsilon(0,0) \neq 0$).

Note that the condition $(\ell, C_t) = (\ell, C_o)$ signifies that $f(0,Y;t) = Y^n$, or -equivalently - that $f_i(0,0;t)$ is identically zero for each of the h irreducible factors in (4) ; thus this property is a consequence of the criterion (B), as we have anticipated in our preceding remarks. We outline the proof of this theorem.

a) The starting point is the following well-known result (see for instance S Abhyankar [1] , Th. 4, p.585, and O. Zariski [7], Theorem 5.p.527) :

If $f(X,Y;t)$ is irreducible in $k[[X,Y;t]]$ and is a monic polynomial in Y, say $f(0,Y;0) = Y^n$, and if condition (B) of the theorem is satisfied, then the roots $Y_1, Y_2, \dots, Y_n$ of f (in the algebraic closure of $k((X,t)))$ belong to the power series ring $k[[X^{1/n};t]]$.

Therefore, the n (conjugate) roots are given by Puiseux expansions

$$(6) \qquad Y_q = \sum_{\nu=0}^{\infty} u_\nu(t)\omega^{q\nu} X^{\nu/n}, \quad (q = 1, 2, \dots, n)$$

where the $u_\nu(t)$ are in $k[[t]]$, $u_0(0) = 0$, and ω is a primitive n^{th} root of unity. We have thus :

$$f(X,Y;t) = \prod_{q=1}^{n} (Y-Y_q),$$

and

$$f(0,Y;t) = [Y - u_o(t)]^n. \tag{7}$$

b) <u>The conclusion</u> (7), <u>i.e., the property that</u> $f(0.Y;t)$ <u>is the</u> n^{th} <u>power of</u> $Y-u_o(t)$, <u>where</u> $u_o(t) \in k[[t]]$, <u>remains true if</u> $f(X,Y;t)$ <u>is reducible</u> [the other assumptions in a) remaining in force]

For, with reference to the factorization (4) of f, it is clear that condition (B) implies that this same condition is satisfied for each irreducible factor f_i of f, i.e., we have

$$\Delta^Y f_i = X^{M_i} \varepsilon_i(X;t) \tag{B_i}$$

where $\varepsilon_i(X;t) \in k[[X;t]]$ and $\varepsilon_i(0;0) \neq 0$. Hence, by a), it follows that if $f_i(0,Y;0) = Y^{n_i}$ then the n_i roots $Y_q^{(i)}$ of f_i are given the Puiseux expansion

$$Y_q^{(i)} = \sum_{\nu=0}^{\infty} u_{\nu,i}(t)\omega_i^{q\nu} X^{\nu/n_i}, \quad (q = 1,2,\dots,n_i)$$

where ω_i is a primitive n_i^{th} proof of unity, and that $f_i(0,Y;t) = [Y - u_{o,i}(t)]^{n_i}$. <u>Our assertion in</u> b) <u>is to the effect that</u> $u_{o,1}(t) = u_{o,2}(t) = \dots = u_{o,h}(t)$ $(=u_o(t))$. For assuming the contrary: $u_{o,i}(t) \neq u_{o,j}(t)$, for some pair of indices i, j in the set $\{1,2,\dots,h\}$, it would follow that if $Y_q^{(i)}$ and $Y_\rho^{(j)}$ denote any root of f_i and f_j respectively, then the difference $Y_q^{(i)}-Y_\rho^{(j)}$ does not divide X^M in $k[[X^{1/n_1 n_2 \dots n_h}; t]]$, contrary to condition (B).

Since we have assumed initially that $f(0,0;t) = 0$, it follows that in our present case we must have $u_o(t) = 0$, i.e., we must have as a consequence of (B).

$$f_i(0, Y; t) = Y^{n_i}, \quad i = 1, 2, \dots, h. \tag{8}$$

c) If we assume (B) and if f is irreducible in $k[[X, Y; t]]$ then $f(X, Y; 0)$ is irreducible in $k[[X, Y]]$.

For, in the notations of a), we must have that the roots $Y_{1,o}, Y_{2,o}, \dots, Y_{n,o}$ of $f(X, Y; 0)$ are

$$Y_{q,o} = \sum_{\nu=o}^{\infty} u_\nu(o)\omega^{q\nu} X^{\nu/n}, \quad q = 1, 2, \dots, n, \tag{9}$$

and thus the irreducibility of $f(X, Y; 0)$ follows from our original assumption that $f(X, Y; 0)$ has no multiple factors.

d) From here on, until further notice, we assume that condition (B) is satisfied. As a consequence of (B) C_t has precisely h branches $\gamma_{t,1}\gamma_{t,2}, \dots, \gamma_{t,h}$, (and not less), where $\gamma_{t,i}$ is defined by the irreducible equation $f_i(X, Y; t) = 0$. If we now apply c) we conclude that the specialization γ_i of $\gamma_{t,i}$, i.e., the algebroid curve $f_{i,o}(X, Y) = 0$, where $f_{i,o}(X, Y) = f_i(X, Y; 0)$, is also irreducible and that therefore C_o also has exactly h branches. We have therefore a particular (1, 1) mapping of the set of branches of C_t onto the set of branches of C_o, namely the mapping π defined by $\pi(\gamma_{t,i}) = \gamma_i$ $(i = 1, 2, \dots, h)$. We call π the specialization pairing between the branches of C_t

and those of C_o

e) We have $m(\gamma_{t,i}) = m(\gamma_i)$, $i = 1, 2, \ldots, h$.

For the proof one may assume that C_t and C_o are irreducible. Let $u_m(t)$ be the first coefficient $u_\nu(t)$ in (6) which is different from zero (necessarily $m > 0$ since $u_o(t) = 0$). Then $m(C_t) = \min. \{m, n\}$. If $m \geqq n$, then it follows from (9) that $m(C_t) = m(C_o) = n$. Assume $m < n$, whence $m(C_t) = m$. Then the proof is completed by observing that we must have in (9) $u_m(0) \neq 0$. For, assuming the contrary, i.e., assuming that $u_m(t)$ is divisible by t, we would find that the difference of the roots

$$Y_1 - Y_n = u_m(t)\ \{\omega^m - 1\}\ X^{m/n} + \ldots$$

does not divide X^M in $k[[X^{1/n}; t]]$, in contradiction with (B).

f) <u>The specialization pairing</u> π <u>is tangentially stable.</u> We have by (3), § 5:

$$(10) \qquad M = \sum_{i=1}^{h} M_i + 2 \sum\sum_{i<j} (\gamma_{t,i}\cdot\gamma_{t,j})$$

and also

$$(11) \qquad M = \sum_{i=1}^{h} M_i + 2 \sum\sum_{i<j} (\gamma_i\cdot\gamma_j),$$

where the M_i are defined in b), (B_i).

Furthermore, we have $(\gamma_{t,i}\cdot\gamma_{t,j}) \leqq (\gamma_i\cdot\gamma_j)$, since π is a specialization pairing. This implies, by (10) and (11), that

$$(12) \qquad (\gamma_{t,i}, \gamma_{t,j}) = (\gamma_i, \gamma_j) ,$$

and this, in view of e), implies that π is tangentially stable.

g) We have already shown in b) [see (8)] that (B) implies the second part of condition (A), namely $(\ell, C_t) = (\ell, C_o)$ $(= n = \sum_{i=1}^{h} n_i)$, where ℓ is the line $X = 0$, We now prove that (B) implies (A) by showing that $C_t \equiv C_o$. More precisely, we shall show that the specialization pairing $\pi : (C_t) \to (C_o)$ is an equivalence. By (12), and in view of Proposition 4.5, it is sufficient to show that corresponding branches $\gamma_{t,i}, \gamma_i$ of C_t and C_o are equivalent. We may therefore assume that C_t and C_o are irreducible. We show now that it is also permissible to assume that the line $\ell : X = 0$ is not a tangent of C_o [and, hence, a fortiori, is not a tangent of C_t, since we know already, by e) that $m(C_t) = m(C_o)$]. For, assume that $X = 0$ is a tangent of C_t (and hence also of C_o). Then the line $Y = 0$ is not a tangent of C_o (since C_o is irreducible, it has only one tangent line). Let v and v_o be the natural valuations of $K((C_t))$ and $k((C_o))$ respectively and let $s = m(C_t) = m(C_o)$. Then $v(x) = v_o(\xi) = n$, while $v_o(y) = v_o(\eta) = s < n$. We have, by (2), § 5 :

$$M_y(C_t) = M_x(C_t) - n + s ,$$

$$M_\eta(C_o) = M_\xi(C_o) - n + s .$$

Since $M_x(C_t) = M_\xi(C_o) = M$, by (B), it follows that $M_y(C) = M_\eta(C_o) =$ = (say) N. This shows that $\Delta^X f$ is of the form $Y^N \varepsilon_1(Y;t)$, where $\varepsilon_1(Y;t) \in k[[Y;t]]$ and $\varepsilon_1(0;0) \neq 0$. Thus (B) is still satisfied when we interchange the roles of X and Y. [In forming the discriminant $\Delta^{X}f$ we must, of course multiply first f be a suitable unit factor in k [[X, Y;t]] so as to make f to be a monic polynomial in X (of degree s)].

We assume therefore that the line ℓ : X = 0 is not tangent to C_o (whence ξ and x are transversal parameters of $\mathfrak{o}$ and $\mathfrak{o}_t$ respectively) . Without loss of generality we may assume that the line Y = 0 is tangent to C_o (replace Y by Y + cX, where c is a suitable element of k). We set $\overline{Y} = Y/X$ and

$$f(X, X\overline{Y}; t) = X^s\overline{f}(X, \overline{Y};t) .$$

Then the quadratic transforms $\overline{C}_t = TC_t)$ and $\overline{C}_o = T(C_o)$ of C_t and C_o are given respectively by the equations

$$\overline{C}_t : \overline{f}(X, \overline{Y};t) = 0 ,$$

$$\overline{C}_o : \overline{f}_o(X, \overline{Y}) = 0, \quad \text{where } \overline{f}_o(X, \overline{Y}) = \overline{f}(X, \overline{Y};0) .$$

The origin of $\overline{C}_o$ is $X = \overline{Y} = 0$. The $\overline{Y}$ - discriminant $\Delta^{\overline{Y}}\overline{f}$ is given by $^{Y}f/X^{s(s-1)}$, and hence

(13) $$\Delta^{\overline{Y}}\overline{f} = x^{M-s(s-1)}\varepsilon(X, t) , \quad \varepsilon(0;0) \neq 0 .$$

Thus (B) is satisfied for the algebroid curve $\overline{C}_t$. We have $\overline{f}(0, 0;0) = \overline{f}_o(0, 0) = 0$, and hence, by b) , we may therefore assert that the polynomial $\overline{f}_o(X, \overline{Y};t)$ (in $\overline{Y}$) has a root of the form $\overline{Y} = \overline{u}_o(t)$, where $\overline{u}_o(t) \in k\,[[t]]$ and $\overline{u}_o(0) = 0$.

We replace $\overline{Y}$ by $\overline{Y} - \overline{u}_o(t)$. This has no effect on $\overline{C}_o$

(since $\overline{u}_o(0) = 0)$, but for the new $\overline{C}_t$ (or for the new embedding of $\overline{C}_t$) we will have $\overline{f}(0,0;t) = 0$. By induction on $\Delta^Y f$, we may therefore assume that $\overline{C}_t \equiv \overline{C}_o$. Hence $C_t \equiv C_o$, as asserted.

h) We now complete the proof of the theorem by showing that (A) implies (B). Since $C_t \equiv C_o$, the two curves C_t and C_o must have the same number of branches. This implies that the number h of irreducible factors of f(X, Y;t) [see (4)] must be the number of branches of C_t [since $f_i(0,0,;0) = 0$ for $i = 1, 2, \ldots, h$, and since f(X, Y;0) has, by assumption, no multiple factors, the number of branches of C_o is at least h) and that $f_{i,o}(X,Y)$ ($=f_i(X,Y;0)$) is irreducible in k[[X, Y,]]. This yields a <u>specialization pairing</u> π between the h branches $\gamma_{t,i} : f_i(X,Y;t) = 0$ of C_t and the h branches $\gamma_i : f_{i,o}(X,Y) = 0$ of C_o. We have, of course, $(\ell, \gamma_{t,i}) \leqq (\ell, \gamma_i)$, for $i = 1, 2, \ldots, h$. However, since $(\ell, C_t) = \sum_{i=1}^{h} (\ell, \gamma_{t,i})$ and $(\ell, C_o) = \sum_{i=1}^{h} (\ell, \gamma_i)$, it follows from our assumption $(\ell, C_t) = (\ell, C_o)$ [included in (A)] that $(\ell, \gamma_{t,i}) = (\ell, \gamma_i)$. If, then, we can prove that the <u>specialization pairing π is an equivalence</u>, then (B) will follow from Proposition 5.2. Now, we prove that <u>the assumption</u> $C_t \equiv C_o$ <u>implies that π is an equivalence.</u> In this proof we may assume that the line X = 0 is not tangent to C_o (and therefore not tangent to C_t). Then $M_x(C_t) = \Delta(C_t)$ and $M_\xi(C_o) = M(C_o)$. Since $C_t \equiv C_o$, it follows from Proposition 5.1 that $M_x(C_t) = M_\xi(C_o) =$ (say) M. This implies (B), and (B) in its turn has been shown already to imply [see g) above] that π is an equivalence.

7. Equisingularity in Codimension 1.

As a matter of interpretation, we can look at the equation $f(X,Y;t) = 0$ in two ways: (1) as in § 6, as an equation defining an analytic family $\mathfrak{F}$ of plane algebroid curves in the (X,Y) plane, with the common origin $X = Y = 0$, and depending on a parameter t; (2) or as the equation of an algebroid surface F in the (X,Y,t)-three space, with origin at $X = Y = t = 0$ and containing the line $W : X = Y = 0$. Then the curves of the family $\mathfrak{F}$ can be looked upon as sections $t = \text{const.}$ of F by planes normal to the line W. To our way of thinking, the property of the family $\mathfrak{F}$ of being equisingular corresponds, in the second way of looking at the equation $f = 0$, to the property of the surface F of being equisingular at the origin $X = Y = t = 0$ along the line W, which is our way of expressing an intuitive statement that the singularity which F has at the special point $X = Y = t = 0$ of the line W is "not worse" than (or is "of the same type" as) the singularity of F at the generic point $(0,0,t)$ of W. We shall now develop this idea through precise definitions and we will add to Theorem 6.1 a number of other criteria of equisingularity, so as to give an adequate picture of our theory of equisingularity in codimension 1. As was pointed out in § 6, the assumption that we have only one parameter t was of no importance whatsoever. We shall, in fact, assume that we have $r - 1$ parameters $t_1, t_2, \ldots, t_{r-1}$ or - equivalently - that we are dealing with an algebroid hypersurface $f(X, Y; t_1, t_2, \ldots, t_{r-1}) = 0$ in an affine $(r+1)$-space.

Let

(1) $$V_r : f(X_1, X_2, \ldots, X_r, X_{r+1}) = 0$$

be an algebroid hypersurface in an affine (r+1)-space A_{r+1}, with the origin Q at $X_1 = X_2 = \ldots = X_{r+1} = 0$. Here f is a formal power series in the X's, with coefficients in our ground field k. We assume that f has no multiple factors in $k[[X_1, X_2, \ldots, X_{r+1}]]$, that $f(0, 0, \ldots, 0) = 0$ and and that the origin Q is a singular point of V. We consider an irreducible analytic subvariety W of V, or codimension 1 on V (hence of dimension r-1), passing through Q and having at Q a simple point. We may then choose the local (analytic) coordinates X_i in the neighborhood of Q in $\mathbb{A}_{r+1}$ in such a way that W is defined by

(2) $$W : X_r = X_{r+1} = 0 .$$

We denote by R the general point $(t_1, t_2, \ldots, t_{r-1}, 0, 0)$ of W, where the t's are analytically independent parameters.

We consider sections C_o of V with algebroid surfaces L_2 passing through Q, having at Q a simple point, and transversal to W at Q. Any such L_2 can be defined by equations of the form

(3) $$L_2 : X_i = A_{i1}(X)X_r + A_{i2}(X) X_{r+1}, \quad i = 1, 2, \ldots, r-1 ,$$

where the A_{ij} are formal power series in $X_1, X_2, \ldots, X_{r+1}$, with coefficients in 1. It is clear that when L_2 is given then we can choose the local coordinates X_i in $\mathbb{A}_{r+1}$ in such a way that L_2 is given by the equations

(4) $$L_2 : X_i = 0 \quad , \qquad i = 1, 2, \ldots, r-1 \ .$$

We require, furthermore, that $L_2 \not\subset V$. Thus, if L_2 is given by the equations (4), then we require that $f(0, 0, \ldots, 0\ X_r, X_{r+1})$ not be identically zero. We call then the section C_o of V with L_2 a W-transversal section of V, at Q. A section C_o may have multiple components [$f(0, 0, \ldots 0, X_r, X_{r+1})$ may have multiple factors], and is thus an algebroid cycle of dimension 1 (a plane algebroid curve, if C_o has no multiple components).

On the contrary, up to an analytical isomorphism there is only one W-transversal section C_t of V, at the general point R of W (see Zariski [8], p. 980); if we set $\mathfrak{o}$ = local ring of V:

$$\mathfrak{o} = k\,[[x_1, x_2, \ldots, x_{r+1}]] = k\,[[X_1, X_2, \ldots, X_{r+1}]]\,(f) \ ,$$

and if $\mathfrak{p}$ denotes the prime ideal of W in $\mathfrak{o}$:

$$\mathfrak{p} = \mathfrak{o} x_r + \mathfrak{o} x_{r+1} \ ,$$

then C_t is a plane algebroid curve, whose local ring is equal to (up to isomorphism) the completion of the local ring $\mathfrak{o}_{\mathfrak{p}}$. A typical C_t is the curve defined by

(5) $$C_t : f(t_1, t_2, \ldots, t_{r-1}; X_r, X_{r+1}) = 0 \ .$$

Note that $k((t_1, t_2, \ldots, t_{r-1}))$ is k-isomorphic to the residue field of the local ring $\mathfrak{o}_{\mathfrak{p}}$ and that C_t is defined over this field, and hence also over the algebraic closure K of $k((t_1, t_2, \ldots, t_{r-1}))$.

Definition 7.1. V is said to be equisingular at Q, along W, if there exists a W-transversal section C_o of W at Q, such that 1) C_o is a curve (i.e., has no multiple components) and such that 2) $C_t \equiv C_o$ (it is implicit in this definition that dim $W = r-1$ and Q is a simple point of W).

Definition 7.2. A singular point Q of V is said to be a singularity of dimensionality type 1 (for V) if there exists an irreducible subvariety W of codimension 1, containing Q, such that V is equisingular at Q along W.

The elements $x_1, x_2, \ldots, x_r, x_{r+1}$ form a minimal basis of the maximal ideal of $\mathfrak{O}$, and they determine the embedding (1) of V. The elements $x_1, x_2, \ldots, x_r$ are parameters of $\mathfrak{O}$ if and only if f is regular in X_{r+1}, and when that is we may assume that f is a monic polynomial in X_{r+1} and that

$$(6) \qquad f(0, 0, \ldots, 0, X_{r+1}) = X_{r+1}^n .$$

If s is the multiplicity of the singular point Q of V, then $n \geqq s$, with equality if and only if the parameters $x_1, x_2, \ldots x_r$ are transversal. Clearly, if $x_1, x_2, \ldots, x_r$ are parameters and if W is defined by (2), then the section C_o of V defined by (4) is transversal to W. Conversely, if the (4) define a transversal section, then there exists a linear combination $x'_r = cx_r + dx_{r+1}$, with $c, d \in k$, such that $x_1, x_2, \ldots, x_{r-1}, x'_r$ are parameters of $\mathfrak{O}$. From now on we deal only with parameters $x_1, x_2, \ldots, x_r$ of $\mathfrak{O}$ which can be extended to a basis $\{x_1, x_2, \ldots, x_r, x_{r+1}\}$ of the maximal ideal of $\mathfrak{O}$

We assume that $x_1, x_2, \ldots, x_r$ are parameters of $\mathfrak{O}$, we denote by $D(X_1, X_2, \ldots, X_r)$ the X_{r+1}-discriminant of the monic polynomial f [of degree n in X_{r+1}; see (6)], by D_o the product of the distinct irreducible factors of D, and by $\Delta_{(x)}$ the algebroid (r-1)-dimensional hypersurface in A_r defined by the equation $D_o(X_1, X_2, \ldots, X_r) = 0$. We call $\Delta_{(x)}$ the critical variety of V associated with the parameters $x_1, x_2, \ldots, x_r$. This variety depends only on $x_1, x_2, \ldots, x_r$ (and is independent of the choice of x_{r+1}).

Definition 7.3. A system of parameters $x_1, x_2, \ldots, x_r$ of $\mathfrak{O}$ is said to be equisingular at Q if the associated critical variety $\Delta_{(x)}$ has a simple point at its origin $X_1 = X_2 = \ldots = X_r = 0$.

The criterion of equisingular analytic families of algebroid curves, derived in § 6, leads (after some additional considerations) to the following criterion of equisingularity in codimension 1 (Zariski [8], Theorems 4.4, 4.5 and 5.2):

Theorem 7.4. The following conditions are equivalent:

(a) The point Q is a singularity of dimensionality type 1.

(b) There exist equisingular systems of parameters at Q

(c) Every system of transversal parameters at Q is equisingular.

Incidental results such as the following can also be proved:

1) If Q is a singular point of dimensionality type 1; then the singular locus S of V has, locally at Q, codimension

1 <u>and a simple point</u> (but not conversely) . Thus, the singular points of the singular locus S and also the singular points of V which do not lie on (r-1)-dimensional components of S are <u>not</u> of dimensionality type 1.

2) <u>If the parameters $x_1, x_2, \ldots, x_r$ form an equisingular system, then the principal ideal generated by $D_o(x_1, x_2, \ldots, x_r)$ in $\mathfrak{O}$ is primary, of codimension 1, and the irreducible subvariety W defined by the associated prime ideal of $\mathfrak{O}\, D_o$ has a simple point at 0, with V equisingular at Q along W.</u> More explicitly: assuming, as we may, that $D_o(x_1, x_2, \ldots, x_r) = x_r$, we can extend $(x_1, x_2, \ldots, x_r)$ to a minimal basis $(x_1, x_2, \ldots, x_r, x_{r+1})$ of the maximal ideal of $\mathfrak{O}$, such that the ideal $\mathfrak{O} x_r + \mathfrak{O} x_{r+1}$ is the prime ideal of the primary ideal $\mathfrak{O}\, x_r$.

3) In defining W-transversal sections C_o of V, at Q, we have not only imposed conditions on the surface L_2 whose intersection with V is C_o [see equation (4) of L_2], but we have also required that $L_2 \not\subset V$. In Definition 7.1 of equisingularity we have required, furthermore, that C_o be a curve (i.e., free from multiple components) . Now, from part c) of Theorem 7.4, it follows easily that if V <u>is equisingular at Q , along W, then:</u> 3a) <u>for every surface L_2 defined as in (3), we have $L_2 \not\subset V$</u> ; 3b) <u>every W-transversal section of V is a curve (i.e., is free from multiple components);</u> 3c) <u>all W-transversal sections of V are equivalent algebroid curves.</u>

4) If V is an algebraic hypersurface then the set of singular points of V which are not of dimensionality type 1 is an algebraic subvariety of V of dimension $\leq r.2$. To see this, let, say, V be projective. We project V onto a projective r-space $\mathbb{P}_r$ using as center of projection various points 0 of $\mathbb{P}_{r+1} - V$, we call π_0 the associated regular map $V \to \mathbb{P}$, we denote by Δ_0 the critical hypersurface (in $\mathbb{P}_r$) of the projection π_0, by S_0 the singular locus of Δ_0, and then we conclude, by (c), that the set of singular points in question is the intersection $\cap\, \pi_0^{-1}\{S_0\}$, as 0 ranges over $\mathbb{P}_{r+1} - V$.

In particular, if V is an algebraic surface, then we call exceptional singularities of V the singular points of V which are not of dimensionality type 1. These points are finite in number, and they include, among others, a) the isolated singular points of V (i.e., the singular points which do not belong to singular curves of V) and b) the singular points of the singular curve of V (if the singular locus of V has dimension 1).

5) The singularities of dimensionality type 1 behave very nicely with respect to monoidal transformations $T: V' \to V$ centered at irreducible (r-1)-dimensional components W of the singular locus S of V. Namely, with any such W we can associate an equivalence class $\{C_W\}$ of plane algebroid curves, where $\{C_W\}$ is the class which contains the W-transversal section C_t of V at the general point R of

W. By Definition 7.1 and by 3), if V is equisingular along W at a point Q of W, then all the W-transversal sections of V at Q belong to the class $\{C_W\}$. Now, let d be the number of distinct tangent lines of a member C of the class $\{C_W\}$, let C' be the quadratic transform of C and let $C_1', C_2', \ldots, C_d'$ be the connected component of C', which correspond to the tangential components $C_1, C_2, \ldots, C_d$ of C. The equivalence classes $\{C_1'\}, \{C_2'\}, \ldots, \{C_d'\}$ depend only on the equivalence class $\{C_W\}$ [see § 3, Definition 1 of (a)-equivalence]. This being said, the following can be shown:

If Q is any point of W such that V is equisingular at Q along W, then the full inverse image $T^{-1}\{Q\}$ of Q on V' consists precisely of d points $Q_1', Q_2', \ldots, Q_d'$. If W' is the full inverse image $T^{-1}\{W\}$ of W on V', then one and only one irreducible component W_i' of W' passes through Q_i' ($i = 1, 2, \ldots, d$; we may have, of course $W_i' = W_j'$ for $i = j$); we have $\dim W_i' = r-1$, and V' is equisingular at Q_i', along W_i'. Furthermore, for a suitable labeling of the points Q_i', the equivalence class $\{C_{W_i}\}$ coincides with the above equivalence class $\{C_i'\}$ ($i = 1, 2, \ldots, d$).

If follows that if V is equisingular at a point Q or W, then the singularity of V at Q is completely determined by the type of singularity of the W-transversal sections of V at Q, i.e., by the equivalence class $\{C_W\}$, and that the singularity of V at Q is resolved by a sequence of consecutive monoidal transformations, with (r-1)-dimensional centers, which runs altogether parallel to the sequence of successive quadratic transformations which are required to desingularize the equivalence class $\{C_W\}$. Since we have just seen that T doesn not blow up such a point Q, the desingularization of the singular point Q of V amounts, in the present case, to the normalization of V at Q.

In particular, if V is a surface which has no exceptional singularities, then the normalization of V is non-singular and can be obtained from V by monoidal transformations centered at curves.

6) If k is the field of complex numbers, there is the following topological property of equisingularity in codimension 1, which is due to Whitney [6] :

If V is equisingular at a point A of W, and if C_o is a W-transversal section of V at Q, then V, as an embedded variety in the complex affine $\mathbb{A}_{r+1}$, is topologically equivalent (more precisely: isotopic), locally at Q, to the direct product $C_o \times W$.

8. A Saturation-Theoretic Criterion of Equivalence of Algebroid Curves.

As an introduction to our new concept of saturated local rings (to be defined later on in this section and in § 9) and to a special type of equisingularity in codimension > 1, based on that concept we shall give here a new criterion of equivalence of algebroid plane curves. For simplicity we shall illustrate our idea on the case of irreducible curves.

Let, then, C be an algebroid branch in the (X, Y)-plane. We go back to § 4 and we consider there the Puiseux expansion (4) of C. We assume that x is a transversal parameter of the local domain $\mathfrak{O}$ of C, hence that

$$(1) \qquad y = \sum_{i=m}^{\infty} a_i x^{1/n} \qquad m \geqq n > 1$$

Here n = m (C) (=multiplicity of the origin 0 of C). We consider the characteristic exponents of the branch C [§ 4, (5)] :

$$\beta_1/n < \beta_2/n < \ldots < \beta_g/n, \quad g \geqq 1 .$$

Let F be the field of fractions of $\mathfrak{O}$:

$$\mathfrak{O} = k[[x]][y],$$
$$F = k((t)), \quad t = x^{1/n} .$$

Let G· be the Galois group of F/K , where K = k((x)) . This is a cyclic group of order n. Let G_i be the subgroup of G which has order equal to the highest common divisor

$$(n, \beta_1, \beta_2, \ldots, \beta_i) ,$$

so that we have now the strictly descending chain

$$(2) \qquad G = G_o > G_1 > G_2 > \ldots > G_g = 1 .$$

The following is easily verified:

"$\tau \in G_{i-1}, \tau \notin G_i$" $\Longrightarrow$ "$y^{\tau} - y = a_{\beta_i}(\omega_{\tau} - 1)x^{\beta_i/n}$ + terms of higher degree".

Here ω_{τ} is a nth root of unity, <u>different from</u> 1. Hence if <u>v</u> is the natural valuation of $F(v(t) = 1,\ v(x) = n)$ then

$$\text{(3)} \qquad v(y^{\tau} - y) = \beta_i,\ \tau \in G_{i-1}, \tau \notin G_i \quad (i=1, 2, \ldots, g) .$$

This yields another characterization of the characteristic exponents β_i/n of the branch C :

$$\{\beta_1/n, \beta_2/n, \ldots, \beta_g/n\} = \{v(y^{\tau} - y)/n \mid \tau \in G, \tau \neq 1\} , \ n = m(C).$$

Now suppose that we have another irreducible plane algebroid branch C', given by

$$\text{(1')} \qquad y' = \sum_{m'}^{+\infty} a_i' x'^{i/n'} , \quad m' \geqq n' > 1 .$$

Let $\mathfrak{o}'$ be the complete local ring of C', and F' the field of fractions of $\mathfrak{o}'$. Let us assume that C and C' have equivalent singularities. Then, in the first place, we must have $n' = n$, since these are the multiplicities of the origin 0 of C and C'. Therefore, if we identify $F = k((t))$ with $F' = k(t'))$ in such a way that $T' = t$ $(t = x^{1/n},\ t' = x'^{1/n'})$, then we have identified x with x'. Let us make this identification. So now

$$\text{(1')} \qquad y' = \sum_{m'}^{+\infty} a_i' x^{i/n} , \quad m' \geqq n > 1 .$$

In the second place, the characteristic exponents of (1) and (1') must be the same. Therefore the chain (2) of subgroups of the Galois group G is the same for C and C', and hence by (3) we have

(4) $v(y^{\tau}-y) = v(y'^{\tau}-y')$, for all $\tau \in G$, $\tau \neq 1$.

Let us set $\bar{\mathfrak{o}} = k[[t]]$ (=integral closure of both $\mathfrak{o}$ and $\mathfrak{o}'$ in F). Then, by (4):

(5) $(y'^{\tau}-y')/(y^{\tau}-y) \in \mathfrak{o}$ and $(y^{\tau}-y)/(y'^{\tau}-y') \in \mathfrak{o}$,
for all $\tau \in G$, $\tau \neq 1$.

Let us consider the following ring contained between $\mathfrak{o}$ and $\bar{\mathfrak{o}}$:

(6) $\tilde{\mathfrak{o}}_x = \{\xi \in \bar{\mathfrak{o}} | (\xi^{\tau}-\xi)/(y^{\tau}-y) \in \bar{\mathfrak{o}}$, all $\tau \in G, \tau \neq 1$.

It is immediate that $\tilde{\mathfrak{o}}_x$ is indeed a ring, hence a complete local ring, of dimension 1 (since both $\mathfrak{o}$ and $\bar{\mathfrak{o}}$ are complete local rings, of dimension 1). The ring $\tilde{\mathfrak{o}}_x$ can be characterized by the following 3 properties :

S_1) $\mathfrak{o} \subset \tilde{\mathfrak{o}}_x \subset \bar{\mathfrak{o}}$.

S_2) If $\zeta \in \tilde{\mathfrak{o}}_x$, $\eta \in \bar{\mathfrak{o}}$ and if $(\eta^{\tau}-\eta)/(\zeta^{\tau}-\zeta)\,\bar{\mathfrak{o}}$ for all $\tau \in G$[if for some τ, $\zeta^{\tau} = \zeta$, this last assumption should read as follows: $\eta^{\tau} = \eta$], then also $\eta \in \mathfrak{o}_x$.

S_3) $\tilde{\mathfrak{o}}_x$ is the smallest ring satisfying conditions S_1) and S_2).

Similarly, we set

(7) $\tilde{\mathfrak{o}}'_x = \{\xi \in \bar{\mathfrak{o}} | (\xi^{\tau}-\xi)/(y'^{\tau}-y') \in \bar{\mathfrak{o}}$, all $\tau \in G, \tau \neq 1$.

We now give the following definition :

Definiton 8.1 Let $\mathfrak{O}$ be the local ring of an irreducible algebroid branch C (not necessarily a plane branch), let x be a parameter of $\mathfrak{O}$, let $\bar{\mathfrak{O}}$ be the integral closure of $\mathfrak{O}$ in the quotient field F of $\mathfrak{O}$ and let G be the Galois group of F/K, where $K = k((x))$. We say that $\mathfrak{O}$ is saturated with respect to x if $\mathfrak{O}$ satisfies condition S_2.

Proposition-Definition 8.2 If $\mathfrak{O}$ and x are as above, there exists a unique ring $\tilde{\mathfrak{O}}_x$ satisfying conditions S_1, S_2, S_3. This ring is called the saturation of $\mathfrak{O}$ with respect to x.

We now go back to the inclusions (5) which hold if C and C' have equivalent singularities. These inclusions imply that $y' \in \tilde{\mathfrak{O}}_x$ and $y \in \tilde{\mathfrak{O}}'_x$, i.e.,

$$\mathfrak{O}' \subset \tilde{\mathfrak{O}}_x \quad \text{and} \quad \mathfrak{O} \subset \tilde{\mathfrak{O}}'_x ,$$

whence

$$\tilde{\mathfrak{O}}_x = \tilde{\mathfrak{O}}'_x . \tag{8}$$

We can express this result, as follows (we drop the identification $F = F'$):

Theorem 8.3 If $C \equiv C'$, then given any transversal parameter x' of $\mathfrak{O}'$, there exists an isomorphism $\varphi : \tilde{\mathfrak{O}}_x \xrightarrow{\sim} \tilde{\mathfrak{O}}'_{x'}$ such that $\varphi(x) = x'$, and conversely.

The truth of the converse can be seen as follows: If φ exists, then we must have in the first place $n=n'$, for in the respective natural valuations v, v' of F and F' we have $v(x) = n$, $v'(x')=n'$. Next, we can identify $\tilde{\mathfrak{O}}_x$ with $\tilde{\mathfrak{O}}'_{x'}$ and x with x'.

So now $\mathfrak{O}$ and $\mathfrak{O}'$ have the same field of fraction $F(=k((x^{1/n})))$, and, of course, the same integral closure $\bar{\mathfrak{O}}$ $(=k[[x^{1/n}]])$. Furthermore, we have $y' \varepsilon \mathfrak{O}' \subset \tilde{\mathfrak{O}}'_x = \tilde{\mathfrak{O}}_x$ and similarly $y \varepsilon \tilde{\mathfrak{O}}'_x$. Therefore, we have the inclusions (5), or-what is the same - the equalities (4), and these imply the equality of the characteristic exponents.

In the special case $C = C'$, Theorem 8.3 tells us that if x and x' are any two transversal parameters of $\mathfrak{O}$, then there exists an isomorphism $\varphi : \tilde{\mathfrak{O}}_x \xrightarrow{\sim} \tilde{\mathfrak{O}}_{x'}$ such that $\varphi(x) = x'$. The following is a far from trivial result and is not covered by what is known from the classical theory :

Theorem 8.4. <u>If</u> x, x' <u>are any two transversal parameters of</u> $\mathfrak{O}$, <u>then</u> $\tilde{\mathfrak{O}}_x = \tilde{\mathfrak{O}}_{x'}$.

The local ring $\tilde{\mathfrak{O}}_{x'}$, x-transversal, which is thus independent of x, shall be denoted by $\tilde{\mathfrak{O}}$ and c-lled <u>the saturation of</u> $\mathfrak{O}$.

The saturation $\tilde{\mathfrak{O}}$ of $\mathfrak{O}$ is the local ring of some irreducible algebroid curve $\tilde{C}$ lying in an affine space $\mathbb{A}_N$ of some high dimension N (N can be computed explicitly in terms of the characteristic exponents). By Theorem 8.3 we have $C \equiv C'$ if and only if $\tilde{C} = \tilde{C}'$. Since $\mathfrak{O} \subset \tilde{\mathfrak{O}}$, C is a projection of $\tilde{C}$. Thus, <u>all irreducible plane algebroid branches in the same equivalence class are projections of one and the same algebroid branch</u> $\tilde{C}$ <u>in some</u> S_N, and - what is important - it can be shown that the <u>generic plane projection of</u> $\tilde{C}$ <u>belongs to that same equivalence class.</u>

We add without proof certain propositions on saturation.

If $\mathfrak{O}$ is the local ring of any irreducible (not necessarily plane) algebroid curve, we say that $\mathfrak{O}$ is <u>saturated</u> if $\mathfrak{O}$ is saturated with respect to some parameter x of $\mathfrak{O}$. Then we have the following propositions :

Proposition 8.5. If $\mathfrak{O}$ is saturated, it is also saturated with respect to any transversal parameter.

Proposition 8.6 If $\mathfrak{O}$ is saturated and if x, x' are any two transversal parameters of $\mathfrak{O}$, then there exists an automorphism of $\mathfrak{O}$ which sends x into x'.

Thus, the automorphisms of a saturated local ring $\mathfrak{O}$ are as abundantly many as they could possibly be.

We also observe that Theorem 8.4 continues to hold true for local rings $\mathfrak{O}$ of algebroid branches which are not plane curves, and thus we can speak of the saturation $\tilde{\mathfrak{O}}$ of any such ring $\mathfrak{O}$, meaning by that the saturation of $\mathfrak{O}$ with respect to any transversal parameter. It can be shown that $\tilde{\mathfrak{O}}$ is the smallest saturated ring between $\mathfrak{O}$ and $\bar{\mathfrak{O}}$. The saturation of $\mathfrak{O}$ with respect to a parameter which is not transversal' may very well depend on the choice of the parameter. It can be proved that various saturations of $\mathfrak{O}$ have one thing in common: they have the same multiplicity as $\mathfrak{O}$. We know that the family of these rings has a lower bound: it is the saturation of $\mathfrak{O}$ (i.e.), the saturation with respect to a transversal parameter of $\mathfrak{O}$). Now, if $n = e(\mathfrak{O})$, the largest local ring between $\mathfrak{O}$ and $\bar{\mathfrak{O}}$ which has multiplicity n is the ring $k + \bar{\mathfrak{m}}^n$, where $\bar{\mathfrak{m}}$ is the maximal ideal of $\bar{\mathfrak{O}}$. It can be shown that this is a saturation of $\mathfrak{O}$ with respect of some parameter x; it is sufficient to take x in $\mathfrak{m}$ so that $v(x)$ and n are relatively prime.

O. Zariski

9. Saturation of Local Rings of Higher Dimension. Equisaturated Families of Hypersurfaces.

In the general theory of saturation we deal with an algebroid variety V of a given dimension r, embedded in some affine space. While in the case of dimension 1 we have restricted ourselves, for simplicity, to irreducible curves, in the general case I shall not assume that V is irreducible. However, I will assume that V is equidimensional. The ground field k is still assumed to be algebraically closed and of characteristic zero. The local ring $\mathfrak{o}$ of V may now have zero-divisors, but will have no nilpotent elements (we are dealing with algebroid varieties, or, if you wish, with reduced local formal schemes) We denote by F the total ring of quotients of $\mathfrak{o}$. Then F is a direct sum of fields :

(1) $$F = F_1 \oplus F_2 \oplus \ldots \oplus F_h ,$$

where h is the number of irreducible components of V. We have that k is a subfield of F. We write

(2) $$1 = \varepsilon_1 + \varepsilon_2 + \ldots + \varepsilon_h, \quad \varepsilon_i = \text{the identity of } F_i .$$

We fix any set of parameters $x_1, x_2, \ldots, x_r$ of $\mathfrak{o}$. It is well known that $\mathfrak{o}$ contains the ring of formal power series $R = k[[x_1, x_2, \ldots, x_r]]$. Our hypothesis that V is equidimensional implies that no element of R, different from zero, is a zero divisor in $\mathfrak{o}$. Hence F contains the field of fractions $K = k((x_1, x_2, \ldots, x_r))$ of R, and is, of course, a finite algebra over K. It is also known that $\mathfrak{o}$ is a finite R-module, hence is integral over R. We denote by $\bar{\mathfrak{o}}$ the integral closure of $\mathfrak{o}$ in F; this need not be a local ring, but is a semi-local ring, having exactly h maximal ideals. This, of course, has to do with the fact that normalization of V separates the irreducible branches of V .

I wish now to define the saturation $\tilde{\mathfrak{o}}_{(x)}$ of $\mathfrak{o}$ with respect to the parameters $x_1, x_2, \ldots, x_r$. In point of fact, the ring $\tilde{\mathfrak{o}}_{(x)}$ which will be thus defined, will depend not so much on the set of parameters x_i as on the field $K(=k(x_1, x_2, \ldots, x_r)))$ of formal meromorphic functions of these parameters. Therefore, it may be just as well to use the notation $\tilde{\mathfrak{o}}_K$. Now, in order to give this definition, it will be simpler to drop the assumption that $\mathfrak{o}$ is a local ring and to define $\tilde{\mathfrak{o}}_K$ for general commutative rings $\mathfrak{o}$ with identity, under the following conditions :

1. $\mathfrak{o}$ has no nilpotent elements :
2. The total ring of quotients F of $\mathfrak{o}$ is a finite direct sum of fields.
3. F contains a field K, and K contains the identity 1 of F.
4. Each field F_i (see (1)) is a finite separable extension of $K\epsilon_i$ (see (2)) .
5. If $R = \mathfrak{o} \cap K$, then $\mathfrak{o}$ is integral over R .

All these conditions are automatically satisfied in the case in which $\mathfrak{o}$ is the local ring of the equidimensional algebroid variety V (in characteristic zero) and K is the field $k((x_1, x_2, \ldots, x_r))$.

We fix an algebraic closure Ω of K and we consider the set M of K-homomorphisms of F into Ω . The set M is finite, since F is a finite algebra over K. If $\psi \in M$, then

$\psi(\varepsilon_i) = 1$ for some i while $\psi(\varepsilon_j) = 0$ for $j \neq i$, and $\psi(F) = \psi(F_i)$ = isomorphic image of the field F_i. The composition of the various fields $\psi(F)$, $\psi \varepsilon M$, is a Galois extension F^* of K. In the special case in which $\mathfrak{o}$ is an integral domain we could have taken for Ω an algebraic closure of the field F, and F^* would then be the least Galois extension of K containing F.

Definition 9.1. If $\zeta, \eta \varepsilon F$, we say that η dominates ζ (with respect to K), in symbols: $\eta > \zeta$, if for any two homomorphisms ψ_1, ψ_2 in the set M we have that

$$[\psi_1(\eta) - \psi_2(\eta)] \;/\; [\psi_1(\zeta) - \psi_2(\zeta)] \text{ is integral over } R \text{ if } \psi_1(\zeta) \neq \psi_2(\zeta)$$

and

$$\psi_1(\eta) = \psi_2(\eta) \text{ if } \psi_1(\zeta) = \psi_2(\zeta).$$

Definition 9.2. We say that $\mathfrak{o}$ is saturated with respect to the field K if $\mathfrak{o}$ contains with any element ζ also every element η of $\bar{\mathfrak{o}}$ which dominates ζ.

It is immediate that any ring between $\mathfrak{o}$ and $\bar{\mathfrak{o}}$ satisfies conditions 1-5 and that the intersection of all the rings between $\mathfrak{o}$ and $\bar{\mathfrak{o}}$ which are saturated with respect to K (there exist such rings; for instance $\bar{\mathfrak{o}}$) is itself saturated with respect to K. We have therefore the following

Proposition-Definition 9.3. There exists a smallest ring between $\mathfrak{o}$ and $\bar{\mathfrak{o}}$ which is saturated with respect to K. This ring is called the saturation of $\mathfrak{o}$ with respect to K and is denoted by $\tilde{\mathfrak{o}}_K$.

A most important property of saturation is given by the following.

Proposition 9.4 . If R(= $\mathfrak{o} \cap K$) *is integrally closed, then over every prime ideal* $\mathfrak{p}$ *of* $\mathfrak{o}$ *there lies one and only prime ideal* $\tilde{\mathfrak{p}}$ *of* $\tilde{\mathfrak{o}}_K$, *and the field of fractions of* $\tilde{\mathfrak{o}}_K/\tilde{\mathfrak{p}}$ *is a purely inseparable extension of the field of fraction of* $\mathfrak{o}/\mathfrak{p}$.

In the Bourbaki terminology one would say that the natural morphism Spec ($\tilde{\mathfrak{o}}_K$) $\longrightarrow$ Spec($\mathfrak{o}$) is *radicial.*

Now, let us go back to the case in which $\mathfrak{o}$ is the local ring of an algebroid variety V . Since $\bar{\mathfrak{o}}$ is a semilocal ring and a finite module over $\mathfrak{o}$, every ring between $\mathfrak{o}$ and $\bar{\mathfrak{o}}$ is semi-local . However, Proposition 9.4, applied just to the maximal ideal $\mathfrak{m}$ of $\mathfrak{o}$ tells us that $\tilde{\mathfrak{o}}_K$ has only one maximal ideal, *and is therefore again a local ring.* It is therefore the local ring of an algebroid variety $\tilde{V}_K$ which dominates V and is dominated by the normalization $\bar{V}$ of V :

$$V \xleftarrow{f} \tilde{V}_K \xleftarrow{g} \bar{V},$$ f *and* g *being morphisms.*

We call the variety $\tilde{V}_K$ the *saturation* on V with respect to K. Saturation, in general, falls short of normalization ; in particular , it *does not separate the branches of V.* Furthermore, it can also be shown that the singular origins of V and $\tilde{V}_K$ *have the same multiplicity*.

Suppose now that we have another equidimensional algebroid variety V' , of dimension r. Let $\mathfrak{o}'$; $\{x'_1, x'_2, \ldots, x'_r\}$; K' have the same meaning for V' as $\mathfrak{o}$; $\{x_1, x_2, \ldots, x_r\}$; K have for V. *Let us assume that there exists a k-isomorphism* $\varphi : \tilde{\mathfrak{o}}_K \xrightarrow{\sim} \tilde{\mathfrak{o}}'_{K'}$, such that $\varphi(x_i) = x'_i$, i = 1, 2, . . . , r. By analogy with the case that V and V' have then equivalent singularities at their respective origins. However, such a conclusion would be meaningless since we do not have a definition of equivalent singularities in higher dimension. But we can come very near such a conclusion if we consider the case of complex-analytic varieties, in which case we can at least inquire after topological equivalence. If V is complex-analytic, the formal morphism $f : \tilde{V}_K \to V$ defines a local holomorphic map of $\tilde{V}_K$ onto V, a map which we continue to denote by f. This map is finite, i.e., for any point Q of V, near 0, the set $f^{-1} Q$ is finite. But Proposition 9.4 tells us that given any irreducible analytic subvariety W of V passing through the origin, the full inverse image $f^{-1}\{W\}$ is an irreducible subvariety W' or $\tilde{V}_K$ passing through the origin $\tilde{0}$, and that, furthermore, W and W' are bimeromorphically equivalent. In such a situation it is easy to show that f is a local homeomorphism.

Thus V and its saturation $\tilde{V}_K$ are locally homeomorphic varieties.

It follows that if there exists a k-isomorphism

$$\varphi : \tilde{\mathfrak{o}}_K \xrightarrow{\sim} \tilde{\mathfrak{o}}'_{K'} \tag{3}$$

such that

$$\varphi(x_i) = x'_i, \quad i = 1, 2, \ldots, r, \tag{4}$$

i.e., if $\tilde{V}_K$ and $\tilde{V}'_{K'}$ are analytically isomorphic complex-analytic varieties (with the additional condition (4)), then V and V' are locally homeomorphic varieties.

In the case in which V and V' are hypersurfaces it is possible to obtain a stronger result, under a mild additional condition. Assume that the set of parameter $x_1, x_2, \ldots, x_r$ can be extended to a minimal base $x_1, x_2, \ldots, x_r, x_{r+1}$ of the maximal ideal $\mathfrak{m}$ of $\mathfrak{O}$; similarly for $x'_1, x'_2, \ldots, x'_r$ and $\mathfrak{O}'$. Then, if there exists k-isomorphism (3) , satisfying (4), the resulting local homeomorphism between V and V' can be extended to a local homeomorphism between the ambient affine space $\mathbb{A}_{r+1}$ and $\mathbb{A}'_{r+1}$ of V and V'. Thus, in this case, the hypersurfaces V and V' are topologically equivalent as embedded varieties. In the case r=1 this gives, in view of theorem 8.3, the classical result that equivalent plane algebroid curves are topologically equivalent (as embedded curves; or as knotted real surfaces in real 4-space).

Using the general concept of saturation and the results stated above, we can obtain a significant partial generalization of our theory of equisingularity in codimension 1 to higher codimension. We shall now indicate this generalization. We begin by analogy with § 6, by considering an analytic family

$$(5) \qquad \mathfrak{F}: f(X_1, X_2, \ldots, X_{s+1}; t) = 0$$

of algebroid hypersurfaces in $\mathbb{A}_{s+1}$ depending on a parameter t and having the same origin $X_1 = X_2 = \ldots = X_{s+1} = 0$ (whence $f(0, 0, \ldots, 0; t)$ is identically zero; without any significant change in what follows, except for more cumbersome notations, we could have assumed that the family $\mathfrak{F}$ depends on a finite number of parameters $t_1, t_2, \ldots, t_\rho$). Here f is a formal power series in $X_1, X_2, \ldots, X_{s+1}$, t, with coefficients in our algebraically closed ground field k of characteristic zero. We make the following assumptions; f is regular if X_{x+1}, has no multiple factors in $k[[X_1, X_2, \ldots, X_{s+1}; t]]$, and if we set

(6) $$f_o(X_1, X_2, \dots, X_{s+1}) = f(X_1, X_2, \dots, X_{s+1}; 0) ,$$

then f_o is not identically zero and is free from multiple factors in $k[[X_1, X_2, \dots, X_{s+1}]]$. We denote by K the algebraic closure of the field $k((t))$, and we regard equation (5) as defining an algebroid hypersurface V_t, defined over K, and with origin at $X_1 = X_2 = \dots = X_{s+1} = 0$; V_t is thus the general member of the family $\mathfrak{F}$. We denote by V_o the special member of $\mathfrak{F}$ corresponding to $t = 0$:

(7) $$V_o : f_o(X_1, X_2, \dots, X_{s+1}) = 0 ,$$

where f_o is defined in (6). Thus V_o is an algebroid hypersurface, with origin at $X_1 = X_2 = \dots = X_{s+1} = 0$ and defined over k ; but we shall regard V_o as defined also over K.

We introduce the following notations :

$\mathfrak{o}_t$ = the (complete) local ring of V_t (over K); $\mathfrak{m}_t$ -its maximal ideal;

$\mathfrak{o}$ = the (complete) local ring of V_o (<u>over</u> K); $\mathfrak{m}$ -its maximal ideal;

$\mathfrak{o}_o$ = the (completed) local ring of V_o (<u>over</u> k) ; $\mathfrak{m}_o$-its maximal ideal ;

(8) $$\mathfrak{O} = k[[x_1, x_2, \dots, x_{s+1}; t]] = k[[X_1, X_2, \dots, X_{s+1}; t]] / (f) ;$$
$\mathfrak{M}$-its maximal ideal.

Then, it is clear that $\mathfrak{o}$ =completion of $K[\mathfrak{o}_o]$ with respect to the maximal ideal $K\mathfrak{m}_o$. If we denote by $\bar{x}_1, \bar{x}_2, \dots, \bar{x}_{s+1}$ the f-residues of $X_1, X_2, \dots, X_{s+1}$, in $K[[X_1, X_2, \dots, X_{s+1}]]$, by $\xi_1, \xi_2, \dots, \xi_{s+1}$

the f_o residues of $X_1, X_2, \ldots, X_{s+1}$ in $k[[X_1, X_2, \ldots, X_{s+1}]]$ and by $x_1, x_2, \ldots, x_{s+1}$ the f-residues of $X_1, X_2, \ldots, X_{s+1}$ in $k[[X_1, X_2, \ldots$ $\ldots, X_{s+1}; t]]$, then

$$\mathfrak{m}_t = \mathfrak{o}_t \cdot (\bar{x}_1, \bar{x}_2, \ldots, \bar{x}_{s+1}), \mathfrak{m} = \mathfrak{o} \cdot (\xi_1, \xi_2, \ldots, \xi_{s+1}),$$
$$\mathfrak{m}_o = \mathfrak{o}_o \cdot (\xi_1, \xi_2, \ldots, \xi_{s+1}) \text{ and } \mathfrak{M} = \mathfrak{D} \cdot (x_1, x_2, \ldots, x_{s+1}, t) .$$

Furthermore, if $\mathfrak{p} = \mathfrak{D} \cdot (x_1, x_2, \ldots, x_{s+1})$ is the prime ideal defined in $\mathfrak{D}$, then $\mathfrak{o}_t$ = completion of $\mathfrak{D}_{\mathfrak{p}}$ and $\bar{x}_i = h(x_i)$, where h is the canonical homomorphism of $\mathfrak{D}$ into $\mathfrak{o}_{\mathfrak{p}}$.
Furthermore, $\{\bar{x}_1, \bar{x}_2, \ldots, \bar{x}_s\}$ and $\{\xi_1, \xi_2, \ldots, \xi_s\}$ are systems of parameters in their respective local rings $\mathfrak{o}_t$ and $\mathfrak{o}$, while $\{x_1, x_2, \ldots, x_s, t\}$ is a system of parameters of $\mathfrak{O}$ (in view of the regularity of f in X_{r+1}).

Remark. The canonical homomorphisms h is an injection if and only if we have $f_i(0, 0, \ldots, 0; t) = 0$ (identically) for each irreducible factor f_i of f in $k[[X_1, X_2, \ldots, X_{s+1}; t]]$.

Definition 9.5. The family $\mathfrak{F}$ is said to be equisaturated if there exists a K-isomorphism of saturated rings

$$(9) \qquad \varphi : (\tilde{\mathfrak{o}}_t)_{K((\bar{x}_1, \bar{x}_2, \ldots, \bar{x}_s))} \xrightarrow{\sim} (\tilde{\mathfrak{o}})_{K((\xi_1, \xi_2, \ldots, \xi_s))} ,$$

such that $\varphi(\bar{x}_i) = \xi_i, \quad i = 1, 2, \ldots, s$.

If such an isomorphism exists, then it can be shown that the above canonical homomorphism h is an injection, and therefore , in that case, the $\bar{x}_i$ can be identified with the x_i .

A more precise way of phrasing the above definition would have been to say that $\mathfrak{F}$ is equisaturated with respect to the system of parameters $x_1, x_2, \ldots, x_s$ of $\mathfrak{o}_t$. For if $\{x'_1, x'_2 \ldots, x'_s, x'_{s+1}, t\}$

is another basis of $\mathfrak{M}$ (the parameter t being unchanged), and $\bar{x}_i'$, ξ_i' having the corresponding meaning for the rings $\mathfrak{o}_t$ and $\mathfrak{o}_o (= \mathfrak{V}/\mathfrak{V}t)$, then the existence of an isomorphism (9) does not imply the existence of a similar isomorphism φ', with $K((\bar{x}))$ and $K((\xi))$ being replaced by $K((x'))$ and $K((\xi'))$.

The following is a generalization of Theorem 6.1 (the equisingularity criterion) and is stated here without proof.

<u>Theorem 9.6.</u> <u>$\mathfrak{F}$ is an equisaturated system if and only if the X_{s+1}-discriminant</u> $D(X_1, X_2, \ldots, X_s;t)$ <u>of f is of the form</u> $D_o(X_1, X_2, \ldots, X_s)\ (X_1, X_2, \ldots, X_s;t)$, <u>where both factors are power series (with coefficients in</u> k) <u>and where</u> $\epsilon(X_1, X_2, \ldots, X_s;t)$ <u>is a unit</u> $k[[X_1, X_2, \ldots, X_s; t]]$.

We now consider, as in § 7, an algebroid hypersurface

$$(10) \qquad V : f(X_1, X_2, \ldots, X_r, X_{r+1}, X_{r+2}, \ldots, X_{r+s}) = 0, \qquad s \geqq 1$$

in affine $\mathbb{A}_{r+s}$, with origin $Q : X_1 = X_2 = \ldots = X_{r+s} = 0$, where we assume that f has no multiple factors in $k[[X_1, X_2, \ldots, X_{r+s}]]$ and the "linear" variety $W : X_r = X_{r+1} = \ldots = X_{r+s} = 0$, of codimension s on V (and hence of dimension r-1), is contained in V. We denote by R the general point $(t_1, t_2, \ldots, t_{r-1}, 0, 0, \ldots, 0)$ of W, where the t's are analytically independent parameters over k, and by V_t the W-transversal section of V at R, defined over $K = k((t_1, t_2, \ldots, t_{r-1})))$ by

$$(11) \qquad V_t : f(t_1, t_2, \ldots, t_{r-1}; X_r, X_{r+1}, \ldots, X_{r+s}) = 0 .$$

Thus, V_t is a hypersurface in affine $\mathbb{A}_{s+1}/K$. Equation (11) also defines an analytic family of hypersurfaces in $\mathbb{A}_{s+1}$, depending on r-q parameters $t_1, t_2, \ldots, t_{r-1}$. We assume that $f(0, 0, \ldots, 0; X_r, X_{r+1}, \ldots, X_{r+s})$ is not identically zero, that it has no multiple factors and that f is regular in X_{r+s} (and hence may be assumed to be a monic polynomial in X_{r+s}).

Definition 9.7. The hypersurface V is said to be equisaturated at the origin Q, along W, if the analytic family (11) is equisaturated.

The main result that we have about equisaturation (in the complex domain) is a generalization of the topological result stated at the very end of § 7, in the special of codimension 1 :

If V is equisaturated at Q along W and if V_o denotes the W-transversal section of V defined by $f(0, 0, \ldots, 0; X_r, X_{r+1}, \ldots, X_{r+s}) = 0$, then V, as an imbedded variety in the complex affine $\mathbb{A}_{r+s}$, is topologically equivalent, locally at Q, to the direct product $V_o \times W$ (see Zariski [9], p. 1019).

Going back to equisaturated systems and using the notations introduced earlier in this section, one derives easily from (9) the following : if $\bar{\mathfrak{O}}$ and $\bar{\mathfrak{o}}_o$ denote the integral closure of $\mathfrak{O}$ and $\mathfrak{o}_o$ respectively, and if the family $\mathfrak{F}$ is equisaturated, then $\bar{\mathfrak{O}}$ can be canonically identified with the ring $\bar{\mathfrak{o}}_o[[t]]$.

Applying this result to the analytic family defined in (11) one finds that if V is equisaturated at Q along W then V and $W \times V_o$ are "bimeromorphically equivalent." However, in general, V and $W \times V_o$ will not be analytically equivalent.

In particular, Theorem 9.6 shows that in the case of codimension s = 1, equisaturation of V at Q along W implies *and is implied by equisingularity of* V *at* Q, *along W.*

10. The Whitney-Thom Conditions

Two differentio-geometric conditions (refered to below as conditions A and B), formulated by Whitney and Thom, concern the behaviour of an algebroid variety V along an irreducible subvariety W, at a simple point of W. These conditions are equivalent, in case of codimension W=1, to our equisingularity condition (§ 7), and will no doubt play an important role in any future general theory of equisingularity (in any codimension).

For simplicity of exposition we shall restrict ourselves to *irreducible algebraic* varieties V and algebraic subvarieties W of V, in order to be in position to use, without any *ad hoc* explanations, the concepts of specialization and valuation.

However, with a few unessential modifications, everything that we state in the sequel is valid also if V is not irreducible (but equidimensional, however) and is algebroid (in particular, if V is a complex-analytic variety). The ground field k is still assumed to be algebraically closed and of characteristic zero.

Let dim V = r, dim W = ρ, and let Q be a *simple* point of W. We denote by P the general point of V and by T(V, P), T(W, Q) the tangent space of V at P and of W at Q respectively. We assume that we are also given a definite embedding of V in an affine n-space $\mathbb{A}_n$ and that Q is, say, the origin.

If R is *any* point of $\mathbb{A}_n$ and q is any positive integer

$< n$, we can represent canonically the set of linear q-space L_q in $\mathbb{A}_n$, passing through R, by the Grassmann variety $G_{n-1,q-1}$ of the linear (q-1)-spaces in the hyperplane at infinity H_∞ of the $\mathbb{A}_n$, by associating with each such L_q the intersection space $L_q \cap H_\infty$. Applying this to the case $q = r$ we consider the product variety $V\ G_{n-1,r-1}$ and the point $P^* = P \times T(V, P)$ of that variety. We can speak of the specializations of P^* (over k) and, in particular, since $Q \in V$, we can speak of the specializations of P^* over the specialization $P \xrightarrow{k} Q$. The point P^* is the general point of an irreducible subvariety V_T of $V\ G_{n-1,r-1}$, birationally equivalent to V, and the birational transformation $V_T \rightarrow V$ is everywhere regular on V^*, while its inverse in biregular outside the singular locus of V.

Definition 10.1. The pair $\{V, W\}$ is said to satisfy condition A at Q if for any specialization $P \times T(V, P) \xrightarrow{k} Q \times T_o$ over $P \rightarrow Q$ we have $T_o \supset T(W, Q)$.

In the complex-analytic domain we could have phrased this definition by considering sequences $\{P_i\}$ of simple points of V which converge to Q and by requiring that for any such sequence and for any accumulation element T_o of the corresponding sequence $\{T(V, P_i)\}$ we should have $T_o \supset T(W, Q)$.

To formulate condition B we consider the Grassmann variety $G_{n,1}$ of the lines of the projective closure $\mathbb{P}_n$ of $\mathbb{A}_n$, we consider a general point R of W such that P and R are k-independent points and we introduce the product variety $V \times W \times G_{n-1,r-1} \times G_{n,1}$. On that product variety we consider the point $P \times R \times T(V, P) \times PR$, where PR is the line joining P and R, and its specializations over the specialization $P \times R \xrightarrow{k} Q \times Q$.

Definition 10.2. The pair $\{V, W\}$ is said to satisfy condition B at Q if for any specialization $P \times R \times T(V, P) \times PR \xrightarrow{k} Q \times Q \times T_o \times \ell_o$ (over $P \times R \xrightarrow{k} Q \times Q$) we have that the line ℓ_o is contained in the r-space T_o

In the complex-analytic domain we could have phrased this definition by considering pairs of sequence $\{P_i\}$, $\{R_i\}$ of simple point P_i of V and simple points R_i of W , which both converge to the point Q , and by demanding that for any two such sequences and for any accumulation element (T_o, ℓ_o) of the corresponding sequence of pairs $\{T(V, P_i), P_iR_i\}$ (assuming $P_i \neq R_i$, for all i) we should have $\ell_o \subset T_o$.

The main reason why the Whitney-Thom conditions A and B have a claim to our attention in connection with a possible general theory of equisingularity is the following property claimed by Thom in [5] (the proof there is very sketchy and we do not claim to having it fully understood) :

If the pair V, W satisfies both conditions A and B at Q then, V , as an imbedded variety in $\mathbf{A}_n$, is topologically equivalent, locally at Q , to the direct product $W \times \Gamma_{n-\rho}$, where $\Gamma_{n-\rho}$ is a W-transversal section of V at Q.

Certainly, if this property is fully proved, it would point to some sort of equisingularity of V at Q, along W , although, of course, topological triviality of along W at Q is in itself no guarantee of equisingularity.

We have a more solid foundation for a connection between the equisingularity and the Whitney-Thom conditions in the special case in which V is a hypersurface and W is of codimension 1 . We have namely the following result (Zariski [8] , Theorem 8.1) :

Theorem 10.3. If V is a hypersurface and if W has codimension 1, then V is equisingular at Q along W if and only if the following two conditions are satisfied:

1) V is equimultiple at Q , along W .
2) The pair {V, W} satisfies conditions A and B , at Q.

Note: In a paper in course of publication [3] , Hironaka has shown, by topological considerations, that for complex-analytic varieties, equimultiplicity of V along W , at Q , is a consequence of conditions A and B in the most general case (i.e. , when V is not necessarily a hypersurface and W is of any codimension) . Thus, in Theorem 10.3, condition 1) can be omitted.

The proof of Theorem 10.3 is based on the following Jacobian criterion of equisingularity for algebroid hypersurfaces (see Zariski [8] Theorem 5.1 and Theorem 5.2) :

Theorem 10.4. In the notations of § 7, let $\bar{\mathfrak{O}}$ be the integral closure of the local ring $\mathfrak{O}$ $(=k[[x_1, x_2, \ldots, x_{r+1}]])$ in the total ring of quotients F of $\mathfrak{O}$, and let J be the ideal generated in $\bar{\mathfrak{O}}$ by the r+1 partial derivatives $\partial f/\partial x_j$, i=1 2, . . . , r+1 . Then Q is a singular point of dimensionality type 1 if and only if the following conditions are satisfied:

(a) J is a principal ideal.
(b) Q is a simple point of the singular locus of V .

Furthermore, if Q is a singular point of dimensionality type 1 then the following conditions are equivalent :

(a') $x_1, x_2, \ldots, x_r$ are transversal parameters of $\mathfrak{o}$.

(b') $J = \mathfrak{o} \cdot \partial f / \partial x_{r+1}$.

Assuming this theorem , we outline the proof of Theorem 10.3; we shall, however, restrict ourselves, for simplicity, to the case in which the algebroid hypersurface V is irreducible at Q .

In the first place we may assume, as in § 7, that $x_1, x_2, \ldots, x_r$ are transversal parameters of $\mathfrak{o}$, and that W is defined by the equations $X_r = X_{r+1} = 0$. In that case, condition 1) of Theorem 10.3 is equivalent to the following condition :

$$x_{r+1}/x_r \in \bar{\mathfrak{o}} . \tag{1}$$

If $\bar{\mathfrak{m}}$ denotes the maximal ideal of $\bar{\mathfrak{o}}$ then it can be shown that condition A is equivalent to

$$\partial x_{r+1} / \partial x_i \in \mathfrak{m} \quad i = 1, 2, \ldots, r-1 \tag{2}$$

[compare (2) with Proposition 10.5 below], while condition B is equivalent to

$$\partial x_{r+1}/\partial x_r - x_{r+1}/x_r \in \mathfrak{m} \tag{3}$$

[compare (3) with Proposition 10.6 below] . Now (1), (2), and (3) together imply that $\partial x_{r+1}/\partial x_i \in \bar{\mathfrak{o}}$, for $i = 1, 2, \ldots, r$, showing that $J = \bar{\mathfrak{o}} \cdot \partial f/\partial x_{r+1}$. Thus, by the first part of Theorem 10.4, it follows that conditions 1) and 2) of Theorem 10.3 imply equisingularity of V at 0, along W. The converse follows easily from the second part of Theorem 10.4. The above relations (1) and (2) are a special case of a valuation-theoretic formulation of conditions A and B.

For illustrative purposes, we shall give here this formulation in the case in which V is an irreducible algebraic hypersurface

$$f(X_1, X_2, \ldots, X_r, X_{r+1}) = 0 ,$$

and W is the linear space $X_{\rho+1} = X_{\rho+2} = \ldots = X_{r+1} = 0$. Let $k(V) = k(x_1, x_2, \ldots, x_{r+1})$ be the function field of V.

Proposition 10.5 A necessary and sufficient condition in order that the pair $\{V, W\}$ satisfy condition A at the origin Q is that the following inequality hold true for any valuation v of k(V) centered at Q :

$$\min. \{v(\partial f/\partial x_1), v(\partial f/\partial x_2), \ldots, v(\partial f/\partial x_\rho)\} > \min\{v(\partial f/\partial d_{\rho+1}), v(\partial f/\partial x_{\rho+2}), \ldots, v(\partial f/\partial x_{r+1})\} .$$

Proposition 10.6 . If $\{V, W\}$ already satisfies condition A at Q , then a necessary and sufficient condition that $\{V, W\}$ satisfy condition B at Q is that the following inequality hold true for any valuation v of k(V) centered at Q :

$$v\left(\sum_{i=1}^{r+1-\rho} x_{\rho+i}\, \partial f/\partial x_{\rho+i}\right) > \min . \{v(x_{\rho+1}), v(x_{\rho+2}), \ldots, v(x_{r+1})\} + \min . \{v(\partial f/\partial x_{\rho+1}), v(\partial f/\partial x_{\rho+2}), \ldots, v(\partial f/\partial x_{r+1})\} .$$

Although conditions A and B have been formulated in terms of a given embedding of V in an affine space, it can be proved that those conditions are intrinsic local properties of V at Q , in the sense that if they hold for one embedding of V they hold for all embedding of V . In the case of condition A it is even possible to give this condition an intrinsic formulation (in which only the local rings of V and W at Q are involved) . This formulation is due to Kunz. (see [4] , unpublished) and is as follows :

Let S denote the set of all divisors of k(V) centered at Q. We set $\mathfrak{O}$ = local ring of V at Q, $\mathfrak{O}_1$ = local ring of W at Q. For any v in S we danote by R the valuation ring of v , by M the maximal ideal of R and by $\Delta = R/M$ the residue field of v (to say that v is centered at Q means therefore that $R \supset \mathfrak{O}$ and that $M \cap \mathfrak{O}$ is maximal ideal of $\mathfrak{O}$) . We denote by $D_k(\mathfrak{O})$ the differential module $\mathfrak{O}$ ' over k, and by, $D_k(\mathfrak{O}_1)$ the differential module of $\mathfrak{O}_1$ over k. Since $\mathfrak{O}_1\ \mathfrak{m}_1 (\cong \mathfrak{O}/\mathfrak{m})$ is canonically a subfield of Δ , we can regard Δ canonically an $\mathfrak{O}_1$-module. We can therefore consider the $\mathfrak{O}$-module.

(4) $$R \otimes_{\mathfrak{O}} D_k(\mathfrak{O})$$

and the $\mathfrak{O}_1$ module

(5) $$\Delta \otimes_{\mathfrak{O}_1} D_k(\mathfrak{O}_1) .$$

In addition to the canonical surjection $R \to \Delta$, with kernel M , we have also a canonical surjection $\mathfrak{O} \to \mathfrak{O}$, with kernel I = prime ideal of W in $\mathfrak{O}$. This yields a canonical homomorphism $D_k(\mathfrak{O}) \to D_k(\mathfrak{O}_1)$, whose kernel is $\mathfrak{O}\, D_k I$, and therefore a canonical homomorphism of the module (4) into the module (5) :

(6) $$\sigma : R \otimes_{\mathfrak{O}} D_k(\mathfrak{O}) \to \Delta \otimes_{\mathfrak{O}_1} D_k(\mathfrak{O}_1) ,$$

whose kernel is

$$M \otimes_{\mathcal{O}} D_k(\mathcal{O}) + R \otimes_{\mathcal{O}} D_k I \, .$$

Let $T(R, \mathcal{O})$ be the torsion submodule of $R \otimes_{\mathcal{O}} D_k(\mathcal{O})$. The Kunz formulation of condition A is as follows :

Proposition 10.7. The pair $\{V, W\}$ *satisfies condition* A *at* Q *if and only if for every* v *in* S *it is true that the homomorphism* (6) *sends* $T(R, \mathcal{O})$ *into zero ; or equivalently -if and only if*

$$T(R, \mathcal{O}) \subset M \otimes_{\mathcal{O}} D_k(\mathcal{O}) + R \otimes_{\mathcal{O}} D_k I \, .$$

Using this proposition it can be shown that $\{V, W\}$ *satisfies condition* A *at the general point of* W.

Much shorter is the proof that (V, W) *satisfies also condition* B at the general point of W . This is practically evident if W is a point (i.e. , if dim $W = 0$) . If dim $W = \rho > 0$ and R is a general point of W/k , then the proof can be easily reduced to the case $\rho = 0$ by a ground field extension $k \to k^{*}$ = algebraic closure of the field $k(R)$.

We shall now conclude with some remarks concerning the global aspects of conditions A and B, say, on affine varieties $V \subset \mathbb{A}_n$.

We consider again the irreducible subvariety V_T of $V \times G_{n-1, r-1}$ defined by its general point $P \times T(V, P)$. Similarly, let W_T be the irreducible subvariety of $W \times G_{n-1, \rho -1}$ defined by its general point $R \times T(W, R)$. Here P and R are k-independent general points of V and W. We set

$$H = V_T \cap (W \times G_{n-1, r-1})$$

and we consider on $W \times G_{n-1, r-1} \times G_{n-1, \rho-1}$ the variety

(7) $$\varphi = (H \times G_{n-1, \rho-1}) \cap (W_T \times G_{n-1, r-I})$$

Let Z_o denote the algebraic subvariety of $G_{n-1, r-1} \times G_{n-1, \rho-1}$ consisting of those pairs $(L_{r-1}, L_{\rho-1})$ of lienar spaces which satisfy the incidence condition $L_{\rho-1} \subset L_{r-1}$, and let

$$Z = W \times Z_o.$$

Both Z and φ are subvarieties of the product variety $W \times G_{n-1, r-1} \times G_{n-1, \rho-1}$. It is easily seen that if Q is a simple point of W, then $\{V, W\}$ satisfies condition A at Q if and only if

(8) $$\varphi \cap (Q \times G_{n-1, r-1} \times G_{n-1, \rho-1}) \subset Z .$$

By the definition (7) of φ, we have a projection map $\psi : \varphi \to W$ (everywhere regular on φ). By (8), the set of simple points Q of W such that $\{V, W\}$ satisfies condition A at Q is the complement, on W, of the set $S_o \cup \psi\{\varphi - (\varphi \cap Z)\}$, where S_o is the singular locus of W. Since $\varphi - (\varphi \cap Z$ is an open subset of φ, it follows that the ψ-transform of this set is a contructible subset of W.

We have thus shown that the set of simple points of W at which $\{V, W\}$ satisfies condition A is constructible. Since we know (as was pointed out above) that this set contains the general point of W , it follows that the set contains a non-empty open subset of W .

Using similar considerations it can be shown that also the set of simple points of W where $\{V, W\}$ satisfies condition B is constructible (and therefore contains a non-empty open subset of W) .

It may very well happen that the set of points of W where condition A is satisfied is not open (or-equivalently - that the set of points of W where condition A is not satisfied is not an algebraic subvariety of W ; we know, however, that this set is contained in some proper algebraic subvariety of W) . The following is an example of this possibility :

Let V be the hypersurface $X_4^2 - (X_1X_2+X_3)X_3^2 = 0$ in A_4 and let W be plane $X_3 = X_4 = 0$. Thus W is the entire singular locus of V, and is a double plane of V. The set of points of W where condition A is not satisfied is the union of the two coordinate axes $X_2 = X_3 = X_4 = 0$ and $X_1 = X_3 = X_4 = 0$, from which, however, the origin $X_1 = X_2 = X_3 = X_4 = 0$ has been removed (at the origin, $\{V, W\}$ does satisfy condition A) .

We know of no example in which the set of points of W at which both conditions A and B are satisfied is not an open subset of W. However, in the special case in which V is a hypersurface and W has codimension 1' and is singular for V, the set of points in question is the set of points of W which are not of dimensionality type 1, and this set is algebraic (see § 7, result 4)) .

O. Zariski

Bibliography

[1] S. Abhyankar, On the ramification of algebraic functions, American American Journal of Mathematics, v.77 (1955).

[2] ________, Inversion and invariance of characteristic pairs, American Journal of Mathematics, v.89 (1967).

[3] H. Hironaka, Normal cones in analytic Whitney stratifications, Institutut des Hautes Etudes scientifiques (1969), Publications Mathématiques, No. 36 (volume dedicated to Oscar Zariski).

[4] Ernst Kunz, Uber gewisse singuläre Punkte algebraischer Manningfaltigkeiten (1969, in course of publication in Crelle's Journal).

[5] R. Thom, Ensembles et Morphismes stratifiés, Bulletin of the American Mathematical Society, v.75, No.2 (1969).

[6] H. Whitney, Local properties of analytic varieties, Differential and Combinatorial Topology, A symposium in honor of Marston Morse Princeton, 1965).

[7] O. Zariski, Studies in Equisingularity I. Equivalent singularities of plane algebroid curves, American Journal of Mathematics, v.87 (1965).

[8] ________, Studies in Equisingularity II. Equisingularity in codimension 1 (and characteristic zero). American Journal of Mathematics, v.87 (1965).

[9] ________, Studies in Equisingularity III. Saturation of local rings and equisingularity, American Journal of Mathematics, v. 90 (1968).

[10] O. Zariski and P. Samuel, Commutative Algebra, v.1 and v.2, (D. Van Nostrand Co., Princeton, USA, 1958 and 1960).

[11] Frédéric Pham et Bernard Teissier, Fractions Lipschitziennes d'une algèbre analytique complexe et saturation de Zariski Centre de Mathématique de l'Ecole Polytechnique, n. M 170669 (June, 1969).

SOME OPEN QUESTIONS IN THE THEORY OF SINGULARITIES[1]

BY OSCAR ZARISKI

ABSTRACT. Three approaches to a theory of equisingularity of complex analytic (or algebraic) hypersurfaces are outlined, based respectively on topology (topological equivalence of embedded varieties), differential geometry (Conditions A and B of Whitney) and algebraic geometry (the author's inductive discriminant criterion). For each of these approaches some unsolved questions and (or) conjectures are formulated, especially in regard to the relationship between these three points of view.

1. If I were giving this address some 7 or 8 years ago, the open question which I would have then given top priority in my talk would have been the problem of reduction of singularities of algebraic varieties of dimension >3, over ground fields of characteristic zero, the case of surfaces and of three-dimensional varieties having been settled in some of my earlier work in the late thirties and the early forties, work which also included the solution of the general problem of local uniformization in characteristic zero, in any dimension (see [10], [11], [12] and [13]). For twenty years, after these earlier papers of mine had appeared in print, no further progress was made in the direction of the solution of the problem of reduction of singularities. Personally I felt that I have devoted enough time and effort to that problem, that I needed a change of pace, and have therefore turned to other questions in my field. Some of my mathematical friends believed, no doubt, that during these twenty years I never gave up trying, and it is quite possible that these friends have drawn the—to me flattering—conclusion that since I am not able to prove the general reduction theorem, that theorem must be false. It is even probable that they were greatly tempted to look for and find a counterexample. Fortunately, Hironaka put a stop to this state of affairs by his

[1] Retiring Presidential address delivered before the Seventy-seventh Annual Meeting of the Society in Atlantic City on January 22, 1971; received by the editors February 11, 1971; the author wishes to thank the NSF for support under the Grant No. GP-9667 to Harvard University.

AMS 1970 *subject classifications*. Primary 14B05, 32C40; Secondary 14H20, 32C45.

Key words and phrases. Singularity, hypersurface, topological equisingularity, differential-geometric equisingularity, algebro-geometric equisingularity, equimultiplicity, blowing-up transformation, equisingularity along a subvariety, Whitney stratification, equisingular projection, equisaturation.

[Reprinted from *Bulletin of the American Mathematical Society*, Vol. 77, No. 4, July 1971, pp. 481–491.]

fundamental paper published in 1964 [4]. Thus that general problem was settled, at last, in the affirmative.

There still remains, however, the unsolved problem of reduction of singularities in characteristic $p\neq 0$. Whatever progress was made in this problem is associated primarily with the name of one single man, and that is Abhyankar. (Also some recent work of Hironaka, some of it unpublished, has contributed significantly to the problem in the case $p\neq 0$.) In his dissertation, published in 1956 [1], Abhyankar solved this problem for surfaces over perfect ground fields. Ten years later he extended his proof to arithmetic surfaces [2], i.e., to surfaces defined over a Dedekind domain, instead of over a field. Using various algorithms devised by him in the case of characteristic $p\neq 0$ he was also able to extend to that case [3] my proof of reduction of singularities of *embedded* surfaces, and therefore also my original proof of the *birational* reduction of singularities of three-dimensional varieties, under the assumption that $p<3!$, i.e., $p\neq 2, 3, 5$. This is the state of the problem at present. Even the problem of local uniformization in characteristic $p\neq 0$ is still unsolved in dimension >3 (and also in dimension 3, if $p=2, 3, 5$).

2. What I wish to discuss here today is not this open question of how to *get rid* of singularities in characteristic $p\neq 0$, but rather the one of how to *classify* singularities *in characteristic zero*, in fact—and more specifically—*in the complex domain*. In recent years there was an upsurge of interest in the study of singularities of algebraic, or—more generally—of complex-analytic varieties. The most substantial contributions here were made by differential topologists rather than by algebraic geometers. The fields of topology and differential geometry have already in their possession a number of powerful tools for the exploration of the structure of the neighborhood of a singular point of a complex-analytic variety. On the other hand, the purely algebraic approach, while still in its infancy, seems to be the most natural approach to the subject, for it is doubtful whether singularities of complex-analytic varieties are purely topological or even differential-geometric phenomena.

I will restrict myself to singular points P of hypersurfaces V_r:

$$V_r: f(x_1, x_2, \cdots, x_r, x_{r+1}) = 0, \qquad P \in V_r \subset A_{r+1}, \tag{1}$$

i.e., to singular points P of r-dimensional varieties V_r such that, locally at P, V_r can be embedded in a complex affine $(r+1)$-space A_{r+1}. Hence, locally at P, V_r can be defined by the single equation (1), where f is a convergent power series and P is the origin. The case of hypersurfaces is the most interesting one, and is, of course, also the one which is relatively easier to handle (see also footnote 2). Now, suppose we have another hypersurface V' of the same dimension r, and a singular point P' on it. The basic question is the following: *what shall we mean by saying that the two singularities P, P' are equivalent?* The relation of equivalence which we are trying to spell out and which we shall designate by the term "*equisingularity*" should formalize our vague and not very intuitive idea of singularities of the same type, of the same degree of complexity. One thing is clear: it must be an equivalence relation which is much weaker than an analytical isomorphism.

Topology provides one possible answer which has the great advantage of being clear-cut and unambiguous: V_r and V'_r are equisingular at P and P' if they are *topologically equivalent, as embedded* varieties, in the neighborhood of P and of P', i.e., *if there exists a local homeomorphism $f: \boldsymbol{A}_{r+1} \rightarrow \boldsymbol{A}'_{r+1}$ of the ambient affine spaces of V_r and V'_r which sends P into P' and V_r into V'_r.*[2]

This definition raises immediately a number of questions. I shall formulate here only two questions, which, as far as their degree of complexity is concerned, are respectively on the relatively shallow and the relatively deep end of a whole spectrum of possible questions. If the answer to the easier question cannot be provided by topologists in a relatively short order, I would be greatly disappointed. On the other hand, I would not blame them if they found it difficult to answer the second question.

The simplest numerical character of a singular point P of V is its *multiplicity* e, i.e., the degree of the leading form of the power series $f: e = e(V, P)$. Here $e = 1$ if and only if V is an analytic manifold at P. Any definition of equisingularity should imply *equimultiplicity*, at the very least. Thus our first question is the following:

A. *Does topological equisingularity of V_r and V'_r at P and P' imply that $e(V_r, P) = e(V'_r, P')$?*

The answer is known to be in the affirmative in the case of curves ($r = 1$). For $r > 1$, I find in Milnor's account on singularities of complex hypersurfaces a general theorem [6, p. 5] from which an affirmative answer can be deduced in the following case: $e(V_r, P) = 1$, and P' is at worst an isolated singular point of V'_r.

The second question is the following:

We apply to the affine space $\boldsymbol{A}_{r+1}$ a locally quadratic transformation with center P, or a blowing-up transformation, whose effect

[2] This definition would definitely be out of order for varieties which are not hypersurfaces. For instance, it is not difficult to see that *any two* algebroid branches V_1 and V'_1 in $\boldsymbol{A}_n$, $n \geq 3$, are topologically equivalent as embedded varieties (of real dimension 2).

can be informally described by saying that we remove the point P from the affine space and we replace it by a projective r-space $\boldsymbol{P}_r$ whose points represent the directions about P. The result is a non-singular $(r+1)$-dimensional manifold $\overline{M}_{r+1}$ and a continuous surjective map

$$T:\overline{M}_{r+1} \to \boldsymbol{A}_{r+1},$$

such that

$$T^{-1}\{P\} = \overline{E}_r \subset \overline{M}_{r+1}, \qquad \overline{E}_r \cong \boldsymbol{P}_r,$$

and such that the restriction of T to $\overline{M}_{r+1}-\overline{E}_r$ is an analytic isomorphism:

$$\overline{M}_{r+1} - \overline{E}_r \overset{\sim}{\to} A_{r+1} - P.$$

There is then a unique r-dimensional variety $\overline{V}_r \subset \overline{M}_{r+1}$ defined by the two conditions

$$T(\overline{V}_r) = V_r, \qquad \overline{E}_r \not\subset \overline{V}_r.$$

The variety $\overline{V}_r$ is sometimes referred to as the *proper transform of* V_r. In view of the inclusion $\overline{V}_r \subset \overline{M}_{r+1}$, this variety is locally a hypersurface at each of its points. Let

$$\bar{\mathcal{E}}_{r-1} = \overline{E}_r \cap \overline{V}_r.$$

The variety $\mathcal{E}_{r-1}$ is pure $(r-1)$-dimensional; it is the inverse image, on $\overline{V}_r$, of the point P, and is called the *exceptional variety* created by our blowing-up transformation of V_r. Again, also $\bar{\mathcal{E}}_{r-1}$ is locally a hypersurface, in view of the inclusion $\bar{\mathcal{E}}_{r-1} \subset \overline{E}_r$. Thus the total effect of our transformation leads from the pair (V_r, P) to the pair $(\overline{V}_r, \bar{\mathcal{E}}_{r-1})$.

Let us operate by a similar transformation on the pair (V'_r, P'), getting a pair $(\overline{V}'_r, \bar{\mathcal{E}}'_{r-1})$ consisting of the proper transform $\overline{V}'_r$ of V'_r and of the exceptional $(r-1)$-dimensional subvariety $\bar{\mathcal{E}}'_{r-1}$ which corresponds to the point P'. Our second question is the following:

B. *Prove that there exists a homeomorphism* $f:\bar{\mathcal{E}}_{r-1} \to \bar{\mathcal{E}}'_{r-1}$ *such that if* $\overline{P}' = f(\overline{P})$, $\overline{P} \in \bar{\mathcal{E}}_{r-1}$ *then* (1) $(\bar{\mathcal{E}}_{r-1}, \overline{P})$ *and* $(\bar{\mathcal{E}}'_{r-1}, \overline{P}')$ *are topologically equisingular and* (2) $(\overline{V}_r, \overline{P})$ *and* $(\overline{V}'_r, \overline{P}')$ *are also topologically equisingular.*

Again the answer to this question is in the affirmative for $r=1$, showing that the classical algebro-geometric notion of equivalence of singularities of plane algebraic (or algebroid) curves, which goes back to Max Noether (see, for instance [**14**]), coincides with the topological concept of equivalence. (Note that if $r=1$ then both $(\bar{\mathcal{E}}_{r-1}$ and $\bar{\mathcal{E}}'_{r-1}$ are finite sets of points.)

3. Let me now take a new tack which promises a better wind. Instead of dealing with a *pair* of hypersurfaces, let us consider *analytic families* of hypersurfaces V_r, all having a singular point at the origin and depending on a set of parameters $(t)=(t_1, t_2, \cdots, t_s)$. Thus, the variable member of this family will be a hypersurface $V_r^{(t)}$ defined by an equation

$$V_r^{(t)}: f(x_1, x_2, \cdots, x_r, x_{r+1}; t_1, t_2, \cdots, t_s) = 0, \tag{2}$$

where f is a convergent power series in $r+s+1$ variables (x), (t), in the neighborhood of $(x)=(t)=0$, and where

$$f(0, 0, \cdots, 0, 0; t_1, t_2, \cdots, t_s) \equiv 0, \quad \text{identically.}$$

Equation (2) can also be interpreted as the equation of a hypersurface

$$V_n: f((x); (t)) = 0, \qquad n = r + s, \; V_n \subset A_{n+1}. \tag{3}$$

This hypersurface carries the nonsingular manifold

$$W: x_1 = x_2 = \cdots = x_{r+1} = 0; \qquad r = \operatorname{cod}_{V_n} W. \tag{4}$$

Our family of r-dimensional hypersurface $V_r^{(t)}$ now appears as a family of sections $V_r^{(\bar{t})}$ of V_n, transversal to W:

$$V_r^{(\bar{t})} = V_n \cap (t_j = \bar{t}_j, \; j = 1, 2, \cdots, r),$$

the singular point of $V_r^{(\bar{t})}$ being the point

$$P^{(\bar{t})}: (0, 0, \cdots, 0, \bar{t}_1, \bar{t}_2, \cdots, \bar{t}_s).$$

We are interested particularly in the initial zero values of the parameters t, and therefore in the particular pair $(V_r^{(0)}, P^{(0)})$, which we shall denote by (V_r, P).

DEFINITION 1. *V_n is topologically equisingular (top. eqs.) at P, along W, if $(V_r^{(\bar{t})}, P^{(\bar{t})})$ and (V_r, P) are topologically equisingular, for all $(\bar{t})$ sufficiently near zero.*

Following Whitney [8] and [9], one introduces two differential-geometric conditions A and B which may or may not be satisfied by the triplet (V_n, W, P):

Condition A. *If $P=\lim P_i$, where the P_i are simple points of V_n, and if $T_0=\lim T(V_n, P_i)$, then $T_0 \supset T(W, P)$.*

Here $T(V_n, P_i)$ denotes the tangent space of V_n at P_i, and similarly for $T(W, P)$.

Condition B. *If $P=\lim P_i$ (as in* A*), if also $P=\lim Q_i$, where the Q_i are simple points of W $(P_i \neq Q_i)$ and if* $\lim P_iQ_i = l_0$, *and* $\lim T(V_n, P_i) = T_0$, *then $l_0 \subset T_0$.*

It can be proved that both conditions are intrinsic, i.e., they are independent of the embedding of V_n in affine $(n+1)$-space.

DEFINITION 2. *V_n is differentially equisingular (diff. eqs.) at P, along W, if both Conditions* A, B *are satisfied by* (V_n, W, P).

Whitney has proved that given any subvariety W of V_n, the set of points

$$\bar{Z}(V_n, W) = \{P \in W \mid V_n \text{ is not diff. eqs. at } P \text{ along } W\},$$

is contained in a *proper* analytic (or algebraic, if V_n and W are algebraic varieties) subvariety of W. This allows him to define a stratification of V_n (called the *Whitney stratification*) with the property that V_n is differentially equisingular along each stratum, at each point of the stratum. If $r>1$ it is not known whether the answer to the following question is in the affirmative:

C. *Is $\bar{Z}(V_n, W)$ a subvariety of W?*

If $r=1$, the answer is known to be in the affirmative (see, for instance, Zariski [15]).

The deepest result obtained so far in the theory of equisingularity is the following result, due to Thom and proved by Thom [7] and later also by Mather:

If S is any stratum of a Whitney stratification of V_n then V_n is topologically equisingular along S, at each point of S.

We note explicitly that this result does not signify that diff. eqs. of V_n at P, along a subvariety W (having P as a simple point) implies top. eqs. of V_n at P, along W. This latter statement may actually be false. For instance, if a, b, c are integers such that $1<a\leqq\min\{b, c\}$, then the singular locus of the surface $V_2: z^a=x^by^c$ consists of the two lines $W_1: x=z=0$ and $W_2: y=z=0$. For each $i=1, 2$ and for each point Q of W_i, the triplet (V, W_i, Q) satisfies both Conditions A and B. In particular, V_2 is therefore diff. eqs. at the origin P along W_i $(i=1, 2)$. However, while *any* section of V_2, transversal to W_i, at *any* point Q of W, *different from P*, has at Q an ordinary double point (i.e., a node), *no* section of V_2, transversal to W_i, at P has a singularity at P as simple as a node. Thus V_2 is not top. eqs. at P, along W_i $(i=1, 2)$. Since in a Whitney stratification any two strata must be disjoint and the boundary of any stratum must be a union of strata, the Whitney stratification of our surface V_2 must consist of the following four strata: $V_2-(W_1\cup W_2)$, W_1-P, W_2-P, P. We thus see that the strata of a Whitney stratification of V_n are not necessarily maximal sets of differential equisingularity of V_n; in other words, it may very well happen for some stratum S that V_n is diff. eqs. along the closure $\bar{S}$ of S at some point P of the boundary $\bar{S}-S$ of S.

One may ask whether the converse of Thom's result is true:

D. *Does topological equisingularity imply differential equisingularity?*

I have proved that the answer is in the affirmative if $r=1$. (We recall that r is the codimension of W on V_n.)

One may cite at this point also the following result, due to Hironaka [5]:

Differential equisingularity of V_n at P, along W, implies equimultiplicity of V_n at P, along W.

This result should be compared with the content of Question A.

4. I now come to the last part of my talk, in which I would like to describe briefly my own attempt to define equisingularity in a purely algebraic fashion. While this attempt was completely successful in codimension 1, in higher codimension I have now, by and large, only questions and conjectures. I may say, however, that my attempt to deal with the general case did pay me some dividends, since it led me to my general theory of saturation of local rings of singular points, a theory which I have initiated in 1968 [16] and which will be further developed in a series of papers, some of which are in course of publication and some in preparation.

Let the parameters $t_1, t_2, \cdots, t_s$ in (3) be now denoted by x_{r+2}, $x_{r+3}, \cdots, x_{n+1}$, where $n=r+s$. Then $V\ (=V_n)$ is defined by

$$V: f(x_1, x_2, \cdots, x_n, x_{n+1}) = 0,$$

and P still denotes the origin. Consider a set of n elements $z_1, z_2, \cdots, z_n$ of the local ring of V at P:

$$z_i = z_i(x_1, x_2, \cdots, x_{n+1}) = z_{i,1} + z_{i,2} + \cdots, \qquad i = 1, 2, \cdots, n,$$

where the z_i are convergent power series in the x's, and $z_{i,\alpha}$ is homogeneous, of degree α.

We shall say that the n elements z_i form a *set of parameters* if the following two conditions are satisfied:

(a) $(x)=0$ *is an isolated solution of the* $n+1$ *equations*

$$z_1(x) = z_2(x) = \cdots = z_n(x) = f(x) = 0.$$

(b) *The n linear forms $z_{i,1}$ are linearly independent.*

If condition (b) is satisfied, then the n linear equations

$$z_{i,1}(x) = 0, \qquad i = 1, 2, \cdots, n,$$

define a line l_z through P. If that line does not belong to the tangent cone of V at P, then condition (a) is automatically satisfied, and the z_i are parameters. We say in this case that the z_i are *transversal parameters.*

If the z_i are parameters, then the n equations $z_i(x)=0$, $i=1, 2,$

$\cdots$, n, define a regular curve Γ through P whose tangent line is l_z, and P is an isolated intersection of Γ and V. Let m be the intersection multiplicity of Γ and V, at P:

$$m = i(V, \Gamma; P).$$

Here $m \geqq e(V, P)$, with equality if and only if the z_i are transversal parameters.

The n parameters z_i define a projection π_z of V onto a neighborhood of the origin of the affine n-space of the n variables z_i:

$$\pi_z : V \to A_n, \qquad \pi_z(\bar{x}_1, \cdots, \bar{x}_{n+1}) = (z_1(\bar{x}), z_2(\bar{x}), \cdots, z_n(\bar{x})).$$

The direction of the line l_z above is called the *direction of the projection*. We call the projection π_z *transversal* if the parameters z_i are transversal.

The full inverse image $\pi_z^{-1}(\bar{z})$ of any point $\bar{z}$, near zero, consists of at most m points. The set of critical points $\bar{z}$, for which this full inverse image consists of less than m points, is a hypersurface in A_n, which we shall denote by Δ_z; it is the *critical variety* of the projection π_z. The equation $D_0(z)=0$ of Δ_z can be obtained by forming the discriminant D of a suitable polynomial, of degree m, with coefficients which are power series in the z_i, and by letting $D_0(z)$ be the product of the distinct irreducible factors of D.

Now, let W be a subvariety of V, of codimension r, having at P a simple point. We shall say that the projection π_z is *permissible*, or—more precisely—*W-permissible*, if the line l_z (the direction of π_z) is not contained in $T(W, P)$. Note that each transversal projection is W-permissible, since $T(W, P) \subset$ tangent cone of V.

Let π_z be a permissible projection. Then $\pi_z(W)$ is a nonsingular variety $\overline{W}$, of the same dimension as W, with a simple point at $\overline{P} = \pi_z(P)$. Since we assume that W is a singular subvariety of V, we have $\overline{W} \subset \Delta_z$, $\text{cod}_{\Delta_z} \overline{W} = r-1$, and we are dealing with a triple $(\Delta_z, \overline{W}, \overline{P})$, *in codimension* $r-1$.

If $r=1$, then $\text{cod}_{\Delta_z} \overline{W} = 0$, i.e., $\overline{W}$ is an irreducible component of Δ_z. In that case, equisingularity of Δ_z, at $\overline{P}$, along $\overline{W}$, means simply that $\overline{P}$ is a simple point Δ_z. I have proved in 1965 the following if $r=1$ (see [**15**]):

(1) *If there exists a permissible projection π_z such that P is simple for Δ_z, then V is equisingular at P, along W.*

(2) *Conversely, if V is equisingular at P, along W, then $\overline{P}$ is simple for Δ_z for any permissible* **transversal** *projection π_z.*

I therefore was led to the following inductive definition of equisingularity:

DEFINITION 3. *V is (algebro-geometrically) equisingular at P, along W, if there exists a permissible projection π_z such that Δ_z is equisingular at $\bar{P}$, along $\bar{W}$.*

With this definition of equisingularity certain statements which are not obvious or even not known to be true for topological or differential equisingularity, become either straightforward consequences or can be proved inductively without too much difficulty. For instance, we can assert the following:

1. *Equisingularity of V at P, along W, implies equimultiplicity of V at P, along W.*

2. *The set* Eqs(V, W) *of points P of W such that V is equisingular at P along W, is the complement of a proper analytic (or algebraic) subvariety of W.*

(Compare Result 2 with the unsolved Question C).

These results allow us to define a stratification of V such that V is equisingular along each stratum, at each point of the stratum.

The following is an open question:

If P is a point on the boundary of a stratum S, it is true then that V is not equisingular at P, along the closure of S.

The basic open question is the following:

E. *Does algebro-geometric equisingularity imply topological equisingularity or even differential equisingularity?*

Of course, in codimension 1 we know that all the 3 types of equisingularity are essentially identical concepts.

Assuming equisingularity of V at P, along W, an even more important question, from an algebro-geometric point of view, is the one which presents itself naturally when we apply to the ambient affine space $\boldsymbol{A}_{n+1}$ of V a blowing-up transformation T *with center* W (rather than with center P, as we have done earlier in connection with Question B). The pair $\boldsymbol{A}_{n+1} \supset W$ is transformed into a pair $M'_{n+1} \supset E'_n$ of nonsingular varieties of dimension $n+1$ and n, where E'_n is the blow-up of W, and E'_n is fibred by a family of r-dimensional projective spaces $F'(P_t)$, where $F'(P_t)$ is the full inverse image of the variable point P_t of W. We have the proper transform V' of V, in M'_{n+1}, the exceptional variety $\mathcal{E}'_{n-1} = E'_n \cap V'$ in V' and the fibres $\mathfrak{F}'(P_t) = F'(P_t) \cap \mathcal{E}'_{n-1}$. Since $V' \subset M'_{n+1}$, $\mathcal{E}'_{n-1} \subset E'_n$ and $\mathfrak{F}'(P_t) \subset F'(P_t)$ and since M'_{n+1}, E'_n and $F'(P_t)$ are nonsingular varieties, all the three varieties $V' \supset \mathcal{E}'_{n-1} \supset \mathfrak{F}'(P_t)$ all hypersurfaces, locally, at each of their points, and we can therefore consider the equisingularity stratification of each of them. We are interested in particular in the initial fibre $\mathfrak{F}'(P_0) = \mathfrak{F}'(P)$. The equisingularity of V at P, along W, must be reflected somehow in the way the strata of V' and $\mathcal{E}'_{n-1}$ are related to the special fibre $\mathfrak{F}'(P)$, and also in the way the special *reduced* fibre $\mathfrak{F}'(P)$ is related to the general reduced fibre $\mathfrak{F}'(P_t)$. To illustrate what I have in mind, I shall state a few conjectures which seem reasonable to me, without claiming, however, that these conjectures, even if these were proved to be true, say all that there is to say about the required relationship.

First we introduce the following notation: if P is any point of V then $S(V, P)$ shall denote the (algebro-geometric) stratum of equisingularity of V which contains P.

F. *If* $P' \in \mathfrak{F}'(P)$ *and* $S' = S(\mathcal{E}_{n-1}, P')$, *then*

$$T(S', P') + T(F'(P), P') = T(E_n', P').$$

G. *If* $P' \in \mathfrak{F}'(P)$, $S_1' = S(V', P')$ *and* $S_2' = S(\mathcal{E}_{n-1}', P')$, *then*

$$T(S_1' \cap S_2', P') + T(F'(P), P') = T(E_n', P').$$

H. *The (reduced) fibre* $\mathfrak{F}'(P)$ *is a specialization of the (reduced) general fibre* $\mathfrak{F}'(P_t)$.

Note that G implies F.

I shall formulate one final question, whose solution seems to be an essential preliminary for attacking the three questions just formulated above.

If V is equisingular at P along W, we call a permissible projection π_z *equisingular* if Δ_z is equisingular at $\bar{P}$ $(=\pi_z(P))$ along $\bar{W}$ $(=\pi_z(W))$. Let

$$z_i = \sum_{(\nu)} a^{(i)}_{\nu_1,\nu_2,\cdots,\nu_{n+1}} x_1^{\nu_1} x_2^{\nu_2} \cdots x_{n+1}^{\nu_{n+1}}, \qquad i = 1, 2, \cdots, n.$$

I. *Prove that there exists a finite set of polynomials* G_μ *in the indeterminate coefficients* $a_\nu^{(i)}$ *such that* $\prod_\mu G_\mu(a_\nu^{(i)}) \neq 0 \Rightarrow \pi_z$ *is equisingular.*

An affirmative answer to this question would imply, in particular, that if there exist equisingular projections then there also exist transversal equisingular projections, and the existence of these latter projections is essential for an inductive approach to Questions F, G, H.

I will conclude with the following remark. A very special case of equisingularity is *analytic triviality* of V at P, along W. That case arises if, for a suitable choice of the local coordinates $x_1, x_2, \cdots, x_n, x_{n+1}$ at P, the equation of W is $x_1 = x_2 = \cdots = x_{r+1} = 0$ (where $r = \text{cod}_V W$) *and the variables* $x_{r+2}, \cdots, x_{n+1}$ *do not occur in the equation* $f=0$ of V. I am now making the following assumption: there exists an equisingular projection π_z of V such that the critical variety Δ_z is *analytically trivial* at $\bar{P}$, along $\bar{W}$ (we express this assumption by saying that

V is *equisaturated at* P, along W). In that special case of equisingularity I was able to prove that V is also topologically equisingular at P, along W (see [15]; this answers, in the affirmative, in this special case, the first part of Question E). If, furthermore, the above equisingular projection π_s is also transversal, then I can also prove that V is differentially equisingular at P, along W (unpublished). I may say, quite generally, that the special case of equisingularity, represented by equisaturation, provided me with a great deal of experimental material which will be useful in the study of the general theory of equisingularity.

REFERENCES

1. S. Abhyankar, *Local uniformization on algebraic surfaces over ground fields of characteristic* $p\neq 0$, Ann. of Math. (2) **63** (1956), 491–526; Correction, ibid. (2) **78** (1963), 202–203. MR **17**, 1134; MR **27** #145.

2. ———, *Resolution of singularities of arithmetical surfaces*, Proc. Conf. Arithmetical Algebraic Geometry (Purdue University, 1963), Harper & Row, New York, 1965, pp. 111–152. MR **34** #171.

3. ———, *Resolution of singularities of embedded algebraic surfaces*, Pure and Appl. Math., vol. 24, Academic Press, New York, 1966. MR **36** #164.

4. H. Hironaka, *Resolution of singularities of an algebraic variety over a field of characteristic zero.* I, II, Ann. of Math. (2) **79** (1964), 109–326. MR **33** #7333.

5. ———, *Normal cones in analytic Whitney stratifications*, Inst. Hautes Études Sci. Publ. Math. No. 36 (1969), 127–138.

6. J. Milnor, *Singular points of complex hypersurfaces*, Ann. of Math. Studies, no. 61, Princeton Univ. Press, Princeton, N. J.; Univ. of Tokyo Press, Tokyo, 1968. MR **39** #969.

7. R. Thom, *Ensembles et morphismes stratifies*, Bull. Amer. Math. Soc. **75** (1969), 240–284. MR **39** #970.

8. H. Whitney, *Local properties of analytic varieties*, Differential and Combinatorial Topology (A Symposium in Honor of Marston Morse), Princeton Univ. Press, Princeton, N. J., 1965, pp. 205–244. MR **32** #5924.

9. ———, *Tangents to an analytic variety*, Ann. of Math. (2) **81** (1965), 496–549. MR **33** #745.

10. O. Zariski, *The reduction of the singularities of an algebraic surface*, Ann. of Math. (2) **40** (1939), 639–689. MR **1**, 26.

11. ———, *Local uniformization on algebraic varieties*, Ann. of Math. (2) **41** (1940), 852–896. MR **2**, 124.

12. ———, *Simplified proof for the resolution of singularities of an algebraic surface*, Ann. of Math. (2) **43** (1942), 583–593. MR **4**, 52.

13. ———, *Reduction of the singularities of algebraic three dimensional varieties*, Ann. of Math. (2) **45** (1944), 472–542. MR **6**, 102.

14. ———, *Studies in equisingularity.* I. *Equivalent singularities of plane algebroid curves*, Amer. J. Math. **87** (1965), 507–536. MR **31** #2243.

15. ———, *Studies in equisingularity.* II. *Equisingularity in codimension* 1 (*and characteristic zero*), Amer. J. Math. **87** (1965), 972–1006. MR **33** #125.

16. ———, *Studies in equisingularity.* III. *Saturation of local rings and equisingularity*, Amer. J. Math. **90** (1968), 961–1023. MR **38** #5775.

GENERAL THEORY OF SATURATION AND OF SATURATED LOCAL RINGS I.*

Saturation of complete local domains of dimension one having arbitrary coefficient fields (of characteristic zero).

By Oscar Zariski.

Contents. page

Introduction. In our paper [2], published in this Journal, we have introduced an operation on rings (and, in particular, on complete local rings) which we have called *saturation,* and we have studied in that paper the saturation of complete equicharacteristic local rings $\mathfrak{o}$, having a coefficient field k which is *algebraically closed and of characteristic zero.* In Part I of that paper we have restricted ourselves to the case in which $\mathfrak{o}$ satisfies the additional conditions of being a local *domain* of dimension *one,* and under this additional assumption we have developed a theory which was essentially complete. In the general case, studied in Part II of that paper (always under the assumption that k is algebraically closed and of characteristic zero),

Received December 30, 1970.

* This work was supported by a National Science Foundation Grant GP-9667.

[Reprinted from *American Journal of Mathematics,* Vol. XCIII, No. 3, pp. 573–648.]

our results were far from being complete. In the last section of [2] we gave a preview of the theory of saturation in that more general case, but have pointed out that the treatment of this general case requires the following two-fold generalization of the contents of Part I of [2]: I). In the first place, in the case of *one-dimensional* complete local *domains*, it is necessary to extend the theory developed in Part I to the case in which the coefficient field k is not algebraically closed (but still of characteristic zero); II) in the second place, it is necessary to extend the theory to complete local *rings* of *dimension one* (not necessarilly domains) free from nilpotent elements and having an arbitrary coefficient field k of characteristic zero.

With this paper we initiate a new series of three papers, dealing with the general theory of saturation. In the present paper we carry out the above generalization (I). The main ingredients of this generalization are the following: a) the structure theorem for saturated local domains of dimension one (§ 4); b) properties of transversal parameters, and automorphisms of saturated local domains of dimension one (§§ 7-8); c) the independence of the operation of saturation of the choice of the coefficient field k (§ 9). In a first reading, the reader may skip section 6 which is devoted almost entirely to the proof of Theorem 6.2 and the contents of which are not used in subsequent sections. While the proofs given by us in Part I of [2] provide the general guidelines for the present study, still the proofs in the present paper are, as a rule, much more difficult, from a technical point of view, than the proofs given in [2], the difficulties being caused, of course, by the fact that the coefficient field k is not assumed to be algebraically closed. This fact also causes, in a number of cases, some significant modifications in the statements of results, when the attempt is made to generalize the results from the special case (k-algebraicaly closed) to the present more general case.

In the next (second) paper of this series we shall deal with the above indicated generalization (II), therefore essentially—and geometrically speaking—with saturation of reducible algebroid curves (defined over an arbitrary coefficient field k, of characteristic zero). Finally, in the last (and third) paper of the series, we shall round up our theory of saturation of algebroid varieties which we have initiated in Part II of our paper [2]. There we shall make essential use of the results established in dimension one in this and in the next paper.

We refer to our paper [2] (Introduction) for our general definition of saturation. The construction of the saturation $\tilde{A}_K$ given in that "Introduction" (see [2], formula (0), p. 964) and the Lemma 1.4 (p. 970), are the only facts from [2] which we shall use here. For the rest, the present paper is self-contained.

1. Some preliminary results on complete local fields. In this section we shall constantly deal with a local field F which 1) is complete with respect to a discrete, rank 1 valuation v and 2) is equicharacteristic and (unless otherwise specified) of characteristic zero.

We fix a coefficient field $\bar{k}$ of F, i.e., a field of representatives of the valuation ring R_v of v (R_v is a complete, equicharacteristic local domain, of characteristic zero and Krull dimension 1). We fix once and for always a subfield k of $\bar{k}$ such that $\bar{k}$ is a finite (algebraic) extension of k. Any element t of F such that $v(t)=1$ will be called a *uniformizing parameter* of F. For any such t, the field F contains the power series ring $\bar{k}[[t]]$, and this ring is the valuation ring of v.

We consider an element x of F such that

$$v(x)=n>0,$$

and we set $K=k((x))=$ field of formal meromorphic functions of x (over k). Then K is a subfield of F, and the power series ring $k[[x]]$ is a subring of R_v. Since $\bar{k}$ is finite over k, the power series ring $\bar{k}[[x]]$ is a finite module over $k[[x]]$. On the other hand, if t is a uniformizing parameter of F, then $\{1,t,t^2,\cdots,t^{n-1}\}$ is a $\bar{k}[[x]]$-module basis of R_v ($=\bar{k}[[t]]$). Hence R_v is a finite module over $k[[x]]$, and F is a finite extension of K. We shall denote by F^* the least Galois extension of K containing F.

LEMMA 1.1. *There exists a uniformizing parameter t of F and an element a in $\bar{k}$, $a\neq 0$, such that*

$$x=at^n. \tag{1}$$

Proof. If u is any uniformizing parameter of F then we can write $x=a(u^n+b_1u^{n+1}+\cdots)$, *where* $a,b_1,b_2,\cdots$ belong to $\bar{k}$ and $a\neq 0$. Since the characteristic of $\bar{k}$ is zero, we can find a power series

$$t=u+c_1u^2+c_2u^3+\cdots,$$

with the c_i in $\bar{k}$, such that $t^n=u^n+b_1u^{n+1}+\cdots$. The coefficients c_i can, in fact, be determined successively from the relations

$$nc_1=b_1,\quad nc_2+\frac{n(n-1)}{2}c_1{}^2=b_2,\cdots$$

$$nc_i+\phi_i(c_1,c_2,\cdots,c_{i-1})=b_i,$$

where ϕ_i is a certain polynomial function of $c_1,c_2,\cdots,c_{i-1}$, with integral coefficients. This yields (1).

We fix an alegbraic closure Ω of F^*. We note that every k-monomorphism of $\bar{k}$ into Ω can be extended to a K-monomorphism of F into Ω (for instance, to the unique K-monomorphism of F which leaves fixed a given uniformizing parameter of F and which induces in $\bar{k}$ the given k-monomorphism of $\bar{k}$). It follows that every k-monomorphism of $\bar{k}$ into Ω maps $\bar{k}$ into F^*, and consequently F^* contains the least Galois extension of k containing $\bar{k}$ (and contained in Ω).

LEMMA 1.2. *Let $\tilde{k}$ be the least Galois extension of k containing $\bar{k}$ and contained in F^*. Let t and a be as in Lemma 1.1, let $a_1(=a), a_2, \cdots, a_\nu$ be the distinct conjugates of a in $\bar{k}$, over k, and let k^* denote the field obtained by adjoining to $\tilde{k}$ all the n-th roots of each of the quotients a_i/a_1 $(i=1,2,\cdots,\nu)$. Then*

1) $F^* = k^*((t))$;
2) *v has a unique extension to F^*, and the extension is unramified;*
3) *v is the only extension to F of the restriction $v \mid K$.*

Proof. If we set $\bar{K} = \bar{k}((x))$, then $F = \bar{K}(t) = \bar{K} + \bar{K}\cdot t + \cdots + \bar{K}t^{n-1}$, and from Lemma 1.1 it follows that the irreducible polynomial $T^n - \dfrac{x}{a}$ in $\bar{K}[T]$ is the minimal polynomial of t over $\bar{K}$. The ν polynomials $T^n - \dfrac{x}{a_i}$ are the conjugates of $T^n - \dfrac{x}{a}$ over K, and hence the following $n\nu$ elements $t_{i,j}$ (which are not necessarily distinct) form a complete set of conjugates of t over K (in F^*):

$$t_{i,j} = t\omega^j\sqrt[n]{a_1/a_i}, \qquad \begin{array}{l} i = 1,2,\cdots,\nu; \\ j = 0,1,\cdots,n-1, \end{array}$$

where ω is a primitive n-th root of unity and where, for each i, $\sqrt[n]{a_1/a_i}$ is some fixed determination of an n-th root of a_1/a_i in Ω. It follows that $k^* \subset F^*$ and that therefore $k^*((t)) \subset F^*$. On the other hand, F^* is the compositum of the fields which are the K-conjugates of F in Ω, i.e., F^* is contained in the compositum of the fields $\bar{k}_\alpha((t_{i,j}))$, where $\bar{k}_\alpha$ ranges over the set of subfields of $\tilde{k}$ which form the complete set of conjugates of $\bar{k}$ over k, and where $i = 1,2,\cdots,\nu$ and $j = 0,1,\cdots,n-1$. All these fields $\bar{k}_\alpha((t_{i,j}))$ are contained in $k^*((t))$, and hence $F^* \subset k^*((t))$. This proves part 1) of the lemma.

The power series ring $k^*[[t]]$ is the integral closure of $\bar{k}[[t]]$ in F^*. Hence, the natural valuation v^* of the local field F^* (the one having $k^*[[t]]$

as valuation ring) is the only extension of v to F^*. Since $v^*(t) = v(t) = 1$, v^* is unramified over F.

The power series ring $\bar{k}[[t]]$ is the integral closure in F of the power series ring $k[[x]]$. These two rings are respectively the valuation rings of v and $v \mid K$. This completes the proof.

In the sequel we shall drop the asterisk in v^* and we shall thus use the same letter v for the natural valuation of F and for the unique extension of that valuation to F^*.

We now state a result which has been communicated to us orally by Ax. Our proof is based, in part, on a key idea of the original proof of Ax [see observations 1) and 2) below]. We state our result in a very general form, although we shall only apply it in the very special case in which $n = 1$ and k has characteristic zero.

PROPOSITION 1.3. *Let $\mathfrak{o} = k[[x_1, x_2, \cdots, x_n]]$ be a power series ring in n indeterminates, over a field k (of arbitrary characteristic) and let F be the field $k((x_1, x_2, \cdots, x_n))$ of fractions of $\mathfrak{o}$. If τ is any automorphism of F (not necessarily a k-automorphism) then $\tau(\mathfrak{o}) = \mathfrak{o}$.*

Proof. If ξ is an element of F, the following statement (referred to as *property* A) may be true or false:

PROPERTY A. *There exist infinitely many positive integers q such that F contains a q-th root of ξ.*

Let $\mathfrak{m}$ be the maximal ideal of $\mathfrak{o}$. We make the following two key observations:

1) *Any element ξ of F of the set $1 + \mathfrak{m}$ has property* A.

2) *If an element ξ of F has property* A *then ξ belongs to $\mathfrak{o}$ and is a unit in $\mathfrak{o}$.*

It is clear that if $\xi \in 1 + \mathfrak{m}$ then A holds true for all positive integers, *if k has characteristic zero.* If, however, the characteristic p of k is $\neq 0$, then A holds for all integers q which are not divisible by p. That takes care of observation 1). We now proceed to the proof of observation 2).

Assume that ξ has property A. We shall show that the assumption that ξ does not belong to $\mathfrak{o}$ leads to a contradiction. We write $\xi = \phi/\psi$, where ϕ and ψ are elements of $\mathfrak{o}$, free from common factors. Since $\xi \notin \mathfrak{o}$, ψ is a non-unit in $\mathfrak{o}$. Let g be an irreducible factor of ψ. If q is any positive integer such that F contains a q-th root of ξ, then we have $\phi/\psi = \phi'^q/\psi'^q$, where ϕ' and ψ' are elements of $\mathfrak{o}$, free from common factors. This implies that ψ

and ψ'^q are associates in $\mathfrak{o}$ and that therefore g must divide ψ'. *Therefore g^q must divide ψ.* This, however, can happen only for a finite number of positive integers q—contrary to assumption.

We have thus shown that $\xi \in \mathfrak{o}$. The proof that ξ is a unit in $\mathfrak{o}$ will be based again on an indirect argument. If ξ is a non-unit in $\mathfrak{o}$, let s be the degree of the leading form of ξ (i.e., let $\xi \in \mathfrak{m}^s$, $\xi \notin \mathfrak{m}^{s+1}$, $s \geqq 1$). If for some positive integer q we have $\xi = \eta^q$ $(\eta \in F)$, then necessarily $\eta \in \mathfrak{o}$ (since $\mathfrak{o}$ is integrally closed in F), and hence *s must be divisible by q.* This can happen only for a finite number of integers q—contrary to our assumption.

It is clear that the set of elements ξ of F which have property A is transformed into itself by τ. In particular, it follows from observations 1) and 2) that $\tau(1+\mathfrak{m}) \subset \mathfrak{o}$. Since $\tau(1) = 1$, it follows that

$$\tau(\mathfrak{m}) \subset \mathfrak{o}. \tag{2}$$

We shall now show that

$$\tau(k) \subset \mathfrak{o}, \tag{3}$$

and this will complete the proof of the proposition, since (2) and (3) imply that $\tau(\mathfrak{o}) \subset \mathfrak{o}$ and since the same argument, applied to τ^{-1}, yields $\tau^{-1}(\mathfrak{o}) \subset \mathfrak{o}$.

The proof of (3) will be again indirect. Assume that (3) is false and let, then, a be an element of k such that $\tau(a) \notin \mathfrak{o}$. Then we can write $\tau(a) = \phi/\psi$, where ϕ and ψ are two elements of $\mathfrak{o}$ which are relatively prime and where ψ is a non-unit in $\mathfrak{o}$. If q is any positive integer, then $a^q\psi \in \mathfrak{m}$ (since $a^q \in k \subset \mathfrak{o}$), and hence by (2) we have that $\phi^q\tau(\psi)/\psi^q \in \mathfrak{o}$. Thus ψ^q must divide the element $\tau(\psi)$ of $\mathfrak{o}$ ($\tau(\psi)$ is an element of $\mathfrak{o}$ in view of (2), since $\psi \in \mathfrak{m}$), and this must be so for all positive integers q, which is absurd.

The following result is an application of the preceding proposition:

PROPOSITSON 1.4. *Let $F = \bar{k}((t))$ be our complete (equicharacteristic) local field F, where this time $\bar{k}$ is of arbitrary characteristic. Let K be a subfield of F such that F is a finite separable extension of K. If v is the natural valuation of F (having $\bar{k}[[t]]$ as valuation ring) and if v_0 is the restriction of v to K, then K is complete with respect to v_0,*[1] *and v is the only extension of v_0 to F.*

Proof. As above, let F^* be the least Galois extension of K containing F. In view of the separability of the extension F^*/F, the integral closure R^* in F^* of the valuation ring $\bar{k}[[t]]$ of v is a finite $\bar{k}[[t]]$-module and is

[1] This proposition is false if F is inseparable over K. For the sake of completeness we include the treatment of the inseparable case in a special section 1 bis.

therefore a *complete* semilocal domain, and therefore necessarily a *local* domain (for otherwise R^* would be the direct sum of its localizations at it maximal ideals). Since R^* is also the intersection of the valuation rings of the various extensions of v to F^*, and since R^* is a local domain, it follows that R^* is itself a valuation ring and that v has a unique extension v^* to F^*. Thus F^* is a local field, complete with respect to the discrete, rank 1 valuation v^*, and is, of course, equicharacteristic. We fix a coefficient field k^* of F^* and a uniformizing parameter u of F^*, whence $R^* = k^*[[u]]$. Now, it is known that the extensions of v_0 to F^* form a complete set of conjugates over K. Since v^* is one of the extensions of v_0, every other extension of v_0 is of the form $v^{*\tau}$, where τ is an automorphism of F^* (in fact, a K-automorphism). By Proposition 1.3, $v^{*\tau}$ and v^* have the same valuation ring, namely R^* $(= k^*[[u]])$. Hence $v^{*\tau} = v^*$, showing that *v^* is the only extension of v_0 to F^**, and that, by a stronger reason, v is the only extension of v_0 to F.

We consider the trace function $T\colon F \to K$ defined on F, relative to K. Since $F \mid K$ is separable, T is not zero, and hence T maps F surjectively onto K. We fix an element ζ in F such that $T(\zeta) = 1$. We show first that *T is a continuous function* (in the topology defined in F by the valuation v). Let $\{\xi_i\}$ be a null sequence in F (i.e., such that $\lim \xi_i = 0$). We have then that $v(\xi_i) \to +\infty$ as $i \to +\infty$. Let $g = [F\colon K]$ and let us fix g K-automorphisms $\tau_1, \tau_2, \cdots, \tau_g$ of F^* such that the restrictions $\tau_i \mid F$ are distinct (and give therefore all the K-monomohphisms of F into F^*) We have $T(\xi_i) = \sum_{\nu=1}^{g} \xi_i^{\tau_\nu}$. Since $v^* = v^{*\tau_\nu}$ and $v^*(\xi_i) \to +\infty$, it follows that also $v^*(T(\xi_i)) \to +\infty$ as $i \to +\infty$. In other words, we have that $T(\xi_i) \to 0$ as $i \to +\infty$, which proves the continuity of T.

Now let $\{\xi_i\}$ be a Cauchy sequence in F *consisting of elements of K*, and let $\xi = \lim \xi_i$. We have to show that $\xi \in K$. We have $\xi\zeta = \lim \xi_i\zeta$, and since $T(\zeta) = 1$, we find, by taking traces and using the K-linearity of T, that $T(\xi\zeta) = \lim \xi$, i.e., $T(\xi\zeta) = \xi$, showing that $\xi \in K$. This completes the proof.

COROLLARY 1.5. *If F is of characteristic zero and K is as in Proposition 1.4, then there exist coefficient fields $\bar{k}$ of F such that if $\bar{k}_0 = \bar{k} \cap K$ then $\bar{k}_0$ is a coefficient field of K (whence $K = \bar{k}_0((u))$, where u is a uniformizing parameter of K) and $[\bar{k}\colon k_0] < \infty$.*

For, let $\bar{k}_0$ be a coefficient field of K (such a field exists, since K is complete and is equicharacteristic) and let $\bar{k}$ be the algebraic closure of $\bar{k}_0$ in F. Since F is of characteristic zero, $\bar{k}$ is contained in some coefficient field Δ of F. Since F is algebraic over K, also Δ will be algebraic over $\bar{k}_0$. Hence $\Delta = \bar{k}$. Since $\bar{k}_0$ is maximally algebraic in K and since we are in the case of characteristic zero, the fields K and $\bar{k}$ are linearly disjoint over $\bar{k}_0$. Since $[F:K] < \infty$, it follows therefore that also $[\bar{k}:\bar{k}_0] < \infty$. Finally, it is clear that $\bar{k} \cap K = \bar{k}_0$ (since $\bar{k}_0$ is a maximal subfield of K).

1 bis. The inseparable case of Proposition 1.4. It is clearly sufficient to consider the case in which F is purely inseparable over K, of degree $=$ characteristic p.

Let e and n denote respectively the reduced ramification index and the relative degree of v with respect to v_0 (see [1], v. 2, pp. 52-53). It is clear that in the present, purely inseparable case, v is the only extension of v_0 to F (since $R_v{}^p \subset R_{v_0}$, whence R_v is the integral closure of R_{v_0} in F). It is known that $en \leqq p$, with equality if and only if R_v if a finite R_{v_0}-module ([1], v. 2, Th. 20 on pp. 60-61 and observation at the bottom of page 33). Since the residue field of v is purely inseparable over the residue field of v_0, the integer n is either 1 or p. Similarly, in the present inseparable case, the reduced ramification index can only be 1 or p. Thus we can have only the following 3 cases:

Case 1: $n = p$, $e = 1$.
Case 2: $n = 1$, $e = p$.
Case 3: $n = 1$, $e = 1$.

We shall show that *Proposition* 1.4 *is still valid in Cases* 1 *and* 2. In Case 3, the proposition is certainly false, for if K is complete with respect to v_0, then $K = \bar{k}_0((u))$, where $\bar{k}_0$ is a coefficient field of R_{v_0} and u is an element of K such that $v_0(u) = 1$. Since $e = 1$, we have also $v(u) = 1$, and since $n = 1$, $\bar{k}_0$ is also a coefficient field of R_v. Therefore, $F = \bar{k}_0((u)) = K$, in contradiction with $[F:K] = p$.

We shall show that *Case* 3 *can actually occur only if* $[\bar{k}:\bar{k}^p] = \infty$. *Hence, if* $[\bar{k}:\bar{k}^p] < \infty$, *then Proposition* 1.4 *is valid also in the inseparable case.*

We treat simultaneously Cases 1 and 2. In Case 1 we fix an element ω in F such that $v(\omega) = 0$ and such that the v-residue of ω does not belong to the residue field of v. In Case 2 we fix an element ω in F such that $v(\omega) = 1$. Then it is immediately seen that $1, \omega, \omega^2, \cdots, \omega^{p-1}$ form an independent basis of F over K. Now let ξ be any element of F, $\xi \neq 0$, and let us write

$$\xi = a_0 + a_1\omega + \cdots + a_{p-1}\omega^{p-1}, \quad a_i \in K.$$

In Case 1, if the index i is so chosen as to have

$$v(a_i) = \min \cdot \{v(a_0), v(a_1), \cdots, v(a_{p-1})\},$$

then the v-residue of ξ/a_i is different from zero (for if $\bar{\omega}$ is the v-residue of ω then $X^p - \bar{\omega}^p$ is the minimal polynomial of $\bar{\omega}$ over the residue field of v_0). Hence

$$(1) \qquad v(\xi) = \min\{v(a_0), v(a_1), \cdots, v(a_{p-1})\}.$$

In Case 2, we have for each index i such that $a_i \neq 0$: $v(a_i\omega^i) \equiv i \pmod{p}$. Therefore

$$(2) \qquad v(\xi) = \min \cdot \{v(a_0), 1 + v(a_1), \cdots, p-1+v(a_{p-1})\}.$$

From (1) and (2) it follows that *in either case it is true that if $v(\xi)$ is very large then also $v(a_0), v(a_1), \cdots, v(a_{p-1})$ are all very large.* Now, let $\{\xi_i\}$ be a Cauchy sequence in F, consisting of elements of K and let $\xi = \lim \xi_i$. Writing ξ as above and observing that $v(\xi - \xi_i) \to +\infty$ as $i \to +\infty$, it follows that $v(a_0 - \xi_i) \to +\infty$ as $i \to +\infty$. Hence $a_0 = \lim \xi_i = \xi$, i.e., $\xi \in K$.

We show next that *if $[\bar{k}: \bar{k}^p] < \infty$ and if $n = e = 1$ then (under the assumption that F/K is a purely inseparable extension of degree $\leqq p$) $F = K$.* We have $F = \bar{k}[[t]]$, and since $e = 1$ we can choose the element t to be in K. We have, at any rate: $F^p \subset K$, hence $\bar{k}^p((t^p)) \subset K$. Since

$$\bar{k}^p((t)) = \sum_{i=0}^{p-1} \bar{k}^p((t^p)) \cdot t^i$$

and $t \in K$, it follows that $\bar{k}^p((t)) \subset K$. Since $[\bar{k}: \bar{k}^p] < \infty$, $\bar{k}$ has a finite p-basis $\omega_1, \omega_2, \cdots, \omega_m$ over $\bar{k}^p$, where m is the integer such that $[\bar{k}: \bar{k}^p] = p^m$. Since $n = 1$, we can find in K m elements $\xi_1, \xi_2, \cdots, \xi_m$ whose v-residues are $\omega_1, \omega_2, \cdots, \omega_m$ respectively. It is immediate now that the p^m monomials $\xi_1^{\nu_1}\xi_2^{\nu_2} \cdots \xi_m^{\nu_m}$ $(0 \leqq \nu_i \leqq p-1)$ are linearly independent over $\bar{k}^p((t))$. Hence $[K: \bar{k}^p((t))] \geqq p^m$. On the other hand, it is obvious that $[F: \bar{k}^p((t))] = p^m$. This implies that $F = K$, as asserted.

Let now $[\bar{k}: \bar{k}^p] = \infty$, and assume for simplicity that $\bar{k}$ has a countable p-basis, say $\omega_1, \omega_2, \cdots, \omega_n, \cdots$. We set

$$\xi_i = \omega_i + t\omega_{i+1}, \quad i = 1, 2, \cdots, n, \cdots$$

and we consider the following subfield K of F:

$$K = \bar{k}^p((t))[\xi_1, \xi_2, \cdots, \xi_n, \cdots].$$

Since $t \in K$ and since the v-residue of ξ_i is ω_i, it follows that $n = e = 1$. We show next that $K \neq F$ by showing that *none of the ω_i is in K.* Suppose the contrary is true: some ω_i is in K. Then there must exist an integer N such that

$$\omega_i \in k^p((t))[\xi_1, \xi_2, \cdots, \xi_N],$$

or, equivalently (since $\xi_j{}^p = \omega_j{}^p + t^p\omega_{j+1}{}^p \in k^p[[t]]$):

$$\omega_i \in k^p[\xi_1, \cdots, \xi_N]((t)).$$

More precisely: ω_i is a power series in t, with coefficients which are polynomials in $\xi_1, \xi_2, \cdots, \xi_N$, with coefficients in k^p and of degree $\leqq p-1$ in each ξ_j $(1 \leqq j \leqq N)$. This power series cannot start with a term αt^{-q}, $q > 0$, $\alpha \neq 0$, since $v(\omega_i) = 0$ and $v(\alpha) = 0$ (the latter equality follows from the fact that the v-residues of $\xi_1, \xi_2, \cdots, \xi_N$ are $\omega_1, \omega_2, \cdots, \omega_N$ and are therefore p-independent over k^p). Hence, $\omega_i = \alpha_0 + \alpha_1 t + \alpha_2 t^2 + \cdots$, where α_0 is necessarily equal to ξ_i. Hence

$$\omega_{i+1} = \alpha_1 + \alpha_2 t^2 + \cdots \in k^p[\xi_1, \xi_2, \cdots, \xi_N][[t]].$$

In a similar way it follows that $\omega_{i+2} \in k^p[\xi_1, \xi_2, \cdots, \xi_N][[t]]$, and so on. Thus all the ω_j's belong to $k^p[\xi_1, \xi_2, \cdots, \xi_N][[t]]$, and this is impossible if $j > N$.

2. Some preliminary properties of saturation in dimension one. From now on, till the end of this paper, $\mathfrak{o}$ will denote a complete noetherian local domain, of dimension 1, equicharacteristic and of characteristic zero, while $\mathfrak{m}$, k, F, $\bar{\mathfrak{o}}$ and $\bar{k}$ will denote respectively the maximal ideal of $\mathfrak{o}$, a field of coefficients of $\mathfrak{o}$, the field of fractions of $\mathfrak{o}$, the integral closure of $\mathfrak{o}$ (in F) and the algebraic closure of k in F. Then necessarily $\bar{\mathfrak{o}}$ is a finite module over $\mathfrak{o}$ and is a power series ring $\bar{k}[[t]]$ over $\bar{k}$. The finiteness of $\bar{\mathfrak{o}}$ over $\mathfrak{o}$ implies that $\bar{k}$ is a finite extension of k. The natural valuation of the complete local field F will be denoted by v.

If x is a parameter of $\mathfrak{o}$, then F contains the field $K = k((x))$ of formal meromorphic functions of x, over k, and since $[\bar{k}: k] < \infty$ it follows that $\bar{\mathfrak{o}}$ is a finite $k[[x]]$-module and that therefore F is a finite algebraic extension of K. Furthermore, $\mathfrak{o} \cap K = k[[x]]$ and $\mathfrak{o}$ is integral over $k[[x]]$. Therefore, the saturation $\tilde{\mathfrak{o}}_K$ of $\mathfrak{o}$ with respect to K is defined (see [2], Introduction). Since the coefficient field k of $\mathfrak{o}$ is fixed throughout this chapter (except in sections 9-10), we shall denote $\tilde{\mathfrak{o}}_K$ by $\tilde{\mathfrak{o}}_x$ ("the saturation of $\mathfrak{o}$ with respect to x"; see [2], Introduction). In the rest of this paper, the least Galois extension of K containing F will be denoted by F^*, and G will denote the Galois group of F^* over K. The unique extension of v to F^* (see Lemma 1.2) will be donoted by the same letter v.

PROPOSITION 2.1. *Let x be a parameter of $\mathfrak{o}$, let $K = k((x))$ and let $\tilde{\mathfrak{o}}_x$ be the saturation of $\mathfrak{o}$ with respect to K. Then $\tilde{\mathfrak{o}}_x$ is a complete domain, and k is also a coefficient field of $\tilde{\mathfrak{o}}_x$.*

Proof. Since $\tilde{\mathfrak{o}}_x$ is a ring between the two complete local domains $\mathfrak{o}$ and $\bar{\mathfrak{o}}$, it is itself a complete local domain which dominates $\mathfrak{o}$ and is dominated by $\bar{\mathfrak{o}}$. We use the construction of the saturation $\tilde{A}_K$ given in [2], Introduction p. 964). We have then

$$(1) \qquad \tilde{\mathfrak{o}}_x = \bigcup_{i=1}^{\infty} \mathfrak{o}_i,$$

where $\mathfrak{o}_{i+1} = \mathfrak{o}_i[L_i]$ (and $\mathfrak{o}_0 = \mathfrak{o}$). Note that in the present case the ascending chain of local rings $\mathfrak{o} = \mathfrak{o}_0 \subset \mathfrak{o}_1 \subset \mathfrak{o}_2 \subset \cdots$ is necessarily finite, since $\tilde{\mathfrak{o}}_x$ is a finite $\mathfrak{o}$-module. To prove that k is a coefficient field of $\tilde{\mathfrak{o}}_x$, it will be sufficient to prove that k is a coefficient field of each ring $\mathfrak{o}_i$. By assumption, this is true for $i = 0$. Assume this to be true for all integers i less than or equal to a given integer q. Since $\mathfrak{o}_{q+1} = \mathfrak{o}_q[L_q]$, the proof will be complete if we prove that the v-residue (in $\bar{k}$) of any element η of the set L_q belongs to k. Now, by definition of L_q, given an element η of L_q there exists an element ζ in $\mathfrak{o}_q$ such that $v(\eta^\tau - \eta) \geqq v(\zeta^\tau - \zeta)$ for all τ in G. Since, by our induction, the v-residue of ζ (in $\bar{k}$) belongs to k, and $k \subset K$, it follows that ζ^τ and ζ have the same v-residue (in $\bar{k}$). Therefore, $v(\zeta^\tau - \zeta) > 0$ for all τ in G. Therefore, also $v(\eta^\tau - \eta) > 0$ for all τ in G. This implies that if c denotes the v-residue of η in $\bar{k}$, then $c^\tau = c$ for all τ in G. Since G induces in $\bar{k}$ all the k-monomorphisms of $\bar{k}$ into the algebraic closure Ω of F, it follows that $c \in k$. This completes the proof.

Note. The above proposition is a special case of a general theorem which we have proved in [2] (see [2], Theorem 4.1).

The result which we prove next is a lemma consisting of two parts, of which only the first will be used by us in this paper; the second part will be used in a subsequent paper, dealing with saturation of varieties of higher dimension.

LEMMA 2.2. *The notation being the same as in Proposition 2.1, let K_0 be a subfield of K such that $[K: K_0]$ is finite. Then:*

1) *We have $\tilde{\mathfrak{o}}_{K_0} \subset \tilde{\mathfrak{o}}_K$; in particular, if $\mathfrak{o}$ is saturated with respect to K, $\mathfrak{o}$ is also saturated with respect to K_0.*

2) *In the special case in which K_0 is a field of the form $k_0((x))$, where k_0 is a subfield of k such that $[k: k_0]$ is finite, we have $\tilde{\mathfrak{o}}_{K_0} = \tilde{\mathfrak{o}}_K$ and, in*

particular, $\mathfrak{o}$ is saturated with respect to K if and only if $\mathfrak{o}$ is saturated with respect to K_0.

Proof. 1) We first show that the saturation of $\mathfrak{o}$ with respect to the field K_0 is defined. For this we have to show that $\mathfrak{o}$ is integral over the ring $\mathfrak{o} \cap K_0$. Now, $\mathfrak{o} \cap K_0 = \mathfrak{o} \cap K \cap K_0 = k[[x]] \cap K_0$. Let v be the natural valuation of F. Then $k[[x]]$ is the valuation ring of $v \mid K$, and hence $\mathfrak{o} \cap K_0$ is the valuation ring of the restriction $v \mid K_0$. By Proposition 1.4, applied to pair of fields (F, K_0) (instead of the pair (F, K)), v is the only extension of $v \mid K_0$ to F. Therefore the valuation ring R_v of v, i.e., the integral closure $\bar{\mathfrak{o}}$ of $\mathfrak{o}$ in F, is integral over $\mathfrak{o} \cap K_0$. By a stronger reason, $\mathfrak{o}$ is integral over $\mathfrak{o} \cap K_0$.

The second part of the statement 1) (beginning with the words "in particular") is a consequence of the first part, i.e., of the assertion $\tilde{\mathfrak{o}}_{K_0} \subset \tilde{\mathfrak{o}}_K$. For if this assertion is true, and if we assume that $\mathfrak{o}$ is saturated with respect to K, i.e., that $\tilde{\mathfrak{o}}_K = \mathfrak{o}$, then we have $\tilde{\mathfrak{o}}_{K_0} \subset \mathfrak{o}$, and since $\mathfrak{o} \subset \tilde{\mathfrak{o}}_{K_0}$, it follows that $\tilde{\mathfrak{o}}_{K_0} = \mathfrak{o}$, i.e., $\mathfrak{o}$ is also saturated with respect to K_0. Conversely, the inclusion $\tilde{\mathfrak{o}}_{K_0} \subset \tilde{\mathfrak{o}}_K$ is a consequence of the second part of the statement 1). For assume that it has been proved that whenever $\mathfrak{o}$ is saturated with respect to a subfield K of F (F = field of fractions of $\mathfrak{o}$, $K = k((x))$, with k a coefficient field of $\mathfrak{o}$ and x a parameter of $\mathfrak{o}$) then $\mathfrak{o}$ is also saturated with respect to any subfield K_0 of K such that $[K : K_0] < \infty$. Since k is also a coefficient field of $\tilde{\mathfrak{o}}_K$ (Proposition 2.1), we can apply this result to the saturated ring $\tilde{\mathfrak{o}}_K$, and we have therefore that $\tilde{\mathfrak{o}}_K$ is saturated with respect to K_0. By the minimality property of $\tilde{\mathfrak{o}}_{K_0}$, the inclusion $\tilde{\mathfrak{o}}_{K_0} \subset \tilde{\mathfrak{o}}_K$ follows.

So we now proceed to prove the second part of statement 1). Let F_0^* be the least Galois extension of K_0 containing F. We have then

$$K_0 \subset K \subset F \subset F^* \subset F_0^*.$$

By part 3) of Lemma 1.2, applied to the pair of fields (K, F_0^*) (instead of to the pair (K, F)), v has a unique extension to F_0^*. We shall use the same letter v to denote that extension.

We assume then that $\mathfrak{o}$ is saturated with respect to K. Let ζ be any element of $\mathfrak{o}$, let η be an element of $\bar{\mathfrak{o}}$, and assume $v(\eta^\sigma - \eta) \geqq v(\zeta^\sigma - \zeta)$ for all σ in the Galois group $\mathfrak{G}(F_0^*/K_0)$. To prove that $\mathfrak{o}$ is also saturated with respect to K_0, we have to show that

$$\eta \in \mathfrak{o}. \tag{2}$$

The assumption implies, in particular, that $v(\eta^\sigma - \eta) \geqq v(\zeta^\sigma - \zeta)$ for all

$\sigma \in \mathfrak{G}(F_0^*/K)$. For any element σ of this latter Galois group it is true that the restriction $\sigma \mid F^*$ belongs to the Galois group $\mathfrak{G}(F^*/K)$. Furthermore, any element τ of $\mathfrak{G}(F^*/K)$ is the restriction to F^* of some element of $\mathfrak{G}(F_0^*/K)$. Our assumption implies therefore that $v(\eta^\tau - \eta) \geqq v(\zeta^\tau - \zeta)$ for all $\tau \in \mathfrak{G}(F^*/K)$ (since both η and ζ are in F). Since $\mathfrak{o}$ is saturated with respect to K, and since $\zeta \in \mathfrak{o}$, we have (2).

2) To prove the second part of the lemma, it is sufficient to prove that if $\mathfrak{o}$ is saturated with respect to K_0 it is also saturated with respect to K.

Let, again, ζ be an element of $\mathfrak{o}$, η an element of $\bar{\mathfrak{o}}$, and assume that

$$(3) \qquad v(\eta^\tau - \eta) \geqq v(\zeta^\tau - \zeta), \text{ for all } \tau \in \mathfrak{G}(F^*/K).$$

We have to prove the inclusion (2).

If c is the v-residue of ζ in $\bar{k}$, then $c \in k$ (since k is a coefficient field of $\mathfrak{o}$), $\zeta - c$ is an element of $\mathfrak{o}$, and we may replace ζ by $\zeta - c$, without affecting $\zeta^\tau - \zeta$. We may therefore assume that $v(\zeta) > 0$. We now fix a primitive element α of k/k_0 and we set $\xi = \alpha + \zeta$. Then $\xi \in \mathfrak{o}$ and we have $\zeta^\tau - \zeta = \xi^\tau - \xi$, for all τ in $\mathfrak{G}(F^*/K)$. To show that $\eta \in \mathfrak{o}$ it will be sufficient to show (since $\mathfrak{o}$ is saturated with respect to K_0) that

$$(4) \qquad v(\eta^\sigma - \eta) \geqq v(\xi^\sigma - \xi), \text{ for all } \sigma \in \mathfrak{G}(F_0^*/K_0).$$

If the restriction $\sigma \mid K$ is the identity, then (4) follows from (3). If $\sigma \mid K$ is not the identity, the necessarily $\sigma \mid k$ is not the identity, since $K = k((x))$ and $\sigma(x) = x$ (in view of the assumption that $K_0 = k_0((x))$. Since $\sigma \mid k_0$ is the identity, it follows that $\alpha^\sigma \neq \alpha$. We have $\xi^\sigma = \alpha^\sigma + \zeta^\sigma$ and $\xi^\sigma - \xi = (\alpha^\sigma - \alpha) + (\zeta^\sigma - \zeta)$. Since $v(\zeta) > 0$, also $v(\zeta^\sigma) > 0$, and since $0 \neq \alpha^\sigma - \alpha \in \bar{k}$, it follows that $v(\xi^\sigma - \xi) = 0$, and (4) follows from the fact that $\eta \in \bar{\mathfrak{o}}$. The proof of the lemma is now complete.

In Section 7 (see end of that section) we shall show by an example that if K_0 is not of the indicated form $k_0((x))$ (or, equivalently, if K is ramified over K_0), then the equality $\tilde{\mathfrak{o}}_{K_0} = \tilde{\mathfrak{o}}_K$ need not be true, even if $k_0 = k$.

LEMMA 2.3. *Let $n = [F : K]$, let η be an element of F, different from zero, and let*

$$\eta_0 = \frac{1}{n} T_{F/K}(\eta),$$

where T denote the trace function. A necessary and sufficient condition that the inequality $v(\eta^\sigma - \eta) > v(\eta)$ hold true for all τ in G is that there exist an element η_0' in K such that $v(\eta - \eta_0') > v(\eta)$, and when that is so then we have already $v(\eta - \eta_0) > v(\eta)$.

Proof. The assertion is trivial if $\eta \in K$. We assume therefore that $\eta \notin K$ and then consequently $\eta \neq \eta_0$. If g denotes the order of the Galois group G, then it is clear that

$$\eta - \eta_0 = \frac{1}{g} \sum_{\tau \in G} (\eta - \eta^\tau).$$

Consequently, if $v(\eta^\tau - \eta) > v(\eta)$ for all $\tau \in G$, then $v(\eta - \eta_0) > v(\eta)$. This proves the necessity of the condition. Conversely, assume that we have $v(\eta - \eta_0') > v(\eta)$ for some element η_0' of K. We have, for all τ in G:

$$\eta^\tau - \eta = (\eta^\tau - \eta_0') - (\eta - \eta_0') = (\eta - \eta_0')^\tau - (\eta - \eta_0').$$

Since $v[(\eta - \eta_0')^\tau] = v(\eta - \eta_0')$, it follows that $v(\eta^\tau - \eta) \geqq v(\eta - \eta_0') > v(\eta)$, and this proves the sufficiency of the condition.

COROLLARY 2.4. *If $T_{F/K}(\eta) = 0$ then $v(\eta^\tau - \eta) = v(\eta)$ for some τ in G. In particular, if we set $\xi = \eta - \eta_0$ then $v(\xi^\tau - \xi) = v(\xi)$ for some τ in G.*

Obvious.

The above lemma (and its Corollary) will be frequently applied in the sequel. A first application is the following result:

PROPOSITION 2.5. *The local domains $\mathfrak{o}$ and $\bar{\mathfrak{o}}_x$ $(= \bar{\mathfrak{o}}_K)$ have the same multiplicity.*

Proof. Let s denote the minimum of the set of positive integers $v(\xi)$, as ξ ranges over the maximal ideal $\mathfrak{m}$ of $\mathfrak{o}$. We shall call s the *reduced multiplicity* [2] of $\mathfrak{o}$; in symbols: $s = e_0(\mathfrak{o})$. Now, since k is also a coefficient field of $\bar{\mathfrak{o}}_x$ (Proposition 2.1), it follows that in order to prove our proposition, we have only to show that s is also the reduced multiplicty of $\bar{\mathfrak{o}}_x$, i.e., that we have $v(\xi) \geqq s$ for every element ξ of the maximal ideal $\mathfrak{m}$ of $\bar{\mathfrak{o}}_x$. We use the representation (1) of $\bar{\mathfrak{o}}_x$ and we prove by induction on i that if $\mathfrak{m}_i$ is the maximal ideal of $\mathfrak{o}_i$ then $v(\xi) \geqq s$ for all ξ in $\mathfrak{m}_i$ (recall that $\mathfrak{o}_0$ is $\mathfrak{o}$). We assume therefore that this is true for all i less than or equal to a given integer $q \geqq 0$. To prove that $v(\xi) \geqq s$ for all ξ in $\mathfrak{m}_{q+1}$ it is sufficient to prove this inequality for all elements ξ of $\mathfrak{m}_{q+1}$ which are of the form $\eta - c$, where $\eta \in L_q$ and c is the v-residue of η (in k). In other words: we have

[2] The *multiplicity* $e(\mathfrak{o})$ of $\mathfrak{o}$ (as distinguished from what we call the *reduced multiplicity* s of $\mathfrak{o}$) is equal to $s.[\bar{k}:k]$. For the proof, apply formula (8) on p. 299 of [1], v. 2, to the case $A = \mathfrak{o}$, $B = \bar{\mathfrak{o}}$, $\mathfrak{q} = \mathfrak{m}$. In that case, $[B:A]$ is 1, $e(\mathfrak{q}) = e(\mathfrak{o})$, the summation on the right hand side of that formula consists of the single term $[\bar{\mathfrak{o}}/\bar{\mathfrak{m}}: \mathfrak{o}/\mathfrak{m}] e(\bar{\mathfrak{o}}\mathfrak{m})$, i.e., $[\bar{k}:k] e(\bar{\mathfrak{m}}^s)$, and this is precisely $[\bar{k}:k]s$.

only to prove the inequality $v(\xi) \geqq s$ for the elements ξ of the set $L_q \cap \mathfrak{m}_{q+1}$ (note that if $\eta \in L_q$ and $c \in k$ then also $\eta - c \in L_q$, by the definition of L_q). Let

$$\xi_0 = \frac{1}{n} T_{F/K}(\xi),$$

and let $\xi' = \xi - \xi_0$. Since $\xi \in \bar{\mathfrak{o}}$, we have $\xi_0 \in k[[x]] \subset \bar{\mathfrak{o}}$, and therefore also ξ' belongs to L_q (since $\xi'^{\tau} - \xi' = \xi^{\tau} - \xi$ for every τ in G). Since $v(\xi^{\tau}) = v(\xi) > 0$ for all τ in G, it follows $v(\xi_0) > 0$, and therefore $\xi_0 \in xk[[x]]$. Since $x \in \mathfrak{m}$, we have $v(\xi_0) \geqq v(x) \geqq s$. Thus in order to prove the inequality $v(\xi) \geqq s$ *it is sufficient to prove that* $v(\xi') \geqq s$. Now, ξ' is an element $L_q \cap \mathfrak{m}_{q+1}$ with the property that $T_{F/K}(\xi') = 0$. Therefore, by Corollary 2.4, we have

$$(5) \qquad v(\xi'^{\tau} - \xi') = v(\xi'), \text{ for some } \tau \text{ in } G.$$

On the other hand, since $\xi' \in L_q$ there exists an element ζ in $\mathfrak{o}_q$ such that

$$(6) \qquad v(\xi'^{\tau} - \xi') \geqq v(\zeta^{\tau} - \zeta), \text{ for all } \tau \text{ in } G.$$

In (6) we may replace ζ by $\zeta - c$, where c is the v-residue of ζ in k. Therefore we may assume that $\zeta \in \mathfrak{m}_q$. By our induction hypothesis, we have then $v(\zeta) \geqq s$. Since also $v(\zeta^{\tau}) = v(\zeta)$, for all τ in G, it follows from (6) that $v(\xi'^{\tau} - \xi') \geqq s$ for all τ in G. This conclusion, together with (5), yields the desired inequality $v(\xi') \geqq s$ and completes the proof of the proposition.

3. Saturators and characteristic of $\mathfrak{o}$ relative to a parameter x.

Definition 3.1. *Let x be a parameter of $\mathfrak{o}$, let y be an element of $\mathfrak{o}$ and let $\mathfrak{o}' = k[[x]][y]$. Then y is called an x-saturator of $\mathfrak{o}$ (more precisely: a K-saturator of $\mathfrak{o}$, where $K = k((x))$), if $\tilde{\mathfrak{o}}_x = \tilde{\mathfrak{o}}_x'$.*

If y is an x-saturator of $\mathfrak{o}$ and if c is the v-residue of y in k, then also $y - c$ is an x-saturator of $\mathfrak{o}$, since $k[[x]][y-c] = k[[x]][y]$. Hence, if there exist x-saturators of $\mathfrak{o}$, there also exist x-saturators of $\mathfrak{o}$ in $\mathfrak{m}$.

Since, by the definition of saturation, the domains $\mathfrak{o}'$ and $\tilde{\mathfrak{o}}_x'$ have the same field of fractions, it follows that if y is an x-saturator of $\mathfrak{o}$ then y is a primitive element of F/K.

COROLLARY 3.2. *If y is an x-saturator of $\mathfrak{o}$ then*

$$\tilde{\mathfrak{o}}_x = \{\eta \in \bar{\mathfrak{o}} \mid v(\eta^{\tau} - \eta) \geqq v(y^{\tau} - y)\}, \textit{ for all } \tau \textit{ in } G.$$

Conversely, if y is an element of $\mathfrak{o}$ slch that $v(\eta^{\tau} - \eta) \geqq v(y^{\tau} - y)$ for all η in $\mathfrak{o}$ and all τ in G, then y is an x-saturator of $\mathfrak{o}$.

The first part of the corollary is merely a re-statement of Lemma 1.4 of our paper [2]. For the converse, we observe first of all that our assumption implies that whenever $y^\tau = y$ (for some τ in G) then $\eta^\tau = \eta$ for all η in $\mathfrak{o}$, and hence also for all η in F. Hence $F = K(y)\,(= K[y])$, and $\bar{\mathfrak{o}}' = \bar{\mathfrak{o}}$, where $\mathfrak{o}' = k[[x]][y]$. Our assumption also implies that $\bar{\mathfrak{o}}_x \subset \bar{\mathfrak{o}}_x'$. The opposite inclusion follows from $\mathfrak{o}' \subset \mathfrak{o}$.

PROPOSITION 3.3. *There exist x-saturators of $\mathfrak{o}$.*

Proof. We fix a finite set of generators $\xi_1, \xi_2, \cdots, \xi_q$ of $\mathfrak{o}$ over $k[[x]]$ and we denote by V the vector space (over k) $\Sigma k \xi_i$. For every τ in G such that τ is not the identity on F we consider the subset V_τ of V consisting of the elements ξ of V such that

$$v(\xi^\tau - \xi) > \min \cdot \{v(\xi_1{}^\tau - \xi_1), v(\xi_2{}^\tau - \xi_2), \cdots, v(\xi_q{}^\tau - \xi_q)\}.$$

It is clear that V_τ is a *proper subspace* of V. Since k is an infinite field, we can find an element y in V which does not belong to any of the subspaces V_τ. We assert that y is an x-saturator of $\mathfrak{o}$. For, if $\mathfrak{o}' = k[[x]][y]$ then $\bar{\mathfrak{o}}_x' \subset \bar{\mathfrak{o}}_x$ (since $\mathfrak{o}' \subset \mathfrak{o}$). On the other hand, if $\tau \in G$ then $v(\xi_i{}^\tau - \xi_i) \geqq v(y^\tau - y)$ for $i = 1, 2, \cdots, q$, since $y \notin V_\tau$ (the above q inequalities also hold formally if τ is the identity on F, for in that case $\xi_i{}^\tau = \xi_i$ and $y^\tau = y$). Hence the ξ_i belongs to $\bar{\mathfrak{o}}_K'$, and consequently $\mathfrak{o} \subset \bar{\mathfrak{o}}_x'$, showing that $\bar{\mathfrak{o}}_x \subset \bar{\mathfrak{o}}_x'$. This completes the proof.

For any τ in G and any η in $\bar{\mathfrak{o}}_x$, $v(\eta^\tau - \eta)$ is either a positive integer (if $\eta^\tau \neq \eta$) or is $+\infty$ (by Proposition 2.1, the v-residue of η in $\bar{k}$ belongs to k and hence η^τ and η have the same v-residue). We set

$$\beta_\tau = \min \cdot \{v(\eta^\tau - \eta), \eta \in \bar{\mathfrak{o}}_x\}, \tag{1}$$

or equivalently, in view of Corollary 3.2:

$$\beta_\tau = v(y^\tau - y), \tag{1'}$$

where y is any x-saturator of $\mathfrak{o}$. Then β_τ is a positive integer, unless $\tau \in \mathfrak{G}(F^*/F)$, in which case $\beta_\tau = +\infty$. We shall only consider those τ in G for which $\beta_\tau \neq \infty$. If $\tau \neq \tau'$, β_τ and $\beta_{\tau'}$ need not be distinct. Let the set of integers β_τ consist of g elements. We arrange these integers in increasing order:

$$\beta_1 < \beta_2 < \cdots < \beta_g, \qquad g \geqq 1. \tag{2}$$

By Proposition 2.5 it follows that if s is the reduced multiplicity of $\mathfrak{o}$ then

$$s \leqq \beta_1. \tag{3}$$

Let $v(x) = n$.

Definition 3.4. *The set of integers* $\{n;\beta_1,\beta_2,\cdots,\beta_g\}$ *is called the x-characteristic of* $\mathfrak{o}$ *(more precisely: the* (x,k)*-characteristic, or the K-characteristic, where* $K=k((x))$*).*

PROPOSITION 3.5. *An element* y *of* $\mathfrak{o}$ *is an x-saturator if and only if we have for all* τ *in* G: $v(y^\tau - y)=\beta_\tau$.

This is an immediate consequence of Corollary 3.2.

We shall now define certain subgroups of the Galois group G of F^*/K (where $K=k((x))$). We set, for $i=1,2,\cdots,g$:

(4) $$G_i=\{\tau\in G \mid v(\eta^\tau-\eta)>\beta_i \text{ for all } \eta\in\bar{\mathfrak{o}}_x\}.$$

Equivalently, we have (in view of the definition (1) of β_τ):

(4′) $$G_i=\{\tau\in G\mid \beta_\tau>\beta_i\},$$

and also (as a mere re-statement of Corollary 3.5):

(4″) $$G_i=\{\tau\in G\mid v(y^\tau-y)>\beta_i\},$$

where y is any x-saturator of $\mathfrak{o}$.

It is easy to see that G_i is a *subgroup* of G. For if $\tau,\sigma\in G_i$, then we write $y^{\sigma\tau}-y=(y^\tau)^\sigma-y^\sigma+y^\sigma-y=(y^\tau-y)^\sigma+(y^\sigma-y)$. Since $v(y^\tau-y)>\beta_i$ and since the valuation v of F^* is invariant under all K-automorphisms of F^* [see part 3) of Lemma 1.2], it follows that $v(y^\tau-y)^\sigma]=v(y^\tau-y)$. Thus, $v[(y^\tau-y)^\sigma]>\beta_i$, and since also $v(y^\sigma-y)>\beta_i$, we see that $v(y^{\sigma\tau}-y)>\beta_i$, i.e., $\sigma\tau\in G_i$. Similarly, if $\tau\in G_i$ then $\tau^{-1}\in G_i$ since $y^{\tau^{-1}}-y=(y-y^\tau)^{\tau^{-1}}$.

Note the following consequences of (4) and of Corollary 3.2:

(5) $\tau\in G_{i-1}$ *and* $\tau\notin G_i$ if and only if $v(y^\tau-y)=\beta_i$; $(i=1,2,\cdots,g)$

(5′) $$\bar{\mathfrak{o}}_x=\{\eta\in\bar{\mathfrak{o}}\mid v(\eta^\tau-\eta)\geqq\beta_\tau, \text{ for all } \tau \text{ in } G\};$$

(5″) $$\bar{\mathfrak{o}}_x=\{\eta\in\bar{\mathfrak{o}}\mid v(\eta^\tau-\eta)\geqq\beta_i, \text{ for all } \tau \text{ in } G_{i-1};\ i=1,2,\cdots,g\}.$$

We set $G_0=G$. We have a strictly descending chain of subgroups:

(6) $$G=G_0>G_1>G_2>\cdots>G_{g-1}>G_g=\mathfrak{G}(F^*/F).$$

We denote by K_i the fixed field of G_i. Then

(7) $$K=K_0<K_1<K_2<\cdots<K_{g-1}<K_g=F.$$

We call the groups G_i *the higher saturation groups of* $\mathfrak{o}$ *relative to* x (more precisely: *relative to* (x,k), *or to* K), and we call the fields K_i *the intermediate saturation fields of* $\mathfrak{o}$ *relative to* x.

2

We shall use consistently the following notations:

$$(8)\qquad \begin{aligned} &e_i = \text{ramification index of } F/K_i, \quad (i = 0, 1, \cdots, g);\\ &n_i = e_{i-1}/e_i, \quad (i = 1, 2, \cdots, g). \end{aligned}$$

In particular, we have therefore

$$(9)\qquad e_0 = n \ (= v(x)); \ e_g = 1,$$

and each n_i is a positive integer.

PROPOSITION 3.6. *For each $i = 1, 2, \cdots, g$, the integer β_i is divisible by e_i.*

Proof. We fix an x-saturator y of $\mathfrak{o}$ and we set

$$y_i' = \frac{1}{[F: K_i]} T_{F/K_i}(y), \qquad i = 0, 1, \cdots, g,$$

and

$$y_i = y_i' - y_{i-1}', \qquad i = 1, 2, \cdots, g;$$

$$y_0 = y_0'.$$

Since $y_g' = y$, it follows that

$$(10)\qquad y = y_0 + y_1 + \cdots + y_g.$$

Since $y \in \bar{\mathfrak{o}}$ and since traces of integral quantities are integral, it follows that

$$(11)\qquad y_i \in \bar{\mathfrak{o}} \cap K_i, \qquad i = 0, 1, \cdots, g.$$

We have

$$y_{i-1}' = \frac{1}{[K_i: K_{i-1}]} \cdot T_{K_i/K_{i-1}}(y_i'),$$

and $y_i = y_i' - y_{i-1}'$. Therefore

$$T_{K_i/K_{i-1}}(y_i) = 0, \qquad i = 1, 2, \cdots, g.$$

This implies, in view of Corollary 2.4, that

$$(12)\qquad \exists \tau \in G_{i-1} \mid v(y_i^\tau - y_i) = v(y_i), \qquad i = 1, 2, \cdots, g.$$

[Note that if K_i^* is the least Galois extension of K_{i-1} which contains K_i, then every K_{i-1}-monomorphism of K_i into K_i^* is induced by some element of G_{i-1}, since $G_{i-1} = \mathfrak{G}(F^*/K_{i-1})$, $G_i = \mathfrak{G}(F^*/K_i)$ and $K_{i-1} \subset K_i$].

For *any* τ in G_{i-1} we have by (10) and (5):

$$(13)\qquad v[(y_i + y_{i+1} + \cdots + y_g)^\tau - (y_i + y_{i-1} + \cdots + y_g)] \geqq \beta_i,$$

with equality if and only if $\tau \notin G_i$. Applying (12) and (13) to the case $i=g$ we find that $v(y_g) \geqq \beta_g$, and applying again (13) to the case $i=g$ and taking τ *not* in G_g, we conclude that

$$(14) \qquad v(y_g) = \beta_g.$$

Assume that it has already been proved that $v(y_j) = \beta_j$, for $j = i+1, i+2, \cdots, g$, where i is some integer such that $1 \leqq i \leqq g-1$. Since $\beta_j > \beta_i$ if $j > i$, it follows from (13) that $v(y_i^\tau - y_i) \geqq \beta_i$ *for all* $\tau \in G_{i-1}$. Hence, in view of (12), it follows that $v(y_i) \geqq \beta_i$. Now, applying (13), with τ *not* in G_i, we find that $v(y_i^\tau - y_i) = \beta_i$, and this, in conjunction with the inequality $v(y_i) \geqq \beta_i$, implies that $v(y_i) = \beta_i$. We have thus proved that

$$(15) \qquad v(y_i) = \beta_i, \qquad i = 1, 2, \cdots, g.$$

Since $y_i \in K_i$ and since e_i is the ramification index of F/K_i, the proof of the proposition is now complete, in view of (15).

We set

$$(16) \qquad \beta_i = m_i e_i, \qquad i = 1, 2, \cdots, g.$$

PROPOSITION 3.7. *The highest common divisor of* β_i *and* e_{i-1} *is* e_i; *in symbols:*

$$(17) \qquad (\beta_i, e_{i-1}) = e_i, \qquad i = 1, 2, \cdots, g,$$

or equivalently

$$(17)' \qquad (m_i, n_i) = 1, \qquad i = 1, 2, \cdots, g,$$

where n_i *and the* e_i *are defined in* (8) *and* (9).

Proof. For any $i = 1, 2, \cdots, g$ let k_i be the algebraic closure of k in K_i. Then $k \subset k_i \subset \bar{k}$, and k_i is a coefficient field of K_i; furthermore $k_{i-1} \subset k_i$ (where we set $k_0 = k$).

Let x_{i-1} be a uniformizing parameter of K_{i-1}; thus $K_{i-1} = k_{i-1}((x_{i-1}))$. By Lemma 1.1, we can find a uniformizing parameter x_i of K_i such that

$$(18) \qquad x_{i-1} = a_i x_i^{n_i}, \qquad a_i \in k_i.$$

We set

$$\Sigma_i = k_i((x_i^{m_i})),$$

and we proceed to prove the following equality:

$$(19) \qquad G_i = G_{i-1} \cap \mathcal{G}(F^*/\Sigma_i).$$

Since $\Sigma_i \subset K_i$ and $G_i = \mathfrak{G}(F^*/K_i)$, the inclusion

$$G_i \subset G_{i-1} \cap \mathfrak{G}(F^*/\Sigma_i)$$

is obvious. So we have only to prove the inclusion

$$(19') \qquad G_i \supset G_{i-1} \cap \mathfrak{G}(F^*/\Sigma_i).$$

Let τ be any element of $G_{i-1} \cap \mathfrak{G}(F^*/\Sigma_i)$. We use the expression (10) of y. Since $\tau \in G_{i-1}$, we have $y_j^\tau = y_j$ for $j = 0, 1, \cdots, i-1$. Hence

$$y^\tau - y = (y_i^\tau - y_i) + (y_{i+1}^\tau - y_{i+1}) + \cdots + (y_g^\tau - y_g).$$

By (15) we have $v(y_j^\tau - y_j) > \beta_i$ if $j > i$. As to $v(y_i^\tau - y_i)$ we observe that y_i is a power series in x_i, with coefficients in k_i, and that this power series begins with a term $cx_i^{m_i}$, $c \in k_i$, since $v(y_i) = \beta_i = m_i e_i$. .Now, $cx_i^{m_i} \in \Sigma_i$, and hence is invariant under τ. It follows that we have also $v(y_i^\tau - y_i) > \beta_i$. This implies by (6) that $\tau \in G_i$, and this proves (19′) and hence also (19).

Since K_i is the fixed field of G_i, it follows from (18) and (19) that $k_i\{\{x_i\}\}$ is the compositum of the two fields $k_{i-1}\{\{x_i^{n_i}\}\}$ and $k_i\{\{x_i^{m_i}\}\}$. This implies at once that $(m_i, n_i) = 1$, and the proof is now complete.

Definition 3.8. *The g pairs of integers (m_i, n_i) $(i = 1, 2, \cdots, g)$ are called the characterstic pairs of $\mathfrak{o}$ with respect to x.*

PROPOSITION 3.9. *Let $\mathfrak{o}'$ be another complete local domain having the same field of fractions F and the same integral closure $\bar{\mathfrak{o}}$ as $\mathfrak{o}$ and having k as cofficient field. Let x be a common parameter of $\mathfrak{o}$ and $\mathfrak{o}'$. We make the following assumptions: the local domains $\mathfrak{o}$, $\mathfrak{o}'$ have the same intermediate saturation fields relative to x and the same x-characteristic. Then $\tilde{\mathfrak{o}}_x = \tilde{\mathfrak{o}}_x'$. In the special case in which $k = \bar{k}$ the assumption that $\mathfrak{o}$ and $\mathfrak{o}'$ have the same intermediate saturation fields with respect to x is already a consequence of the assumption that $\mathfrak{o}$ and $\mathfrak{o}'$ have the same x-characteristic.*

Proof. The assumptions imply that $\mathfrak{o}$ and $\mathfrak{o}'$ have the same saturation groups G_i, relative to x. Since also the x-characteristic $(n; \beta_1, \beta_2, \cdots, \beta_g)$ is the same for both $\mathfrak{o}$ and $\mathfrak{o}'$, the equality $\tilde{\mathfrak{o}}_x = \tilde{\mathfrak{o}}_x'$ follows from (5″). If $k = \bar{k}$, then the intermediate saturation fields K_i and K_i' of $\mathfrak{o}$ and $\mathfrak{o}'$ with respect to x have the same coefficient field, namely k. Therefore the relative desgree $[K_i : K_{i-1}]$ is the ramification index of K_i over K_{i-1}, and similarly for $[K_i' : K_{i-1}']$. If we now assume that $\mathfrak{o}$ and $\mathfrak{o}'$ have the same x-characteristic $(n; \beta_1, \beta_2, \cdots, \beta_g)$, we conclude from Proposition 3.7 that $e_i' = e_i$ $(i = 1, 2, \cdots, g)$, since $e_0' = e_0 = n$ $(= v(x))$ [see (9)]. Therefore $[K_i : K_{i-1}] = [K_i' : K_{i-1}'] = n_i$ [see (8)]. Thus $K_1, K_2, \cdots, K_g$ and also

$K_1', K_2', \cdots, K_g'$ are successive extensions of K $(= k(x))$, of degrees $n_1, n_2, \cdots, n_g$, and since they all have the same coefficient field k as K, these extensions are *cyclic*. They are therefore uniquely determined by K and by the ramification indices (= relative degrees) n_i. This completes the proof.

We shall conclude this section with two results which we shall have occasion to use later on.

First we give the following definition:

Definition 3.10. *If s is the reduced multiplicity of $\mathfrak{o}$, then any parameter x of $\mathfrak{o}$ such that $v(x) = s$ is called a transversal parameter of $\mathfrak{o}$.*

PROPOSITION 3.11. *We have always*

$$\min \cdot \{n, \beta_1\} = s, \tag{20}$$

where $n = v(x)$. In particular, if x is not a transversal parameter, i.e., if $n > s$, then $\beta_1 = s$ (and thus the equality sign holds in (3)).

Proof. Since, by (3), we have $\beta_1 \geqq s$, (20) certainly holds if $n = s$, i.e., if x is a transversal parameter. Now assume that $n > s$. By Proposition 2.5 and by definition of an x-saturator we have that if y is an x-saturator of $\mathfrak{o}$, then s is also the reduced multiplicity of the local domain $k[[x]][y]$. We may assume that $v(y) > 0$. Then the fact that $v(x) > s$ and that s is the reduced multiplicity of $k[[x]][y]$ implies that $v(y) = s$. We now use the decomposition (10) of y. Since $y_0 \in xk[[x]]$, we have $v(y_0) > s$. Therefore $v(y) = v(y_1 + y_2 + \cdots + y_g)$. In view of (15), and since $\beta_1 < \beta_2 < \cdots < \beta_g$, we conclude that $v(y) = v(y_1)$, i.e., $s = \beta_1$, and this establishes (20).

Let y be an x-saturator of $\mathfrak{o}$. By (15), and in the notations of the proof of Proposition 3.7, we have that $y_i \in k_i[[x_i]]$ and, namely:

$$y_i = \alpha_i x_i^{m_i} + \text{terms of degree} > m_i, \qquad \alpha_i \neq 0. \tag{21}$$

PROPOSITION 3.12. *We have $k_i = k_{i-1}(a_i, \alpha_i)$, where a_i is defined in* (18) *and α_i is defined in* (21). *In particular, if $a_i \in k_{i-1}$ then α_i is a primitive element of k_i/k_{i-1}.*

Proof. Let K_i' denote the field $K_{i-1}(\alpha_i x_i^{m_i})$, so that we have $K_{i-1} \subset K_i' \subset K_i$. If $\tau \in \mathfrak{G}(F^*/K_i')$ then, by (21), we have $v(y_i^\tau - y_i) > m_i n_i = \beta_i$, and hence $v(y^\tau - y) > \beta_i$, showing that $\tau \in G_i$ [see (4″)]. Therefore $K_i = K_i' = K_{i-1}(\alpha_i x_i^{m_i}) = K_{i-1}[\alpha_i x_i^{m_i}]$ and from this it follows that $K_i = K_{i-1}[\alpha_i, x_i]$, and hence, by (18), we have $K_i = k_{i-1}(a_i, \alpha_i)((x_i))$. This completes the proof.

Note. The condition $a_i \in k_{i-1}$ can always be satisfied if $v(x_{i-1}) = v(x_i)$, i.e., if $e_{i-1} = e_i$, for in that case we have $n_i = 1$ and we can take for x_i the element x_{i-1}, in which case a_i is 1.

4. The structure theorem for saturated local domains. We assume generally in this section (unless the contrary is explicitly specified) that our local domain $\mathfrak{o}$ is saturated. Thus $\mathfrak{o}$ will be saturated with repect to some parameter x. We fix a parameter x such that $\mathfrak{o}$ is saturated with respect to x and we use the notations of the preceding section as to the higher saturation groups G_i of $\mathfrak{o}$, the intermediate saturation fields K_i of $\mathfrak{o}$, etc. Let $\mathfrak{M}_i = \bar{\mathfrak{m}} \cap K_i$ be the prime ideal of the valuation induced in K_i by the valuation v (here $\bar{\mathfrak{m}}$ denotes the maximal ideal of the integral closure $\bar{\mathfrak{o}}$ of $\mathfrak{o}$). Then the set $v(\mathfrak{M}_i)$ is the set of all positive multiples of e_i, and for any integer q the ideal $\mathfrak{M}_i^q$ is the set of elements ξ of K_i such that $v(\xi) \geqq qe_i$. In particular, $\mathfrak{M}_i^{m_i}$ is the set of elements ξ of K_i such that $v(\xi) \geqq \beta_i$ [see (16), § 3].

THEOREM 4.1. (*the structure theorem*). *If* $\mathfrak{o}$ *is saturated with respect to* x *then*

$$\mathfrak{m} = \mathfrak{M}_0 + \mathfrak{M}_1^{m_1} + \mathfrak{M}_2^{m_2} + \cdots + \mathfrak{M}_g^{m_g}. \tag{1}$$

Proof. Clearly $\mathfrak{M}_0 = xk[[x]] \subset \mathfrak{m}$ since $k[[x]] \subset \mathfrak{o}$ and $x \in \mathfrak{m}$. We shall show that $\mathfrak{M}_i^{m_i} \subset \mathfrak{m}$ for $i = 1, 2, \cdots, g$, thus establishing the inclusion

$$\mathfrak{M}_0 + \mathfrak{M}_1^{m_1} + \mathfrak{M}_2^{m_2} + \cdots + \mathfrak{M}_g^{m_g} \subset \mathfrak{m}. \tag{2}$$

Let ξ be any element $\mathfrak{M}_i^{m_i}$ and let τ be any element of G. We fix an x-saturator y of $\mathfrak{o}$. We must show that $v(\xi^\tau - \xi) \geqq v(y^\tau - y)$. Now, if $\tau \in G_i$ then $\xi^\tau = \xi$, and there is nothing to prove. Assume that $\tau \in G_{j-1}$, $\tau \notin G_j$ for some $j \leqq i$. Then, using (5) of § 3, we find

$$v(y^\tau - y) = \beta_j \leqq \beta_i \leqq v(\xi) \leqq v(\xi^\tau - \xi).$$

This proves inclusion (2). We now proceed to the proof of the opposite inclusion. Quite generally, if η is any element of $\bar{\mathfrak{o}}$, we decompose η into a sum of $g+1$ elements:

$$\eta = \eta_0 + \eta_1 + \cdots + \eta_g, \qquad \eta_i \in \bar{\mathfrak{o}} \cap K_i, \tag{3}$$

by the same method used in § 3 to obtain the decomposition (10) of y. The individual terms η_i are therefore defined as follows:

$$\eta_i = \eta_i' - \eta_{i-1}', \qquad i = 1, 2, \cdots, g;$$
$$\eta_0 = \eta_0',$$

where

$$\eta_i' = \frac{1}{[F:K_i]} T_{F/K_i}(\eta), \qquad i = 0, 1, \cdots, g.$$

We have

$$\eta_{i-1}' = \frac{1}{[K_i:K_{i-1}]} T_{K_i/K_{i-1}}(\eta_i'),$$

and therefore

$$T_{K_i/K_{i-1}}(\eta_i) = 0.$$

This last relation implies, by Corollary 2.4, that

$$(4) \qquad \exists \tau \in G_{i-1} \mid v(\eta_i^\tau - \eta_i) = v(\eta_i), \qquad i = 1, 2, \cdots, g.$$

Now, assume that $\eta \in \mathfrak{m}$. Then, by (4) of Section 3, and by (3) above we find that

$$(5) \quad v(\eta^\tau - \eta) = v[(\eta_i + \eta_{i+1} + \cdots + \eta_g)^\tau - (\eta_i + \eta_{i+1} + \cdots + \eta_g)] \geqq \beta_i, \quad \text{for all } \tau \text{ in } G_{i-1}.$$

Applying (4) and (5) to the case $i = g$ we find that $v(\eta_g) \geqq \beta_g$, whence $\eta_g \in \mathfrak{M}_g^{m_g}$. Assume that it has already been proved that $\eta_j \in \mathfrak{M}_j^{m_j}$ (whence $v(\eta_j) \geqq \beta_j$) for $j = i+1, i+2, \cdots, g$, where i is some integer such that $1 \leqq i \leqq g-1$. Since $\beta_j > \beta_i$ if $j > i$, it follows from (5) that $v(\eta_i^\tau - \eta_i) \geqq \beta_i$ for *all* τ in G_{-1}. Hence, by (4), we find that $v(\eta_i) \geqq \beta_i$, i.e., $\eta_i \in \mathfrak{M}_i^{m_i}$. We have thus shown that $\eta_i \in \mathfrak{M}_i^{m_i}$ for $i = 1, 2, \cdots, g$. Clearly $\eta_0 \in \mathfrak{M}_0$ (since $\eta \in \mathfrak{m}$).

This completes the proof.

COROLLARY 4.2. *An element η of $\bar{\mathfrak{o}}$ is a x-saturator of $\mathfrak{o}$ if and only if*

$$(6) \qquad v(\eta_i^\tau - \eta_i) = \beta_i, \textit{ for all } \tau \textit{ such that } \tau \in G_{i-1}, \tau \notin G_i; \quad (i = 1, 2, \cdots, g)$$

and when that is so then $v(\eta_i) = \beta_i$ (or equivalently: $\eta_i \in \mathfrak{M}_i^{m_i}$, $\eta_i \notin \mathfrak{M}_i^{m_i+1}$) and $K_i = K_{i-1}(\eta_i)$.

We first note that no element τ of G_{i-1} such that (4) holds can belong to G_i, unless $\eta_i = 0$; for if $\tau \in G_i$ then $\eta_i^\tau = \eta_i$. Now, if (6) holds, then necessarily $\eta_i \neq 0$, and therefore, using (4), we find that $\eta_i \in \mathfrak{M}_i^{m_i}$, for $i = 1, 2, \cdots, g$. Since $\eta_0 \in k[[x]] = k + \mathfrak{M}_0$, we conclude that $\eta \in \mathfrak{o}$. Furthermore, from (6) and (5) follows that $v(\eta^\tau - \eta) = \beta_i$ for $\tau \in G_{i-1}$, $\tau \notin G_i$, and this implies that η is an x-saturator of $\mathfrak{o}$ (see Proposition 3.5 and (4′) of Section 3). The converse is an immediate consequence of (5), Section 3 (with y replaced by η) and of (1) and (5), this section.

If $\eta \in \mathfrak{o}$ then $\eta_i \in \mathfrak{M}_i^{m_i}$, i.e., $v(\eta_i) \geqq \beta_i$. If, then, (6) holds, we must have $v(\eta_i) = \beta_i$. Furthermore, if $\tau \in G_{i-1}$ and $\eta_i^\tau = \eta_i$, then, by (6), we must have $\tau \in G_i$. Since $\eta_i \in K_i$, this implies that the subfield $K_{i-1}[\eta_i]$ of K_i coincides, in fact, with the entire field K_i.

Going back to the decomposition (1) of $\mathfrak{m}$ (it being assumed, of course, that $\mathfrak{o}$ is saturated with respect to x), we note that if n $(=v(x))$ is greater than or equal to β_1, then $\mathfrak{M}_0$ is redundant in (1), i.e.,

$$\mathfrak{M}_0 \subset \mathfrak{M}_1^{m_1} + \mathfrak{M}_2^{m_2} + \cdots + \mathfrak{M}_g^{m_g}$$

(in fact, in that case $\mathfrak{M}_0$ is already contained in $\mathfrak{M}_1^{m_1}$); and conversely. We now give the following definition:

Definition 4.3. *If $n < \beta_1$ then* (1) *is called the standard decomposition of $\mathfrak{m}$ relative to x. If $n \geqq \beta_1$ then*

$$\mathfrak{m} = \mathfrak{M}_1^{m_1} + \mathfrak{M}_2^{m_2} + \cdots + \mathfrak{M}_g^{m_g} \tag{7}$$

is the standard decomposition of $\mathfrak{m}$ relative to x.

We note that by Proposition 3.11 the inequality $n < \beta_1$ implies that $n = s$, i.e., that x is a transversal parameter, while if $n \geqq \beta_1$ then $\beta_1 = s$. At any rate, if x is *not* a transversal parameter then (7) is the standard decomposition of $\mathfrak{o}$ relative to x. If x is a transversal parameter, both cases $s < \beta_1$ and $s = \beta_1$ are possible. Precise information on these two cases will be given later on (see Proposition 4.10).

We shall now add to the above structure theorem 4.1 a strong converse, which together with Theorem 4.1 will provide us with a *structural characterization* of our saturater local domains (of dimension 1).

THEOREM 4.4. *Given any ascending finite sequence of fields*

$$K_1 \subset K_2 \subset \cdots \subset K_h = F$$

such that $k \subset K_1$, $[\bar{k}: k]$ and $[F: K_1]$ are finite, given any sequence of positive integers $m_1, m_2, \cdots, m_h$ and setting $\mathfrak{M}_i = \bar{\mathfrak{m}} \cap K_i$,

$$\mathfrak{m} = \mathfrak{M}_1^{m_1} + \mathfrak{M}_2^{m_2} + \cdots + \mathfrak{M}_h^{m_h}, \tag{8}$$

the set $\mathfrak{o} = k + \mathfrak{m}$ is a complete saturated local doman (having k as coefficient field and F as field of fractions).

Proof. We set e_i = ramification index of F/K_i and $\beta_i = e_i m_i$. It is clear that the set $k + \mathfrak{m}$ is a ring, for if $i \leqq j$ then $\mathfrak{M}_i^{m_i}\mathfrak{M}_j^{m_j} \subset \mathfrak{M}_j^{m_j}$. We have $\mathfrak{M}_h = \bar{\mathfrak{m}}$, whence $\mathfrak{o} \supset \bar{\mathfrak{m}}^{m_h}$, showing that $\mathfrak{o}$ is a complete local domain having F as field of fractions, and that the valuation ring $\bar{\mathfrak{o}}$ of v is the integral closure of $\mathfrak{o}$.

The theorem is a consequence of the follwoing lemma:

LEMMA 4.5. *The assumptions and notations being the same as in Theorem 4.4, let x be any element of $\mathfrak{M}_1^{m_1}$. Then $\mathfrak{o}$ is saturated with respect to the parameter x.*

Proof of the lemma. If for some pair of indices i, j we have $i < j$ and $\beta_i \geqq \beta_j$, then $\mathfrak{M}_i^{m_i} \subset \mathfrak{M}_j^{m_j}$ and $\mathfrak{M}_i^{m_i}$ is superfluous in (8). Thus, by omitting, if necessary, some of the fields K_i and the corresponding ideals $\mathfrak{M}_i^{m_i}$, we obtain a new representation of m:

$$(8') \qquad \mathfrak{m} = \mathfrak{M}_{i_1}^{m_{i_1}} + \mathfrak{M}_{i_2}^{m_{i_2}} + \cdots + \mathfrak{M}_{i_\tau}^{m_{i_\tau}},$$

in which we have $\beta_{i_1} < \beta_{i_2} < \cdots < \beta_{i_\tau}$, with $1 \leqq i_1 < i_2 < \cdots < i_\tau = h$. If $i_1 = 1$, the statement of the lemma is not affected if we replace the decomposition (8) by (8′). If $i_1 > 1$, i.e., if $\mathfrak{M}_1^{m_1}$ has been omitted in the process, then necessarily β_1 must be greater than or equal to some of the β_{i_j} $(j = 1, 2, \cdots, \tau)$. Hence $\beta_1 \geqq \beta_{i_1}$, $\mathfrak{M}_1^{m_1} \subset \mathfrak{M}_{i_1}^{m_{i_1}}$, $x \in \mathfrak{M}_{i_1}^{m_{i_1}}$, and thus, if we prove the lemma with reference to the decomposition (8′) of $\mathfrak{m}$ it will follows *a fortiori* that the lemma is true with reference to our original decomposition (8) of $\mathfrak{m}$. We may therefore assume that

$$(9) \qquad \beta_1 < \beta_2 < \cdots < \beta_h.$$

We set $K = K_0 = k((x))$. Since we have assumed that $[\bar{k} : k]$ is finite, and since $x \in \bar{\mathfrak{m}}$ (in view of $m_1 > 0$), it follows that $[F : K]$ is finite. We denote, as usual, by G the Galois group of F^*/K, where F^* is the least Galois extension of K containing F. Let ζ, η be elements of $\bar{\mathfrak{o}}$ such that $\zeta \in \mathfrak{m}$, and such that

$$(10) \qquad v(\eta^\tau - \eta) \geqq v(\zeta^\tau - \zeta), \text{ for all } \tau \text{ in } G.$$

We have to prove that $\eta \in \mathfrak{o}$. Since $\zeta \in \mathfrak{m}$, we can write

$$\zeta = \zeta_1 + \zeta_2 + \cdots + \zeta_h, \quad \zeta_i \in \mathfrak{M}_i^{m_i}.$$

We proceed as in the proof of Theorem 4.1. Using the sequence of fields $K = K_0 \subset K_1 < K_2 < \cdots < K_h = F$, we write the decomposition (3) of η, using the successive traces η_i'. Then, upon setting $G_i = \mathfrak{G}(F^*/K_i)$ $(i = 0, 1, \cdots, h)$, we obtain (4) (with g replaced by h). We then show by induction from $i+1$ to i that $\eta_i \in \mathfrak{M}_i^{m_i}$ $(i = 1, 2, \cdots, h)$. That part of the proof of Theorem 4.1 was based entirely on (4) and on the inequalities (5), and these inequalties were valid then because it was *given* that η belongs to $\mathfrak{o}$. In the present case, however, the inequalities (5) are still valid, but for a different reason: they simply follow directly from (9) and (10), from the fact that $v(\zeta_i) \geqq \beta_i$ (since $\zeta_i \in \mathfrak{M}_1^{m_i}$) and from the equality

$$\xi^\tau - \xi = (\zeta_i^\tau - \zeta_i) + (\zeta_{i+1}^\tau - \zeta_{i+1}) + \cdots + (\zeta_h^\tau - \zeta_h),$$

which holds for all τ in G_{i-1}.

As to η_0 we observe that since $\eta \in \bar{\mathfrak{o}}$ we have

$$\eta \in k[[x]] \subset k + \mathfrak{M}_0 \subset k + \mathfrak{M}_1^{m_1},$$

since $x \in \mathfrak{M}_1^{m_1}$. This completes the proof.

We note the following consequence of Theorem 4.1 and Lemma 4.5:

COROLLARY 4.6. *If a local domain* $\mathfrak{o}$ *is saturated, it is also saturated with respect to some transversal parameter.*

For, by Theorem 4.1 we have a decomposition of $\mathfrak{m}$ of type (8), with the inequalities (9) being valid. It is clear that the reduced multiplicity of $\mathfrak{o}$ is then β_1. It is then sufficient to take an element x of $\mathfrak{m}$ such that $x \in \mathfrak{M}_1^{m_1}$, $x \notin \mathfrak{M}_1^{m_1+1}$ and apply Lemma 4.5.

In Section 7 we will derive a much stronger result (see Theorem 7.2).

Theorems 4.1 and 4.4 together yield the following necessary and sufficient condition for our local domain $\mathfrak{o}$ to be saturated: $\mathfrak{o}$ *is saturated if and only if the maximal ideal* $\mathfrak{m}$ *of* $\mathfrak{o}$ *has a decomposition of the form* (8) *described in Theorem* 4.4. In addition, Theorem 4.1 says that if $\mathfrak{o}$ is saturated then $\mathfrak{m}$ has such decompositions which enjoy certain special properties; these decompositions are obtained by choosing any parameter x of $\mathfrak{o}$ such that $\mathfrak{o}$ is saturated with respect to x and considering the *standard decomposition* of $\mathfrak{m}$ relative to x (Definition 4.3). The special properties which these decompositions possess are spelled out in the following definition (compare with (2), Section 3 and Proposition 3.7).

Definition 4.7. *A decomposition* (8) *of* $\mathfrak{m}$ *is called standard if* $K_i \neq K_{i+1}$ *for* $i = 1, 2, \cdots, h-1$, *and if the following two conditions are satisfied* (*we use the notations of Theorem* 4.4 *and of the proof of that theorem*):

(a) $\beta_1 < \beta_2 < \cdots < \beta_h$;

(b) $(\beta_i, e_{i-1}) = e_i, \quad i = 2, 3, \cdots, h.$

In the next section we shall study properties of standard decompositions and we shall also show that every standard decomposition of $\mathfrak{m}$ is in fact the standard decomposition of $\mathfrak{m}$ relative to some parameter x of $\mathfrak{o}$ with respect to which $\mathfrak{o}$ is saturated.

We introduce some notations which we will be using consistently throughout the remainder of this section, and in the next section as well.

Whether $\mathfrak{o}$ is or is not saturated, we consider the set $\mathfrak{m}v$ (= set of all integers $v(\xi)$, $\xi \in \mathfrak{m}$) and for any integer γ in $\mathfrak{m}v$ we denote by $k^{(\gamma)}$ the subfield of $\bar{k}$ which is generated over k by the v-residue (in $\bar{k}$) of those elements of F which are of the form ξ/η, where $\xi, \eta \in \mathfrak{m}$ and $v(\xi) = v(\eta) = \gamma$.

We note that $\mathfrak{m}v$ is a semigroup of positive integers and that the reduced multiplicty s of $\mathfrak{o}$ is the smallest integer in $\mathfrak{m}v$. The field $k^{(s)}$ will be of particular interest to us.

If $\mathfrak{o}$ is saturated and we are given a decomposition (8) of $\mathfrak{m}$, then we set $k_i = K_i \cap \bar{k}$ $(i = 1, 2, \cdots, h)$. For each i, k_i is therefore the algebraic closure of k in K_i and is the coefficient field of K_i containing k.

Definition 4.8. *The fields* $k^{(\gamma)}$ $(\gamma \in \mathfrak{m}v)$ *will be called the higher residue fields of* $\mathfrak{o}$. *If* $\mathfrak{o}$ *is saturated and* (8) *is a standard decomposition of* $\mathfrak{m}$, *then the* h *fields* $K_1, K_2, \cdots, K_h$ *and the* h *fields* $k_1, k_2, \cdots, k_h$ *will be called respectively the intermediate fields and the characteristic coefficient fields of* $\mathfrak{o}$, *relative to the given decomposition* (8). *Furthermore, the set of integers* $\{e_1; \beta_1, \beta_2, \cdots, \beta_h\}$ *will be called the characteristic of the standard decomposition* (8).

PROPOSITION 4.9. *Let* $\mathfrak{o}$ *be saturated and let* (8) *be a standard decomposition of* $\mathfrak{m}$. *Then* $\beta_1 = s$ (= *reduced multiplicity of* $\mathfrak{o}$) *and* $k^{(s)} = k_1$.

Proof. The assertion $\beta_1 = s$ is obvious (and follows exclusively from the inequalities $\beta_1 < \beta_i$, $i = 2, 3, \cdots, h$). If $0 \neq c \in k_1$, we fix an element ξ in $\mathfrak{M}_1^{m_1}$ which does not belong to $\mathfrak{M}_1^{m_1+1}$. Then $\xi, c\xi \in \mathfrak{m}$, $v(\xi) = v(c\xi) = s$, and c is the v-residue of $\dfrac{c\xi}{c}$ (in $\bar{k}$). Thus $k_1 \subset k^{(s)}$.

Conversely, if ξ, η are two elements of $\mathfrak{m}$ such that $v(\xi) = v(\eta) = s$ (in other words, if ξ and η are transversal parameters of $\mathfrak{o}$), we can write: $\xi = \xi_1 + \xi_2 + \cdots + \xi_h$, $\eta = \eta_1 + \eta_2 + \cdots + \eta_h$, where $\xi_i, \eta_i \in \mathfrak{M}_i^{m_i}$ $(i = 1, 2, \cdots, h)$ and where necessarily $v(\xi_1) = v(\eta_1) = s$, while $v(\xi_i) \geqq \beta_i > s$ and $v(\eta_i) \geqq \beta_i > s$ for $i = 2, 3, \cdots, h$. Therefore the v-residue of ξ/η is the same as the v-residue of ξ_1/η_1 and thus belongs to k_1. This shows that $k^{(s)} \subset k_1$ and completes the proof.

Let $\mathfrak{o}$ be saturated with respect to a parameter x, let $v(x) = n$, and let $\{n; \beta_1, \beta_2, \cdots, \beta_g\}$ be the x-characteristic of $\mathfrak{o}$ (Definition 3.4). We have seen earlier in this section (see Definition 4.3) that if x is not transversal then $n > \beta_1$, and (7) (rather than (1)) is the standard decomposition of $\mathfrak{m}$ relative to x. If x is transversal (whence $n = s$), then both cases

$s < \beta_1$ and $s = \beta_1$ are possible, and the standard decomposition of $\mathfrak{m}$ is then respectively (1) or (7). Which one of these two cases arises is a question which is answered by the following proposition:

PROPOSITION 4.10. *Let $\mathfrak{o}$ be saturated with respect to a transversal parameter x, let $\{s; \beta_1, \beta_2, \cdots, \beta_g\}$ be the x-characteristic of $\mathfrak{o}$ and let $k, k_1, k_2, \cdots, k_g$ be the characteristic coefficient fields of $\mathfrak{o}$ relative to the decomposition* (1) *of $\mathfrak{m}$. Then $s < \beta_1$ (and hence* (1) *is a standard decomposition of $\mathfrak{m}$) if and only if $k^{(s)} = k$. If, however, $s = \beta_1$ (in which case* (1) *is a standard decomposition of $\mathfrak{m}$), then $k^{(s)} = k_1 > k$, and the exponent m_1 of $\mathfrak{M}_1$ in* (7) *is equal to* 1.

Proof. If $s < \beta_1$ then the assertion that $k^{(s)} = k$ follows directly from the preceding Proposition 4.9 and from the fact that (1) is the standard decomposition of $\mathfrak{m}$. Assume now that $s = \beta_1$. In that case, (7) is a standard decomposition of $\mathfrak{m}$, and hence, again by Proposition 4.9, we have $k^{(s)} = k_1$. We observe now that we have $(\beta_1, s) = e_1$ (see Proposition 3.7 for $i = 1$ and take into account (9) of Section 3), and therefore (since we have assumed that $s = \beta_1$) $e_0 = e_1 = s$. From $\beta_1 = e_1$ $(= s)$ follows the equality $m_1 = 1$ (see (16), Section 3). From $e_0 = e_1$ and K_0 $(= K) < K_1$ follows that $k < k_1$, which in conjunction with the above established equality $k_1 = k^{(s)}$ yields $k < k^{(s)}$. This completes the proof.

It is of interest to note here that whether $\mathfrak{o}$ is saturated or not, the field $k^{(s)}$ (the first of the higher residue fields of $\mathfrak{o}$) appears also in a different connection. We have namely the following result:

PROPOSITION 4.11. *Without assuming that $\mathfrak{o}$ is saturated, let $\mathfrak{o}'$ denote the quadratic transform of $\mathfrak{o}$. Then $k^{(s)}$ is a coefficient field of the* (*complete*) *local domain $\mathfrak{o}'$.*

Proof. Let $\{x_1, x_2, \cdots, x_N\}$ be a basis of $\mathfrak{m}$ and let, say,

$$v(x_1) = s \quad (= \min \cdot \{v(x_1), v(x_2), \cdots, v(x_N)\}).$$

Then we have

$$\mathfrak{o}' = \mathfrak{o}\left[\frac{x_2}{x_1}, \frac{x_3}{x_1}, \cdots, \frac{x_N}{x_1}\right],$$

and $\mathfrak{o}'$ is complete (since $\mathfrak{o} \subset \mathfrak{o}' \subset \bar{\mathfrak{o}}$). If c_i is the v-residue of x_i/x_1 in $\bar{k}$, then the field $k' = k[c_2, c_3, \cdots, c_N]$ is a coefficient field of $\mathfrak{o}'$. Now if $v(x_i) > s$ then $c_i = 0$; and if $v(x_i) = s$ then $c_i \in k^{(s)}$. Hence $k' \subset k^{(s)}$. On the other hand, if $\xi, \eta \in \mathfrak{m}$ and $v(\xi) = v(\eta) = s$, then we can write

$$\xi = a_1x_1 + a_2x_2 + \cdots + a_Nx_N,$$
$$\eta = b_1x_1 + b_2x_2 + \cdots + b_Nx_N,$$

where $a_i, b_i \in \mathfrak{o}$. Let $\bar{a}_i, \bar{b}_i$ be the v-residues of a_i, b_i in $\bar{k}$. Then

$$0 \neq v\text{-residue of } \frac{\xi}{x_1} = \bar{a}_1 + \bar{a}_2c_2 + \cdots + \bar{a}_Nc_N,$$

$$0 \neq v\text{-residue of } \frac{\eta}{x_1} = \bar{b}_1 + \bar{b}_2c_2 + \cdots + \bar{b}_Nc_N.$$

Since $\bar{a}_i, \bar{b}_i \in k$, this shows that the v-residue of ξ/η in $\bar{k}$ belongs to k'. Hence $k^{(s)} \subset k'$, and thus $k^{(s)} = k'$.

We shall now discuss another point concerning the field $k^{(s)}$. We know that the coefficient field k of $\mathfrak{o}$ is also a coefficient field of any saturation of $\mathfrak{o}$ (Proposition 2.1). We may now ask what is the effect of saturation on the field $k^{(s)}$. Let $\tilde{\mathfrak{o}}_x$ be the saturation of $\mathfrak{o}$ with respect to some parameter x of $\mathfrak{o}$, let $k^{(s)}$ be the first of the higher residue fields of $\mathfrak{o}$ and let $\tilde{k}^{(s)}$ be the first of the higher residue fields of $\tilde{\mathfrak{o}}_x$ (here s = reduced multiplicity of $\mathfrak{o}$ = reduced multiplicity of $\tilde{\mathfrak{o}}_x$; see Proposition 2.5). It is clear, of course, that $k^{(s)} \subset \tilde{k}^{(s)}$.

PROPOSITION 4.12. *If x is a transversal parameter then $k^{(s)} = \tilde{k}^{(s)}$, while if x is not transversal it may very well happen that $k^{(s)}$ is a proper subfield of $\tilde{k}^{(s)}$.*

Proof. We denote by $\tilde{\mathfrak{m}}$ the maximal ideal of $\tilde{\mathfrak{o}}_x$. Assume x transversal. We may also assume that $\tilde{k}^{(s)} \neq k$, for otherwise $\tilde{k}^{(s)} = k^{(s)} = k$. By Proposition 4.10, the standard decomposition of $\mathfrak{m}$ with respect to x is

$$\tilde{\mathfrak{m}} = \mathfrak{M}_1 + \mathfrak{M}_2^{m_2} + \cdots + \mathfrak{M}_g^{m_g},$$

and therefore, by Proposition 4.9, we have $\tilde{k}^{(s)} = k_1$. Let now y be an x-saturator of $\mathfrak{o}$ and let us consider the first term y_0 in the decomposition (10), Section 3, of y. Let $\xi = y - y_0 = y_1 + y_2 + \cdots + y_g$. Since $y_0 \in k[[x]] \subset \mathfrak{o}$, also $\xi \in \mathfrak{o}$. By (15) and (2) of Section 3 it follows that $v(\xi) = \beta_1$ $(= s)$. Hence, if c is the v-residue of $\dfrac{\xi}{x}$ in $\bar{k}$ then $c \in k^{(s)}$. On the other hand, by (5), Section 3, we have that $v(\xi^\tau - \xi) = \beta_1$ $(= s)$ for all τ in G, such that $\tau \notin G_1$. This implies that $c^\tau \neq c$ for all such τ. Since $c \in k^{(s)} \subset \tilde{k}^{(s)} = k_1 \subset K_1$, we have also $c^\tau = c$ for all τ in G_1. Therefore G_1 is the Galois group of $F^*/K(c)$. This implies that $K_1 = K(c)$, whence $k_1 = k(c) \subset k^{(s)}$, i.e., $\tilde{k}^{(s)} \subset k^{(s)}$. This establishes the equality $k^{(s)} = \tilde{k}^{(s)}$.

Consider now the following example. Let k be the field of rational numbers, let $F = k(\sqrt{2})((t))$, let $x = \sqrt{2}\cdot t^2$ and let $\mathfrak{o} = k[[x]][t] = k[[x,t]]$. In this case we have $s = 1$, x is not a transversal parameter of $\mathfrak{o}$, and $k^{(1)} = k$. The field F is an algebraic extension of K, of degree 4, and we have $F = K(t)$, the minimal polynomial of t over K being $T^4 - \frac{x^2}{2}$. The least Galois extension F^* of K containing F is the field $k^*((t))$, where $k^* = k(\sqrt{2}, \sqrt{-1})$, and the Galois group G of F^*/K has order 8. The saturation $\tilde{\mathfrak{o}}_x$ is the set of all elements η of $\bar{\mathfrak{o}}$ $(= k(\sqrt{2})[[t]]$ such that $v(\eta^\tau - \eta) \geqq v(t^\tau - t)$, for all τ in G for which $t^\tau \neq t$. But for any such τ we have either $t^\tau = \pm\sqrt{-1}\cdot t$ or $t^\tau = -t$, and hence $v(t^\tau - t) = 1$. Therefore the only condition on η is $v(\eta^\tau - \eta) \geqq 1$. This implies that $\tilde{k}^{(1)} = k(\sqrt{2})$.

The above proposition can be related to a question which we shall discuss in more detail in a subsequent paper, namely to the question of permutability of the following two operations: saturation and quadratic transformation. Let x be a parameter of our complete (this time not necessarly saturated) local domain $\mathfrak{o}$, let $\mathfrak{o}'$ be the quadratic transform of $\mathfrak{o}$, $(\tilde{\mathfrak{o}}_x)'$ the quadratic transform of the saturation $\tilde{\mathfrak{o}}_x$ of $\mathfrak{o}$ with respect to x. Let $k^{(s)}$ and $\tilde{k}^{(s)}$ have the same meaning as in Proposition 4.12. We know that $k^{(s)}$ is a coefficient field of $\mathfrak{o}'$ (Proposition 4.11). We denote by $(\tilde{\mathfrak{o}}')_x$ the saturation of $\mathfrak{o}'$ with respect to x; more precisely: the saturation of $\mathfrak{o}'$ with respect to the field $k^{(s)}((x))$. We know that $k^{(s)}$ is also a coefficient field of $(\tilde{\mathfrak{o}}')_x$ (Proposition 2.1), and that $\tilde{k}^{(s)}$ is a coefficient field of $(\tilde{\mathfrak{o}}_x)'$ (Proposition 4.11). If, then, $k^{(s)} \neq \tilde{k}^{(s)}$, then $(\tilde{\mathfrak{o}}')_x \neq (\tilde{\mathfrak{o}}_x)'$.

5. Properties of standard decompositions. We begin by proving a proposition which can be regarded in some sense, as a complement to our definition (Definition 4.7) of a standard decomposition.

PROPOSITION 5.1. *The notation being the same as in Theorem* 4.4, *a decomposition of* $\mathfrak{m}$ *(as given by* (8), *Section* 4):

$$\mathfrak{m} = \mathfrak{M}_1^{m_1} + \mathfrak{M}_2^{m_2} + \cdots + \mathfrak{M}_h^{m_h}, \tag{1}$$

is standard if and only if the following two conditions are satisfied.

(a_1) *no term* $\mathfrak{M}_i^{m_i}$ *is redundant in* (1);

(b_1) $(\beta_i, e_{i-1}) = e_i$, *for* $i = 2, 3, \cdots, h$.

Proof. We shall refer to conditions (a) and (b) of Definition 4.7. If (a) is not satisfied, i.e., if we have $\beta_i \geqq \beta_j$ for some pair of indices i, j such that $i < j$, then $\mathfrak{M}_i^{m_i} \subset \mathfrak{M}_j^{m_j}$ and hence $\mathfrak{M}_i^{m_i}$ is redundant in (1).

Since the conditions (b) and (b_1) are identical, this shows that if (a_1) and (b_1) are satisfied then (1) is a standard decomposition.

Conversely, assume that (1) is a standard decomposition. We consider a specific term $\mathfrak{M}_i^{m_i}$ in (1) and we prove that it is not redundant in (1). This is obvious if $i=1$, for $\mathfrak{M}_1^{m_1}$ contains elements whose value in v is exactly β_1 (namely the elements of $\mathfrak{M}_1^{m_1}$ which do not belong to $\mathfrak{M}_1^{m_1+1}$), while every element in $\mathfrak{M}_2^{m_2}+\mathfrak{M}_3^{m_3}+\cdots+\mathfrak{M}_h^{m_h}$ has value $\geqq \beta_2 > \beta_1$ [see condition (a) of Definition 4.7]. We shall therefore assume that $i>1$. We then consider separately two cases, according as $e_i < e_{i-1}$ or $e_i = e_{i-1}$.

Case 1. $e_i < e_{i-1}$. Again we observe that $\mathfrak{M}_i^{m_i}$ contains elements whose value in v is exactly β_i. By condition (b) of Definition 4.7 and by our assumption that $e_i < e_{i-1}$ it follows that *β_i is not divisible by e_{i-1}*. On the other hand, we can easily see that *if ξ is any element of*

$$\mathfrak{M}_1^{m_1}+\mathfrak{M}_2^{m_2}+\cdots+\mathfrak{M}_{i-1}^{m_{i-1}}+\mathfrak{M}_{i+1}^{m_{i+1}}+\cdots+\mathfrak{M}_h^{m_h}$$

such that $v(\xi) < \beta_{i+1}$ then $v(\xi)$ is divisible by e_{i-1} (this will prove that $\mathfrak{M}_i^{m_i}$ is not redundant, since $\beta_i < \beta_{i+1}$). For, if we write $\xi = \eta + \zeta$, where $\eta \in \mathfrak{M}_1^{m_1}+\mathfrak{M}_2^{m_2}+\cdots, \mathfrak{M}_{i-1}^{m_{i-1}}$ and $\zeta \in \mathfrak{M}_{i+1}^{m_{i+1}}+\cdots+\mathfrak{M}_h^{m_h}$, then $v(\zeta) \geqq \beta_{i+1} > v(\xi)$, whence $v(\xi) = v(\eta)$. Now, $\eta \in K_{i-1}$ and hence $v(\eta)$ is divisible by e_{i-1}, and the assertion is proved.

Case 2. $e_i = e_{i-1}$. Let $k_1, k_2, \cdots, k_h$ be the characteristic coefficient field of $\mathfrak{o}$ relative to the given standard decomposition (1). Let x_j be a uniformizing parameter of K_j (whence $K_j = k_j((x_j))$). Since $e_{i-1} = e_i$, x_{i-1} is also a uniformizing parameter of K_i, and we have therefore $K_{i-1} = k_{i-1}((x_{i-1}))$, $K_i = k_i((x_{i-1}))$. Since $K_{i-1} < K_i$, it follows that $k_{i-1} < k_i$. We fix an element c in k_i such that $c \notin k_{i-1}$ and we consider the element $z = cx_{i-1}^{m_i}$ of $\mathfrak{M}_i^{m_i}$. We have $v(z) = \beta_i$, and the v-residue (in $\bar{k}$) of the quotient $z/x_{i-1}^{m_i}$ is the element c which does not belong to k_{i-1}. On the other hand, we can easily see that if *ξ is any element of*

$$\mathfrak{M}_1^{m_1}+\mathfrak{M}_2^{m_2}+\cdots+\mathfrak{M}_{i-1}^{m_{i-1}}+\mathfrak{M}_{i+1}^{m_{i+1}}+\cdots+\mathfrak{M}_h^{m_h},$$

then the v-residue (in $\bar{k}$) of the quotient $\xi/x_{i-1}^{m_i}$ belongs to k_{i-1} or is infinite (this will prove that $\mathfrak{M}_i^{m_i}$ is not redundant in (1)). For if we write $\xi = \eta + \zeta$, where $\eta \in \mathfrak{M}_1^{m_1}+\mathfrak{M}_2^{m_2}+\cdots+\mathfrak{M}_{i-1}^{m_{i-1}}$ and

$$\zeta \in \mathfrak{M}_{i+1}^{m_{i+1}}+\mathfrak{M}_{i+2}^{m_{i+2}}+\cdots+\mathfrak{M}_h^{m_h},$$

then

$$v(\xi/x_{i-1}^{m_i}-\eta/x_{i-1}^{m_i}) = v(\zeta/x_{i-1}^{m_i}) \geqq \beta_{i+1}-\beta_i > 0,$$

showing that the v-residues (in $\bar{k}$) of $\xi/x_{i-1}^{m_i}$ and $\eta/x_{i-1}^{m_i}$ are equal. Since $\eta/x_{i-1}^{m_i} \in K_{i-1}$, the assertion is proved.

Before we proceed with the derivation of other properties of standard decompositions, we wish to observe that by Theorems 4.1 and 4.4 our saturated local domains $\mathfrak{o}$ could also be defined by the property that the maximal ideal $\mathfrak{m}$ of $\mathfrak{o}$ admits a decomposition of the type specified in Definition 4.7. If such a decomposition exists, then we needed Theorem 4.1 and Lemma 4.5 in order to derive the existence of a *standard* decomposition. It is therefore of interest to give a *direct* method which leads from any decomposition of the type specified in Definition 4.7 to a *standard* decomposition ("direct" in the sense that it does not go via the concept of saturation). We shall prove the following result:

PROPOSITION 5.2. *If the maximal ideal* $\mathfrak{m}$ *of* v *admits a decomposition of type* (1), *relative to a sequence of fields* $K_1 < K_2 < \cdots < K_h = F$ (*where we assume, as in Theorem* 4.4, *that* $k \subset K_1$ *and that* $[\bar{k}: k]$ *and* $[F: K_1]$ *are finite*), *then by repeated deletions of redundant terms* $\mathfrak{M}_i^{m_i}$ *and insertions of new fields in the given sequence of fields* K_i, *it is possible to obtain a standard decomposition of* $\mathfrak{m}$.

Proof. First of all we delete the redundant terms in (1). We may therefore assume that no term in (1) is redundant (and that therefore $\beta_1 < \beta_2 < \cdots < \beta_h$). Assume, for a moment, that for some index i $(1 \leqq i \leqq h-1)$ we have $k_i = k_{i+1}$ *and* $\beta_{i+1} \equiv 0 \pmod{e_i}$. Then, if we set $\beta_{i+1} = e_i m_{i+1}'$, it is clear that we have $\mathfrak{M}_{i+1}^{m_{i+1}} = \mathfrak{M}_i^{m_{i+1}'} + \mathfrak{M}_{i+1}^{m_{i+1}+1}$. Since $m_{i+1}' > m_i$, we have $\mathfrak{M}_i^{m_i} + \mathfrak{M}_{i+1}^{m_{i+1}} = \mathfrak{M}_i^{m_i} + \mathfrak{M}_{i+1}^{m_{i+1}+1}$. This gives a new decomposition of $\mathfrak{m}$, obtained by replacing in (1) the sum $\mathfrak{M}_i^{m_i} + \mathfrak{M}_{i+1}^{m_{i+1}}$ by the sum $\mathfrak{M}_i^{m_i} + \mathfrak{M}_{i+1}^{m_{i+1}+1}$. The new integer analogous to β_{i+1} is $\beta_{i+1}' = \beta_{i+1} + e_{i+1}$. The term $\mathfrak{M}_{i+1}^{m_{i+1}+1}$ in the new composition cannot be redundant, because in the contrary case the term $\mathfrak{M}_{i+1}^{m_{i+1}}$ in the original decomposition would be redundant. It follows that no term in the new decomposition of $\mathfrak{m}$ is redundant. This time, however, we have $\beta_{i+1}' \not\equiv 0 \pmod{e_i}$ since $\beta_{i+1} \equiv 0 \pmod{e_i}$ and $e_{i+1} < e_i$ (in view of $k_i = k_{i+1}$ and $K_i < K_{i+1}$). We may therefore assume that the decomposition (1) is not only irredundant but has also the following property: *if for a given* i $(1 \leqq i \leqq h-1)$ *we have* $k_i = k_{i+1}$ *then* $\beta_{i+1} \not\equiv 0 \pmod{e_i}$.

Let us assume that the sequence (1) is standard up to and including its i-th term, i.e., that we have $(\beta_{j+1}, e_j) = e_{j+1}$ for $j = 1, 2, \cdots, i-1$, but that $(\beta_{i+1}, e_i) = e_{i+1}\epsilon$, with $\epsilon > 1$ (the first part of this assumption is vacuous if $i = 1$). We shall now construct an irredundant decomposition of $\mathfrak{o}$, of the same type as (1), which is standard up to and including the $(i+1)$-th term and which is relative either to the sequence of fields

$$K_1 < K_2 < \cdots < K_i < K_{i+1}' < K_{i+1} < \cdots < K_h$$

or to the sequence of fields

$$K_1 < K_2 < \cdots < K_i < K_{i+1}' < K_{i+2} < \cdots < K_h,$$

where K_{i+1}' is some field between K_i and K_{i+1}. This procedure must ultimately lead to a standard decomposition of $\mathfrak{m}$, since the number of terms in any decomposition such as (1) is bounded by the relative degree $[F: K_1]$.

Let $\beta_{i+1} = e_{i+1}\epsilon m_{i+1}'$, $e_i = e_{i+1}\epsilon n_{i+1}'$, where $(m_{i+1}', n_{i+1}') = 1$. We may choose the uniformizing parameters x_i and x_{i+1} of K_i and K_{i+1} in such a manner as to have $x_i = a x_{i+1}^{n_{i+1}}$, where $n_{i+1} = e_i/e_{i+1}$ and $a \in k_{i+1}$ (see Lemma 1.1). We have $n_{i+1} = \epsilon n_{i+1}'$. We set $x_{i+1}' = x_{i+1}^{\epsilon}$ and $K_{i+1}' = k_{i+1}((x_{i+1}'))$. Then $x_i = a x_{i+1}'^{n_{i+1}'}$, and therefore $K_i \subset K_{i+1}' \subset K_{i+1}$. We cannot have $K_i = K_{i+1}'$, because in the contrary case we would have $k_i = k_{i+1}$ and $n_{i+1}' = 1$, i.e., $\beta_{i+1} \equiv 0 \pmod{e_i}$, contrary to the assumed property of the decomposition (1). We also have $K_{i+1}' < K_{i+1}$, since $\epsilon > 1$. Thus $K_i < K_{i+1}' < K_{i+1}$. After inserting the new field K_{i+1}', the pair of ramification indices e_i, e_{i+1} and the coefficient fields k_i, k_{i+1} are replaced by the triplets

$$e_i,\ e_{i+1}'\ (= e_{i+1}\epsilon),\ e_{i+1};$$

$$k_i,\ k_{i+1}'\ (= k_{i+1}),\ k_{i+1}.$$

Since $k_{i+1}' = k_{i+1}$ and $\beta_{i+1} \equiv 0 \pmod{e_{i+1}'}$, we can write (see the first part of the proof):

$$\mathfrak{M}_{i+1}^{m_{i+1}} = \mathfrak{M}_{i+1}'^{m_{i+1}'} + \mathfrak{M}_{i+1}^{m_{i+1}+1},$$

where $\mathfrak{M}_{i+1}' = \bar{\mathfrak{m}} \cap K_{i+1}'$. Hence we have the following new decomposition of $\mathfrak{m}$ (as a sum of $h+1$ terms):

$$(2)\qquad \mathfrak{m} = \mathfrak{M}_1^{m_1} + \mathfrak{M}_2^{m_2} + \cdots + \mathfrak{M}_i^{m_i} + \mathfrak{M}_{i+1}'^{m_{i+1}'} + \mathfrak{M}_{i+1}^{m_{i+1}+1} + \mathfrak{M}_{i+2}^{m_{i+2}} + \cdots + \mathfrak{M}_h^{m_h}.$$

The pair of integers β_i, β_{i+1} is now replaced by the triplet

$$\beta_i,\ \beta_{i+1}'\ (= \beta_{i+1}),\ \beta_{i+1} + e_{i+1}.$$

If $\mathfrak{M}_{i+1}^{m_{i+1}+1}$ is redundant in (2) we omit it, and we get an irredundant decomposition. If not, then already (2) is irredundant. We now have $(\beta_{i+1}', e_i) = e_{i+1}'$, showing that the new decomposition of $\mathfrak{m}$ is standard up to and including its $(i+1)$-th term. This completes the proof.

3

Throughout this section we assume (unless the contrary is explicitly specified) tha our local domain $\mathfrak{o}$ is saturated. We consider a *standard* decomposition (1) of the maximal ideal $\mathfrak{m}$ of $\mathfrak{o}$. We shall use the notations of the preceding section (such as $\beta_i, e_i, n_i, k_i, k^{(\gamma)}$ where $\gamma \in \mathfrak{m}v$; see Definition 4.8). We shall denote the elements of the set $\mathfrak{m}v$ by $\gamma_1 < \gamma_2 < \cdots < \gamma_q < \cdots$; here $\gamma_1 = \beta_1 =$ reduced multiplicity s of $\mathfrak{o}$.

PROPOSITION 5.3. *Let γ be an element of $\mathfrak{m}v$, let the index i $(1 \leqq i \leqq h)$ be determined by the condition*

$$\beta_i \leqq \gamma < \beta_{i+1},$$

where we agree to set $\beta_{h+1} = \infty$. Then γ is divisible by e_i [and therefore the set of integers $\mathfrak{m}v$, which by (1) certainly contains the set of integers $\beta_i + \lambda e_i$ $(i = 1, 2, \cdots, h$; λ-any integer $\geqq 0)$, coincides with this last set]. Furthermore $k^{(\gamma)} = k_i$, and consequently $k^{(\gamma)} \subset k^{(\gamma')}$ if $\gamma < \gamma'$. (Thus the set of distinct higher residue fields of $\mathfrak{o}$ coincides with the set of distinct characteristic fields of $\mathfrak{o}$ relative to any standard decomposition of $\mathfrak{m}$).

Proof. Fix an element ξ in $\mathfrak{m}$ such that $v(\xi) = \gamma$. Using (1) we write ξ in the form

$$\xi = \xi_\nu + \xi_{\nu+1} + \cdots + \xi_h, \quad \xi_j \in \mathfrak{M}_j^{m_j},$$

and we assume that this decomposition of ξ is one in which ν is maximum $(1 \leqq \nu \leqq g)$. We have then $v(\xi) \geqq \beta_\nu$, and we assert that $v(\xi) < \beta_{\nu+1}$. For in the contrary case we would have that $v(\xi_\nu + \xi_{\nu+1}) \geqq \beta_{\nu+1}$, whence $\xi_\nu + \xi_{\nu+1} = \xi_{\nu+1}' \in \mathfrak{M}_{\nu+1}^{m_{\nu+1}}$, contrary to our choice of ν. It follows therefore in the first place that $\nu = i$, and in the second place that $v(\xi) = v(\xi_\nu)$. Since $\xi_\nu \in K_\nu$, the assertion that γ is divisible by e_i is proved.

Let ξ, η be any two elements of m such that $v(\xi) = v(\eta) = \gamma$. We have just shown that $\xi = \xi_i + \xi_{i+1} + \cdots + \xi_h$, $\eta = \eta_i + \eta_{i+1} + \cdots + \eta_h$, with $\xi_j, \eta_j \in \mathfrak{M}_j^{m_j}$ and $\gamma = v(\xi_i) = v(\eta_i)$. Since $\gamma < \beta_{i+1}$, the two quotients ξ/η and ξ_i/η_i have the same v-residue in $\bar{k}$. Thus the v-residue of ξ/η is in k_i, and this shows that $k^{(\gamma)} \subset k_i$. On the other hand, since γ is an integer of the form $\beta_i + \lambda e_i$, $\lambda \geqq 0$, any element ξ of $\mathfrak{M}_i^{m_i+\lambda}$ which does not belong to $\mathfrak{M}_i^{m_i+\lambda+1}$ has value γ. If then c is an element of k_i and if we set $\eta = c\xi$, then the fact that $c = \eta/\xi$ shows that $c \in k^{(\gamma)}$. Hence $k_i \subset k^{(\gamma)}$, and this completes the proof.

COROLLARY 5.4. *For any γ in $\mathfrak{m}v$, the field $k^{(\gamma)}$ coincides with the set of v-residues, in $\bar{k}$, of quotients of the form ξ/η, where $\xi, \eta \in \mathfrak{m}$ and $v(\xi) = v(\eta) = \gamma$.*

This follows directly from the equality $k^{(\gamma)}=k_i$ and from the above proof. Note that if $\mathfrak{o}$ is not saturated, the above set only *generates* $k^{(\gamma)}$ over k (see Definition 4.8) and is not, in general, even a field.

The integers $\beta_1, \beta_2, \cdots, \beta_h$ belong to the semigroup $\mathfrak{m}v$ (and, in particular, β_1 is the smallest element of $\mathfrak{m}v$). The following proposition characterizes the integers β_i in terms of the following three entities: (1) the semigroup $\mathfrak{m}v$, (2) the (finite) ascending chain of the higher residue fields $k^{(\gamma)}$ of $\mathfrak{o}$ and (3) the ramification exponent e_1 of F/K_1. Note that (1) and (2) are intrinsically related to the ring $\mathfrak{o}$ and that only (3) depends on the choice of the standard decomposition of $\mathfrak{m}$.

PROPOSITION 5.5. *An element γ_ν of $\mathfrak{m}v$ coincides with one of the integers $\beta_1, \beta_2, \cdots, \beta_h$ if and only if one of the following three conditions issatisfied:* (a) $\nu=1$ *(in which case $\gamma_1=\beta_1=s=$ reduced multiplicity of $\mathfrak{o}$)*; (b) $\nu>1$ *and* $(e_1, \gamma_2, \gamma_3, \cdots, \gamma_{\nu-1})>(e_1, \gamma_2, \gamma_3, \cdots, \gamma_\nu)$; (c) $\nu>1$ *and* $k^{(\gamma_{\nu-1})}<k^{(\gamma_\nu)}$.

Proof. Assume that γ_ν does not belong to the set $\{\beta_1, \beta_2, \cdots, \beta_h\}$. Then, of course, condition (a) is not satisfied. To prove that also conditions (b) and (c) are not satisfied we shall first consider the case in which $\gamma_\nu>\beta_h$. In that case all the β_i belong to the set $\{\gamma_1, \gamma_2, \cdots, \gamma_{\nu-1}\}$, and hence $(e_1, \gamma_2, \gamma_3, \cdots, \gamma_{\nu-1}) \leqq (e_1, \beta_2, \beta_3, \cdots, \beta_h)=e_h=1$ (see Definition 4.7). Therefore $(e_1, \gamma_2, \gamma_3, \cdots, \gamma_{\nu-1})=1$ and condition (b) is not satisfied. Furthermore, $k^{(\gamma_{\nu-1})} \geqq k^{(\beta_h)}$, by Proposition 5.3; i.e., $k^{(\gamma_{\nu-1})}=\bar{k}$ (since $K_h=F$), and thus condition (c) is not satisfied.

We consider now the case $\gamma_\nu<\beta_h$. Let say, $\beta_i<\gamma_\nu<\beta_{i+1}$, $i<h$. By Proposition 5.3 we have then $\gamma_j \equiv 0 \pmod{e_i}$ for all $j \leqq \nu$, and since $(e_1, \beta_2, \beta_3, \cdots, \beta_i)=e_i$ and $\beta_i<\gamma_\nu$, it follows that

$$(e_1, \gamma_2, \gamma_3, \cdots, \gamma_{\nu-1})=(e_1, \gamma_2, \gamma_3, \cdots, \gamma_\nu)=e_i.$$

Thus condition (b) is not satisfied. Finally, since $\beta_i \leqq \gamma_{\nu-1}<\beta_{i+1}$ and $\beta_i<\gamma_\nu<\beta_{i+1}$, we have, by Proposition 5.3: $k^{(\gamma_{\nu-1})}=k^{(\gamma_\nu)}=k_i$. Hence condition (c) is not satisfied.

Conversely, assume now that $\gamma_\nu=\beta_i$ for some $i>1$ (whence $\nu>1$) and that condition b) is *not* satisfied. We have by Proposition 5.3 that $(e_1, \gamma_2, \gamma_3, \cdots, \gamma_\nu)=(e_1, \beta_2, \beta_3, \cdots, \beta_i)=e_i$, and we have also $\beta_{i-1} \leqq \gamma_{\nu-1}<\beta_i$, whence $(e_1, \gamma_2, \gamma_3, \cdots, \gamma_{\nu-1})=e_{i-1}$ (always by Proposition 5.3). Since we have assumed that condition (b) is not satisfied, it follows that $e_i=e_{i-1}$ This implies that $k_{i-1}<k_i$ (since $K_{i-1}<K_i$). By Proposition 5.3 we have

however, $k_{i-1} = k^{(\gamma_{\nu-1})}$ and $k_i = k^{(\gamma_\nu)}$. Thus condition (c) is satisfied, and this completes the proof.

PROPOSITION 5.6. *The first ramification exponent e_1 determines uniquely the characteristic $(e_1; \beta_1, \beta_2, \cdots, \beta_h)$ of a standard decomposition of $\mathfrak{m}$ (and hence also the remaining ramification exponents $e_2, e_3, \cdots, e_h$, the integers $m_1, m_2, \cdots, m_h$ and the characteristic residue field $k_1, k_2, \cdots, k_h$). Furthermore, the field K_1 determines uniquely the remaining fields $K_2, K_3, \cdots, K_h$ (and hence also the entire standard decomposition of $\mathfrak{m}$).*

Proof. We have first of all that β_1 is the same for all standard decompositions of $\mathfrak{m}$; namely $\beta_1 = \gamma_1 = s =$ reduced multiplicity of $\mathfrak{o}$. The integer β_2 is either the first element γ_ν of $\mathfrak{m}v$ which is not divisible by e_1 or is the first element γ_ν of $\mathfrak{m}v$ such that $k(^{(\beta_1)} < k^{(\gamma_\nu)}$, whichever is smaller. Next we have $e_2 = (e_1, \beta_2)$ (= highest common divisor of e_1 and β_2) and $m_2 = \beta_2/e_2$. Similarly, β_3 is either the first element γ_ν of $\mathfrak{m}v$ which is not divisible by e_2 or is the first element γ_ν of $\mathfrak{m}v$ such that $k^{(\gamma_\nu)} < k^{(\beta_2)}$, whichever is smaller. Next, we have $e_3 = (e_2, \beta_3)$ and $m_3 = \beta_3/e_3$. This process stops when we reach a pair $\{e_h, \beta_h\}$ such that $e_h = 1$ *and* $k^{(\beta_h)} = \bar{k}$. Since $k_i = k^{(\beta_i)}$, the first part of the proposition is proved.

We note that the procedure described above shows again that, in addition to the inequalities $\beta_1 < \beta_2 < \cdots < \beta_h$ which are part of the *definition* of standard decomposition (see Definition 4.7), we also have

$$e_1 \geqq e_2 \geqq \cdots \geqq e_h \quad (=1);$$

$$1 \leqq m_1 < m_2 < \cdots < m_h.$$

To prove the second part of the proposition, let

$$\mathfrak{m} = \mathfrak{M}_1{}^{m_1} + \mathfrak{M}_2{}'^{m_2} + \cdots + \mathfrak{M}_h{}'^{m_h}$$

be another standard decomposition of $\mathfrak{m}$ in which the associated chain of intermediate fields $K_1' < K_2' < \cdots < K_h' = F$ is such that $K_1' = K_1$ (whence $e_1' = e_1$, and consequently the number h of terms of the two decompositions and the exponents $m_1, m_2, \cdots, m_h$ are the same for both decompositions). By the first part of the proof we know that the decompositions have the same characteristic $(e_1; \beta_1, \beta_2, \cdots, \beta_h)$ and the same characteristic coefficient fields $k_1, k_2, \cdots, k_h$. Our claim is that the two decompositions are identical, i.e., that $K_i' = K_i$, $i = 2, 3, \cdots, h$. By Lemma 1.1 we can find uniformizing parameters x_1, x_2, x_2' of K_1, K_2 and K_2' respectively such that

$$x_1 = a_2 x_2{}^{n_2} = a_2' x_2'{}^{n_2}, \tag{3}$$

where $n_2 = e_1/e_2$ and $a_2, a_2' \in k_2$. From (3) it follows that the quotient $\alpha = x_2'/x_2$ belongs to $\bar{k}$, since

$$(4) \qquad \alpha^{n_2} = a_2/a_2' \in k_2 \subset \bar{k}.$$

Now $x_2^{m_2}$ and $x_2'^{m_2}$ belong to $\mathfrak{M}_2^{m_2}$ and $\mathfrak{M}_2'^{m_2}$ respectively, hence $x_2^{m_2}$ and $x_2'^{m_2}$ belong to $\mathfrak{m}$, and we have $v(x_2^{m_2}) = v(x_2'^{m_2}) = e_2 m_2 = \beta_2$. Therefore α^{m_2} $(= x_2'^{m_2}/x_2^{m_2})$ belongs to $k^{(\beta_2)}$, i.e., to k_2. Since, by (4), also $\alpha^{n_2} \in k_2$ and since $(m_2, n_2) = 1$, we conclude that $\alpha \in k_2$. Since $x_2' = \alpha x_2$, it follows that K_2' $(= k_2((x_2')))$ and K_2 $(= k_2((x_2)))$ coincide. In a similar way one proves the equalities $K_i' = K_i$ for $i > 2$.

If x is a parameter of $\mathfrak{o}$ such that $\mathfrak{o}$ is saturated with respect to x, then we have defined in Section 3 (see Definition 3.4) the x-characteristic $\{n; \beta_1, \beta_2, \cdots, \beta_g\}$ of $\mathfrak{o}$ (here $n = v(x)$), and in Section 4 (see Definition 4.3) we have also defined the standard decomposition of $\mathfrak{m}$ relative to x, such a decomposition being indeed standard in the sense of Definition 4.7. The characteristic of the standard decomposition of $\mathfrak{m}$ relative to x does not necessarily coincide with the x-characteristic $\{n; \beta_1, \beta_2, \cdots, \beta_g\}$ of $\mathfrak{m}$, but can easily be recovered from the x-characteristic by using Proposition 4.10. One finds at once the following:

Case 1. *x is transversal and $k^{(s)} = k$* ($s =$ reduced multiplicity of $\mathfrak{o}$). In that case the characteristic of the standard decomposition of $\mathfrak{m}$ relative to x is $\{s; s, \beta_1, \beta_2, \cdots, \beta_g\}$.

Case 2. *x is transversal and $k^{(s)} > k$.* In that case the characteristic of the standard decomposition of $\mathfrak{m}$ relative to x is $\{s; s, \beta_2, \cdots, \beta_g\}$.

Case 3. *x is not transversal.* The characteristic of the standard decomposition of $\mathfrak{m}$ relative to x is in this case $\{e_1'; s, \beta_2, \cdots, \beta_g\}$, where e_1' is the highest common divisor of n and s (whence $e_1' < n$).

Note that in cases 2 and 3 we have $\beta_1 = s$. We can therefore summarize by saying that the characteristic of the standard decomposition of $\mathfrak{m}$ relative to x is in all cases given either by $\{e'; s, \beta_1, \beta_2, \cdots, \beta_g\}$ or by $\{e'; \beta_1, \beta_2, \cdots, \beta_g\}$, according as $\beta_1 > s$ or $\beta_1 = s$; here $e' = (n, s)$ (whence $e' = s$ if $n = s$, i.e., if x is transversal). The following is therefore an immediate consequence of Proposition 5.6:

COROLLARY 5.7. *If $\mathfrak{o}$ is saturated with respect to a parameter x then the x-characteristic of $\mathfrak{o}$ depends only on $v(x)$.*

In particular, the x-characteristic of $\mathfrak{o}$ is the same for all transversal parameters x such that $\mathfrak{o}$ is saturated with respect to x (and we know, by

Corollary 4.6, that there exist transversal parameter with respect to which $\mathfrak{o}$ is saturated; see, however, Theorem 7.2). We therefore give the following definition:

Definition 5.8. *If $\mathfrak{o}$ is saturated with respect to a parameter x and if x is transversal, then the x-characteristic of $\mathfrak{o}$ is called the characteristic of $\mathfrak{o}$.*

The number g of the β_i which occur in the x-characteristic of $\mathfrak{o}$ (x-*transversal*) is thus independent of x. The standard decomposition of $\mathfrak{m}$ relative to x (x-transveral) has $g+1$ or g terms, according as $k^{(s)}=k$ or $k^{(s)}>k$.

We shall conclude this section with an analysis of the following question: *find all standard decompositions of* $\mathfrak{m}$.

PROPOSITION 5.9. *If $\mathfrak{o}$ is saturated then given any standard decomposition of* $\mathfrak{m}$:

$$\mathfrak{m}=\mathfrak{M}_1^{m_1}+\mathfrak{M}_2^{m_2}+\cdots+\mathfrak{M}_h^{m_h}, \tag{5}$$

there exists a parameter z of $\mathfrak{o}$ such that $\mathfrak{o}$ is saturated with respect to z and such that (5) *is the standard decomposition of $\mathfrak{m}$ relative to z (in the sense of Definition* 4.3).

Proof. Let $K_1, K_2, \cdots, K_h$ be the intermediate fields of the standard decomposition (5) and let x_1 be a uniformizing parameter of K_1 (i.e., let $K_1=k_1((x_1))$, where k_1 is the first characteristic coefficient field of $\mathfrak{o}$, relative to the decomposition (5); see Definition 4.8). With the same notations which have ben used throughout this section, we fix an integer n' such that $n'\geqq m_1$ and $(n', m_1)=1$. If we set $z=x_1^{n'}$, then $z\in\mathfrak{M}_1^{m_1}$, and hence $\mathfrak{o}$ is saturated with respect to the parameter z (Lemma 4.5). We shall prove that (5) is the standard decomposition of $\mathfrak{m}$ relative to this parameter z.

Let

$$\mathfrak{m}=\mathfrak{M}_1'^{m_1'}+\mathfrak{M}_2'^{m_2'}+\cdots+\mathfrak{M}'^{m'_{h'}}_{h'} \tag{6}$$

be the standard decomposition of $\mathfrak{m}$ relative to z. Let $K_1', K_2', \cdots, K'_{h'}$ be the associated intermediate fields. We shall prove that $K_1'=K_1$, and this will show that the decompositions (5) and (6) of $\mathfrak{m}$ are identical (see the last assertion in Proposition 5.6). Let $K_1'=k_1'((x_1'))$, $e_1'=v(x_1')$, $e_1=v(x_1)$. We have $k_1'=k_1=k^{(s)}$. We also have $e_1'=e_1$ since $e_1'=(s, v(z))=(e_1m_1, e_1n')$ and $(m_1, n')=1$, by our choice of the integer n'. Since $z\in k_1((x_1'))$, we may assume that $z=ax_1'^{n'}$, $a\in k_1$. From $e_1'=e_1$ and

$s = e_1m_1 = e_1'm_1'$ it follows that $m_1' = m_1$. Thus $x_1^{m_1}$ and $x_1'^{m_1}$ are elements of $\mathfrak{m}$, having value s. Therefore $(x_1/x_1')^{m_1} \in k_1$ $(= k^{(s)})$. Thus we have

$$(7) \qquad (x_1/x_1')^{n'} \in k_1 \text{ and } (x_1/x_1')^{m_1} \in k_1,$$

and since $(n', m_1) = 1$, the (7) imply that $x_1/x_1' \in k$, and therefore $K_1' = K_1$.

The above proposition shows that in order to find all standard decompositions of $\mathfrak{m}$ ($\mathfrak{o}$ being assumed to be saturated and the coefficient field k being fixed) it is sufficient to find the set S of all parameters z of $\mathfrak{o}$ such that $\mathfrak{o}$ is saturated with respect to z—a question which is of some interest in itself. We shall prove in Section 7 that *the set S in question includes all transversal parameters* (see Theorem 7.2). We shall assume this result in this section. Then Lemma 4.5 shows that this set includes also all the (positive integral) powers x^q of any transversal parameter x. The complete description of the set S will involve the consideration of *fractional powers* of transversal parameters, as can be seen from the following (preliminary) proposition:

PROPOSITION 5.10. *Assume that $\mathfrak{o}$ is saturated with respect to a parameter x (not necessarily a transveral one). Let $v(x) = n$ and let γ be any element of the semi-group $\mathfrak{m}v$. Then there exists an element z in $\mathfrak{m}$ and an element a in k such that $z^n = ax^\gamma$ (whence $v(z) = \gamma$).*

Proof. Let $\mathfrak{m} = \mathfrak{M}_1^{m_1} + \mathfrak{M}_2^{m_2} + \cdots + \mathfrak{M}_h^{m_h}$ be the standard decomposition of $\mathfrak{m}$ relative to the parameter x, let $K_i = k_i((x_i))$, $i = 1, 2, \cdots, h$, be the associated intermediate fields, and let $(e_1; \beta_1, \beta_2, \cdots, \beta_h)$ be the associated characteristic of $\mathfrak{o}$. Here $\beta_1 = s$, $e_1 = v(x_1) = (s, n)$. We assume that the x_i have been chosen so as to have $x_{i-1} = a_ix_i^{n_i}$, $a_i \in k_i(x_0 = x)$, where $n_i = e_{i-1}/e_i$, with $e_i = v(x_i)$ $(e_0 = n)$. We determine the index j by the condition $\beta_j \leqq \gamma < \beta_{j+1}$, unless $\beta_h \leqq \gamma$ in which case we set $j = h$. We know then that $\gamma \equiv 0 \pmod{e_j}$ (Proposition 5.3). If $\gamma = qe_j$, we set $z = x_j^q$. Then $z \in \mathfrak{m}$, since $\gamma \geqq \beta_j = e_jm_j$ implies that $q \geqq m_j$, whence $z \in \mathfrak{M}_j^{m_j} \subset \mathfrak{m}$. We have $z^n = x_j^{qn}$ and we also have $x = bx_j^{n/e_j}$, where $b = a_1a_2^{n_1} \cdots a_j^{n_1n_2\cdots n_{j-1}}$. Hence $x^\gamma = x^{qe_j} = b^\gamma x_j^{qn}$, and this, in conjunction with $z^n = x_j^{qn}$, implies that $z^n = ax^\gamma$, where $a = 1/b^\gamma$.

Anticipating the results proved in Section 7, to the effect that $\mathfrak{o}$ is saturated with respect to any transversal parameter (Proposition 7.2), we can state the following corollary:

COROLLARY 5.11. *If x is a transversal parameter of $\mathfrak{o}$ and γ is any element of the semi-group $\mathfrak{m}v$, there exists an element z in $\mathfrak{m}$ and an element a in $\bar{k}$ such that $z^s = ax^\gamma$ (whence $v(z) = \gamma$).*

6. Characterization of saturation parameters.

Definition 6.1. *If $\mathfrak{o}$ is saturated, a parameter z of $\mathfrak{o}$ is called a saturation parameter if $\mathfrak{o}$ is saturated with respect to z.*

We shall assume in this section a theorem proved in the next section, to the effect that every transverse parameter x of $\mathfrak{o}$ is a saturation parameter (see Theorem 7.2). As was pointed out already in the preceding section, this implies, in view of Lemma 4.5, that also every positive integral power of a transversal parameter is a saturation parameter. Our object in this section is to characterize all saturation parameters of $\mathfrak{o}$.

Let z be a parameter of the saturated ring $\mathfrak{o}$ and let

$$v(z) = \gamma, \tag{1}$$

whence $\gamma \geqq s$ (= reduced multiplicity of $\mathfrak{o}$). We know from Proposition 5.10 (where we must now interchange the letters x and z and where we set $\gamma = s$), that *a necessary condition* for z to be a saturation parameter is that z^s have an expression of the form bx^γ, where $b \in \bar{k}$ (= algebraic closure of k in F) and where $x \in \mathfrak{o}$ (whence x is necessarily a *transversal* parameter of $\mathfrak{o}$). So from now on our search for the full set of saturation (non-transversal) parameters of $\mathfrak{o}$ will begin as follows: we start with a *transversal* parameter x of $\mathfrak{o}$ and with a given element γ of the semi-group $\mathfrak{m}v$ such that $\gamma > s$. Then by Proposition 5.10, part (a), there will exist an element z in $\mathfrak{m}$ such that $v(z) = \gamma$ and such that $z^s/x^\gamma \in \bar{k}$. We wish to find a necessary and sufficient condition for z to be a saturation parameter.

We consider the standard decomposition of $\mathfrak{m}$ relative to the transversal parameter x:

$$\mathfrak{m} = \mathfrak{M}_1 + \mathfrak{M}_2^{m_2} + \cdots + \mathfrak{M}_h^{m_h}. \tag{2}$$

We denote, as usual, by $K_1, K_2, \cdots, K_h$ the intermediate fields of the decomposition (2). Here

$$\begin{aligned} &K_i = k_i((x_i)), && i = 1, 2, \cdots, h; \\ &x_{i-1} = a_i x_i^{n_i},\ a_i \in k_i, && i = 2, 3, \cdots, h; \\ &x_1 = x; \quad K_h = F. && \end{aligned} \tag{3}$$

We denote by $e_1, e_2, \cdots, e_h$ the ramification exponents of $K_1, K_2, \cdots, K_h$. Here

$$\begin{gathered} e_1 = s; \quad e_i = v(x_i), \quad e_h = 1, \\ e_{i-1}/e_i = n_i. \end{gathered} \tag{4}$$

We also set

$$(5)\qquad \begin{aligned} \beta_i &= e_i m_i, && i=2,3,\cdots,h;\\ \beta_1 &= e_1 m_1 = s, && (m_1=1). \end{aligned}$$

Thus $\{s;\beta_1,\beta_2,\cdots,\beta_h\}$ is the characteristic of the standard decomposition (2). We introduce the following h integers ϵ_i:

$$(6_1)\qquad \epsilon_1 = (\beta_1,\gamma) \quad [= (s,\gamma)].$$

$$(6_i)\qquad \epsilon_i = (\beta_i,\epsilon_{i-1}), \qquad i=2,3,\cdots,h.$$

We observe that

$$(7)\qquad \epsilon_i \textit{ divides } e_i, \qquad i=1,2,\cdots,h.$$

For $i=1$ this follows directly from (6_1) and from $e_1 = s$. By induction, assuming that ϵ_{i-1} divides e_{i-1}, we note that by (6_i) ϵ_i divides ϵ_{i-1}, and hence ϵ_i divides e_{i-1}. Since ϵ_i also divides β_i and $(e_{i-1},\beta_i)=e_i$, (7) is proved.

We also observe that ϵ_i is the highest common divisor of γ and e_i:

$$(8)\qquad \epsilon_i = (e_i,\gamma), \qquad i=1,2,\cdots,h.$$

For $i=1$ this follows directly from (6_1). By induction, assuming that $\epsilon_{i-1}=(e_{i-1},\gamma)$, we have, by (6_i): $\epsilon_i=(\beta_i,e_{i-1},\gamma)=(e_i,\gamma)$ [since the decomposition (2) is standard and since therefore $(\beta_i,e_{i-1})=e_i$].

We now set

$$(9)\qquad e_i=\epsilon_i\mu_i, \qquad \gamma=\epsilon_i\nu_i, \qquad i=1,2,\cdots,h,$$

whence, by (6_i), the positive integers μ_i and ν_i are relatively prime:

$$(\mu_i,\nu_i)=1, \qquad i=1,2,\cdots,h.$$

THEOREM 6.2. *The parameter z is a saturation parameter if and only if the following two conditions are satisfied:*

$$\text{(a)}\qquad \beta_i+\epsilon_i\in \mathrm{mv}, \qquad i=1,2,\cdots,h;$$

$$\text{(b)}\qquad z^{\mu_i}/x_i^{\nu_i}\in k_i, \qquad i=1,2,\cdots,h.$$

[*Note* I. Since $v(z)=\gamma$ and $v(x_i)=e_i$, it follows from (9) that $v(z^{\mu_i}/x_i^{\nu_i})=0$. Since, on the other hand, $z^s/x^\gamma\in\bar{k}$ and since by (3) we have $x=c_ix_i^{s/e_i}$, for $c_i\in k_i$, it follows that the quotient $z^{\mu_i}/x_i^{\nu_i}$ at any rate belongs to $\bar{k}$.]

Proof. Since $e_h=1$ and ϵ_i divides e_i, by (7), it follows that also ϵ_h is 1.

Thus $e_h = \epsilon_h$. We denote by l the smallest integer such that $0 \leqq l \leqq h-1$ and such that $\epsilon_{l+1} = e_{l+1}$. If $l < h-1$ then

$$\epsilon_{l+2} = (\beta_{l+2}, \epsilon_{l+1}) = (\beta_{l+2}, e_{l+1}) = e_{l+2};$$

and more generally, we find that

$$(10) \qquad \epsilon_j = e_j, \qquad j = l+1, l+2, \cdots, h.$$

while, by definition of l, we have, *if l is positive*:

$$(11) \qquad \epsilon_i < e_i, \qquad i = 1, 2, \cdots, l.$$

[*Note* II. If $i > l$, then $\epsilon_i = e_i$, $\mu_i = 1$ in (9), and condition (b), for $i > l$, is simply this: $z/x_i^{\nu_i} \in k_i$, $i = l+1, l+2, \cdots, h$. These $h-l$ conditions are all consequences of the first: $z/x_{l+1}^{\nu_{l+1}} \in k_{l+1}$. In fact, we have $\gamma = \nu_i e_i = \nu_{i+1} e_{i+1}$, if $i \geqq l+1$, and, by (3), we have $x_i = a_{i+1} x_{i+1}^{n_{i+1}}$, $a_{i+1} \in k_{i+1}$, where $n_{i+1} = e_i/e_{i+1} = \nu_{i+1}/\nu_i$. Therefore,

$$x_i^{\nu_i} = a_{i+1}^{\nu_i} x_{i+1}^{\nu_{i+1}} \quad \text{and} \quad z/x_{i+1}^{\nu_{i+1}} = a_{i+1}^{\nu_i} z/x_i^{\nu_i}.$$

Since $a_{i+1}^{\nu_i} \in k_{i+1}$ and $k_i \subset k_{i+1}$, the inclusion $z/x_{i+1}^{\nu_{i+1}} \in k_{i+1}$ follows from the inclusion $z/x_i^{\nu_i} \in k_i$. Note also that *we have $l = 0$ if and only if γ is divisible by s*, and in that case condition (b) reduces to one condition only, namely to the following: $z/x^{\nu_1} \in k_1$; here $\nu_1 = \gamma/s$ and $x = x_1$.]

Part 1. *Necessity*. We assume then that $\mathfrak{o}$ is saturated with respect to z and we consider the standard decomposition of $\mathfrak{m}$ with respect to z:

$$(12) \qquad \mathfrak{m} = \mathfrak{M}_1'^{m_1'} + \mathfrak{M}_2'^{m_2'} + \cdots + \mathfrak{M}_q'^{m_q'}.$$

We denote by $K_1', K_2', \cdots, K_q'$ the intermediate fields of the decomposition (12):

$$(13) \qquad \begin{aligned} &K_i' = k_i'((x_i')), && i = 1, 2, \cdots, q; \\ &x_{i-1}' = a_i' x_i'^{n_i'} && i = 2, 3, \cdots, q; \\ &K_q' = F. \end{aligned}$$

Let $e_1', e_2', \cdots, e_q'$ be the ramification exponents of the fields $K_1', K_2', \cdots, K_q'$:

$$(14) \qquad \begin{aligned} &e_i' = v(x_i'); \\ &e_q' = 1; \\ &e_{i-1}'/e_i' = n_i'. \end{aligned}$$

We also set

$$(15) \qquad \beta_i' = e_i' m_i', \qquad i = 1, 2, \cdots, q.$$

We have always $\beta_1' = s$, and $\{e_1'; \beta_1', \beta_2', \cdots, \beta_q'\}$ is the characteristic of the standard decomposition (12).

By (2), we have $\beta_i + e_i \in \mathfrak{m}v$, $i = 1, 2, \cdots, h$. Therefore, by (10), the relations (a) are trivially satisfied for $i = l+1, l+2, \cdots, h$. We now observe that by (12), we have $\beta_i' + e_i' \in \mathfrak{m}v$, for $i = 1, 2, \cdots, q$. Assuming now that $l > 0$, relations (a) for $i = 1, 2, \cdots, l$, will be a consequence of the following stronger result which we proceed to establish:

$$(16) \qquad \epsilon_i = e_i', \quad \beta_i = \beta_i', \qquad i = 1, 2, \cdots, l.$$

For $i = 1$, (16) follows from $\beta_1 = \beta_1' = s$, from the definition (6_1) of ϵ_1 and from $e_1' = (\gamma, s)$ [see § 3, (9) and (17) for $i = 1$]. We now use induction on the index i. We assume therefore that (16) is true, for $i \leqq \rho$, where ρ is some integer such that $1 \leqq \rho < l$, and we proceed to prove (16) for $i = \rho + 1$.

We shall prove first that

$$(17) \qquad \beta_{\rho+1} < \beta_\rho + \epsilon_\rho, \qquad (1 \leqq \rho < l).$$

We give an indirect proof of (17). Assume first that $\beta_{\rho+1} > \beta_\rho + \epsilon_\rho$. By our induction hypothesis we know already that $\beta_\rho + \epsilon_\rho$ $(= \beta_\rho' + e_\rho')$ belongs to $\mathfrak{m}v$. From the characterization of the β_i, given in Proposition 5.5, follows that any element of $\mathfrak{m}v$ which is strictly *less* than $\beta_{\rho+1}$ is divisible by e_ρ. Hence $\beta_\rho + \epsilon_\rho$ must be divisible by e_ρ. This, however, is impossible, since e_ρ divides β_ρ and since, by (11), we have $\epsilon_\rho < e_\rho$. This contradiction shows then that we cannot have $\beta_{\rho+1} > \beta_\rho + \epsilon_\rho$. Assume now that $\beta_{\rho+1} = \beta_\rho + \epsilon_\rho$. In that case we have $e_{\rho+1} = (\beta_{\rho+1}, e_\rho) = (\beta_\rho + \epsilon_\rho, e_\rho) = \epsilon_\rho$ [since ϵ_ρ divides e_ρ, by (7), and since e_ρ divides β_ρ], and therefore $\epsilon_{\rho+1} = (\beta_{\rho+1}, \epsilon_\rho) = (\beta_{\rho+1}, e_{\rho+1}) = e_{\rho+1}$, in contradiction with (11) (since $\rho ; 1 \leqq l$). Thus (17) is proved.

From (17), in conjunction with $\beta_\rho < \beta_{\rho+1}$ it follows that

$$(18) \qquad \beta_{\rho+1} \not\equiv 0 \pmod{\epsilon_\rho},$$

since ϵ_ρ divides e_ρ, by (17), and since e_ρ divides β_ρ. From (18) follows *a fortiori* that $\beta_{\rho+1} \not\equiv 0 \pmod{e_\rho}$. From Proposition 5.5 we know that every element of $\mathfrak{m}v$ which is *less* than $\beta_{\rho+1}$ is divisible by e_ρ, and hence also by ϵ_ρ (since ϵ_ρ divides e_ρ). Therefore, by (18), $\beta_{\rho+1}$ is the smallest element of $\mathfrak{m}v$ which is *not* divisible, by ϵ_ρ. Since, by our induction hypothesis, we have $\epsilon_\rho = e_\rho'$, we see that *$\beta_{\rho+1}$ is the smallest element of $\mathfrak{m}v$ which is not divisible by e_ρ'.* Now, by Proposition 5.5 we know also that $k^{(\alpha)} \subset k^{(\beta_\rho)}$, for all α in mv which are *less* than $\beta_{\rho+1}$. Again, by our induction hypothesis, we have $\beta_\rho = \beta_\rho'$, and thus we now also know that *$k^{(\alpha)} \subset k^{(\beta_\rho')}$ for all α in mv which are less than $\beta_{\rho+1}$.* The two italicized statements, just made above, imply, by Proposition 5.5 [applied to the standard decomposition

(12)], that $\beta_{\rho+1} = \beta_{\rho+1}'$. We have next: $e_{\rho+1}' = (\beta_{\rho+1}', e_\rho') = (\beta_{\rho+1}, \epsilon_\rho) = \epsilon_{\rho+1}$ [this last equality is simply the definition $(6_{\rho+1})$ of $\epsilon_{\rho+1}$]. This completes the proof of (16) and of condition (a) of the theorem. We note that the part of the proof of (17) which consisted in showing that $\beta_{\rho+1}$ cannot be *greater* than $\beta_\rho + \epsilon_\rho$, depended only on the inductive assumption that (16) is true for $i = \rho$ and on the fact that $\epsilon_\rho < e_\rho$ [see (11)]. Since we have now shown that (16) is valid also if $i = l$, and since we have also $\epsilon_l < e_l$ [by (11)], we can now conclude that the inequality $\beta_{\rho+1} \leqq \beta_\rho + \epsilon_\rho$ is also true for $\rho = l$, provided $l > 0$. Thus, in addition to (17), we now have also

$$(19) \qquad \beta_{l+1} \leqq \beta_l + \epsilon_l, \text{ if } l > 0.$$

Note the following consequence of (16):

$$(20) \qquad k_i' = k_i, \qquad i = 1, 2, \cdots, l.$$

This is so, since $k_i = k^{(\beta_i)}$ and $k_i' = k^{(\beta_i')}$.

To complete the proof of the necessity of our conditions (a) and (b), it remains to prove condition (b) of the theorem, and for that purpose we first proceed to derive the following relations:

$$(21) \qquad \beta_i' \geqq \beta_i, \qquad i = 1, 2, \cdots, q;$$

$$(22) \qquad K_i' \supset K_i, \qquad i = 1, 2, \cdots, q.$$

For $i = 1$, (21) is trivial (and is, in fact, a strict equality: $\beta_1' = \beta_1 = s$). We first prove (22) for $i = 1$.

As usual, we set $K' = k((z))$ and $F'^* =$ least Galois extension of K' containing F. Let $G' = \mathfrak{G}(F'^*/K')$ and let $G_i' = \mathfrak{G}(F'^*/K_i')$. We use the results established in § 3. We have $G_1' = \{\tau \in G' \mid v(\zeta^\tau - \zeta) > s$, for all ζ in $\mathfrak{m}\}$. Now, since z^s/x^γ is algebraic over k and since $z^\tau = z$ for all $\tau \in G'$, it follows that $x^\tau = c_\tau x$, for $\tau \in G'$; here c_τ belongs to the algebraic closure of k in F'^*. We have thus: $(x^\tau - x) = (c_\tau - 1)x$. Since $v(x) = s$, it follows that $c_\tau = 1$ for all $\tau \in G_1'$, i.e., $x^\tau = x$ for all $\tau \in G_1'$. Therefore $x \in K_1'$. Furthermore, if b is any element of k_1, then $(bx)^\tau - bx = (b^\tau - b)x$, for all τ in G_1'. But since $bx \in \mathfrak{m}$ and $v(x) = s$, we must have $b^\tau = b$, for all $\tau \in G_1'$. Hence $k_1 \subset k_1'$ (here $k_1 =$ algebraic closure of k in K_1 and $k_1' =$ algebraic closure of k on K_1'), and $K_1 = k_1((x)) \subset K_1'$.

We now use induction on i. We assume that (21) and (22) have already been proved for $i = 1, 2, \cdots, \rho$, where $1 \leqq \rho < q$. We know from § 3 that $\beta_{\rho+1}'$ is the minimum of the set of integers $v(\zeta^\tau - \zeta)$ obtained by letting ζ range over $\mathfrak{m}$ and letting τ' range over all K_ρ'-isomorphisms of F (into a fixed algebraic closure Ω of F). Similarly, $\beta_{\rho+1}$ is the minimum of the set of integers $v(\zeta^{\tau}-\zeta)$ obtained by letting ζ range over $\mathfrak{m}$ and letting τ range over all the K_{ρ}-isomorphisms of F into Ω. Since, by our induction we have $K_{\rho}' \supset K_{\rho}$ every τ' is also a τ, and therefore $\beta_{\rho+1}' \geqq \beta_{\rho+1}$.

To prove the inclusion $K_{\rho+1}' \supset K_{\rho+1}$, we have to show, for each τ in $G_{\rho+1}'$, that $x_{\rho+1}{}^{\tau} = \tau$ and that $\tau \mid k_{\rho+1}$ is the identity. Now, by our usual choice of the uniformizing parameters x_i and x_i' in K_i and K_i', x is of the form $ax_i^{\alpha_i}$ and z is of the form $a'x_i'^{\alpha_i'}$, where $a \in k_i$, $a' \in k_i'$ $(\alpha_i = s/e_i$, $\alpha_i' = \gamma/e_i')$. Since z^s/x^{γ} is algebraic over x, it follows that $x_{\rho+1}'$ is a fractional power of $x_{\rho+1}$ (multiplied by some factor which is algebraic over k). Therefore, we have for each $\tau \in G_{\rho+1}'$: $x_{\rho+1}{}^{\tau} = \rho_{\tau} x_{\rho+1}$, ρ_{τ}-algebraic over k, and hence

$$(x_{\rho+1}{}^{m_{\rho+1}})^{\tau} - x_{\rho+1}{}^{m_{\rho+1}} = (\rho_{\tau}{}^{m_{\rho+1}} - 1) x_{\rho+1}{}^{m_{\rho+1}}. \tag{23}$$

We have $v(\zeta^{\tau}-\zeta) > \beta_{\rho+1}'$ for all $\zeta \in \mathfrak{m}$ and all $\tau \in G_{\rho+1}'$. Since $\beta_{\rho+1}' \geqq \beta_{\rho+1}$ and since $v(x_{\rho+1}{}^{m_{\rho+1}}) = \beta_{\rho+1}$, it follows from (23) that we must have $\rho_{\tau}{}^{m_{\rho+1}} = 1$ for all τ in $G_{\rho+1}'$. In other words, we have: $x_{\rho+1}{}^{m_{\rho+1}}$ is invariant under all τ in $G_{\rho+1}'$. Therefore

$$x_{\rho+1}{}^{m_{\rho+1}} \in K_{\rho+1}'. \tag{24}$$

Let now b be any element of $k_{\rho+1}$. If we apply the above argument to the product $bx_{\rho+1}{}^{m_{\rho+1}}$ (this is an element of $\mathfrak{m}$), we conclude that also $bx_{\rho+1}{}^{x_{\rho+1}}$ is invariant under $G_{\rho+1}'$. This shows that $b \in K_{\rho+1}'$, whence

$$k_{\rho+1} \subset K_{\rho+1}'. \tag{25}$$

Now, we observe that $x_{\rho} = a_{\rho+1} x_{\rho+1}{}^{n_{\rho+1}}$, with $a_{\rho+1} \in k_{\rho+1}$. Since, by our induction, we have $x_{\rho} \in K_{\rho}'$, we conclude, by (25) that

$$x_{\rho+1}{}^{n_{\rho+1}} \in K_{\rho+1}'. \tag{26}$$

Since $m_{\rho+1}, n_{\rho+1}$ are relatively prime, we see that $x_{\rho+1} \in K_{\rho+1}'$. This completes the proof of (21) and (22).

We next prove the following lemma (which will complement relations (21) and (22):

LEMMA 6.3. *Assuming that z is a saturation parameter, we have the following:*

1) *If either $l = 0$, or $l > 0$ and $k_{l+1} > k_l$, or $l > 0$ and $\beta_{l+1} < \beta_l + \epsilon_l$ [see (19)], then $q = h$ and*

$$K_j' = K_j, \quad \beta_j' = \beta_j, \qquad j = l+1, l+2, \cdots, h. \tag{27}$$

2) *If* $l>0$, $k_{l+1}=k_l$ *and* $\beta_{l+1}=\beta_l+\epsilon_l$ *[see* (19)*], then* $q=h-1$ *and*

$$K_j'=K_{j+1}, \qquad j=l,l+1,\cdots,q \tag{28}$$

and

$$\beta_j'=\beta_{j+1}, \qquad j=l+1,\cdots,q.$$

Proof of the lemma. Assume first $l=0$. Since $\beta_1'=s=\beta_1$, the equality $\beta_j'=\beta_j$ is true for $j=l+1$. Since in this case we have $v(z)=\gamma\equiv 0 \pmod s$, we have, by Proposition 3.7, $e_1'=(\beta_1',\gamma)=(s,\gamma)=s=e_1$. Thus K_1 and K_1' has some coefficient field k_1 $(=k^{(s)})$ and the same ramification exponent s. Since $K_1\subset K_1'$, by (22), it follows that $K_1=K_1'$. This proves (27) for $j=1$, in this case (i.e., in the case $l=0$), and the lemma follows in this case from the last part of Proposition 5.6.

Now assume $l>0$. By Propositions 5.3 and 5.5 applied to the standard decomposition (2), it follows that for any element α in $\mathfrak{m}v$, such that $\beta_l\leqq\alpha<\beta_{l+1}$, we have $k^{(\alpha)}=k_l$ and $\alpha\equiv 0 \pmod{e_l}$. Since ϵ_l divides e_l, since, by (16), $\epsilon_l=e_l'$ and $\beta_l=\beta_l'$, and since, by (20), $k_l=k_l'$, we see that for any element α in $\mathfrak{m}v$ such that $\beta_l'\leqq\alpha<\beta_{l+1}$ we have $k^{(\alpha)}=k_l'$ and $\alpha\equiv 0 \pmod{e_l'}$. *Now assume that* $k_{l+1}>k_l$, i.e., $k^{(\beta_{l+1})}>k_l=k_l'$. Then the validity of the equality $k^{(\alpha)}=k_l'$ for all α such that $\beta_l'\leqq\alpha<\beta_{l+1}$ implies, in view of Proposition 5.5, that $\beta_{l+1}'\leqq\beta_{l+1}$, whence $\beta_{l+1}'=\beta_{l+1}$, in view of (21). *Assume on the other hand that* $\beta_{l+1}<\beta_l+\epsilon_l$. Since ϵ_l divides β_l and since $\beta_{l+1}>\beta_l$, it follows that ϵ_l does not divide β_{l+1}, i.e., e_l' *does not divide* β_{l+1}. Then the validity of the relation $\alpha\equiv 0 \pmod{e_l'}$ for all α such that $\beta_l'\leqq\alpha<\beta_{l+1}$ implies, in view of Proposition 5.5, that $\beta_{l+1}'\leqq\beta_{l+1}$, whence again $\beta_{l+1}'=\beta_{l+1}$, in view of (21). From $\beta_{l+1}'=\beta_{l+1}$, follows $k_{l+1}'=k_{l+1}$ since $k_{l+1}'=k^{(\beta_{l+1}')}$ and $k_{l+1}=k^{(\beta_{l+1})}$. Furthermore, using (16) and (10), we now have $e_{l+1}'=(\beta_{l+1}',e_l')=(\beta_{l+1},\epsilon_l)=\epsilon_{l+1}=e_{l+1}$. We have found thus that K_{l+1} and K_{l+1}' have the same coefficient field k_{l+1} and the same ramification exponent e_{l+1}. Since $K_{l+1}\subset K_{l+1}'$, by (22), we conclude that $K_{l+1}=K_{l+1}'$. This proves (27) for $j=l+1$, in all cases listed under 1).

Let F_{l+1}^* be the least Galois extension of K_{l+1} which contains F and let $G^*=\mathfrak{G}(F_{l+1}^*/K_{l+1})$. By (1) and (4) of § 3, we find that the two sets of integers $\{\beta_{l+2},\beta_{l+3},\cdots,\beta_h\}$ and $\{\beta_{l+2}',\beta_{l+3}',\cdots,\beta_q'\}$ must coincide, since both sets coincide with the set of integers β_τ, $1\neq\tau\in G^*$, where

$$\beta_\tau=\min\cdot\{v(\eta^\tau-\eta),\eta\in\mathfrak{o}\}. \tag{29}$$

Thus $q=h$ and $\beta_j'=\beta_j$, $j=l+1,l+2,\cdots,h$. The argument given above, which led from $\beta_{l+1}'=\beta_{l+1}$ to the conclusion $K_{l+1}=K_{l+1}'$ [via (16), (10), and (22)], can now be repeated without change, and compltes the proof of (27) and of part 1) of the lemma.

Assume now that $l > 0$, that $k_{l+1} = k_l$ and that $\beta_{l+1} = \beta_l + \epsilon_l$. This last equality implies that $e_{l+1} = (\beta_{l+1}, e_l) = (\beta_l + \epsilon_l, e_l) = \epsilon_l$ (since ϵ_l divides e_l and e_l divides β_l). Since $\epsilon_l = e_l'$, by (16), we conclude that

$$\epsilon_l = e_l' = e_{l+1} \quad (= v(x_l') = v(x_{l+1}). \tag{30}$$

Since $k_{l+1} = k_l = k_l'$, we see that the field K_l' and K_{l+1} have the same coefficient field k_{l+1} $(= k_l')$ and the same ramification exponent (as subfields of F). The equality $K_l' = K_{l+1}$ will follow *if we prove that* $K_{l+1} \subset K_l'$.

We have $K_l' = k_l((x_l'))$ and $K_{l+1} = k_{l+1}((x_{l+1})) = k_l((x_{l+1}))$. By (30), we have $v(x_{l+1}/x_l') = 0$. By (3) and (13) we have that the quotients $x/x_l^{s/e_l}$, $z/x^{\gamma/e_l'}$ belong to $\bar{k}$, and we know also that z^s/x^γ is in $\bar{k}$. Therefore $x_l/x_l'^{n_{l+1}} \in \bar{k}$, where $n_{l+1} = e_l/e_{l+1} = e_l/e_l'$. But $x_l \in K_l \subset K_l' = k_l((x_l'))$. Therefore $x_l = cx_l'^{n_{l+1}}$, with $c \in k_l$. On the other hand, we have also $x_l = a_{l+1}x_{l+1}^{n_{l+1}}$, with $a_{l+1} \in k_{l+1} = k_l$. Therefore,

$$x_{l+1}/x_l' = b, \text{ and } b^{n_{l+1}} \in k_l. \tag{31}$$

On the other hand, we have $x_{l+1}^{m_{l+1}} \in M_{l+1}^{m_{l+1}} \in \mathfrak{m}$, and also

$$x_l'^{m_l'+1} \in \mathfrak{M}_l'^{m_l'+1} \subset \mathfrak{m}.$$

From $\beta_{l+1} = \beta_l + \epsilon_l$ and $\beta_l' = \beta_l$ follows that $m_{l+1}e_{l+1} = m_l'e_l' + \epsilon_l$, and therefore, by (30): $m_{l+1} = m_l' + 1$. Thus $x_l'^{m_{l+1}} \in \mathfrak{m}$ and $x_{l+1}^{m_{l+1}} \in \mathfrak{m}$ with $v(x_l'^{m_{l+1}}) = v(x_{l+1}^{m_{l+1}}) = \beta_{l+1}$. Since $k_l = k_{l+1} = k^{(\beta_{l+1})}$, it follows that

$$b^{m_{l+1}} \in k_l. \tag{32}$$

From (31), (32), and from $(m_{l+1}, n_{l+1}) = 1$, follows that $b \in k_l$ and therefore that $x_{l+1} \in k_l((x_l')) = K_l'$. This shows that $K_{l+1} \subset K_l'$ and establishes the equality $K_{l+1} = K_l'$.

Let F_{l+1}^* denote, as before, the least Galois extension of K_{l+1} which contains F and let $G^* = \mathfrak{G}(F_{l+1}^*/K_{l+1})$. From $K_{l+1} = K_l'$ and from (1) and (4) of § 3 we deduce that the two sets of integers $\{\beta_{l+2}, \beta_{l+3}, \cdots, \beta_h\}$ and $\{\beta_{l+1}', \beta_{l+2}', \cdots, \beta_q'\}$ are the same, since they both coincide with the set of integers β_τ, $1 \neq \tau \in G^*$, where β_τ is as in (29). Thus $\beta_j' = \beta_{j+1}$, $j = l+1, l+2, \cdots, q$, and now the equalities $K_j' = K_{j+1}$ follow for $j = l+1, \cdots, q$ by an argument similar to the one used in the proof that $K_l' = K_{l+1}$. Namely, from $\beta_j' = \beta_{j+1}$ follows that k_j' $(= k^{(\beta_j')}) = k_{j+1}$ $(= k^{(\beta_{j+1})})$. Furthermore, from $e_l' = e_{l+1}$ [see (30)] and from $\beta_{l+1}' = \beta_{l+2}$, follows that $e_{l+1}' = (e_l', \beta_{l+1}') = (e_{l+1}, \beta_{l+2}) = e_{l+2}$. In a similar fashion, by induction, it follows that

(33) $e_j' = e_{j+1} \, (= v(x_j') = v(x_{j+1})), \qquad j = l+1, l+2, \cdots, q.$

(33) generalizes relation (30). From $K_j \subset K_j'$ follows that $x_j = dx_j'^{n_j}$, where $n_j = e_j/e_j' = e_j/e_{j+1}$ [by (33)] and $d \in k_j'$. On the other hand, $x_j = a_{j+1}x_{j+1}^{n_{j+1}}$, with $a_{j+1} \in k_{j+1} = k_j'$. Therefore, if we set $x_{j+1}/x_j' = u$, then $u^{n_{j+1}} \in k_j'$. On the other hand we find that $v(x_{j+1}^{m_{j+1}}) = v(x_j'^{m_j'}) = \beta_{j+1}$ $(= \beta_j')$, and therefore $u^{m_{j+1}} \in k_j'$. Since $(n_{j+1}, m_{j+1}) = 1$, it follows that $u \in k_j'$, showing that $x_{j+1} \in K_j'$. Thus $K_{j+1} \subset K_j'$ and therefore $K_{j+1} = K_j'$ (since $k_{j+1} = k_j'$ and $e_{j+1} = e_j'$).

This completes the proof of the lemma.

We shall now complete the proof of the necessity of conditions (a) and (b) by proving (b).

If we are in the case 1) of the above lemma, then $k_i = k_i'$ for $i = 1, 2, \cdots, h$, and we have

$$z = c_i x_i'^{\nu_i}, \quad x_i = d_i x_i'^{\mu_i},$$

with $c_i, d_i \in k_i' = k_i$. Therefore $z^{\mu_i}/x_i^{\nu_i} \in k_i$.

Now assume that we are in the case 2) of the lemma. Then we still have $k_i' = k_i$ for $i = 1, 2, \cdots, l$, and condition (b) follows as in the preceding case, for $\nu = 1, 2, \cdots, l$. Now, let $l+1 \leqq i \leqq q$. We observe that in this case we have $\mu_i = 1$ in (9), since $\epsilon_i = e_i$ for $i > l$. We have that $v(z) = \gamma = \epsilon_i \nu_i = e_{i-1}'\nu_i$ [by (33)], and hence $z = c_i x_{i-1}'^{\nu_i}$, $c_i \in k_{i-1}' = k_i$. We have also shown above that $x_i = u x_{i-1}'$, $u \in k_{i-1}' = k_i$. Hence $z/x_i^{\nu_i} \in k_i$. This completes the proof of the necessity of conditions (a) and (b).

Part 2. Sufficiency.

We assume now that conditions (a) and (b) are satisfied. Let, then

$$(34_i) \qquad z^{\mu_i}/x_i^{\nu_i} = c_i \in k_i, \qquad i = 1, 2, \cdots, h.$$

Since μ_i and ν_i are positive, relatively prime integers, we can find positive integers A_i, B_i such that

$$(35_i) \qquad A_i\nu_i - B_i\mu_i = 1, \qquad i = 1, 2, \cdots, h.$$

We set

$$(36_i) \qquad x_i' = z^{A_i}/x_i^{B_i}.$$

Then $x_i' \in F$, and we have $v(x_i') = A_i\gamma - B_i e_i = (A_i\nu_i - B_i\mu_i)\epsilon_i = \epsilon_i$, by (35_i) and (9). Thus

$$(37_i) \qquad v(x_i') = \epsilon_i, \qquad i = 1, 2, \cdots, h.$$

We now observe that by (36_i) we have $x_i'^{\mu_i} = z^{A_i\mu_i}/x_i^{B_i\mu_i}$, whence, by (34_i) and (35_i): $x_i'^{\mu_i} = c_i^{A_i}x_i^{A_i\nu_i}/x_i^{B_i\mu_i} = c_i^{A_i}x_i$. Similarly, we have

$$x_i'^{\nu_i} = z^{A_i\nu_i}/x_i^{B_i\nu_i} = c_i^{B_i}z^{A_i\nu_i}/z^{B_i\mu_i} = c_i^{B_i}z.$$

Thus

$$(38_i) \qquad x_i = \frac{1}{c_i^{A_i}}\, x_i'^{\mu_i},$$

$$i = 1, 2, \cdots, h.$$

$$(39_i) \qquad z = \frac{1}{c_i^{B_i}}\, x_i'^{\nu_i}.$$

We now set

$$(40_i) \qquad K_i' = k_i((x_i')), \qquad i = 1, 2, \cdots, h.$$

We know that $x_{i-1} = a_i x_i^{n_i}$, with $a_i \in k_i$, $i = 1, 2, \cdots, h$ [see (3)]. Hence, it follows from (36_{i-1}), (39_i) and (38_i) that

$$(41_i) \qquad x_{i-1}' = a_i' x_i'^{n_i'},\ a_i' \in k_i,\ n_i' = \epsilon_{i-1}/\epsilon_i, \qquad i = 1, 2, \cdots, h.$$

This proves that

$$(42) \qquad K_1' \subset K_2' \subset \cdots \subset K_h'.$$

From the definition (40_i) of K_i' and from (38_i) it also follows that

$$(43_i) \qquad K_i \subset K_i', \qquad i = 1, 2, \cdots, h.$$

Let $\mathfrak{M}_i' = \bar{\mathfrak{m}} \cap K_i' =$ prime ideal of the valuation induced in K_i' by v. We set

$$(44_i) \qquad m_i' = \beta_i/\epsilon_i, \qquad i = 1, 2, \cdots, h.$$

We proceed to prove that

$$(45) \qquad \mathfrak{m} = \mathfrak{M}_1'^{m_1'} + \mathfrak{M}_2'^{m_2'} + \cdots + \mathfrak{M}_h'^{m_h'}.$$

First of all, we observe that $v(\mathfrak{M}_i'^{m_i}) = m_i \cdot v(x_i') = \beta_i$, by (40_i), (37_i), and (44_i). Since $K_i' \supset K_i$ [see (43_i)], it follows that $\mathfrak{M}_i'^{m_i'} \supset \mathfrak{M}_i^{m_i}$, and hence, by (2):

$$(46) \qquad \mathfrak{m} \subset \mathfrak{M}_1'^{m_1'} + \mathfrak{M}_2'^{m_2'} + \cdots + \mathfrak{M}_h'^{m_h'}.$$

We shall now prove the opposite inclusion by showing that

$$(47_i) \qquad \mathfrak{M}_i'^{m_i'} \subset \mathfrak{m}, \qquad i = 1, 2, \cdots, h.$$

We first observe that, by (10) and (37_i), we have $v(x_i') = v(x_i) = e_i$, for $i = l+1, l+2, \cdots, h$. Thus K_i' and K_i, which, by definition (40_i) of K_i', have already in common the coefficient field k_i, have also the same ramifica-

4

tion exponent e_i, as subfields of F. Therefore, by (43_i), we have $K_i = K_i'$, $\mathfrak{M}_i' = \mathfrak{M}_i$, for $i = l+1, l+2, \cdots, h$. Thus, (47_i) is true for $i \geqq l+1$. We shall now use induction from $i+1$ to i, where we assume that we have

$$(48) \qquad 1 \leqq i \leqq l$$

and that

$$(49) \qquad \mathfrak{M}_{i+1}'^{m_{i+1}'} \subset \mathfrak{m}.$$

We observe that

$$(50) \qquad \beta_{i+1} \leqq \beta_i + \epsilon_i.$$

For, we have that e_i divides β_i, but e_i is strictly greater than ϵ_i, by (11) and (48). Hence e_i does *not* divide $\beta_i + \epsilon_i$. Since, by condition (a) of the theorem, we have that $\beta_i + \epsilon_i \in \mathfrak{m}v$, (50) follows from Proposition 5.3.

We have by (38_i): $x_i'^{m_i'} = b_i x_i^{m_i}$, where $b_i = c_i^{A_i m_i} \in k_i$. Thus

$$(51) \qquad x_i'^{m_i'} \in \mathfrak{M}_i^{m_i} \subset \mathfrak{m}.$$

If λ is any integer $> m_i'$, then $v(x_i'^{\lambda}) = \beta_i + (\lambda - m_i')\epsilon_i \geqq \beta_i + \epsilon_i \geqq \beta_{i+1}$, by (50). Since $x_i' \in K_{i+1}'$ and since $\beta_{i+1} = v(\mathfrak{M}_{i+1}'^{m_{i+1}'})$, it follows by (49), that

$$(52) \qquad x_i'^{\lambda} \in \mathfrak{M}_i'^{m_{i+1}'} \subset \mathfrak{m}, \qquad \text{if } \lambda > m_i'.$$

From (51) and (52), we conclude that $\mathfrak{M}_i'^{m_i} \subset \mathfrak{m}$. Thus the (47_i) are proved for all i, and the equality (45) is proved.

By (39_i), we have that $z \in K_1'$, and since $v(z) = \gamma \geqq s$, it follows that $z \in \mathfrak{M}_1'^{m_1'}$. Hence, by Lemma 4.5, $\mathfrak{o}$ is saturated with respect to z.

This completes the proof of the theorem.

Corollary 6.4. *Let $\{s; \beta_1, \beta_2, \cdots, \beta_h\}$ be the characteristic of the standard decomposition of $\mathfrak{m}$ with respect to a transversal parameter (here $\beta_1 = s$, and we know, by Corollary 5.7, that the above characteristic is the same for all transversal parameters), and let $\{e_1'; \beta_1', \beta_2', \cdots, \beta_q'\}$ be the characteristic of any standard decomposition of $\mathfrak{m}$. Let $\epsilon_1 = e_1'$ and let $\epsilon_2, \epsilon_3, \cdots, \epsilon_h$ and l be defined as in the (6_i) and (10). Then:*

(1) *if $l > 0$, $k_{l+1} = k_l$ and $\beta_{l+1} = \beta_l + \epsilon_l$, we have $q+1 = h$, $\beta_i' = \beta_i$, $i = 1, 2, \cdots, l$; $\beta_j' = \beta_{j+1}$, $j = l+1, l+2, \cdots, q$.*

(2) *In all other cases we have $q = h$, $\beta_i' = \beta_i$, $i = 1, 2, \cdots, h$.*

Remark. Assume that k is algebraically closed. Then conditions (b) of Theorem 6.2 are trivially satisfied, and we are then left only with conditions (a) of that theorem. Conditions (a) are always satisfied if $l = 0$ (in which case we have $\epsilon_i = e_i$, $i = 1, 2, \cdots, h$). The case $l = 0$ (or, equivalently, the case $\epsilon_1 = e_1 = s$) is the one in which γ is divisible by s, and our conclusion that in this case z is always a saturation parameter is also directly obvious, for if k is algebraically closed, the equality $z^s/x^\gamma = a \in k = \bar{k}$ implies that $z = x_1^\mu$, where $x_1 = \sqrt[\mu]{a \cdot x}$ is a transverse parameter and where $\mu = \gamma/s$. Thus, if there exist saturation parameters which are not powers of transversal parameters, we must have $l > 0$, i. e. $\epsilon_1 < e_1$ (ϵ_1-a proper division of e_1). From this it follows easily that *a necessary condition for the existence of saturation parameters z which are not of the form x^μ, x-transversal, is that the multiplicity s' of the quadratic transform $\mathfrak{o}'$ of $\mathfrak{o}$ be smaller than the multiplicity s of $\mathfrak{o}$.* For, since ϵ_1 is a proper divisor of e_1 $(= s = \beta_1)$, we have $\beta_1 + \epsilon_1 \not\equiv 0 \pmod{e_1}$. Since, on the other hand, condition (a) requires that $\beta_1 + \epsilon_1$ belong to $\mathfrak{m}v$, it follows from Proposition 5.3 that $\beta_2 \leqq \beta_1 + \epsilon_1$. Since $\beta_1 + \epsilon_1 = s + \epsilon_1 < 2s$, we have $s' = \beta_2 - s < s$, as asserted.

As an illustration, let us consider the case in which the integer h is 2, i. e. the case in which β_2 and s are relatively prime. Let $\beta_2 = s + s'$, $0 < s' < s$. The semi-group $\mathfrak{m}v$ is there generated over $\mathbf{Z}_+$ by s and β_2. From $s + \epsilon_1$ $(= \beta_1 + \epsilon_1) \in \mathfrak{m}v$ and from $s + \epsilon_1 < 2s$ follows then that we must have $s + \epsilon_1 = \beta_2$, i. e. $\epsilon_1 = s'$. Hence in this case, a *necessary and sufficient condition that a parameter z be a saturation parameter in that* either $\gamma \equiv 0 \pmod{s}$ *or that* $(\gamma, s) = s'$, where $\gamma = v(z)$.

It would be interesting to analyze the full implications of Theorem 6.2 and thus obtain explicitly a description of the set of all saturation parameters of $\mathfrak{o}$ also in the general case $h > 2$.

7. Saturation-theoretic properties of transversal parameters. In this section a coefficient field k of $\mathfrak{o}$ is fixed once and for always, and $\bar{k}$ denotes the algebraic closure of k in F, whence $\bar{\mathfrak{o}} = \bar{k}[[t]]$ and $F = \bar{k}((t))$.

The following simple lemma will be of prime importance for the proofs of a number of theorems in this and in later sections.

LEMMA 7.1. *Let f_1, f_2 be two automorphisms of F such that $f_1 \mid \bar{k} = f_2 \mid \bar{k}$. Let t be an element of F such that $v(t) = 1$ (whence $F = \bar{k}((t))$). Assume that*

$$(1) \qquad v(f_2(t) - f_1(t)) = 1 + \lambda, \lambda > 0.$$

Then we have for all elements η of F such that $\eta \neq 0$ and $v(\eta) \neq 0$:

$$(2) \qquad v(f_2(\eta) - f_1(\eta)) = v(\eta) + \lambda.$$

Proof. The lemma is trivial if $f_1 = f_2$, for then (1) and (2) are satisfied, λ being $+\infty$ in that case. We shall therefore assume that $f_1 \neq f_2$.

We set $f = f_1^{-1}f_2$, whence f is a non-identical $\bar{k}$-automorphism of F. By Proposition 1.3, we have $vf_1 = v$, and hence $v(f_2(\eta) - f_1(\eta)) = vf_1(f(\eta) - \eta) = v(f(\eta) - \eta)$. Thus condition (1) now reads as follows:

$$v(f(t) - t) = 1 + \lambda, \lambda > 0, \tag{3}$$

while the assertion (2) is equivalent to

$$v(f(\eta) - \eta) = v(\eta) + \lambda. \tag{4}$$

Thus, it is sufficient to prove the lemma under the assumption that f_1 is the indentity. We assume this and we write f for f_2. We are given that (3) holds for some element t of F such that $v(t) = 1$ and we have to prove (4) for any element η of F such that $\eta \neq 0$ and $v(\eta) \neq 0$.

Relation (3) implies that $f(t) = t\xi$, where $\xi = 1 + at^\lambda + \cdots, 0 \neq a \in \bar{k}$. If n is any positive integer then

$$f(t^n) - t^n = t^n(\xi^n - 1) = t^n(\xi - 1)(\xi^{n-1} + \xi^{n-2} + \cdots + 1).$$

Now, $v(\xi - 1) = \lambda > 0$, and therefore the v-residue of ξ is 1. Consequently, the v-residue of $\xi^{n-1} + \xi^{n-2} + \cdots + 1$ is $n \neq 0$, and hence (4) holds for $\eta = t^n$. Now, let η be any element of F such that $v(\eta) = n =$ a positive integer:

$$\eta = c_0 t^n + c_1 t^{n+1} + \cdots, \qquad c_i \in \bar{k}, c_0 \neq 0.$$

We have then (since f is a $\bar{k}$-automorphism):

$$f(\eta) - \eta = c_0[f(t^n) - t^n] + c_1[f(t^{n+1}) - t^{n+1}] + \cdots.$$

Since $v[f(t^{n+i}) - t^{n+i}] = n + i + \lambda$ and $c_0 \neq 0$, (4) follows. If now η is an element of F such that $v(\eta) < 0$, then we set $\zeta = \frac{1}{\eta}$ and we have

$$v(f(\eta) - \eta) - v(\eta) = v\left(\frac{1}{f(\zeta)} - \frac{1}{\zeta}\right) + v(\zeta)$$
$$= v(\zeta - f(\zeta)) - v(f(\zeta)) = v(f(\zeta) - \zeta) - v(\zeta) = \lambda.$$

This completes the proof.

THEOREM 7.2. *If $\mathfrak{o}$ is saturated (i.e., k-saturated) then it is saturated with respect to any transversal parameter x (more precisely: $\mathfrak{o}$ is k-saturated with respect to x, i.e., $\mathfrak{o}$ is saturated with respect to the field $K = k((x))$).*

Proof. To say that $\mathfrak{o}$ is saturated means to say that $\mathfrak{o}$ is k-saturated with respect to some parameter. Let then ξ be a parameter of $\mathfrak{o}$ such that $\mathfrak{o}$ is k-saturated with respect to ξ. By Lemma 4.5, $\mathfrak{o}$ is also saturated with respect to the parameter ξ^q, for any positive integer q. However, here is a short direct proof of this consequence of Lemma 4.5. Let $L = k((\xi))$, $L_0 = k((\xi^q))$ and let Σ^*, $\Sigma_0{}^*$ denote respectively the least Galois extensions of L and L_0 containing F. Any element τ of the Galois group Γ of Σ^*/L can be extended to an automorphism τ_0 of $\Sigma_0{}^*/L_0$. Therefore, if $\zeta \in \mathfrak{o}$, $\eta \in \bar{\mathfrak{o}}$ and $v(\eta^\sigma - \eta) \geqq v(\zeta^\sigma - \zeta)$ for all σ in the Galois group of $\Sigma_0{}^*/L_0$, then we have, in particular, $v(\eta^\tau - \eta) \geqq v(\zeta^\tau - \zeta)$ for all τ in Γ. Hence, since $\mathfrak{o}$ is saturated with respect to ξ, it follows that $\eta \in \mathfrak{o}$, showing that $\mathfrak{o}$ is also saturated with respect to ξ^q.

Now if $q > 1$, the ξ^q is certainly not a transversal parameter. We have therefore shown that if $\mathfrak{o}$ is saturated, it is also saturated with respect to some non-transversal parameter. We may therefore assume that *ξ is not a transversal parameter.*

Let now x be any *transversal* parameter of $\mathfrak{o}$. We have to show that $\mathfrak{o}$ is saturated with respect to x.

We shall use the above notation L $(= k((\xi)))$, Σ^* and Γ. We begin by making a number of observations.

1. *There exists an element τ in Γ such that $v(x^\tau - x) = v(x)$.* For, let

$$x_0 = \frac{1}{[F:L]} T_{F/L}(x).$$

Then $x_0 \in \xi k[[\xi]]$, $v(x_0) \geqq v(\xi) > v(x)$ (since x is a transversal parameter while ξ is not), $v(x - x_0) = v(x)$, and the assertion follows directly from Lemma 2.3.

2. Let

$$s = v(x) \quad (= e_0(\mathfrak{o}) = \text{reduced multiplicity of } \mathfrak{o}),$$

and let

$$\Gamma_1 = \{\tau \in \Gamma \mid v(\eta^\tau - \eta) > s \text{ for } all\ \eta \text{ in } \mathfrak{m}\}.$$

It is immediate that Γ_1 is a subgroup of Γ and that $\Gamma_1 \supset \mathfrak{G}(\Sigma^*/F)$; by 1, Γ_1 is a *proper* subgroup of Γ. Let L_1 be the fixed field of Γ_1. Thus $L < L_1 \subset F$.

3. Let

$$x' = \frac{1}{[F:L_1]} \cdot T_{F/L_1}(x).$$

Since $v(x^\tau - x) > v(x)$ for all τ in Γ_1 (by definition of Γ_1) it follows from Lemma 2.3 that

$$(5) \qquad v(x - x') > v(x)$$

and that consequently

$$(5') \qquad v(x) = v(x').$$

4. Let y be a ξ-saturator of $\mathfrak{o}$ in $\mathfrak{m}$ (Proposition 3.3). We assert that

$$v(y^\tau - y) = s \textit{ for all } \tau \textit{ in } \Gamma \textit{ such that } \tau \notin \Gamma_1.$$

For, if $\tau \notin \Gamma_1$ then there exists an element ζ in $\mathfrak{m}$ such that $v(\zeta^\tau - \zeta) = s$ (by definition of Γ_1). From Corollary 3.2 and from the fact that $\mathfrak{o}$ is saturated with respect to ξ follows that $v(\zeta^\tau - \zeta) \geqq v(y^\tau - y)$, i. e., $v(y^\tau - y) \leqq s$. Since $y \in \mathfrak{m}$, we have $v(y) \geqq s$, whence $v(y^\tau - y) \geqq s$, and our assertion is proved.

5. *The element x' introduced in 3 is an element of* $\mathfrak{o}$. For, let τ be any element of Γ. If $\tau \in \Gamma_1$, then $x'^\tau = x'$, since $x' \in L_1$ and $\Gamma_1 = \mathfrak{G}(F^*/L_1)$. If $\tau \notin \Gamma_1$ then $v(x'^\tau - x') \geqq v(y^\tau - y)$, since $v(x'^\tau - x') \geqq s$ [by (5')], while, by 4, $v(y^\tau - y) = s$. Thus, in all cases, we have $v(x'^\tau - x') \geqq v(y^\tau - y)$, and this shows that $x' \in \mathfrak{o}$, as asserted. It follows now from (5') that *also x' is a transversal parameter of* $\mathfrak{o}$.

6. *The ring $\mathfrak{o}$ is saturated with respect to x'.* Let ζ be any element of $\mathfrak{o}$, let η be an element of $\bar{\mathfrak{o}}$ and assume that $v(\eta^\sigma - \eta) \geqq v(\zeta^\sigma - \zeta)$ for any $k((x'))$-isomorphism of F into Σ^*. Consider any element τ of Γ. If $\tau \in \Gamma_1$, then τ induces a $k((x'))$-isomorphism of F and hence $v(\eta^\tau - \eta) \geqq v(y^\tau - y)$, since $\eta^\tau = \eta^\sigma$, $\zeta^\tau = \zeta^\sigma$ and $v(\zeta^\tau - \zeta) \geqq v(y^\tau - y)$ (see Corollary 3.2). If $\tau \notin \Gamma_1$ and if we denote by c the v-residue of η $(c \in k)$, then

$$v(\eta^\tau - \eta) \geqq v((\eta - c)^\tau - (\eta - c)) \geqq s,$$

since $\eta - c$ belongs, by assumption, to the maximal ideal of the saturated ring $\bar{\mathfrak{o}}_{x'}$, whence $v(\eta - c) \geqq s =$ reduced multiplicity of $\bar{\mathfrak{o}}_{x'}$; while, by 4, $v(y^\tau - y) = s$. Thus again $v(\eta^\tau - \eta) \geqq v(y^\tau - y)$. This holds for any element τ of Γ, showing that $\eta \in \mathfrak{o}$, and $\mathfrak{o}$ is therefore saturated with respect to x'.

As a result of the preceding observations we have now found a *transversal* parameter x' such that $\mathfrak{o}$ *is saturated with respect to x'* and which is related to the intial transversal parameter x by the relation (5). *That relation implies that the v-residue of x'/x is* 1.

The proof of the theorem is thus reduced to the proof of the following statement:

Let x, x' be two transversal parameters of $\mathfrak{o}$ such that 1) $\mathfrak{o}$ *is saturated with respect to x' and* 2) *the v-residue of x'/x is* 1. *Then $\mathfrak{o}$ is also saturated with respect to x.*

We now proceed to prove this statement. We fix elements t and t' in F such that $v(t) = v(t') = 1$ and such that (see (1), §1)

$$x = t^s/a, \qquad a \in \bar{k};$$
$$x' = t'^s/a', \qquad a' \in \bar{k}.$$

Let α denote the v-residue of t'/t ($\alpha \in \bar{k}$). Then we will have (since the v-residue of x'/x is 1):

$$a'/a = \alpha^s. \tag{6}$$

We shall now use Lemma 1.2. If $\tilde{k}$ denotes the least Galois extension of k containing $\bar{k}$ and Γ denotes the Galois group of $\tilde{k}/k$, then (6) shows that the field k^* obtained by adjoining to $\tilde{k}$ all the s-th roots of the quotients a^σ/a, $\sigma \in \Gamma$, coincides with the field k'^* obtained by adjoining to $\tilde{k}$ all the s-th roots of the quotient a'^σ/a', $\sigma \in \Gamma$. Since $t' \in \bar{k}[[t]] \subset k^*[[t]]$ and $t \in \bar{k}[[t']] \subset k'^*[[t']]$, it follows from Lemma 1.2 that the least Galois extension F^* of K $(= k((x)))$ containing F coincides with the least Galois extension of K' $(= k((x')))$ containing F. Here $F^* = k^*((t)) = k^*((t'))$.

For convenience we shall now replace the uniformizing parameter t' by the quotient t'/α. This quotient is still a uniformizing parameter of F, and if we change our notation and use the letter t' to denote this new uniformizing parameter of F, then we will have, by (6):

$$\begin{aligned} x &= t^s/a, \\ x' &= t'^s/a, \end{aligned} \tag{7}$$

and t'/t has v-residue 1.

We denote by G and G' respectively, the Galois groups $\mathfrak{G}(F^*/K)$ and $\mathfrak{G}(F^*/K')$, where $K = k((x))$ and $K' = k((x'))$. If τ is any element of G and τ' is any element of G' then it follows from (7) and from $x^\tau = x$, $x'^{\tau'} = x'$ that

$$t^\tau = \gamma_\tau t, \qquad \gamma_\tau{}^s = \frac{a^\tau}{a}; \tag{8}$$

$$t'^{\tau'} = \gamma_{\tau'} t', \qquad \gamma_{\tau'}{}^s = \frac{a^{\tau'}}{a}. \tag{8'}$$

Clearly, the element τ of G is uniquely determined by γ_τ and by the restriction of τ to k^* (this restriction is an element of the Galois group of k^*/k). Similarly, the element τ' of G' is uniquely determined by $\gamma_{\tau'}$ and by the restriction of τ' to k^*. It is also clear that we obtain a group isomorphism

$$\phi\colon G \xrightarrow{\sim} G'$$

by setting $\phi(\tau) = \tau'$ $(\tau \in G, \tau' \in G')$ if $\gamma_\tau = \gamma_{\tau'}$ and $\tau \mid k^* = \tau' \mid k^*$.

Since the v-residue of t'/t is 1, we can write

$$t' = t + \text{terms of degree} > 1 \text{ in } t.$$

Therefore, if $\tau' = \phi(\tau)$ $(\tau \in G)$ then

$$t'^{\tau'} = \gamma_{\tau'} t' = \gamma_{\tau'} t + \text{terms of degree} > 1 \text{ in } t;$$
$$t'^{\tau} = t^{\tau} + \cdots = \gamma_\tau t + \text{terms of degree} > 1 \text{ in } t.$$

Since $\gamma_{\tau'} = \gamma_\tau$, it follows that

$$v(t'^{\tau'} - t'^{\tau}) > 1.$$

Since, moreover, $\tau' \mid k^* = \tau \mid k^*$, we are in a position to apply Lemma 7.1 to the two automorphisms τ', τ of F^*. We have therefore

$$(9) \qquad v(\eta^{\tau'} - \eta^{\tau}) = v(x^{\tau'} - x) - v(x) + v(\eta), \text{ if } \tau' = \phi(\tau),$$

for all elements η of F^* such that $\eta \neq 0$ and $v(\eta) \neq 0$.

Let now ζ and η be elements of F such that $\zeta \in \mathfrak{o}$, $\eta \in \bar{\mathfrak{o}}$ and such that

$$(10) \qquad v(\eta^{\tau} - \eta) \geqq v(\zeta^{\tau} - \zeta), \text{ for all } \tau \text{ in } G.$$

To prove that $\mathfrak{o}$ is saturated with respect to x we have to prove that $\eta \in \mathfrak{o}$. We fix an x'-saturator y of $\mathfrak{o}$ in $\mathfrak{m}$. Then, in view of Corollary 3.2, we see that what we have to prove is that

$$(11) \qquad v(\eta^{\tau'} - \eta) \geqq v(y^{\tau'} - y), \text{ for all } \tau' \text{ in } G'.$$

In proving (11) we may assume (as we have done repeatedly on previous occasions) that $v(\zeta) > 0$ and $v(\eta) > 0$ (observe that the v-residue of ζ is in k, since $\zeta \in \mathfrak{o}$, and that (10) implies that the v-residue of η is invariant under all automorphism of k^*/k, hence also belongs to k). We may even assume (by substracting from η a suitable element of $k[[x]]$) that $T_{F/k((x))}(\eta) = 0$ and that consequently (see Corollary 2.4)

$$(12) \qquad v(\eta^{\tau} - \eta) = v(\eta) \text{ for some } \tau \text{ in } G.$$

Since $\zeta \in \mathfrak{m}$, we have

$$(13) \qquad v(\zeta) \geqq s,$$

and since $v(\zeta^\tau - \zeta) \geqq v(\zeta)$, it follows from (10), (12), and (13) that

$$v(\eta) \geqq s. \tag{14}$$

Given τ' in G', we set $\tau = \phi^{-1}(\tau')$, we use (9) for the pair (τ, τ') and the relation similar to (9) obtained upon replacing η by ζ:

$$v(\zeta^{\tau'} - \zeta^\tau) = v(x^{\tau'} - x) - v(x) + v(\zeta), \qquad \tau' = \phi(\tau). \tag{15}$$

Since $v(x) = s$, it follows from (9), (14), (15), and (13) that $v(\eta^{\tau'} - \eta^\tau) \geqq v(x^{\tau'} - x)$ and $v(\zeta^{\tau'} - \zeta^\tau) \geqq v(x^{\tau'} - x)$. Since $x \in \mathfrak{o}$ and y is an x'-saturator of $\mathfrak{o}$, it follows from Corollary 3.2 that $v(x^{\tau'} - x) \geqq v(y^{\tau'} - y)$. Thus

$$v(\eta^{\tau'} - \eta^\tau) \geqq v(y^{\tau'} - y), \tag{16}$$

$$v(\zeta^{\tau'} - \zeta^\tau) \geqq v(y^{\tau'} - y). \tag{17}$$

Now, since $\zeta \in \mathfrak{o}$, we have $v(\zeta^{\tau'} - \zeta) \geqq v(y^{\tau'} - y)$, and consequently, by (17): $v(\zeta^\tau - \zeta) \geqq v(y^{\tau'} - y)$. This inequality, in conjuction with (10), yield the inequality $v(\eta^\tau - \eta) \geqq v(y^{\tau'} - y)$, and from this the desired inequality (11) follows in view of (16). This completes the proof.

COROLLARY-DEFINITION 7.3. *The saturation of $\mathfrak{o}$ with respect to a transversal parameter is a ring which is independent of the choice of that parameter. This ring is called the saturation of $\mathfrak{o}$ and will be denoted by $\tilde{\mathfrak{o}}$.*

For, let x and x' be two transversal parameters of $\mathfrak{o}$ and let $\tilde{\mathfrak{o}} = \tilde{\mathfrak{o}}_x$, $\tilde{\mathfrak{o}}' = \tilde{\mathfrak{o}}_{x'}$ (the saturations being relative to the given coefficient field k of $\mathfrak{o}$). Since $\mathfrak{o}$ and $\tilde{\mathfrak{o}}'$ have the same reduced multiplicity (Propositions 2.1 and 2.5), x is also a transverse parameter of $\tilde{\mathfrak{o}}'$. Therefore, by the above theorem, $\tilde{\mathfrak{o}}'$ is saturated with respect to x. By the minimality property of $\tilde{\mathfrak{o}}_x$ it follows that $\tilde{\mathfrak{o}} \subset \tilde{\mathfrak{o}}'$. Similary, $\tilde{\mathfrak{o}}' \subset \tilde{\mathfrak{o}}$, and hence $\tilde{\mathfrak{o}}' = \tilde{\mathfrak{o}}$.

COROLLARY 7.4. *The ring $\tilde{\mathfrak{o}}$ is the smallest saturated ring between $\mathfrak{o}$ and $\bar{\mathfrak{o}}$.*

First of all, if z is any parameter of $\mathfrak{o}$ and x is a transversal parameter of $\mathfrak{o}$, then x is also a transversal parameter of $\tilde{\mathfrak{o}}_z$, and $\tilde{\mathfrak{o}}_z$ is therefore also saturated with respect to x. Thus $\tilde{\mathfrak{o}}_z \supset \tilde{\mathfrak{o}}_x$ $(= \tilde{\mathfrak{o}})$; in other words, $\tilde{\mathfrak{o}}$ *is the smallest saturation of* $\mathfrak{o}$. Now, let $\mathfrak{o}'$ be any saturated local ring between $\mathfrak{o}$ and $\bar{\mathfrak{o}}$ (not necessarily the saturation of $\mathfrak{o}$ with respect to some parameter; note that any ring between $\mathfrak{o}$ and $\bar{\mathfrak{o}}$ is necessarily a complete local domain, of dimension 1). We fix a transversal parameter x' of $\mathfrak{o}'$ and we denote by k' the algebraic closure of k in $\mathfrak{o}'$. Then k' is a coefficient field of $\mathfrak{o}'$, and $\mathfrak{o}'$ is saturated with respect to the field $k'((x'))$. Since k' is a finite extension of k, it follows from Lemma 2.2 that $\mathfrak{o}'$ is also saturated with respect to the field $k((x'))$. It follows at once (see the reasoning given in the beginning of the proof of Theorem 7.2) that if q is any positive integer then $\mathfrak{o}'$ is saturated with respect to the field $k((x'^q))$. Now if q is sufficiently large then x'^q is contained in the conductor of $\mathfrak{o}$ with respect to $\bar{\mathfrak{o}}$ and is thus an element of $\mathfrak{o}$. Thus $\mathfrak{o}' \supset \tilde{\mathfrak{o}}_{x'^q}$ (since $\mathfrak{o}' \supset \mathfrak{o}$), and on the other hand we have just established the fact that $\tilde{\mathfrak{o}}_{x'^q} \supset \tilde{\mathfrak{o}}$. Hence $\mathfrak{o}' \supset \tilde{\mathfrak{o}}$, as asserted.

COROLLARY 7.5. *If x is a transversal parameter of $\mathfrak{o}$ and x' is any element of the maximal ideal of $k[[x]]$, then the saturation of $\mathfrak{o}$ with respect to $k((x'))$ is equal to the saturation $\tilde{\mathfrak{o}}$ of $\mathfrak{o}$ (with respect to $k((x))$).*

By Corollary 7.4 we have $\tilde{\mathfrak{o}} \subset \tilde{\mathfrak{o}}_{k((x'))}$. On the other hand, if F'^* is the least Galois extension of $k((x'))$ which contains F, then the Galois group of $F'^*/k((x))$ is a subgroup of the Galois group of $F'^*/k((x'))$ and from this it follows immediately that $\tilde{\mathfrak{o}}_{k((x))}$ is also saturated with respect to $k((x'))$, whence $\tilde{\mathfrak{o}}_{k((x'))} \subset \tilde{\mathfrak{o}}$.

The saturation of $\mathfrak{o}$ with respect to a parameter which is not transversal may very well depend on the choice of the parameter. The question naturally arises whether the family of rings which are saturations of $\mathfrak{o}$ has an upper bound, and if so what is that upper bound. If $\mathfrak{o}'$ is any saturation of $\mathfrak{o}$, then we know that k is also a coefficient field of $\mathfrak{o}'$ and that $\mathfrak{o}$ and $\mathfrak{o}'$ have the same multiplicity (and therefore also the same reduced multiplicity). If, then, s is the reduced multiplicity of $\mathfrak{o}$, and $\mathfrak{m}'$ is the maximal ideal of $\mathfrak{o}'$, we must have $v(\xi) \geqq s$ for every ξ in $\mathfrak{m}'$, with equality for some elements ξ in $\mathfrak{m}'$. If we denote by $\bar{\mathfrak{m}}$ the maximal ideal of the integral closure $\bar{\mathfrak{o}}$ of $\mathfrak{o}$ ($\bar{\mathfrak{o}} =$ valuation ring of v), it follows then that $\mathfrak{m}' \subset \bar{\mathfrak{m}}^s$ and therefore $\mathfrak{o}' \subset k + \bar{\mathfrak{m}}^s$. We observe now that the set $k + \bar{\mathfrak{m}}^s$ is a ring between $\mathfrak{o}$ and $\bar{\mathfrak{o}}$, and we proceed to prove the following result.

PROPOSITION 7.6. *There exists a parameter x of $\mathfrak{o}$ such that the local ring $k + \bar{\mathfrak{m}}^s$ is the k-saturation of $\mathfrak{o}$ with respect to x. This ring $k + \bar{\mathfrak{m}}^s$ is therefore the upper bound of the family of all saturation of $\mathfrak{o}$.*

Proof. We fix a transversal parameter z of $\mathfrak{o}$ and we write $z = t^s/a$, where $0 \neq a \in \bar{k}$ and t is a suitable element of $\bar{\mathfrak{o}}$ such that $v(t) = 1$ [see (1), §1]. The conductor of $\mathfrak{o}$ in $\bar{\mathfrak{o}}$ is a power $\bar{m}^N$ of the maximal ideal $\bar{\mathfrak{m}} = \bar{\mathfrak{o}} \cdot t$ of $\bar{\mathfrak{o}}$. Hence, for any $n \geqq N$ and for any element b in $\bar{k}$, the element t^n/b belongs to $\mathfrak{m}$. We take for b a *primitive* element of $\bar{k}$ over k. Let $b_1, b_2, \cdots, b_q$ be the distinct conjugates of b over k, where we assume that $b_1 = b$. Upon replacing b, if necessary, by $c + b$, where c is a suitable element of k, we may assume that

$$(18) \qquad (b_i/b_1)^s \neq 1, \qquad i = 2, 3, \cdots, q.$$

Having thus fixed b, we now choose a positive integer n satisfying the following conditions:

$$(19) \qquad n \geqq N;$$

$$(20) \qquad (n, s) = 1;$$

$$(21) \qquad (b_i/b_1)^s \neq (a_j/a_1)^n, \quad \text{for } i = 2, 3, \cdots, q, \quad j = 2, 3, \cdots, r,$$

where $a_1, a_2, \cdots, a_r$ the distinct conjugates of a, and $a_1 = a$. To see that such integers n exist, let us denote by l_j the following non-negative integer: if a_j/a_1 is a root of unity, then we assume that a_j/a_1 is a *primitive* l_j-th root of unity; in the contrary case we set $l_j = 0$. We have then that

$$(22) \qquad 0 \leqq l_j \neq 1, \qquad j = 2, 3, \cdots, q.$$

If for a given pair of indices (i, j) $(2 \leqq i \leqq q; 2 \leqq j \leqq r)$ there exists an integer n satisfying the equation

$$(23) \qquad (b_i/b_1)^s = (a_j/a_1)^n,$$

we denote by $n_{i,j}$ any such integer. Then, to satisfy the conditions (19), (20), and (21), we have only to choose an integer n satisfying (19), (20) and the following conditions:

$$(24) \qquad n \not\equiv n_{i,j} \pmod{l_j},$$

for *each* pair (i, j) for which (23) has a solution in n (and $n_{i,j}$ being *one* such solution, fixed in advance). In view of (22), there exist infinitely many integers n satisfying (19), (20), and (24).

We now fix such an integer n and we set

$$(25) \qquad x = t^n/b,$$

and we show that $\tilde{\mathfrak{o}}_x = k + \bar{m}^s$. Let τ be any $k((x))$-automorphism of the least Galois extension F^* of $k((x))$ containing F, such that τ is not the identity on F. Since $x^\tau = x$, we must have, by (25): $t^\tau = \epsilon_\tau t$, where ϵ_τ is algebraic over k:

$$(26) \qquad \epsilon_\tau^n = b^\tau/b.$$

W- have $z^\tau = \epsilon_\tau{}^s t^s / a^\tau$ and hence

$$(27) \qquad z^\tau - z = \frac{1}{a^\tau}(\epsilon_\tau{}^s - \frac{a^\tau}{a}) t^s.$$

We assert that $z^\tau \neq z$. For, in the contrary case, we would have

$$(28) \qquad \epsilon_\tau{}^s = \frac{a^\tau}{a}$$

and hence by (26)

$$(30) \qquad (\frac{b^\tau}{b})^s = (\frac{a^\tau}{a})^n.$$

In view of (18) and (21), (30) implies $b^\tau = b$. Since $\bar{k} = k(b)$ it follows that τ *is the identity on* $\bar{k}$. Since $a \in \bar{k}$, we have also $a^\tau = a$. By (26) and (28) we now have $\epsilon_\tau{}^n = \epsilon_\tau{}^s = 1$, whence $\epsilon_\tau = 1$, by (20). Thus $t^\tau = t$. This, in conjunction with the fact that τ is the identity on $\bar{k}$, implies that τ is the identity on F $(= \bar{k}((t)))$, contrary to assumption. We have thus shown that $z^\tau \neq z$. Hence, by (27), $v(z^\tau - z) = s$, for all $\tau \in G = \mathfrak{G}(F^*/k((x)))$, such that τ is not the identity on F. If then η is any element of $\bar{\mathfrak{m}}^s$, we have $v(\eta^\tau - \eta) \geqq v(\eta) \geqq s = v(z^\tau - z)$ for all τ as above. This implies that $\bar{\mathfrak{m}}^s \subset \tilde{\mathfrak{m}}_x$, and the proposition is proved.

Note that $k + \bar{\mathfrak{m}}^s$ is the greatest local ring between $\mathfrak{o}$ and $\bar{\mathfrak{o}}$ which has k as coefficient field and has reduced multiplicity s. The consideration of the *maximal* saturation of $\mathfrak{o}$ is thus of scarce interest, since the only invariants of that saturation are the reduced multiplicity s of $\mathfrak{o}$ and the coefficient field k. On the contrary, the minimal saturation $\tilde{\mathfrak{o}}$ of $\mathfrak{o}$ (see Corollary 7.4) is of considerable interest for the classification of singularities of plane algebroid curves (see [2], Theorem 2.1. In our next paper, this theorem will be extended to reducible algebroid curves, in the plane).

The proof of the above proposition allows one to show, at least in the case in which k is algebraically closed (or—more generally—if $k = \bar{k}$), that in Corollary 7.5 the condition that x be transversal may be essential. For, if $k = \bar{k}$, we have in the notations of the above proof, that $a \in k$. So we may assume $z = t^s$. We can then take $x = t^n$, with $(n, s) = 1$, and we will then have $\tilde{\mathfrak{o}}_x = k + \bar{\mathfrak{m}}^s$. Now, let $x' = x^s$. Then $x' = z^n$, and by Corollary 7.5, we have therefore $\tilde{\mathfrak{o}}_{x'} = \tilde{\mathfrak{o}}_z = \tilde{\mathfrak{o}}$ (since z is transversal). But $x' = x^s$, and unless $\mathfrak{m} = \bar{\mathfrak{m}}^s$, we will have $\tilde{\mathfrak{o}} < k + \bar{\mathfrak{m}}^s = \tilde{\mathfrak{o}}_x$.

If we identify in Lemma 2.2, the fields K and K_0 with $k((x))$ and $k((x'))$ respectively, the above counterexample shows that it may very well happen that $\tilde{\mathfrak{o}}_{K_0} < \tilde{\mathfrak{o}}_K$.

8. The automorphisms of $\mathfrak{o}$ in the saturated case. In this section we continue to deal once and for always with a preassigned field k of representatives of $\mathfrak{o}$, while $\bar{k}$ will denote, as before, the algebraic closure of k in F. *We assume in this section that* $\mathfrak{o}$ *is saturated* (i.e., k-saturated).

The fact that F is a complete local field $\bar{k}((t))$ has as a consequence that F has a big group of $\bar{k}$-automorphisms. If, namely, t' is any element of F such that $v(t')=1$, then there exists a unique $\bar{k}$-automorphism ϕ of F such that $\phi(t)=t'$. More generally, given any two elements x, x' of F, such that $v(x)=v(x')=n\neq 0$, then *a necessary and sufficient condition that there exist a* $\bar{k}$-*automorphism* ϕ *of* F *such that* $\phi(x)=x'$, *is that the* v-*residue of* x'/x *in* $\bar{k}$ *be an* n-*th power of an element of* $\bar{k}$. For if we write $x=at^n+a_1t^{n+1}+\cdots$, $a_i\in\bar{k}$ and if ϕ is any $\bar{k}$-automorphism of F, then—writing $\phi(t)=t'$—we have $\phi(x)=at'^n+a_1t'^{n+1}+\cdots$, and hence the v-residue of $\dfrac{\phi(x)}{x}$ in $\bar{k}$ is the n-th power of the v-residue of $\dfrac{t'}{t}$ in $\bar{k}$. This proves the necessity of the condition. On the other hand, we can choose the uniformizing parameters t, t' so as to have [see (1), § 1]:

$$x=t^n/a,\quad x'=t'^n/a'; \qquad a,a'\in\bar{k}.$$

Upon replacing t' by ct', where c is a suitable element of $\bar{k}$, we may assume that the v-residue of t'/t is 1. When that is so, then the v-residue of x'/x is $\dfrac{a}{a'}$. If we assume that $\dfrac{a}{a'}=\alpha^n$, $\alpha\in\bar{k}$, then upon replacing t' by $\alpha t'$ we can write $x=t^n/a$, $x'=t'^n/a$, and thus the (unique) $\bar{k}$-automorphism of F which takes t into t' will send x into x'.

Remark. If the residue of x'/x is 1 then we can set $\alpha=1$ in the preceding considerations, and we thus obtain a $\bar{k}$-automorphism ϕ of F such that $\phi(x)=x'$ and such that the residue of $\phi(t)/t$ $(=t'/t)$ is 1. It then follows from Lemma 7.1 that $v(\phi(\xi)-\xi)>v(\xi)$, for all ξ in F, $\xi\neq 0$.

Each $\bar{k}$-automorphism of F transforms, of course, the integral closure $\bar{\mathfrak{o}}$ $(=\bar{k}[[t]])$ of $\mathfrak{o}$ onto itself, and thus, given any two elements x, x' of $\bar{\mathfrak{o}}$ such that $v(x)=v(x')=n>0$, there exists a $\bar{k}$-automorphism of $\bar{\mathfrak{o}}$ which sends x into x', provided only the v-residue of x'/x be the n-th power of an element of $\bar{k}$ (a condition which is automatically satisfied if $\bar{k}$ is algebraically closed). This is a sort of restricted transitivity property of the group of $\bar{k}$-automorphism of the ring $\bar{\mathfrak{o}}$.

In this section we shall show that also the group of automorphisms of any of our saturated local domains $\mathfrak{o}$ enjoys a suitably restricted transitivity property, so that—at any rate—this group is very "big." As an immediate—and most important—illustration of what we have in mind, we cite the following result.

THEOREM 8.1. *If* $\mathfrak{o}$ *is saturated and if* x, x' *are two transversal parameters of* $\mathfrak{o}$ *such that the* v*-residue of* x'/x *is of the form* c^s, *where* $s = v(x)$ $=$ *reduced multiplicity of* $\mathfrak{o}$ *and* c *is an element of* k, *then there exists a* k*-automorphism of* F *which sends* x *into* x' *and which induces an automorphism of* $\mathfrak{o}$.

Actually, we shall prove the following, more general theorem:

THEOREM 8.2. *Let* $\mathfrak{o}$ *be saturated, let* x *and* x' *be two parameters of* $\mathfrak{o}$ *such that* 1) $\mathfrak{o}$ *is saturated with respect to* x; 2) $v(x) = v(x')$ $(= n$, *say*$)$: 3) *the* v*-residue of* x'/x *in* $\bar{k}$ *is an element of the form* c^n, $c \in k$. *Then a necessary and sufficient condition that there exist a* $\bar{k}$*-automorphism of* F *which sends* x *into* x' *and which induces an automorphism of* $\mathfrak{o}$, *is that the following inequality*

$$v(x'^{\tau} - x') \geqq v(y^{\tau} - y) + n - s \tag{1}$$

hold for all τ *in the Galois group* G *of* $F^*/k((x))$; *here* F^* *is the least Galois extension of* $k((x))$ *containing* F, y *is an* x*-saturator of* $\mathfrak{o}$ *and* s *is the reduced multiplicity of* $\mathfrak{o}$.

Note. Theorem 8.1 is an immediate consequence of Theorem 8.2. For, if x and x' are transversal parameters then $n = s$ and the inequality (1), which now reduces to $v(x'^{\tau} - x') \geqq v(y^{\tau} - y)$, is satisfied, since y is an x-saturator of $\mathfrak{o}$ and since $x' \in \mathfrak{o}$ (see Corollary 3.2).

Proof. There certainly exists a $\bar{k}$-automorphism f of $\bar{F}$—*and only one*—such that $f(x) = x'$ and

$$f(t) = ct + \text{terms of degree} > 1 \text{ in } t, \tag{2}$$

where t is some uniformizing parameter of F. *To prove our theorem is tantamount to proving that* $f(\mathfrak{o}) = \mathfrak{o}$ *if and only if inequality* (1) *holds.* We now observe that the equality $f(\mathfrak{o}) = \mathfrak{o}$ can be considered as a consequence of the inclusion $f(\mathfrak{o}) \subset \mathfrak{o}$. For, assume that we have $f(\mathfrak{o}) \subset \mathfrak{o}$. The $\mathfrak{o}$-module $\bar{\mathfrak{o}}/\mathfrak{o}$ has a finite composition series. Since $f(\bar{\mathfrak{o}}) = \bar{\mathfrak{o}}$, it follows that the $f(\mathfrak{o})$-module $\bar{\mathfrak{o}}/f(\mathfrak{o})$ and the $\mathfrak{o}$-module $\bar{\mathfrak{o}}/\mathfrak{o}$ have composition series of the same length. Now, since $f(\mathfrak{o}) \subset \mathfrak{o}$, the $\mathfrak{o}$-module $\bar{\mathfrak{o}}/\mathfrak{o}$ can also be regarded as an $f(\mathfrak{o})$-module, and any composition series of the $\mathfrak{o}$-module $\bar{\mathfrak{o}}/\mathfrak{o}$ is also a normal sequence (without repetitions) of the $f(\mathfrak{o})$-module $\bar{\mathfrak{o}}/\mathfrak{o}$. Since the

$f(\mathfrak{o})$-module $\bar{\mathfrak{o}}/\mathfrak{o}$ is a factor module of the $f(\mathfrak{o})$-module $\bar{\mathfrak{o}}/f(\mathfrak{o})$, the equality of the lengths of the $\mathfrak{o}$-module $\bar{\mathfrak{o}}/\mathfrak{o}$ and the $f(\mathfrak{o})$-module $\bar{\mathfrak{o}}/f(\mathfrak{o})$ implies at once that we must have $f(\mathfrak{o})=\mathfrak{o}$.

Summarizing, we can say now that *to prove our theorem is tantamount to proving that the inclusion*

$$f(\mathfrak{o})\subset\mathfrak{o} \tag{3}$$

holds if and only if inequality (1) *holds.* This we proceed to prove.

We extend f to a k^*-automorphism of F^* and we continue to denote by f this extended automorphism. Let η be any element of the maximal ideal $\mathfrak{m}$ of $\mathfrak{o}$ and let $\eta'=f(\eta)$. Then $\eta'\in\mathfrak{o}$ if and only if

$$v(\eta'^{\tau}-\eta')\geqq v(y^{\tau}-y), \qquad \text{for all } \tau \text{ in } G. \tag{4}$$

We fix any τ in G, $\tau\neq 1$, and we set

$$f_1=\tau f\tau^{-1}.$$

Then f_1 is also a k^*-automorphism of F^*, and it follows from (2) that

$$f_1(t)=ct+\text{terms of degree}>1 \text{ in } t.$$

Thus $v[f_1(t)-f(t)]>1$ and we can apply Lemma 7.1 (where F and $\bar{k}$ are to be replaced by F^* and k^* respectively). We have therefore, in particular, for the above element η of $\mathfrak{m}$ the following equality

$$v(f_1(\eta)-f(\eta))=v(\eta)+v(f_1(x)-f(x))-n. \tag{5}$$

Now, $f(x)=x'$ and $f_1(x)=x'^{\tau}$. Hence we can re-write (5) as follows:

$$v[f_1(\eta)-f(\eta)]=v(\eta)+v(x'^{\tau}-x')-n. \tag{6}$$

Now, observe that $f_1(\eta^{\tau})=\eta'^{\tau}$ and

$$\eta'^{\tau}-\eta'=f_1(\eta^{\tau})-f(\eta)=f_1(\eta^{\tau}-\eta)+f_1(\eta)-f(\eta).$$

Since $\eta\in\mathfrak{o}$, we have $v(\eta^{\tau}-\eta)\geqq v(y^{\tau}-y)$, and therefore also $v[f_1(\eta^{\tau}-\eta)]\geqq v(y^{\tau}-y)$. Therefore (4) is equivalent to

$$v[f_1(\eta)-f(\eta)]\geqq v(y^{\tau}-y),$$

which, by (6), is in its turn equivalent to

$$v(x'^{\tau}-x')\geqq v(y^{\tau}-y)+n-v(\eta). \tag{7}$$

What we have shown is that $f(\mathfrak{o})\subset\mathfrak{o}$ if and only if (7) holds for all elements η of $\mathfrak{o}$. That is equivalent to saying that $f(\mathfrak{o})\subset\mathfrak{o}$ if and only if

(1) holds, since $s = \min \cdot \{v(\eta) \mid \eta \in \mathfrak{o}\}$. This completes the proof of the theorem.

COROLLARY 8.3. *Under the assumptions of Theorem 8.2 and assuming furthermore that* (1) *holds,* $\mathfrak{o}$ *is also saturated with respect to* x'.

Obvious.

THEOREM 8.4. *Let* $\mathfrak{o}$ *be saturated, let* x, x' *be two parameters of* $\mathfrak{o}$ *such that* $\mathfrak{o}$ *is saturated with respect to* x *and* $v(x) = v(x')$ $(= n$, *say*) (*in other words, the assumptions are the same as in Theorem* 8.2, *except that we do not assume any longer that the* v-*residue of* x'/x *is of the form* c^n, $c \in k$). *If inequality* (1) *holds for all* τ *in* G (y *being again an* x-*saturator of* $\mathfrak{o}$) *then* $\mathfrak{o}$ *is also saturated with respect to* x'.

Proof. Let c be the residue of $\frac{x'}{x}$ in $\bar{k}$ and let $\bar{x} = cx$. Assume that (1) holds. *We first assert that* $\bar{x} \in \mathfrak{o}$ *and that* $\mathfrak{o}$ *is saturated with respect to* $\bar{x}$ (this assertion is, of course, trivial if $c \in k$, for then $k((x)) = k((\bar{x}))$; however, we only know that $c \in \bar{k}$). Let τ be any element of G. If $\tau(c) = c$, then $\bar{x}^\tau = \bar{x}$. If $\tau(c) \neq c$, then, the residue of $\frac{x'^\tau}{x}$ in $\bar{k}$ being $\tau(c)$ (since $\tau(x) = x$), it follows that the residue of x'^τ/x' is $\frac{\tau(c)}{c}$, i.e., that residue is different from 1, and then $v(x'^\tau - x') = v(x') = n$. This implies by (1), that $v(y^\tau - y) \leqq s$, and since $\bar{x}^\tau - \bar{x} = (\tau(c) - c)x$, it follows that $v(\bar{x}^\tau - \bar{x}) = n \geqq s$, i.e., $v(\bar{x}^\tau - x) \geqq v(y^\tau - y)$. We have shown that $\bar{x}$ dominates y and hence $\bar{x} \in \mathfrak{o}$.

To prove now that $\mathfrak{o}$ is k-saturated with respect to $\bar{x}$, we consider the least Galois extension $\bar{F}^*$ of $k((\bar{x}))$ (in the algebraic closure of F^*) which contains F, and we denote by $\bar{G}$ the Galois group of $\bar{F}^*/k((\bar{x}))$. We consider any pair of elements ξ, η of F such that $\xi \in \mathfrak{o}$, $\eta \in \bar{\mathfrak{o}}$ and such that

(8) $$v/(\eta^{\bar{\tau}} - \eta) \geqq v(\xi^{\bar{\tau}} - \xi) \text{ for all } \bar{\tau} \text{ in } \bar{G}.$$

We have to prove that $\eta \in \mathfrak{o}$. Since k is also a coefficient field of $\bar{\mathfrak{o}}_{k((x))}$ (Proposition 2.1) and since $\eta \in \bar{\mathfrak{o}}_{k((x))}$, we may assume that $v(\eta) > 0$. Let τ be any element of G. If $\tau(c) = c$ then $\tau(\bar{x}) = \bar{x}$, and thus $\tau \mid F$ is a $k((\bar{x}))$-isomorphism of F. This isomorphism can be extended to a $k((\bar{x}))$-automorphism $\bar{\tau}$ of $\bar{F}^*$, and we have thus $\bar{\tau} \in \bar{G}$ and $\bar{\tau} \mid F = \tau \mid F$. Hence, by (8), we have $v(\eta^\tau - \eta) \geqq v(\xi^\tau - \xi)$, and consequently

(9) $$v(\eta^\tau - \eta) \geqq v(y^\tau - y),$$

since $\xi \in \mathfrak{o}$. Assume now that $\tau(c) \neq c$. Then, as we have just seen above, we have $v(y^\tau - y) \leqq s$. On the other hand, s is still the reduced multiplicity of $\tilde{\mathfrak{o}}_{k((\bar{x}))}$, and $\eta \in \tilde{\mathfrak{o}}_{k((\bar{x}))}$. Since $v(\eta) > 0$, we have $v(\eta) \geqq s$, and hence $v(\eta^\tau - \eta) \geqq s$. Thus (9) holds true also in the case in which $\tau(c) \neq c$. This proves that $\eta \in \mathfrak{o}$, as asserted.

We now note that $v(x') = v(\bar{x})$ $(= v(x))$ and that the v-residue of $x'/\bar{x}$ is 1 (since $\bar{x} = cx$). Hence, in order to provs that $\mathfrak{o}$ is also saturated with respect to x', it will be sufficient to show (in view of Corollary 8.3) that the following inequality (similar to (1))

$$(10) \qquad v(x'^{\bar{\tau}} - x') \geqq v(\bar{y}^{\bar{\tau}} - \bar{y}) + n - s$$

holds for all $\bar{\tau}$ in $\bar{G}$; here $\bar{y}$ denotes an $\bar{x}$-saturator of $\mathfrak{o}$. This we proceed to prove.

Let, then, $\bar{\tau}$ be a given element of $\bar{G}$. For the proof of (10) we shall consider separately two cases, according as $\bar{\tau}(c) = c$ or $\bar{\tau}(c) \neq c$.

First case. $\bar{\tau}(c) = c$. In this case, we have $\bar{\tau}(x) = x$, since $x = \bar{x}/c$ and $\bar{\tau}(\bar{x}) = \bar{x}$. Thus, the restriction of $\bar{\tau}$ to F is a $k((x))$-isomorphism of F, and this $k((x))$-isomorphism can be extended to $k((x))$-automorphism τ of F^*. Then τ and $\bar{\tau}$ have the same restriction to F, and (10) follows from (1), since x' and $\bar{y}$ belong to F and since, futhermore, $y^\tau - y = y^{\bar{\tau}} - y$ (in view of $y \in F$) and $v(y^{\bar{\tau}} - y) \geqq v(\bar{y}^{\bar{\tau}} - \bar{y})$ (in view of the fact that $y \in \mathfrak{o}$ and $\bar{y}$ is an $\bar{x}$-saturator of $\mathfrak{o}$).

Second case. $\bar{\tau}(c) \neq \bar{c}$. Since inequality (10) is trivial if $n = s$, we may assume that $n > s$, i. e., that both x and $\bar{x}$ are non-transversal parameters. Therefore, in the notations of § 3 and by Proposition 3.11 we have $\beta_1 = \bar{\beta}_1 = s$, where $\bar{\beta}_1$ is the integer analogous to β_1, obtained upon replacing x by $\bar{x}$ (recall that we have shown that $\mathfrak{o}$ is also saturated with respect to $\bar{x}$). Let

$$\bar{G} = \bar{G}_0 < \bar{G}_1 < \bar{G}_2 < \cdots$$

be the sequence of higher saturation groups of $\mathfrak{o}$ relative to $\bar{x}$.

We first show that

$$(11) \qquad c\mathfrak{m} \subset \mathfrak{o}.$$

For, if $\eta \in \mathfrak{m}$ and $\tau \in G$, and if we set $\bar{\eta} = c\eta$, then we have $v(\bar{\eta}^\tau - \bar{\eta}) = v(\eta^\tau - \eta)$ if $\tau(c) = c$, and $v(\bar{\eta}^\tau - \bar{\eta}) \geqq v(\bar{\eta}) \geqq s \geqq v(y^\tau - y)$, if $\tau(c) \neq c$ [we note that we have shown earlier in the proof that $\tau(c) \neq c \Rightarrow v(y^\tau - y) \leqq s$]. In both bases we find that $v(\bar{\eta}^\tau - \bar{\eta}) \geqq v(y^\tau - y)$, whence $\bar{\eta} \in \mathfrak{o}$.

5

We next show that if $\bar{\tau} \in \bar{G}_1$ *then* $\bar{\tau}(c) = c$. For if $\bar{\tau} \in \bar{G}_1$ then $v(\eta^{\bar{\tau}} - \eta) > \bar{\beta}_1 = s$ for all η in $\mathfrak{o}$. In particular, $v[(cy)^{\bar{\tau}} - cy] > s$. Now, $(cy)^{\bar{\tau}} - cy = c^{\bar{\tau}}(y^{\bar{\tau}} - y) + y(c^{\bar{\tau}} - c)$, and $v(y^{\bar{\tau}} - y) > s$ while $v(y) = s$ (x being a non-transversal parameter). Hence $v(c^{\bar{\tau}} - c) > 0$, and this shows that $c^{\bar{\tau}} = c$.

Now, finally, we can prove (10) in the case under consideration (i.e., when $\bar{\tau}(c) \neq c$). From what we have just proved it follows that $\bar{\tau} \notin G_1$, and hence, by the definition of the integer $\bar{\beta}_1$, we have $v(\bar{y}^{\tau} - \bar{y}) = \bar{\beta}_1 = s$, and thus (10) holds since $v(x'^{\bar{\tau}} - x') \geqq v(x') = n$. This completes the proof.

Note that inequality (1) certainly holds true if $n = s$, i.e., if x and x' are transversal parameters. Thus Theorem 8.4 tells us, in particular, that if $\mathfrak{o}$ is saturated with respect to one transversal parameter x it is also saturated with respect to every transversal parameter x'. This fact, however, is covered by the much stronger result formulated in Theorem 7.2.

Remark. For the existence of a $\bar{k}$-automorphism of F which sends x into x' it is at any rate necessary that the following condition be satisfied: *the v-residue of x'/x in $\bar{k}$ is the n-th power of an element γ of $\bar{k}$.* The following example will show that this additional condition $\gamma \in k$ in Theorem 8.2 may be quite necessary.

Let k be the field of rational numbers, let n be an integer greater than 1, let β_1 be an integer greater than n, such that $(\beta_1, n) = 1$, and let $\bar{k}$ be any algebraic extension of k containing a n-th root $\sqrt[n]{2}$ of 2. Let $F = \bar{k}((t))$, $x = t^n$, $K = k((x))$, $K_1 = k((t))$, $K_2 = F$. We set

$$\mathfrak{o} = k + \mathfrak{m} = k + \mathfrak{M} + \mathfrak{M}_1^{\beta_1} + \mathfrak{M}_2^{\beta_2},$$

where $\mathfrak{M}_2 = \bar{\mathfrak{m}} = t \cdot \bar{k}[[t]]$, $\mathfrak{M}_1 = \mathfrak{M}_2 \cap K_1$, $\mathfrak{M} = \mathfrak{M}_2 \cap K$ and β_2 is any integer greater than β_1. By Lemma 4.5, $\mathfrak{o}$ is saturated with respect to x. Now, let $x' = 2x$. Then $\mathfrak{o}$ is also saturated with respect to x', and the v-residue of x'/x in $\bar{k}$ is γ^n, where $\gamma = \sqrt[n]{2} \in \bar{k}$. The ring $\mathfrak{o}$ is the set of all power series in t, of the form

$$a_0 + a_1 t^n + a_2 t^{2n} + \cdots + b_1 t^{\beta_1} + b_2 t^{\beta_1+1} + \cdots + c_1 t^{\beta_2} + c_2 t^{\beta_2+1} + \cdots,$$

where the a_i and b_i are arbitrary elements of k, while the coefficients $c_1, c_2, \cdots$ are arbitrary elements of $\bar{k}$. Now, let f be *any* $\bar{k}$-automorphism of F which sends x into x'. Then $f(t) = \sqrt[n]{2}\, t$, where $\sqrt[n]{2}$ is one of the determinations of the n-th roots of 2. Therefore $f(\mathfrak{o})$ is the set of all power series in t, which are of the form

$$a_0 + a_1 t^n + a_2 t^n + \cdots + b_1(\sqrt[n]{2})^{\beta_1} t^{\beta_1} + b_2(\sqrt[n]{2})^{\beta_1+1} t^{\beta_1+1} + \cdots$$
$$+ c_1 t^{\beta_2} + c_2 t^{\beta_2+1} + \cdots,$$

where the coefficients a_i, b_j and c_ν are as above. Since $\beta_1 \not\equiv 0 \pmod{n}$, it is clear that $f(\mathfrak{o}) \neq \mathfrak{o}$.

9. Change of coefficient field; intrinsic nature of the saturation concept. Let k and k' be two coefficient fields of our one-dimensional local domain. We denote by θ the *canonical* isomorphism of k onto k', defined by the condition: $\theta(c)\text{-}c \in \mathfrak{m}$, for all c in k. Our main object in this section is to prove the following result:

THEOREM 9.1. *If $\mathfrak{o}$ is k-saturated, then given any transversal parameter x of $\mathfrak{o}$ there exists an automorphism ϕ of $\mathfrak{o}$ such that $\phi \mid k = \theta$ and $\phi(x) = x$, and hence (see Theorem 7.2) $\mathfrak{o}$ is also k'-saturated.*

Proof. Let $\bar{k}$ and $\bar{k}'$ denote respectively the algebraic closure of k and of k' in the quotient field F of $\mathfrak{o}$. Then $\bar{k}$ and $\bar{k}'$ are coefficient fields of F, and if t is any uniformizing parameter of F then $F = \bar{k}((t)) = \bar{k}'((t))$, while the integral closure $\bar{\mathfrak{o}}$ of $\mathfrak{o}$ is given by both $\bar{k}[[t]]$ and $\bar{k}'[[t]]$. We denote by $\bar{\theta}$ the canonical isomorphism of $\bar{k}$ onto $\bar{k}'$, defined by the condition: $v(\bar{\theta}(\bar{c}) - \bar{c}) > 0$, for any c in $\bar{k}$; here v denotes the natural valuation of F. It is clear that $\bar{\theta}$ is an extension of θ.

We choose the uniformizing parameter t so as to have (see Lemma 1.1): $x = at^n$, where $a \in \bar{k}$ and $n = s = v(x)$ ($=$ reduced multiplicity of $\mathfrak{o}$). We use the notations of Lemma 1.2. Let F^* be the least Galois extension of K $(= k((x)))$ containing F, let $\tilde{k}$ be the least Galois extension of k containing $\bar{k}$ and contained in F^*, and let $\tilde{F} = \tilde{k}((t))$. Then

$$F \subset \tilde{F} \subset F^* = k^*((t)),$$

where k^* is the field obtained by adjoining to $\tilde{k}$ all the n-th roots of each of the quotients a_i/a_1; here a_1 $(= a)$, $a_2, \cdots, a_\nu$ are the distinct conjugates of a, over k, in $\tilde{k}$.

If $\tilde{k}'$ denotes the algebraic closure of k' in $\tilde{F}$, then both $\tilde{k}'$ and $\tilde{k}$ are coefficient fields of $\tilde{F}$, and we can therefore extend the isomorphism $\bar{\theta}\colon \bar{k} \xrightarrow{\sim} \bar{k}'$ to the canonical isomorphism $\tilde{\theta}$ of $\tilde{k}$ onto $\tilde{k}'$. It is clear that $\tilde{k}'$ is a least Galois extension of k' containing $\bar{k}'$.

If we set $a' = \bar{\theta}(a)$, then $a = a' + a_1't + a_2't^2 + \cdots$, where the a_i' are in $\bar{k}'$. Hence $x = at^n = a'(t^n + b_1't^{n+1} + b_2't^{n+2} + \cdots)$, where $b_i' = a_i'/a' \in \bar{k}'$. Let t' be the unique element of $\bar{k}'[[t]]$ $(= \bar{\mathfrak{o}})$ which is of the form

$$t' = t + c_1't^2 + c_2't^3 + \cdots, c_i' \in \bar{k}',$$

and which is such that $t'^n = t^n + b_1't^{n+1} + b_2't^{n+2} + \cdots$. Then we have:

$$(1) \qquad x = at^n = a't'^n, \qquad a' = \bar{\theta}(a),$$

$$(2) \qquad v((t'/t) - 1) > 0,$$

and both t, t' are uniformizing parameters of F.

We now pass to the field $F^* = k^*((t))$ and we denote by k'^* the algebraic closure of $\bar{k}'$ in F^*. Then we have

$$F^* = k^*((t)) = k^{*\prime}((t')).$$

We extend $\bar{\theta}: \bar{k} \xrightarrow{\sim} \bar{k}'$ to the canonical isomorphism $\theta^*: k^* \xrightarrow{\sim} k'^*$ defined by the condition $v(\theta^*(c) - c) > 0$, $c \in k^*$; here v continues to denote the canonical valuation of F^*, an unramified extension of the canonical valuation v of F. Finally, we denote by ϕ the unique automorphism of F^* which is determined by the conditions $\phi(t) = t'$, $\phi \mid k^* = \theta^*$. It is obvious that ϕ induces automorphisms of F and of $\bar{F}$, that $\phi(x) = x$ and that $\phi \mid k = \theta$, $\phi \mid \bar{k} = \bar{\theta}$, $\phi \mid \bar{k} = \bar{\theta}$. Furthermore, it is also clear that F^* is the least Galois extension of K' $(= k'((x)))$ which contains F. We observe that in view of (2) and of $v(\theta^*(c) - c) > 0$ for all $c \in k^*$, we have for any element η in F^* $(\eta \neq 0)$:

$$(3) \qquad v(\phi(\eta)/\eta - 1) > 0,$$

or—equivalently:

$$(4) \qquad v(\phi(\eta) - \eta) > v(\eta).$$

We denote by $\mathfrak{o}'$ the ring $\phi(\mathfrak{o})$. It is then clear that $\mathfrak{o}'$ is k'-saturated with respect to x. Our theorem will be proved if we show that

$$(5) \qquad \mathfrak{o}' = \mathfrak{o}.$$

It will be sufficient to show that

$$(6) \qquad \mathfrak{o}' \subset \mathfrak{o},$$

for the equality $\mathfrak{o} = \mathfrak{o}'$ can then be derived from the inclusion (6) by observing that $\phi(\bar{\mathfrak{o}}) = \bar{\mathfrak{o}}$ and by applying the reasoning which was used in § 8 in deriving from the inclusion (3) of § 8, the equality $\mathfrak{o} = f(\mathfrak{o})$.

Let $G = \mathfrak{G}(F^*/K)$, $G' = \mathfrak{G}(F^*/K')$ be the Galois group of F^* over K and K' respectively, where

$$K = k((x)), \qquad K' = k'((x)). \tag{7}$$

It is clear that $G' = \phi G \phi^{-1}$.

Let

$$G = G_0 > G_1 > G_2 > \cdots > G_g = \mathfrak{G}(F^*/F)$$

and

$$K = K_0 < K_1 < K_2 < \cdots < K_g = F$$

be respectively the higher saturation groups of $\mathfrak{o}$ relative to $k((x))$ $(= K)$ and the associated intermediate saturation fields (see (6) and (7) of § 3). Let, similarly,

$$G' = G_0' > G_1' > G_2' > \cdots > G_g' = \mathfrak{G}(F^*/F)$$

and

$$K' = K_0' < K_1' < K_2' < \cdots < K_g' = F$$

be the higher saturation groups and the intermediate saturation fields of $\mathfrak{o}'$ relative to $k'((x))$ $(= K')$. We have, of course:

$$\begin{aligned} G_i' &= \phi G_i \phi^{-1}, \\ K_i' &= \phi(K_i). \end{aligned} \qquad i = 0, 1, 2, \cdots, g. \tag{8}$$

Let k_i be the algebraic closure of k in K_i and let k_i' be the algebraic closure of k' in K_i'. If $\mathfrak{m}$ and $\mathfrak{m}'$ are the maximal ideals of $\mathfrak{o}$ and $\mathfrak{o}'$, then we have [see § 4, Theorem 4.1]:

$$\mathfrak{m} = \mathfrak{M}_0 + \mathfrak{M}_1^{m_1} + \mathfrak{M}_2^{m_2} + \cdots + \mathfrak{M}_g^{m_g}, \tag{9}$$

$$\mathfrak{m}' = \mathfrak{M}_0' + \mathfrak{M}_1'^{m_1} + \mathfrak{M}_2'^{m_2} + \cdots + \mathfrak{M}_g'^{m_g}, \tag{9'}$$

and $k, k_1, k_2, \cdots, k_g$ (resp., $k', k_1', k_2', \cdots, k_g'$) are the characteristic coefficient fields of $\mathfrak{o}$ (resp., of $\mathfrak{o}'$) relative to the decomposition (9) (resp. (9')) of $\mathfrak{m}$ (resp., of $\mathfrak{m}'$) [see Definition 4.8]. We have obviously:

$$\begin{aligned} \mathfrak{M}_i' &= \phi(\mathfrak{M}_i), \\ k_i' &= \phi(k_i), \end{aligned} \qquad (i = 0, 1, 2, \cdots, g)$$

and $k_g = \bar{k}$, $k_g' = \bar{k}'$. We use the notations of § 3: e_i = ramification index of F/K_i = ramification index of F/K_i', $\beta_i = m_i e_i$ [see (8) and (16) of § 3]

$= \min \cdot \{v(\xi) \mid \xi \in \mathfrak{M}_i^{m_i}\} = \min \cdot \{v(\xi') \mid \xi' \in \mathfrak{M}_i'^{m_i}\}$; furthermore, $K_i = k_i((x_i))$, $K_i' = k_i'((x_i'))$, where

$$x_{i-1} = a_i x_i^{n_i}, \quad (a_i \in k_i, n_i = e_{i-1}/e_i) \tag{10}$$

$$x_{i-1}' = a_i' x_i'^{n_i}, \quad (a_i' = \phi(a_i) \in k_i', x_i' = \phi(x_i)) \tag{10'}$$

$i = 1, 2, \cdots, g,$

and where $x_0 = x_0' = x$ and $e_i = v(x_i) = v(x_i')$ [see (18), § 3].

The proof of the inclusion (6) (and hence also of the theorem) consists of a number of steps, each of which is given below as a separate lemma.

LEMMA 9.2. *If $c' \in k_i'$ and $\tau \in G_i$, then $v(c'^\tau - c') \geqq \beta_{i+1}$ ($i = 0, 1, \cdots, g-1$; note that if $\tau \in G_g$, then $c'^\tau = c'$ since $c' \in \bar{k}' \subset F$ and since $G_g = \mathfrak{G}(F^*/F)$).*

Proof. Let $f(c') = \sum a_\nu' c'^\nu = 0$ be the irreducible equation which c' satisfies over the field k'. We have

$$f(c'^\tau) = f(c'^\tau) - f(c') = (c'^\tau - c')\{f'(c') + \tfrac{1}{2}(c'^\tau - c)f''(c') + \cdots\}. \tag{11}$$

We also have

$$0 = f(c') = f^\tau(c'^\tau) = \sum_\nu a_\nu'^\tau (c'^\tau)^\nu$$
$$= \sum_\nu (c'^\tau)^\nu [a_\nu'^\tau - a_\nu'] + \sum_\nu a_\nu'[(c'^\tau)^\nu - c'^\nu], \text{ i.e.,}$$

$$\sum_\nu (c'^\tau)^\nu [a_\nu'^\tau - a_\nu'] + f(c'^\tau) - f(c') = 0. \tag{12}$$

Since $a_\nu' \in k' \subset \mathfrak{o}$, we have $v(a_\nu'^\tau - a_\nu') \geqq \beta_{i+1}$, and since $v(c'^\tau) \geqq 0$, it follows from (12) that

$$v[f(c'^\tau) - f(c')] \geqq \beta_{i+1}. \tag{13}$$

Now, let $c = \phi^{-1}(c')$. Then $c \in k_i$, $v(c' - c) > 0$, and hence also $v(c'^\tau - c^\tau) > 0$. Since $\tau \in G_i$ and $c \in k_i \subset K$, we have $c^\tau = c$, and thus we find that $v(c'^\tau - c) > 0$. This, together with the inequality $v(c' - c) > 0$, implies that $v(c'^\tau - c') > 0$. We have $0 \neq f'(c') \in \bar{k}'$, whence $v(f'(c')) = 0$. This, together with the inequality $v(c'^\tau - c') > 0$, implies that the expression in the curly brackets on the right-hand side of (11), has v-value zero. Therefore $f(c'^\tau) - f(c')$ and $c'^\tau - c$ have the same value, and the lemma now follows from (13)

LEMMA 9.3. *If $\eta' \in \mathfrak{o}'$ and $v(\eta') \geqq \beta_i$, then $v(\eta'^\tau - \eta') > \beta_i$ for all $\tau \in G_i$.*

Proof. Let $\eta' = \phi(\eta)$, where $\eta \in \mathfrak{o}$ and also $v(\eta) \geqq \beta_i$. We can write:

$$\eta'^\tau - \eta' = (\eta'^\tau - \eta^\tau) + (\eta^\tau - \eta) + (\eta - \eta'). \tag{14}$$

By (4) we have $v(\eta-\eta')=v(\eta'^{\tau}-\eta^{\tau})>v(\eta)\geqq\beta_i$. Since $\tau\in G_i$ and $\eta\in\mathfrak{o}$, we also have $v(\eta^{\tau}-\eta)>\beta_i$, and from this the lemma follows, in view of (14).

LEMMA 9.4. *For any* $i=0,1,\cdots,g$ *we have*

$$v\left(\frac{x_i'^{\tau}}{x_i'}-1\right)\geqq\beta_{j+1} \tag{15}$$

if

$$j\geqq i \text{ and } \tau\in G_j. \tag{16}$$

Proof. The lemma is trivial for $i=0$, for in that case we have $x_0'=x_0'^{\tau}=x$, for all τ in G $(=G_0)$. We shall assume that $i>0$, that the lemma is true for $i-1$ and we prove it for i. We have, by (10′):

$$x_{i-1}'^{\tau}/x_{i-1}'=\frac{a_i'^{\tau}}{a_i'}\cdot\left(\frac{x_i'^{\tau}}{x_i'}\right)^{n_i}. \tag{17}$$

By the induction hypothesis we have that $v(x_{i-1}'^{\tau}/x_{i-1}'-1)\geqq\beta_{j+1}>0$, and by Lemma 9.2 we have $v(a_i'^{\tau}/a_i'-1)\geqq\beta_{i+1}>0$ (since $a_i'\in k_i'$ and $\tau\in G_j\subset G_i$). Hence, it follows from (17) that

$$v\left[\left(\frac{x_i'^{\tau}}{x_i}\right)^{n_i}-1\right]>0. \tag{18}$$

Since $x_i'^{m_i}\in\mathfrak{o}'$ and $v(x_i'^{m_i})=\beta_1$, we have by Lemma 9.3:

$$v\left[\left(\frac{x_i'^{\tau}}{x_i'}\right)^{m_i}-1\right]>0. \tag{19}$$

Since m_i and n_i are relatively prime, it follows from (18) and (19) that

$$v\left(\frac{x_i'^{\tau}}{x_i'}-1\right)>0 \tag{20}$$

For convenience let us set $\zeta=x_{i-1}'^{\tau}/x_{i-1}'$, $\eta=x_i'^{\tau}/x_i'$ and $\alpha=a_i'^{\tau}/a_i'$. By our induction hypothesis, we have $v(\zeta-1)\geqq\beta_{j+1}$, and by Lemma 9.2 we have $v(\alpha-1)\geqq\beta_{j+1}$. By (17) we have $\eta^{n_i}=\zeta/\alpha$, whence

$$\eta^{n_i}-1=\frac{(\zeta-1)-(\alpha-1)}{\alpha}.$$

Since $v(\alpha)=0$, we conclude that $v(\eta^{n_i}-1)\geqq\beta_{j+1}$. By (20) we have $v(\eta-1)>0$, and hence $\eta^{n_i-1}+\eta^{n_i-2}+\cdots+1$ has v-residue $n_i\neq0$. Therefore $v(\eta-1)\geqq\beta_{j+1}$, and this completes the proof of the lemma.

We now complete the proof of Theorem 9.1 by proving the inclusion (6). We have to prove that $\mathfrak{M}_i'^{m_i}\subset\mathfrak{o}$. For that it is sufficient to prove that

$c'x_i'^{\lambda} \in \mathfrak{o}$ for all $c' \in k_i'$ and all $\lambda \geqq m_i$. In other words, we have to prove that if $\tau \in G_j$, then $v(c'^{\tau}(x_i'^{\tau})^{\lambda} - c'x_i'^{\lambda}) \geqq \beta_{j+1}$, for $j = 0, 1, \cdots, g-1$. Now, this assertion is obvious if $j < i$, for in that case $\beta_{j+1} \leqq \beta_i \leqq v(c'x_i'^{\lambda})$. If, however, $j \geqq i$, our assertion is an immediate consequence of Lemma 9.4 and Lemma 9.2 (in applying Lemma 9.2 we must observe that if $j \geqq i$ then $c' \in k_i' \subset k_j'$).

COROLLARY 9.5. *If the local domain* $\mathfrak{o}$ *is not saturated, and* k, k' *are two coefficient fields of* $\mathfrak{o}$, *then the k-saturation of* $\mathfrak{o}$ *coincides with the* k'-*saturation of* $\mathfrak{o}$.

Let $\tilde{\mathfrak{o}}$ and $\tilde{\mathfrak{o}}'$ denote respectively the k-saturation and the k'-saturation of $\mathfrak{o}$ (see Corollary-Definition 7.3). Proposition 2.1 and the consideration of the canonical isomorphism $\theta: k \longrightarrow k'$ show at once that k and k' are coefficients fields of both $\tilde{\mathfrak{o}}$ and $\tilde{\mathfrak{o}}'$. From Therorem 9.1 follows that $\tilde{\mathfrak{o}}'$ is k-saturated and therefore, by Corollary 7.4, we have $\tilde{\mathfrak{o}}' \supset \tilde{\mathfrak{o}}$. Similarly it follows that $\tilde{\mathfrak{o}} \supset \tilde{\mathfrak{o}}'$, and hence $\tilde{\mathfrak{o}} = \tilde{\mathfrak{o}}'$.

Theorem 9.1 refers to the *canonical* isomorphism between k and k'. The question arises whether an *arbitrary* isomorphism $\theta: k \longrightarrow k'$ can also be extended to an automorphism ϕ of $\mathfrak{o}$. We have shown in [2] that the answer is affirmative if k is algebraically closed (see [2], Thtorem 1.16). In view of Theorem 9.1, the question is equivalent to the following one: *can any automorphism* θ *of* k *be extended to an automorphism of* $\mathfrak{o}$? That the answer to this question is in general negative is obvious, for if there exists an extension of θ to an automorphism ϕ of $\mathfrak{o}$, ϕ has an extension to an automorphism ψ of the field of fractions F of $\mathfrak{o}$, and the restriction of ψ to the algebraic closure $\bar{k}$ of k in F is an automorphism $\bar{\theta}$ of $\bar{k}$ which is an extension of θ. Thus, a *necessary* condition that θ have an extension to an automorphism of $\mathfrak{o}$ is that θ have an extension to an automorphism $\bar{\theta}$ of $\bar{k}$, and this condition is not always satisfied. For instance, if $\mathfrak{Q}$ is the field of rational number and u is a transcendental over $\mathfrak{Q}$, we set $k = \mathfrak{Q}(u)$, $\bar{k} = k(\sqrt[3]{u})$, $F = \bar{k}((t))$, and $\mathfrak{o} = k + \bar{\mathfrak{m}}$, where $\bar{\mathfrak{m}} = t \cdot \bar{k}[[t]]$. Then $\mathfrak{o}$ is a saturated local domain, with k as coefficient field and F as field of fractions. There is a $\mathfrak{Q}$-automorphism θ of k determined by the condition $\theta(u) = 2u$, but this automorphism θ of k has no extension to an automorphism of $\bar{k}$, for $\sqrt[3]{2u}$ $(= \sqrt[3]{u} \cdot \sqrt[3]{2})$ does not belong to $k(\sqrt[3]{u})$.

However, the existence of an automorphism $\bar{\theta}$ of $\bar{k}$ which is an extension of θ is in itself far from being sufficient for the existence of an extension of θ to an automorphism of $\mathfrak{o}$. The theorem proved below gives a necessary and sufficient condition in order that a *given* automorphism $\bar{\theta}$ of $\bar{k}$ *which is an extension of* θ have an extension to an automorphism ψ of F such that $\phi = \psi \mid \mathfrak{o}$ is an automorphism of $\mathfrak{o}$.

THEOREM 9.6. *Let x be a transversal parameter of the saturated local domain $\mathfrak{o}$, let $K = K_0 < K_1 < \cdots < K_g = F$ be the intermediate saturation fields of $\mathfrak{o}$ relative to K $(= k((x)))$, let $K_i = k_i((x_i))$, where x_{i-1} and x_i are related by* (10). *Let θ be an automorphism of k and let $\bar{\theta}$ be a given extension of θ to an automorphism of the algebraic closure $\bar{k}$ of k in F (assuming that such an extension exists). Then $\bar{\theta}$ has an extension to an automorphism ψ of F such that $\psi \mid \mathfrak{o}$ is an automorphism of $\mathfrak{o}$ if and only if the following conditions are satisfied:*

a) $\bar{\theta}(k_i) = k_i$, $i = 1, 2, \cdots, g$;

b) *For each $i = 0, 1, \cdots, g$, there exists an element γ_i in k_i such that*

$$\bar{\theta}(a_{i+1})/a_{i+1} = \frac{\gamma_i}{\gamma_{i+1}^{n_{i+1}}}, \qquad i = 0, 1, 2, \cdots, g-1. \tag{21}$$

Here $k_0 = k^{(s)}$, and thus $k_0 = k$ if $\mathfrak{M}_0$ is not redundant in (9), *and $k_0 = k_1$ in the contrary case (in which case $m_1 = 1$ and $x = x_1$; see Proposition* 4.10).

Proof.

A) *Proof of necessity.* Assume that there exists an automorphism ψ of F which is an extension of $\bar{\theta}$ and such that the restriction of ψ to $\mathfrak{o}$ is an automorphism of $\mathfrak{o}$. Let c be an arbitrary element of k_i and let us fix an element η of $\mathfrak{o}$ such that $v(\eta) = \beta_i$ (where $\beta_i = m_i e_i$; we use the notations of the proof of Theorem 9.1), for instance $\eta = x_i^{m_i}$. Then η and $c\eta$ belong to $\mathfrak{o}$, and hence also η' and $c'\eta'$ belong to $\mathfrak{o}$, where $\eta' = \psi(\eta)$ and $c' = \psi(c)$. Thus c' appears as the residue in $\bar{k}$, of the quotient of two elements $c'\eta'$ and η' of $\mathfrak{o}$ which have v-value β_i. Therefore $c' \in k_i$ (see Proposition 5.3), and we thus see that $\bar{\theta}(k_i) \subset k_i$. Then, either by considering ψ^{-1}, or by observing that $\bar{\theta}(k) = k$ and that therefore $[k_i : k] = [\psi(k_i) : k]$, we conclude that $\bar{\theta}(k_i) = k_i$.

Let $x_i' = \phi(x_i)$ and let γ_i be the residue of x_i'/x_i in $\bar{k}$. We shall prove by induction on i that $\gamma_i \in k_i$ and that the (21) hold true. For $i = 0$, we have $x_0 = x \in \mathfrak{o}$, $x_0' \in \mathfrak{o}$, and since $v(x) = v(x_0') = s$, we have $\gamma_0 \in k^{(s)} = k_0$. We have $x = a_1 x_1^{n_1}$, $x' = a_1' x_1'^{n_1}$, where $x' = \phi(x)$ and $a_1' = \psi(a_1)$ $(= \bar{\theta}(a_1))$. Hence $\gamma_0 = a_1'/a_1 \cdot \gamma_1^{n_1}$, which is (21) for $i = 0$. Assume that the inclusions $\gamma_i \in k_i$ and the relations (21) have already been proved for $i = 0, 1, \cdots, j$, where j is a given integer such that $0 \leqq j < g$. The relation (21), for $i = j$, yields the inclusion $\gamma_{j+1}^{n_{j+1}} \in k_{j+1}$ [since $a_{i+1} \in k_{i+1}$ and therefore, by a), also

$\theta(a_{i+1}) \in k_{i+1}]$. Since $\gamma_{j+1}{}^{m_{j+1}}$ is the residue of $x'_{j+1}{}^{m_{j+1}}/x_{j+1}{}^{m_{j+1}}$, and both numerator and denominator of this quotient are elements of $\mathfrak{o}$ having v-value β_{j+1}, it follows that $\gamma_{j+1}{}^{m_{j+1}} \in k_{j+1}$. This inclusion, together with the above inclusion $\gamma_{j+1}{}^{n_{j+1}} \in k_{j+1}$, implies that $\gamma_{j+1} \in k_{j+1}$, since $(m_{j+1}, n_{j+1}) = 1$. If $j = g - 1$, there is nothing more to be proved. If, however, $j < g - 1$, then we can write $x_{j+1} = a_{j+2}x_{j+2}{}^{n_{j+2}}$, $x'_{j+1} = a'_{j+2}x'_{j+2}{}^{n_{j+2}}$, where $a'_{j+2} = \bar{\theta}(a_{j+2})$, and hence $\gamma_{j+1} = a'_{j+2}/a_{j+2} \cdot \gamma_{j+2}{}^{n_{j+2}}$, which is (21) for $i = j + 1$.

B) *Proof of sufficiency.* Assuming that conditions a) and b) are satisfied, we recall that x_g is a uniformizing parameter of F, i.e., that $F = \bar{k}((x_g))$. We set $x_g' = \gamma_g x_g$. Then we have also $F = \bar{k}((x_g'))$. We denote ψ the automorphism of F which is uniquely determined by the conditions $\psi \mid \bar{k} = \bar{\theta}$ and $\psi(x_g) = x_g'$. We shall prove that $\psi(\mathfrak{o}) = \mathfrak{o}$. First of all we prove by induction from $i+1$ to i that $\psi(x_i) = \gamma_i x_i$. In fact, we have $x_i = a_{i+1}x_{i+1}{}^{n_{i+1}}$, and hence $\psi(x_i) = \bar{\theta}(a_{i+1})\gamma_{i+1}{}^{n_{i+1}}x_{i+1}{}^{n_{i+1}} = a_{i+1}\gamma_i x_{i+1}{}^{n_{i+1}}$ (by 21), and this proves that $\psi(x_i) = \gamma_i x_i$. Since $\gamma_i \in k_i$ and $\psi(k_i) = k_i$, it follows at once that $\psi(K_i) = K_i$, $\psi(\mathfrak{M}_i) = \mathfrak{M}_i$, and hence, by (9), $\psi(\mathfrak{m}) = \mathfrak{m}$, which proves that $\psi(\mathfrak{o}) = \mathfrak{o}$. This completes the proof of the theorem.

10. Characteristic pairs and characteristic parameters of a saturated local domain; an isomorphism criterion. If $\mathfrak{o}$ is a saturated local domain with maximal ideal $\mathfrak{m}$, and x is a transveral parameter of $\mathfrak{o}$, then we have introduced in § 4 the standard decomposition of $\mathfrak{m}$ with respect to x (the coefficient field k of $\mathfrak{o}$ being fixed). We have, as in (1), § 4:

$$\mathfrak{m} = \mathfrak{M}_0 + \mathfrak{M}_1{}^{m_1} + \mathfrak{M}_2{}^{m_2} + \cdots + \mathfrak{M}_g{}^{m_g}, \tag{1}$$

and this is the standard decomposition of $\mathfrak{m}$ relative to x if and only if $k^{(s)} = k$, where s is the reduced multiplicity of $\mathfrak{o}$ (see Proposition 4.10). If, however, $k^{(s)} > k$, then $k^{(s)} = k_1$, we have $m_1 = 1$ in (1), and

$$\mathfrak{m} = \mathfrak{M}_1 + \mathfrak{M}_2{}^{m_2} + \cdots + \mathfrak{M}_g{}^{m_g} \tag{2}$$

is the standard decomposition of $\mathfrak{m}$ relative to x (see again Proposition 4.10).

In our usual notations, we have the intermediate saturation field of $\mathfrak{o}$:

$$k((x)) = K = K_0 < K_1 < K_2 < \cdots < K_g = F,$$

where $K_i = k_i((x_i))$, and

$$x_{i-1} = a_i x_i{}^{n_i}, \qquad i = 1, 2, \cdots, g, \qquad (x_0 = x) \tag{3}$$

with $a_i \in k_i$. We have $(m_i, n_i) = 1$, and the g pairs $\{m_i, n_i\}$ were called by *the characteristic pairs of* $\mathfrak{o}$ (see Definition 3.8; see also Corollary 5.7 and Definition 5.8). If (1) is the standard decomposition of $\mathfrak{m}$ relative to x, then $m_1 > n_1$ (since $m_1 e_1 = \beta_1 > s = n_1 e_1$; see Proposition 4.10). If (2) is the standard decomposition, then $m_1 = n_1 = 1$, and we may choose x_1 so as to have $x = x_1$ (whence $a_1 = 1$).

Definition 10.1. *The g elements $a_1, a_2, \cdots, a_g$ of $\bar{k}$ are called characteristic parameters of* $\mathfrak{o}$.

Given the transversal parameter x, the characteristic parameters $a_1, a_2, \cdots, a_g$ are not uniquely determined. They depend on the choice of the uniformizing parameters $x_1, x_2, \cdots, x_g$ of $K_1, K_2, \cdots, K_g$. If $x_1', x_2', \cdots, x_g'$ is another set of uniformizing parameters of these fields, such that

$$x_{i-1}' = a_i' x_i'^{n_i}, \qquad i = 1, 2, \cdots, g; \qquad (x_0' = x),$$

then we have, in the first place, $a_1' x_1'^{n_1} = a_1 x_1^{n_1}$, whence $a_1' = a_1 (\frac{x_1}{x_1'})^{n_1}$, and here $\frac{x_1}{x_1'}$ must be an element of k_1, say b_1. Thus

$$(4) \qquad x_1' = \frac{x_1}{b_1}, \qquad b_1 \in k_1,$$

$$(5) \qquad a_1' = a_1 b_1^{n_1}.$$

From (4) and in view of $x_1' = a_2' x_2'^{n_2}$, it follows that $x_1 = b_1 a_2' x_2'^{n_2} = a_2 x_2^{n_2}$, whence we have by a similar argument:

$$(6) \qquad x_2' = \frac{x_2}{b_2}, \qquad b_2 \in k_2,$$

$$(7) \qquad a_2' = a_2 \cdot b_2^{n_2}/b_1.$$

More generally, we find that

$$(8) \qquad x_i' = \frac{x_i}{b_i}, \qquad b_i \in k_i,$$

$$(9) \qquad a_i' = a_i \cdot b_i^{n_i}/b_{i-1}, \qquad i = 1, 2, \cdots, g.$$

Conversely, if $b_1, b_2, \cdots, b_g$ are arbitrary elements of $k_1, k_2, \cdots, k_g$, respectively, *all different from zero,* and if we then define the a_i' as in (9), then $a_1', a_2', \cdots, a_g'$ are also characteristic parameters of $\mathfrak{o}$; they correspond to the choice of $x_1', x_2', \cdots, x_g'$ as uniformizing parameters of $K_1, K_2, \cdots, K_g$ respectively; here the x_i' are defined by (8). We may also replace x by $\frac{x}{b_0}$, where b_0 is any element of $k^{(s)}$, and then we get a more general set of characteristic parameters $a_1', a_2', \cdots, a_g'$:

$$(10) \qquad a_1' = \frac{a_1 b_1^{n_1}}{b_0}, \quad a_2' = \frac{a_2 b_2^{n_2}}{b_1}, \cdots, a_g' = \frac{a_g b_g^{n_g}}{b_{g-1}},$$

where b_i is an arbitrary element of k_i, $b_i \neq 0$ $(i = 0, 1, \cdots, g;\ k_0 = k^{(s)})$.

PROPOSITION 10.2. *The set* (10) *of characteristic parameters is independent of the choice of transversal parameters x of $\mathfrak{o}$.*

Proof. If ξ is any other transversal parameter, then the residue of $\frac{x}{\xi}$ in $\bar{k}$ is an element b_0 of $k^{(s)}$. Upon replacing x by x/b_0, we may assume that the residue of $\frac{x}{\xi}$ is 1. Then, by Proposition 8.2, there exists a $\bar{k}$-automorphism ϕ of F which induces an automorphism of $\mathfrak{o}$ and such that $\phi(x) = \xi$. It is then obvious (since ϕ is a $\bar{k}$-automorphism) that if $a_1, a_2, \cdots, a_g$ are characteristic parameters of $\mathfrak{o}$, relative to x, then $a_1, a_2, \cdots, a_g$ will also be characteristic parameters of $\mathfrak{o}$, relative to ξ.

Let now $\mathfrak{o}'$ be another saturated local domain, with coefficient field k', characteristic pairs (m_j', n_j) $(j = 1, 2, \cdots, g')$, and characteristic residue fields $k_1', k_2', \cdots, k'_{g'}$. Then, the following is an immediate consequence of Proposition 10.2 and Theorem 9.6:

THEOREM 10.2. *If $\bar{k}$ (resp. $\bar{k}'$) denotes the algebraic closure of k (resp. of k'), in the field of fractions of $\mathfrak{o}$ (resp. of $\mathfrak{o}'$), and if θ is an isomorphism $\bar{k} \xrightarrow{\sim} \bar{k}'$ such that $\theta(k) = k'$, then θ can be extended to an isomorphism $\mathfrak{o} \xrightarrow{\sim} \mathfrak{o}'$ if and only if the following conditions are satisfied:*

a) *The two rings $\mathfrak{o}$, $\mathfrak{o}'$ have the same characteristic pairs, i.e., we have* $g = g'$, $m_j = m_j'$, $n_j = n_j'$ $(j = 1, 2, \cdots, g)$.

b) $\theta(k_j) = k_j'$, $j = 1, 2, \cdots, g$.

c) $\theta(a_1), \theta(a_2), \cdots, \theta(a_g)$ *are characteristic parameters of* $\mathfrak{o}'$.

HARVARD UNIVERSITY.

REFERENCES.

[1] O Zariski and P. Samuel, *Commutative Algebra.*

[2] O. Zariski, "Studies in equisingularity III. Saturation of local rings and equisingularity," *American Journal of Mathematics*, vol. 90 (1968), pp. 961-1023.

GENERAL THEORY OF SATURATION AND OF SATURATED LOCAL RINGS II.*

Saturated local rings of dimension 1.

By Oscar Zariski.

To Wei-Liang Chow on his 60th birthday.

Contents. page

Introduction. This is the second paper in a series of three papers on the general theory of saturation and saturated (complete) local rings [the first paper was published in an earlier issue of this journal (see [6]); it will be referred to as GTS I]. In this paper we develop the theory of saturation of complete local rings of dimension 1, the emphasis being on rings with zero divisors (but free from nilpotent elements). Throughout this paper we deal only with local rings $\mathfrak{o}$ which are equicharacteristic, of characteristic zero (or equivalently: complete local rings having a coefficient field k, of characteristic zero). A coefficient field k of $\mathfrak{o}$ is fixed throughout the paper, but it is shown in the very last section (§ 8) that the whole theory is independent of the choice of the coefficient field (see Theorem 8.1 and Corollary 8.2). The coefficient field k is not assumed to be algebraically closed. It is,

Received April 5, 1971.

* This work was supported by a National Science Foundation Grant GP-9667.

[Reprinted from *American Journal of Mathematics,* Vol. XCIII, No. 4, pp. 872-964.]

in fact, essential for our program to develop the theory of saturation over coefficient fields k which are not algebraically closed, in view of the manner in which we plan to develop, in our third paper of this series, the theory of saturation of complete local rings of dimension > 1. At least in the case of Cohen-Macaulay rings $\mathfrak{o}$, and—in particular—in the case of the local ring $\mathfrak{o}$ of a point of an algebroid hypersurface, we shall obtain the bulk of our theory by localizing $\mathfrak{o}$ at its various prime ideals $\mathfrak{p}$ of height one and applying the results of the present paper to the one-dimensional local rings $\mathfrak{o}_{\mathfrak{p}}$ thus obtained [for it will be proved—and this is the key of this particular approach to hypersurfaces—that the saturation of $\mathfrak{o}$ is the intersection of the saturations of the local rings $\mathfrak{o}_{\mathfrak{p}}$ (with respect to suitable parameters $x_{\mathfrak{p}}$)] as $\mathfrak{p}$ ranges over the set of all prime ideals $\mathfrak{p}$ of $\mathfrak{o}$ of height one.

The special case in which k is algebraically closed is, of course, of particular geometric interest, and we devote a good deal of space to a discussion and a geometric presentation of our general results in this special case. The most important topic studied exclusively in the case in which k is algebraically closed is developed in § 7. In that section we generalize to *reducible* plane algebroid curves C, defined over k, the saturation—theoretic criterion of equisingularity, which we have obtained in our paper [5] for irreducible plane algebroid curves. Also in § 6 we deal exclusively with the case of an algebraically closed coefficient field k, obtaining in that case a suggestive geometric interpretation of the structure theorem for saturated rings (proved in §§ 2, 3). In particular we obtain a very simple geometric characterization of what we call the *saturation components* of C, relative to a transversal parameter x (see Theorem 6.5; for the definition of saturation components of C, relative to a parameter x, see Definition 4.11). Note that the cited Theorem 6.5 shows that the saturation components of C, relative to a transversal parameter x, are independent of the choice of x; we refer to them simply as *the* saturation components of C (see Definition 6.7).

If the branches of C do not all have the same tangent then the saturation components of C are simply the tangential components of C (see Corollary 4.8), as defined in § 2 of our paper [4]). If the branches of C do have all the same tangent, then the saturation components Γ_ρ of C are to a *large extent* (i.e., under additional, very mild condition) characterized by the condition that the branches of Γ_ρ have in common some infinitely near points outside the base set B of points common to all branches of C (see Proposition 6.6). For convenience of the reader we have included in our paper a self-contained (and semi-original) account of the theory of infinitely near points in the plane (for arbitrary characteristic).

2

The rest of the paper, in particular Sections 2 and 3, is devoted to what is perhaps the central result of this work, namely the structure theorem for saturated complete local rings of dimension 1. This result is an extension of our structure theorem 1.12 proved in our paper [5] for local domains having algebraically closed coefficient fields k and generalized in [6] (Theorem 4.1) to the case of arbitrary coefficient fields. The saturation components $\Gamma_1, \Gamma_2, \cdots, \Gamma_r$ of a plane algebroid curve C are of importance, because our structure theorem describes the saturation of the local ring $\mathfrak{o}$ of C in terms of the saturations of the local rings of the curves Γ_ρ. This is the basis of a number of inductive proofs of properties of saturated rings, the induction being with respect to the number h of irreducible components of C.

Theorems, lemmas, or definitions from our paper GTS I will be referred to by the number which they carry in that paper, preceded by the Latin numeral I.

1. Generalities and preliminaries. In this chapter we shall deal constantly with a complete (noetherian) local ring $\mathfrak{o}$ of dimension 1 which 1) is definitely not a domain, 2) has no nilpotent elements different from zero, 3) is equidimensional and 4) is equicharacteristic, of characteristic zero. The maximal ideal of $\mathfrak{o}$ will be denoted by $\mathfrak{m}$. A coefficient field k of $\mathfrak{o}$ will be fixed once and for always. Beginning with § 2 of this chapter, the ring $\mathfrak{o}$ will be assumed to be saturated (i.e., k-saturated with respect to some parameter x), unless the contrary is explicitly specified.

We denote by F the total ring of quotients of $\mathfrak{o}$. By condition 1), 2) above, and in view of the noetherian character of $\mathfrak{o}$, F will be a direct sum of h field, $h > 1$:

$$F\epsilon_1 + F\epsilon_2 + \cdots + F\epsilon_h.$$

Here $\epsilon_i^2 = \epsilon_i$, $\epsilon_i\epsilon_j = 0$ if $i \neq j$, ϵ_i is the identity element of the field $F\epsilon_i$, and

$$1 = \epsilon_1 + \epsilon_2 + \cdots + \epsilon_h.$$

For each i, the ring $\mathfrak{o}\epsilon_i$ is a complete local domain, having $\mathfrak{m}\epsilon_i$ as maximal ideal, $k\epsilon_i$ as coefficient field and $F\epsilon_i$ as field of fractions. If N is the embedding dimension of $\mathfrak{o}$ ($N =$ number of elements in a minimal basis of $\mathfrak{m}$), then $\mathfrak{o}$ is the local ring of a reducible algebroid curve Γ in the affine N-space A_N (over k) ($\Gamma =$ a reducible and reduced formal local scheme). The curve Γ has h irreducible components (or *branches*) $\Gamma_1, \Gamma_2, \cdots, \Gamma_h$, all having center at the origin of A_N, and $\mathfrak{o}\epsilon_i$ is the local domain of Γ_i (at the origin).

More generally, if I is any non-empty subset of the set of indices $\{1,2,\cdots,h\}$, and if we set

$$(1) \qquad \epsilon_I = \sum_{i \in I} \epsilon_i,$$

then $\mathfrak{o}\epsilon_I$ is a complete local ring, having $\mathfrak{m}\epsilon_I$ as maximal ideal, $k\epsilon_I$ as coefficient field (with ϵ_I as identity element) and $F\epsilon_I$ as total ring of quotients. The ring $\mathfrak{o}\epsilon_I$ is the local ring of the algebroid curve $\Gamma_I = \bigcup_{i \in I} \Gamma_i$. If

$$(2) \qquad \{1,2,\cdots,h\} = \bigcup_{\rho=1}^{r} I_\rho$$

is a partition of the set of indices $1,2,\cdots,h$ into r non-empty, mutually disjoint subsets I_ρ and if we set

$$(3) \qquad \epsilon_\rho' = \sum_{i \in I_\rho} \epsilon_i, \qquad \rho = 1,2,\cdots,r,$$

then we have $1 = \epsilon_1' + \epsilon_2' + \cdots + \epsilon_r'$, $\epsilon_\rho'^2 = \epsilon_\rho'$, $\epsilon_\rho'\epsilon_\sigma' = 0$ if $\rho \neq \sigma$, and $F = F\epsilon_1' + F\epsilon_2' + \cdots + F\epsilon_r'$, where the sum is direct.

We denote by $\bar{\mathfrak{o}}$ the integral closure of $\mathfrak{o}$ in F and by $\overline{\mathfrak{o}\epsilon_I}$ the integral closure of $\mathfrak{o}\epsilon_I$ in $F\epsilon_I$, where ϵ_I is defined by (1). It is easy to see that

$$(4) \qquad \overline{\mathfrak{o}\epsilon_I} = \bar{\mathfrak{o}}\epsilon_I$$

and that if $\epsilon_1', \epsilon_2', \cdots, \epsilon_r'$ are as in (3), then $\bar{\mathfrak{o}}$ is the (direct) sum of the local rings $\bar{\mathfrak{o}}\epsilon_\rho'$:

$$(5) \qquad \bar{\mathfrak{o}} = \bar{\mathfrak{o}}\epsilon_1' + \bar{\mathfrak{o}}\epsilon_2' + \cdots + \bar{\mathfrak{o}}\epsilon_r'.$$

For, since $\epsilon_\rho'^2 - \epsilon_\rho' = 0$, it folows $\epsilon^{\rho'} \in \bar{\mathfrak{o}}$, whence $\sum_{\rho=1}^{r} \overline{\mathfrak{o}\epsilon^{\rho'}} \subset \bar{\mathfrak{o}}$. On the other hand, since $1 = \epsilon_1' + \epsilon_2' + \cdots + \epsilon_r'$, we have $\bar{\mathfrak{o}} \subset \bar{\mathfrak{o}} \cdot 1 \subset \sum_{\rho=1}^{r} \bar{\mathfrak{o}}\epsilon_\rho'$, and this proves (5). If $\eta \in \bar{\mathfrak{o}}$, then $\eta\epsilon_I$ is integral over $\mathfrak{o}\epsilon_I$, whence $\bar{\mathfrak{o}}\epsilon_I \subset \overline{\mathfrak{o}\epsilon_I}$. On the other hand, we have $\mathfrak{o}\epsilon_I \subset \bar{\mathfrak{o}}\epsilon_I \subset \bar{\mathfrak{o}}$ (since $\epsilon_I \in \bar{\mathfrak{o}}$), and hence $\overline{\mathfrak{o}\epsilon_I} = \overline{\mathfrak{o}\epsilon_I} \cdot \epsilon_I \subset \bar{\mathfrak{o}}\epsilon_I$ (since $F\epsilon_I \subset F$), and this establishes (4).

Let x be a parameter of $\mathfrak{o}$ and let K be the formal power series field $k((x))$. Then K is a field contained in F, and the saturation $\tilde{\mathfrak{o}}_x$ of $\mathfrak{o}$ with respect to x (i.e., with respect to K) is defined (see SES III, Introduction). For each $i = 1,2,\cdots,h$, the field $F\epsilon_i$ is a finite algebraic extension of the subfield $K\epsilon_i$. We set

$$(6) \qquad [F\epsilon_i : K\epsilon_i] = n_i, \qquad i = 1,2,\cdots,h.$$

The ring F is a finite-dimensional vector space over K:

(7) $$\dim_K F = n = n_1 + n_2 + \cdots + n_h.$$

To study the saturation $\tilde{\mathfrak{o}}_x$ we have to deal with the set H of all K-homomorphisms ψ of F into a fixed algebraic closure Ω of K. We set

(8) $$H_i = \{\psi \in H \mid \psi(\epsilon_i) = 1\}, \quad i = 1, 2, \cdots, h.$$

Then we have for any ψ in H_i: $\psi(\epsilon_j) = 0$ if $j \neq i$, the restriction of ψ to $F\epsilon_i$ is an isomorphism, and we have $\psi(K\epsilon_i) = K$, with $\psi(\xi\epsilon_i) = \xi$, $\xi \in K$. As ψ ranges over H_i, the restriction of ψ to $F\epsilon_i$ ranges over all isomorphisms ϕ of $F\epsilon_i$ into Ω such that $\phi(\xi\epsilon_i) = \xi$ for all ξ in K, and distinct ψ's in H_i have distinct restrictions to $F\epsilon_i$, since $\psi(F\epsilon_j) = 0$ for all $j \neq i$. Hence, by (6), H_i has exactly n_i elements:

(9) $$H_i = \{\psi_i^{(1)}, \psi_i^{(2)}, \cdots, \psi_i^{(n_i)}\}, \quad i = 1, 2, \cdots, h,$$

and H consists of n elements, since

(10) $$H = \bigcup_{i=1}^{h} H_i.$$

More generally, if I is any non-empty subset of the set $\{1, 2, \cdots, h\}$ and if ϵ_I is defined by (1), we set

(11) $$H_I = \{\psi \in H \mid \psi(\epsilon_I) = 1\}.$$

It is clear that

$$H_I = \bigcup_{i \in I} H_i,$$

that

$$\dim_{K\epsilon_I} F\epsilon_I = \sum_{i \in I} n_i,$$

and that as ψ ranges over H_I, the restriction of ψ to $F\epsilon_I$ ranges over the set of all n' homomorphism ϕ of $F\epsilon_I$ into Ω such that $\phi(\xi\epsilon_I) = \xi$ for all $\xi \in K$; here $n' = \sum_{i \in I} n_i$.

A partition (2) of the set of indices $1, 2, \cdots, h$ yields a corresponding partition of H:

(12) $$H = \bigcup_{\rho=1}^{r} H_{I_\rho}.$$

For each element $\psi_i^{(\nu_i)}$ of H_i $(\nu_i = 1, 2, \cdots, n_i)$ we set

(13) $$\psi_i^{(\nu_i)}(F) = F_i^{(\nu_i)} \quad (= \psi_i^{(\nu_i)}(F\epsilon_i)),$$
$$i = 1, 2, \cdots, h,$$
$$\nu_i = 1, 2, \cdots, n_i.$$

The n_i fields $F_i^{(\nu_i)}$ form a complete set of K-conjugate (not necessarily distinct) subfields of Ω. We shall denote by F_i^* the compositum of the n_i fields $F_i^{(\nu_i)}$ and by F^* the compositum of the h fields F_i. The fields F_i^*, F^* are power series fields and are Galois extensions of the field K $(=k((x)))$; more precisely: F_i^* is the least Galois extension of K which contains any of the n_i fields $F_i^{(\nu_i)}$. Note that these n_i fields are isomorphic to $F\epsilon_i$.

We shall denote by G the Galois group of F^*/K. For expository reasons we shall adopt from now on the following notational procedure: we shall *fix* for each $i=1,2,\cdots,h$ an element in H_i and we shall denote that element by ψ_i. The corresponding field $\psi_i(F)$, in the set of n_i K-conjugate fields $F_i^{(\nu_i)}$, shall be denoted by F_i. As τ ranges over the Galois group G, the fields F_i^τ $(=\tau(F_i))$ range over the set of n_i fields $F_i^{(\nu_i)}$ (here $F_i^\tau = F_i$ if $\tau \in \mathfrak{G}(F^*/F_i)$), and the product $\tau\cdot\psi_i$ ranges over the set H_i (here $\tau\cdot\psi_i = \psi_i$ if and only if $\tau \in \mathfrak{G}(F^*/F_i)$). Later on (and especially in § 2) we shall find it convenient to make an appropriate choice of the ψ_i (in their respective sets H_i). Actually, what we shall have to do—after having fixed arbitrarily one of the ψ_i, say ψ_1—is to choose the other $h-1$ homomorphisms ψ_i in an appropriate manner.

We shall denote by v and v_i the canonical valuation of the power series field F^* and the power series field F_i^* respectively. Here v may very well be ramified over v_i. We shall denote by e_i the ramification index of v over v_i, or—equivalently—the ramification index of F^* over F_i; this is the same as the ramification index of F^* over F_i^*, since F_i^* is unramified over F_i (I, Lemma 1.2).

For any element η of F we set

$$\eta_i = \psi_i(\eta), \quad i=1,2,\cdots,h. \tag{14}$$

Thus η_i is an element of F_i, and each of the K-conjugates $\psi_i^{(\nu_i)}(\eta)$ of η_i is of the form η_i^τ, $\tau\in G$. We note that *given any set of h elements $\eta_1,\eta_2,\cdots,\eta_h$, belonging to $F_1,F_2,\cdots,F_h$ respectively, there exists one and only one element η of F such that* (14) *holds.* For, since the map $\psi_i\colon F\epsilon_i \rightarrow F_i$ is surjective, we can find an element η_i' in $F\epsilon_i$ such that $\psi_i(\eta_i')=\eta_i$. If we then set $\eta=\eta_1'+\eta_2'+\cdots+\eta_h'$, we will have $\psi_i(\eta)=\eta_i$ since $\psi_i(\eta_j)=0$ if $i\neq j$. Now, the kernel of ψ_i in F is the ideal $\mathfrak{A}_i=\sum_{j\neq i} F\epsilon_j$, and the intersection of the h ideals $\mathfrak{A}_i$ is zero. Hence η is uniquely determined by the relations (14). We also note the obvious fact that $\eta\in\bar{\mathfrak{o}}$ *if and only if each of the η_i is integral over* $k[[x]]$.

The following definition is identical with Definition I, 3.1:

Definition 1.1. *An element y of $\mathfrak{o}$ is called an x-saturator of $\mathfrak{o}$ if*

$$\tilde{\mathfrak{o}}_x = \widetilde{k[[x]][y]}_x. \tag{15}$$

Let $\xi_1, \xi_2, \cdots, \xi_N$ be a set of ring generators of $\mathfrak{o}$ over $k[[x]]$.

PROPOSITION 1.2. *An element y of $\mathfrak{o}$ is an x-saturator of $\mathfrak{o}$ if*

$$v(\psi(\xi_\alpha) - \psi'(\xi_\alpha)) \geqq v(\psi(y) - \psi'(y)), \quad \text{for } \alpha = 1, 2, \cdots, N \text{ and all } \psi, \psi' \text{ in } H. \tag{16}$$

Furthermore, if y is an x-saturator of $\mathfrak{o}$ then $F = K[y]$.

Proof. The inequalities (16) signify that each of the generators ξ_α, and hence also every element of $\mathfrak{o}$, dominates y (with respect to the parameter x). Furthermore, the (16) imply that $\psi(y) \neq \psi'(y)$ if $\psi \neq \psi'$. Thus the n elements $\psi(y)$, $\psi \in H$, are distinct. Since these elements are the roots of the minimal polynomial of y over K (the degree of that polynomial being at most n, since $n = \dim_K F$), we see that $\dim_K F = \dim_K K[y] = n$, i.e., $K[y] = F$. Thus if the (16) hold then the two rings $\mathfrak{o}$, $k[[x]][y]$ have the same total ring of quotients. Since $\mathfrak{o}$ is integral over $k[[x]]$ and since the set of elements of $\overline{k[[x]][y]}$ which dominate y is precisely the saturation of $k[[x]][y]$ with respect to x, we conclude that (16) is equivalent to the following set-theoretic inclusion: $\mathfrak{o} \subset \widetilde{k[[x]][y]}_x$. If (15) holds, then this inclusion is certainly satisfied. Conversely, if the above inclusion holds, then $\tilde{\mathfrak{o}}_x \subset \widetilde{k[[x]][y]}_x$ by the minimality property of $\tilde{\mathfrak{o}}_x$, and since the opposite inclusion follows trivially from $k[[x]][y] \subset \mathfrak{o}$, we have (15). The second part of the proposition follows from the fact, just established above, that the (16) imply the equality $F = K[y]$.

PROPOSITION 1.3. *There exist x-saturators of $\mathfrak{o}$.*

Proof. We shall proceed along lines similar to those of the proof of Proposition I, 3.3. If $\{\xi_1, \xi_2, \cdots, \xi_N\}$ is a set of generators of $\mathfrak{o}$ over $k[[x]]$, let V be the vector space, over k, spanned by the N elements ξ_α. We can now find, already in the vector space V, an element y satisfying (16), as follows:

For any pair of distinct elements ψ, ψ' in H, let $V_{\psi,\psi'}$ denote the set of elements ξ of V for which the following inequality holds:

$$v(\psi(\xi) - \psi'(\xi)) > \min \cdot \{v(\psi(\xi_1) - \psi'(\xi_1)), \\ v(\psi(\xi_2) - \psi'(\xi_2)), \cdots, v(\psi(\xi_N) - \psi'(\xi_N))\}.$$

It is immediate that if $\psi \neq \psi'$ then $V_{\psi,\psi'}$ is a *proper subspace* of V. Since the number of K-homomorphisms of F into Ω is finite and since k is an infinite field, it follows that the union of the subspaces $V_{\psi,\psi'}(\psi \neq \psi')$ is a proper subset V_0 of V. Any element y of the complement $V - V_0$ satisfies (16) and is therefore an x-saturator.

Let y be an x-saturator of $\mathfrak{o}$. In agreement with the notational procedure introduced above, we have h elements $y_1, y_2, \cdots, y_h$ in $F_1, F_2, \cdots, F_h$, respectively, where $y_i = \psi_i(y)$ [see (14)]. An element η of $\bar{\mathfrak{o}}$ belongs to $\tilde{\mathfrak{o}}_x$ if and only if for any pair of distinct K-homomorphisms ψ, ψ' of F into Ω we have

$$(17) \qquad v(\psi(\eta) - \psi'(\eta)) \geqq v(\psi(y) - \psi'(y)).$$

Now, if, say, $\psi \in H_i$ and $\psi' \in H_j$ [see (8); here we *may* have $i = j$], then $\psi(\eta) = \eta_i^\tau$ and $\psi'(\eta) = \eta_j^\sigma$, while $\psi(y) = y_i^\tau$ and $\psi'(y) = y_j^\sigma$, where τ and σ are suitable elements of the Galois group G of F^*/K. Therefore, *if y is an x-saturator of $\mathfrak{o}$ then*

$$(18) \qquad \tilde{\mathfrak{o}}_x = \{\eta \in \bar{\mathfrak{o}} \mid v(\eta_i^\tau - \eta_i^\sigma) \geqq v(y_i^\tau - y_j^\sigma)\}, \text{ for all } i, j = 1, 2, \cdots, h \text{ and all } \tau, \sigma \text{ in } G.$$

Now, let I be any non-empty subset of the set of indices $\{1, 2, \cdots, h\}$ and le ϵ_I be defined as in (1). The element $x\epsilon_I$ is then a parameter of $\mathfrak{o}\epsilon_I$, and the saturation of $\mathfrak{o}\epsilon_I$ with respect to $x\epsilon_I$, i.e., with respect to the field $K\epsilon_I$ $(= k\epsilon_I((x\epsilon_I)))$ is defined. According to our general definition of saturation, the saturation of $\mathfrak{o}\epsilon_I$ with respect to $K\epsilon_I$ is defined in terms of the various $K\epsilon_I$-homomorphisms ϕ of $F\epsilon_I$ into some preassigned algebraic closure Ω' of $K\epsilon_I$. However, after fixing a definite isomorphism $\theta: \Omega' \xrightarrow{\sim} \Omega$ such that $\theta(\xi\epsilon_I) = \xi$, for all ξ in K, we see that we can replace in the definition of $(\widetilde{\mathfrak{o}\epsilon_I})_{K\epsilon_I}$, the $K\epsilon_I$-homomorphisms ϕ by the homomorphisms ψ of $F\epsilon_I$ into Ω such that $\psi(\xi\epsilon_I) = \xi$ for all ξ in K; or—equivalently—we can replace the ϕ's by the K-homomorphisms of F into Ω such that $\psi(\epsilon_I) = 1$. *These homomorphisms form the set H_I, defined in* (11). From this we can conclude with the following result:

PROPOSITION 1.4. *If y is an x-caturator of $\mathfrak{o}$, then $y\epsilon_I$ is an $x\epsilon_I$-saturator of $\mathfrak{o}\epsilon_I$, and we have*

$$(19) \qquad \widetilde{(\mathfrak{o}\epsilon_I)}_{K\epsilon_I} = \{\eta\epsilon_I \mid \eta \in \bar{\mathfrak{o}} \textit{ and } v(\eta_i^\tau - \eta_j^\sigma) \geqq v(y_i^\tau - y_j^\sigma) \textit{ for all } i, j \in I \textit{ and all } \tau, \sigma \textit{ in } G.$$

Proof. If $\xi_1, \xi_2, \cdots, \xi_N$ are ring generators of $\mathfrak{o}$ over $k[[x]]$, the elements $\xi_1\epsilon_I, \xi_2\epsilon_I, \cdots, \xi_N\epsilon_I$ are ring generators of $\mathfrak{o}\epsilon_I$ over $k\epsilon_I[[x\epsilon_I]]$. If y is an x-saturator of $\mathfrak{o}$, the inequalities (15) are valid for any two distinct K-homomorphism ψ, ψ' of F into Ω. Since $\psi(\epsilon_I) = 1$ whenever $\psi \in H_I$, it follows that the (15) remain valid if we replace each ξ_α by $\xi_\alpha\epsilon_I$ and y by $y\epsilon_I$, provided we let ψ, ψ' range only over the subset H_I of H. What the resulting inequalities tell us is precisely that $y\epsilon_I$ is an $x\epsilon_I$-saturator of $\mathfrak{o}\epsilon_I$. The expression (18) of the saturation of $\mathfrak{o}\epsilon_I$ with respect to $K\epsilon_I$ now follows from the definition of x-saturators, from the equality (4) and from the fact that an element ψ of H belongs to H_I if and only if there exists an index i in I and an element τ in G such that for any η in F we have $\psi(\eta) = \eta_i^\tau$.

From now on we fix an x-saturator y of $\mathfrak{o}$. Upon subtracting from y a suitable element of k we may assume that y belongs to the maximal ideal $\mathfrak{m}$ of $\mathfrak{o}$. The minimal polynomial $f(x;y)$ of y over K is then a monic polynomial in Y, of degree n, whose coefficients belong to the power series ring $k[[x]]$ and which vanishes at the origin $x = y = 0$. Furthermore, we have $F(0;Y) = Y^n$. The polynomial $f(x;Y)$ factors, over K, into h distinct irreducible factors $f_1(x;Y), f_2(x;Y), \cdots, f_h(x;Y)$; here $f_i(x;Y)$ is a monic polynomial in Y, of degree n_i, and coefficients in $k[[x]]$. The roots of $f_i(x;Y)$ in Ω are the n_i quantities $\psi_i^{(\nu_i)}(y)$, where the $\psi^{(\nu_i)}$ range over the set H_i [see (8)]. We have h irreducible plane algebroid curves

$$C_i\colon f_i(X;Y) = 0, \quad i = 1, 2, \cdots, h,$$

which all have center at the origin $X = Y = 0$ and are the branches of the plane algebroid curve

$$C\colon f(X;Y) = 0.$$

The local domain of C_i is the subring $k\epsilon_i[[x\epsilon_i]][y\epsilon_i]$ of $\mathfrak{o}\epsilon_i$, or also the isomorphic ring $k[[x]][y_i]$, while the local ring of C is the subring $k[]x[][y]$ of our original local ring $\mathfrak{o}$. The assumption that y is an x-saturator of $\mathfrak{o}$ signifies that the saturation, with respect to x, of the local ring of the plane algebroid curve C is equal to the saturation of $\mathfrak{o}$ with respect to x.

We observe that if I is any non-empty subset of $\{1, 2, \cdots, h\}$ and ϵ_I is defined as in (1), then we have by (18) and (19):

$$\tilde{\mathfrak{o}}_x\epsilon_I \subset \widetilde{(\mathfrak{o}\epsilon_I)}_{K\cdot\epsilon_I}. \tag{20}$$

It will be a far from trivial result, proved in Section 3 (see Corollary 3.6) as a consequence of a structure theorem for saturated rings (see §§ 2, 3), that in (20) *the inclusion sign* $\subset$ *can, in fact, be replaced by the equality sign* in the following two cases: (1) I consists of a single index; (2) I is an element of what we call the *saturation partition of* $\{1,2,\cdots,h\}$ (see § 3).

We note that (20) implies at any rate the (saturated) ring $\widetilde{(\mathfrak{o}\epsilon_I)}_{K\epsilon_I}$ contains the saturation of $\tilde{\mathfrak{o}}_x\cdot\epsilon_I$ with respect to the field $K\epsilon_I$. On the other hand, since $\mathfrak{o}\epsilon_I\subset\tilde{\mathfrak{o}}_x\epsilon_I$, the ring $\widetilde{(\mathfrak{o}\epsilon_I)}_{K\epsilon_I}$ is contained in the saturation of $\tilde{\mathfrak{o}}_x\epsilon_I$, with respect to $K\epsilon_I$. Therefore we have

$$\widetilde{(\tilde{\mathfrak{o}}_x\epsilon_I)}_{K\epsilon_I}=\widetilde{(\mathfrak{o}\epsilon_I)}_{K\epsilon_I}. \tag{21}$$

We shall denote by $\mathfrak{o}_i$ the saturation of the ring $k[[x]][y_i]$ with respect to K:

$$\mathfrak{o}_i=\widetilde{k[[x]][y_i]}_K,\quad i=1,2,\cdots,h. \tag{22}$$

Proposition 1.4 says that $\widetilde{(\mathfrak{o}\epsilon_i)}_{K\epsilon_i}=\widetilde{k\epsilon_i[[x\epsilon_i]][y\epsilon_i]}_{K\epsilon_i}$. Since $\psi_i(y\epsilon_i)=y_i$ and $\psi_i(K\epsilon_i)=K$, it follows from (21) and (22), for $I=\{i\}$, that

$$\mathfrak{o}_i=\psi_i(\widetilde{(\tilde{\mathfrak{o}}_x\epsilon_i)}_{K\epsilon_i}),\quad i=1,2,\cdots,h. \tag{23}$$

Thus the rings $\mathfrak{o}_1,\mathfrak{o}_2,\cdots,\mathfrak{o}_h$ depend only on $\tilde{\mathfrak{o}}_x$ and on the choice of the h homomorphisms $\psi_1,\psi_2,\cdots,\psi_h$ within the h sets $H_1,H_2,\cdots,H_h$. The field of fractions of $\mathfrak{o}_i$ is F_i $(=K[y_i]=\psi_i(F)=\psi_i(F\epsilon_i))$. If, for a given i $(i=1,2,\cdots,h)$, ψ_i' is any other element of H_i and $y_i'=\psi_i'(y)$, then $\psi_i'=\alpha\cdot\psi_i$, where $\alpha\colon F_i\xrightarrow{\sim}F_i'=\psi_i'(F)$ is a K-isomorphism between F_i and the K-conjugate field $F_i'=K[y_i']$. The ring $\mathfrak{o}_i'$ associated with ψ_i' is the ring $\widetilde{k[[x]][y_i']}_K$, and we have, of course, $\mathfrak{o}_i'=\alpha(\mathfrak{o}_i)$. Thus, each ring $\mathfrak{o}_i$ is uniquely determined by $\tilde{\mathfrak{o}}_x$ and by the index i, up to K-conjugacy within the field F_1^*.

Let

$$K=K_{i,0}<K_{i,1}<\cdots<K_{i,g_i}=F_i \tag{24}$$

be the sequence of intermediate saturation fields of $\mathfrak{o}_i$, relative to x [see I, § 3, (1)], and let

$$\mathfrak{m}_i=\mathfrak{M}_{i,0}+\mathfrak{M}_{i,1}{}^{m_{i,1}}+\cdots+\mathfrak{M}_{i,g_i}{}^{m_{i,g_i}} \tag{25}$$

be the associated standard decomposition of the maximal ideal $\mathfrak{m}_i$ of $\mathfrak{o}_i$ (Thecrem I, 4.1). The integers g_i, $m_{i,1},\cdots,m_{i,g_i}$ depend only on $\mathfrak{o}$, x and the index i, while the chain of fields (24) is uniquely determined, for a given i, up to an arbitrary K-isomorphism of F_i into F_i^*. In the special case in which $\mathfrak{o}_i=k[[x]]$ (and hence $F_i=K$) we agree to set $g_i=0$. In all other cases we have $g_i > 0$. The integers m_{i,α_i} and the integers $\beta_{i,\alpha_i}, e_{i,\alpha_i}$ defined below in (27) and (28) are defined only if $g_i > 0$.

Applying to the x-saturator y_i of $\mathfrak{o}_i$ the decomposition (10) of I, § 3, we can write

$$(26) \qquad y_i = y_{i,0} + y_{i,1} + \cdots + y_{i,g_i}, \qquad i = 1, 2, \cdots, h.$$

According to Definition I, 3.4, the set of integers

$$\{v_i(x), v_i(y_{i,1}), \cdots, v_i(y_{i,g_i})\}$$

is the x-characteristic of $\mathfrak{o}_i$; here—we recall—v_i is the natural valuation of the power series field F_i. However, we prefer the natural valuation v of F^* (= the least Galois extension of K containing the h fields $F_1, F_2, \cdots, F_h$). We set therefore

$$(27) \qquad v(y_{i,\alpha_i}) = \beta_{i,\alpha_i}, \quad i = 1, 2, \cdots, h;$$

$$(28) \qquad e_{i,\alpha_i} = e(F^*/K_{i,\alpha_i}), \quad \alpha_i = 1, 2, \cdots, g_i;$$

and

$$(29) \qquad e_0 = e_{i,0} = e(F^*/K) = v(x).$$

It will be the set of integers

$$(30) \qquad \{e_0; \beta_{i,1}, \beta_{i,2}, \cdots, \beta_{i,g_i}\}$$

that will be referred to in the sequel, without further ado, as the x-characteristic of $\mathfrak{o}_i$. It is obtained from the original x-characteristic of $\mathfrak{o}_i$ by multiplying each element of the latter by the integer $e_i = e(F^*/F_i)$, and it is independent of the choice of the fixed homomorphism ψ_i in the set H_i. All the properties of the usual x-characteristic of $\mathfrak{o}_i$, as established in I, § 3, continue to hold true for the modified x-characteristic (30). Thus we have:

$$(31) \qquad \beta_{i,1} < \beta_{i,2} < \cdots < \beta_{i,g_i};$$

$$(32) \qquad \beta_{i,\alpha_i} = e_{i,\alpha_i} \cdot m_{i,\alpha_i}, \quad i = 1, 2, \cdots, h;$$

$$(33) \qquad (\beta_{i,\alpha_i}, e_{i,\alpha_i - 1}) = e_{i,\alpha_i}, \quad \alpha_i = 1, 2, \cdots, g_i;$$

and if we set

$$(34) \qquad G_{i,\alpha_i} = \mathfrak{G}(F^*/K_{i,\alpha_i}),$$

then

$$(35) \qquad v(y_i^\tau - y_i) = v(y_{i,\alpha_i}^\tau - y_{i,\alpha_i}) = \beta_{i,\alpha_i}, \text{ for all } \tau \text{ such that } \tau \in G_{i,\alpha_i - 1}, \tau \notin G_{i,\alpha_i} \quad (1 \leqq \alpha_i \leqq g_i).$$

Furthermore, we have (see Corollary I, 4.2)

$$(36) \qquad K_{i,\alpha_i} = K_{i,\alpha_i-1}(y_{i,\alpha_i}), \quad \alpha_i = 1, 2, \cdots, g_i.$$

We also wish to stress the property of y_{i,α_i} stated in Proposition I, 3.11. Thus, in our present notations, if we denote by k_{i,α_i} the coefficient field of K_{i,α_i} which contains k and if we choose the uniformizing parameters x_{i,α_i} of the fields K_{i,α_i} in such a way as to have [see (18), I, § 3]

$$(37) \qquad x_{i,\alpha_i-1} = a_{i,\alpha_i} x_{i,\alpha_i}{}^{n_{i,\alpha_i}} \quad a_{i,\alpha_i} \in k_{i,\alpha_i},$$

then the v-residue c_{i,α_i} of $y_{i,\alpha_i}/x_{i,\alpha_i}{}^{m_{i,\alpha_i}}$ in k_{i,α_i} satisfies the condition

$$(38) \qquad k_{i,\alpha_i} = k_{i,\alpha_i-1}(a_{i,\alpha_i}, c_{i,\alpha_i}), \quad \alpha_i = 1, 2, \cdots, g_i.$$

Since $\mathfrak{o}_i$ depends ony on $\mathfrak{o}$, x and the index i, up to an arbitrary K-isomorphism of F_i, *the x-characteristic* (30) *depends only on the index i and does not depend on the choice of the x-saturator y of $\mathfrak{o}$.*

The decomposition (26) of y_i was obtained by us in I, § 3 by a well-defined process of taking successive traces (see the proof of Proposition I, 3.6). However, we shall often want to modify the decomposition (26) of y_i, subject to the condition that the modified decomposition still satisfy (27), (35), (36) and (38). Any such decomposition of y_i will still be called by us *a standard decomposition of y_i relative to K.* The following lemma describes a particular mode of modifying a standard decomposition in such a way that the new decomposition of y_i is still standard:

LEMMA 1.5. *Assume that* (26) *is a standard decomposition of y_i relative to K, and let, for a given α $(0 \leqq \alpha < g_i)$, ζ_α be an element of $K_{i,\alpha}$ such that $v(\zeta_\alpha) \geqq \beta_{i,\alpha+1}$. If we set*

$$y'_{i,\alpha} = y_{i,\alpha} - \zeta_\alpha, \quad y'_{i,\alpha+1} = y_{i,\alpha+1} + \zeta_\alpha,$$

then the decomposition

$$(39) \qquad y_i = y_{i,0} + y_{i,1} + \cdots + y_{i,\alpha-1} + y'_{i,\alpha} + y'_{i,\alpha+1} + y_{i,\alpha+2} + \cdots + y_{i,g_i}$$

of y_i is still standard relative to K.

Proof. We first check the validity of (27) for the new decomposition (39), i.e., we prove that $v(y'_{i,\alpha+1}) = \beta_{i,\alpha+1}$ *and also that* $v(y'_{i,\alpha}) = \beta_{i,\alpha}$ *if* $\alpha > 0$. If $\alpha > 0$, then $v(\zeta_\alpha) \geqq \beta_{i,\alpha+1} > \beta_{i,\alpha}$, while $v(y_{i,\alpha}) = \beta_{i\,\alpha}$. Thus $v(y'_{i,\alpha}) = \beta_{i,\alpha}$. Similarly, if $v(\zeta_\alpha) \geqq \beta_{i,\alpha+1}$ then $v(y'_{i,\alpha+1}) = \beta_{i,\alpha+1}$. Assume, however, that $v(\zeta_\alpha) = \beta_{i,\alpha+1}$. This implies that $\beta_{i,\alpha+1}$ is an integral multiple of $e_{i,\alpha}$, and therefore, by (33), we have $e_{i,\alpha+1} = e_{i,\alpha}$. Thercfore we may assume that for

$\alpha_i = \alpha + 1$ we have in (37): $x_{i,\alpha} = x_{i,\alpha+1}$, whence $a_{i,\alpha+1} = 1$. But then, by (38), we have $k_{i,\alpha+1} = k_{i,\alpha}(c_{i,\alpha+1})$. Since $K_{i,\alpha} < K_{i,\alpha+1}$, we must have $k_{i,\alpha} < k_{i,\alpha+1}$, whence

$$c_{i,\alpha+1} \notin k_{i,\alpha}. \tag{40}$$

Now, the residue $\bar{c}$ of $\zeta_\alpha / x_{i,\alpha+1}{}^{m_{i,\alpha+1}} (= \zeta_\alpha / x_{i,\alpha}{}^{m_{i,\alpha+1}})$ belongs to $k_{i,\alpha}$. Therefore, by (40), the v-residue $\bar{c} + c_{i,\alpha+1}$ of $y'_{i,\alpha+1}/x_{i,\alpha+1}{}^{m_{i,\alpha+1}}$ is different from zero, and this shows that $v(y'_{i,\alpha+1}) = \beta_{i,\alpha+1}$.

We have $y'_{i,\alpha}{}^\tau - y'_{i,\alpha} = (y_{i,\alpha}{}^\tau - y_{i,\alpha}) - (\zeta_\alpha{}^\tau - \zeta_\alpha)$. If $\alpha > 0$, and $\tau \in G_{i,\alpha-1}$, $\tau \notin G_{i,\alpha}$, then $v(y_{i,\alpha}{}^\tau - y_{i,\alpha}) = \beta_{i,\alpha}$, while

$$v(\zeta_\alpha{}^\tau - \zeta_\alpha) \geqq \beta_{i,\alpha+1} > \beta_{i,\alpha}.$$

Hence $v(y'_{i,\alpha}{}^\tau - y'_{i,\alpha}) = \beta_{i,\alpha}$. On the other hand, if $\tau \in G_{i,\alpha}$, then $\zeta_\alpha{}^\tau = \zeta_\alpha$, and thus $y'_{i,\alpha+1}{}^\tau - y'_{i,\alpha} = y_{i,\alpha+1}{}^\tau - y_{i,\alpha+1}$. This establishes the validity of (35) for the decomposition (39).

The (36) are consequences of (34) (see end of proof of Corollary I, 4.2). As to (38), let $c'_{i,\alpha+1}$ by the v-residue of $y'_{i,\alpha+1}/x_{i,\alpha+1}{}^{m_{i,\alpha+1}}$ and let, if $\alpha > 0$, $c'_{i,\alpha}$ be the v-residue of $y'_{i,\alpha}/x_{i,\alpha}{}^{m_i}$. If $\alpha > 0$, then $c'_{i,\alpha} = c_{i,\alpha}$ (since $v(\zeta_\alpha) \geqq \beta_{i,\alpha+1} > \beta_{i,\alpha}$). If also $v(\zeta_\alpha) > \beta_{i,\alpha+1}$ then $c'_{i,\alpha+1} = c_{i,\alpha+1}$. If, however, $v(\zeta_\alpha) = \beta_{i,\alpha+1}$ then the above proof of the validity of the (27) for the decomposition (39) shows that $c'_{i,\alpha+1} - c_{i,\alpha+1}$ is an element of $k_{i,\alpha}$ and that $a_{i,\alpha+1}$ may be assumed to be 1. Then from $k_{i,\alpha+1} = k_{i,\alpha}(c_{i,\alpha+1})$ follows also that $k_{i,\alpha+1} = k_{i,\alpha}(c'_{i,\alpha+1})$. This completes the proof.

We now go back to Proposition 1.4. If we apply (19) to the case $i = j \in I$, we see that if $\eta\epsilon_I$ belongs to $\widehat{(\mathfrak{o}\epsilon_I)_{K\epsilon_I}}$ then, for any i in I, η_i must belong to $\mathfrak{o}_i$, i.e., η_i must have a decomposition of the form

$$\eta_i = \eta_{i,0} + \eta_{i,1} + \cdots + \eta_{i,g_i}; \eta_{i,\alpha_i} \in \mathfrak{M}_{i,\alpha_i}{}^{m_{i,\alpha_i}}, \quad (\alpha_i = 1, 2, \cdots, g_i) \tag{41}$$
$$\eta_{i,0} \in k + \mathfrak{M}_{i,0} = k + \mathfrak{M} = k[[x]].$$

We now state the following lemma, which is an easy consequence of (19):

LEMMA 1.6. *Let I be any non-empty subset of the set of indices $\{1, 2, \cdots, h\}$ and let*

$$\epsilon_I = \sum_{i \in I} \epsilon_i.$$

If η is an element of F, then $\eta\epsilon_I \in \widehat{(\mathfrak{o}\epsilon_I)_{K\epsilon_I}}$ if and only if the following conditions are satisfied:

a) *For each $i \in I$, the element η_i of F_i must belong to $\mathfrak{o}_i$, i.e., η_i must be of the form indicated in* (41).

b) *For any pair of distinct indices i, j in I we most have:*

(42) $$v(\eta_i{}^\tau - \eta_j{}^\sigma) \geqq v(y_i{}^\tau - y_j{}^\sigma), \text{ for all } \tau, \sigma \text{ in } G.$$

Proof. Condition a) is contained in (19), in the special case $j = i$, while b) gives the rest of (19) (i. e., when i and j are distinct elements of the set of indices I). It remains to prove that conditions a) and b) imply that $\eta\epsilon_I \subset \bar{\mathfrak{o}}\epsilon_I$. Now, condition a) implies that η_i is integral over $k[[x]]$, i. e., that $\eta_i \in \psi_i(\bar{\mathfrak{o}}\epsilon_i)$, for all i in I. Let, say, $\eta_i = \psi_i(\xi_i \cdot \epsilon_i)$, $\xi_i \in \bar{\mathfrak{o}}$, $i \in I$. Then $\psi_i(\eta - \sum_{i \in I} \xi_i \cdot \epsilon_i) = 0$ for all i in I, i. e., $\eta - \sum_{i \in I} \xi_i \cdot \epsilon_i \in \sum_{j \notin I} F\epsilon_j$. Since $\xi_i \xi_j = 0$ for all $j \notin I$, we see that $\eta\epsilon_I = \xi\epsilon_I$, where $\xi = \sum_{i \in I} \xi_i \cdot \epsilon_i \in \bar{\mathfrak{o}}$.

In the sequel we shall often have to deal with partitions (2) of the set of indices $\{1, 2, \cdots, h\}$ into mutually disjoint non-empty subset $I_1, I_2, \cdots, I_r$. The following is an immediate consequence of (18) and the preceding Lemma 1.6:

COROLLARY 1.7. *Given a partition* (2) *of* $\{1, 2, \cdots, h\}$ *and setting* $\epsilon_\rho' = \sum_{i \in I_\rho} \epsilon_i$, *an element* η *of* F *belongs to* $\tilde{\mathfrak{o}}_x$ *if and only if the following two conditions are satisfied:*

(a) $\eta\epsilon_\rho' \in \widetilde{(\mathfrak{o}\epsilon_\rho')}_{K\epsilon_\rho'}$, $\rho = 1, 2, \cdots, r$.

(b) $v(\eta_i{}^\tau - \eta_j{}^\sigma) \geqq v(y_i{}^\tau - y_j{}^\sigma)$, *for all pairs of indices* i, j *which do not belong to one and the same set* I_ρ, *and for all* τ, σ *in* G.

2. A structure theorem for saturated rings. Beginning with this section it will be assumed (unless the contrary is explicitly stated) that our local ring $\mathfrak{o}$ is saturated with respect to the parameter x. Our main object in this and in the next section will be to derive a *structure theorem* which will be a generalization to saturated local ring of the structure theorem I, 4.1 proved in GTS I for saturated local *domains.*

We fix an x-saturator y of $\mathfrak{o}$ and we use the notations of the preceding section. We shall assume that $y \in \mathfrak{m}$. We have then $y_i \in \mathfrak{m}_i$, $i = 1, 2, \cdots, h$, and therefore $v(y_i{}^\tau) > 0$ for all τ in G $(= \mathcal{G}(F^*/K))$. It will be convenient to define the integers β_{i,α_i} [see § 1, (27) or (32)] also for $\alpha_i = 0$ and $\alpha_i > g_i$. We set

(1) $\beta_{i,0} = 0$;

(2) $\beta_{i,\alpha_i} = +\infty$ if $\alpha_i > g_i$.

In particular, the integers β_{i,α_i} are now defined also if $g_i = 0$, i.e., if $\mathfrak{o}_i = k[[x]]$.

We next define an integer γ as follows:

$$(3) \qquad \gamma = \min_{\substack{i \neq j \\ 1 \leqq i,j \leqq h}} \max_{\tau,\sigma \in G} [v(y_i^\tau - y_j^\sigma)].$$

Then γ is a positive integer, whence, in view of (1), we have

$$(4) \qquad \gamma > \min. \{\beta_{1,0}, \beta_{2,0}, \cdots, \beta_{h,0}\}.$$

We now define an integer q by the following propedty: q is the greatest integer such that

$$(5) \qquad \gamma > \min. \{\beta_{1,q-1}, \beta_{2,q-1}, \cdots, \beta_{h,q-1}\}.$$

In view of (4), q is a *positive* integer, and by (2) we have

$$q - 1 \leqq \max. \{g_1, g_2, \cdots, g_h\}.$$

We also note that from the definition of x-saturators it follows that *the integer γ, and hence also the integer q, is independent of the choice of the x-saturator y.*

Let $\{\psi_1, \psi_2, \cdots, \psi_h\}$ be a set of h K-homomorphisms of F into Ω, such that $\psi_i \in H_i$ for each i, i.e., such that $\psi_i(\epsilon_i) = 1$ and $\psi_i(\epsilon_j) = 0$ if $j \neq i$. Let $y_i = \psi_i(y)$, $i = 1, 2, \cdots, h$, and let C be the plane algebroid curve defined by the ring $k[[x]][y]$ (see § 1).

Definition. We say that the set $\{\psi_1, \psi_2, \cdots, \psi_h\}$ *is a normal set of homomorphisms for the pair* $\{\mathfrak{o}, x\}$ *or for the pair* $\{C, x\}$, *if we have*

$$(6) \qquad v(y_i - y_j) \geqq \gamma, \text{ for all } i, j = 1, 2, \cdots, h, i \neq j.$$

It is a simple matter to show the existence of normal sets of homomorphisms; in fact, we can easily show that any homomorphism ψ in H belongs to some normal set of homomorphisms. To see this, we start with any set of h homomorphisms $\psi_1, \psi_2, \cdots, \psi_h$ such that $\psi_i \in H_i$, and we show that, for instance, ψ_1 belongs to some normal set of homomorphisms. Let $y_i = \psi_i(y)$, as above. By (3) there exist pairs of elements (τ_i, σ_i) in the Galois group G of F^*/K, $i = 2, 3, \cdots, h$, such that $v(y_1^{\tau_i} - y_i^{\sigma_i}) \geqq \gamma$. If we then set $\sigma_i' = \tau_i^{-1} \cdot \sigma_i$, then we find that $v(y_1 - y_i^{\sigma_i'}) \geqq \gamma$, and hence also $v(y_i^{\sigma_i'} - y_j^{\sigma_j'}) \geqq \gamma$, for all $i \neq j$. Therefore, if we set $\bar{\psi}_1 = \psi_1$, $\bar{\psi}_i = \sigma_i' \cdot \psi_i$, $i = 2, 3, \cdots, h$, then it folows that $\{\bar{\psi}_1, \bar{\psi}_2, \cdots, \bar{\psi}_h\}$ is a normal set of homomorphisms.

We shall now prove the following theorem:

THEOREM 2.1.

(a) *We have* $\beta_{1,\alpha}=\beta_{2,\alpha}=\cdots=\beta_{h,\alpha}$ $(=\beta_\alpha$, *say*), *for* $\alpha=0,1,\cdots,q-1$. *Furthermore, if* $\{\psi_1,\psi_2,\cdots,\psi_h\}$ *is a normal set of K-homomorphisms for the pair* $\{C,x\}$, *and if we set* $y_i=\psi_i(y)$, *then the following is true:*

(b) *There exist standard decompositions of the* y_i *such that*

$$y_{1,\alpha}=y_{2,\alpha}=\cdots=y_{h,\alpha}\quad(=y^{(\alpha)},\ say),$$

for $\alpha=0,1,\cdots,q-2$;

(c) $K_{1,\alpha}=K_{2,\alpha}=\cdots=K_{h,\alpha}$ $(=K_\alpha$, *say*), *and hence* $e_{1,\alpha}=e_{2,\alpha}=\cdots=e_{h,\alpha}$ $(=e_\alpha$, *say*) *and* $m_{1,\alpha}=m_{2,\alpha}+\cdots=m_{h,\alpha}=m_\alpha=\beta_\alpha/e_\alpha$, *for* $\alpha=0,1,\cdots,q-1$.

We note immediately the following consequence:

COROLLARY 2.2. *If the set* $\{\psi_1,\psi_2,\cdots,\psi_h\}$ *is normal and if the standard decompositions of the* y_i *are so chosen as to satisfy condition* (b), *then*

$$(7)\qquad v(y_{i,q-1}-y_{j,q-1})\geqq\gamma,\ \ for\ all\ i,j=1,2,\cdots,h,i\neq j,$$

and hence, if we set $y^{(q-1)}=y_{1,q-1}$, *then we have*

$$(8)\qquad y_{i,q-1}=y^{(q-1)}+\bar{y}_{i,q-1},$$

where

$$(9)\qquad v(\bar{y}_{i,q-1})\geqq\gamma.$$

For, we have then $y_i-y_j=(y_{i,q-1}-y_{j,q-1})+(y_{i,q}-y_{j,q})+\cdots$, and since, for $\alpha\geqq q$, we have

$$v(y_{i,\alpha}-y_{j,\alpha})\geqq\min.\{\beta_{i,\alpha},\beta_{j,\alpha}\}\geqq\min\{\beta_{i,q},\beta_{j,q}\}$$

and $\min.\{\beta_{i,q},\beta_{j,q}\}\geqq\gamma$, by the definition of q, (7) follows from (6).

Proof of the theorem. The case $q=1$ is trivial, for in this case (a) and (c) are trivially true, in view of (1) and in view of $K_{i,0}=K$ [see § 1, (24)], while (b) is vacuous.

We shall now deal with the general case $q\geqq 2$. We observe that (a) and (c) are trivially true for $\alpha=0$, in view of (1), while (b) is vacuous for $\alpha=-1$. Furthermore, we have $v(y_{i,0}-y_{j,0})>\beta_0$ $(=\beta_{i,0})$, since $\beta_0=0$. We shall therefore proceed by induction on α. Namely, we shall assume that there exists an integer ρ, $0\leqq\rho<q-1$, with the property that (a) and (c) are true for all $\alpha\leqq\rho$, that (b) is true for all $\alpha=0,1,2,\cdots,\rho-1$, and that

we have furthermore $v(y_{i,\rho}-y_{j,\rho})>\beta_\rho$, for all $i\neq j$. (This last inequality already follows from the preceding induction hypothesis, in view of (6) and by the definition of the integers γ and q). We shall then prove that (a) and (c) are also true for $\alpha=\rho+1$, that (b) is true for $\alpha=0,1,2,\cdots,\rho$, and that we have

$$(10) \qquad v(y_{i,\rho+1}-y_{j,\rho+1})>\beta_{\rho+1}.$$

For the sake of clarity of exposition we first consider the case $\rho=0$. Let, say, $\beta_{1,1}=\min.\{\beta_{1,1},\beta_{2,1},\cdots,\beta_{h,1}\}$. Since $q\geqq 2$ we have

$$(11) \qquad \gamma>\beta_{1,1}.$$

Since the set $\{\psi_1,\psi_2,\cdots,\psi_h\}$ is normal, it follows from (11) and (6) that

$$(12) \qquad v(y_i-y_j)>\beta_{1,1}, \text{ for all } i,j=1,2,\cdots,h, i\neq j.$$

Now, we have for all $\alpha\geqq 1$:

$$v(y_{i,\alpha}-y_{j,\alpha})\geqq \min.\{\beta_{i,\alpha},\beta_{j,\alpha}\}\geqq \min.\{\beta_{i,1},\beta_{j,1}\}\geqq\beta_{1,1}.$$

Therefore we can conclude, by (12), that

$$(13) \qquad v(y_{i,0}-y_{j,0})\geqq\beta_{1,1}, \text{ for all } i,j=1,2,\cdots,h, i\neq j.$$

Now, we first prove (a) for $\alpha=1$ $(=\rho+1)$, i.e., we prove that $\beta_{1,1}=\beta_{2,1}=\cdots=\beta_{h,1}$. Assume the contrary, and let, say, $\beta_{i,1}>\beta_{1,1}$ for some i. Omitting in the difference y_1-y_i the terms $y_{1,\alpha}$ $(\alpha>1)$ and $y_{i,\alpha}$ $(\alpha>0)$ which have definitely value greater than $\beta_{1,1}$, we find by (12), for $j=1$:

$$(14) \qquad v(y_{1,1}+y_{1,0}-y_{i,0})>\beta_{1,1}.$$

If we now set $\zeta_0=y_{1,0}-y_{i,0}$ then $\zeta_0\in K$ and it follows from Lemma 1.5, in view of (13), that the following decomposition of y_1:

$$y_1=y'_{1,0}+y'_{1,1}+y_{1,2}+\cdots,$$

where $y'_{1,0}=y_{1,0}-\zeta_0$ $(=y_{i,0})$ and $y'_{1,1}=y_{1,1}+\zeta_0$, is still standard. Hence we must have $v(y'_{1,1})=\beta_{1,1}$, in contradiction with (14) (since $y'_{1,1}=y_{1,1}+y_{1,0}-y_{i,0}$). This proves the equalities $\beta_{1,1}=\beta_{2,1}=\cdots=\beta_{h,1}$ $(=\beta_1$, say).

We next prove (b) for $\alpha=0$ $(=\rho)$. More precisely, we have to prove that the standard decompositions of the y_i can be so chosen as to have $y_{1,0}=y_{2,0}=\cdots=y_{h,0}$. This, however, follows immediately from (13), and from $\beta_{i,1}=\beta_1$ for all i, for by Lemma 1.5 we can replace $y_{i,1}$ by

$y'_{i,1} = y_{i,1} + y_{i,0} - y_{1,0}$ and $y_{i,0}$ by $y'_{i,0} = y_{i,0} - (y_{i,0} - y_{1,0}) = y_{1,0}$, and $y'_{i,0}$ is now independent of i.

Finally we prove (c) for $\alpha = 1$ $(= \rho + 1)$. Since $\beta_{i,1} = \beta_1$, the (12) take the form: $v(y_i - y_j) > \beta_1$. Since we have already arranged to have $y_{1,0} = y_{2,0} = \cdots = y_{h,0}$, and since $\beta_{i,\alpha} > \beta_{i,1} = \beta_1$ for $\alpha > 1$, it follows, from the above new form of (12), that

$$(15) \qquad v(y_{i,1} - y_{j,1}) > \beta_1, \quad i, j = 1, 2, \cdots, h,$$

Let τ be any element of the Galois group $G_{j,1} = \mathfrak{G}(F^*/K_{j,1})$. Since $y_{j,1} \in K_{j,1}$ it follows from (15) that $v(y_{i,1}{}^\tau - y_{j,1}) > \beta_1$, and hence, again using (15), we have $v(y_{i,1}{}^\tau - y_{i,1}) > \beta_1$ $(= \beta_{i,1})$. The property of standard decompositions expressed by (35), § 1, in the case $\alpha_i = 1$, tells us therefore that $G_{j,1} \subset G_{i,1}$. Similarly it follows that $G_{i,1} \subset G_{j,1}$ and hence $G_{i,1} = G_{j,1}$ and $K_{i,1} = K_{j,1}$.

We now deal with the general case in which ρ is positive. Let, say, $\beta_{1,\rho+1} = \min. \{\beta_{1,\rho+1}, \beta_{2,\rho+1}, \cdots, \beta_{h,\rho+1}\}$. Since $\rho + 1 \leqq q - 1$, it follows from the definition of the integer q that

$$(16) \qquad \gamma > \beta_{1,\rho+1}.$$

Since the set $\{\psi_1, \psi_2, \cdots, \psi_h\}$ is normal, it follows from (16) and (6) that

$$(17) \qquad v(y_i - y_j) \geqq \beta_{1,\rho+1}.$$

By our induction hypothesis we have standard decompositions

$$(18) \qquad y_i = y^{(0)} + y^{(1)} + \cdots + y^{(\rho-1)} + y_{i,\rho} + y_{i,\rho+1} + \cdots,$$

where $y^{(\alpha)} = y_{1,\alpha} = y_{2,\alpha} = \cdots = y_{h,\alpha} \in K_\alpha = K_{1,\alpha} = K_{2,\alpha} = \cdots = K_{h,\alpha}$, for $\alpha = 0, 1, 2, \cdots, \rho - 1$. By our induction hypothesis we also have $\beta_{1,\alpha} = \beta_{2,\alpha} = \cdots = \beta_{h,\alpha} = \beta_\alpha$ for $\alpha = 1, 2, \cdots, \rho$, and $K_{1,\rho} = K_{2,\rho} = \cdots = K_{h,\rho}$ $(= K_\rho$, say). By (18) and (17) we have

$$(19) \qquad v[y_{i,\rho} - y_{j,\rho}) + (y_{i,\rho+1} - y_{j,\rho+1}) + \cdots] > \beta_{1,\rho+1}.$$

Now, we have for all $\alpha \geqq \rho + 1$:

$$v(y_{i,\alpha} - y_{j,\alpha}) \geqq \min. \{\beta_{i,\alpha}, \beta_{j,\alpha}\} \geqq \min. \{\beta_{i,\rho+1}, \beta_{j,\rho+1}\} \geqq \beta_{1,\rho+1}.$$

Therefore we conclude, by (19), that

$$(20) \qquad v(y_{i,\rho} - y_{j,\rho}) \geqq \beta_{1,\rho+1}, \text{ for all } i, j = 1, 2, \cdots, h.$$

We now prove part (a) of the theorem for $\alpha = \rho + 1$, i.e., we prove that

$$(21) \qquad \beta_{1,\rho+1} = {}_{2,\rho+1} = \cdots = \beta_{h,\rho+1}.$$

3

Assume the contrary, and let, say, $\beta_{i,\rho+1} > \beta_{1,\rho+1}$ for some i. We consider (19) for that particular index i and for $j=1$. If we delete those terms on the left-hand side of (19) which definitely have value $> \beta_{1,\rho+1}$ in the valuation v, we obtain the inequality

$$(22) \qquad v(y_{1,\rho+1} + y_{1,\rho} - y_{i,\rho}) > \beta_{1,\rho+1}.$$

If we now set $\zeta_\rho = y_{1,\rho} - y_{i,\rho}$, then $\zeta_\rho \in K_\rho$, and, by (20), we have $v(\zeta_\rho) \geqq \beta_{1,\rho+1}$. Therefore, by Lemma 1.5 (for $i=1$ and $\alpha=\rho$), the following decomposition of y_1:

$$y_1 = y^{(0)} + y^{(1)} + \cdots + y^{(\rho-1)} + y'_{1,\rho} + y'_{1,\rho+1} + y_{1,\rho+2} + \cdots,$$

where $y'_{1,\rho} = y_{1,\rho} - \zeta_\rho$ $(= y_{i,\rho})$ and

$$y'_{1,\rho+1} = y_{1,\rho+1} + \zeta_\rho \quad (= y_{1,\rho+1} + y_{1,\rho} - y_{i,\rho}),$$

is still standard. Hence we must have $v(y'_{1,\rho+1}) = \beta_{1,\rho+1}$ in contradiction with (22). This proves the equalities (21). We denote by $\beta_{\rho+1}$ the common value of the $\beta_{i,\rho+1}$:

$$(23) \qquad \beta_{\rho+1} = \beta_{1,\rho+1} = \beta_{2,\rho+1} = \cdots = \beta_{h,\rho+1}.$$

We next prove part (b) of the theorem for $\alpha = 0, 1, 2, \cdots, \rho$, i.e., we prove that we can find a standard decomposition of each y_i in such a way as to have, in addition to the equalities $y_{1,\alpha} = y_{2,\alpha} = \cdots = y_{h,\alpha}$ $(= y^{(\alpha)})$, for $\alpha = 0, 1, \cdots, \rho - 1$ [see (18)], also the equalities

$$y_{1,\rho} = y_{2,\rho} = \cdots = y_{h,\rho}.$$

To see this, we observe that, by (20), we have $v(y_{1,\rho} - y_{i,\rho}) \geqq \beta_{\rho+1}$, where $\beta_{\rho+1}$ was defined in (23). This being so, it follows that if we set $y'_{i,\rho} = y_{1,\rho}$ and $y'_{i,\rho+1} = y_{i,\rho+1} + y_{i,\rho} - y_{1,\rho}$, then, for each i, the decomposition

$$y_i = y^{(0)} + y^{(1)} + \cdots + y^{(\rho-1)} + y'_{i,\rho} + y'_{i,\rho+1} + y_{i,\rho+2}, \cdots$$

is standard and that $y'_{1,\rho} = y'_{2,\rho} = \cdots = y'_{h,\rho}$ $(= y_{1,\rho})$. This completes the proof of part (b) of the theorem.

We set $y^{(\rho)} = y_{1,\rho}$, whence

$$(24) \quad y_i = y^{(0)} + y^{(1)} + \cdots + y^{(\rho)} + y_{i,\rho+1} + \cdots, i = 1, 2, \cdots, h.$$

We next prove part (c) of the theorem for $\alpha = \rho + 1$, i.e., we show that $K_{1,\rho+1} = K_{2,\rho+1} = \cdots = K_{h,\rho+1}$. We shall use the inequalities (17). In view of (24) and (23), these inequalities imply that

$$(25) \qquad v(y_{i,\rho+1} - y_{j,\rho+1}) > \beta_{\rho+1},$$

which—incidentally—establishes (10). Let τ be any element of $G_{j,\rho+1}$ $(=\mathfrak{G}(F^*/K_{j,\rho+1})$. Then we obtain from (25) that $v(y_{i,\rho+1}{}^{\tau}-y_{j,\rho+1})>\beta_{\rho+1}$, and that consequently, again by (25), $v(y_{i,\rho+1}{}^{\tau}-y_{i,\rho+1})>\beta_{\rho+1}$. By the property (35), § 1, of standard decompositions we conclude from this that $\tau\in G_{i,\rho+1}$. We have therefore shown that $G_{j,\rho+1}\subset G_{i,\rho+1}$. In a similar way the opposite inclusion follows, and hence $G_{j,\rho+1}=G_{i,\rho+1}$ and $K_{j,\rho+1}=K_{i,\rho+1}$, as asserted. We set

$$K_{\rho+1}=K_{1,\rho+1}=K_{2,\rho+1},\cdots,K_{h,\rho+1}.$$

This completes the proof of the theorem.

3. The structure theorem (continuation) and some consequences. We define a partition of the set of indices $\{1,2,\cdots,h\}$ into disjoint non-empty subsets $I_1,I_2,\cdots,I_r$ as follows: two indices i, j belong to the same set I_ρ if there exist automorphisms σ, τ in the Galois group G of F^*/K such that $v(y_i^\sigma-y_j^\tau)>\gamma$, where γ is the integer defined in (3), § 2. Since the relation between pair of indices i, j defined by this condition is obviously an equivalence, the condition defines indeed a partition of $\{1,2,\cdots,h\}$ into disjoint non-empty subsets. By the definition of γ, there exists a pair of distinct indices i, j such that $\max_{\sigma,\tau} v(y_i^\sigma-y_j^\tau)=\gamma$. Therefore the number r of sets I_ρ is $\geqq 2$. We call $\{I_1,I_2,\cdots,I_r\}$ *the saturation partition of* $\{1,2,\cdots,h\}$ *relative to the pair* $\{\mathfrak{o},x\}$ ($\mathfrak{o}$ being saturated with respect to x).

In accordance with Theorem 2.1 we assume that the h homomorphisms $\psi_1,\psi_2,\cdots,\psi_h$ of F into Ω form a normal set and that consequently conditions (b), (c) of that theorem are satisfied. We will have then (using also Corollary 2.2):

$$(1)\qquad y_i=y^{(0)}+y^{(1)}+\cdots+y^{(q-2)}+y^{(q-1)}+\bar{y}_{i,q-1}+y_{i,q}+\cdots+y_{i,g_i},$$

where the $\bar{y}_{i,q-1}$ satisfy (9), § 2. Note that since $\gamma>\beta_{q-1}$ and $v(y_{i,q-1})=\beta_{q-1}$, it follows from (8) and (9) of § 2 that $v(y^{(q-1)})=\beta_{q-1}$. Thus, by parts (a) and (b) of Theorem 2.1 we have

$$(2)\qquad v(y^{(\alpha)})=\beta_\alpha,\quad \alpha=0,\ 1,2,\cdots,q-1,$$

while

$$(3)\qquad v(\bar{y}_{i,q-1})\geqq\gamma>\beta_{q-1},\quad i=1,2,\cdots,h.$$

Note also that if $q=1$, then (1) becomes

$$(4)\qquad y_i=y^{(0)}+\bar{y}_{i,0}+y_{i,1}+\cdots$$

and $\beta_0=0$.

Given any element η of F we set, as usual, $\psi_i(\eta)=\eta_i$, $i=1,2,\cdots,h$. If $\eta\in\mathfrak{m}$, then we know [see (41), § 1] that each η_i must admit a decomposition of the form

$$(5)\qquad \eta_i=\eta_{i,0}+\eta_{i,1}+\cdots+\eta_{i,g_i},$$

where

$$(6)\qquad \eta_{i,\alpha_i}\in\mathfrak{M}_{i,\alpha_i}{}^{m_{i,\alpha_i}},\quad i=1,2,\cdots,h;\ \alpha_i=1,2,\cdots,g_i;$$

$$(7)\qquad \eta_{i,0}\in\mathfrak{M}_{i,0}=\mathfrak{M},$$

or equivalently:

$$(6')\qquad \eta_{i,\alpha_i}\in K_{i,\alpha_i}\text{ and } v(\eta_{i,\alpha_i})\geqq\beta_{i,\alpha_i},\quad \alpha_i=1,2,\cdots,g_i;$$

$$(7')\qquad \eta_{i,0}\in K\text{ and } v(\eta_{i,0})>0.$$

In view of Theorem 2.1, we may re-write the (6) and (7), for $\alpha_i=1,2,\cdots,q-1$, as follows:

$$(8)\qquad \eta_{i,\alpha}\in\mathfrak{M}_\alpha{}^{m_\alpha},\quad \alpha=0,1,\cdots,q-1,$$

where we have set $\mathfrak{M}_0=\mathfrak{M}$, $m_0=1$. We shall now use the notations of Corollary 1.7, where—we recall—ϵ'_ρ stands for $\sum_{i\in I_\rho}\epsilon_i$.

THEOREM 3.1. (*The structure theorem: first formulation*). *We have* $\eta\in\mathfrak{m}$ *if and only if the* η_i *admit decompositions of the form* (5), *satisfying conditions* (8) (*for* $\alpha=0,1,\cdots,q-1$) *and* (6) (*for* $\alpha_i=q,q+1,\cdots,g_i$), *and satisfying furthermore the following conditions:*

(a) $\eta_{1,\alpha}=\eta_{2,\alpha}=\cdots,\eta_{h,\alpha}$ ($=\eta^{(\alpha)}$, *say*), $\alpha=0,1,\cdots,q-2$;

(b) $v(\eta_{i,q-1}-\eta_{j,q-1})\geqq\gamma$, *for all* $i,j=1,2,\cdots,h$;

or—equivalently:

there exists an element $\eta^{(q-1)}$ *in* K_{q-1} *such that if we set*

$$(9)\qquad \bar\eta_{i,q-1}=\eta_{i,q-1}-\eta^{(q-1)}$$

then

$$(10)\qquad v(\bar\eta_{i,q-1})\geqq\gamma,\quad i=1,2,\cdots,h;$$

(c) $\eta\epsilon_\rho'\in\overline{(\mathfrak{o}\epsilon_\rho')_{K\epsilon_\rho'}}$, $\rho=1,2,\cdots,r$.

Note. It is clear that if (b) is satisfied, then we can take for $\eta^{(q-1)}$ any of the elements $\eta_{i,q-1}$, say $\eta^{(q-1)}=\eta_{1,q-1}$. Conversely, it is clear that if

(9) and (10) are satisfied for some element $\eta^{(q-1)}$ in K_{q-1}, then the inequalities (b) are satisfied. Moreover, since, by (6′), we have $v(\eta_{i,q-1}) \geqq \beta_{q-1}$, while $\gamma > \beta_{q-1}$, it follows from (9) and (10 that $v(\eta^{(q-1)}) \geqq \beta_{q-1}$. Thus the structure theorem says that $\eta \in \mathfrak{m}$ *if and only if the following is true*: 1) *the* η_i *have decompositions of the form*

$$(11) \qquad \eta_i = \eta^{(0)} + \eta^{(1)} + \cdots + \eta^{(q-1)} + \bar{\eta}_{i,q-1} + \eta_{i,q} + \cdots + \eta_{i,g_i},$$

where

$$(12) \qquad \eta^{(\alpha)} \in K_\alpha \text{ and } v(\eta^{(\alpha)}) \geqq \beta_\alpha, \quad \alpha = 0, 1, \cdots, q-1,$$

$$(13) \qquad \bar{\eta}_{i,q-1} \in K_{q-1},$$

and (10) *and the* (6′) (*for* $\alpha_i = q, q+1, \cdots, g_i$) *hold*; 2) *condition* (c) *of the theorem is satisfied for* $\rho = 1, 2, \cdots, r$.

Proof. Condition (c) is merely a re-statement of condition (a) of Corollary 1.7. We also know that conditions (5), (6) and (7) are necessary. We shall begin now with the proof of the necessity of conditions (a) and (b) of the theorem, and we first consider the case $q = 1$. In this case, condition (a) is vacuous. If i and j are distinct indices in the set $\{1, 2, \cdots, h\}$, then we have, by (17), §1, $v(\eta_i - \eta_j) \geqq v(y_i - y_j)$. Since the homomorphisms $\psi_1, \psi_2, \cdots, \psi_h$ form a normal set, we have the inequalities $v(y_i - y_j) \geqq \gamma$. Thus, $v(\eta_i - \eta_j) \geqq \gamma$, and this implies, in view of (5), (6′) and (7′), that $v(\eta_{i,0} - \eta_{j,0}) \geqq \gamma$. Therefore condition (b) is satisfied.

We now assume that $q \geqq 2$. We have $v(y_i - y_j) \geqq \gamma > \beta_{q-1} \geqq \beta_1$. Hence $v(\eta_i - \eta_j) > \beta_1$, and this implies, as in the above case $q = 1$, that $v(\eta_{i,0} - \eta_{j,0}) \geqq \beta_1$. We set $\eta^0 = \eta_{1,0}$ and $\eta'_{i,1} = \eta_{i,1} + \eta_{i,0} - \eta_{1,0}$. Then the new decompositions $\eta_i = \eta^0 + \eta'_{i,1} + \eta_{i,2} + \cdots$ of the η_i still satisfy (6) and (7), while condition (a) is satisfied for this new decomposition in the case $\alpha = 0$. If $q = 2$, this means that ((a) is satisfied. Assume $q > 2$ and assume, by induction, that we have already found decompositions (5) of the η_i satisfying (6) and (7) and such that (a) is satisfied for $\alpha = 0, 1, \cdots, a$, where a is some integer such that $0 \leqq a < q-2$. We set $\eta^{(\alpha)} = \eta_{i,\alpha}$, for $\alpha = 0, 1, \cdots, a$. Thus,

$$\eta_i = \eta^{(0)} + \eta^{(1)} + \cdots + \eta^{(a)} + \eta_{i,a+1} + \cdots.$$

Since $a + 2 \leqq q - 1$, we have $\gamma > \beta_{a+2}$, and $v(\eta_i - \eta_j) > \beta_{a+2}$ since $v(y_i - y_j) \geqq \gamma$. From this it follows that $v(\eta_{i,a+1} - \eta_{j,a+1}) \geqq \beta_{a+2}$. If we now set $\eta^{(a+1)} = \eta_{1,a+1}$ and $\eta'_{i,a+2} = \eta_{i,a+2} + \eta_{i,a+1} - \eta_{1,a+1}$, then the following new decompositions of the η_i:

$$\eta_i = \eta^{(0)} + \eta^{(1)} + \cdots + \eta^{(a)} + \eta^{(a+1)} + \eta'_{i,a+2} + \eta_{i,a+3} + \cdots$$

still satisfy the conditions analogous to (6) and (7), but these new decompositions satisfy condition (a) for $\alpha = 0, 1, \cdots, a+1$. This establishes the necessity of condition (a). We can therefore write

$$\eta_i = \eta^{(0)} + \eta^{(1)} + \cdots + \eta^{(q-2)} + \eta_{i,q-1} + \eta_{i,q} + \cdots,$$

and we have $\eta_i - \eta_j = (\eta_{i,q-1} - _{j,q-1}) + (\eta_{i,q} - \eta_{j,q}) + \cdots$. Since

$$v(\eta_i - \eta_j) \geqq v(y_i - y_j) \geqq \gamma \text{ and } \gamma \leqq \min. \{\beta_{1,q}, \beta_{2,q}, \cdots, \beta_{h,q}\},$$

it follows that $v(\eta_{i,q-1} - \eta_{j,q-1}) \geqq \gamma$. Thus, condition (b) is satisfied.

We now proceed to the proof of sufficiency of the conditions stated in the theorem. Since condition (c) of the theorem is identical with condition (a) of Corollary 1.7, it follows from that Corollary that we have only to show that condition (b) of Corollary 1.7 is satisfied. It will be sufficient to prove that condition (b) of Corollary 1.7 is satisfied for $\sigma = 1$, i.e., that we have $v(\eta_i^\tau - \eta_j) \geqq v(y_i^\tau - y_j)$ for all τ in G and for all pairs of indices i, j which does not belong to one and the same set I_ρ. We shall consider three cases, according as 1) τ does not belong to G_{q-2}, 2) τ belongs to G_{q-2} but does not belong to G_{q-1}, and 3) $\tau \in G_{q-1}$. Here—we recall—for $\alpha = 0, 1, 2, \cdots, q-1$, G_α denoted the Galois group of F^*/K_α, where

$$K_\alpha = K_{1,\alpha} = K_{2,\alpha} = \cdots = K_{h,\alpha}.$$

Case 1. $\tau \notin G_{q-2}$. Since $\tau \in G_0$, let α be the integer such that $1 \leqq \alpha \leqq q-2$ and $\tau \in G_{\alpha-1}$, $\tau \notin G_\alpha$. We have

$$y_i^\tau - y_j = [y^{(\alpha)\tau} - y^{(\alpha)}] + [y_{i,\alpha+1}{}^\tau - y_{j,\alpha+1}] + \cdots,$$

and since $\tau \notin G_\alpha$, we have $v[y^{(\alpha)\tau} - y^{(\alpha)}] = \beta_\alpha$ [see (35), § 1]; it follows, therefore, by (27) and (31), § 1, that $v(y_i^\tau - y_j) = \beta_\alpha$. On the other hand, we have $\eta_i^\tau - \eta_j = [\eta^{(\alpha)\tau} - \eta^{(\alpha)}] + [\eta_{i,\alpha+1}{}^\tau - \eta_{j,\alpha+1}] + \cdots$, and since $v(\eta^{(\alpha)}) \geqq \beta_\alpha$, it follows by (6′) that $v(\eta_i^\tau - \eta_j) \geqq v(y_i^\tau - y_j)$.

We note that in this case no use has been made of the assumption that i and j belong to distant sets I_ρ.

Case 2. $\tau \in G_{q-2}$, $\tau \notin G_{q-1}$. In this case, we have

$$y_i^\tau - y_j = (y_{i,q-1}{}^\tau - y_{j,q-1}) + (y_{i,q}{}^\tau - y_{j,q}) + \cdots.$$

Now,

$$y_{i,q-1}{}^\tau - y_{j,q-1} = (y_{i,q-1}{}^\tau - y_{i,q-1}) + (y_{i,q-1} - y_{j,q-1}),$$

and since $\tau \notin G_{q-1}$, the v-value of the expression in the first parenthesis is equal exactly to β_{q-1}. As to the expression in the second parenthesis, (7) of § 2 implies that its v-value is greater than β_{q-1} (since $\gamma > \beta_{q-1}$). Hence $v(y_i^\tau - y_j) = \beta_{q-1}$ (since $v(y_{i,\alpha})$ and $v(y_{j,\alpha})$ are greater than β_{q-1}, if $\alpha \geqq q$). On the other hand, we have $\eta_i^\tau - \eta_j = (\eta_{i,q-1}{}^\tau - \eta_{j,q-1}) + (\eta_{i,q}{}^\tau - \eta_{j,q}) + \cdots$, and this impies that $v(\eta_i^\tau - \eta_j) \geqq \beta_{q-1}$ [see (6′)]. Thus we find again $v(\eta_i^\tau - \eta_j) \geqq v(y_i^\tau - y_j)$.

We note that also in this case no use has been made of the assumption that i and j belong to distinct sets I_ρ.

Case 3. $\tau \in G_{q-1}$. In this case we have

$$\eta_i^\tau - \eta_j = (\eta_{i,q-1} - \eta_{j,q-1}) + (\eta_{i,q}{}^\tau - \eta_{j,q}{}^\tau) + \cdots.$$

Since $\gamma \leqq \min. \{\beta_{1,q}, \beta_{2,q}, \cdots, \beta_{h,q}\}$, the expressions in each parenthesis on the right hand side, except the first, have v-value $\geqq \gamma$. By condition (b) of the theorem, we have also $v(\eta_{i,q-1} - \eta_{j,q-1}) \geqq \gamma$. On the other hand, *since i, j belong to distinct sets of our partition* $\{I_1, I_2, \cdots, I_r\}$ we have $v(y_i^\tau - y_j) \leqq \gamma$, for *all* $\tau \in G$. This completes the proof of sufficiency.

We shall now add a proposition the proof of which is practically contained in the considerations just developed in the second (sufficiency) part of the above proof and which we shall have occasion to use shortly. We first observe that we know from § 1 that given any set of h elements $\eta_1, \eta_2, \cdots, \eta_h$ in $F_1, F_2, \cdots, F_h$ respectively, there exists one and only one element η of F such that the equations (14) of § 1, i.e., the relations $\psi_i(\eta) = \eta_i$, $i = 1, 2, \cdots, h$, hold true. Now, for $\alpha = 0, 1, \cdots, q-1$, the field K_α ($= K_{i,\alpha}$, $i = 1, 2, \cdots, h$; see Theorem 2.1, part (c)) is contained in each of the fields $F_1, F_2, \cdots, F_h$. Therefore, applying the above remark to the case in which $\eta_1 = \eta_2 = \cdots = \eta_h = \eta^{(\alpha)} \in K_\alpha$ $(0 \leqq \alpha \leqq q-1)$, we see that there exists a unique element η of F such that $\psi_i(\eta) = \eta^{(\alpha)}$ for $i = 1, 2, \cdots, h$. This defines, therefore, a bijection θ_α of the field K_α into the ring F $(\alpha = 0, 1, \cdots, q-1)$, and we have

$$(14) \qquad \theta_\alpha(\eta^{(\alpha)}) = \eta \Rightarrow \psi_i(\eta) = \eta^{(\alpha)}, \text{ for } i = 1, 2, \cdots, h.$$

Since the ψ_i are K-homomorphisms, it follows that if we set $L_\alpha = \operatorname{Im} \theta_\alpha$, then L_α *is a field* ($L_0 = K$), *containing K, and θ_α is a K-isomorphism of K_α onto L_α* (θ_0 is the identity map of K). Since $\theta_\alpha^{-1} = \psi_i \mid L_\alpha$ $(i = 1, 2, \cdots, h)$ *and* $K_0 < K_1 < \cdots < K_{q-1}$, it follows that

$$(15) \qquad K = L_0 < L_1 < \cdots < L_{q-1} \subset F$$

and that

$$(16) \qquad \theta_\alpha^{-1} = \theta_{q-1}^{-1} \mid L_\alpha, \quad \alpha = 0, 1, \cdots, q-1.$$

We shall show later on that the fields L_α depend only on $\mathfrak{o}$ and L_0, being independent of the choice of the *normal* set of h homomorphisms ψ_i.

Each of the fields L_α is a complete local field, i.e., a power series field, and we denote by $\mathfrak{N}_\alpha$ the maximal ideal of the power series ring (discrete valuation ring) of which L_α is the field of fractions. We have therefore $\mathfrak{N}_\alpha = \theta_\alpha(\mathfrak{M}_\alpha)$.

PROPOSITION 3.2. *The integers $m_1, m_2, \cdots, m_{q-1}$ being those introduced in part* (c) *of Theorem* 2.1, *the saturated ring* $\mathfrak{o}$ *contains the sets* $\mathfrak{N}_0, \mathfrak{N}_1^{m_1}, \cdots, \mathfrak{N}_{q-1}^{m_{q-1}}$ *(and hence* $\mathfrak{o}$ *contains the saturated local domain* $k + \mathfrak{N}_0 + \mathfrak{N}_1^{m_1} + \cdots + \mathfrak{N}_{q-1}^{m_{q-1}}$, *which has L_{q-1} as field of fractions).*

Proof. Let η be any element of $\mathfrak{N}_0 + \mathfrak{N}_1^{m_1} + \cdots + \mathfrak{N}_{q-1}^{m_{q-1}}$. Then $\psi_i(\eta)$ is independent of i $(i = 1, 2, \cdots, h)$, namely

$$\psi_i(\eta) = \eta' = \eta^{(0)} + \eta^{(1)} + \cdots + \eta^{(q-1)},$$

where $\eta^{(0)} \in \mathfrak{M}_0$ and $\eta^{(\alpha)} \in \mathfrak{M}_\alpha^{m_\alpha}$, $\alpha = 1, 2, \cdots, h$. We have to show that $v(\eta'^\tau - \eta') \geqq v(y_i^\tau - y_j)$ for all pairs of indices i, j $(= 1, 2, \cdots, h)$ and for all τ in G. The inclusion $\eta^{(\alpha)} \in M_\alpha^{m_\alpha}$ implies (and is equivalent to) the inequality $v(\eta^{(\alpha)}) \geqq \beta_\alpha$, and hence if $\tau \notin G_{q-1}$ the considerations developed in the cases 1 and 2 of the sufficiency proof of Theorem 3.1 are applicable, and the inequality $v(\eta'^\tau - \eta') \geqq v(y_i^\tau - y_j)$ is valid if $\tau \notin G_{q-1}$. If, however, $\tau \in G_{q-1}$ then $\eta'^\tau = \eta'$ and $v(\eta'^\tau - \eta') = +\infty$. This completes the proof of the proposition.

We know that the integers m_{i,α_i} and q are uniquely determined by $\mathfrak{o}$ and x, and hence also the integers $m_1, m_2, \cdots, m_{q-1}$ are uniquely determined by $\mathfrak{o}$ and x. We shall now prove the following:

PROPOSITION 3.3. *The field $L_1, L_2, \cdots, L_{q-1}$ are also uniquely determined by* $\mathfrak{o}$ *and x (and therefore also the saturated local domain*

$$k + \mathfrak{N}_0 + \mathfrak{N}_1^{m_1} + \cdots + \mathfrak{N}_{q-1}^{m_{q-1}}$$

is uniquely determined by $\mathfrak{o}$ *and x).*

Proof. Let $\psi_1', \psi_2', \cdots, \psi_h'$ be another normal set of h homomorphisms. Let K'_{i,α_i} be the fields analogous to the fields K_{i,α_i} but associated with the new homomorphisms ψ_i'. We have $\psi_i' = \tau_i \cdot \psi_i$, where τ_i is a K-isomorphism $F_i \xrightarrow{\sim} F_i'$ of F_i into F_i^*, and $K'_{i,\alpha_i} = K_{i,\alpha_i}{}^{\tau_i}$. In particular, we have then

$K'_{1,\alpha}=K'_{2,\alpha}=\cdots=K'_{h,\alpha}=K'_{\alpha}=K_{\alpha}^{\tau_i}$, for $\alpha=0,1,2,\cdots,q-1$ and $i=1,2,\cdots,h$. *We assert that the isomorphism*

$$\tau_i \mid K_\alpha\colon K_\alpha \xrightarrow{\sim} K'_\alpha$$

is independent of i. This is trivial for $\alpha=0$, since $K_0=K_0'=K$ and since the τ_i are K-isomorphisms. Assume the assertion is true for $\alpha\leqq a$, where a is an integer such that $0\leqq a<q-1$. We have $K_{a+1}=K_a(y_{i,a})$, $K'_{a+1}=K'_a(y'_{i,a})$ [see (36), § 1] and $\psi'_i(y)=y'_i$. This implies $\tau_i(y_i)=y'_i$ and $\tau_i(y_{i,a})=y'_{i,a}$ (the $y_{i,\alpha}$ having been obtained from y_i by a canonical process of successive trace formations: similarly for the $y'_{i,\alpha}$ and the y'_i). Since by our induction hypothesis we have that $\tau_i \mid K_\alpha$ is independent of i, our assertion is proved. Let us then denote the isomorphism $\tau_i \mid K_\alpha$ (which depends only on α) by $\tau^{(\alpha)}$ $(\alpha=0,1,\cdots,q-1)$, and let $\eta^{(\alpha)}$, $\eta'^{(\alpha)}$ be corresponding elements of K_α and K'_α: $\eta'^{(\alpha)}=\tau^{(\alpha)}(\eta^{(\alpha)})$. If we set $\eta=\theta_\alpha(\eta^{(\alpha)})$ (whence $\eta\in L_\alpha$), we find:

$$\psi'_i(\eta)=(\tau_i\cdot\psi_i)(\eta)=\tau_i(\eta^{(\alpha)})=\tau^{(\alpha)}(\eta^{(\alpha)})=\eta'^{(\alpha)},\ i=1,2,\cdots,h.$$

This shows that if θ'_α denote the K-isomorphism $K'_\alpha \xrightarrow{\sim} L'_\alpha$ analogous to θ_α and associated with the ψ'_i, then

$$\theta'_\alpha=\theta_\alpha\cdot\tau^{(\alpha)-1},\qquad L'_\alpha=L_\alpha,$$

which completes the proof of the proposition.

We shall now give a second formulation of the structure theorem, which bypasses the homomorphisms $\psi_1,\psi_2,\cdots,\psi_h$ and describes directly and intrinsically the structure of the saturated ring $\mathfrak{o}$, *in terms of the* r *saturated rings* $(\widetilde{\mathfrak{o}\epsilon_\rho'})_{K\epsilon_\rho'}$. Since the number of the direct field summands of $F\epsilon_\rho'$ is *less* than h (being equal to the number of indices in the set I_ρ), this formulation of the structure theorem is inductive in nature, the induction being with respect to h [the same is essentially true also of our first formulation of the structure theorem (Theorem 3.1); see part (c) of that theorem].

We denote by $\mathfrak{o}_\rho'$ the saturated ring $(\widetilde{\mathfrak{o}\epsilon_\rho'})_{K\epsilon_\rho'}$:

(17) $$\mathfrak{o}_\rho'=(\widetilde{\mathfrak{o}\epsilon_\rho'})_{K\epsilon_\rho'},\quad \rho=1,2,\cdots,r,$$

and by $\mathfrak{A}_\rho'$ the following ideal of $\mathfrak{o}_\rho'$:

(18) $$\mathfrak{A}_\rho'=\{\xi'\in\mathfrak{o}_\rho' \mid v(\xi_i')\geqq\gamma,\text{ for all } i \text{ in } I_\rho\}.$$

We note that if $i\notin I_\rho$ then $\xi_i'=0$. It is immediately seen that $\mathfrak{A}_\rho'$ is indeed an ideal in $\mathfrak{o}_\rho'$. It is obvious that $\mathfrak{A}_\rho'$ contains some power of the maximal ideal of $\mathfrak{o}_\rho'$, and hence the local rings $\mathfrak{o}_\rho'$ and $k + \mathfrak{A}_\rho'$ have the same total ring of quotents, namely $F\epsilon_\rho'$. By Proposition 1.4, we have $\mathfrak{o}_\rho' = \widetilde{k\epsilon_\rho'[[x\epsilon_\rho']][y\epsilon_\rho']}_{K\epsilon\rho'}$. Since $\psi_i(y\epsilon_\rho') = y_i$ for $i \in I_\rho$, it follows that

$$\mathfrak{o}_i = \psi_i(\mathfrak{o}_\rho'),\ \text{if } i \in I_\rho, \tag{19}$$

where $\mathfrak{o}_i$ was defined in (22), § 1. Using the standard decomposition of $\mathfrak{o}_i$:

$$\mathfrak{o}_i = k + \mathfrak{M}_{i,0} + \mathfrak{M}_{i,1}{}^{m_{i,1}} + \cdots + \mathfrak{M}_{i,q-1}{}^{m_{i,q-1}} + \cdots,$$

and using the relation (a) of Theorem 2.1, the definition of the integer q [(5), § 2] and the inequality $\gamma > \beta_{q-1}$, we see at once that the ideal $\mathfrak{A}_\rho'$ can also be described as follows:

$$\mathfrak{A}_\rho' = \{\xi' \in \mathfrak{o}_\rho' \mid \xi_i' \in \bar{\mathfrak{A}}_{q-1} + \mathfrak{M}_{i,q}{}^{m_{i,q}} + \mathfrak{M}_{i,q+1}{}^{m_{i,q+1}} + \cdots\},\ i \in I_\rho, \tag{20}$$

where

$$\bar{\mathfrak{A}}_{q-1} = \{\bar{\xi}^{(q-1)} \in K_{q-1} \mid v(\bar{\xi}^{(q-1)}) \geqq \gamma\}. \tag{21}$$

THEOREM 3.4. (*The structure theorem: second formulation*). *The saturated local ring* $\mathfrak{o}$ *admits the following decomposition:*

$$\mathfrak{o} = k + \mathfrak{N}_0 + \mathfrak{N}_1{}^{m_1} + \mathfrak{N}_2{}^{m_2} + \cdots + \mathfrak{N}_{q-1}{}^{m_{q-1}} + \sum_{\rho=1}^{r} \mathfrak{A}_\rho'. \tag{22}$$

Proof. That the ring $\mathfrak{o}$ is contained in the right-hand side of (22) is a direct consequence of Theorem 3.1. Since we also know that $\mathfrak{N}_0 + \mathfrak{N}_1{}^{m_1} + \cdots + \mathfrak{N}_{q-1}{}^{m_{q-1}} \subset \mathfrak{o}$ (see Proposition 3.2), to complete the proof we have only to show that $\mathfrak{A}_\rho' \subset \mathfrak{o}$, $\rho = 1, 2, \cdots, r$. Let, then, η be any element of $\mathfrak{A}_\rho'$. We have $\eta\epsilon_i = 0$ if $i \notin I_\rho$, and by (20) it follows that conditions (a) and (b) of Theorem 3.1 are satisfied, with $\eta_{i,\alpha} = 0$, $\alpha = 0, 1, \cdots, q-2$, and also $\eta^{(q-1)} = 0$. Conditions (c) of Theorem 3.1 are trivially satisfied, since $\eta\epsilon_\rho' = \eta \in \mathfrak{A}_\rho' \subset \mathfrak{o}_\rho'$, while $\eta\epsilon_\sigma' = 0$ if $\sigma \neq \rho$. This completes the proof.

We note that the sum $\sum_{\rho=1}^{r} \mathfrak{A}_\rho'$ is direct.

As a consequence of Theorem 3.4 we now have the following:

PROPOSITION 3.5. *We have*

$$\mathfrak{o}_\rho' = \mathfrak{o}\epsilon_\rho',\quad \rho = 1, 2, \cdots, r, \tag{23}$$

where $\mathfrak{o}_\rho'$ *is defined in* (17), *and*

$$\widetilde{(\mathfrak{o}\epsilon_i)}_{K\epsilon_i} = \mathfrak{o}\epsilon_i,\quad i = 1, 2, \cdots, h. \tag{24}$$

Proof. For all indices i in the (non-empty) set I_ρ we have $\psi_i(\mathfrak{o}_\rho')=\mathfrak{o}_i$, by (19). Let q' and γ' be the integers analogous to q and γ and associated with the saturated ring $\mathfrak{o}_\rho'$. It is clear that $\gamma'>\gamma$ (by the definition of the partition $\{I_1,I_2,\cdots,I_r\}$). Since $\beta_{i,\alpha}=\beta_\alpha<\gamma$, for $\alpha=0,1,\cdots,q-1$, it follows that $q'\geqq q$ and that the chain of fields $K=K_0'<K_1'<\cdots<K'_{q'-1}$ associated with $\mathfrak{o}_\rho'$ and analogous to the chain of fields $K=K_0<K_1<\cdots<K_{q-1}$ associated with $\mathfrak{o}$ (see Theorem 2.1, part c), includes this later chain, i.e., we have $K'_\alpha=K_\alpha$, $\alpha=0,1,\cdots,q-1$. Therefore, the chain of subfields $L'<L_1'<\cdots<L'_{q'-1}$ of $F\epsilon_\rho'$ associated with $\mathfrak{o}_\rho'$ and analogous to the chain $L_0<L_1<\cdots<L_{q-1}$, includes the chain

$$L_0\epsilon_\rho'<L_1\epsilon_\rho'<\cdots<L_{q-1}\epsilon_\rho',$$

i.e., we have $L'_\alpha=L_\alpha\epsilon_\rho'$. Relation (22) tells us that every element η of $\mathfrak{o}$ is of the form $\eta=\xi+\zeta$, where $\eta\in k+\mathfrak{N}_0+\mathfrak{N}_1^{m_1}+\cdots+\mathfrak{N}_{q-1}^{m_{q-1}}$ and ζ is an element of $\mathfrak{o}$ such that $v(\zeta_i)\geqq\gamma$, $i=1,2,\cdots,h$. Applying this result to the saturated ring $\mathfrak{o}_\rho'$ and observing that

$$(k+\mathfrak{N}_0+\mathfrak{N}_1^{m_1}+\cdots+\mathfrak{N}_{q-1}^{m_{q-1}})\epsilon_\rho'\subset\mathfrak{o}_\rho'$$

and $\mathfrak{A}_\rho'\subset\mathfrak{o}_\rho'$, we conclude that

$$\mathfrak{o}_\rho'=(k+\mathfrak{N}_0+\mathfrak{N}_1^{m_1}+\cdots+\mathfrak{N}_{q-1}^{m_{q-1}})\epsilon_\rho'+\mathfrak{A}_\rho'. \tag{25}$$

Sine $\mathfrak{A}_\sigma'\epsilon_\rho'=0$ if $\sigma\neq\rho$, while $\mathfrak{A}_\rho'\epsilon_\rho'=\mathfrak{A}_\rho'$, we see that (23) follows from (25).

To prove (24), we use induction on h. If $i\in I_\rho$, we have, by applying our induction hypothesis to the ring $\mathfrak{o}_\rho'$ (which now replaces the ring $\mathfrak{o}$), that $\widetilde{(\mathfrak{o}\epsilon_i)}_{K\epsilon_i}=\mathfrak{o}_\rho'\epsilon_i$ (note that $\widetilde{(\mathfrak{o}\epsilon_i)}_{K\epsilon_i}=\widetilde{(\mathfrak{o}_\rho'\epsilon_i)}_{K\epsilon_\rho'\epsilon_i}$). Since, by (23), we have $\mathfrak{o}_\rho'\epsilon_i=\mathfrak{o}\epsilon_i$, the proof of the proposition is now complete.

COROLLARY 3.6. *Without assuming that $\mathfrak{o}$ is saturated with respect to the parameter x, the following is true:*

$$\widetilde{(\mathfrak{o}\epsilon_i)}_{K\epsilon_i}=\tilde{\mathfrak{o}}_K\cdot\epsilon_i,\qquad i=1,2,\cdots,h. \tag{26}$$

$$\widetilde{(\mathfrak{o}\epsilon_\rho')}_{K\epsilon_\rho'}=\tilde{\mathfrak{o}}_K\cdot\epsilon_\rho',\qquad \rho=1,2,\cdots,r. \tag{27}$$

Here K, as always, denotes the power series field $k((x))$.

For, by (24), the local ring $\tilde{\mathfrak{o}}_K\cdot\epsilon_i$ is saturated with respect to the field $K\epsilon_i$, and hence $\widetilde{(\mathfrak{o}\epsilon_i)}_{K\epsilon_i}\subset\tilde{\mathfrak{o}}_K\cdot\epsilon_i$ (since $\mathfrak{o}\epsilon_i\subset\tilde{\mathfrak{o}}_K\cdot\epsilon_i$). Since the opposite inclusion is trivial [see (20), § 1], (26) follows. In a similar way, (27) follows from (23).

By the *reduced multiplicity* s of $\mathfrak{o}$ we mean the sum of the reduced multiplicities $s_1, s_2, \cdots, s_h$ of the h local domains $\mathfrak{o}\epsilon_1, \mathfrak{o}\epsilon_2, \cdots, \mathfrak{o}\epsilon_h$. We know (see footnote 2 in I, 2) that the multiplicity of $\mathfrak{o}\epsilon_i$ is equal to $s_i \cdot [\bar{k}\epsilon_i : k\epsilon_i]$. Now, if $\bar{\mathfrak{o}}$ is the integral closure of $\mathfrak{o}$, and if $\bar{\mathfrak{m}}_1, \bar{\mathfrak{m}}_2, \cdots, \bar{\mathfrak{m}}_h$ are the h maximal ideal of $\bar{\mathfrak{o}}$, then $\bar{\mathfrak{o}}\epsilon_i$ is isomorphic to the localization $\bar{\mathfrak{o}}_{\mathfrak{m}_i}$ of $\bar{\mathfrak{o}}$ at $\bar{\mathfrak{m}}_i$, and $\bar{\mathfrak{o}}$ is the direct sum of the h local domains $\bar{\mathfrak{o}}\epsilon_i$, while each of these local domains $\bar{\mathfrak{o}}\epsilon_i$ *is* a power series ring over the field $\bar{k}\epsilon_i$. From the definition of s_i it follows at once that $\bar{\mathfrak{o}}\mathfrak{m} = \bar{\mathfrak{m}}_1^{s_2} \cap \bar{\mathfrak{m}}_2^{s_2} \cap \cdots \cap \bar{\mathfrak{m}}_h^{s_h}$, and since the ideal $\bar{\mathfrak{m}}_i^{s_i}$ in $\bar{\mathfrak{o}}$ has length s_i, it follows, by the well-known projection formula (see [7], v. p. 299, Corollary 1) that

$$(28) \qquad e(\mathfrak{o}) = \sum_{i=1}^{h} [\bar{k}\epsilon_i : k\epsilon_i] s_i = \sum_{i=1}^{h} e(\mathfrak{o}\epsilon_i).$$

We can now state the following generalization of Proposition I, 2.5:

PROPOSITION 3.7. *If x is any parameter of $\mathfrak{o}$ then $\mathfrak{o}$ and $\tilde{\mathfrak{o}}_x$ have the same multiplicity and the same reduced multiplicity.*

Proof. By Proposition I, 2.5, the local domains $\mathfrak{o}\epsilon_i$ and $(\widetilde{\mathfrak{o}\epsilon_i})_{x\epsilon_i}$ have the same multiplicity and the same reduced multiplicity. The proposition now follows from the definition of the reduced multiplicity, from the fact that $(\widetilde{\mathfrak{o}\epsilon_i})_{x\epsilon_i}$ coincides with $\tilde{\mathfrak{o}}_x \cdot \epsilon_i$ and from the formula (26) as applied to both $\mathfrak{o}$ and $\tilde{\mathfrak{o}}_x$.

4. Properties of transversal parameters (generalization of Theorems I, 7.2 and I, 8.1). We shall denote by $\bar{k}_i$ the algebraic closure of the field $k\epsilon_i$ in the field $F\epsilon_i$. It is easily seen that the algebraic closure $\bar{k}$ of k in F is the direct sum of the h fields $\bar{k}_i$ and that $\bar{k}_i = \bar{k}\epsilon_i$ $(i = 1, 2, \cdots, h)$. Each field $F\epsilon_i$ is a power series field over $\bar{k}_i$, and given any element ξ_i of $F\epsilon_i$, which belongs to the valuation ring of the canonical valuation of $F\epsilon_i$, this element has a definite residue in $\bar{k}_i$. If ξ_i does not belong to that valuation ring, we say that the residue of ξ_i is infinite.

Definition 4.1. *Let $\xi \in F$, and assume that for each $i = 1, 2, \cdots, h$, the element $\xi\epsilon_i$ of $F\epsilon_i$ has a finite residue c_i $(\in \bar{k}_i)$. Then we call the element $\sum_{i=1}^{h} c_i$ of $\bar{k}$ the residue of ξ.*

In the case of complete local fields $F = \bar{k}((t))$ we have made certain remarks in the beginning of I, § 8. We shall summarize and complete these demarks in the following lemma:

LEMMA 4.2. *Let* $F = \bar{k}((t))$ *bs a power series field over a field* $\bar{k}$, *and let* x, x' *be elements of* $\bar{k}[[t]]$ *such that* $v(x) = v(x') = n > 0$ *(here* v *denotes the natural valuation of* F*).*

(a) *A necessary and sufficient condition that there exist* $\bar{k}$*-automorphisms of* F *which send* x *into* x' *is that the residue of* x'/x *in* $\bar{k}$ *be the* n*-th power of an element of* k.

(b) *If* c *is any element of* $\bar{k}$ *such that the residue of* x'/x *in* $\bar{k}$ *is equal to* c^n *then there exists one and only one* $\bar{k}$*-automorphism* f_c *of* F *sending* x *into* x', *which satisfies the following condition: if* ξ *is any element of* F *such that* $v(\xi) = s \neq 0$, *then the residue of* $f_c(\xi)/\xi$ *in* $\bar{k}$ *is* c^s.

(c) *Any* $\bar{k}$*-automorphism of* F *sending* x *into* x' *is of the form* f_c *for a suitable choice of the element* c *in* $\bar{k}$ *having the property that the residue of* x'/x *in* $\bar{k}$ *is* c^n.

Proof. Part (a) of the lemma has already been proved in the beginning of I, § 8.

(b) Let c be an element of $\bar{k}$ such that the residue of x'/x in $\bar{k}$ is equal to c^n. We can find uniformizing parameters t, t' in F such that $x = at^n$, $x' = a't'^n$, with $a, a' \in \bar{k}$. If b denotes the residue of t'/t, then we have clearly: $a'/a = c^n/b^n$. If we then replace t' by $t'c/b = t''$, we will have $x' = at''^n$ and t''/t has residue c. So we may assume that we have $x = at^n$, $x' = at'^n$ and $c =$ residue t'/t. Let now f_c be the $\bar{k}$-automorphism of F such that $f(t) = t'$. Then clearly $f_c(x) = x'$. We write $\xi = bt^s +$ terms of degree $> s$ and we introduce the $\bar{k}$-automorphism g of F defined by the condition $g(t) = ct$. We have then $f_c(t) - ct = t' - ct$, and since c is the residue of t'/t, it follows that $v(f_c(t) - g(t)) \geqq 2$. Hence, by Lemma I, 7.1, we have $v[f_c(\xi) - g(\xi)] > s$, or equivalently—the residue of $(f_c(\xi) - g(\xi))/\xi$ is zero. But since $g(\xi) = bc^st^s +$ terms of degree $> s$, the residue of $g(\xi)/\xi = c^s$. Therefore also the residue of $f_c(\xi)/\xi$ is c^s.

If f is any other $\bar{k}$-automorphism of F sending x into x' and satisfying the condition: "residue of $f(\xi)/\xi$ in $\bar{k}$ is c^s, if $s = v(\xi) > 0$," then from $f(at^n) = a[f(t)]^n$ and $f(at^n) = f(x) = x' = at'^n$, follows that $f(t) = \omega t'$, where ω is an n-th root of unity. Now for $\xi = t$ the above condition on f tells us that the residue $f(t)/t$ is equal to c. On the other hand, this residue must also be equal to ωc, since $f(t) = \omega t'$ and since t'/t has residue c. Hence $\omega = 1$ and $f = f_c$, which establishes part (b) of the lemma.

(c) Let now f be any $\bar{k}$-automorphism of F such that $f(x) = x'$. We can find a uniformizing parameter t of F such that $x = at^n$, $a \in \bar{k}$, $a \neq 0$. Then $x' = at'^n$, where $f(t) = t'$. If then we denote by c the residue of t'/t in $\bar{k}$, it will follow that the residue of x'/x is c^n. On the other hand, since the residue of $f(t)/t$ is c, the proof of part (b) of the lemma shows that the residue of $f(\xi)/\xi$ is c^s for any element ξ of F such that $v(\xi) = s \neq 0$. This shows that f is the automorphism f_c introduced in part (b) of the lemma.

COROLLARY 4.3. *Let F, x, x' be as in Lemma 4.2 and assume that the residue of x'/x in $\bar{k}$ is of the form c^n, $c \in \bar{k}$. Let L be a power series field which contains F as a subfield and which is finite over F. Let $\tilde{k}$ be the algebraic closure of k in L (whence L is a power series field over $\tilde{k}$; we remind the reader that all our fields are of characteristic zero). Let e be the ramification exponent of L/F, and let v, $\tilde{v}$ denote the natural valuation of F and L respectively (both having the additive group of integers as value group; thus $\tilde{v}(x) = ev(x) = ne$ and $v(x') = ev(x') = ne$). Assume that the residue of x'/x in $\tilde{k}$ is of the form $\tilde{c}^{en}$, $\tilde{c} \in \tilde{k}$, and that $\tilde{c}^e \in \bar{k}$. Then there exists one and only one $\tilde{k}$-automorphism $g_{\tilde{c}}$ of L satisfying the following conditions: (1) $g_{\tilde{c}}(F) = F$; (2) $g_{\tilde{c}}(x) = x'$; (3) if ξ is any element of L such that $\tilde{v}(\xi) = \tilde{s} \neq 0$, then the residue of $g_{\tilde{c}}(\xi)/\xi$ is equal to $\tilde{c}^{\tilde{s}}$. Furthermore, given any $\tilde{k}$-automorphism of F, there exists an element $\tilde{c}$ in $\tilde{k}$ such that $c^e \in \bar{k}$, the residue of x'/x in $\tilde{k}$ is c^{ne} and f is the automorphism $g_{\tilde{c}}$ described above.*

For the proof we first constduct the $\bar{k}$-automorphism f_c of F referred to in Lemma 4.2, where $c = \tilde{c}^e$. W have that $f_c(t) = t'$ and the residue of t'/t is c. Here t, t' are suitable uniformizing parameters of F. Since $v(t) = v(t')$ $= e$ and since $c = c^e$, we can apply Lemma 4.2 again, replacing F, x, x' by L, t, t' respectively.

If $\mathfrak{o}$ is our local (not necessarily saturated) ring and x is a parameter of $\mathfrak{o}$, then $x\epsilon_i$ is a parameter of $\mathfrak{o}\epsilon_i$, for $i = 1, 2, \cdots, h$; and conversely. We say that x *is a transversal parameter of* $\mathfrak{o}$ if for $i = 1, 2, \cdots, k$, $x\epsilon_i$ is a transversal parameter of $\mathfrak{o}\epsilon_i$. Our next objectice is to generalize Theorem I, 8.1 from local *domains* to local *rings*. We prove, namely, the following theorem:

THEOREM 4.4. *Assume that our local (one-dimensional) ring $\mathfrak{o}$ is saturated with respect to a transversal parameter x, and let x' be another transversal parameter of $\mathfrak{o}$ such that the residue of x'/x is 1 (note that since*

$x\epsilon_i$ and $x'\epsilon_i$ are both transversal parameters of $\mathfrak{o}\epsilon_i$, for $i=1,2,\cdots,h$, the quotient $x'/x\cdot\epsilon_i$ has finite residue in $F\epsilon_i$). Then there exists a $\bar{k}$-automorphism of F such that: (1) $f(\mathfrak{o})=\mathfrak{o}$ and (2) $f(x)=x'$ (and hence $\mathfrak{o}$ is also saturated with respect to x').

Proof. By Proposition 3.5, $\mathfrak{o}\epsilon_i$ is saturated with respect to $x\epsilon_i$, and since the residue of $x'\epsilon_i/x\epsilon_i$ is ϵ_i, it follows from Theorem I, 8.1 that there exists a $\bar{k}_i$-automorphism f_i of $F\epsilon_i$ such that (1) $f_i(\mathfrak{o}\epsilon_i)=\mathfrak{o}\epsilon_i$ and (2) $f_i(x\epsilon_i)=x'\epsilon_i$. Furthermore, by part (b) of Lemma 4.2 it follows that such an automorphism f_i exists which satisfies the additional condition: $f_i(\xi_i)/\xi_i$ has residue ϵ_i, for any ξ_i in $F\epsilon_i$, $\xi_i\neq 0$. Since F is the direct sum of the h fields $F\epsilon_i$, the sum $\sum f_i$ defines a $\bar{k}$-automorphism f of F, such that $f(x)=x'$ and such that, furthermore, $f(\xi)/\xi$ has residue 1, for any element ξ of F which is not a zero-divisor. If $\bar{\mathfrak{o}}$ is the integral closure of $\mathfrak{o}$ in F, we know that $\bar{\mathfrak{o}}$ is ths direct sum of the integral closures $\overline{\mathfrak{o}\epsilon_i}$ of the rings $\mathfrak{o}\epsilon_i$ in their respective fields of fractions $F\epsilon_i$ [see (5), § 1]. Since $f_i(\mathfrak{o}\epsilon)=\mathfrak{o}\epsilon_i$, it follows that $f_i(\overline{\mathfrak{o}\epsilon_i})=\overline{\mathfrak{o}\epsilon_i}$, and hence $f(\bar{\mathfrak{o}})=\bar{\mathfrak{o}}$. We shall now prove that $f(\mathfrak{o})=\mathfrak{o}$ (and this will establish our theorem). It will be sufficient to prove that

$$(1) \qquad f(\mathfrak{o})\subset\mathfrak{o},$$

for once this is established, the equality $f(\mathfrak{o})=\mathfrak{o}$ will follow by the same argument on the lengths of the $\mathfrak{o}$-module $\bar{\mathfrak{o}}/\mathfrak{o}$ and the $f(\mathfrak{o})$-module $\bar{\mathfrak{o}}/f(\mathfrak{o})$ which was used in the proof of Theorem I, 8.1 (see the reasoning employed just before the inclusion (8) of I, § 8).

Let then η be any element of $\mathfrak{o}$. To prove that $f(\eta)\in\mathfrak{o}$ we have to show the following: if ψ_1, ψ_2 are any two K-homomorphisms of F into Ω (here $K=k((x))$, and Ω is an algebraic closure of K) then the following quantity

$$(2) \qquad [\psi_1\cdot f)(\eta)-\psi_2\cdot f)(\eta)]/(\psi_1(y)-\psi_2(y))$$

is integral over $k[[x]]$; here y is an x-saturator of $\mathfrak{o}$. In showing this we may assume that $\eta\in\mathfrak{m}$.

Now, quite generally, given any K-homomorphism ψ of F into Ω and assuming that $\psi\in H_i$ (i.e., that $\psi(\epsilon_i)=1$), then $\operatorname{Ker}\psi=\sum_{j\neq i}F\epsilon_j$, and since $f(F\epsilon_j)=F\epsilon_j$, $\operatorname{Ker}\psi$ is invariant under f. Hence f determines an automorphism f_ψ of the field $\psi(F)$ such that the following diagram is commutative:

$$\begin{array}{ccc} F & \xrightarrow[\sim]{\;f\;} & F \\ \Big\downarrow \psi & & \Big\downarrow \psi \\ \psi(F) & \xrightarrow[f_\psi]{\sim} & \psi(F). \end{array}$$

Since the restriction f_i of f to F_{ϵ_i} is a $\bar{k}_i$-automorphism, where $\bar{k}_i$ is the algebraic closure of k_{ϵ_i} in F_{ϵ_i}, it follows that f_ψ is a $\bar{k}'$-automorphism, where $\bar{k}'$ is the algebraic closure of k in $\psi(F)$. Applying these remarks to the two K-homomorphisms ψ_1, ψ_2 and setting $f_{\psi_1} = f'_1$, $f_{\psi_2} = f'_2$, we have the two automorphisms

$$f'_1 : \psi_1(F) \xrightarrow{\sim} \psi_1(F),$$

$$f'_2 : \psi_2(F) \xrightarrow{\sim} \psi_2(F),$$

and from the above commutative diagram follows that

$$\psi_i \cdot f = f'_i \cdot \psi_i, \quad i = 1, 2. \tag{3}$$

Furthermore, f'_i is an automorphism over the algebraic closure $\bar{k}'_i$ of k in $\psi_i(F)$ $(i = 1, 2)$, and for any element ξ of $\psi_i(F)$, different from zero, the residue of $f_i(\xi)/\xi$ is equal to 1.

In view of (3), we can rewrite the quantity (2) as follows:

$$[(f'_1 \cdot \psi_1)(\eta) - (f_2' \cdot \psi_2)(\eta)]/(\psi_1(y) - \psi_2(y)). \tag{4}$$

Let $\psi_i(x') = x'_i$ $(i = 1, 2)$. Then, by (3) and since $f(x) = x'$ and $\psi_i(x) = x$, we have

$$f'_i(x) = x'_i, \quad i = 1, 2. \tag{5}$$

Let F^* be the least Galois extension of K $(= k((x)))$ which contains all the fields $\psi(F)$, $\psi \in H$. By Corollary 4.3, the automorphisms f'_1 and f'_2 can be extended (in a unique way) to automorphisms of F^* (which we shall continue to denote by f'_1, f'_2) such that the condition: "residue of $f'_i(\xi)/\xi$ is equal to 1," is still satisfied for any element ξ of F^*, different from zero. Therefore, for any such element ξ we have that the residue of $f'_1(\xi)/f'_2(\xi)$ is equal to 1. It follows therefore from Lemma I, 7.1, that for any element ξ of F^*, different from zero, we have that *the difference* $v[f'_1(\xi) - f'_2(\xi)] - v(\xi)$ *is positive and is independent of* ξ; here v denotes the natural valuation of the power series field F^*. Now, we can write

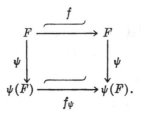

$$(6)\quad \begin{aligned}&(f'_1\cdot\psi_1)(\eta)-(f'_2\cdot\psi_2)(\eta)\\ &\qquad =f'_1(\psi_1(\eta)-\psi_2(\eta))+[f'_1(\psi_2(\eta))-f'_2(\psi_2(\eta))].\end{aligned}$$

Since $\eta\in\mathfrak{o}$, we have $v[\psi_1(\eta)-\psi_2(\eta)]\geqq v[\psi_1(y)-\psi_2(y)]$, and hence

$$(7)\qquad v[f'_1(\psi_1(\eta)-\psi_2(\eta)]\geqq v[(\psi_1(y)-\psi_2(y))].$$

Applying the above cited Lemma I, 7.1 to the two automorphisms f'_1, f'_2 of F^*, we have

$$\begin{aligned}&v[f'_1(\psi_2(\eta))-f'_2(\psi_2(\eta))]\\ &\qquad =v[f'_1(\psi_2(x))-f'_2(\psi_2(x))]+v(\psi_2(\eta))-v(\psi_2(x)),\end{aligned}$$

and hence, since $\psi_2(x)=x$ and in view of (5):

$$(8)\qquad v[f'_1(\psi_2(\eta))-f'_2(\psi_2(\eta))]=v(x'_1-x'_2)+v(\psi_2(\eta))-v(x).$$

Since $\eta\in\mathfrak{m}$ and x is a transversal parameter of $\mathfrak{o}$, we have $v(\psi(\eta))\geqq v(x)$, for any ψ is H (and, in particular, for ψ_2). Hence by (8), we have

$$(9)\qquad v[f'_1(\psi_2(\eta))-f'_2(\psi_2(\eta))]\geqq v(x'_1-x'_2).$$

Now, if we write (6), with $\eta=x$, and apply the commutativity relations (3) we find (observing that $f(x)=x'$):

$$\psi_1(x')-\psi_2(x')=f'_1[\psi_1(x)-\psi_2(x)]+[f'_1(\psi_2(x))-f'_2(\psi_2(x))],$$

or, since $\psi_1(x)=\psi_2(x)=x$:

$$(10)\qquad \psi_1(x')-\psi_2(x')=x'_1-x'_2.$$

Since $x'\in\mathfrak{o}$, (10) implies that $v(x'_1-x'_2)\geqq v(\psi_1(y)-\psi_2(y))$. From this last inequality and from (6), (7) and (9) follows that the v-value of the quotient (4) is non-negative. This completes the proof of the theorem.

We shall now proceed with our main objective, and that is the generalization of Theorem I, 7.2 from the case of a local *domain* $\mathfrak{o}$ to the general case of a local ring $\mathfrak{o}$:

THEOREM 4.5. *If the local ring $\mathfrak{o}$ is saturated, then it is saturated with respect to any transversal parameter of $\mathfrak{o}$.*

Proof. We shall use induction with respect to the number h of direct field summands F_{ϵ_i} of the total ring of quotients F of $\mathfrak{o}$, since the theorem is true if $h=1$, i.e., if $\mathfrak{o}$ is a local domain (Theorem I, 7.2). Our first step will be to establish the following preliminary result (an exact replica of Corollary I, 4.6):

4

A) *If* $\mathfrak{o}$ *is saturated, then there exists a transversal parameter of* $\mathfrak{o}$ *with respect to which* $\mathfrak{o}$ *is saturated.*

For the proof of assertion A) we fix some parameter x with respect to which $\mathfrak{o}$ is saturated. If x is a transversal parameter of $\mathfrak{o}$, there is nothing to prove. Assume then that x *is not a transversal parameter of* $\mathfrak{o}$. We use the noattions of the preceding sections, and we first consider the case $q = 1$. In this case we shall prove in fact that the following stronger assertion is valid:

A_1) *If* $\mathfrak{o}$ *is saturated with respect to a non-transversal parameter* x *and if* $q = 1$, *then* $\mathfrak{o}$ *is saturated with respect to any transversal parameter.*

In other words, in the special case of x non-transversal and $q = 1$, we will prove without further ado the validity of Theorem 4. 5.

Since $q = 1$, we have

$$\gamma \leqq \min\{\beta_{1,1}, \beta_{2,1}, \cdots, \beta_{h,1}\}. \tag{11}$$

Consider the saturated ring $\mathfrak{o}'_\rho = \mathfrak{o}\epsilon'_\rho$ $(\rho = 1, 2, \cdots, r)$ and the ideal $\mathfrak{A}'_\rho$ in $\mathfrak{o}'_\rho$ (defined in (18), § 3). *We assert that* $\mathfrak{A}'_\rho$ *is the entire maximal ideal* $\mathfrak{m}'_\rho$ *of* $\mathfrak{o}'_\rho$. To see this, we first observe that since $x\epsilon_i$ is a non-transversal parameter of the saturated ring $\mathfrak{o}\epsilon_i$, for some $i = 1, 2, \cdots, h$, we must have $\beta_{i,1} < v(x)$ for scme i. Hence, by (11), we have

$$\gamma < v(x). \tag{12}$$

Now, if ξ' is any element of the maximal ideal of $\mathfrak{o}'_\rho$, and i is any index in the set I_ρ, then we have $\xi'_i = \xi_{i,0} + \xi_{i,1} + \xi_{i,2} + \cdots$, with $v(\xi_{i,0}) \geqq v(x) > \gamma$ and $v(\xi_{i,\alpha}) \geqq \beta_{i,\alpha} \geqq \beta_{i,1} \geqq \gamma$. Hence $v(\xi_i') \geqq \gamma$ for all $i \in I_\rho$, and this proves our assertion: $\mathfrak{A}'_\rho = \mathfrak{m}'_\rho$. Since $\mathfrak{m}'_\rho = \mathfrak{m} \cdot \epsilon'_\rho$ (in view of the equality $\mathfrak{o}'_\rho = \mathfrak{o} \cdot \epsilon'_\rho$), it follows from (22), § 3, that *the maximal ideal* $\mathfrak{m}$ *of* $\mathfrak{o}$ *is the direct sum of the* r *maximal ideals* $\mathfrak{m}'_\rho$:

$$\mathfrak{m} = \bigoplus_{\rho=1}^{r} \mathfrak{m}'_\rho. \tag{13}$$

Now, let z be any transversal parameter of $\mathfrak{o}$, and let $z_\rho = z\epsilon'_\rho$. Then z_ρ is a transversal parameter of $\mathfrak{o}'_\rho$, and by our induction hypothesis the saturated ring $\mathfrak{o}'_\rho$ is saturated with respect to the field $\Sigma_\rho = k\epsilon'_\rho((z_\rho))$. It is then practically obvious from (13) that $\mathfrak{o}$ is saturated with respect to z (and that is precisely what A_1 asserts); nevertheless, we shall give the details of the proof.

Let $\Sigma = k((z))$, let ζ be any element of $\mathfrak{o}$, let η be an element of $\bar{\mathfrak{o}}$, let Ω denote the algebraic closure of the field Σ, and let us assume that for any two distinct Σ-homomorphisms ϕ_1, ϕ_2 of F into Ω, the quotient $[\phi_1(\eta)-\phi_2(\eta)]/[\phi_1(\zeta)-\phi_2(\zeta)]$ is integral over $k[[z]]$. If we set $\eta_\rho=\eta\epsilon'_\rho$, $\zeta_\rho=\zeta\epsilon'_\rho$, then by (13) we have $\zeta_\rho\in\mathfrak{o}'_\rho$, and clearly $\eta_\rho\in\overline{\mathfrak{o}'_\rho}$ (= integral closure of $\mathfrak{o}'_\rho$ in $F\epsilon'_\rho$). Furthermore, we have

$$(14)\qquad \eta=\eta_1+\eta_2+\cdots+\eta_r.$$

Now let Ω_ρ be an algebraic closure of the field $\Sigma_\rho=k\epsilon'_\rho((z_\rho))=\Sigma\epsilon'_\rho$, let θ be an isomorphism $\Omega_\rho\xrightarrow{\sim}\Omega$ such that $\theta(z_\rho)=z$ and $\theta(c\epsilon'_\rho)=c$ for all $c\in k$. Let ψ_1, ψ_2 be any two Σ_ρ-homomorphisms of $F\epsilon'_\rho$ into Ω_ρ, and let us define two Σ-homomorphisms ϕ_1, ϕ_2 of F into Ω as follows: $\phi_j\mid F\epsilon'_\rho=\theta\cdot\psi_j$, and $\phi_j\mid F\epsilon'_\sigma=0$ if $\sigma\neq\rho$ $(j=1,2)$. From our hypothesis that

$$[\phi_1(\eta)-\phi_2(\eta)]/[\phi_1(\zeta)-\phi_2(\zeta)]$$

is integral over $k[[z]]$, and from the fact that $\phi_j(\eta)=\theta(\psi_j(\eta_\rho))$, $\phi_j(\zeta)=\theta(\psi_j(\zeta_\rho))$, follows at once that the quotient

$$[\psi_1(\eta_\rho)-\psi_2(\eta_\rho)]/[\psi_1(\zeta_\rho)-\psi_2(\zeta_\rho)]$$

is integral over $k\epsilon'_\rho[[z_\rho]]$. This being true for any two Σ_ρ-homomorphism of $F\epsilon'_\rho$ into Ω_ρ, and ζ_ρ being an element of $\mathfrak{o}'_\rho$, we conclude that $\eta_\rho\in\mathfrak{o}'_\rho$ (since $\mathfrak{o}'_\rho$ is saturated with respect to Σ_ρ). By (13) and (14), this implies that $\eta\in\mathfrak{o}$, which proves that $\mathfrak{o}$ is saturated with respect to the transversal parameter z.

We now deal with the case $q>1$. Again we may assume that x is not a transversal parameter of $\mathfrak{o}$. To say that for some $i=1,2,\cdots,h$, $x\epsilon_i$ is not a transversal parameter of $\mathfrak{o}\epsilon_i$ is the same as saying that $\beta_{i,1}<v(x)$. But all the $\beta_{i,1}$ are equal: $\beta_{i,1}=\beta_1$. It follows therefore that $x\epsilon_i$ is a non-transversal parameter of $\mathfrak{o}\epsilon_i$, for all $i=1,2,\cdots,h$, and we conclude, by (22), § 3, that any element z of the field L_1 such that $z\in\mathfrak{N}_1^{m_1}$, $z\notin\mathfrak{N}_1^{m_1+1}$, is a transversal parameter of $\mathfrak{o}$. We shall prove that $\mathfrak{o}$ is saturated with respect to any such parameter z. We note that in the present case, we have $\mathfrak{N}_0\subset\mathfrak{N}_1^{m_1}$, whence

$$(15)\qquad \mathfrak{o}=k+\mathfrak{N}_1^{m_1}+\mathfrak{N}_2^{m_2}+\cdots+\mathfrak{N}_{q-1}^{m_{q-1}}+\sum_{\rho=1}^{r}\mathfrak{A}'_\rho.$$

We set $z_\rho=z\epsilon'_\rho$. We have the representation (25) (§ 3) of the saturated ring $\mathfrak{o}'_\rho$, z_ρ is a transversal parameter of $\mathfrak{o}'_\rho$, and hence, by our induction hypothesis, $\mathfrak{o}'_\rho$ is saturated with respect to the field $\Sigma_\rho=k\epsilon'_\rho((z_\rho))$. Let now $\Sigma=k((z))$, let ζ be any element of $\mathfrak{o}$, η an element of $\bar{\mathfrak{o}}$, and assume that $[\phi_1(\eta)-\phi_2(\eta)]/[\phi_1(\zeta)-\phi_2(\zeta)]$ is integral over $k[[z]]$, for any two Σ-homomorphisms of F into some fixed algebraic closure of $k((z))$. If we set $\eta'_\rho=\eta\epsilon'_\rho$, $\zeta'_\rho=\zeta\epsilon'_\rho$, we will have $\zeta'_\rho\in\mathfrak{o}'_\rho$, $\eta'_\rho\in\bar{\mathfrak{o}}'_\rho$ and our assumption implies that η'_ρ dominates ζ'_ρ (with respect to the field Σ_ρ). Hence $\eta'_\rho\in\mathfrak{o}'_\rho$ (since $\mathfrak{o}'_\rho$ is saturated with respect to the field Σ_ρ), and what we have to prove is that

$$\eta\in\mathfrak{o}. \tag{16}$$

Since $\eta\in\tilde{\mathfrak{o}}_\Sigma$, and since k is also a coefficient field of the local ring $\tilde{\mathfrak{o}}_\Sigma$, we may assume, upon replacing η by $\eta-c$, where c is a suitable element of k, that η belong to the maximal ideal of $\tilde{\mathfrak{o}}_\Sigma$. Since $\mathfrak{o}'_\rho$ $(=\mathfrak{o}\epsilon'_\rho)$ is saturated with respect to the field Σ_ρ $(=\Sigma\epsilon'_\rho)$, it follows from Corollary 3.6 that $\mathfrak{o}'_\rho=\tilde{\mathfrak{o}}_\Sigma\cdot\epsilon'_\rho$. Therefore η'_ρ belongs to the maximal ideal of $\mathfrak{o}'_\rho$, for $\rho=1,2,\cdots,r$:

$$\eta'_\rho\in(\mathfrak{N}_1^{m_1}+\mathfrak{N}_2^{m_2}+\cdots+\mathfrak{N}_{q-1}^{m_{q-1}})\epsilon'_\rho+\mathfrak{A}'_\rho, \tag{17}$$

and if (16) is to be true, we will have to show that

$$\eta\in\mathfrak{m}. \tag{18}$$

To prove (18) we use the structure theorem 3.1. Condition (c) of that theorem is satisfied, since $\eta'_\rho\in\mathfrak{o}'_\rho$. So we have only to show that conditions (a) and (b) of Theorem 3.1 are satisfied. We recall that we have $\psi_1\mid L_1=\psi_2\mid L_1=\cdots=\psi_h\mid L_1$ $(=\theta_1^{-1}$; see (14), § 3), and $\theta_1^{-1}(L_1)=K_1$. We shall identify L_1 with K_1, using the isomorphism $\theta_1\colon K_1\longrightarrow L_1$. Then $\psi_1,\psi_2,\cdots,\psi_h$ become K_1-homomorphisms of F into Ω, and Ω is, of course, also an algebraic closure of K_1. Since $\Sigma=k((z))\subset L_1=K_1$, the ψ_i *are also Σ-homomorphisms* of F into Ω, and Ω is also an algebraic closure of Σ. Since η dominates ζ with respect to Σ, we have

$$v(\eta_i-\eta_j)\geqq v(\zeta_i-\zeta_j),$$

where $\eta_i=\psi_i(\eta)$, $\zeta_i=\psi_i(\zeta)$ $(i=1,2,\cdots,h)$. Since $\zeta\in\mathfrak{o}$ we have, by Theorem 3.1,

$$\zeta_i=\zeta^{(0)}+\zeta^{(1)}+\cdots+\zeta^{(q-1)}+\bar{\zeta}_{i,q-1}+\zeta_{i,q}+\cdots$$

where $\zeta^{(\alpha)}\in K_\alpha$ $(0\leqq\alpha\leqq q-1)$, $\bar{\zeta}_{i,q-1}\in K_{q-1}$, $v(\zeta_{i,q-1})\geqq\gamma$ and $v(\zeta_{i,\alpha})\geqq\beta_{i,\alpha}$, for $\alpha\geqq q$. Therefore $v(\zeta_i-\zeta_j)\geqq\gamma$ and $v(\eta_i-\eta_j)\geqq\gamma$. Since the derivation of conditions (a) and (b) of Theorem 3.1 was based exclusively on the inequalities $v(\eta_i-\eta_j)\geqq\gamma$ and on the fact that $\eta_i\in\mathfrak{o}_i$, and since the inclusion $\eta_i\in\mathfrak{o}_i$ follows in our present case from the inclusion $\eta\epsilon'_\rho\in\mathfrak{o}'_\rho$, the inclusion (18) is established, and this proves assertion A).

We now proceed to the proof of our Theorem 4.5, by using Theorem 4.4 and the just established assertion A). Given the saturated local ring $\mathfrak{o}$ and given any transversal parametere z of $\mathfrak{o}$, we fix some *transversal* parameter x of $\mathfrak{o}$ such that $\mathfrak{o}$ is saturated with respect to x. Let c be the residue of z/x in $\bar{k}$. Then $c \neq 0$ and also $c^{\epsilon_i} \neq 0$, for $i = 1, 2, \cdots, h$. We set $x' = cx$. *We shall prove that* 1) $x' \in \mathfrak{o}$ *and that* 2) $\mathfrak{o}$ *is saturated with respect to the* (*transversal*) *parameter* x'. From this Theorem 4.5 will follow in view of Theorem 4.4, since the residue of z/x' is 1.

Let H be the set of all K-homomorphisms of F into an algebraic closure Ω of K, where $K = k((x))$. To prove 1), we have to show (in the notation of § 1) that

$$v(\psi_1(x') - \psi_2(x')) \geqq v(\psi_1(y) - \psi_2(y)), \tag{19}$$

for any two homomorphisms ψ_1, ψ_2 in H; here y is an x-saturator of $\mathfrak{o}$. We have $\psi_i(x') = \psi_i(c) \cdot x$, since $\psi_i(x) = x$. Hence

$$\psi_1(x') - \psi_2(x') = [\psi_1(c) - \psi_2(c)]x.$$

If $\psi_1(c) = \psi_2(c)$, (19) is trivially satisfied. Assume that $\psi_1(c) \neq \psi_2(c)$. Writing

$$\frac{\psi_1(z)}{\psi_2(z)} = \frac{\psi_1(z)}{\psi_1(x)} \cdot \frac{\psi_2(x)}{\psi_2(z)},$$

we find that the residue of $\psi_1(z)/\psi_2(z)$ (in the valuation v) is equal to $\psi_1(c)/\psi_2(c)$, and hence is different from 1. On the other hand, since $z \in \bar{\mathfrak{o}}$, both $\psi_1(z)$ and $\psi_2(z)$ are integral over $k[[x]]$, and thus $v(\psi_i(z)) \geqq 0$. Therefore $v[\psi_1(z) - \psi_2(z)] = v(\psi_1(z)) = v(\psi_2(z))$. Furthermore, since $c^{\epsilon_i} \neq 0$, the v-residue of $\psi_i(z/x)$ is $\neq 0$. Thus $v(\psi_1(z)) = v(x)$. We thus have: $v[\psi_1(z) - \psi_2(z)] = v(x)$, and since $z \in \mathfrak{o}$, it follows that

$$v(x) \geqq v(\psi_1(y) - \psi_2(y)).$$

Since $\psi_1(x') - \psi_2(x') = [\psi_1(c) - \psi_2(c)]x$, the inequality (19) is established also in the case $\psi_1(c) \neq \psi_2(c)$. Thus the inclusion $x' \in \mathfrak{o}$ is proved.

We next prove that $\mathfrak{o}$ is saturated with respect to the parameter x'. We denote by Ω' an algebraic closure of the field $k((x'))$. Let ζ be any element of $\mathfrak{o}$ and let η be an element of $\bar{\mathfrak{o}}$ which dominates ζ with respect to the field $k((x'))$. We have to show that η belongs to $\mathfrak{o}$. For that it will be sufficient to show that if y is an x-saturator of $\mathfrak{o}$ then η dominates y with respect to the field $k((x))$, i.e., that we have

$$v[\psi_1(\eta) - \psi_2(\eta)] \geqq v[\psi_1(y) - \psi_2(y)], \tag{20}$$

for any two $k((x))$-homomorphisms ψ_1, ψ_2 of F into the algebraic closure Ω of $k((x))$. We consider two cases, according as $\psi_1(c) = \psi_2(c)$ or $\psi_1(c) \neq \psi_2(c)$.

Case I. $\psi_1(c) = \psi_2(c) = c'$ (say). Then

$$\psi_1(x') = \psi_2(x') = c'x. \tag{21}$$

Let θ be a given k-isomorphism of Ω onto Ω' such that $\theta(c'x) = x'$ and $\theta(k((c'x))) = k((x'))$. Let $\phi_i = \theta \cdot \psi_i$. Then ϕ_1, ϕ_2 are $k((x'))$-homomormorphisms of F into Ω'. Since η dominates ζ with respect to the field $k((x'))$, it follows that the quotient

$$[\phi_1(\eta) - \phi_2(\eta)]/[\phi_1(\zeta) - \phi_2(\zeta)]$$

is integral over $k[[x']]$. Applying the isomorphism θ^{-1} and observing that c' is algebraic over k, we conclude that the quotient

$$[\psi_1(\eta) - \psi_2(\eta)]/[\psi_1(\zeta) - \psi_2(\zeta)]$$

is integral over $k[[x]]$, and from this (20) follows since $\zeta \in \mathfrak{o}$ and since therefor ζ dominates the x-saturator y.

Case II. $\psi_1(c) \neq \psi_2(c)$. The preceding part of the proof, where we have shown that $x' \in \mathfrak{o}$, shows that in the present case we have $v(x) \geqq v(\psi_1(y - \psi_2(y))$. Hence, in order to prove (20) it will be sufficient to show that $v[\psi_1(\eta) - \psi_2(\eta)] \geqq v(x)$—equivalently—that $v[\psi_1(\eta) - \psi_2(\eta)] \geqq v(c'x)$. Since c' is algebraic over k, the ring $k(c')[[x]]$ is integral over each of its two subrings $k[[x]]$ and $k[[c'x]]$. Hence, it is sufficient to show that the quotient $[\psi_1(\eta) - \psi_2(\eta)]/c'x$ is integral over $k[[c'x]]$. By applying the isomorphism θ, we see that this is equivalent to showing *that the quotient* $[\phi_1(\eta) - \phi_2(\eta)]/x'$ *is integral over* $k[[x']]$. Now, η is in $\tilde{\mathfrak{o}}_{x'}$, and k is still a coefficient field of $\tilde{\mathfrak{o}}_{x'}$. Hence, after subtracting from η a suitable element of k, we may assume that η belongs to the maximal ideal $\tilde{\mathfrak{m}}_{x'}$ of $\tilde{\mathfrak{o}}_{x'}$. Now, the transforms $\mathfrak{o}'_1 = \phi_1(\tilde{\mathfrak{o}}_{x'})$ and $\mathfrak{o}_2 = \phi_2(\tilde{\mathfrak{o}}_{x'})$ are saturated local domains, with maximal ideals $\phi_1(\tilde{\mathfrak{m}}_{x'})$ and $\phi_2(\tilde{\mathfrak{m}}_{x'})$ respectively. Thus, $\phi_i(\eta)$ belongs to the maximal ideal of the saturated local domain $\mathfrak{o}_i'$ $(i = 1, 2)$. Now, if, say, $\phi_1(\epsilon_i) = 1$ $\phi_2(\epsilon_j) = 1$ $[i, j \in \{1, 2, \cdots, h\}]$, then $\mathfrak{o}_1' = \phi_1(\tilde{\mathfrak{o}}_x \cdot \epsilon_i)$ and $\mathfrak{o}_2' = \phi_2(\tilde{\mathfrak{o}}_x \cdot \epsilon_j)$ [see (23), § 1 and Corollary 3.6]. Since x' is a transversal parameter of $\mathfrak{o}$, $x'\epsilon_i$ is a transversal parameter of the local domain $\mathfrak{o}\epsilon_i$ and therefore $x'\epsilon_i$ is also a transversal parameter of $\widehat{(\mathfrak{o}\epsilon_i)}_{k((x'))\cdot\epsilon_i}$, i.e., of $\tilde{\mathfrak{o}}_x \cdot \epsilon_i$ [see Corollary 3.6]. Consequently, *x' is a transversal parameter of* $\mathfrak{o}_1'$. Similarly, x' is a transversal parameter of $\tilde{\mathfrak{o}}_2'$. It follows that both quotients $\phi_1(\eta)/x'$ and $\phi_2(\eta)/x'$ are integral over $k[[x']]$. This shows that the quotient

$[\phi_1(\eta) - \phi_2(\eta)]/x'$ is integral over $k[[x']]$, and the proof of our theorem is now complete.

The following proposition is almost an immediate consequence of the preceding theorem and is a generalization of an identical result proved by us elsewhere (see Corollary-Definition I, 7.3) for local domains.

PROPOSITION 4.6. *If* $\mathfrak{o}$ *is one of our local, not necessarily saturated, rings, then the saturation* $\tilde{\mathfrak{o}}_x$ *of* $\mathfrak{o}$ *with respect to a transversal parameter* x *of* $\mathfrak{o}$ *is independent of the choice of the transversal parameter (this ring* $\tilde{\mathfrak{o}}_x$ *will be denoted from now on by* $\tilde{\mathfrak{o}}$*).*

Proof. We first observe that if z is any parameter of $\mathfrak{o}$ and x is a transversal parameter of $\mathfrak{o}$, then x is also a transversal parameter of the caturation $\tilde{\mathfrak{o}}_z$. For, if x is transversal for $\mathfrak{o}$, $x\epsilon_i$ is transversal for $\mathfrak{o}\epsilon_i$, and therefore, by Proposition I, 2.5, $x\epsilon_i$ is also transversal for $\widetilde{(\mathfrak{o}\epsilon_i)}_{z\epsilon_i}$. This latter ring is equal to $\tilde{\mathfrak{o}}_z \cdot \epsilon_i$, by Corollary 3.6. Thus, $x\epsilon_i$ is a transversal parameter of $\tilde{\mathfrak{o}}_z \cdot \epsilon_i$, for $i = 1, 2, \cdots, h$, showing that x is transversal for $\tilde{\mathfrak{o}}_z$. Now, let x, x' be any two transversal parameters of $\mathfrak{o}$. Then x' is still a transversal parameter of $\tilde{\mathfrak{o}}_x$, and hence, by Theorem 4.5, $\tilde{\mathfrak{o}}_x$ is saturated with respect to x'. Therefore $\tilde{\mathfrak{o}}_{x'} \subset \tilde{\mathfrak{o}}_x$. Similarly we must have $\tilde{\mathfrak{o}}_x \subset \tilde{\mathfrak{o}}_{x'}$, and hence $\tilde{\mathfrak{o}}_{x'} = \tilde{\mathfrak{o}}_x$.

We shall conclude this section with the presentation of some special cases of the structure theorem, which we shall occasion to apply later on (see § 7).

Let $\mathfrak{o}$ be our saturated local ring, $\mathfrak{m}$ its maximal ideal, let x be any transversal parameter of $\mathfrak{o}$, and let $\{I_1, I_2, \cdots, I_r\}$ be the saturation-partition of the set $\{1, 2, \cdots, h\}$ relative to the pair $\{\mathfrak{o}, x\}$ (note that we know already that $\mathfrak{o}$ is saturated with respect to x). Let $\mathfrak{m}_\rho'$ denote the maximal ideal of the saturated ring $\mathfrak{o}\epsilon_\rho'$ $(\rho = 1, 2, \cdots, r)$.

PROPOSITION 4.7. *We have* $\mathfrak{m} = \sum_{\rho=1}^{r} \mathfrak{m}_\rho'$ *if and only if* $\gamma = v(x)$, *and when that is so then* $q = 1$.

Proof. Assume $\gamma = v(x)$. Since x is transversal, we have $\beta_{i,1} > v(x)$ for $i = 1, , \cdots, h$. Hence $q = 1$. By Theorem 3.4 we have then

$$\mathfrak{m} = \mathfrak{M} + \sum_{\rho=1}^{r} \mathfrak{N}_\rho'. \tag{22}$$

Since x is a transversal parameter, we have for any $\rho = 1, 2, \cdots, r$, and for any element ξ in $\mathfrak{m}_\rho'$: $v(\xi_i) \geqq v(x)$, for all $i \in I_\rho$; here ξ_i stands for $\psi_i(\xi)$.

This shows, in view of the definition (18), § 3, of $\mathfrak{A}_\rho'$, that if $\gamma = v(x)$, then $\mathfrak{A}_\rho' = \mathfrak{m}_\rho'$. Since $\mathfrak{M} \subset \Sigma\mathfrak{M}\epsilon_\rho' \subset \Sigma\mathfrak{m}_\rho'$, the equality $\mathfrak{m} = \sum_{\rho=1}^{r} \mathfrak{m}_\rho'$ follows from (22).

Conversely, assume that we have $\mathfrak{m} = \sum_{\rho=1}^{r} \mathfrak{m}_\rho'$. Fix two indices i, j which belong to different set I_ρ. The rings $\mathfrak{o}\epsilon_i$ and $\mathfrak{o}\epsilon_j$ are saturated, and their maximal ideals are $\mathfrak{m}\epsilon_i$ and $\mathfrak{m}\epsilon_j$ respectively. If $i \in I_\rho$ and $j \in I_\sigma$ $(\rho \neq \sigma)$, we have $\mathfrak{m}\epsilon_i = \mathfrak{m}_\rho'\epsilon_i$ and $\mathfrak{m}\epsilon_j = \mathfrak{m}_\sigma'\epsilon_j$. It follows that given any element $\xi^{(i)}$ in $\mathfrak{m}\epsilon_i$ and any element $\xi^{(j)}$ in $\mathfrak{m}\epsilon_j$, there exists an element ξ in $\mathfrak{m}$ such that $\xi\epsilon_i = \xi^{(i)}$ and $\xi\epsilon_j = \xi^{(j)}$. We take for $\xi^{(i)}$ any element of $\mathfrak{m}\epsilon_i$ such that $v(\psi_i(\xi^{(i)})) > v(x)$, and we take for $\xi^{(j)}$ the element $x\epsilon_j$. Then $\psi_i(\xi) - \psi_j(\xi) = \psi_i(\xi^{(i)}) - x$, and hence $v(\psi_i(\xi) - \psi_j(\xi)) = v(x)$. Since $v(\psi_i(\xi) - \psi_j(\xi)) \geqq v(y_i - y_j)$, and since, by the transversality of x, we have $x(y_{i'}^{\tau} - y_{j'}^{\sigma}) \geqq v(x)$ for any pair of distinct indices in $\{1, 2, \cdots, h\}$ and for any τ and σ in the Galois group G, it follows that $v(y_i - y_j) = v(x)$ and that $\gamma = v(x)$. This completes the proof.

COROLLARY 4.8. *Assume k algebraically closed. Then the have $\mathfrak{m} = \sum_{\rho=1}^{r} \mathfrak{m}_\rho'$ if and only if the plane irreducible curves $C_1, C_2, \cdots, C_h$ do not all have the same tangent at the origin O, and when that is so then two indices i, j belong to the same set I_ρ if and only if the curves C_i, C_j have the same tangent at O.*

For, we have $\gamma = v(x)$ if and only if there exists a pair of indices i, j such that $v(y_i - y_j) = v(x)$, and this last equality is valid if and only if C_i and C_j have distinct tangents at O (always under our assumption that x is a transversal parameter of the reducible curve $C = C_1 + C_2 + \cdots + C_h$). Therefore $\gamma = v(x)$ if and only if the curves $C_1, C_2, \cdots, C_h$ do not all have the same tangent, and the rest of the corollary follows immediately.

COROLLARY 4.9. *Assume that k is algebraically closed and that for some partition $\{\bar{I}_1, \bar{I}_2, \cdots, \bar{I}_s\}$ of $\{1, 2, \cdots, h\}$ we have $\mathfrak{m} = \sum_{\nu=1}^{s} \mathfrak{m}\bar{\epsilon}_\nu$, where $\bar{\epsilon}_\nu = \sum_{i \in \bar{I}_\nu} \epsilon_i$. Then for any pair of indices i, j in the set $\{1, 2, \cdots, h\}$ which belong to distinct sets $\bar{I}_\sigma$, it is true that C_i and C_j have distinct tangents at O; or—equivalently: each set $\bar{I}_\nu$ is the union of sets I_ρ, $\rho = 1, 2, \cdots, r$.*

For the proof we have only to repeat the reasoning given in the second part of the proof of Proposition 4. 7.

PROPOSITION 4.10. *Assume that k is algebraically closed and that the h irreducible components $C_1, C_2, \cdots, C_h$ of C have the same tangent at O. Then for any two transversal parameters x, x' of $\mathfrak{o}$ it is true that the residue of x'/x in $\bar{k}$ belongs to k.*

Proof. Let $c_i\epsilon_i$ be the residue of x'/x in $k\epsilon_i$, where $c_i \in k$. Then the residue c of x'/x in $\bar{k}$ $(= k\epsilon_1 + k\epsilon_2 + \cdots + k\epsilon_h)$ is $c_1\epsilon_1 + c_2\epsilon_2 + \cdots + c_h\epsilon_h$. We have to prove that $c_1 = c_2 = \cdots = c_h$, which will show that $c \in k$. Assume the contrary, and let, say, $c_1 \neq c_2$. Let ψ_1, ψ_2 be two $k((x))$-homomorphisms of F into the algebraic closure of $k((x))$ such that $\psi_1(\epsilon_1) = \psi_2(\epsilon_2) = 1$. Then it is clear that the residue $[\psi_1(x') - \psi_2(x')]/x$ in the integral closure of $k[[x]]$ in F^* is equal to $c_1 - c_2 \neq 0$. Therefore $v(\psi_1(x') - \psi_2(x')) = v(x)$, which implies that $v(\psi_1(y) - \psi_2(y)) = v(x)$, where y is an x-saturator of $\mathfrak{o}$. This shows that $\gamma = v(x)$ and hence, by Proposition 4.7, the C_i do not all have the same tangent (in fact, already C_1 and C_2 have distinct tangents). This is contrary to our assumption, and thus the proposition is proved.

Let C be a plane algebraic curve having h irreducible components $C_1, C, \cdots, C_h$, and defined over k. Let $k[[x]][y]$ be the local ring of C (where x is then a parameter of that ring) and let $\mathfrak{o}$ be the saturation of $k[[x]][y]$ with respect to $k((x))$.

Definition 4.11. *The saturation partition $\{I_1, I_2, \cdots, I_r\}$ of $\{1, 2, \cdots, h\}$, relative to the pair $\{\mathfrak{o}, x\}$, will also be referred to as the saturation partition of $\{1, 2, \cdots, h\}$, relative to the pair $\{C, x\}$. If we set*

$$\Gamma_\rho = \bigcup_{i \in I_\rho} C_i, \quad \rho = 1, 2, \cdots, r,$$

then the r curves Γ_ρ will be called the saturation components of C, relative to the parameter x.

COROLLARY 4.12. *If k is algebraically closed and if the branches C_i of C do not all have the same tangent, then the saturation components of C, relative to any transversal parameter, coincide with the tangential components of C.*

This is simply a reformulation of Corollary 4.8.

5. A parenthesis: the theory of infinitely near points in the plane (for arbitrary characteristic). In this section we shall develop briefly some classical concepts from the theory of infinitely near points in the plane, or—more generally—in the neighborhood of a simple point O of an analytic surface F. The ground field k will be assumed algebraically closed and *of arbitrary characteristic,* except at the end of the section (following the proof of Theorem 5.15), where the characteristic of k will be again assumed to be zero. The material developed in this section will be applied in the next section towards a geometric intepretation of the structure theorem proved in Sections 2 and 3. In developing this material, we refer the reader, for further details, to the following sources: Enriques-Chisini [2], Zariski [3] (Chapter I) and SES I.

A locally quadratic transformation T, with center O, blows up the point O into an irreducible, non-singular, rational curve E_1 (the exceptional curve of T) and transforms F into a non-singular surface F_1 containing E_1. Each point O_1 of E_1 (together with its local ring on F_1) is called *a quadratic transform of O.* We shall consider an infinite sequence $OO_1O_2\cdots O_i\cdots$ of points such that each point O_i is a quadratic transform of its predecessor O_{i-1} $(i \geqq 1, O_0 = O)$. We say then that O_i is a point *infinitely near O,* in *the i-th order neighborhood of O.* More generally, if $i < j$ then O_j is a point infinitely near O_i, in the $(j-i)$-th order neighborhood of O_i. Note that given O_i, the sequence $OO_1\cdots O_{i-1}$ is uniquely determined. We shall denote by F_i the surface which carries the point O_i (we set $F = F_0$) and by $T_{i,j}$ $(i < j)$ the transformation from F_i *to* F_j. Thus $T_{i,i+1}$ is a locally quadratic transformation of F_i, with center O_i, while $T_{i,j}$ $(i < j)$ is a product of the $j-i$ locally quadratic transformations $T_{i,i+1}, T_{i+1,i+2}, \cdots, T_{j-1,j}$.

If C_i is an algebroid (not necessarily irreducible) curve in F_i, with origin at O_i, we denote by $T_{i,j}[C_i]$ $(i < j)$ the *proper* transform of C_i in F_j, and by $T_{i,j}\{C_i\}$ the *total* transform of C_i, i.e., the union of the proper transform of C_i and the exceptional curves in F_j created by $T_{i,j}$.

We denote by $\mathcal{E}_{i,j}$ the total exceptional curve of $T_{i,j}$, on F_j, i.e., the total transform on F_j, of the point O_i. Thus, in particular, $\mathcal{E}_{i,i+1}$ is irreducible and non-singular; this irreducible curve will also be denoted by $E_{i,i+1}$. We set

$$E_{i,j} = T_{i+1,j}[E_{i,i+1}], \quad j > i+1.$$

It is easy to see that $\mathcal{E}_{i,j}$ is the union of the $j-i$ irreducible (non-singular) curves $E_{i,j}, E_{i+1,j}, \cdots, E_{j-1,j}$, that $\mathcal{E}_{i,j}$ is a connected curve, and that if $j > i+1$, then $\mathcal{E}_{i,j}$ has singular points, which are, however, only ordinary double points (normal crossings of two irreducible components).

If C_i is as above and $i < j$, then

$$T_{i,j}\{C_i\} = T_{i,j}[C_i] \cup \mathcal{E}_{i,j}. \tag{1}$$

In particular, if $i < j < q$, then

$$\mathcal{E}_{i,q} = T_{j,q}[\mathcal{E}_{i,j}] \cup \mathcal{E}_{j,q}. \tag{2}$$

Definition 5.1. *If $i < j$, then O_j is said to be free with respect to O_i, if O_j is a simple point of $\mathcal{E}_{i,j}$; O_j is called a satellite of O_i if it is a singular point (hence a double point) of $\mathcal{E}_{i,j}$.*

We proceed to derive a number of properties of free points and of satellites.

PROPOSITION 5.2. *O_{i+1} is free with respect to O_i.*

Proof. Trivial, since $E_{i,i+1}$ is a non-singular curve.

PROPOSITION 5.3. *Given O_i and O_j, with $i < j$. Then:*

(a) *Every satellite of O_j is also a satellite of O_i.*

(b) *If, furthermore, O_{j+1} is free with respect to O_i, then, conversely, every satellite O_q of O_i, $q > j + 1$, is also a satellite of O_j. (It is tacitly understood that O_q is also a point which is infinitely near O_{j+1}.)*

Proof. (a) If q is any integer greater than j, then, by (2), we have $\mathcal{E}_{j,q} \subset \mathcal{E}_{i,q}$. Hence if O_q is a singular point of $\mathcal{E}_{j,q}$, it is *a fortiori* a singular point of $\mathcal{E}_{i,q}$.

(b) By assumption, O_{j+1} is a simple point of $\mathcal{E}_{i,j+1}$, and, by (2), we have $\mathcal{E}_{i,j+1} = T_{j,j+1}[\mathcal{E}_{i,j}] \cup \mathcal{E}_{j,j+1}$, with $O_{j+1} \in \mathcal{E}_{j,j+1}$. Therefore $O_{j+1} \notin T_{j,j+1}[\mathcal{E}_{i,j}]$. This implies that for any $q > j + 1$, the two curves $\mathcal{E}_{j+1,q}$ and $T_{j,q}[\mathcal{E}_{i,j}]$ $(= T_{j+1,q}[T_{j,j+1}[\mathcal{E}_{i,j}]])$ have no common points. Since $O_q \in \mathcal{E}_{j+1,q}$, it follows that

$$O_q \notin T_{j,q}[\mathcal{E}_{i,j}]. \tag{3}$$

If then O_q is a singular point of $\mathcal{E}_{i,q}$, it follows from (2) and (3) that O_q is also a singular point of $\mathcal{E}_{j,q}$, and this completes the proof.

PROPOSITION 5.4. *Given O_i and O_j, with $i < j$, the first order neighborhood of O_j contains exactly one satellite of O_i or exactly two satellies of O_i, according as O_j is free with respect to O_i or is a satellite of O_i.*

Proof. If O_j is a simple point of $\mathcal{E}_{i,j}$, then the irreducible exceptional curve $E_{j,j+1}$ in F_{j+1} meets $T_{j,j+1}[\mathcal{E}_{i,j}]$ in exactly one point, say O'_{j+1}. Since $\mathcal{E}_{i,j+1} = T_{j,j+1}\ \mathcal{E}_{i,j}] \cup E_{j,j+1}$, it follows that O'_{j+1} is the only point of $E_{j,j+1}$ which is singular for $\mathcal{E}_{i,j+1}$. If, however, O_j is a double point of $\mathcal{E}_{i,j}$, then $E_{j,j+1}$ meets $T_{j,j+1}[\mathcal{E}_{i,j}]$ in exactly two points, say O'_{j+1}, O''_{j+1} (corresponding to the two tangential directions of $\mathcal{E}_{i,j}$ at O_j), and these two points are therefore the only points of $E_{j,j+1}$ which are singular for $\mathcal{E}_{i,j+1}$. Q. E. D.

We shall now deal with an infinite sequence of infinitely near points $OO_1O_2 \cdots$ *which lie on a given irreducible algebroid curve* C *in* F, *with origin* O. When we say that the points O_i all lie on C we mean of course that, for any i, the point O_i is the origin of the proper transform $C_i = T_{0,i}[C]$ of C under $T_{0,i}$. Since any irreducible algebroid curve has a unique tangential direction at its origin, the sequence $OO_1O_2 \cdots$ is uniquely determined by C. We shall denote by s_i the multiplicity of the point O_i of C_i (in other words, s_i is the multiplicity of the local domain of O_i on C_i). We shall constantly make use of the following well known facts (see Zariski, SES I, p. 515):

$$s_i = (C_{i+1}, E_{i,i+1}) = (C_{i+1}, E_{i,i+1})_{O_{i+1}}, \tag{4}$$

where the right-hand side stands for the intersection multiplicity of C_{i+1} and $E_{i,i+1}$ at O_{i+1} (this is also the intersection number of C_{i+1}, $E_{i,i+1}$, since these two curves meet only in O_{i+1});

$$(C_i, D_i)_{O_i} = (C_{i+1} + s_i E_{i,i+1}, D_{i+1} + \sigma_i E_{i,i+1}), \tag{5}$$

where D_i is another irreducible algebroid curve through O_i, and σ_i is the multiplicity of O_i on D_i;

$$(E_{i,i+1}, E_{i,i+1}) = -1. \tag{6}$$

From (4) it follows that

$$s_i \geqq s_{i+1}, \quad i = 0, 1, 2, \cdots. \tag{7}$$

Since the singularity of C at O can be resolved by a finite number of locally quadratic transformations it follows that

$$s_i = 1 \text{ for } i \text{ large}. \tag{8}$$

From (4), (5) and (6) follows easily that

$$(C_i, D_i)_{O_i} = (C_{i+1}, D_{i+1}) + s_i\sigma_i. \tag{9}$$

Note that if O'_{i+1} denotes the point on D which immediately follows O_i, then C_{i+1}, D_{i+1} meet if and only if $O_{i+1} = O'_{i+1}$, and when that is so then $(C_{i+1}, D_{i+1}) = (C_{i+1}, D_{i+1})_{O_{i+1}}$.

PROPOSITION 5.5. *Let* $s_i = hs_{i+1} + s'$, *where* $h \geqq 1$ *and* $0 \leqq s' < s_{i+1}$. *Then the following is true:*

(a) $s_{i+1} = s_{i+2} = \cdots = s_{i+h}$.

(b) *If* $h > 1$, *then* $O_{i+2}, O_{i+3}, \cdots, O_{i+h}$ *are satellites of* O_i [*hence also of* O; *see Proposition* 5.3, *part* (a)]

(c) *If* $s' \neq 0$, *then* $s_{i+h+1} = s'$, *and* O_{i+h+1} *is also a satellite of* O_i (*and hence also of* O).

(d) *If* $s' = 0$, *then* O_{i+h+1} *is free with respect to* O_i. *If furthermore* $s' = 0$ *and* $h > 1$, *then* O_{i+h+1} *is free even with respect to* O.

Proof. (a) This part of the theorem is vacuous if $h = 1$. We shall assume therefore that $h > 1$. Applying (9) to the case in which C_i, D_i and O_i are replaced by C_{i+1}, $E_{i,i+1}$ and O_{i+1}, in which case the left hand side of (9) is equal to s_i, by (4), while s_i and σ_i are to be replaced by s_{i+1} and 1 respectively, we find:

$$(10) \qquad (C_{i+2}, E_{i,i+2})_{O_{i+2}} = s_i - s_{i+1} \geqq s_{i+1} \text{ (since } h > 1\text{)},$$

whence, at any rate, $O_{i+2} \in E_{i,i+2}$. Since $(C_{i+2}, E_{i+1,i+2})_{O_{i+2}} = s_{i+1}$ [by (4)], and since $E_{i+1,i+2}$ and $E_{i,i+2}$ have distinct tangents at O_{i+2}, it follows from this last equality and from (10) that $s_{i+2} = s_{i+1}$. If $h = 2$, the proof of (a) is complete. If $h > 2$, then, using (10) and applying (9), with C_i, D_i, O_i replaced by C_{i+2}, $E_{i,i+2}$ and O_{i+2}, respectively, we have, in view of the equality $s_{i+2} = s_{i+1}$ just established above, that $(C_{i+3}, E_{i,i+3})_{O_{i+3}} = s_i - 2s_{i+1} \geqq s_{i+1}$ (since $h > 2$), and from this it follows, in a similar fashion, that $s_{i+3} = s_{i+1}$ [in view of the equality $(C_{i+3}, E_{i+2,i+3})_{O_{i+3}} = s_{i+2} = s_{i+1}$]. If $h = 3$, the proof of (a) is complete. In this fashion the proof of (a) can be completed for arbitrary h.

(b) The proof of (a) shows that we have

$$(11) \qquad (C_{i+q}, E_{i,i+q})_{O_{i+q}} = s_i - (q-1)s_{i+1} \geqq s_{i+1} > 0, \quad \text{for } q = 2, 3, \cdots, h.$$

Hence $O_{i+q} \in E_{i,i+q}$ $(2 \leqq q \leqq h)$, and since O_{i+q} also belongs to $E_{i+q-1,i+q}$, and $E_{i,i+q} \neq E_{i+q-1,i+1}$ (since $q \geqq 2$), it follows that O_{i+q} is a double point of $\mathcal{E}_{i,i+q}$, for $2 \leqq q \leqq h$, and this establishes part (b) of the theorem.

(c) The reasoning employed in the proof of part (a) shows that we have (also if $h = 1$)

$$(12) \qquad (C_{i+h+1}, E_{i,i+h+1})_{O_{i+h+1}} = s_i - hs_{i+1} = s' < s_{i+1}.$$

Since $(C_{i+h+1}, E_{i+h,i+h+1}) = s_{i+1}$ [in view of $s_{i+h} = s_{i+1}$ and in view of (4), as applied to the case in which i is replaced by $i + h$)], and since we are now dealing with ease in which $s' > 0$, it follows from (12) that $s_{i+h+1} = s'$ (always using the fact that $E_{i,i+h+1}$ and $E_{i+h,i+h+1}$ have distinct tangents at O_{i+h+1}). Furthermore, the inequality $s' > 0$ implies that O_{i+h+1} lies on *both* curves $E_{i,i+h+1}$ and $E_{i+h,i+h+1}$, whence O_{i+h+1} *is* double point of $\mathcal{E}_{i,i+h+1}$.

(d) If $s' = 0$, then (11) implies that

$$(C_{i+h}, E_{i,i+h})_{O_{i+h}} = s_{i+1}, \tag{13}$$

where we note that by (4), (13) is valid also if $h = 1$, since we have then $s_{i+1} = s_i$. On the other hand, if we apply (4) to C_{i+h} instead of to C_{i+1}, we find

$$(C_{i+h}, E_{i+h-1,i+h})_{O_{i+h}} = s_{i+h-1}. \tag{14}$$

If $h > 1$, then $s_{i+h-1} = s_{i+1}$, by part (a) of the proposition. In this case, (13) and (14) imply therefore that the two tangents of $\mathcal{E}_{0,i+h}$ at O_{i+h}, i.e., the tangents of $E_{i,i+h}$ and $E_{i+h-1,i+h}$ at O_{i+h}, are distinct from the tangent of C_{i+h} at O_{i+h} (note that since $h > 1$, $E_{i,i+h}$ and $E_{i+h-1,i+h}$ are *distinct* curves). But this implies that O_{i+h+1} lies only one one irreducible component of $\mathcal{E}_{0,i+h+1}$, namely on $E_{i+h,i+h+1}$, and is therefore a simple point of $\mathcal{E}_{0,i+h+1}$.

Assume now $h = 1$ (and s' still equal to zero), whence $s_i = s_{i+1}$. In this case, (4) implies that C_{i+1} is not tangent to $E_{i,i+1}$ at O_{i+1}. This signifies that the origin O_{i+2} of C_{i+2} does not lie on the proper transform $E_{i,i+2}$ of $E_{i,i+1}$. Therefore O_{i+2} is a simple point of $\mathcal{E}_{i,i+2}$ $(= E_{i+1,i+2} \cup E_{i,i+2})$.

This completes the proof of the theorem.

COROLLARY 5.6. *We have* $s_i = s_{i+1}$ *if and only if* O_{i+2} *is free with respect to* O_i.

For, if $s_i = s_{i+1}$, then we are dealing with the case $h = 1$, $s' = 0$ of Proposition 5.5, and hence O_{i+2} is free with respect to O_i, by part (d) of the proposition. If, however, $s_i > s_{i+1}$, then we have either $h > 1$ or $h = 1$ and $s' > 0$, and in either case O_{i+2} is a satellite of O_i, by parts (b) and (c) of the proposition.

COROLLARY 5.7. *If* $s = 1$ *(and therefore all* s_i *are* 1*), all the points* O_i *are free with respect to* O.

O_1 is free with respect to O, by the definition of free points, and O_2 is free with respect to O, by Corollary 5.6 (since $s_1 = s_2 = 1$). We now use induction on i. Assume that we know already that $O_1, O_2, \cdots, O_i$ are free with respect to O. If we apply part (b) of Proposition 5.3, with i, j replaced respectively by 0 and $i-1$, we see that the assumption of part (b) of that proposition is satisfied: namely, O_i is free with respect to O. Therefore, every point which is free with respect to O_{i-1} is also free with respect to O.

But O_{i+1} is free with respect to O_{i-1}, by Corollary 5.6 (since $s_{i-1} = s_i = 1$). Hence O_{i+1} is free with respect to O.

COROLLARY 5.8. *If $s > 1$ then all the points O_i, for i sufficiently large, are free with respect to* 0. *More precisely, if q is the integer such that $s_q > 1$ and $s_i = 1$ for all $i > q$, then all the points O_i, $i \geqq q + s_q + 1$, are free with respect to O, while $O_{q+2}, O_{q+3}, \cdots, O_{q+s_q}$ are satellies of O.*

We have $s_q > s_{q+1} = 1$, and hence, if we apply Proposition 5.5 to the case $i = q$, we find $h = s_q > 1$ and $s' = 0$, and thus, by part (b) of that proposition it follows that $O_{q+2}, O_{q+3}, \cdots, O_{q+s_q}$ are satellites of O, while part (d) of the proposition implies that (a) *O_{q+s_q+1} is free with respect to O.* We also have that (b) *every point O_i, $i > q + s_q$ is free with respect to O_{q+s_q}*, in view of Corollary 5.7, since $s_i = 1$ for all $i \geqq q + s_q$. The assertion that O_i is free with respect to O also if $i > q + s_q + 1$ follows now from (a) and (b) above and from Proposition 5.3, part (b), as applied to the case $i = 0$, $j = s + s_q$.

For any given point O_i there exists a positive integer ρ (depending on i) such that $s_i = s_{i+1} + s_{i+2} + \cdots + s_{i+\rho}$. Namely, if $s_i = s_{i+1}$, then $\rho = 1$; if $s_i > s_{i+1}$ and if, say, $s_i = hs_{i+1} + s'$, $0 \leqq s' < s_{i+1}$, then, by Proposition 5.5, we find that $\rho = h + 1$ if $s' \neq 0$ (in which case $s_{i+1} = s_{i+2} = \cdots = s_{i+\rho-1}$, $s' = s_{i+\rho}$) and $\rho = h$ if $s' = 0$ (in which case $s_{i+1} = s_{i+2} = \cdots = s_{i+\rho}$). This observation leads to the classical concept of *proximate points of O_i*:

Definition 5.9. *Each of the points $O_{i+1}, O_{i+2}, \cdots, O_{i+\rho}$ (such that $s_i = s_{i+1} + s_{i+2} + \cdots, s_{i+\rho}$) is called a proximate point of O_i.*

Every point O_j $(j > 0)$ is a proximate point of at least one point, namely of O_{j-1}. If there is another point O_i, different from O_{j-1}, such that O_j is a proximate point of O_i, then Proposition 5.5 shows that we must have $s_i > s_{i+1} = s_{i+2} = \cdots = s_{j-1}$. Thus, the index i (and hence also the point O_i) is uniquely determined: *i is the greatest integer such that $s_i > s_{j-1}$.* Of course, if $s = s_{j-1}$ (and only in that case), such an integer i does not exist, and in that case therefore O_{j-1} is the only point which has O_j as a proximate point. But also if $s > s_{j-1}$, it may very well happen that O_j is not a proximate point of the point O_i defined above. At any rate, we see that any point O_j $(j > 0)$ is a proximate point of at *most* two points, namely of O_{j-1} and the above point O_i. The precise situation is an immediate consequence of Proposition 5.5 and is described in the following corollary:

COROLLARY 5.10. (a) *If $s_{j-1} = s$ then O_j is a proximate point of only one point, namely of O_{j-1}.*

(b) *If* $s_{j-1} < s$, *let* i *be the integer defined above, namely the greatest integer* i *such that* $s_i > s_{j-1}$ (*whence* $s_{i+1} = s_{j-1}$). *If we write* $s_i = hs_{j-1} + s'$, *with* $0 \leqq s' < s_{j-1}$, *then* O_j *is a proximate point of* O_i *if and only if either* $s' \neq 0$ *or* $s' = 0$ *and* $j \leqq i + h$. *Furthermore,* O_{j-1} *is the only point different from* O_i *which has* O_j *as proximate point.*

The only part of the corollary that needs a word of explanation is the assertion that O_j is a proximate point of O_i if $s' \neq 0$. Now, it follows from Proposition 5.5 and from the equality $s_{j-1} = s_{i+1}$, that if $s' \neq 0$ then the set of proximate points of O_i includes *all* the points O_q such that $s_q = s_{i+1}$ and *one* additional point, of multiplicity $s' < s_{i+1}$. Thus, the *last* proximate point of O_i cannot have multiplicity s_{j-1} $(= s_{i+1})$; in particular, O_{j-1} is *not* the lsast proximate point of O_i, and this implies that O_j is still in the set of proximate points of O_i.

We are now in a position to prove the following characterization of the points O_j which are *free* with respect to O:

PROPOSITION 5.11. *A point* O_j *is free with respect to* O *if and only if* O_j *is a proximate point of the point* O_{j-1} *only.*

Proof. Assume first that O_j is also a proximate point of the point O_i defined in Corollary 5.10. Part (b) and (c) of Proposition 5.5 show that *all* the proximate points of a point O_i, except perhaps O_{i+1}, are satellites of O. Since $i < j - 1$, i.e., $i + 1 < j$, we conclude that O_j is a satellite of O.

Assume now that O_{j-1} is the only point which has O_j as a proximate point. We must prove that O_j is then free with respect to O. We shall consider separately the two cases (which are the only possible ones, according to Corollary 5.10):

(a) $s_{j-1} = s$; (b) $s_{j-1} < s$ and $s' = 0$, $j > i + h$ (the notations being those of Corollary 5.10).

Case (a). We have in this case $s = s_1 = s_2 = \cdots = s_{j-1}$. If $j = 2$, then O_j $(= O_2)$ is free with respect to O, by Corollary 5.6. If $j > 2$ we use induction from $j - 1$ to j. We assume therefore that O_{j-1} is known to be free with respect to O. It follows then, by part (b) of Proposition 5.3, that every point which is free with respect to O_{j-2} is also free with respect to O. But O_j is free with respect to O_{j-2}, by Corollary 5.6 (since $s_{j-2} = s_{j-1}$). Hence O_j is free with respect to O.

Case (b). We have in this case $s_i = hs_{i+1}$, with $h > 1$ (since $s_i > s_{j-1}$ $= s_{i+1}$). Hence, by part (d) of Proposition 5.5, we have that O_{i+h+1} is free with respect to O. Therefore, by part (b) of Proposition 5.3, every point which is free with respect to O_{i+h} is also free with respect to O. Since $j > i + h$, it remains to prove that O_j is free with respect to O_{i+h}. But this follows from the case (a) of the proof, as applied to the sequence of points $O_{i+h}, O_{i+h+1}, \cdots, O_j$, since we have $s_{i+h} = s_{i+h+1} = \cdots = s_{j-1}$.

We shall now introduce a concept which will be useful in our forthcoming geometric interpretation of the structure theorem (§ 6).

Definition 5.12. *A satellite O_i of O os called a terminal satellite of O if O_{i+1} is free with respect to O. The point O_{i+1} is then called a leading free point of the algebroid curve C. (In other words, a point O_j, $j > 1$, of the sequence $OO_1O_2 \cdots$ is a leading free point on C (with respect to O) if it is free with respect to O and if its immediate predecessor O_{j-1} is a satellite of O).*

If C is a regular curve, i.e., if $s = 1$, there are no satellites in the sequence $OO_1O_2 \cdots$ and no leading free points. If, however, $s > 1$, there exist terminal satellites, and the number of terminal satellites (which is finite, by Corollary 5.8) is the same as the number of leading free points.

The following proposition deals mainly with the concept of leading free points and terminal satellites.

PROPOSITION 5.13. (a) *If a point O_{j-1} is free with respect to O (in particular, if O_j is a terminal satellite of O), then the set of terminal satellites of O_j coincides with the set of terminal satellites of O which follow O_j (or equivalently: the set of leading free points with respect to O_j coincides with the set of leading free points with respect to O which follow O_{j+1}).*

(b) *If a point O_{j+1} is free with respect to a point O_q and if $s_j \neq s_q$ (whence $q < j$), then O_{j+1} is also free with respect to O. In particular, if O_{j+1} is a leading free point of some point O_q, then O_{j+1} is also a leading free point of O.*

Proof. (a) This is a direct consequence of Proposition 5.3.

(b) We have, by assumption, $s_q > s_j$. Let i be the greatest integer such that $s_i > s_j$ (whence $q \leqq i < j$). By Proposition 5.11, as applied to O_{j+1} and O_q instead of to O_j and O, we have that O_{j+1} is *not* a proximate point of O_i. Therefore, again by Proposition 5.11 and part (b) of Corollary 5.10, O_{j+1} is also free with respect to O. If O_{j+1} is a leading free point of O_q, then O_j is a satellite of O_q, and therefore $s_q > s_j$ (by part (a) of Corollary 5.10 and by Proposition 5.11). On the other hand, O_j, being a satellite of O_q, is *a fortiori* a satellite of O. Thus O_{j+1} is a leading free point with respect to O.

5

By Proposition 5.11 there is close connection between the concept of satellites and the concept of proximate points. We establish below some results which further illustrate that connection and justify to some extent the term "satellite."

PROPOSITION 5.14. *Let O_i be a point in the i-th order neighborhood of O and let O_{i+1} be a point in the first order neighborhood of O_i. Let C, C′ be two irreducible algebraic curves, both passing through O_i and O_{i+1}, and let P, P′ be the set of proximate points of O_i on C and C′ respectively. Then at least one of the sets P, P′ is always a subset of the other. More precisely, if s_i, s_{i+1} are the multiplicites of C at O_i and O_{i+1} and s'_i, s'_{i+1} are the multiplicities of C′ at O_i and O_{i+1}, then $P \subset P'$ if either $[s_i/s_{i+1}] < [s'_i/s'_{i+1}]$ (where [ ] denotes the integral part) or $[s_i/s_{i+1}] = [s'_i/s'_{i+1}]$ and*

$$[(s_i - 1)/s_{i+1}] \leqq [(s'_i - 1)/s'_{i+1}].$$

Proof. Let $h = [s_i/s_{i+1}]$, $h' = [s'_i/s'_{i+1}]$ and let $s_i = hs_{i+1} + \sigma$, where $0 \leqq \sigma < s_{i+1}$. The first point of the (naturally ordered) sets P and P' is O_{i+1}, and P consists of h points $O_{i+1}, O_{i+2}, \cdots, O_{i+h}$ or of $h+1$ points $O_{i+1}, O_{i+2}, \cdots, O_{i+h}, O_{i+h+1}$, according as $\sigma' = 0$ or $\sigma' > 0$. We use the relations (11), and we assume that $h \leqq h'$. For $q = 2$ (and assuming $h \geqq 2$) relation (11) shows that O_{i+2} is uniquely determined by O_i and O_{i+1}: it is the common point of the exceptional curve $E_{i+1,i+2}$ (the blow up of O_{i+1}) and the exceptional curve $E_{i,i+2}$ (the proper transform of $E_{i,i+1}$ by $T_{i+1,i+2}$). Therefore $O'_{i+2} = O_{i+2}$. Similarly, if $h \geqq 3$, then the point O_{i+3} is uniquely determined: it is the common point of $E_{i+2,i+3}$ and of the proper transform $E_{i,i+3}$ of $E_{i,i+2}$ by $T_{i+2,i+3}$. Hence $O'_{i+3} = O_{i+3}$. In this fashion, we find successively that $O'_{i+q} = O_{i+q}$, for $q = 1, 2, \cdots, h$ (since $h \leqq h'$). *Thus the inclusion $P \subset P'$ is proved if $h \leqq h'$ and $\sigma = 0$.* So from now on we may assume that $\sigma > 0$. Let now $h < h'$. Since $\sigma > 0$, P has an extra point O_{i+h+1} (of multiplicity σ for C), and (12) shows that O_{i+h+1} is the common point of $E_{i+h,i+h+1}$ and $E_{i,i+h+1}$. Since $h < h'$, P' contains a point O'_{i+h+1}, and (11), applied to C' and to $q = h+1$ ($\leqq h'$), shows that O'_{i+h+1} is also the common point of $E_{i+h,i+h+1}$ and $E_{i,i+h+1}$. Thus $O'_{i+h+1} = O_{i+h+1}$, and the inclusion $P \subset P'$ is proved.

Assume now that $h = h'$ and that $[(s_i - 1)/s_{i+1}] \leqq [(s'_i - 1)/s'_{i+1}]$. Since $\sigma > 0$, we have $[(s_i - 1)/s_{i+1}] = h$. From $[(s'_i - 1)/s'_{i+1}] \geqq h$ and $[s'_i/s'_{i+1}] = h' = h$, follows that also $\sigma' > 0$, and this implies that the point O_{i+h+1} also belongs to P' (and that, in this case, we have in fact $P = P'$). This completes the proof.

THEOREM 5.15. *Let O_i be a point in the i-th neighborhood of O and let O_{i+1} be a point in the first neighborhood of O_i. Let C, C' be two irreducible algebroid curves passing through both points O_i, O_{i+1}, and let s_i, s_{i+1} (resp., s'_i, s'_{i+1}) be the multiplicities of O_i, O_{i+1} for C (resp., for C'). Assume that $s_i/s_{i+1} = s'_i/s'_{i+1}$, say $s_i/s_{i+1} = s'_i/s'_{i+1} = n/n_1$, where $(n, n_1) = 1$. We develop n/n_1 into a continued fraction:*

$$n/n_1 = h_1 + \cfrac{1}{h_2 + \cfrac{}{\ddots + \cfrac{1}{h_q}}}, \tag{15}$$

and we write

$$\begin{aligned} n &= h_1 n_1 + n_2, \\ n_1 &= h_2 n_2 + n_3, \qquad n_1 > n_2 > \cdots > n_q > 0, \\ &\vdots \qquad\qquad\qquad\qquad q \geqq 1. \\ n_{q-1} &= h_q n_q, \end{aligned} \tag{16}$$

Then C, C' have also in common the $h_1 + h_2 + \cdots + h_{q-1}$ points $O_{i+2}, O_{i+3}, \cdots, O_{i+h_1+h_2+\cdots+h_q}$, all these points are satellites with respect to O_i, and $O_{i+h_1+h_2+\cdots+h_q}$ is a terminal satellite with respect to O_i, on both C and C'. Furthermore, if we set

$$s_{i+1} = \rho n_1, \quad s'_{i+1} = \rho' n_1,$$

then the first h_1 of the $h_1 + h_2 + \cdots + h_q$ points $O_{i+1}, O_{i+2}, \cdots, O_{i+h_1+h_2+\cdots+h_q}$ are of multiplicity ρn_1 (resp., $\rho' n_1$) on C (resp. on C') ;the next h_2 points are of multiplicity ρn_2 (resp., $\rho' n_2$) on C (resp., on C'), etc.; the last h_q points are of multiplicity ρn_q (resp., $\rho' n_q$) on C (resp., on C').

Proof. Since $s_i/s_{i+1} = n/n_1$ and $s_{i+1} = \rho n_1$, we have $s_i = \rho n$, and thus we have relations similar to the (16), with the integers $n, n_1, n_2, \cdots, n_q$ replaced by $\rho n, \rho n_1, \rho n_2, \cdots, \rho n_q$. From the first of these relations, namely $s_i = h_1 s_{i+1} + \rho n_2$, follows that $O_{i+1}, O_{q+2}, \cdots, O_{i+h_1+1}$ are the proximate points of O_i on C, and that their multiplicities are $s_{i+1} = s_{i+2} = \cdots = s_{i+h_1}$ $(= \rho n_1)$ and $s_i + h_1 + 1$ $(= \rho n_2)$. [Here, we assume naturally that $n_2 > 0$, otherwise the set of proximate points is $O_{i+1}, O_{i+2}, \cdots, O_{i+h}$, all of the same multiplicity ρn_1.] The second of the relations (16) yields $\rho n_1 = h_2 \cdot \rho n_2 + \rho n_3$, and this implies (assumes that $n_3 > 0$, i. e., that $q > 2$) that $O_{i+h_1+1}, O_{i+h_1+2}, \cdots, O_{i+h_1+h_2+1}$ are the proximate points of O_{i+h_1}, and their multiplicities on C are all equal to ρn_2, except for the last point $O_{i+h_1+h_2+1}$ which has multiplicity ρn_3. At the last q-th step, we find that the point $O_{i+h_1+h_2+\cdots+h_{q-1}}$, of multiplicity ρn_{q-1}, has the h_q proximate points $O_{i+h_1+h_2+\cdots+h_{q-1}+1},\cdots,O_{i+h_1+h_2+\cdots+h_q}$, and that these all have multiplicity ρn_q on C. This analysis shows that each of the $h_1+h_2+\cdots+h_q-1$ points O_j in the set $\{O_{i+2},O_{i+3},\cdots,O_{i+h_1+h_2+\cdots+h_q}\}$ is a proximate point of *two* points (namely of its immediate predecessor O_{j-1} and of one of the points $O_{i+h_1+h_2+\cdots+h_\nu}$, different from O_{j-1}), and hence is a satellite of O_i (Proposition 5.11, with O replaced by O_i). Since for each point O_j in the set $\{O_i,O_{i+1},\cdots,O_{i+h_1+h_2+\cdots+h_q}\}$ it is true that all the proximate points of O_j are contained in that set, it follows that the point $O_{i+h_1+h_2+\cdots+h_q+1}$ is the proximate point only of its immediate predecessor, and therefore $O_{i+h_1+h_2+\cdots+h_q}$ is a terminal satellite with respect to O_i (Proposition 5.11). Similar considerations apply to C', and the assertion that all the points

$$O_{i+2},O_{i+3},\cdots,O_{i+h_1+h_2+\cdots+h_q}$$

belong also to C' follows from Proposition 5.14.

In the remainder of this section we shall assume that our algebraically closed ground field k is of characteristic zero. In that case, our irreducible plane algebroid curve C will be represented by a Puiseux expansion

$$(17)\quad C: y=a_1x+a_2x^2+\cdots+a_hx^h+b_1x^{\beta_1/n}+\cdots+b_2x^{\beta_2/n}+\cdots+b_gx^{\beta_g/n}+\cdots,$$

where $\beta_1/n,\beta_2/n,\cdots,\beta_g/n$ are the characteristic exponents of the above expansion $y=y(x^{1/n})$. We shall assume that x is a transversal parameter of the local ring $\mathfrak{o}$ of C. In that case we have $\beta_1/n>1$. We set

$$(18)\quad e_0=n,\quad e_1=(\beta_1,e_0),\cdots,e_{\nu+1}=(\beta_{\nu+1},e_\nu),\quad \nu=0,1,\cdots,g-1.$$

We recall the definition of the characteristic exponent: β_1/n is the first exponent which is not an integer, β_2/n is the first exponent such that $\beta_2\not\equiv 0 \pmod{e_1}$, etc.; β_g is the first exponent such that $\beta\not\equiv 0 \pmod{e_{g-1}}$, and $e_g=(\beta_g,e_{g-1})=1$. Thus

$$e_0>e_1>\cdots>e_{g-1}>e_g=1.$$

The set of integers $(n;\beta_1,\beta_2,\cdots,\beta_g)$ is called *the x-characteristic of C*. It is known that if x is a transversal parameter then the x-characteristic of C is independent of x; we call then the set of $g+1$ integers $(n;\beta_1,\beta_2,\cdots,\beta_g)$ *the characteristic of C*. For all this we refer to our paper SES III. The exponent h which appears in (17) is the integral part $[\beta_1/n]$ of β_1/n.

It is clear from (17) that the multiplicity s of O on C is equal to n. For $i=1,2,\cdots,h-1$, the i-th quadratic transform C_i of C is given by

(19) $$C_i\colon y_i = a_{i+1}x + \cdots + a_h x^{h-i} + b_1 x^{(\beta_1 - in)/n} + \cdots + b_g x^{(\beta_g - in)/n} + \cdots,$$

while the transform C_h of C is given by

(20) $$C_h\colon y_h = b_1 x^{(\beta_1 - hn)/n} + \cdots + b_g x^{(\beta_g - hn)/n} + \cdots,$$

where

(21) $$y_i = [y - (a_1 x + a_2 x^2 + \cdots + a_i x^i)]/x^i, \quad i = 1.2, \cdots, h.$$

If we set

(22) $$\beta_1 = hn + n_1, \quad \text{o} < n_1 < n,$$

then it follows from (19) and (20) that

$$s = s_1 = s_2 = \cdots = s_{h-1} = n;\ s_h = n_1,$$

and that the point $O_1, O_2, \cdots, O_h$ are therefore free with respect to O. Furthermore, from (21) it follows that the position of the h free points $O_1, O_2, \cdots, O_h$ depends on the values of the h coefficients $a_1, a_2, \cdots, a_h$ in (17).

We now expand the quotient n/n_1 into a continued fraction (15) and we use the notations of (16). Since $s_{h-1} = n$ and $s_h = n_1$, we can apply the information, supplied by Theorem 5.15, to the curce C and the points O_i, O_{i+1}, where i is now equal to $h-1$ and where also the integer ρ which appears in the statement of that theorem is now equal to 1. Thus, we conclude that the $h_1 + h_2 + \cdots + h_{q-1}$ points $O_{h+1}, O_{h+2}, \cdots, O_{h+h_1+\cdots+h_{q-1}}$ which follow O_h on C are satellies of O, that the last of these points is a terminal satellite of O, and that

(23) $$\begin{aligned} &s_h = s_{h+1} = \cdots = s_{h+h_1-1} = n_1, \\ &s_{h+h_1} = s_{h+h_1+1+\cdots+} = s_{h+h_1+h_2-1} = n_2, \\ &s_{h+h_1+\cdots+h_{q-1}} = s_{h+h_1+\cdots+h_{q-1}+1} = \cdots = s_{h+h_1+\cdots+h_q-1} = n_q. \end{aligned}$$

The positions of these satellitss as well as their number and their multiplicities are completely determined by O_h, and by the first characteristic exponent β_1/n, *and are independent of the alue of the coefficient* b_1.

We shall denote the first terminal satellite $O_{h+h_1+\cdots+h_q-1}$ of O on C by $\bar{O}_1$, and its immediate successor (the first leading free point on C) by $O_1{}^*$. The following theorem gives, in principle, complete information concerning the composition of the sequence of points following the first terminal satellite $\bar{O}_1$.

Theorem 5.16. *There are exactly g terminal satellies of O on C. If*

$\bar{O}_1, \bar{O}_2, \cdots, \bar{O}_g$ denote these terminal satellites, and if $\bar{C}_\nu$ denotes the transform of C which has origin at $\bar{O}_\nu$, then the characteristic of $\bar{C}_\nu$ is equal to $(e_\nu; \beta_{\nu+1} - \beta_\nu + e_\nu, \cdots, \beta_g - \beta_\nu + e_\nu)$ $(\nu = 1, 2, \cdots, g)$, where the e_ν are the integers defined in (18), and where it is tacitly understood that when $\nu = g$ we have a regular curve $\bar{C}_g$ ($e_g = 1$, while the remaining integers of the characteristic of $\bar{C}_g$ are not defined).

Proof. By induction on g, it will be sufficient to show that the characteristic of $\bar{C}_1$ is $(e_1; \beta_2 - \beta_1 + e_1, \cdots, \beta_g - \beta_1 + e_1)$, since from (18) it follows that

$$e_2 = (\beta_2 - \beta_1 + e_1, e_1), \cdots, e_{\nu+1} = (\beta_{\nu+1} - \beta_1 + e_1, e_\nu), \quad \nu = 1, 2, \cdots, g-1.$$

Now, from (20), it follows that x is *not* a transversal parameter of the local ring of C_h (since $\beta_1 - hn = n_1 < n$) and the x-characteristic of C_h is $(n; n_1, \beta_2 - \beta_1 + n_1, \cdots, \beta_g - \beta_1 + n_1)$. It follows from the inversion formula (see Abhyankar [1]; see also Zariski, SES III, p. 996; we must use case (b) of the inversion formula, since $n \not\equiv 0 \pmod{n_1}$) that the characteristic of C_h is equal to

$$(n_1; n, \beta_2 - \beta_1 + n, \cdots, \beta_g - \beta_1 + n).$$

Applying the same argument to C_h (with $n; \beta_1, \beta_2, \cdots, \beta_g$ replaced by $n_1; n, \beta_2 - \beta_1 + n, \cdots, \beta_g - \beta_1 + n$), and recalling that $n = h_1 n_1 + n_2$, we find that the characteristic of C_{h+h_1} is $(n_2; n_1, \beta_2 - \beta_1 + n_1, \cdots, \beta_g - \beta_1 + n_1)$. Proceeding in this fashion we find that the characteristic of $C_{h+h_1+\cdots+h_{q-2}}$ is $(n_{q-1}; n_{q-2}, \beta_2 - \beta_1 + n_{q-2}, \cdots, \beta_g - \beta_1 + n_{q-2})$. We have $n_{q-2} = h_{q-1} n_{q-1} + n_q$, and thus, after h_{q-1} successive quadratic transformations, we get the curve $C_{h+h_1+\cdots+h_{q-2}+h_{q-1}}$, whose characteristic with respect to a suitable *non-transversal* parameter is $(n_{q-1}; n_q, \beta_2 - \beta_1 + n_q, \cdots, \beta_g - \beta_1 + n_q)$ (the transition from $C_{h+h_1+\cdots+h_{q-2}}$ to $C_{h+h_1+\cdots+h_{q-2}+h_{q-1}}$ is similar to that of from C to C_h). But this time we have $n_{q-1} \equiv 0 \pmod{n_q}$, and it is the case (a) of the inversion formula (loc. cit.) that is applicable. We thus find that the characteristic of $C_{h+h_1+\cdots+h_{q-1}}$ is

$$(n_q; \beta_2 - \beta_1 + n_{q-1}, \cdots, \beta_g - \beta_1 + n_{q-1}).$$

Applying to this curve other $h_q - 1$ successive quadratic transformations, we get finally the curve $\bar{C}_1$, having origin at the first terminal satellite $\bar{O}_1$ of O, and the characteristic of $\bar{O}_1$ is thus equal to

$$(n_q;) \beta_2 - \beta_1 + n_q, \beta_3 - \beta_1 + n_q, \cdots, \beta_g - \beta_1 + n_q).$$

This proves our assertion, since n_q is clearly the highest common divisor e_1 of β_1 and n.

6. Geometric interpretation of the structure theorem. We maintain the assumption that the coefficitnt field k is algebraically closed and we first study the infinitely near points common to two given irreducible plane algebroid curves C, C', having the same origin O. We assume that C and C' are defined by given Puiseux expansions in $x^{1/n}$ and $x^{1/n'}$ respectively:

$$C: y = \sum_{i=1}^{\infty} a_i x^{i/n}, \tag{1}$$

$$C': y' = \sum_{j=1}^{\infty} a_j' x^{i/n'}. \tag{1'}$$

We shall assume that x is a transversal parameter for both C and C'. We shall denote by $\beta_1/n, \beta_2/n, \cdots, \beta_g/n$ the characteristic exponents of the Puiseux expansion (1), and by $\beta_1'/n', \beta_2'/n', \cdots, \beta_{g'}/n'$ the characteristic exponents in (1'). We shall make the following conventions (similar to (1) and (2), § 2):

$$\beta_0 = 0, \qquad \beta_l = +\infty \text{ if } l > g; \tag{2}$$

$$\beta_0' = 0, \qquad \beta_l' = +\infty \text{ if } l > g'. \tag{2'}$$

We also agree that $g = 0$ if and only if $n = 1$, and, similarly, that $g' = 0$ if and only if $n' = 1$.

We denote by ω a primitive n-th root of unity and by ω' a primitive n'-th root of unity. The n conjugates of y over the field $k((x))$ will be denoted by $y_1, y_2, \cdots, y_n$:

$$y_\nu = \sum_i a_i \omega^{\nu i} x^{i/n}, \qquad \nu = 1, 2, \cdots, n.$$

Thus $y_n = y$. Similarly, we have the n' conjugates of y':

$$y_\mu' = \sum a_j' \omega'^{\mu i} x^{j/n'}, \qquad \mu = 1, 2, \cdots, n',$$

with $y'_{n'} = y'$.

We denote by v the natural valuation of the field $k((x^{1/nn'}))$ and we normalize v by setting $v(x) = 1$. We set

$$\gamma = \max_{\substack{1 \leqq \nu \leqq n \\ 1 \leqq \mu \leqq n}} \{v(y_\nu - y_\mu')\}, \tag{3}$$

and we define an integer q by the following condition:

$$(4) \qquad \gamma > \min\cdot\left\{\frac{\beta_{q-1}}{n}, \frac{\beta_{q-1}'}{n'}\right\},$$

$$(5) \qquad \gamma \leqq \min\cdot\left\{\frac{\beta_q}{n}, \frac{\beta_q'}{n'}\right\}.$$

Since $\gamma > 0$, it follows from (2) that $1 \leqq q$ and that if both C and C' are regular curves (i. e., if $n = n' = 1$) then certainly q is equal to 1. The definition of the number γ signifies that γ is the greater number such that for a suitable choice of μ the two series y and y_μ' differ only in the terms of degree $\geqq \gamma$. By choosing conveniently our notations we may assume that $y_\mu' = y'$. We have then the following: if $i/n < \gamma$ and $a_i \neq 0$ (resp., if $j/n' < \gamma$ and $a_j' \neq 0$) then there exists an integer j (resp., an integer i) such that $i/n = j/n'$, and for any such pair of integers $\{i, j\}$ we have $a_i = a_j'$. This implies, in particular, that the set of characteristic exponents of (1) which are less than γ coincides with the set of characteristic exponents of (1′) which are less than γ:

(6) $\qquad \beta_l/n = \beta_l'/n'$ *if either* β_l/n *or* β_l'/n' *is less than* γ.

We now show that

$$(7) \qquad q - 1 \leqq \min\cdot\{g, g'\}.$$

This is obvious if $g = g'$, for $\beta_l = \infty$ if $l > g$ and $\beta_l' = \infty$, if $l > g'$, and hence the inequality $q - 1 > \min\cdot\{g, g'\}$ would imply in this case that $\gamma = \infty$. Now assume that $g \neq g'$, say $g < g'$. We have to show then than $q - 1 \leqq g$. We assume therefore that $q - 1 \geqq g$ and we prove that in that case we must have $q - 1 = g$. Since $q - 1 \geqq g$, it follows from (4) that $\gamma > \min\cdot\{\beta_g/n, \beta_g'/n'\}$. Hence, by (6), we have $\beta_l/n = \beta_l'/n'$, for $l = 1, 2, \cdots, g$. We know that the highest common divisor of $n, \beta_1, \beta_2, \cdots, \beta_g$ is 1. On the other hand, since $g' > g$, the highest common divisor of $n', \beta_1', \cdots, \beta_g'$ is different from 1, say is $e' > 1$. It follows that $n' = e'n$, and thus all the exponents in the series (1) are of the form $\dfrac{e'i}{n'}$. Now $\beta_1'/n', \beta_2'/n', \cdots, \beta_g'/n'$ are exponents which occur in the series (1), but the highest common divisor of $n', \beta_1', \beta_2', \cdots, \beta_{g+1}'$ is *less* than e'. Hence the term $x^{\beta'_{g+1}/n'}$ cannot occur in the series (1), showing that $\gamma < \beta_{g+1}'/n'$. Then

$$\gamma < \min\cdot\{\infty, \beta_{g+1}'/n'\} = \min\cdot\{\beta_{g+1}/n, \beta_{g+1}'/n'\},$$

showing that $q = g + 1$, as asserted. This establishes (7).

From (6) and (7) follows now that

$$(8)\qquad \beta_l/n = \beta_l'/n', \qquad l = 1, 2, \cdots, q-1.$$

We set

$$e_0 = n,\ e_l = (\beta_l, e_{l-1}), \qquad l = 1, 2, \cdots, g;$$
$$e_0' = n',\ e_l' = (\beta_l', e_{l-1}'), \qquad l = 1, 2, \cdots, g'.$$

It follows from (8) that

$$(9)\qquad e_l'/e_l = \beta_l'/\beta_l = n'/n, \qquad l = 1, 2, \cdots, q-1.$$

We may assume (interchanging, if necessary, the roles of the curves C, C') that

$$(10)\qquad \beta_q/n \leqq \beta_q'/n'.$$

Note that this implies, in particular, that $g \geqq g'$ if the equality sign holds in (7). By the definition (3) of the number γ and in view of (5), it follows that

$$(11)\qquad \gamma \leqq \beta_q/n.$$

All the exponents of the Puiseux expansion $y = y(x^{1/n})$ which are *greater* than β_{q-1}/n and *less* than β_q/n are of the form $\beta_{q-1}/n + i \cdot e_{q-1}/n$, where i is a positive integer. Similarly, all the exponents in the Puiseux expansion $y'(x^{1/n'})$ which are *greater* than β_{q-1}'/n' $(= \beta_{q-1}/n)$ and *less* than β_q'/n' are of the form $\beta_{q-1}'/n' + i \cdot e_{q-1}'/n'$, or—what is the same—of the form $\beta_{q-1}/n + i \cdot e_{q-1}/n$, since, by (9), we have $e_{q-1}/n = e_{q-1}'/n'$. If γ is *less* than β_q/n, then γ must be equal to one of the exponents just indicated above. Therefore, γ is always a rational number which is both of the form c/n and of the form $\dfrac{c'}{n'}$:

$$(12)\qquad \gamma = c/n = c'/n',$$

where

$$(13)\qquad \beta_{q-1} < c = \gamma n \leqq \beta_q, \qquad \beta_{q-1}' < c' = \gamma n' \leqq \beta_q',$$

and

$$(14)\qquad c \equiv 0 \pmod{e_{q-1}}, \text{ if } c < \beta_q,$$

and similarly $c' \equiv 0 \pmod{e_{q-1}'}$ if $c < \beta_q'$.

We shall now prove the following:

PROPOSITION 6.1. *The intersection multiplicity* $(C, C')_0$ *of the two* *curves C, C' at their common origin O is given by the following expression:*

$$(15)\qquad \begin{aligned}(C,C')_O &= \beta_1(e_0'-e_1')+\beta_2(e_1'-e_2')+\cdots\\ &\qquad +\beta_{q-1}(e_{q-2}'-e_{q-1})+\gamma n e_{q-1}'\\ &= \beta_1'(e_0-e_1)+\beta_2'(e_1-e_2)+\cdots\\ &\qquad +\beta_{q-1}'(e_{q-2}-e_{q-1})+\gamma n' e_{q-1}.\end{aligned}$$

(*Note.* That the two expressions on the right-hand side are equal is a direct consequence of (9)).

Proof. The expression

$$\prod_{1\leqq\nu\leqq n}\prod_{1\leqq\mu\leqq n'}(y_\nu-y_\mu')$$

is an integral power series in x, and the intersection multiplicity $(C,C')_O$ is the order of vanishing of this power series at the origin $x=0$. It is clear that this order is equal to n times the order of vanishing of the (fractional) power series

$$(16)\qquad \prod_{1\leqq\mu\leqq n'}(y-y_\mu'),$$

because any two of the n partial products $\prod_{\mu=1}^{n'}(y_\nu-y_\mu')$ (ν-fixed) are conjugate over $k((x))$.

We shall now prove the following two assertions:

(a) If for some l, such that $1\leqq l\leqq q-1$, we have

$$(17)\qquad \mu\equiv 0(n'/e_{l-1}'),\qquad \mu\not\equiv 0(n'/e_l'),$$

then

$$(18)\qquad v(y-y_\mu')=\beta_l/n\quad(=\beta_l'/n').$$

(b) If

$$(19)\qquad \mu=0(n'/e_{q-1}'),$$

then

$$(20)\qquad v(y-y_\mu')=\gamma.$$

Assume (17). The two series $y(x^{1/n})$ and $y'(x^{1/n'})$ $(=y_n'(x^{1/n'}))$ differ only in terms of degree $\geqq\gamma$. In particular, the terms of degree $\leqq\beta_l/n$ $(=\beta_l'/n')$ are the same in both series. Let $a_ix^{i/n}=a_j'x^{j/n'}$, with $i/n=j/n'$ $\leqq\beta_l/n$ and $a_i=a_j$, be any such terms. Then the corresponding term in the series y_μ' is $a_j'\omega'^{\mu j}x^{j/n'}$, and therefore the coefficient of the term $x^{i/n}$ in the series $y-y_\mu'$ is $a_i(1-\omega'^{\mu j})$. Now, if $j<\beta_l'$, then $j\equiv 0 \pmod{e_{l-1}'}$, and hence, by the fiirst of the two conditions (17), μj is divisible by n', and $1-\omega'^{\mu j}$ is zero. Let now $j=\beta_l'$. We have $\mu e_{l-1}'=\lambda n'$, where λ is an integer (by the first of the two conditions (17)), and by the second of the conditions (17) we have $\mu e_l'\not\equiv 0 \pmod{n'}$, or—equivalently—$\lambda e_l'\not\equiv 0$ (e_{l-1}'). Since $(\beta_l', e_{l-1}')=e_l'$, this implies that also $\beta_l'\lambda$ is not divisible by e_{l-1}', or—equivalently—$\beta_l'\mu\not\equiv 0 \pmod{n'}$ (since $\mu=\lambda\cdot n'/e_{l-1}'$). Hence $\omega^{\mu\beta_l'}\neq 1$, and this proves that the leading term of $y-y_\mu'$ has exponent β_l'/n' $(=\beta_l/n)$.

Now assume (19). Since the terms of degree $<\gamma$ are the same in both series y, y' and since we have $j\equiv 0 \pmod{e_{q-1}'}$ for every term $a_j'\cdot x^{j/n'}$ in y' such that $a_j'\neq 0$ and $j/n'<\gamma$ (since $\gamma\leqq\beta_q/n\leqq\beta_q'/n'$, and hence $j<\beta_q'$), it follows at once that for every such j the coefficient of $x_j{}^{/n'}$ in $y-y_\mu'$ is zero, by (19). However, the coefficient of $x^{\gamma/n}$ cannot be zero in the series $y-y_\mu'$, in view of the definition (3) of γ. This proves (20).

We thus see that there are exactly $e_{l-1}-e_l$ *factors* $y-y_\mu'$ such that $v(y-y_\mu')=\beta_l/n$ $(l=1,2,\cdots,q-1)$, and that there are exactly e_{q-1}' factors $y-y_\mu'$ such that $v(y-y_\mu')=\gamma$. This takes care of all the n' factors $y-y_\mu'$ since $(e_0'-e_1')+e_1'-e_2'+\cdots+(e_{q-2}'-e_{q-1}')+e_{q-1}'=e_0'=n'$. It follows that the power series (16) has order equal to

$$\frac{1}{n}[\beta_1(e_0'-e_1')+\beta_2(e_1'-e_2')+\cdots+\beta_{q-1}(e_{q-2}'-e_{q-1}')+\gamma n(e_{q-1}')].$$

This completes the proof of the proposition.

We must add to Proposition 6.1 a complementary result which says essentially that equality (15) is satisfied only for *one* pair of integer q and c $(=\gamma n)$. More precisely, we have the following result:

PROPOSITION 6.2. *Let $\bar q$, $\bar c$ be positive integers such that $\beta_{q-1}<\bar c\leqq\beta_{\bar q}$ and such that*

$$(C,C')_O=\beta_1(e_0'-e_1')+\beta_2(e_1'-e_2')+\cdots+\beta_{\bar q-1}(e_{\bar q-2}'-e_{\bar q-1}')+\bar c\cdot e_{\bar q-1}'. \tag{21}$$

Then necessarily $\bar q=q$ and $\bar c=c$.

Proof. It will be sufficient to show that $\bar q=q$, for then it will follow from (15) and (21) that $ce_{q-1}'=\bar c e_{q-1}'$, whence $c=\bar c$. Let us show, for instance, that the assumption $\bar q<q$ leads to a contradiction (the assumption that $q<\bar q$ can be disposed of in exactly the same fashion). If $\bar q<q$ then we find from (15) and (21) that

$$\bar c e_{\bar q-1}'=\beta_{\bar q}(e_{\bar q-1}'-e_{\bar q}')+\cdots+\beta_{q-1}(e_{q-2}'-e_{q-1}')+ce_{q-1}'.$$

The right-hand side of this last equality can also be written in the form

$$\beta_{\bar{q}}e_{\bar{q}-1}' + e_{\bar{q}}'(\beta_{\bar{q}+1} - \beta_{\bar{q}}) + \cdots + e_{q-2}'(\beta_{q-1} - \beta_{q-2}) + e_{q-1}'(c - \beta_{q-1})$$

and hence is *greater* than $\beta_{\bar{q}}e_{\bar{q}-1}'$ (since $c > \beta_{q-1}$). On the other hand, $\bar{c}e_{\bar{q}-1}'$ is *not greater* than $\beta_{\bar{q}}e_{\bar{q}-1}'$, since $\bar{c} \leqq \beta_{\bar{q}}$. This is the contradiction.

We denote by $OO_1O_2 \cdots$ the sequence of infinitely near points on C, and by $OO_1'O_2' \cdots$ the similar sequence on C'. Furthermore, we denote (as in Theorem 5.16) by $\bar{O}_1, \bar{O}_2, \cdots, \bar{O}_g$ the terminal satellites on C, and by $\bar{O}'_1, \bar{O}'_2, \cdots, \bar{O}'_{g'}$ the terminal satellites on C'. Similarly we denote by $O_1^*, O_2^*, \cdots, O_g^*$ the leading free points (with respect to O) on C, and by $O_1'^*, O_2'^*, \cdots, O'^*_{g'}$ the leading free points on C'. Note that if $g = 0$, i. e., if C is a regular curve (or equivalently: if $n = 1$), then all the points O_i are free, there are no satellites, and, in particular, there are no terminal satellites, and therefore there are also no *leading* free points (by our definition 5.12, O is not regarded as a leading free point). Similar remarks apply to C' if $g' = 0$.

Proposition 6.3. *The integer q and the number γ being those defined in* (3), (4), (5), *and assuming that we have, say, $\beta_q/n \leqq \beta_q'/n'$ (see* (10)), *the following is true:*

(a) $O_\alpha^* = O_\alpha'^*$, *for $\alpha \leqq q-1$, but $O_q^* \neq O_q'^*$ (if $q = 1$ this signifies that C and C' have no leading free points in common; this includes the case in which either C or C' is a regular curve and therefore carries no leading free point at all).*

(b) *If $\gamma < \beta_q/n$ [see* (13) *and* (14)], *the curves C, C' have in common exactly $(\gamma n - \beta_{q-1})/e_{q-1} - 1$ points infinitely near (and following) their last common leading free point O_{q-1}^* (infinitely near O if $q = 1$). These points are all free with respect to O and have multiplicity e_{q-1} for C and multiplicity e_{q-1}' for C'.*

(c) *If $\gamma = \beta_q/n$, then the set of common points of C and C' which follow O_{q-1}^* (O, if $q = 1$) begins exactly with $[(\beta_q - \beta_{q-1})/e_{q-1}]$ points which are free with respect to O; all remaining common points (if any) are satellites of O.*

Proof. We shall use induction on q, and we shall first prove the proposition in the case $q = 1$. In this case we consider separately the two subcases: 1) $\gamma < \beta_1/n$; 2) $\gamma = \beta_1/n$.

Case 1). $\gamma < \beta_1/n$. In this case, γ is necessarily an integer, $\gamma \geqq 1$, and the expansions for C and C' are of the form:

$$C: y = a_n x + \cdots + a_{(\gamma-1)n} x^{\gamma-1} + a_{\gamma n} x^{\gamma} + \text{terms of degree} > \gamma;$$

$$C': y' = a_n x + \cdots + a_{(\gamma-1)n} x^{\gamma-1} + a_{\gamma n}' x^{\gamma} + \text{terms of degree} > \gamma,$$

with $a_{\gamma n} \neq a_{\gamma n}'$. This implies at once $O_i = O_i'$, $i = 1, 2, \cdots, \gamma - 1$, while $O_\gamma \neq O_\gamma'$, and that furthermore the $\gamma - 1$ points $O_1, O_2, \cdots, O_{\gamma-1}$ are free with respect to O and that $s_i = n$, $s_i' = n'$, $i = 1, 2, \cdots, \gamma - 1$. This proves the proposition in this case, since we are now in the case (b) of the proposition (in the present case $q = 1$ we have $\beta_{q-1} = \beta_0 = 0$ and $e_{q-1} = e_0 = n$, $e_0' = n'$).

Case 2). $\gamma = \beta_1/n$. In this case we have necessarily $g \geqq 1$ (i.e., $\beta_1 \neq \infty$). We set $\beta_1 = hn + n_1 = h'n' + n_1'$, where we agree to set $h' = \infty$ if $\beta_1' = \infty$ (i.e., if $g' = 0$). Since $\beta_1/n \leqq \beta_1'/n'$, we have $h \leqq h'$. The expansion for C is of the form:

$$(22) \qquad C: y = a_n x + \cdots + a_{hn} x^h + a_{\beta_1} x^{\beta_1/n} + \text{terms of degree} > \beta_1/n.$$

As to the expansion for C', it is either of the form

$$(23) \qquad C': y' = n_n x + \cdots + a_{hn} x^h + a'_{(h+1)n} x^{h+1} + \text{terms of degree} > h + 1, \quad \text{if } h < h',$$

or is of the form

$$(24) \qquad C': y' = a_n x + \cdots + a_{hn} x^h + a'_{\beta_1'} x^{\beta_1'/n'} + \text{terms of degree} > \beta_1'/n', \quad \text{if } h = h',$$

and in the latter case we must have $a_{\beta_1/n} \neq a'_{\beta_1'/n'}$ if $\beta_1/n = \beta_1'/n'$. From (22) it follows that $O_1 O_2 \cdots O_h$ are free points on C, while O_{h+1} is a satellite of O on C. If the expansion of C' is given by (23), then it is clear that $O_i' = O_i$, $i = 1, 2, \cdots, h$, while $O_{h+1}' \neq O_{h+1}$ since O_{h+1}' is still free with respect to O. This proves the proposition in this case, since $h = [\beta_1/n]$. If the expansion of C' is given by (24), then we still have $O_i' = O_i$, $i = 1, 2, \cdots, h$; there may be other common points of C, C' *after* O_h, but they can only be satellites of O, since O_{h+1} is a satellite of O and since $O_1{}^*$ cannot lie on C' (this follows either from the inequality $\beta_1/n \neq \beta_1'/n'$, or—if $\beta_1/n = \beta_1'/n'$—from the inequality $a_{\beta_1/n} \neq a'_{\beta_1'/n'}$). This completes the proof of the proposition in the case $q = 1$.

Let now $q \geqq 2$ and assume that the proposition is true for $q - 1$. The expansions for C and for C' are then given by (22) and (24) respectively, with $\beta_1/n = \beta_1'/n'$ *and* $a_{\beta_1/n} = a'_{\beta_1'/n'}$. It follows that the curves C, C' have in common the above free points $O_1, O_2, \cdots, O_h$, and that the multiplicities of these points on C and C' are respectively $s = s_1 = \cdots = s_{h-1} = n$, s_h

$= n_1 < n$, and $s' = s_1' = \cdots = s_{h-1}' = n'$, $s_h' = n' < n'$. Now, from $\beta_1/n = \beta_1'/n'$ follows from $n/n_1 = n'/n_1'$, and thus, by Theorem 5.15, the two curves C, C' have in common the first terminal satellite: $\bar{O}_1 = \bar{O}_1'$. We consider the transforms $\bar{C}_1$, $\bar{C}_1'$ of C, C', having $\bar{O}_1$ as origin. We write:

$$(25) \qquad n = h_1 n_1 + n_2,\ n_1 = h_2 n_2 + n_3, \cdots, n_{t-1} = h_t n_t,\ 0 < n_\rho < n_{\rho-1},$$

$$(26) \qquad n' = h_1 n_1' + n_2',\ n_1' = h_2 n_2' + n_3', \cdots, n_{t-1}' = h_t n_t',\ 0 < n_\rho' < n_{\rho-1}'.$$

Theorem 5.15 gives then complete information on the multiplicities of C, C' at the points O_i preceding $\bar{O}_1$. If we now apply this information and also use formula (9) of § 5, we find the following:

$$(27) \qquad (C, C')_O = hnn' + h_1 n_1 n_1' + \cdots + t_{t-1} n_{t-1} n_{t-1}' + (h_t - 1) n_t n_t' + (\bar{C}_1, \bar{C}_1')_{\bar{O}_1}.$$

We have $hn = \beta_1 - n_1$, $h_i n_i = n_{i-1} - n_{i+1}$ (with $n_0 = n$), $1 \leqq i \leqq t-1$, and $h_t n_t = n_{t-1}$. Therefore

$$hnn' + \sum_{i=1}^{t} h_i n_i n_i' = \beta_1 n' - n_1 n' + \sum_{i=1}^{t-1} (n_{i-1} n_i' - n_{i+1} n_i') + n_{t-1} n_t'.$$

Since $n_{i-1}/n_i = n_{i-1}'/n_i'$ $(i = 0, 1, \cdots, t; n_0 = n, n_0' = n')$, it follows that the above sum $hnn' + \sum_{i=1}^{t} h_i n_i n_i'$ is equal to $\beta_1 n'$. Thus, by (27), and observing that n_t is the highest common divisor of β_1 and n, i. e., $n_t = e_1$, and that similarly $n_t' = e_1'$, we have

$$(C, C')_O = \beta_1 n' - e_1 e_1' + (\bar{C}_1, \bar{C}_1')_{O_1},$$

and, by Proposition 6.1, we now find that

$$(28) \qquad (\bar{C}_1, \bar{C}_1')_{O_1} = -\beta_1 e_1' + e_1 e_1' + \beta_2 (e_1' - e_2') + \cdots + \beta_{q-1}(e_{q-2}' - e_{q-1}') + \gamma n e_{q-1}', \text{ if } q > 2,$$

and

$$(29) \qquad (\bar{C}_1, \bar{C}_1')_{O_1} = -\beta_1 e_1' + e_1 e_1' + \gamma n e_1', \text{ if } q = 2.$$

Now, by Theorem 5.16, we know that the characteristics of $\bar{C}_1$ and $\bar{C}_1'$ are respectively $(e_1, \bar{\beta}_1, \bar{\beta}_2, \cdots, \bar{\beta}_{g-1})$ and $(e_1'; \bar{\beta}_1', \bar{\beta}_2', \cdots, \bar{\beta}'_{g'-1})$, where

$$(30) \qquad \bar{\beta}_i = \beta_{i+1} - \beta_1 + e_1, \quad i = 1, 2, \cdots, g-1;$$

$$(30') \qquad \bar{\beta}_j' = \beta_{j+1}' - \beta_1' + e_1', \quad j = 1, 2, \cdots, g'-1.$$

Note that relation (30) does not hold of $i = 0$ (in particular, relations (30) are not applicable if $g = 1$, i. e., if $\bar{C}_1$ is a regular curve), for $\bar{\beta}_0$ is, by definition, equal to zero [see (2)], while for $i = 0$ the right-hand side of (30) is equal to s_1. Similarly, (30′) is false for $j=0$. If we now set $\bar{e}_0'=e_1'$, $\bar{e}_1'=(\bar{\beta}_1',\bar{e}_0'),\cdots,\bar{e}'_{g'-1}=(\bar{\beta}_{g'-1},\bar{e}'_{g'-2})$ (the $\bar{e}_j'$ not being defined for $j>0$ if $\bar{C}_1'$ is a regular curve), then we find that

$$\bar{e}_j'=e_{j+1}',\quad (j=0,1,\cdots,g'-1).$$

If $q>2$, we substitute in (28) for the β_{i+1} $(i=1,2,\cdots,q-1)$ their expressions $\bar{\beta}_i+\beta_1-e_1$, obtained from (30). If $q=2$ we use (29) instead. We obtain then easily the following expression for $(\bar{C}_1,\bar{C}_1')$, valid for $q>2$ and *also* for $q=2$:

$$(31)\qquad (\bar{C}_1,\bar{C}_1')_{O_1}=\bar{\beta}_1(\bar{e}_0'-\bar{e}_1')+\cdots+\bar{\beta}_{q-2}(\bar{e}_{q-3}'-\bar{e}_{q-2}')+\bar{c}\bar{e}_{q-2},$$

where

$$(32)\qquad \bar{c}=\gamma n-\beta_1+e_1,$$

and where therefore, by (13) and (30):

$$(33)\qquad \beta_{q-2}<\bar{c}\leqq\bar{\beta}_{q-1}$$

Note that for $q=2$ it is understood that relation (31) reduces to the following:

$$(34)\qquad (\bar{C}_1,\bar{C}_1')=\bar{c}\bar{e}_0',$$

where—we recall—$\bar{e}_0'=e_1'=$ multiplicity of the origin $\bar{O}_1$ of $\bar{C}_1'$.

By (31) and (33) it follows, applying Proposition 6.2 to the two curves $\bar{C}_1$, $\bar{C}_1'$, that the integers similar to q and c, but associated with the curves $\bar{C}_1$ and $\bar{C}_1'$, are $q-1$ and $\bar{c}$ respectively. Thus the quantity $\bar{\gamma}$ analogous to γ and associated with the pair of curves $\bar{C}_1$, $\bar{C}_1'$ is $\bar{c}/\bar{e}_0$ $(=\bar{c}/e_1)$. Therefore, by (32):

$$(35)\qquad \bar{\gamma}=\frac{\gamma n}{e_1}-\frac{\beta_1}{e_1}+1.$$

By our induction hypothesis, the curves $\bar{C}_1$, $\bar{C}_1'$ have in common precisely $q-2$ leading free points. We note that, by the definition of γ and q, we have $\gamma n>\beta_{q-1}$ Since we are dealing with the case $q\geqq 2$, it follows that $\gamma n>\beta_1$, and hence, by (32), we have $\bar{c}>e_1=$ multiplicity of $\bar{C}_1$ at $\bar{O}_1$. Therefore, by (34), we have $(\bar{C}_1,\bar{C}_1')>e_1e_1'=$ product of the multiplicities of $\bar{C}_1$ and $\bar{C}_1'$ at $\bar{O}_1$. Thus, $\bar{C}_1$ and $\bar{C}_1'$ have the same tangent at $\bar{O}_1$, showing that $O_1{}^*=O_1'^*$. This proves that C and C' have exactly $q-1$ leading free points in common, which establishes part (a) of the proposition.

We now observe that since we have assumed that $\beta_q/n\leqq\beta_q'/n'$, it follows from (30) and (30′) and from $n/n'=\beta_1/\beta_1'=e_1/e_1'$ [see (9), where

q is $\geqq 2$], that $\bar{\beta}_{q-1}/\bar{e}_0 \geqq \bar{\beta}_{q-1}'/\bar{e}_0'$. We also observe that by (35), and by (30) for $i=q-1$, we have

$$\bar{\gamma}-\frac{\bar{\beta}_{q-1}}{\bar{e}_0}=(\gamma-\frac{\beta_q}{n})\frac{n}{e_1}, \tag{36}$$

and hence $\bar{\gamma}<\bar{\beta}_{q-1}/\bar{e}_0$ if $\gamma<\beta_q/n$, and $\bar{\gamma}=\bar{\beta}_{q-1}/\bar{e}_0$ if $\gamma=\beta_q/n$.

Assume first that $\gamma<\beta_q/n$, so that we are now dealing with the case (b) of the proposition. We have then $\bar{\gamma}<\bar{\beta}_{q-1}/\bar{e}_0$, and hence, by our induction, the curves $\bar{C}_1$, $\bar{C}_1'$ have in common exactly $(\bar{\gamma}\bar{e}_0-\bar{\beta}_{q-2})/\bar{e}_{q-2}-1$ points infinitely near (and following) their last common leading free point (infinitely near $\bar{O}_1$ if $q=2$). These points are all free with respect to $\bar{O}_1$ and have multiplicity $\bar{e}_{q-2}$ $(=e_{q-1})$ for $\bar{C}_1$ and multiplicty $\bar{e}_{q-2}'$ $(=e_{q-1}')$ for $\bar{C}_1'$. Since the immediate successor O_1^* of $\bar{O}_1$ is free with respect to O, it follows from Proposition 5.3, part (b), that the above common points of $\bar{C}_1$, $\bar{C}_1'$, following $\bar{O}_1$, are also free with respect to O. Now, if $q=2$, then $\bar{\beta}_{q-2}=0$, and thus in this case the number of common points of $\bar{C}_1$, $\bar{C}_1'$, following $\bar{O}_1$, is equal to $\bar{\gamma}-1$, i.e., by (35), this number is equal to $(\gamma n-\beta_1)/e_1$. Hence the number of common points of C_1 and C_1', *following* O_1^*, is 1 less, i.e., is equal to $(\gamma n- \ _1)/e_1-1$, which proves the proposition in this case $q=2$, $\gamma<\beta_q/n$). If $q>2$, then the last common leading free point of $\bar{C}_1$ and $\bar{C}_1'$ is the point O_{q-1}^*. In this case, however, we have from (35), and from (30) for $i=q-2$:

$$(\bar{\gamma}\bar{e}_0-\bar{\beta}_{q-2})/\bar{e}_{q-2}-1=(\gamma n-\beta_{q-1})/e_{q-1}-1,$$

and this completes the proof of part (b) of the proposition.

Assume now that $\gamma=\beta_q/n$, whence $\bar{\gamma}=\bar{\beta}_{q-1}/\bar{e}_0$. If $q>2$, then, by our induction, $\bar{C}_1$, $\bar{C}_1'$ have exactly

$$[\bar{\beta}_{q-1}-\bar{\beta}_{q-2})/\bar{e}_{q-2-}] \tag{37}$$

common free points (with respect to $\bar{O}_1$), following O_{q-1}^*. Since $q>2$, it follows from the relations (30) that the integer (37) is equal to $[(\beta_q-\beta_{q-1})/e_{q-1}]$. If, however, $q=2$, then the integer (37) is equal to $[\bar{\beta}_1/\bar{e}_0]=[(\beta_2-\beta_1+e_1)/e_1]=[(\beta_2-\beta_1)/e_1]+1$, and this time this gives the number of common free points *following* $\bar{O}_1$ (*not* O_1^*). Hence the number of common free points following O_1^* is $[(\beta_2-\beta_1)/e_1]$. This completes the proof of the proposition.

We now go back to the structure theorem proved in sections 2 and 3. Our object is to give a geometric description of the saturation partition $\{I_1, I_2, \cdots, I_r\}$ of the set of irreducible components $C_1, C_2, \cdots, C_h$ of C, relative to the pair (C, x) (see Definition 4.11), in the case of an algebraically closed ground field k, and under the assumption that the parameter x is transversal for the curve C. Since every saturated local ring is saturated with respect to every transversal parameter, the assumption that x is transversal implies no loss of generality. We shall find easily the geometric significance of the integer q which appears in the structure theorem.

When dealing with the structure theorem we dealt with the valuation v of F^*, normalized by the condition that $v(x) = e_0 = e(F^*/K)$, where $K = k((x))$ [see (29), § 1]. In the present section we used a different normalization of the valuation, namely the one in which x has value 1. Let us denote this last valuation of F^* by $\tilde{v}$; thus v and v are equivalent valuations of F^*, but $v(x) = e_0$, while $\tilde{v}(x) = 1$.

We now set, for $i, j \in \{1, 2, \cdots, h\}$, $i \neq j$:

$$\gamma_{i,j} = \max_{\tau, \sigma \in G} \{v(y_i^\tau - y_j^\sigma)\}, \tag{38}$$

and

$$\tilde{\gamma}_{i,j} = \max_{\tau, \sigma} \{\tilde{v}(y_i^\tau - y_j^\sigma)\}. \tag{39}$$

We have, by (3) and (5) of § 2:

$$\gamma_{i,j} > \min \cdot \{\beta_{i,q-1}, \beta_{j,q-1}\} \tag{40}$$

and by the definition of the integer q given in § 2, we have

$$\gamma_{i,j} \leqq \min \cdot \{\beta_{i,q}, \beta_{j,q}\} \text{ for } \textit{some } i \neq j. \tag{41}$$

In (30), § 1, we have introduced a *modified* x-characteristic

$$\{e_0, \beta_{i,1}, \beta_{i,2}, \cdots, \beta_{i,g_i}\}$$

of C_i. If we denote by $\{\tilde{e}_{i,0}; \tilde{\beta}_{i,1}, \tilde{\beta}_{i,2}, \cdots, \tilde{\beta}_{i,g_i}\}$ the original x-characteristic of C_i (here $\tilde{e}_{i,0} = e(F_i/K) =$ multiplicity of C_i, since x is transversal), then

$$\tilde{\beta}_{i,\alpha_i} = \beta_{i,\alpha_i} \cdot \frac{\tilde{e}_{i,0}}{e_0}, \quad i = 1, 2, \cdots, h, \quad \alpha_i = 1, 2, \cdots, g_i. \tag{42}$$

By (38) and (39) we have

$$\tilde{\gamma}_{i,j} = \gamma_{i,j}/e_0, \tag{43}$$

and hence, using (40), (41) and (42) we find that

$$\gamma_{i,j} > \min \cdot \{\tilde{\beta}_{i,q-1}/\tilde{e}_{i,0}, \tilde{\beta}_{j,q-1}/\tilde{e}_{j,0}\}, \tag{44}$$

6

and that

$$\tilde{\gamma}_{i,j} \leqq \min \cdot \{\tilde{\beta}_{i,q}/\tilde{e}_{i,0}, \tilde{\beta}_{j,q}/\tilde{e}_{j,0}\}, \text{ for some } i \neq j. \tag{45}$$

Now, the number $\tilde{\gamma}_{i,j}$ plays here, for the two curves C_i, C_j, precisely the role that the number γ defined in (3) played, for the two curves C, C'. It follows therefore from part (a) of Proposition 6.3 that the integer q which occurs in the structure theorem has the following geometric significance:

PROPOSITION 6.4. *The h irreducible plane curves $C_1, C_2, \cdots, C_h$ have precisely $q-1$ leading free points in common.*

Our saturation partition $\{I_1, I_2, \cdots, I_r\}$ of the set of indices $(1, 2, \cdots, h)$ was defined as follows: two indices i and j belong to the same subset I_ρ if and only if $\gamma_{i,j} > \gamma$, where

$$\gamma = \min_{i \neq j} \cdot \gamma_{i,j}. \tag{46}$$

Let $\beta_1, \beta_2, \cdots, \beta_{q-1}$ be the integers defined in part (a) of Theorem 2.1, and let

$$e_i = (\beta_i, e_{i-1}), \quad i = 1, 2, \cdots, q-1. \tag{47}$$

We set

$$J = \beta_1(e_0 - e_1) + \beta_2(e_1 - e_2) + \cdots + \beta_{q-1}(e_{q-2} - e_{q-1}) + \gamma e_{q-1}. \tag{48}$$

A geometric description of the partition $\{I_1, I_2, \cdots, I_r\}$ is given by the following theorem:

THEOREM 6.5. *If i and j are arbitrary distinct indices in the set $\{1, 2, \cdots, h\}$ then*

$$(C_i, C_j)_O \geqq \frac{\tilde{e}_{i,0}\tilde{e}_{j,0}}{e_0^2} \cdot J, \tag{49}$$

with strict inequality $>$ if and only if the indices i, j belong to one and the same subset I_ρ.

Proof. We first note, with reference to formula (15), that if t is any positive integer *less* than q then we can write (15) in the following form:

$$\begin{aligned}(C, C')_O = \beta_1(e_0' - e_1') &+ \cdots + \beta_{t-1}(e_{t-2}' - e_{t-1}') \\ &+ \beta_t e_{t-1}' + (\beta_{t+1} - \beta_t)e_t' + \cdots \\ &+ (\beta_{q-1} - \beta_{q-2})e_{q-2}' + (\gamma n - \beta_{q-1})e_{q-1}',\end{aligned}$$

and hence, since $\gamma n > \beta_{q-1}$ [see (13)], we have the following strict inequality:

$$(50)\qquad (C,C')_O > \beta_1(e_0'-e_1') + \cdots + \beta_{t-1}(e_{t-2}'-e_{t-1}') + \beta_t e_{t-1}', \quad 0<t<q.$$

Now let us apply formula (15) to the pair of curves C_i, C_j. We shall denote the integer analogous to q, and associated with these two curves C_i, C_j, by q_{ij}. By (41) we have

$$(51)\qquad q \leqq q_{ij},$$

where now q denotes the integer introduced in the beginning of § 2.

If $q < q_{ij}$, then applying the inequality (50) to the curve (C_i, C_j) (with the role of q and t now being played by $q_{i,j}$ and q respectively) we get the inequality

$$(52)\qquad (C_i,C_j)_O > \tilde{\beta}_{i,1}(\tilde{e}_{j,0}-\tilde{e}_{j,1}) + \cdots + \beta_{i,q-1}(\tilde{e}_{j,q-2}-\tilde{e}_{j,q-1}) + \beta_{i,q}\tilde{e}_{j,q-1},$$

where we have set

$$(53)\qquad \tilde{e}_{j,\alpha} = (e_{j,\alpha-1}, \tilde{\beta}_{j,\alpha}).$$

It follows at once from (42) that

$$(54)\qquad \tilde{e}_{j,\alpha} = e_{j,\alpha}\cdot\frac{\tilde{e}_{j,0}}{e_0},$$

where $e_{j,\alpha} = (e_{j,\alpha-1}, \beta_{j,\alpha})$.

Now, since $\beta_{j,\alpha} = \beta_j$ for $\alpha = 1, 2, \cdots, q-1$, the inequality (52) takes the following form (taking into account (42) and (54)):

$$(55)\quad (C_i,C_j)_O > \frac{\tilde{e}_{i,0}\cdot\tilde{e}_{j,0}}{e_0{}^2}[\beta_1(e_0-e_1) + \cdots + \beta_{q-1}(e_{q-2}-e_{q-1}) + \beta_{i,q}e_{q-1}].$$

By the definition of the integer q, given in § 2, we have that

$$\gamma \leqq \min\cdot\{\beta_{1,q}, \beta_{2,q}, \cdots, \beta_{h,q}\}.$$

Thus $\beta_{i,q} \geqq \gamma$, and the inequailty (55) implies therefore the strict inequality in (49). We note that our assumption $q < q_{ij}$ implies that

$$\gamma_{i,j} > \min\cdot\{\frac{\tilde{\beta}_{i,q}}{e_{i,0}}, \frac{\tilde{\beta}_{j,q}}{e_{j,0}}\},$$

whence $\gamma_{i,j} > \min\cdot\{\beta_{i,q}, \beta_{j,q}\} \geqq \gamma$. Thus in this case, C_i and C_j belong to one and the same subset I_ρ.

Now assume that we have in (51) the equality sign: $q = q_{ij}$. Then, formula (15) yield the equality

$$(56)\qquad (C_i,C_j)_O = \tilde{\beta}_{i,1}(\tilde{e}_{j,0}-\tilde{e}_{j,1}) + \cdots + \tilde{\beta}_{i,q-1}(\tilde{e}_{j,q-2}-\tilde{e}_{j,q-1}) + \tilde{\gamma}_{i,j}\tilde{e}_{i,0}\tilde{e}_{j,q-1},$$

which, in view of (42), (54) and by $\beta_{j,\alpha}=\beta_j$ and $e_{j,\alpha}=e_\alpha$, for $\alpha=1,2,\cdots,q-1$, is the same as the equality

$$(57)\quad (C_i,C_j)_O=\frac{\bar{e}_{i,0}\cdot\bar{e}_{j,0}}{e_0^2}[\beta_0(e_0-e_1)+\cdots+\beta_{q-1}(e_{q-2}-e_{q-1})+\gamma_{i,j}e_{q-1}].$$

Since $\gamma_{i,j}\geqq\gamma$, with strict inequality if and only if the curves C_i, C_j belong to the same subset I_ρ of our partition, the proof of the theorem is complete.

We can present Theorem 6.5 in another, somewhat less explicit, form, by observing that by that theorem, the quantity J/e_0^2 is equal to $\min_{i\neq j}\cdot\{(C_i,C_j)_O/\bar{e}_{i,0}\bar{e}_{j,0}\}$. Hence, we may say that i and j *belong to one and same subset* I_ρ *of the partition* $\{I_1,I_2,\cdots,I_r\}$ *if and only if* $(C_i,C_j)_O/\bar{e}_{i,0}\bar{e}_{j,0}$ *is greater than that minimum.* Here, we recall, $\bar{e}_{i,0}$ denotes the multiplicity of C_i at O. Incidentally, this yields another proof of the second part of Corollary 4.8, for we have always $(C_i,C_j)_O\geqq\bar{e}_{i,0}\bar{e}_{j,0}$, with equality if and only if the two curves C_i, C_j have distinct tangents at 0. Therefore, if the h curves $C_1,C_2,\cdots,C_h$ do not all have the same tangent, then

$$\min_{i\neq j}\cdot\{(C_i,C_j)_O/\bar{e}_{i,0}\bar{e}_{j,0}\}=1.$$

In the proof of Proposition 6.4 we have used only part (a) of Proposition 6.3. If we make use of Proposition 6.3 to the fullest extent, we can derive another, more expressive, geometric characterization of the partition $\{I_1,I_2,\cdots,I_r\}$, *provided that we assume that in the inequality*

$$\gamma\leqq\min\cdot\{\beta_{1,q},\beta_{2,q},\cdots,\beta_{h,q}\}$$

[see the definition of the integer q in § 2, namely inequality (5)] the strict inequality sign $<$ holds true. Let us, namely, denote by B *the set of base points of the* h *curves* $C_1,C_2,\cdots,C_h$, i.e., the set of infinitely near points $OO_1\cdots$ which lie on each of these h curves. We know that B includes a $(q-1)$-th leading free point O_{q-1}^* (and of course, all the points preceding O_{q-1}^*). It may possibly include some satellites of O, *after* O_{q-1}^*, but certainly no points after O_{q-1}^* which are free with respect to O (see Proposition 6.4). We prove now the following result:

Proposition 6.6. *If* $\gamma<\min\cdot\{\beta_{1,q},\beta_{2,q},\cdots,\beta_{h,q}\}$ *then two indices* i, j *belong to one and the same set* I_ρ *if and only if* C_i *and* C_j *have common points outside the base set* B.

Proof. Let C_i and C_j be two curves which do *not* belong to one and the same set I_ρ. We have then $\gamma_{i,j}=\gamma$, and hence, by our assumption on γ, $\gamma_{i,j}<\min\cdot\{\beta_{i,q},\beta_{j,q}\}$. Let, say, $\beta_{i,q}\leqq\beta_{j,q}$. Then $\gamma_{i,j}<\beta_{i,q}$, and hence $\bar{\gamma}_{i,j}<\bar{\beta}_{i,q}/\bar{e}_{i,0}$, while $\beta_{i,q}\leqq\beta_{j,q}$ implies $\bar{\beta}_{i,q}/\bar{e}_{i,0}\leqq\bar{\beta}_{j,q}/\bar{e}_{j,0}$. If we then apply part

(b) of Proposition 6.3, we find that C_i and C_j have precisely $(\gamma-\beta_{q-1})e_{q-1}-1$ common points, beyond the leading free point $O_{q-1}{}^*$. Let then B denote the set of common (infinitely near) points of C_i and C_j. If C_ν is any other of the curves $C_1, C_2, \cdots, C_h$, then either C_i and C_ν or C_j and C_ν belong to different sets I_σ of our partition $\{I_1, I_2, \cdots, I_h\}$. Let, say, C_i and C_ν belong to different sets I_σ. Then $\gamma_{i,\nu}=\gamma$, and hence also C_i and C_ν have precisely $(\gamma-\beta_{q-1})/e_{q-1}-1$ points in common, beyond the point $O_{q-1}{}^*$. This shows that $B\subset C_\nu$, for every $\nu=1,2,\cdots,h$, and B is thus the required base set. The above proof shows that if two curves C_i, C_j belong to different subsets I_σ, then B is exactly the set of common points of C_i and C_j. Suppose now that C_i and C_j belong to one and the same set I_ρ, i.e., that $\gamma_{i,j}>\gamma$. We have to show that the set of common points of C_i and C_j is bigger than B. We consider several cases.

Case 1. $\gamma_{i,j}>\min\cdot\{\beta_{i,q},\beta_{j,q}\}$, i.e., $q_{i,j}>q$. In that case the two curves C_i and C_j have in common at least q *leading* free points, and since B contains only $q-1$ leading free points, the set of common points of C_i and C_j is bigger than B.

Case 2. $\gamma_{i,j}<\min\cdot\{\beta_{i,q},\beta_{j,q}\}$, say $\gamma_{i,j}<\beta_{i,q}\leqq\beta_{j,q}$. In that case, part (b) of Proposition 6.3 is applicable, and we find that C_i, C_j have precisely $(\gamma_{i,j}-\beta_{q-1})/e_{q-1}-1$ common points, beyond $O_{q-1}{}^*$. Since both $\gamma_{i,j}$ and γ are divisible by e_{q-1} and $\gamma_{i,j}>\gamma$, we conclude again that the set of common points of C_i and C_j is bigger than B.

Case 3. $\gamma_{i,j}=\min\cdot\{\beta_{i,q},\beta_{j,q}\}$, say $\gamma_{i,j}=\beta_{i,q}\leqq\beta_{j,q}$. In this case, part (c) of Proposition 6.3 is applicable and we find that C_i, C_j have at least $[(\beta_{i,q}-\beta_{q-1})/e_{q-1}]$ common points, beyond $O_{q-1}{}^*$. Since this integer is obviously greater than $(\gamma-\beta_{q-1})/e_{q-1}-1$, we reach the same conclusion as in the previous two cases. This completes the proof.

We shall conclude this section with a remark related to Theorem 6.5. That theorem shows that the saturation partition of the set of indices $\{1,2,\cdots,h\}$, relative to the pair $\{C,x\}$ (see Definition 4.11), where x is a *transversal* parameter, is independent of the choice of x, depending only on C. We therefore give the following definition:

Definition 6.7. *The saturation partition of the set of indices* $\{1,2,\cdots,h\}$, *relative to the pair* $\{C,x\}$, *x-transversal, is called the saturation partition of* $\{1,2,\cdots,h\}$, *relative to* C, *and the saturation components* $\Gamma_1,\Gamma_2,\cdots,\Gamma_r$ *of* C, *relative to* x (*see Definition* 4.11), *are called the saturation components of* C.

7. A saturation-theoretic criterion of equisingularity of plane algebroid curves (k-algebraically closed). In our paper SES I we have given an inductive definition of the relation of algebro-geometric equivalence of two algebroid curves C, D, defined over one and the same algebraically closed ground field k, of characteristic zero. This relation, which we shall now call *equisingularity,* was defined by the condition that there exist (1,1) correspondences between the branches of C and D, which satisfy certain conditions and which we have caled *equivalences* [actually, we have given in SES I three different definitions of equivalences (see SES I, definitions 2, 3 and 4), but have proved (SES I, Theorem 1 and Corollary of Lemma 3) that these definitions are equivalent]. If we have a (1,1) correspondence π between the branches C_i of C and the branches D_i of D, we shall write $\pi\colon C \to D$ and we shall say that π is a pairing between the branches of C and the branches of D. If C_i, D_i are corresponding branches, we shall write $\pi(C_i) = D_i$. If furthermore, π is an equivalence, then we shal write $\pi\colon C \overset{\rightarrow}{=} D$. We note that our definition of equivalence $C \overset{\rightarrow}{=} D$ is meaningful also if C and D are defined (not over the same field k but) over isomorphic, algebraically closed fields $k(C)$, $k(D)$.

In another paper we have established a saturation-theoretic criterion of equisingularity in the case in which C and D are irreducible curves (see SES III, Theorem 2.1). In this section we shall extend that criterion to the case of reducible curves C, D, *defined over one and the same algebraically closed ground field k, of characteristic zero* (or—more generally—defined over isomorphic ground fields, which are algebraically closed and are of characteristic zero).

For C and D to be equisingular, it is necessary at any rate that they have the same number of irreducible components (or *branches*). Let us then assume that we are dealing with two algebroid curves C, D, each having h branches. Let $\pi\colon C \to D$ be a pairing between the branches $C_1, C_2, \cdots, C_h$ of C and the branches $D_1, D_2, \cdots, D_h$ of D. We assume that $\pi(C_i) = D_i$, $i = 1, 2, \cdots, h$. We begin with a lemma which essentially does the following: it reduces the test of equisingularity of reducible curves to that of irreducible curves, provided one adds the condtion that the (local) intersection numbers of corresponding pairs of branches be equal.

LEMMA 7.1. *The pairing $\pi\colon C \to D$ is an equivalence $C \overset{\sim}{=} D$ if and only if the following two conditions are satisfied:*

(a) *If* $D_i = \pi(C_i)$, *then* $D_i \equiv C_i$ (*i. e.,* D_i *and* C_i *are equisingular branches*).

(b) $(C_i, C_j) = (D_i, D_j)$, *for* $i, j = 1, 2, \cdots, h$, $i \neq j$.

Proof. Let $\bar{C}_1, \bar{C}_2, \cdots, \bar{C}_t$ be the tangential components of C; here t denotes the number of distinct tangents $p_1, p_2, \cdots, p_t$ of C, at the origin P of C, and for each $\nu = 1, 2, \cdots, t$, $\bar{C}_\nu$ denotes the union of the branches of C which have p_ν as tangent. Let T_P be the locally quadratic transformation with center P, and let $\bar{C}_\nu' = T_P[\bar{C}_\nu]$ be the proper transform of $\bar{C}_\nu$. Then $\bar{C}_1', \bar{C}_2', \cdots, \bar{C}_t'$ are plane algebroid curves with distinct origins. Since the lemma is trivial if C is a regular curve, we shall proceed by induction, i. e., we shall assume that the lemma is true if C is replaced by any of the t curves $\bar{C}_1', \bar{C}_2', \cdots, \bar{C}_t'$.

Assume first that π is an equivalence. Then π is tangentially stable (see SES I, § 2, Definition 1), and we also have

$$(1) \qquad m_P(C_i) = m_Q(D_i), \quad i = 1, 2, \cdots, h,$$

where Q denotes the origin of D. Since π is tangentially stable, also D has exactly t tangential components, say $\bar{D}_1, \bar{D}_2, \cdots, \bar{D}_t$, and we may assume that the tangential component of D which corresponds, under π, to $\bar{C}_\nu$ is $\bar{D}_\nu$. Thus π induces a pairing between the branches of $\bar{C}_\nu$ and the branches of $\bar{D}_\nu$ $(\nu = 1, 2, \cdots, t)$. Let T_Q be the locally quadratic transformation with center Q, and let $\bar{D}_\nu' = T_Q[\bar{D}_\nu]$ $(\nu = 1, 2, \cdots, t)$, $C_i' = T_P[C_i]$, $D_i' = T[_Q D_i]$ $(i = 1, 2, \cdots, h)$. By the definition of equivalences, the pairing π_ν' between the branches of $\bar{C}'_\nu$ and the branches of $\bar{D}_\nu'$, induced by π, is an equivalence between the two curves $\bar{C}_\nu'$, $\bar{D}_\nu'$. If C_i is any of the brances of $\bar{C}_\nu$, then D_i is a branch of $\bar{D}_\nu$, C_i' is a branch of $\bar{C}_\nu'$ and D_i' is the corresponding branch $\pi_\nu'(C_i')$ of $\bar{D}_\nu'$. By our induction hypothesis we have therefore $D_i' \equiv C_i'$ for $i = 1, 2, \cdots, h$, and $(C_i', C_j') = (D_i', D_j')$, for $i, j = 1, 2, \cdots, h$, $i \neq j$; here, the equality $(C_i', C_j') = (D_i', D_j')$ follows from our induction hypothesis, if C_i' and C_j' are branches of one and the same curve $\bar{C}_\nu'$; in the contrary case, both intersection numbers (C_i', C_j') and (D_i', D_j') are zero, since in that case C_i', C_j' have distinct origins, and also D_i', D_j' have distinct origins. From (1) and from $C_i' \equiv D_i'$ follows that $C_i \equiv D_i$ (apply Definition 2 of § 3, in SES I, to the case in which the two curves C, D are the irreducible curves C_i, D_i respectively). Since we have

$$(2) \qquad (C_i, C_j) = m_P(C_i) m_P(C_j) + C_i', C_j'),$$

and similarly

$$(3) \qquad (D_i, D_j) = m_Q(D_i) m_Q(D_j) + (D_i', D_j'),$$

for $i, j = 1, 2, \cdots, h$ and $i \neq j$, the equalities (b) of the lemma now follow from (2), (3), (1) and from the equalities $(C_i', C_j') = (D_i', D_j')$.

Conversely, let us assume that the pairing π satisfies conditions (a) and (b) of the lemma. By (a), we have the equalities (1), showing that condition 2) of our Definition 2 of equivalence given in SES I, § 3, is satisfied. Condition b) and the equalities (1) imply that we have $(C_i, C_j) > m_P(C_i) m_P(C_j)$ if and only if $(D_i, D_j) > m_Q(D_i) m_Q(D_j)$, i.e., C_i, C_j have the some tangent if and only if D_i, D_j have the same tangent. Thus π is tangentially stable, and this is condition 1) of the above cited definition of equivalence.

From $C_i \equiv D_i$ follows that $C_i' \equiv D_i'$. Furthermore, by (1), (2) and (3) and using condition (b) of the lemma, we find that $(C_i', C_j') = (D_i', D_j')$, $i, j = 1, 2, \cdots, h$, $i \neq j$. Thus conditions (a) and (b) of the lemma are satisfied by the pairing $\pi_\nu' \colon \bar{C}_\nu' \rightarrow \bar{D}_\nu'$, for each $\nu = 1, 2, \cdots, t$, where $\bar{C}_\nu$, $\bar{D}_\nu$ are corresponding tangential compoentnts of C and D under our *tangentially stable* pairing π. We therefore have, by our induction hypothesis, that the pairing π_ν' is an equivalence for $\nu = 1, 2, \cdots, t$. Thus, also condition 3) of the above cited definition of equivalence is satisfied. This completes the proof of the lemma.

COROLLARY 7.2. *Let $\pi \colon C \rightarrow D$ be a tangentially stable pairing between the branches of C and the branches of D, and let $\bar{C}_\nu$, $\bar{D}_\nu$ be corresponding tangential components of C and D, under π $(\nu = 1, 2, \cdots, t)$. Let $\pi_\nu \colon \bar{C}_\nu \rightarrow \bar{D}_\nu$ be the pairing between the branches of $\bar{C}_\nu$ and the branches of $\bar{D}_\nu$, induced by π. Then π is an equivalence if and only if each π_ν $(\nu = 1, 2, \cdots, t)$ is an equivalence.*

For, if π is an equivalence, then Lemma 7.1 implies at once that each π_ν is an equivalence. On the other hand, if each π_ν is an equivalence, the condition (a) of the lemma is obviously satisfied, while condition (b) of the lemma is satisfied whenever C_i and C_j are branches of one and the same tangential component Γ_ν, i.e., whenever C_i and C_j have the same tangent. If, however, C_i and C_j have distinct tangents, then also D_i and D_j have distinct tangents, and therefore $(C_i, C_j) = m_P(C_i) m_P(C_j)$ and $(D_i, D_j) = m_Q(D_i) m_Q(D_j)$. We have therefore $(C_i, C_j) = (D_i, D_j)$, in view of the equations (1) (which are consequences of condition (a) of the lemma).

We now fix some notations which we shall use in the remainder of this section and which will differ to some extent from the notations used in the earlier sections, due to the fact that we now have to deal with *two* plane curves C and D. We shall use the symbols $\mathfrak{o}(C)$ and $F(C)$ to denote respectively the local ring of C and the total ring of quotients of $\mathfrak{o}(C)$. The symbols $\mathfrak{o}(D)$, $F(D)$ will have a similar meaning for the curve D. We assume, of course, from the very beginning that C and D havs the same number h of irreducible components C_i, D_i $(i=1,2,\cdots,h)$. We denote by $\epsilon_1,\epsilon_2,\cdots,\epsilon_h$ the mutually orthogonal idempotents in $F(C)$ such that $1=\epsilon_1+\epsilon_2+\cdots+\epsilon_h$. Similarly, let $\delta_1,\delta_2,\cdots,\delta_h$ be the mutually orthogonal idempotents in $F(D)$ such that $1=\delta_1+\delta_2+\cdots+\delta_h$.

The local ring $\mathfrak{o}(C)$ has exactly h minimal prime ideals $\mathfrak{P}_1,\mathfrak{P}_2,\cdots,\mathfrak{P}_h$, where

$$\mathfrak{P}_i=\mathfrak{o}(C)\cap\sum_{j\neq i}F(C)\cdot\epsilon_j,\qquad i=1,2,\cdots,h.$$

In other words, $\mathfrak{P}_i$ is the set of element ξ of $\mathfrak{o}(C)$ such that $\xi\epsilon_i=0$. The irreducible components $C_1,C_2,\cdots,C_h$ of C are associated, in one-to-one fashion, with the prime ideals $\mathfrak{P}_1,\mathfrak{P}_2,\cdots,\mathfrak{P}_h$, and we may assume that C_i is associated with $\mathfrak{P}_i$, i.e., that $C_i=\mathfrak{V}(\mathfrak{P}_i)$. Thus, each irreducible component C_i of C is associated canonically with the idempotent ϵ_i, and we have

$$\mathfrak{o}(C_i)=\mathfrak{o}(C)/\mathfrak{P}_i\cong\mathfrak{o}(C)\cdot\epsilon_i.$$

We shall identify the local *domain* $\mathfrak{o}(C_i)$ with the local domain $\mathfrak{o}(C)\cdot\epsilon_i$ (this identification is canonical: the $\mathfrak{P}_i$-residue of any element ξ of $\mathfrak{o}(C)$ is identified with the product $\xi\epsilon_i$). Thus the *field* $F(C_i)$ is now identified with the field $F(C)\cdot\epsilon_i$, and the total ring of quotients $F(C)$ is now the direct sum of the h fields $F(C_i)$. Similarly, we may assums that each irreducible component D_i of D is associated with the idempotent δ_i, and we identify $\mathfrak{o}(D_i)$ and $F(D_i)$ with $\mathfrak{o}(D)\cdot\delta_i$ and $F(D)\cdot\delta_i$ respectively.

From now on we shall assume that $\mathfrak{o}(C)$ and $\mathfrak{o}(D)$ have in common one and the same coefficient field k (k-algebraically closed and of characteristic zero). We shall denote by $\widetilde{\mathfrak{o}(C)}$ the k-saturation of $\mathfrak{o}(C)$, i.e., the k-saturation of $\mathfrak{o}(C)$ with respect to any transversal parameter (see Proposition 4.6). Similarly, $\widetilde{\mathfrak{o}(D)}$ will denote the k-saturation of $\mathfrak{o}(D)$. Our main object in this section will be, generally speaking, to prove that *C and D are equisingular if and only if the ring $\widetilde{\mathfrak{o}(C)}$ and $\widetilde{\mathfrak{o}(D)}$ are k-isomorphic.* We shall split this result into two separate theorems, since the proof of the "if" part is much simpler than the proof of the "only if" part of the above statement, and since furthermore, each of the two separate theorems which we intend to prove below contains additional items of information which constitute different refinements of the above general statement.

Before we state the two basic theorems which we propose to prove in this section, we must first give a definition and prove a lemma.

If x is a transversal parameter of $\mathfrak{o}(C)$ and Ω is an algebraic closure of the field $K = k((x))$, then we have introduced in § 1 the set H of all K-homomorphisms of $F(C)$ into Ω and we have denoted by H_i the set of all elements ψ in H such that $\psi(\epsilon_i) = 1$ (and hence $\psi(\epsilon_j) = 0$, $j \neq i$). Now suppose we have a field K' isomorphic to K and that we have fixed an isomorphism $\Theta\colon K \xrightarrow{\sim} K'$. Let Ω' be an algebraic closure of K'. We can then consider the set H' of homomorphisms ψ' of $F(C)$ into Ω' such that the restriction of ψ' to K is the given isomorphism $\Theta\colon K \xrightarrow{\sim} K'$. If F^* is the compositum of the fields $\psi(F(C))$, $\psi \in H$, and F'^* is the compositum of the fields $\psi'(F(C))$, $\psi' \in H'$, then there exists an isomorphism $\phi\colon F^* \xrightarrow{\sim} F'^*$ such that $\phi \mid K = \Theta$, and is clear that H' is the set of the homomorphism $\phi \cdot \psi$, where ψ ranges over H.

In § 2 we have defined *normal sets* of homomorphisms $\{\psi_1, \psi_2, \cdots, \psi_h\}$ in H, relative to the pair $\{C, x\}$ (where $\psi_i \in H_i$, $i = 1, 2, \cdots, h$). It is clear that we can define, in a similar way, normal sets of homomorphisms $\{\psi_1', \psi_2', \cdots, \psi_h'\}$ in H', *relative to* $\{C, x\}$ *and to the given isomorphism* $\Theta\colon K \xrightarrow{\sim} K'$. Namely, if we denote by H_i' the set of all elements ψ' in H' such that $\psi'(\epsilon_i)$ is the element 1 of K' and if we set $\psi_i'(y) = y_i'$, then the set $\{\psi_1', \psi_2', \cdots, \psi_h\}$ is normal if the following conditions are satisfied: (1) $\psi_i' \in H_i'$, $i = 1, 2, \cdots, h$; (2) $v'(y_i' - y_j') \geqq \gamma$, for all $i, j = 1, 2, \cdots, h$. Here v' is the natural valuation of F'^* (whence $v' = v \cdot \phi^{-1}$, where v is the natural valuation of F^*) and γ is the integer defined in (3), § 2. It is clear that the normal sets of homomorphisms in H', relative to $\{C, x\}$ and to the given isomorphism $\Theta\colon K \xrightarrow{\sim} K'$ are the sets of the form $\{\phi \cdot \psi_1, \phi \cdot \psi_2, \cdots, \phi \cdot \psi_h\}$, where $\{\psi_1, \psi_2, \cdots, \psi_h\}$ is any normal set of K-homorphisms of $F(C)$ into Ω, relative to the pair $\{C, x\}$.

This being said, let I be any subset of the set of indices $\{1, 2, \cdots, h\}$, let $\epsilon' = \sum_{i \in I} \epsilon_i$, and let $\Gamma = \bigcup_{i \in I} C_i$. The local ring $\mathfrak{o}(\Gamma)$ can be identified with the ring $F(C) \cdot \epsilon'$. If x is a transversal parameter of $\mathfrak{o}(C)$ then $x' = x\epsilon'$ is a transversal parameter of $\mathfrak{o}(\Gamma)$, and $F(\Gamma)$ contains the field $K' = K\epsilon' = k\epsilon'((x\epsilon'))$, which is isomorphic to K. There is, furthermore, the canonical isomorphism $\Theta\colon K' \xrightarrow{\sim} K$, defined by $\Theta(\xi\epsilon') = \xi$, for ξ in K. If then H_I denotes the set of homomorphisms ψ in H such that $\psi(\epsilon') = 1$, then the set of homomorphisms induced in $F(\Gamma)$ by the elements ψ of H_I coincides with the set of all homomorphisms ψ' of $F(\Gamma)$ into Ω such that $\psi' \mid K' = \Theta$, and we can therefore speak of normal sets of homomorphisms in H_I, relative to the pair $\{\Gamma, x_{\epsilon'}\}$ and to the above canonical isomorphism $\Theta\colon K' \xrightarrow{\sim} K$.

We shall now define, by induction on h, *strictly normal sets* of homomorphisms $\{\psi_1', \psi_2', \cdots, \psi_h'\}$ of $F(C)$ into Ω', relative to $\{C, x\}$ and to a given isomorphism $\Theta\colon K \longrightarrow K'$ (where x is a transversal parameter of $\mathfrak{o}(C)$, and $K = k((x))$).

Definition 7.3.

(a) *If $h = 1$ and H' is as above, then any homomorphism ψ' in H' constitutes a strictly normal set (consisting of the single element ψ').*

(b) *If $h > 1$, and assuming that strictly normal sets of homomorphisms have already been defined for all plane curves with less than h branches, let $\{I_1, I_2, \cdots, I_r\}$ be the saturation partition of $\{1, 2, \cdots, h\}$, relative to C, and let $\Gamma_1, \Gamma_2, \cdots, \Gamma_r$ be the saturation components of C (see Definition 6.7). Then $\{\psi_1', \psi_2', \cdots, \psi_h'\}$ is a strictly normal set of homomorphisms of $F(C)$ into Ω (relative to $\{C, x\}$ and Θ) if the following two conditions are satisfied:*

1) *$\{\psi_1', \psi_2', \cdots, \psi_h'\}$ is a normal set, relative to the pair $\{C, x\}$ and Θ (in the sense of the definition given above).*

2) *For each $\rho = 1, 2, \cdots, r$, the set $\{\psi_i' \mid I_\rho\}$ is strictly normal, relative to the pair $\{\Gamma_\rho, x_{\epsilon_\rho'}\}$ and to the canonical isomorphism $\Theta_\rho\colon K_{\epsilon_\rho'} \xrightarrow{\rightarrow} K'$, defined by $\Theta_\rho(\xi_{\epsilon_\rho'}) = \Theta(\xi)$, $\xi \in K$ (here $\epsilon_\rho' = \sum_{i \in I_\rho} \epsilon_i$).*

It is easy to see that condition 1) of the above definition can be replaced by the following condition:

1′) *$v'(y_i' - y_j') \geqq \gamma$, for all pairs of indices i, j belonging to distinct sets of the partition $\{I_1, I_2, \cdots, I_r\}$* (here $y_i' = \psi_i'(y($ where $\mathfrak{o}(C) = k[[x]][y]$, and γ is the integer defined in (3), § 2).

For, 1) certainly implies 1′), by definition of normal sets of homomorphisms $\{\psi_1', \psi_2', \cdots, \psi_h'\}$. On the other hand, if we set

$$\gamma_\rho = \min_{\substack{i \neq j \\ i, j \in I_\rho}} \cdot \{ \max_{\tau, \sigma \in G} (v(y_i^\tau - y_j^\sigma)) \}, \qquad \rho = 1, 2, \cdots, r, \tag{4}$$

then we have, by the definition of the saturation partition $\{I_1, I_2, \cdots, I_r\}$:

$$\gamma_\rho > \gamma, \qquad \rho = 1, 2, \cdots, r. \tag{5}$$

If condition 2) of Definition 7.3 is satisfied, then we have, for any pair of indices i, j, belonging to one and the same set I_ρ: $v'(y_i'-y_j')\geqq\gamma_\rho$ (since the set $\{\psi_i' \mid i\in I_\rho\}$ is strictly normal, hence normal). Thus $v'(y_i'-y_j')>\gamma$, if i and j belong to one and the same set I_ρ. This, combined with 1′), shows that conditions 1′) and 2) imply condition 1).

If in the above definition we have $K'=K$ and Θ is the identity, then we speak of strictly normal sets of K-homomorphisms of $F(C)$ into Ω, relative to C.

LEMMA 7.4. *The notations being the same as in Definition 7.3, any homomorphism ψ' of $F(C)$ into Ω' ($\psi'\in H'$) is contained in some strictly normal set of homomorphisms of $F(C)$ into Ω' relative to $\{C,x\}$ and Θ.*

Proof. We shall proceed by induction on h, since for $h=1$ the lemma is trivial. We choose our notations in such a way as to have $\rho\in I_\rho$ ($\rho=1,2,\cdots,r$) and $\psi'=\psi_1'\in H_1'$. Here $\{I_1,I_2,\cdots,I_r\}$ denotes the saturation partition of $\{1,2,\cdots,h_r\}$ relative to C. We also fix homomorphisms $\psi_2',\psi_3',\cdots,\psi_r'$ in $H_2',H_3,\cdots,H_r'$ respectively. Let $y_\rho'=\psi_\rho'(y)$, $\rho=1,2,\cdots,r$. By our choice of notations it follows that for each $\rho\neq 1$, we have

$$\max_{\tau\in G}\{v(y_1'-y_\rho'^{\tau})\}=\gamma.$$

We fix, for each $\rho\neq 1$, an element τ_ρ' of the Galois group G' of F'^*/K', such that $v'(y_1'-y_\rho'^{\tau_\rho})=\gamma$. Then $\tau_\rho'\cdot\psi_\rho'$ is still an element of H_ρ', and we may replace ψ_ρ' by $\tau_\rho'\cdot\psi_\rho'$. We may therefore assume that

$$v'(y_1'-y_\rho')=\gamma,\qquad \rho=2,3,\cdots,r. \tag{6}$$

Now, by our induction hypothesis, each ψ_ρ' is contained in some strictly normal set $\{\psi_i' \mid i\in I_\rho\}$ of homomorphisms of $F(\Gamma_\rho)$ into Ω' relative to Γ_ρ and to the canonical isomorphism $\Theta_\rho\colon K\epsilon_\rho'\xrightarrow{\sim}K'$ (see Definition 7.3; here $\Gamma_1,\Gamma_2,\cdots,\Gamma_r$ are saturation components of C). Consider the union $\{\psi_1',\psi_2',\cdots,\psi_h'\}$ of these r sets $\{\psi_i' \mid i\in I_\rho\}$. Condition 2) of Definition 7.3 is certainly satisfied by this set, by construction. To prove that also condition 1) of Definition 7.3 is satisfied, we have only to prove that we have $v'(y_i'-y_j')\geqq\gamma$, whenever i and j belong to different sets of the partition $\{I_1,I_2,\cdots,I_r\}$. Let, say, $i\in I_\rho$, $j\in I_{\rho'}$, $\rho\neq\rho'$. By (6), it is sufficient to consider the case in which both ρ and ρ' are different from 1. We have:

$$y_i'-y_j'=(y_i'-y_\rho')+(y_\rho'-y_1')+(y_1'-y_{\rho'}')+(y_{\rho'}'-y_j').$$

By (6), we have $v'(y_\rho' - y_1') = v'(y_1' - y_{\rho'}') = \gamma$ (since $\rho \neq 1$ and $\rho' \neq 1$). Since $i \in I_\rho$ and $j \in I_{\rho'}$, we have $v'(y_i' - y_\rho') \geqq \gamma_\rho$ and $v'(y_{\rho'}' - y_j') \geqq \gamma_{\rho'}$, showing, by (5), that $v'(y_i' - y_\rho') > \gamma$ and $v'(y_{\rho'}' - y_j') > \gamma$. This proves the inequality $v'(y_i' - y_j') \geqq \gamma$.

We now proceed with the main object of this section.

THEOREM 7.5. *Assume that there exists a k-isomorphism ϕ between the two saturated rings $\widetilde{\mathfrak{o}(C)}$ and $\widetilde{\mathfrak{o}(D)}$:*

$$\phi\colon \widetilde{\mathfrak{o}(S)} \xrightarrow{\sim} \widetilde{\mathfrak{o}(D)}. \tag{7}$$

Assume also that the idempotents ϵ_i and δ_i have been indexed in such a way that the following is true:

$$\phi(\epsilon_i) = \delta_i, \qquad i = 1, 2, \cdots, h. \tag{8}$$

Then the pairing $\pi\colon C \to D$ defined by setting $\pi(C_i) = D_i$ is an equivalence (and hence C and D are equisingular).

Proof. We shall proceed by induction on the number h of irreducible components C_i of C, since the theorem is known to be true in the case $h - 1$ (see SES III, Theorem 2.1).

We first consider the case in which the h branches C_i of C do not all have the same tangent. Let then $\bar{C}_1, \bar{C}_2, \cdots, \bar{C}_t$ be the tangential components of C, $t > 1$. To simplify the notational aspects of our proof we use the given isomorphism ϕ to identify the two saturadted rings $\widetilde{\mathfrak{o}(C)}$ and $\widetilde{\mathfrak{o}(D)}$, and we denote by $\mathfrak{o}$ these identified rings:

$$\mathfrak{o} = \widetilde{\mathfrak{o}(C)} = \widetilde{\mathfrak{o}(D)}. \tag{9}$$

This identification also induces (uniquely) an identification of the total quotient rings $F(C)$ and $F(D)$, and we therefore set $F = F(C) = F(D)$. Furthermore, we have now, by (8): $\epsilon_i = \delta_i$, $i = 1, 2, \cdots, h$.

If $\{I_1, I_2, \cdots, I_r\}$ is the saturation partition of $\{1, 2, \cdots, h\}$, relative to C, then by Corollary 4.8, the integer r is equal to t, and for a suitable indexing of the tangential components $\bar{C}_\nu$ we have: $\bar{C}_\nu = \bigcup_{i \in I_\nu} C_i$ ($\nu = 1, 2, \cdots, t$). We set

$$\bar{\epsilon}_\nu = \sum_{i \in I_\nu} \epsilon_i, \qquad \nu = 1, 2, \cdots, t \ (= r).$$

If we denote by $\mathfrak{m}$ the maximal ideal of the saturated local ring $\mathfrak{o}$ [see

(9)], then, again by Corollary 4.8, we have:

$$\mathfrak{m} = \mathfrak{m}\bar{\epsilon}_1 + \mathfrak{m}\bar{\epsilon}_2 + \cdots + \mathfrak{m}\bar{\epsilon}_t. \tag{10}$$

We now set $\bar{D}_\nu = \bigcup_{i \in I_\nu} D_i$, $\nu = 1, 2, \cdots, t$, and we apply Corollary 4.9, after viewing (10) as a direct decomposition of the maximal ideal of the saturated ring $\mathfrak{o} = \widetilde{\mathfrak{o}(D)}$. It follows then that each curve $\bar{D}_\nu$ is the union of a certain number of tangential components of D. Hence the number of tangential components of D is not less than the number t of tangential components of C. Interchanging the roles of C and D we conclude that C and D have the same number of tangential components. Consequently the t curves $\bar{D}_1, \bar{D}_2, \cdots, \bar{D}_t$ are themselves the tangential components of D. Thus the pairing $\pi: C \to D$ is tangentially stable, and $\bar{C}_\nu$, $\bar{D}_\nu$ are corresponding tangential components under π. By Proposition 3.5 we have

$$\widetilde{\mathfrak{o}(\bar{C}_\nu)} = \widetilde{\mathfrak{o}(\bar{D}_\nu)} = \mathfrak{o} \cdot \bar{\epsilon}_\nu, \tag{11}$$

where the saturations are relative to the coefficient field $k\bar{\epsilon}_\nu$. Since $\bar{C}_\nu$ has less than h branches, it follows from (11) and from our induction hypothesis, that the pairing $\pi_\nu: \bar{C}_\nu \to \bar{D}_\nu$, defined by $\pi_\nu(C_i) = D_i$, $i \in I_\nu$, is an equivalence. Therefore, by Corollary 7.2, also $\pi: C \to D$ is an equivalence.. This completes the proof in the case $t > 1$.

We shall now assume that $t = 1$. Let

$$\mathfrak{o}(C) = k[[x]][y], \tag{12}$$

$$\mathfrak{o}(D) = [[x']][y'], \tag{12'}$$

where we assume that x and x' are transversal parameters of $\mathfrak{o}(C)$ and $\mathfrak{o}(D)$ respectively. Therefore

$$\widetilde{\mathfrak{o}(C)} = \widetilde{\mathfrak{o}(C)_{k((x))}},$$

and

$$\widetilde{\mathfrak{o}(D)} = \widetilde{\mathfrak{o}(D)_{k((x'))}}.$$

Let $\phi^{-1}(x') = \xi$, where ϕ is the given isomorphism (7). Then ξ is also a transversal parameter of $\mathfrak{o}(C)$. Since all the branches of C have the same tangent, it follows from Proposition 4.10 that the residue of ξ/x in $k\epsilon_1 + k\epsilon_2 + \cdots + k\epsilon_h$ is actually an element of k, say c, $c \in k$. We can replace in (12) the element x by cx, without affecting the ring $\mathfrak{o}(C)$. We may therefore assume that the residue of ξ/x is 1. By Theorem 4.4 there exist then a $\bar{k}$-automorphism f of $F(C)$ which induces an automorphism of $\widetilde{\mathfrak{o}(C)}$ and which is such that $f(x) = \xi$; here $\bar{k} = k\epsilon_1 + k\epsilon_2 + \cdots + k\epsilon_h$. The product $\phi \cdot f$ is a new k-isomorphism of $\widetilde{\mathfrak{o}(C)}$ onto $\widetilde{\mathfrak{o}(D)}$, but this new isomorphism takes x into x', and it also sends ϵ_i into δ_i, for $i = 1, 2, \cdots, h$ (since $f(\epsilon_i) = \epsilon_i$, $i = 1, 2, \cdots, h$). *We can therefore assume that* $\phi(x) = x'$.

As in the preceding part of the proof, we again use the isomorphism ϕ to identify the saturated rings $\widetilde{\mathfrak{o}(C)}$ and $\widetilde{\mathfrak{o}(D)}$. Thus we have again (9) and also $\epsilon_i = \delta_i$, $i = 1, 2, \cdots, h$, and $x = x'$, while in view of (12′) we now have

$$\mathfrak{o}(D) = k[[x]][y'].$$

We have furthermore

$$\mathfrak{o} = \widetilde{k[[x]][y]_{k((x))}} = \widetilde{k[[x]][y']_{k((x))}}.$$

In § 1, formula (29), we have introduced the integer e_0, which depends only on $F(C)$ and the subfield $k((x))$ of $F(C)$. In § 2 we have introduced the integers γ and q [see (3) and (5), § 2], the integers $\beta_0, \beta_1, \beta_2, \cdots, \beta_{q-1}$ [part (a) of Theorem 2.1] and the integers $e_1, e_2, \cdots, e_{q-1}$ [part (c) of Theorem 2.1]. All these integers depend only on the saturated ring $\mathfrak{o}$, on the coefficient field k and on the parameter x. Let $\{I_1, I_2, \cdots, I_r\}$ be the saturation partition of the set of indices $\{1, 2, \cdots, h\}$, relative to the pair $\{\mathfrak{o}, x\}$. Since x is a transversal parameter, the saturation components of C and D (see Definition 6.7) are respectively the curves

$$\Gamma_\rho = \bigcup_{i \in I_\rho} C_i, \qquad \Delta_\rho = \bigcup_{i \in I_\rho} D_i, \qquad \rho = 1, 2, \cdots, r.$$

We set

$$\epsilon_\rho' = \sum_{i \in I_\rho} \epsilon_i.$$

We can identify the local rings $\mathfrak{o}(\Gamma_\rho)$ and $\mathfrak{o}(\Delta_\rho)$ with the rings $\mathfrak{o}(C)\epsilon_\rho'$ and $\mathfrak{o}(D)_\rho'$ respectively. These two local rings have in common the coefficient field k_ρ'. If we denote by $\widetilde{\mathfrak{o}(\Gamma_\rho)}$ and $\widetilde{\mathfrak{o}(\Delta_\rho)}$ the $k\epsilon_\rho'$-saturations of these two local rings and if we observe that $x\epsilon_\rho'$ is a transversal parameter of both $\mathfrak{o}(\Gamma_\rho)$ and $\mathfrak{o}(\Delta_\rho)$, we deduce from Corollary 3.6 (where we set first $\mathfrak{o} = \mathfrak{o}(C)$ and next $\mathfrak{o} = \mathfrak{o}(D)$, while K is $k((x))$) that

$$\mathfrak{o}(\Gamma_\rho) = \widetilde{\mathfrak{o}(\Delta_\rho)} = \mathfrak{o}\epsilon_\rho', \qquad \rho = 1, 2, \cdots, h, \tag{13}$$

where $\mathfrak{o}$ is the saturated ring $\widetilde{\mathfrak{o}(C)}$ $[=\widetilde{\mathfrak{o}(D)}$; see (9)]. Since each Γ_ρ has less than h branches, it follows from (13) and from our induction hypothesis, that we have equivalences

$$(14) \qquad \pi_\rho: \Gamma_\rho \overset{\rightarrow}{=} \Delta_\rho, \qquad \rho = 1, 2, \cdots, r,$$

where

$$(15) \qquad \pi_\rho(C_i) = D_i, \qquad i \in I_\rho.$$

It follows that

$$(16) \qquad C_i \equiv D_i, \qquad i = 1, 2, \cdots, h,$$

i.e., C_i and D_i are equisingular curves. Furthermore, in view of (14) and (15) and by Lemma 7.1 (as applied to the two curves Γ_ρ, Δ_ρ), we have

$$(17) \qquad (C_i, C_j) = (D_i, D_j), \qquad i \neq j, i, j \in I_\rho \quad (\rho = 1, 2, \cdots, r).$$

Now assume that i and j belong to *different* sets of the saturation partition $\{I_1, I_2, \cdots, I_r\}$. Then by Theorem 6.5 we have

$$(18) \qquad (C_i, C_j) = (D_i, D_j) = \frac{m_i m_j}{e_0^2} J,$$

where $m_i = m_P(C_i) = m_Q(D_i)$ and $m_j = m_P(C_j) = m_Q(D_j)$ (the equalities $m_P(C_i) = m_Q(D_i)$ follow from (16); we also recall that the integers $e_{i,0}$ which appear in Theorem 6.5 are the multiplicities of the branches C_i of C, at their common origin P, since x is a transversal parameter). Our theorem now follows from (16), (17) and (18), in view of Lemma 7.1.

We now proceed to formulate and prove a strong converse of Theorem 7.5. Basically, the result which we wish to prove is represented by Theorem 7.6. However, we found that in order to prove Theorem 7.6 we must prove at the same time a somewhat stronger result. We shall formulate and prove this stronger result (which will be Theorem 7.9) after we have prepared the ground and have established a few simple preliminary facts (see Lemma 7.7 and Corollary 7.8).

THEOREM 7.6. *Let C, D be equisingular plane algebroid curves (both defined over the algebraically closed field k, of characteristic zero) and let $\pi: C \overset{\rightarrow}{=} D$ be a given equivalence between these two curves. We assume that $\pi(C_i) = D_i$, $i = 1, 2, \cdots, h$. Let x and x' be transversal parameters of the local ring $\mathfrak{o}(C)$ and the local ring $\mathfrak{o}(D)$ respectively. Then there exists a* *k-isomorphism* $\phi\colon \widetilde{\mathfrak{o}(D)} \xrightarrow{\sim} \widetilde{\mathfrak{o}(C)}$ *such that*

$$\phi(\delta_i) = \epsilon_i, \qquad i = 1, 2, \cdots, h; \tag{19}$$

$$\phi(x') = x. \tag{20}$$

Before we proceed to the proof of this theorem, we make some preliminary remarks, concerning saturated local *domains* $\mathfrak{o}$ of dimension 1, having our algebraically closed field k (of characteristic zero) as coefficient field. Let F be the field of fractions of $\mathfrak{o}$, let x be any parameter of $\mathfrak{o}$ such that $\mathfrak{o}$ is saturated with respect to the field $K = k((x))$ (for instance, let x be a transversal parameter of $\mathfrak{o}$), let v be the natural valuation of $\mathfrak{o}$ and $n = v(x)$. Then F is a cyclic extension of K, of degree n. Let G be the Galois group of F/K. There is then one very simple property of the saturated local domain $\mathfrak{o}$ which is a direct consequence of a result (Corollary 1.14) of our paper SES III. It is the following:

LEMMA 7.7. *The saturated local domain $\mathfrak{o}$ is invariant under all the K-automorphisms τ of F (thus τ is any element of the Galois group G).*

For the proof, apply Corollary 1.14 of SES III to the special case $x = x'$.

COROLLARY 7.8. *Let Ω be a field containing the field K $(= k((x)))$, and let $\mathfrak{o}$, $\mathfrak{o}'$ be two saturated local domains of dimension 1, contained in Ω, both having k as cofficient field. Assume that x belongs to both $\mathfrak{o}$ and $\mathfrak{o}'$ and that it is a transversal parameter of both rings. If $\mathfrak{o}$ and $\mathfrak{o}'$ are k-isomorphic rings, then $\mathfrak{o} = \mathfrak{o}'$.*

For, if ϕ is a k-isomorphism between $\mathfrak{o}$ and $\mathfrak{o}'$, we may assume that $\phi(x) = x$, since there always exists a k-automorphism of $\mathfrak{o}'$ which sends $\phi(x)$ into x (see SES III, Corollary 1.14). Thus, if F and F' are the fields of fractions of $\mathfrak{o}$ and $\mathfrak{o}'$ respectively, they are conjugate fields over K, and hence $F = F'$ [since they are Galois (cyclic) extensions of K]. Therefore, the isomorphism ϕ induces a K-automorphism of F, and hence, by Lemma 7.7, $\phi\mathfrak{o}) = \mathfrak{o}$, i.e., $\mathfrak{o}' = \mathfrak{o}$.

We go back to Theorem 7.6 and we observe first of all that for any $i = 1, 2, \cdots, h$, the fields $k\epsilon_i$ and $k\delta_i$ are canonically isomorphic, and that therefore also the direct sums $k\epsilon_1 + k\epsilon_2 + \cdots + k\epsilon_h$ and $k\delta_1 + k\delta_2 + \cdots + k\delta_h$ are canonically isomorphic, the isomorphism f in question being defined by $f(\sum_{i=1}^{h} c_i\epsilon_i) = \sum_{i=1}^{k} c_i\delta_i$, where $c_1, c_2, \cdots, c_h \in k$. We use this canonical isomorphism f to identify the two direct sums above. We therefore have now $\epsilon_i = \delta_i$, $i = 1, 2, \cdots, h$. We set $\bar{k} = k\epsilon_1 + k\epsilon_2 + \cdots + k\epsilon_h$ and we observe that both x and x' are analytically independent over $\bar{k}$, i.e., $\bar{k}[[x]]$ and $\bar{k}[[x']]$ are power series rings, over $\bar{k}$, contained in $F(C)$ and $F(D)$ respectively. We identify x with x', and therefore also (canonically) $\bar{k}[[x]]$ with $\bar{k}[[x']]$. So, from now on, x will be an element of both local rings $\mathfrak{o}(C)$ and $\mathfrak{o}(D)$, and the field $K = k((x))$ is a subfield of both total quotient rings $F(C)$ and $F(D)$. Furthermore, we can write:

$$\mathfrak{o}(C) = k[[x]][y],$$
$$\mathfrak{o}(D) = k[[x]][y'].$$

If Ω denotes, as usual, an algebraic closure of the field K, let H be the set of K-homomorphisms of $F(C)$ into Ω and let H' be the set of K-homomorphisms of $F(D)$ into Ω. As in § 1, so also here we denote by H_i the set of all ψ in H such that $\psi(\epsilon_i) = 1$. Similarly, we denote by H_i' the set of all ψ' in H' such that $\psi'(\epsilon_i) = 1$ $(i = 1, 2, \cdots, h)$. For each i, the local rings $\mathfrak{o}(C_i)$ and $\mathfrak{o}(D_i)$ can be identified with $\mathfrak{o}(C) \cdot \epsilon_i$ and $\mathfrak{o}(D) \cdot \epsilon_i$ respectively, i.e., we have:

$$\mathfrak{o}(C_i) = k\epsilon_i[[x\epsilon_i]][y\epsilon_i],$$
$$\mathfrak{o}(D_i) = k\epsilon_i[[x\epsilon_i]][y'\epsilon_i], \qquad i = 1, 2, \cdots, h.$$

The local *domains* $\mathfrak{o}(C_i)$ and $\mathfrak{o}(D_i)$ have $k\epsilon_i$ as coefficient field and $x\epsilon_i$ as transversal parameter. By the assumptions made in Theorem 7.6, the curves C_i and D_i are equisingular. Hence, by the saturation-theoretic criterion proved by us elsewhere (see SES III, Theorem 2.1), the $k\epsilon_i$-saturations $\widetilde{\mathfrak{o}(C_i)}$ and $\widetilde{\mathfrak{o}(D_i)}$ are isomorphic over $k\epsilon_i[[x\epsilon_i]]$:

$$\widetilde{\mathfrak{o}(C_i)} \xrightarrow{\sim} \widetilde{\mathfrak{o}(D_i)}, \text{ over } k\epsilon_i[[x\epsilon_i]]. \tag{21}$$

This implies, of course, that the fields $F(C_i)$ $(= K\epsilon_i[y\epsilon_i])$ and $F(D_i)$ $(= K\epsilon_i[y'\epsilon_i])$ are $K\epsilon_i$-isomorphic. Therefore, if we set $n_i = [F(C_i) : K\epsilon_i]$, then we also have $n_i = [F(D_i) : K\epsilon_i]$, and both sets H, H' consist of n_i homomorphisms.

If, for a given $i = 1, 2, \cdots, h$, ψ is any element of H_i and ψ' is any element of H_i', then the restrictions $\psi \mid F(C_i)$, $\psi' \mid F(D_i)$ are isomorphisms, and we have

$$\psi(F(C_i)) = \psi'(F(D_i)) = F_i,$$

where F_i, the cyclic extension of K, of degree n_i, in Ω, depends only on i, being independent of the choice of ψ in H_i and of ψ' in H_i'. The saturated local domain $\psi(\widetilde{\mathfrak{o}(C_i)})$, $\psi'(\widetilde{\mathfrak{o}(D_i)})$ are k-isomorphic, in fact $k[[x]]$-isomorphic, by (21). It follows from Lemma 7.7 that these saturated local domains coincide and are thus independent of the choice of ψ in H_i and of ψ' in H_i'. We set therefore

$$\mathfrak{o}_i = \psi(\widetilde{\mathfrak{o}(C_i)}) = \psi'(\widetilde{\mathfrak{o}(D_i)}), \text{ for all } \psi \text{ in } H_i \text{ and all } \psi' \text{ in } H_i'. \quad i = 1, 2, \cdots, h. \tag{22}$$

By (22), any pair of homomorphisms ψ, ψ' such that $\psi \in H_i$ and $\psi' \in H_i'$ defines a $k_{\epsilon_i}[[x_{\epsilon_i}]]$-*isomorphism* $\psi^{-1} \cdot \psi' \colon \widetilde{\mathfrak{o}(D_i)} \xrightarrow{\sim} \widetilde{\mathfrak{o}(C_i)}$, where for simplicity we have written ψ' when what we actually have in mind is the restriction of ψ' to $\widetilde{\mathfrak{o}(D_i)}$, and similarly for ψ and $\widetilde{\mathfrak{o}(C_i)}$. The isomorphism $\psi^{-1} \cdot \psi'$ can, of course, be extended (uniquely) to a K_{ϵ_i}-isomorphism between the fields $F(D_i)$ and $F(C_i)$.

With the identifications $\epsilon_i = \delta_i$ and $x = x'$ made above, the assertion of Theorem 7.6 is to the effect that there exists a K-isomorphism ϕ: $F(D) \xrightarrow{\sim} F(C)$ such that $\phi(\widetilde{\mathfrak{o}(C)}) = \widetilde{\mathfrak{o}(D)}$. We shall prove this assertion under the following stronger form:

THEOREM 7.9. *Let* $\{\psi_1, \psi_2, \cdots, \psi_h\}$ *and* $\{\psi_1', \psi_2', \cdots, \psi_h\}$ *be two strictily normal sets of K-homomorphisms in H in H′, relative to C and D respectively. We assume (as in Theorem 7.6) that there exists an equivalence* $\pi\colon C = D$ *such that* $\pi(C_i) = D_i$, $i = 1, 2, \cdots, h$, *and that consequently (with the identifications* $\epsilon_i = \delta_i$, $x = x'$, *and by what has just been shown above) there exists, for each* $i = 1, 2. \cdots, h$, *a* $k_{\epsilon_i}((x_{\epsilon_i}))$-*isomorphism*

$$\phi_i = \psi_i^{-1} \cdot \psi_i' \colon F(D_i) \xrightarrow{\sim} F(C_i). \tag{23}$$

If we set

$$\phi = \phi_1 + \phi_2 + \cdots + \phi_h \colon F(D) \xrightarrow{\sim} F(C), \tag{24}$$

then the K-isomorphism ϕ *induces an isomorphism between* $\widetilde{\mathfrak{o}(D)}$ *and* $\widetilde{\mathfrak{o}(C)}$.

Proof. We shall use induction on h, for in the case $h = 1$ the theorem says nothing more than any K-isomorphism of the *fields* $F(D)$ and $F(C)$ induces an isomorphism of the saturated local domains $\widetilde{\mathfrak{o}(C)}$ and $\widetilde{\mathfrak{o}(D)}$, and this assertion is a direct consequence of Lemma 7.7 and of the saturation-theoretic criterion of equisingularity proved by us in SES III.

A priori, we have now two saturation partitions of the set of indices

$\{1, 2, \cdots, h\}$: one, relative to C, and the other relative to D. We show now that *these two partitions are identical.* For that purpose, we use Theorem 6.5, observing that the integers $\tilde{e}_{i,0}$ which appear in that theorem are the multiplicities $m_P(C_i)$, since x is a transversal parameter. It follows from Theorem 6.5 that two distinct indices i, j belong to one and the same set of the saturation partition relative to C, if and only if

$$(25) \qquad (C_i, C_j)/m_P(C_i)m_P(C_j) > \min_{\substack{\alpha,\beta \\ \alpha\neq\beta}}.\{(C_\alpha, C_\beta)/m_P(C_\alpha)m_P(C_\beta)\}.$$

Similarly, two distinct indices i, j belong to one and the same set of the saturation partition relative to D, if and only if

$$(26) \qquad (D_i, D_j)/m_Q(D_i)m_Q(D_j) > \min_{\substack{\alpha,\beta \\ \alpha\neq\beta}}.\{(D_\alpha, D_\beta)/m_Q(D_\alpha)m_Q(D_\beta)\}.$$

Now, since we have the equivalence $\pi\colon C = D$, with $\pi(C_i) = D_i$, $i = 1, 2, \cdots, h$, it follows from Lemma 7.1 that $(C_i, C_j) = (D_i, D_j)$ for all $i \neq j$, and also that $m_P(C_i) = m_Q(D_i)$. Hence, for any pair of distinct indices i, j, the inequality (25) holds if and only if the inequality (26) holds, and this proves that the two saturation partitions are identical. Let $\{I_1, I_2, \cdots, I_r\}$ be this common saturation partition of the set of indices $\{1, 2, \cdots, h\}$.

We next consider the two integers γ and q introduced in § 2, in relation to the pair $\{\mathfrak{o}, x\}$ where $\mathfrak{o}$ is now $\widetilde{\mathfrak{o}(C)}$. Similarly, we denote by γ', q' the analogous integers, relative to the pair $\{\widetilde{\mathfrak{o}(D)}, x\}$. Thus, while γ and q are defined by (3) and (5) of § 2, the integers γ', q' are defined in a similar fashion, upon replacing y by y':

$$\gamma' = \min_{\substack{i\neq j \\ 1\leqq i,j\leqq h}}.\ \max_{\tau,\sigma\in G}.\{v(y_i'^{\tau} - y_j'^{\sigma})\},$$

where $y_i' = \psi_i'(y')$, and $\psi_1', \psi_2, \cdots, \psi_h'$ can be any h elements of H' belonging respectively to $H_1', H_2', \cdots, H_h'$, while q' is the greatest integer such that

$$\gamma' > \min.\{\beta_{1,q'-1}, \beta_{2,q'-1}, \cdots, \beta_{h,q'-1}\}.$$

Here $\{e_{i,0}; \beta_{i,1}, \beta_{i,2}, \cdots, \beta_{i,g_i}\}$ is the (modified) x-characteristic of the ring $\mathfrak{o}_i$ defined in (22). *We assert now that*

$$(27) \qquad \gamma' = \gamma, \quad q' = q.$$

To see this, let, say, $q' \geqq q$. By the structure theorem (Theorem 2.1), applied to $\widetilde{\mathfrak{o}(D)}$ (which now replaces the ring $\mathfrak{o}$ of that theorem), we have $\beta_{1,\alpha} = \beta_{2,\alpha} = \cdots = \beta_{h,\alpha}$ $(= \beta_\alpha$, say) for $\alpha = 0, 1, \cdots, q' - 1$. Hence

$$\beta_0 < \beta_1 < \cdots < \beta_{q-1} < \gamma \leqq \min.\{\beta_{1,q}, \beta_{2,q}, \cdots, \beta_{h,q}\};$$

$$\beta_0 < \beta_1 < \cdots < \beta_{q'-1} < \gamma' \leqq \min.\{\beta_{1,q'}, \beta_{2,q'}, \cdots, \beta_{h,q'}\}.$$

For any two distinct indices i, j in the set $\{1, 2, \cdots, h\}$ we set

$$\gamma_{i,j} = \max_{\tau,\sigma \in G}. \{v(y_i^{\tau} - y_j^{\sigma})\},$$

$$\gamma_{i,j}' = \max_{\tau,\sigma \in G}. \{v(y_i'^{\tau} - y_j'^{\sigma})\},$$

and we denote by $q_{i,j}$ the greatest integer such that

$$\gamma_{i,j} > \min.\{\beta_{i,q_{i,j}-1}, \beta_{j,q_{i,j}-1}\}.$$

Similarly, let $q_{i,j}'$ be the greatest integer such that

$$\gamma_{i,j}' > \min.\{\beta_{i,q_{i,j}'-1}, \beta_{j,q_{i,j}'-1}\}.$$

We have obviously, $\gamma \leqq \gamma_{i,j}$, $\gamma' \leqq \gamma_{i,j}'$, and also $q \leqq q_{i,j}$ and $q' \leqq q_{i,j}'$. By the definition of saturation partitions, we have $\gamma_{i,j} = \gamma$ if and only if i and j belong to distinct sets of the partition $\{I_1, I_2, \cdots, I_r\}$, and when that is so we have necessarily $q_{i,j} = q$, since from $\gamma_{i,j} = \gamma \leqq \min.\{\beta_{1,q}, \beta_{2,q}, \cdots, \beta_{h,q}\}$ follows that $\gamma_{i,j} \leqq \min.\{\beta_{i,q}, \beta_{j,q}\}$, whence $q \geqq q_{i,j}$, which in conjunction with $q \leqq q_{i,j}$ yields the equality $q = q_{i,j}$. Similarly, we have $\gamma_{i,j}' = \gamma'$ if and only if the indices i *and* j belong to one and the same set of the partition $\{I_1, I_2, \cdots, I_r\}$, and when that is so we have necessarily $q_{i,j}' = q'$

We fix two indices i, j belonging to *distinct* sets of the partition $\{I_1, I_2, \cdots, I_r\}$. By Theorem 6.5 we have then

$$(28) \qquad (C_i, C_j) = \frac{m_P(C_i) m_P(C_j)}{e_0^2} \cdot \{\beta_1(e_0 - e_1) + \beta_2(e_1 - e_2) + \cdots + \beta_{q-1}(e_{q-2} - e_{q-1}) + \gamma e_{q-1}\},$$

and

$$(D_i, D_j) = \frac{m_Q(D_i) m_Q(D_j)}{e_0^2} \cdot \{\beta_1(e_0 - e_1) + \beta_2(e_1 - e_2) + \cdots + \beta_{q'-1}(e_{q'-2} - e_{q'-1}) + \gamma' e_{q'-1}\}.$$

Here the integers $e_1, e_2, \cdots, e_{q'-1}$ are defined as in (47) of § 6, or—equivalently—as in part (c) of Theorem 2.1 (i.e., $e_\alpha = [F^* : K_\alpha]$, $\alpha = 0, 1, \cdots, q' - 1$; see (28) of § 1). Since $(C_i, C_j) = (D_i, D_j)$ and $m_P(C_i) = m_Q(D_i)$, it follows that in addition to (28) we have also the following expression of (C_i, C_j):

$$(29) \quad (C_i, C_j) = \frac{m_P(C_i) m_P(C_j)}{e_0^2} \cdot \{\beta_1(e_0 - e_1) + \beta_2(e_1 - e_2) + \cdots + \beta_{q'-1}(e_{q'-2} - e_{q'-1}) + \gamma' e_{q'-1}\}.$$

From (28) and (29). the equalities (27) follow by applying Proposition 6.2 to the case $C = C_i$ and $C' = C_j$. However, since our notations in the present case are somewhat different from those of Proposition 6.2, it will be simpler to repeat the little calculation used in the proof of that proposition. We show therefore again that the assumption $q' > q$ leads to the inequality $\gamma > \beta_q$, in contradiction with the definition of γ (note that $q' > q$ implies, at any rate, that $\beta_{1,q} = \beta_{2,q} = \cdots = \beta_{hq} = \beta_q$). From $q' = q$ will follow, of course, that $\gamma' e_{q-1} = \gamma e_{q-1}$, in view of (28) and (29), whence $\gamma' = \gamma$.

We have, by (28) and 29):

$$\beta_q(e_{q-1} - e_q) + \cdots + \beta_{q'-1}(e_{q'-2} - e_{q'-1}) + \gamma' e_{q'-1} = \gamma e_{q-1},$$

or—equivalently:

$$e_{q-1}\beta_q + e_q(\beta_{q+1} - \beta_q) + \cdots + e_{q'-1}(\gamma' - \beta_{q'-1}) = \gamma\, e_{q-1}.$$

Hence, since we have assumed $q' > q$, we have $e_{q-1}\beta_q < e_{q-1}\gamma.$, i.e., $\beta_q < \gamma$, a contradiction.

Let Γ_ρ and Δ_ρ, $\rho = 1, 2, \cdots, r$, be corresponding saturation components of C and D. By definition of strictly normal sets of homomorphisms, we have for each $\rho = 1, 2, \cdots, r$, that the set $\{\psi_i \mid i \in I_\rho\}$ is a strictly normal set of homomorphisms of $F(\Gamma_\rho)$ into Ω, relative to the curve Γ_ρ and to the canonical isomorphism $K\epsilon_\rho' \xrightarrow{\sim} K$. Similarly, the set $\{\psi_i' \mid i \in I_\rho\}$ is a strictly normal set of homomorphisms of $F(\Delta_\rho)$ into Ω, relative to the curve Δ_ρ and the canonical isomorphism $K\epsilon_\rho' \xrightarrow{\sim} K$. It follows therefore, by our induction hypothesis, that if we define the ϕ_i as in (23) and set

$$\phi_\rho^* = \sum_{i \in I_\rho} \phi_i, \quad \rho = 1, 2, \cdots, r,$$

then ϕ_ρ^* induces a $k\epsilon_\rho'[[x\epsilon_\rho']]$-isomorphism between $\mathfrak{o}(\Delta_\rho)$ and $\mathfrak{o}(\Gamma_\rho)$.

We now show that isomorphism ϕ, defined in (24), induces an isomorphism between $\mathfrak{o}(\widetilde{D})$ and $\mathfrak{o}(\widetilde{C})$. To show this it will be sufficient to show that $\phi(\mathfrak{o}(\widetilde{D})) \subset \mathfrak{o}(\widetilde{C})$. For that it will be sufficient to show that

$$\phi(y') \in \mathfrak{o}(\widetilde{C}). \tag{30}$$

We set

$$\eta = \phi(y') \tag{31}$$

and we use the structure theorem 3.1 to prove the inclusion

$$\eta \in \mathfrak{m}. \tag{32}$$

If we set $\eta_i = \psi_i(\eta)$, $i = 1, 2, \cdots, h$, then we have, by the definition (24) of ϕ and the definition (31) of η: $\eta_i = \psi_i'(y') = y_i'$. Thus

$$(33) \qquad \eta_i = y_i' = y_i'^{(0)} + y'^{(1)} + \cdots + y'^{(q-2)} + y_{i,q-1}' + \cdots,$$

where $y'^{(\alpha)} \in K_\alpha$, $y_{i,q-1}' \in K_{q-1}$ and where we have

$$(34) \qquad y'^{(\alpha)} \in \mathfrak{M}_\alpha^{m_\alpha}, \quad \alpha = 1, 2, \cdots, q-2; y_{i,\alpha}' \in \mathfrak{M}_{i,\alpha}^{m_{i,\alpha}},$$

and

$$(35) \qquad v(y_{i,q-1}' - y_{j,q-1}) \geqq \gamma, \text{ for all } i, j = 1, 2, \cdots, h.$$

Thus the (33) give decompositions of the η_i, which are of the type (5) of § 3, and are such that the condition (6) and (8) of § 3 are satisfied (in view of (34)) and such that also condition (a) of Theorem 3.1 is satisfied. Furthermore, also condition (b) of Theorem 3.1 is satisfied, in view of (35). Finally, we have $\eta\epsilon_\rho' = \phi(y') \cdot \epsilon_\rho' = \phi_\rho^*(y'\epsilon_\rho')$, and since

$$\phi_\rho^*(\widetilde{\mathfrak{o}(\Delta_\rho)}) = \widetilde{\mathfrak{o}(\Gamma_\rho)} = (\widetilde{\mathfrak{o}(C)}) \cdot \epsilon_\rho',$$

we see that also condition (c) of Theorem 3.1 is satisfied (with $\mathfrak{o}$ being replaced by $\mathfrak{o}(C)$). This proves the inclusion (32) and completes the proof of the theorem.

8. Change of coefficient field. In this section we shall generalize to the case of reducible curves our theorem on the intrinsic nature of the saturation of local domains of dimension 1, proved by us elsewhere (see SES III, Theorem 1.16). Again we shall restrict ourselves to the case of algebraically closed coefficient fields. Our object, then, is to prove the following result:

THEOREM 8.1. *Let k and k' be two coefficient fields of a complete local ring $\mathfrak{o}$, of dimension 1, and let x be a parameter of $\mathfrak{o}$. We assume that k (and hence also k') is algebraically closed, of characteristic zero, and that $\mathfrak{o}$ is k-saturated with respect to x. Let Θ be any isomorphism $k \longrightarrow k'$ between k and k'. Then Θ can be extended to an automorphism f of $\mathfrak{o}$ such that $f(x) = x$ (and consequently $\mathfrak{o}$ is also k'saturated with respect to x).*

Proof. We use the structure theorem, as given by Theorem 3.4. We have then the following decomposition of the maximal ideal $\mathfrak{m}$ of $\mathfrak{o}$ [see 22), § 3]:

$$(1) \qquad \mathfrak{m} = \mathfrak{N}_0 + \mathfrak{N}_1^{m_1} + \cdots + \mathfrak{N}_{q-1}^{m_{q-1}} + \sum_{\rho=1}^{r} \mathfrak{A}_\rho'.$$

The sequence of prime ideals $\mathfrak{N}_0, \mathfrak{N}_1, \cdots, \mathfrak{N}_{q-1}$ is associated with the chain of fields $K = L_0 < L_1 < \cdots < L_{q-1}$ contained in F, which have been introduced in § 3. We recall that these fields are uniquely determined by $\mathfrak{o}$ and the field $K = k((x))$ (see Proposition 3.3). For the purposes of the proof we shall have to prove that there exists an automorphism f which in addition to satisfying the condition stated in the theorem satisfies also the following additional condition:

A. *Let $\mu_\alpha = [L_\alpha : K]$, and let, for a given $\alpha = 1, 2, \cdots, q-1$, t_α be a uniformizing parameter of the local field L_α such that $x = t_\alpha^{\mu_\alpha}$ (we recall that L_α is then the power series field in t_α, over k, and that there always exist such uniformizing parameters t_α for which $x = t_\alpha^{\mu_\alpha}$, since k is algebraically closed). Then there exists an automorphism f of $\mathfrak{o}$, as stated in the theorem, such that $f(t_\alpha) = t_\alpha$ (and hence, a fortiori, $f(x) = x$).*

We shall prove our theorem, strengthened by the above condition A, by induction on the number h of direct field components of the total ring of quotients F of $\mathfrak{o}$. Let us recall briefly the proof in the case $h = 1$, which was essentially implicit in our proof of Theorem 1.16 in SES III.

If $h = 1$, we have $L_{q-1} = F$, and

$$\mathfrak{m} = \mathfrak{N}_0 + \mathfrak{N}_1^{m_1} + \mathfrak{N}_2^{m_2} + \cdots, \mathfrak{N}_g^{m_g}, \qquad g = q-1.$$

Since $[F : K] = \mu_{q-1}$, we have $[F : L_\alpha] = \mu_{q-1}/\mu_\alpha$. We can then find a uniformizing parameter t_{q-1} of F such that $t_\alpha = t_{q-1}^{\mu_{q-1}/\mu_\alpha}$. Then $x = t_{q-1}^{\mu_{q-1}}$, and the proof of Theorem 1.16 in SES III tells us that the isomorphism $\Theta : k \xrightarrow{\sim} k'$ can be extended to an automorphism f of $\mathfrak{o}$ such that $f(t_{q-1}) = t_{q-1}$. We will have then, *a fortiori*, $f(t_\alpha) = t_\alpha$.

We now consider the case $h > 1$. We fix a uniformizing parameter t_{q-1} of L_{q-1} such that $t_\alpha = t_{q-1}^{\mu_{q-1}/\mu_\alpha}$. It will be sufficient to prove the existence of an automorphism f of $\mathfrak{o}$ such that $f(t_{q-1}) = t_{q-1}$ and such that $f \mid k = \Theta$.

For each $\rho = 1, 2, \cdots, r$, we consider the local ring $\mathfrak{o}\epsilon_\rho'$, where $\epsilon_\rho' = \sum_{i \in I_\rho} \epsilon_i$. This local ring has two coefficient fields $k\epsilon_\rho'$, $k'\epsilon_\rho'$, and we have the isomorphism

$$\Theta_\rho' : k\epsilon_\rho' \xrightarrow{\sim} k'\epsilon_\rho', \tag{2}$$

defined by $\Theta_\rho'(c\epsilon_\rho') = \Theta(c) \cdot \epsilon_\rho'$, $c \in k$. The local ring $\mathfrak{o}\epsilon_\rho'$ is saturated with respect to the field $k\epsilon_\rho'((x\epsilon_\rho'))$ $(= K\epsilon_\rho')$ [see Proposition 3.5 and the formula

(17) of § 3]. The decomposition of the maximal ideal $\mathfrak{m}\epsilon_\rho'$ of $\mathfrak{o}\epsilon_\rho'$, analogous to (1), will begin with the terms

$$\mathfrak{m}\epsilon_\rho' = \mathfrak{N}_0\epsilon_\rho' + \mathfrak{N}_1\epsilon_\rho' + \cdots + \mathfrak{N}_{q-1}\epsilon_\rho' + \cdots,$$

in the sense that the integer q' analogous to q but associated with $\mathfrak{o}\epsilon_\rho'$ and the field $K\epsilon_\rho'$ is always $\geqq q$, and the chain of q' fields

$$L_{0,\rho}' < L_{1,\rho}' < \cdots < L_{q'-1,\rho}',$$

analogous to the chain $L_0 < L_1 < \cdots < L_{q-1}$, will be such that $L_{\alpha,\rho}' = L_\alpha \cdot \epsilon_\rho'$, for $\alpha = 0, 1, \cdots, q-1$. All this follows without any difficulty from the way in which the fields $L_0, L_1, \cdots, L_{q-1}$ were defined in § 3, in terms of the pair $\{\mathfrak{o}, k((x))\}$.

We now apply our induction hypothesis to each pair $\{\mathfrak{o}\epsilon_\rho', k\epsilon_\rho'(x_\rho')\}$, and we take for α the integer $q-1$, while as uniformizing parameter of $L_{q-1} \cdot \epsilon_\rho'$ we take the element $t_{q-1}\epsilon_\rho'$, for we have then $x\epsilon_\rho' = (t_{q-1}\epsilon_\rho')^{\mu_{q-1}}$. As an isomorphism analogous to $\Theta \colon k \xrightarrow{\sim} k'$ we take the isomorphism Θ_ρ' defined by (2). It follows then, by our induction hypothesis, that Θ_ρ' can be extended to an automorphism f_ρ' of $\mathfrak{o}\epsilon_\rho'$, such that

$$f_\rho'(t_{q-1}\epsilon_\rho') = t_{q-1}\epsilon_\rho', \qquad \rho = 1, 2, \cdots, r. \tag{3}$$

The automorphism f_ρ' can be extended (uniquely) to an automorphism of the total ring of quotients $F\epsilon_\rho'$ of $\mathfrak{o}\epsilon_\rho'$, and we shall contiuue to denote by f_ρ' this extended automorphism. The ring F being the direct sum of the rings $F\epsilon_\rho'$, we define an automorphism f of F by

$$f = f_1' + f_2' + \cdots + f_r'. \tag{4}$$

Since the restriction of f_ρ' to $k\epsilon_\rho'$ is the isomorphism Θ_ρ' defined by $\Theta_\rho'(c\epsilon_\rho') = \Theta(c)\epsilon_\rho'$, it follows that the restriction of f to k is the given isomorphism $\Theta \colon k \xrightarrow{\sim} k'$. From (3), it follows that $f(t_{q-1}) = t_{q-1}$. Hence, to complete the proof of the theorem we have only to show that $f(\mathfrak{o}) = \mathfrak{o}$. It will be sufficient to show that

$$f(\mathfrak{o}) \subset \mathfrak{o}, \tag{5}$$

for if $\bar{\mathfrak{o}}$ denotes the integral closure of $\mathfrak{o}$ in F, we have $f(\bar{\mathfrak{o}}) = \bar{\mathfrak{o}}$ (since $\bar{\mathfrak{o}}$ is also the integral closure of both $k[[x]]$ and $k'[[x]]$, and since $f(k[[x]])$

$= k'[[x]])$, and hence the $\mathfrak{o}$-module $\bar{\mathfrak{o}}/\mathfrak{o}$ and the $f(\mathfrak{o})$-module $\bar{\mathfrak{o}}/f(\mathfrak{o})$ have the same length.

To prove (4), we set $t_\alpha = t_{q-1}^{\mu_{q-1}/\mu_\alpha}$, so that $\mathfrak{N}_\alpha = k[[t_\alpha]]\cdot t_\alpha$, $\alpha = 0, 1, \cdots, q-1$. From $f(t_{q-1}) = t_{q-1}$ follows $f(t_\alpha) = t_\alpha$, and $f(t_\alpha^\lambda) = t_\alpha^\lambda$ for all integers λ. If $\xi = \sum_{\lambda \geqq m_\alpha} c_\lambda t_\alpha^\lambda$ is any element of $\mathfrak{N}_\alpha^{m_\alpha}$, then $f(\xi) = \sum_{\lambda \geqq m_\alpha} c_\lambda' t_\alpha^\lambda$, where $c_\lambda' = f(c_\lambda) \in k' \subset \mathfrak{o}$. The infinite series representing $f(\xi)$ is convergent in the *complete* local ring $\mathfrak{o}$, since for $\lambda \geqq m_\alpha$ we have $t_\alpha^\lambda \in (\mathfrak{N}_\alpha^{m_\alpha})^{[\lambda/m_\alpha]}$, where [ ] stands for the integral part of a number, whence $t_\alpha^\lambda \in \mathfrak{m}^{\nu(\lambda)}$ with $\nu(\lambda) \to +\infty$ as $\lambda \to +\infty$. Hence $f(\xi) \in \mathfrak{m}$. We have thus shown that $f(\mathfrak{N}_\alpha^{m_\alpha}) \subset \mathfrak{m}$ for $\alpha = 0, 1, \cdots, q-1$. In view of (1), the inclusion (5) will be proved if we show that $f(\mathfrak{A}_\rho') \subset \mathfrak{A}_\rho'$, for $\rho = 1, 2, \cdots, r$. Now, from the definition (4) of f and from the fact that $\mathfrak{A}_\rho' \subset F \cdot \epsilon_\rho'$ folows that $f(\mathfrak{A}_\rho') = f_\rho'(\mathfrak{A}_\rho')$. Thus, we have only to show that

$$f_\rho'(\mathfrak{A}_\rho') \subset \mathfrak{A}_\rho'.$$

Let ξ' be any element of $\mathfrak{A}_\rho'$, let $\{\psi_1, \psi_2, \cdots, \psi_h\}$ be a normal set of K-homomorphisms of F into the algebraic closure Ω of K (see § 2) and let $\xi_i' = \psi_i(\xi')$, $i \in I_\rho$. We have (by the definition of the ideal $\mathfrak{A}_\rho'$; see (18), § 3): $v(\xi_i') \geqq \gamma$, for all $i \in I_\rho$. Let $\eta' = f_\rho'(\xi')$. We have to prove that

$$\eta' \in \mathfrak{A}_\rho'.$$

Since f_ρ' is an automorphism of $\mathfrak{o}\epsilon_\rho'$ and since $\mathfrak{A}_\rho' \subset \mathfrak{o}\epsilon_\rho'$, we have, at any rate: $\eta' \in \mathfrak{o}\epsilon_\rho'$. So we have only to prove that if we set

$$\eta_i' = \psi_i(\eta'), \quad i \in I_\rho,$$

then

$$v(\eta_i') \geqq \gamma, \quad i \in I_\rho. \tag{6}$$

Let $\psi_i(F\epsilon_\rho') = K_i$. The kernel of the restriction of ψ_i to $F\epsilon_\rho'$ is the sum $\sum_j F\epsilon_j$, where j ranges over the set of indices in I_ρ, other than i. It is clear that this kernel is invariant under f_ρ', since $f_\rho'(F\epsilon_j) = F\epsilon_j$ for all $j \in I_\rho$. It follows that there is an induced automorphism $f_{\psi_i}\colon K_i \xrightarrow{\sim} K_i$ such that

$$\psi_i \cdot f_\rho' = f_{\psi_i} \cdot \psi_i. \tag{7}$$

We have therefore: $\eta_i' = \psi_i(\eta') = (\psi_i \cdot f_\rho')(\xi') = (f_{\psi_i} \cdot \psi_i)(\xi') = f_{\psi_i}(\xi_i')$.

Since $v(\xi_i') \geqq \gamma$, and the natural valuation of F_i is invariant by the automorphism f_{ψ_i}, it follows that also $(v(f_{\psi_i}(\xi'))) \geqq \gamma$. We have therefore (6), and this completes the proof of the theorem.

COROLLARY 8.2. *If k, k' are two algebraically closed coefficient fields (of characteristic zero) of a complete local ring $\mathfrak{o}$, of dimension 1, and if x is any parameter of $\mathfrak{o}$, then the k-saturation of $\mathfrak{o}$, with respect to x, coincides with the k'-saturation of $\mathfrak{o}$ with respect to x. In particular, the k-saturation of $\mathfrak{o}$ (i.e., the k-saturation of $\mathfrak{o}$ with respect to any transversal parameter; see Proposition 4.6) coincides with the k'-saturation of $\mathfrak{o}$, and therefore is independent of the choice of the coefficient field.*

We have shown in [GTS. I, § 9] that Theorem 8.1 is definitely false, in that generality, if k is not algebraically closed, already in the case of local *domains*. However, we have proved, in the case of local *domains* that even if k is *not algebraically closed*, the *canonical isomorphism* $\Theta: k \xrightarrow{\sim} k'$ (i.e., the isomorphism Θ defined by $\Theta(c) - c \in \mathfrak{m}$, for all c in k) can be extended to an automorphism of the saturated local domain $\mathfrak{o}$ [see GTS I, Theorem 9.1]. It would be interesting to prove, in the more general case in which $\mathfrak{o}$ is not a domain, that Theorem 8.1 continues to hold true for pairs of coefficient fields k, k' which are *not* algebraically closed, provided Θ stands for the canonical isomorphism $k \xrightarrow{\sim} k'$ referred to above.

HARVARD UNIVERSITY.

REFERENCES.

[1] S. Abhyankar, "Inversion and invariance of characteristic pairs," *American Journal of Mathematics*, vol. 89 (1967), pp. 363-372.

[2] F. Enriques and O. Chisini, *Lezioni sulla teoria geometrica delle equazioni e delle funzioni algebriche*, vol. 2, pp. 327-458 (Bologna, 1915-1924).

[3] O. Zariski, *Algebraic Surfaces*, Ergebnisse der Mathematik und iher Grenzgebiete, vol. 3 (Berlin, Springer, 1935), or second supplemented edition, vol. 61 (Springer, Berlin-Heidelberg New York, 1971).

[4] ———, "Studies in equisingularity I. Equivalent singularities of plane algebroid curves," *American Journal of Mathematics*, vol. 87 (1965), pp. 507-536. (This paper will be referred to as SES I.)

[5] ———, "Studies in equisingularity III. Saturation of local rings and equisingularity," *American Journal of Mathematics*, vol. 90 (1968), pp. 961-1023. (This paper will be referred to as SES III.)

[6] ———, "General theory of saturation and of saturated local rings I. Saturation of complete local domains of dimension one having arbitrary coefficient fields (of characteristic zero)," *American Journal of Mathematics*, vol. 93 (1971), pp. 573-648. (This paper will be referred as GTS I.)

[7] O. Zariski and P. Samuel, *Commutative Algebra*, volumes 1 and 2 (Princeton, D. Van Nostrand, 1958 and 1960).

QUATRE EXPOSES SUR LA SATURATION

Oscar ZARISKI

(Notes prises par J.J. RISLER)

Nous noterons S.E.S. I, II ou III (resp G.T.S. I,II ou III), les articles de O. Zariski "Studies in equisingularity" I, II ou III parus dans l'American Journal of Mathematics, vol 87 (1965) et vol 90 (1968) (resp : les articles "General theory of saturation" I, II parus dans la même revue, vol 93 (1971), ou III, en cours de publication dans la même revue) .

Tous les anneaux seront commutatifs, avec éléments unités ; les morphismes d'anneaux étant supposés envoyer l'élément unité sur l'élément unité.

§ 1.- DEFINITION GENERALE DE LA SATURATION

Soient A un anneau, F son anneau total des fractions, K un sous-corps de F.

Nous allons définir la saturation de A par rapport à K : ce sera un sous-anneau de la clôture intégrale $\bar{A}$ de A dans F.

Soit Ω une clôture algébrique (fixée une fois pour toutes) de K ; notons H l'ensemble de K-homomorphismes de F dans Ω, A^* le plus petit sous-anneau de Ω contenant les anneaux $\psi(A)$ (pour tout $\psi \in H$), et $\bar{A}^*$ la clôture intégrale de A^*.
(Remarquons que A^* et $\bar{A}^*$ sont des anneaux intègres).

Définition 1.1

Soient η et ξ deux éléments de F. On dit que η domine ξ par rapport à K (ce que nous noterons $\eta \underset{K}{\succ} \xi$ ou $\eta \succ \xi$ s'il n'y a pas de confusion possible sur le corps K), si, quels que soient ψ_1 et ψ_2 dans H, on a :

$$\frac{\psi_1(\eta) - \psi_2(\eta)}{\psi_1(\xi) - \psi_2(\xi)} \in \bar{A}^* \qquad \text{si} \qquad \psi_1(\xi) \neq \psi_2(\xi)$$

$$\psi_1(\eta) = \psi_2(\eta) \qquad \text{si} \qquad \psi_1(\xi) = \psi_2(\xi)$$

Définition 1.2

Soit B un sous-anneau de $\bar{A}$ contenant A. On dit que B est saturé par rapport à K si B est fermé dans $\bar{A}$ pour la relation de domination, c'est-à-dire si, pour deux éléments ξ et η de $\bar{A}$, les conditions $\xi \in B$ et $\eta \underset{K}{>} \xi$ impliquent $\eta \in B$.

Remarquons que cette définition implique que $\bar{A}$ est saturé par rapport à K : l'ensemble des sous-anneaux de $\bar{A}$ contenant A et saturés n'est donc pas vide. Il est clair, d'autre part, que l'intersection d'une famille d'anneaux saturés est un anneau saturé.

Définition 1.3

Le saturé de A par rapport à K, noté $\tilde{A}_K$, est le plus petit sous-anneau saturé entre A et $\bar{A}$.

Remarque

Dans S.E.S. III, les hypothèses suivantes sont faites :

a) A est réduit,

b) F est de dimension finie sur K (ce qui implique que F est noetherien et de plus isomorphe à une somme directe finie de corps F_i),

c) les F_i sont des extensions séparables de K.

Ces hypothèses ne sont pas indispensables pour donner une définition de la saturation, mais sont probablement nécessaires pour pouvoir en dire quelque chose d'intéressant.

§ 2-CAS DES VARIETES ALGEBROÏDES

Nous appellerons variété algébroïde (ou variété algébrique formelle) le spectre V d'un anneau $\mathcal{O}$ satisfaisant aux conditions suivantes :

a) $\mathcal{O}$ est un anneau local noetherien, complet pour la topologie définie par l'idéal maximal et équicaractéristique,

b) $\mathcal{O}$ est réduit (ce qui est équivalent à dire que l'idéal (O) est intersection finie d'idéaux premiers minimaux : $(0) = \mathcal{P}_1 \cap \ldots \cap \mathcal{P}_n$),

c) $\mathcal{O}$ est équidimensionnel

(i.e. les dimensions des anneaux $\frac{\mathcal{O}}{\mathcal{P}_i}$ sont égales).

Faisons d'abord quelques rappels d'algèbre commutative ; soit $\varphi : \mathcal{O} \to \mathcal{O}/\underline{m}$ le morphisme canonique ($\underline{m}$ désignant l'idéal maximal de l'anneau $\mathcal{O}$). Un sous-corps k de $\mathcal{O}$ est appelé <u>corps de coefficients</u> si $\varphi|k$ est un isomorphisme de k sur $\mathcal{O}/\underline{m}$. Un tel corps est évidemment maximal parmi les sous corps de $\mathcal{O}$. On peut montrer ("Théorèmes de Cohen") que sous les hypothèses faites plus haut, il existe toujours un corps de coefficients, et que si la caractéristique de $\mathcal{O}$ (et donc de $\mathcal{O}/\underline{m}$) est 0, les corps de coefficients sont exactement les sous-corps maximaux de $\mathcal{O}$.

Si k et k' sont deux corps de coefficients dans $\mathcal{O}$, il existe un isomorphisme canonique $\varphi_{k,k'} : k \tilde{\to} k'$; il suffit de poser :

$$\varphi_{k,k'} = (\varphi|k')^{-1} \circ (\varphi|k)$$

On peut aussi définir $\varphi_{k,k'}$ par la condition :

$$\varphi_{k,k'}(c) - c \in \underline{m} \qquad \forall c \in k.$$

N.B.- $\varphi_{k,k'}$ n'est pas en général le seul isomorphisme de k sur k'.

<u>Définition 2.1</u>

Soit r la dimension de $\mathcal{O}$. Un ensemble d'éléments $x_1, x_2, \ldots, x_r$ de $\underline{m}$ tel que l'idéal $(x_1, x_2, \ldots, x_r)$ soit primaire pour $\underline{m}$ est appelé un <u>système de paramètres</u> pour $\mathcal{O}$.

Il existe toujours des systèmes de paramètres dans un anneau local.

<u>Interprétation géométrique</u> :

Ecrivons $\mathcal{O}$ comme quotient d'un anneau de séries formelles : $\mathcal{O} \tilde{\to} k[[y_1, \ldots, y_n]]/I$, ce qui géométriquement s'interprète comme un plongement local de la variété V dans un espace affine (sur lequel les coordonnées sont $y_1, \ldots, y_n$). Relevons les x_i en des éléments X_i de $k[[y_1, \ldots, y_n]]$. Le fait que $(x_1, x_2, \ldots, x_r)$ soit un système de paramètres pour $\mathcal{O}$ s'interprète

- 23 -

alors en disant que l'intersection de V avec la variété d'équations : $X_1 = \dots = X_r = 0$ est réduite à l'origine.

Fixons maintenant un corps de coefficients k, et un système de paramètres $x_1, \dots, x_r$. On a alors les propriétés suivantes :

1) Les éléments $x_1, \dots, x_r$ sont analytiquement indépendants sur k (cela veut dire que le sous-anneau $R = k[[x_1, \dots, x_r]]$ de $\mathcal{O}$ est isomorphe à l'anneau des séries formelles en r indéterminées).

2) Aucun élément de R n'est diviseur de 0 dans $\mathcal{O}$ (ceci parce que $\mathcal{O}$ est supposé équidimensionnel).

3) $\mathcal{O}$ est un R-module de type fini.

Ces propriétés entraînent que si K est le corps des fractions de R, K est canoniquement contenu dans l'anneau total des fractions F de $\mathcal{O}$ et que F est un K-espace vectoriel de dimension finie.

<u>Définition 2.2</u>

L'anneau $\tilde{\mathcal{O}}_K$ (défini au paragraphe précédent) s'appelle la k-saturation de $\mathcal{O}$ par rapport à $(x_1, \dots, x_r)$. On le note aussi $\tilde{\mathcal{O}}_k\,(x_1, \dots, x_r)$.

On peut montrer que si k est donné, l'anneau $\tilde{\mathcal{O}}_K$ est le même pour tous les choix suffisamment "génériques" des paramètres dans les cas suivants:

- V est une hypersurface (cf. note (1) à la fin du texte)

- V est une courbe (i.e. r = 1).

Précisons un peu ce qui se passe dans le cas r = 1 (le cas d'une hypersurface sera traité au § 5) :

Supposons que k soit de caractéristique 0 et soit x un paramètre de $\mathcal{O}$. Alors $\tilde{\mathcal{O}}_{k,x}$ est un anneau local noetherien de dimension 1, de corps résiduel k et de même multiplicité que $\mathcal{O}$ (dans le cas où $\mathcal{O}$ est intègre, cf. G.T.S. I , proposition 2.5) .

Un paramètre x de $\mathcal{O}$ est dit <u>transverse</u> si on a l'égalité :

- 24 -

$\dim_k(\frac{\mathcal{O}}{(x)}) = e(\mathcal{O})$, $e(\mathcal{O})$ désignant la multiplicité de l'anneau local $\mathcal{O}$ (l'inégalité $\dim_k(\frac{\mathcal{O}}{(x)}) \geqslant e(\mathcal{O})$ est toujours satisfaite). On a alors :

Théorème 2.3

Si k et k' sont deux corps de coefficients de $\mathcal{O}$, et x et x' deux paramètres transverses tels que le résidu de x'/x soit égal à 1 , alors l'isomorphisme canonique $\varphi_{k,k'} : k \overset{\sim}{\to} k'$ peut s'étendre en un automorphisme de $\tilde{\mathcal{O}}_{k,x}$ qui envoie x sur x' (G.T.S. II, théorème 4.4 et G.T.S. III, Appendice A, théorème A.1) .

On voit qu'un anneau saturé possède beaucoup d'automorphismes, un peu comme un anneau de séries formelles.

Il résulte de ce théorème que, dans le cas des courbes, la saturation $\tilde{\mathcal{O}}_{k,x}$ est la même pour tout choix du corps de coefficients et pour tout paramètre x supposé transverse.

§ 3.- LA SATURATION DE PHAM-TEISSIER

(cf. : fractions lipschitziennes d'une algèbre analytique complexe et saturation de Zariski, par Frédéric Pham et Bernard Teissier, Centre de Mathématiques de l'Ecole Polytechnique (Juin 1969)).

Nous allons en donner une définition due à J. Lipman, un peu plus générale que celle de l'article original.

Donnons d'abord quelques rappels algébriques sur la clôture intégrale d'un idéal (cf. par exemple M. Lejeune et B. Teissier : Quelques calculs utiles à la résolution des singularités, séminaire publié par le Centre de Mathématiques de l'Ecole Polytechnique).

Définition 3.1

Soient A un anneau, R un sous-anneau de A et I un R-module contenu dans A.

Alors la clôture intégrale $\overline{I}$ de I dans A, est par définition l'ensemble des éléments x de A qui vérifient une équation de la forme :

$$x^n + a_1 x^{n-1} + \dots + a_n = 0 \qquad \text{avec} \qquad a_i \in I^i \quad (1 \leqslant i \leqslant n)$$

- 25 -

Exemple :

Soit F l'anneau total des fractions de A, et soit x un élément de F qui s'écrira : $x = \frac{a}{b}$ avec a, b $\in$ A, et b $\neq$ 0. Alors x est entier (au sens classique) sur l'anneau R, si, et seulement si, a est entier sur le sous R-module de A engendré par b.

Soient maintenant R un anneau, A et A' deux R-algèbres, g : A $\rightarrow$ A' un R-homomorphisme. On peut considérer A' comme une A-algèbre, et on a un homomorphisme canonique f :

$$A' \underset{R}{\otimes} A' \rightarrow A' \underset{A}{\otimes} A'$$

Nous poserons $I_{R,A,A'}$ = Ker f ; c'est l'idéal de $A' \underset{R}{\otimes} A'$ engendré par les éléments $g(a) \underset{R}{\otimes} 1 - 1 \underset{R}{\otimes} g(a)$ $(\forall a \in A)$.

Définition 3.2

La saturation de Pham-Teissier de A dans A' relativement à R est l'anneau :

$$A^*_{R,A'} = \{x \in A' \mid x \underset{R}{\otimes} 1 - 1 \underset{R}{\otimes} x \in \overline{I_{R,A,A'}}\}$$

Reprenons maintenant les notations du début, et posons R = A $\cap$ K (rappelons que K est un sous-corps de l'anneau total des fractions F de A).

Nous faisons les hypothèses suivantes :

a) A est entier sur R,

b) K est le corps des fractions de R.

a) implique que F (et donc aussi K) est algébrique sur le corps des fractions de R.

Nous envisageons aussi l'hypothèse suivante, plus forte que l'hypothèse b) :

b^*) R est intégralement clos dans K.

- 26 -

Ω étant une clôture algébrique de K, soit H l'ensemble des R-homomorphismes : $\bar{A} \to \Omega$, et A^* le plus petit sous-anneau de Ω contenant les anneaux $\psi(A)$ $(\forall \psi \in H)$.

L'hypothèse a) implique que $\bar{A}^*$ est entier sur R, puisque $\psi(A)$ est entier sur R $\forall \psi \in H$. On voit donc que si $\xi, \eta \in \bar{A}$, la relation $\eta \underset{K}{\succ} \xi$ équivaut à dire que $\psi_1(\eta) - \psi_2(\eta)$ est entier (au sens défini plus haut) sur le R-module engendré dans Ω par l'élément $\psi_1(\xi) - \psi_2(\xi)$ $(\forall \psi_1, \psi_2 \in H)$.

Traduisons ce fait en terme de produit tensoriel : si ψ_1 et $\psi_2 \in H$, soit $f = \psi_1 \otimes \psi_2$ l'homomorphisme : $\bar{A} \underset{R}{\otimes} \bar{A} \to \Omega$ défini par $f(\xi \underset{R}{\otimes} 1) = \psi_1(\xi)$ et $f(1 \underset{R}{\otimes} \xi) = \psi_2(\xi)$.

Il est, d'autre part, clair que tout homomorphisme de $\bar{A} \underset{R}{\otimes} \bar{A}$ dans Ω est de cette forme.

La remarque que nous venons de faire donne alors :

<u>Lemme 3.3</u>

Soient ξ et η deux éléments de $\bar{A}$. Alors $\eta \underset{K}{\succ} \xi$ si, et seulement si, pour tout R-homomorphisme $f : \bar{A} \underset{R}{\otimes} \bar{A} \to \Omega$, $f(\eta \underset{R}{\otimes} 1 - 1 \underset{R}{\otimes} \eta)$ est entier sur le sous R-module de Ω engendré par $f(\xi \underset{R}{\otimes} 1 - 1 \underset{R}{\otimes} \xi)$.

Enonçons maintenant une proposition due à J. Lipman :

<u>Proposition 3.4</u>

Soient ξ et η deux éléments de $\bar{A}$.

1) si $\eta \underset{R}{\otimes} 1 - 1 \underset{R}{\otimes} \eta$ est entier sur le R-module engendré dans $\bar{A} \underset{R}{\otimes} \bar{A}$ par $\xi \underset{R}{\otimes} 1 - 1 \underset{R}{\otimes} \xi$, alors $\eta \underset{K}{\succ} \xi$

2) Si R est intégralement clos dans son corps des fractions K et si F est séparable sur K (i.e. si F est une somme directe de corps F_i, extensions séparables de K), alors la réciproque est vraie.

1) est évident, et nous allons donner des indications de démonstration de 2) (démonstration valable sous l'hypothèse plus faible que R est

géométriquement unibranche, i.e. que $\bar{R}$ est radiciel sur R).

Supposons donc que $\eta \succ_K \xi$ et posons $S = R[\xi, \eta]$ (S est un sous-anneau de $\bar{A}$). Comme S est de type fini sur R, $S \otimes_R S$ est noetherien et n'a qu'un nombre fini d'idéaux premiers minimaux $\mathcal{P}_1, \ldots, \mathcal{P}_m$.

D'autre part, R est intégralement clos, et $S \otimes_R S$ est entier sur R ; on a donc $\mathcal{P}_i \cap R = (0)$ par le "going down" de Cohen-Seidenberg. (c'est ici que l'hypothèse R intégralement clos, ou géométriquement unibranche, intervient) .

Notons φ_i le composé de l'application canonique : $S \otimes_R S \to S \otimes_R S / \mathcal{P}_i$ et d'un plongement $S \otimes_R S / \mathcal{P}_i \to \Omega$; φ_i est de la forme $\varphi_{i,1} \otimes \varphi_{i,2}$ où $\varphi_{i,1}$ et $\varphi_{i,2}$ sont des R-homomorphismes : $S \to \Omega$.

D'après le lemme 3.3, $\varphi_i(\eta \otimes_R 1 - 1 \otimes_R \eta)$ est entier sur le R-module engendré par $\varphi_i(\xi \otimes_R 1 - 1 \otimes_R \xi)$; il existe donc un polynôme homogène $G_i \in R[X,Y]$, de degré n_i, tel que $G_i(X,0) = X^{n_i}$ et $G_i(\eta \otimes_R 1 - 1 \otimes_R \eta, \xi \otimes 1 - 1 \otimes \xi) \in \mathcal{P}_i$.

Posons $G(X,Y) = \prod G_i(X,Y)$: c'est un polynôme homogène de degré $n = \Sigma n_i$, tel que $G(X,0) = X^n$ et que $G(\eta \otimes 1 - 1 \otimes \eta, \xi \otimes 1 - 1 \otimes \xi) \in \cap \mathcal{P}_i = (0)$ (ceci à cause de l'hypothèse de séparabilité) ; $\eta \otimes 1 - 1 \otimes \eta$ est donc bien entier sur le R-module engendré par $\xi \otimes 1 - 1 \otimes \xi$.

<u>Théorème 3.5</u>

Avec les notations précédentes, notons $\tilde{A}_K$ la saturation de Zariski de A (définition 1.3) et $A^*_{R,\bar{A}}$ la saturation de Pham-Teissier (définition 3.2). Supposons que les hypothèses a) et b^*) soient vérifiées. On a alors :

1) $\tilde{A}_K \subset A^*_{R,\bar{A}}$

2) S'il existe un élément $y \in \bar{A}$ tel que $\tilde{A}_K = \widetilde{R[y]}_K$ (un tel élément s'appelle un K-saturateur pour A relativement à R dans la terminologie de Zariski) on a l'égalité :

$$\tilde{A}_K = A^*_{R,\bar{A}} \; .$$

- 28 -

La démonstration de ce théorème est simple : pour 1), il suffit de montrer que $A^*_{R,\overline{A}}$ est saturé par rapport à K dans le sens de Zariski ; soient donc $\xi \in A^*_{R,\overline{A}}$ et $\eta \in \overline{A}$, et supposons que $\eta \underset{K}{>} \xi$: la proposition 3.4 et la transitivité de la dépendance intégrale impliquent immédiatement que $\eta \in A^*_{R,\overline{A}}$.

Pour 2), remarquons que l'on a l'égalité $R[y]^*_{R,\overline{A}} = \widetilde{R[y]_K}$ puisque $x \in R[y]^*_{R,\overline{A}}$ implique que $x \underset{K}{>} y$, d'après la proposition 3.4 (l'idéal $I_{R,A,\overline{A}}$ est principal, engendré par $y \otimes 1 - 1 \otimes y$).

D'autre part, on a :

$$A^*_{R,\overline{A}} \subset (\widetilde{A}_K)^*_{R,\overline{A}} = (\widetilde{R[y]_K})^*_{R,\overline{A}} \qquad \text{(par hypothèse)}$$

$$\subset (R[y]^*_{R,\overline{A}})_{R,\overline{A}} \qquad \text{(d'après la démonstration de 1))}$$

$= R[y]^*_{R,\overline{A}}$. La remarque précédente montre donc que $A^*_{R,\overline{A}} \subset \widetilde{R[y]_K} = \widetilde{A}_K$ ce qui achève la démonstration.

<u>Corollaire 3.6</u>

Soit $\mathcal{O}$ l'anneau local d'une variété algébroïde V de dimension r (les hypothèses sont celles du § 2) ; soit k un corps de coefficients de $\mathcal{O}$ et $(x_1,\ldots,x_r)$ un système de paramètres de $\mathcal{O}$. Alors on a l'égalité :

$$\widetilde{\mathcal{O}_{k,(x_1,\ldots,x_r)}} = \mathcal{O}^*_{k[(x_1,\ldots,x_r)],\overline{\mathcal{O}}}$$

dans les deux cas suivants :

a) V est une hypersurface, et $(x_1,\ldots,x_r)$ peut être prolongé en une base (i.e. un système minimal de générateurs) $x_1,\ldots,x_r, x_{r+1}$ de $\underline{m}$.

b) r = 1 (V est alors une courbe algébroïde) et k est de caractéristique 0.

<u>Démonstration</u>

Dans le cas a), $\mathcal{O}$ est une extension monogène de $R = k[[x_1,\ldots,x_r]]$ ($\mathcal{O} = R[x_{r+1}]$) et l'hypothèse du théorème 3.5 est évidemment vérifiée.

Dans le cas b), Zariski a montré qu'il existait toujours un K-saturateur pour une courbe algébroïde (cf. G.T.S. II, proposition 1.3 , p. 878)

§ 4.- SATURATION ET EQUISINGULARITE

Définition 4.1

Soient S et T deux anneaux commutatifs, $f : S \to T$ un morphisme. On dit que f est radiciel (ou que T est une S-algèbre radicielle) si, pour tout idéal premier $\mathfrak{P}$ de S, il y a au plus un idéal premier $\mathfrak{Q} \subset T$, tel que $f^{-1}(\mathfrak{Q}) = \mathfrak{P}$, et si, de plus, pour tout idéal premier $\mathfrak{Q} \subset T$, le corps des fractions de $T/\mathfrak{Q}$ est une extension purement inséparable du corps des fractions de $S/\mathfrak{P}$ (avec $\mathfrak{P} = f^{-1}(\mathfrak{Q})$).

Exemple : Si on suppose que T est entier sur S, alors f est radiciel si, et seulement si, f induit un homéomorphisme (universel) spec T $\to$ spec S.

Soient V et V' deux variétés analytiques, O un point de V, O' un point de V', $\mathcal{O}$ l'anneau local de V en O et $\mathcal{O}'$ celui de V' en O'.

Proposition 4.2

Supposons que $\mathcal{O}$ soit contenu dans $\mathcal{O}'$ et que $\mathcal{O}'$ soit fini sur $\mathcal{O}$.

Alors si $\mathcal{O}'$ est radiciel sur $\mathcal{O}$, l'application continue $f : V' \to V$ (définie localement par l'inclusion $\mathcal{O} \subset \mathcal{O}'$) est un homéomorphisme local.

La démonstration se fait par récurrence sur la dimension de V, en utilisant l'existence du conducteur de $\mathcal{O}$ dans $\mathcal{O}'$: cf. S.E.S III, prop. 5.1.

Remarque

Soit T une S-algèbre ; pour que T soit radicielle sur S, il faut et il suffit que pour $t \in T$, l'élément $t \otimes_S 1 - 1 \otimes_S t$ soit nilpotent (i.e. appartienne à tout idéal premier minimal) dans l'anneau $T \otimes_S T$.

Cela signifie que le "morphisme diagonal", correspondant au morphisme d'anneaux : $T \otimes_S T \to T$, est surjectif, ce qui entraîne facilement que T est radiciel sur S (le lecteur intéressé pourra se reporter à E.G.A.I).

Théorème 4.3

Soient $R \subset A \subset A'$ trois anneaux, et supposons A' entier sur A. Alors la saturation de Pham-Teissier $A^* = A^*_{R,A'}$ de A dans A' par rapport à R est un anneau radiciel sur A.

Il suffit, d'après la remarque précédente, de montrer que si $x^* \in A^*$, l'élément $x^* \otimes_A 1 - 1 \otimes_A x^*$ est nilpotent dans l'anneau $A^* \otimes_A A^*$. La définition de A^* entraîne immédiatement que $x^* \otimes_A 1 - 1 \otimes_A x^*$ est nilpotent dans l'anneau $A' \otimes_A A'$, car l'élément $x^* \otimes_R 1 - 1 \otimes_R x^*$ est, par définition, entier sur le noyau I de l'application:

$$f : A' \otimes_R A' \to A' \otimes_A A', \text{ d'où une relation :}$$

$$(x^* \otimes_R 1 - 1 \otimes_R x^*)^n - \Sigma a_\nu (x^* \otimes_R 1 - 1 \otimes_R x^*)^{n-\nu} = 0$$

avec $a_\nu \in I^\nu$, d'où :

$$f((x^* \otimes_R 1 - 1 \otimes_R x^*)^n) = (x^* \otimes_A 1 - 1 \otimes_A x^*)^n = 0$$

Soit maintenant $\mathfrak{P}^*$ un idéal premier minimal de $A^* \otimes_A A^*$: il faut montrer que $x^* \otimes_A 1 - 1 \otimes_A x^* \in \mathfrak{P}^*$; il suffit pour cela de montrer qu'il existe un idéal premier $\mathfrak{P}'$ dans $A' \otimes_A A'$ au dessus de $\mathfrak{P}^*$ (puisque, $x^* \otimes_A 1 - 1 \otimes_A x^*$ étant nilpotent dans cet anneau, il appartient à tous les idéaux premiers).

L'idéal $\mathfrak{P}^*$ est le noyau d'un morphisme $\varphi : A^* \otimes_A A^* \to K$ où K est un corps que l'on peut supposer algébriquement clos ; φ définit deux applications : $A^* \to K$ qui coïncident sur A et qui s'étendent en deux applications de A' dans K, puisque A' est supposé entier sur A. On obtient donc un morphisme $\varphi' : A' \otimes_A A' \to K$, et le noyau $\mathfrak{P}'$ de φ' est un idéal premier de $A' \otimes_A A'$ au dessus de $\mathfrak{P}^*$. c.q.f.d.

- 31 -

Corollaire 4.4

Supposons R intégralement clos dans son corps des fractions K, et A entier sur R ; alors la saturation $\widetilde{A}_K$ est radicielle sur A.

En effet, le théorème 3.5 montre que $\widetilde{A}_K \subset A^*_{R,\overline{A}}$, ce qui implique immédiatement le corollaire.

Remarque

L'utilisation de la saturation de Pham-Teissier simplifie beaucoup la démonstration de ce corollaire (cf. S.E.S. III, théorème 4.1, pour la démonstration originale).

Théorème 4.5

Soit V une variété analytique complexe, soit $\mathcal{O}$ l'anneau local de V en un point O et soit $\widetilde{V}$ la saturation de V en O par rapport à un système de paramètres ($\underline{x}$) ($\widetilde{V}$ est défini localement en O par l'anneau $\widetilde{\mathcal{O}}_{\mathbb{C},\{\underline{x}\}}$).

Alors l'application canonique f : $\widetilde{V} \to V$ est un homéomorphisme local.

Cela résulte du corollaire 4.4 et de la proposition 4.2.

Ce théorème implique en particulier que si V et V' sont des variétés analytiques complexes, telles que $\widetilde{V}_{(x)}$ et $\widetilde{V}'_{(x')}$ soient analytiquement isomorphes, alors elles sont localement homéomorphes.

Dans le cas des hypersurfaces, on a un résultat beaucoup plus fort :

Théorème 4.6

Soient V et V' deux hypersurfaces complexes, O (resp O') un point de V (resp V'), $\mathcal{O}$ et $\mathcal{O}'$ les anneaux locaux de V et V' en O et O' ;

Soit $\{x_1,\ldots,x_n\}$ (resp $\{x'_1,\ldots,x'_n\}$) un système de paramètres de $\mathcal{O}$ (resp de $\mathcal{O}'$) qui se prolonge en une base $\{x_1,\ldots,x_{n+1}\}$ de $\underline{m}$ (resp en une base $\{x'_1,\ldots,x'_{n+1}\}$ de $\underline{m}'$).

Supposons qu'il existe un $\mathbb{C}$-isomorphisme f : $\widetilde{\mathcal{O}}_{\mathbb{C},\{x\}} \to \widetilde{\mathcal{O}}'_{\mathbb{C},\{x'\}}$, tel que $f(x_i) = x'_i$ $(1 \leq i \leq n)$; alors l'homéomorphisme local ; $V \to V'$ peut être étendu en un homéomorphisme local des espaces ambiants : $\mathbb{C}^{n+1} \to \mathbb{C}^{n+1}$.

La démonstration de ce théorème, assez longue, se trouve dans S.E.S. III (théorème 6.1).

Examinons maintenant le cas des courbes (sur un corps de caractéristique 0).

Soit $\mathcal{O}$ l'anneau local d'une courbe algébroïde en un point O, k un corps de coefficients et x un paramètre : nous avons vu que l'anneau $\tilde{\mathcal{O}}_{k,x}$ était indépendant du choix de k, et du choix de x si x était choisi transverse (théorème 2.3). Notons cet anneau $\tilde{\mathcal{O}}$; il est facile de voir que $\tilde{\mathcal{O}}$ est la plus petite des saturations de $\mathcal{O}$ (pour tous les paramètres de $\mathcal{O}$) (cf. **note** (2) à la fin du texte) .

Inversement, si l'on suppose $\mathcal{O}$ intègre, la limite supérieure de toutes ces saturations est l'anneau $k + \bar{\underline{m}}^s$, $\bar{\underline{m}}$ étant l'idéal maximal de la clôture intégrale $\bar{\mathcal{O}}$ de $\mathcal{O}$ et s un entier que nous allons définir. $\bar{\mathcal{O}}$ est alors un anneau de valuation discrète, isomorphe à $\bar{k}[[t]]$, $\bar{k}$ étant la clôture algébrique de k dans $\bar{\mathcal{O}}$. Notons s la multiplicité réduite de $\mathcal{O}$: $s = \inf [v(\xi) \mid \xi \in \underline{m}]$; alors l'anneau $k + \bar{\underline{m}}^s$ peut se caractériser ainsi : c'est le plus gros anneau compris entre $\mathcal{O}$ et $\bar{\mathcal{O}}$ ayant s pour multiplicité réduite et k pour corps résiduel ; cf. G.T.S. I, prop. 7.6, p. 630.

<u>Théorème 4.7</u>

Supposons que k (toujours de caractéristique zéro) soit algébriquement clos. Soient C et C' deux courbes algébroïdes planes sur le corps k. Alors une condition nécessaire et suffisante pour que C et C' aient des singularités équivalentes (dans le sens algébro-géométrique comme dans S.E.S. I) est que les saturations $\tilde{\mathcal{O}}$ et $\tilde{\mathcal{O}}'$ soient analytiquement isomorphes.

Ceci est montré dans S.E.S. III (th. 2.1) dans le cas irréductible et dans G.T.S. II dans le cas général.

On peut se poser une question analogue dans le cas des hypersurfaces : soient V et V' deux hypersurfaces algébroïdes complexes ayant des normalisations analytiquement isomorphes (ce qui est bien réalisé pour deux cour-

- 33 -

bes planes ayant le même nombre de composantes irréductibles).

Soient $\tilde{V}$ et $\tilde{V}'$ les saturations de V et V' (pour des choix génériques des paramètres ; cf. le § 2 et le § 5 plus loin) : est-ce que l'équivalence topologique de V et V' implique que $\tilde{V}$ et $\tilde{V}'$ sont isomorphes ? (Nous avons vu que la réciproque était vraie : th. 4.6).

Envisageons maintenant "l'équisaturation" d'une famille d'hypersurfaces ; nous définirons une famille $V^{(t)}$ d'hypersurfaces algébroïdes à l'aide d'une équation $f \in k[[\underline{X},\underline{t}]]$ où $\underline{X} = (X_1,\ldots,X_{r+1})$, $\underline{t} = (t_1,\ldots,t_s)$, satisfaisant aux conditions suivantes :

1) k est algébriquement clos, de caractéristique 0.

2) f est sans facteurs multiples dans $k[[\underline{X},\underline{t}]]$.

3) $f(\underline{0},\underline{t}) = 0$

4) $f(0,X_{r+1},\underline{t}) = X_{r+1}^n$.

Posons $k^* = k((t_1,\ldots,t_s))$ (corps de séries formelles à s indéterminées sur k) ; on peut considérer $V^{(t)}$ comme une hypersurface de dimension r définie sur k^*, dont nous noterons :

$$\mathcal{O}^*_{(t)} = k^*[[x_1,\ldots,x_r,x_{r+1}]] \text{ l'anneau local.}$$

Nous noterons d'autre part V_o l'hypersurface définie sur k définie par l'équation :

$$f_o(\underline{X}) = f(X_1,\ldots,X_{r+1},0) \quad \text{et} \quad \mathcal{O}_o = k[[\xi_1,\ldots,\xi_{r+1}]]$$

son anneau local ; on peut aussi considérer V_o comme une hypersurface définie sur k^*, son anneau local étant alors $\mathcal{O}_o^* = k^*[[\xi_1,\ldots,\xi_{r+1}]]$.

Remarquons que sans hypothèse supplémentaire, l'anneau $\mathcal{O}_o^*$ n'est pas nécessairement réduit.

<u>Définition 4.8</u>

On dit que $V^{(t)}$ est une famille équisaturée d'hypersurfaces (ou que $V^{(t)}$ est une déformation équisaturée de V_o) si :

1) V_o est réduite (i.e. l'équation $f_o(\underline{X})$ n'a pas de facteur multiple).

2) Il existe un k^*-isomorphisme φ d'anneaux saturés :

$$\widetilde{(\mathcal{O}_t^*)}_{k^*((x_1,\ldots,x_r))} \xrightarrow{\sim} \widetilde{(\mathcal{O}_o^*)}_{k^*((\xi_1,\ldots,\xi_r))}$$

tel que $\varphi(x_i) = \xi_i \quad (i = 1,\ldots,r)$.

Si l'on considère f comme l'équation d'une hypersurface V de dimension r+s définie sur k, et si l'on note W l'espace linéaire défini par les équations : $X_1 = \ldots = X_{r+1} = 0$ (qui est par hypothèse contenu dans V), on dira que V est équisaturée à l'origine, le long de W, par rapport aux paramètres $(x_1,\ldots,x_r)$ de la fibre générique $V^{(t)}$, lorsque les deux conditions de la définition 4.8 sont satisfaites.

Remarque

Il n'y a aucune hypothèse de transversalité sur les paramètres $(x_1,\ldots,x_r)$ dans la définition 4.8 ; en particulier, la direction $X_1 = X_r = \ldots = X_{r+1} = t_1 = \ldots = t_s$ peut être tangente à V à l'origine.

Voici maintenant un critère d'équisaturation (cf. S.E.S. III, th. 7.4) :

Théorème 4.9

Conservons les notations précédentes ; soit D le discriminant (par rapport à la variable X_{r+1}) de l'équation $f(X_1,\ldots,X_{r+1},t)$. V est équisaturée à l'origine le long de W par rapport aux paramètres $(x_1,\ldots,x_r)$, si et seulement si

$$D = D_o\ (X_1,\ldots,X_r)\ \varepsilon\ (X_1,\ldots,X_r,\ t_1,\ldots,t_s)$$

où ε est une unité (i.e. $\varepsilon(0) \neq 0$).

Le fait que D soit le produit d'une unité avec une équation ne dépendant que des variables $X_1,\ldots,X_r$ s'interprète géométriquement en disant que l'hypersurface "discriminant" (d'équation D = 0) est analytiquement triviale le long de l'espace W_o d'équations $X_1 = \ldots = X_r = 0$.

On voit que pour une famille de courbes planes ($r = 1$) la notion d'équisaturation est équivalente à la notion d'équisingularité habituelle (cf. S.E.S I).

Voici maintenant un théorème qui prouve que la notion d'équisaturation fournit une bonne notion d'équisingularité dans le cas général :

Théorème 4.10

Supposons V équisaturée à l'origine le long de W par rapport aux paramètres $(x_1,\ldots,x_r)$ de la fibre générique. Soient $\mathcal{O}$ l'anneau local de V (sur k) et $\mathcal{O}_o$ l'anneau local de V_o (sur k). Alors les propriétés suivantes sont réalisées :

1) La clôture intégrale $\overline{\mathcal{O}}$ de $\mathcal{O}$ est canoniquement isomorphe à $\overline{\mathcal{O}}_o[[t_1,\ldots,t_s]]$ (i.e. la normalisation $\overline{V}$ de V est analytiquement triviale le long de l'image réciproque de W dans $\overline{V}$).

2) $\widetilde{\mathcal{O}}_{k((x_1,\ldots,x_r,t_1,\ldots t_s))} \xrightarrow{\sim} (\widetilde{\mathcal{O}}_o)_{k((x_1,\ldots x_r))}[[t_1,\ldots,t_s]]$

3) Si l'on se place dans le cas analytique complexe, V est topologiquement triviale le long de W.

La propriété 1) est une conséquence immédiate de 2) et la propriété 3) résulte facilement du théorème 4.6, énoncé plus haut (S.E.S. III, § 7). La propriété 2) est plus difficile : c'est un théorème du type de celui de Seidenberg dans son article publié à l'I.H.E.S. (volume en l'honneur de O. Zariski).

§ 5.- SATURATION "ABSOLUE" D'UNE HYPERSURFACE (cf. S.E.S. III, § 8)

Soient k un corps algébriquement clos de caractéristique 0, V une hypersurface algébroïde équidimensionnelle définie sur k, d'anneau local $\mathcal{O}$. Soit $(x_1,\ldots,x_r)$ un système de paramètres de $\mathcal{O}$ qui puisse se prolonger en une base (i.e. un système minimal de générateurs) $(x_1,\ldots,x_r,x_{r+1})$ de $\underline{m}$: nous allons définir une notion de transversalité pour $(x_1,\ldots,x_r)$.

Posons $R = k[[x_1,\ldots,x_r]]$: c'est un sous-anneau de $\mathcal{O}$ isomorphe à un anneau de séries formelles en r indéterminées ; on a $\mathcal{O} = R[x_{r+1}]$, ce qui implique que $\mathcal{O}$ est fini (et donc entier) sur R.

Soit $\mathfrak{P}$ un idéal premier de hauteur 1 dans $\mathfrak{O}$; $\mathfrak{P}$ est l'idéal d'une sous-variété irréductible, de codimension 1, W de V.

Comme $\mathfrak{O}$ est entier sur R et que R est intégralement clos, $\mathfrak{P} \cap R$ est un idéal premier de R qui est aussi de hauteur 1. Il est donc principal engendré par un élément $\xi_{\mathfrak{P}}$.

Soit $h_{\mathfrak{P}}$ l'homomorphisme canonique :

$\mathfrak{O} \to \mathfrak{O}_{\mathfrak{P}}$; $h_{\mathfrak{P}}$ restreint à R est injectif, car l'équidimensionnalité de $\mathfrak{O}$ implique que $\mathfrak{O}$ est un R-module sans torsion : on peut donc identifier R à un sous-anneau de $\mathfrak{O}_{\mathfrak{P}}$, et $K = k((x_1,\ldots,x_r))$ à un sous-corps de l'anneau total de fractions de $\mathfrak{O}_{\mathfrak{P}}$; $\xi_{\mathfrak{P}}$ est alors un paramètre de l'anneau $\mathfrak{O}_{\mathfrak{P}}$ (qui est de dimension 1) car $\xi_{\mathfrak{P}}$ est régulier dans $\mathfrak{O}_{\mathfrak{P}}$.

<u>Définition 5.1</u>

On dit que le système de paramètres $(x_1,\ldots,x_r)$ est transverse à $\mathfrak{P}$ (ou à W) si $\xi_{\mathfrak{P}}$ est un paramètre transverse de $\mathfrak{O}_{\mathfrak{P}}$.

Si le lieu singulier de V est de codimension plus grande que 1, V est normale (il suffit d'appliquer le "critère de Serre" à $\mathfrak{O}$) et donc saturée.

Supposons donc que le lieu singulier V_{sing} de V soit de codimension 1, et soient $W_1,\ldots,W_q$ les composantes inrréductibles de dimension r-1 de V_{sing} ; les W_i correspondent à des idéaux premiers $\mathfrak{P}_i$ de hauteur 1 de l'anneau $\mathfrak{O}$. On a alors :

<u>Théorème 5.2</u>

La saturation $\widetilde{\mathfrak{O}}_{k((x_1,\ldots,x_r))}$ est indépendante du choix des paramètres $(x_1,\ldots,x_r)$ pourvu qu'ils fassent partie d'une base $(x_1,\ldots,x_r,x_{r+1})$ de $\underline{m}$ et qu'ils soient transverses à chacune des variétés $W_1,\ldots,W_q$ (au sens de la définition 5.1).

La démonstration complète de ce théorème se trouve dans G.T.S. III cf. aussi S.E.S III, théorème 8.2.

Remarques

1) Le théorème 5.2. implique que pour "presque tous" les systèmes de paramètres de la forme $x'_i = \sum_{j=1}^{r+1} a_{i_j} x_j$ $(i = 1,\dots,r$, $a_{i_j} \in k)$, la saturation $\widetilde{\mathcal{O}_{k((x'_1,\dots,x'_r))}}$ est la même, "presque tous" signifiant qu'il suffit que les a_{i_j} appartiennent à un ouvert de Zariski de l'espace des coefficients.

2) Le théorème 5.2 n'est pas vrai en général pour un système de paramètres transverse au sens habituel.

Indiquons brièvement sur quoi repose la démonstration de ce théorème : soit $\mathfrak{P}$ un idéal premier de hauteur 1, et soient $\hat{R}_{\mathfrak{P}}$ et $\hat{\mathcal{O}}_{\hat{\mathfrak{P}}}$ les complétés des anneaux $R_{\mathfrak{P}}$ et $\mathcal{O}_{\mathfrak{P}}$. Soit $\hat{K}_{\mathfrak{P}}$ le corps des fractions de $R_{\mathfrak{P}}$;

Lemme 5.3

On a l'égalité :

$$\widetilde{\mathcal{O}}_K = F \cap \Big(\bigcap_{\mathfrak{P} \in S} \widetilde{\big(\widehat{\mathcal{O}_{\mathfrak{P}}} \big)_{\widetilde{K}}} \Big)$$

(avec $K = k((x_1,\dots,x_r))$).

Ce lemme permet de se ramener au cas de la dimension 1 ; il faut alors étudier sous quelles conditions les opérations de saturation et de complétion commutent.

Soit donc $\mathcal{O}$ un anneau noetherien semi-local de dimension 1, équidimensionnel, équicaractéristique et de caractéristique 0.

Supposons que $\mathcal{O}$ contienne un sous-anneau R régulier de dimension 1 et que :

1) $\mathcal{O}$ soit un R-module fini,

2) $\mathcal{O}$ soit sans torsion sur R,

3) R soit pseudo-géométrique (au sens de Nagata)

3) signifie que pour tout idéal premier $\mathfrak{P}$ de R, la clôture intégrale de $R/\mathfrak{P}$

dans une extension finie de son corps des fractions est finie sur $R/\mathfrak{p}$ (cf. Nagata, Local Rings, p. 131).

Soient $\hat{R}$ le complété de R, et $\hat{K}$ le corps des fractions de $\hat{R}$; on a alors :

Lemme 5.4

Sous les hypothèses précédentes, il y a un isomorphisme canonique :

$$\widehat{(\widetilde{\mathcal{O}}_K)} \xrightarrow{\sim} \widetilde{(\widehat{\mathcal{O}})}_{\hat{K}}$$

On déduit de là que si $\widehat{\mathcal{O}}$ est saturé par rapport à $\hat{K}$, $\mathcal{O}$ est saturé par rapport à K.

***

NOTES .

Note 1 (p. 24) - (cf. S.E.S. III, théorème 8.2, énoncé sans démonstration pour le cas où k est supposé algébriquement clos et de caractéristique 0 ; la démonstration complète, dans le cas plus général, où k est supposé seulement d'être de caractéristique 0 , est donnée dans G.T.S. III, Corollaire 3.5)

Note 2 (p. 33) - (cf. G.T.S. I, th. 7.2, p. 624, dans le cas où $\mathcal{O}$ est intègre ; dans le cas général cf. G.T.S. II, th. 4.5, p 905 . Pour un théorème plus fort cf G.T.S. I, cor. 7.4 et G.T.S. III, Appendice A, Lemme A9) .

GENERAL THEORY OF SATURATION AND OF SATURATED LOCAL RINGS. III.* SATURATION IN ARBITRARY DIMENSION AND, IN PARTICULAR, SATURATION OF ALGEBROID HYPERSURFACES.

By OSCAR ZARISKI.

Contents

Introduction. This is the third (and last) paper in a series of three papers on the general theory of saturation and saturated local rings, of which the first two were published in earlier issues of this *Journal* in 1971 (see [15] and [16], referred to respectively as GTS I and GTS II). All local or semi-local rings considered in this paper are assumed to be equidimensional, equicharacteristic zero, free from nonzero nilpotent elements, and—unless specified otherwise—noetherian. The principal objective of this paper is the full development of our theory of saturation in higher dimension, as previewed by us in the last section of our 1968 paper SES III (see [14]). This objective is *essentially* achieved by us in Section 3 of the present paper, after some preparatory (but indispensable) work has been carried out in Sections 1 and 2. We say "essentially" for the following reason: While Theorem 8.3 of SES III was announced by us under the

*This work was supported by National Science Foundation Grant GP 30854 X.
Manuscript received March 12, 1973.

[Reprinted from *American Journal of Mathematics*, Vol. 97, No. 2, pp. 415-502.]

sole assumption that the (complete) local ring in question satisfies the so-called condition (S) [see SES III, pp. 1021 and 1023; see also Section 1 of the present paper, where condition (S) is introduced immediately before the statement of Lemma 1.7], in this paper we prove this theorem only for local rings of algebroid hypersurfaces. Whether that theorem holds for all local rings satisfying condition (S) remains an open question.

Theorem 8.3 of SES III, which in the present paper (and for algebroid hypersurfaces only) appears as Theorem 3.1, is the principal result of this paper. It enables us to prove the absolute character of the saturation of the local ring $\mathfrak{o}$ of an algebroid hypersurface, with respect to "generic" parameters. More precisely, we prove in Section 3 that if *k is any coefficient field of* $\mathfrak{o}$ *and* $x_1, x_2, \ldots, x_r$ *are parameters of* $\mathfrak{o}$ *which are k-transversal in codimension* 1 (see Definitions 3.2 and 3.3), *then the k-saturation* $\tilde{\mathfrak{o}}^k_{\{x_1, x_2, \ldots, x_r\}}$ *of* $\mathfrak{o}$ *with respect to the parameters* $x_1, x_2, \ldots, x_r$ *is a local ring* $\tilde{\mathfrak{o}}$ *which depends only on* $\mathfrak{o}$ (see Theorem 3.4 and Corollary 3.5). This ring $\tilde{\mathfrak{o}}$, which may be called *the absolute saturation of* $\mathfrak{o}$, is also characterized *a posteriori* by the *minimality property* of being a *subring* of the k-saturation of $\mathfrak{o}$ with respect to *any* set of parameters of $\mathfrak{o}$ (Proposition 3.6). All these results are obtained as applications of Theorem 3.1, by a reduction to local rings of dimension 1 and by making use of results proved by us for one-dimensional local rings in GTS I and GTS II.

We have already mentioned above that Sections 1 and 2 deal with some essential preliminary results. In either section we do not restrict ourselves to hypersurfaces, as far as our general considerations are concerned, but the principal results in these sections refer either to hypersurfaces or to local rings of dimension 1. Thus, in Section 1 we prove, under very general conditions, the permutability of saturation and localization (Proposition 1.2), but the principal result of that section is Theorem 1.8, which refers exclusively to the case in which the ring A, which is being saturated with respect to its subring R, *is a simple ring extension of* R. Thus, geometrically speaking, Theorem 1.8 is basically a theorem on saturation of algebroid hypersurfaces and, as such, it is used in the proof of Theorem 3.1.

In Section 2 our principal result is Theorem 2.7, which asserts that *for semi-local rings of dimension* 1 *the two operations of saturation and completion commute*. This result is all we need for the applications to hypersurfaces given in Section 3. That saturation and completion permute also in the case of local rings of hypersurfaces of any dimension follows as a consequence of the following two general theorems proved by Lipman in [3] which appears in this same issue of the *American Journal of Mathematics*:

(1) *If* A, F *and* R *satisfy conditions* (A), (B) *and* (C) *of Section* 2 *and if* A *is a simple ring extension of* R, *then the Lipschitz saturation* $A^*_{A,R}$ *coincides*

with our saturation $\tilde{A}_K$ (here K stands for the field of fractions of R and A is the integral closure of A in its total ring of quotients).

(2) *If a subring R of a ring A is a noetherian excellent local ring, if A is a finite R-module and if the total ring of quotients of A is a finite direct sum of fields, then the Lipschitz saturation commutes with the completion, i.e., we have* (*in the notations of Lipman's paper* [3]):

$$(\hat{A})^*_{\hat{\bar{A}},\,\hat{R}} = (A^*_{\bar{A},R}) \otimes_R \hat{R}.$$

[Actually, Lipman proves a more general result in which R is only assumed to be noetherian and $\hat{R}$ is replaced by any normal noetherian R-algebra R'].

As to the first of the above two theorems of Lipman, it may be mentioned here that in a letter to Pham and Teissier, dated July 14, 1970, the author of the present paper gave an example of an algebroid surface in affine 4-space (actually an algebraic cone) for which the Lipschitz (or Pham-Teissier) saturation and our saturation of the local ring $\mathfrak{o}$ of the origin of the algebroid surface (actually, of the local ring of the vertex of the cone), relative to one and the same set of parameters of $\mathfrak{o}$, *lead to different rings*. I shall give this example at the end of this Introduction. At any rate, it remains an open question whether completion commutes with saturation in our sense, when we deal with algebraic varieties which are not hypersurfaces.

It may be pointed out at this stage that our Corollary 2.3 (which essentially restates an old result of ours, to the effect that, under suitable conditions, the two operations of completion and normalization commute; see [12]) can be now viewed as a special case of a more general result proved in Grothendieck EGA IV, (4.14.1).

In Section 4 we prove quite generally that *the saturation* $\tilde{\mathfrak{o}}^k_{\{x_1, x_2, \ldots, x_r\}}$ *of a complete local ring* $\mathfrak{o}$ *with respect to* ***transversal*** *parameters* $x_1, x_2, \ldots, x_r$ *has the same multiplicity as* $\mathfrak{o}$ (Theorem 4.1). When we first presented this theorem in a seminar lecture at Purdue University in 1970, our proof dealt only with hypersurfaces. At that time, Lipman called our attention to a result of Rees [8] which made it possible to extend our proof to the non-hypersurface case.

In the proof of Theorem 3.1 essential use is made of a result concerning one-dimensional local rings $\mathfrak{o}$, namely that *the k-saturation of* $\mathfrak{o}$ *with respect to a* ***transversal*** *parameter x depends only on* $\mathfrak{o}$. (See Appendix A, Corollary A.7). For one-dimensional local *domains* this result was proved by us in GTS I (Corollary 9.5), and for local rings $\mathfrak{o}$ of dimension 1, with zero divisors, we have proved this result in GTS II, but only under the assumption that the residue field of $\mathfrak{o}$ is algebraically closed (see GTS II, Corollary 8.2). For the purposes of the proof of Theorem 3.1 we have to have that result also in the case in which the residue field is not algebraically closed. Thus, the *necessity of* proving the result stated in Corollary A7 is the immediate reason for our having added Appendix A to the present paper. However, in doing so, we were also motivated by a wish to extend to one-dimensional local rings, with zero divisors, a theorem proved by us in GTS I for one-dimensional local *domains* (Theorem 9.1 in GTS I) to the effect that saturated local domains *have "many" automorphisms*. Here the term "many" is used in a well-defined sense which results from combining Theorems 9.1 and 8.1 of GTS I. This extension is represented by Theorem A1 in Appendix A, and most of the Appendix A is essentially dedicated to the proof of Theorem A1, a theorem which is much deeper than its consequence—Corollary A7—which is needed in the proof of Theorem 3.1. Nevertheless, even Corollary A7 is a nontrivial statement. A simpler proof of our Corollary A7, a proof which bypasses the question of existence of "many" automorphisms of saturated one-dimensional local rings ("many"—in the sense which results from combining Theorem A1 with Theorem 4.4 of GTS II) can be found in the second paper [4] of J. Lipman which appears in this same issue of the *Journal.*

The notion of *k-transversality of a set of parameters* $x_1, x_2, \ldots, x_r$ *of a complete local ring* $\mathfrak{o}$, *with respect to a prime ideal* $\mathfrak{p}$ *of* $\mathfrak{o}$, *of height* 1 (*k being a coefficient field of* $\mathfrak{o}$) plays an essential role in Section 3. In Appendix B we make a general study of that notion, and we deal with prime ideals $\mathfrak{p}$ of arbitrary height (see Definition B9). For arbitrary (not necessarily complete) local rings $\mathfrak{o}$ and for any condition C on sets of ρ elements of $\mathfrak{m}^s$ ($\mathfrak{m}$ = the maximal ideal of $\mathfrak{o}$; s, a positive integer) we define what we mean by saying that "*C is an open condition*" (of order s; see Definition B7). The principal result of Appendix B (Theorem B11) asserts that *k-transversality of parameters* $x_1, x_2, \ldots, x_r$ *of* $\mathfrak{o}$, *with respect to a given prime ideal* $\mathfrak{p}$ *of* $\mathfrak{o}$, *is an open condition of order* 1. This result insures that "generic" parameters are k-transversal in codimension 1 (see Definitions 3.2 and 3.3) and that therefore the k-saturation of the local ring of a hypersurface, with respect to "generic" parameters, is independent of the choice of the parameters (and of the coefficient field k; see Corollary 3.5).

We show by examples, at the end of Appendix B, that if $\mathfrak{p}$ is a prime ideal of $\mathfrak{o}$ and k, k' are two coefficient fields of $\mathfrak{o}$, a set of parameters may be k-transversal with respect to $\mathfrak{p}$ without being k'-transversal with respect to $\mathfrak{p}$. Thus, k-transversality of a set of parameters, with respect to a given prime ideal $\mathfrak{p}$ of $\mathfrak{o}$, is not necessarily an intrinsic condition of the parameters [except when $\mathfrak{p} = \mathfrak{m}$ or when $\mathfrak{p}$ is a minimal prime ideal of $\mathfrak{o}$; in the first case k-transversality of parameters $x_1, x_2, \ldots, x_r$, with respect to $\mathfrak{m}$, signifies simply that $x_1, x_2, \ldots, x_r$ are transversal parameters of $\mathfrak{o}$ (see Definition B3). In the second case, any set of parameters is k-transversal with respect to $\mathfrak{p}$].

The principal result of Appendix B, i.e., Theorem B11, is preceded by some preliminary results concerning superficial elements of a given order s (Lemma B1 and Corollary B2) and transversal parameters of $\mathfrak{o}$ (Propositions B4, B5, B6, B8). Some of these results (but not all) are probably either known or are implicitly contained in known facts on local rings. We thought it advisable, nevertheless, to gather together all the relevant facts on transversality, in the precise form in which we needed them. An important role in the proof of the main result (Theorem B11) is the characterization of open condition C of order 1, given in Lemma B10 (under the additional assumption that C is—what we call—*a projectively k-stable condition*).

We shall end this introduction by giving an example of a (non-normal) local domain $\mathfrak{o}$ of dimension 2 (the local ring of the vertex of an algebraic cone in 4-space) and of a pair of parameters x_1, x_2 of $\mathfrak{o}$, such that $\mathfrak{o}$ is k-saturated with respect to x_1, x_2 in our sense (k, a coefficient field of $\mathfrak{o}$), while the relative Pham-Teissier (or Lipschitz) saturation of $\mathfrak{o}$ with respect to the parameters x_1, x_2 is the integral closure $\bar{\mathfrak{o}}$ of $\mathfrak{o}$ (and hence is different from $\mathfrak{o}$, since $\mathfrak{o}$ is not normal).

Let us consider a four-dimensional projective space $\mathbf{P}_4$, over an algebraically closed field k of characteristic zero, and let $\mathbf{P}_3$ be a linear subspace of $\mathbf{P}_4$, of dimension 3, defined over k. Let Γ be a non-singular *rational* quartic curve in the above projective three-space $\mathbf{P}_3$, let O be a point of $\mathbf{P}_4$, not in $\mathbf{P}_3$, and let V be the cone which projects Γ from the vertex O. We take for $\mathfrak{o}$ *the completion* of the local ring of V at O. Since Γ is not arithmetically normal, $\mathfrak{o}$ is not a normal ring. We now proceed to exhibit the parameters x_1, x_2 which we need for our example.

We may assume that Γ is defined in $\mathbf{P}_3$ by the parametric equations

$$z_1 = t^4, \quad z_2 = t^3\tau, \quad z_3 = t\tau^3, \quad z_4 = \tau^4.$$

Then the integral closure $\bar{\mathfrak{o}}$ of $\mathfrak{o}$ is the subring of the formal power series ring $k[[t,\tau]]$ consisting of all the power series of the form $\varphi = \sum_{i=0}^{\infty} \varphi_{4i}(t,\tau)$, where φ_{4i} is a (binary) homogeneous form of degree $4i$ in t,τ, while $\mathfrak{o}$ consists of those elements φ of $\bar{\mathfrak{o}}$ in which the term $t^2\tau^2$ is missing. Thus $\mathfrak{o}$ is a maximal subring of $\bar{\mathfrak{o}}$, and we have $\dim_k \bar{\mathfrak{o}}/\mathfrak{o} = 1$. The local parameters x_1, x_2 which we need will be the following: $x_1 = t^4$, $x_2 = \tau^4$.

We shall now briefly outline the proof of our following two assertions:

(a) *$\mathfrak{o}$ is k-saturated with respect to x_1, x_2.*

(b) *The Lipschitz k-saturation of* $\mathfrak{o}$ *with respect to the parameters* x_1, x_2 *is the integral closure of* $\mathfrak{o}$.

Proof of a. Let $\xi = at^4 + bt^3\tau + ct\tau^3 + d\tau^4 + \varphi_8(t,\tau) + \varphi_{12}(t,\tau) + \cdots$ be any element of $\mathfrak{o}$ (where $a,b,c,d \in k$) and let $\eta = a_1t^4 + b_1t\tau^3 + c_1t\tau^3 + d_1\tau^4 + e_1t^2\tau^2 + \psi_8(t,\tau) + \psi_{12}(t,\tau) + \cdots$ be an element of $\bar{\mathfrak{o}}$. Assume that η dominates ξ with respect to the ring $R = k[[x_1,x_2]]$. We will show that η necessarily belongs to $\mathfrak{o}$, i.e., *that the coefficient* e_1 *of* $t^2\tau^2$ *in* η *is zero*. The field of fractions F of $\mathfrak{o}$ is a cyclic extension of $k((x_1,x_2))$ of degree 4; namely, $F = k((x_1,x_2))[z_2]$, $z_2^4 = x_1^3x_2$. The automorphisms σ of $F/k((x_1,x_2))$ are induced by the particular k-automorphism σ^* of $k((t,\tau))$ such that $\sigma^*(t) = \epsilon_1 t$ and $\sigma^*(\tau) = \epsilon_2\tau$, where ϵ_1 and ϵ_2 are arbitrary 4th roots of unity. Let us take, in particular, $\epsilon_1 =$ a primitive 4th root of unity and $\epsilon_2 = -1$. Then

$$\sigma(\eta) - \eta = b_1(\epsilon_1 - 1)t^3\tau - c_1(\epsilon_1 + 1)t\tau^3 - 2e_1t^2\tau^2 + \text{sum of terms}$$

$$\psi'_{4i}(t,\tau), \qquad i \geqslant 2.$$

Similarly,

$$\sigma(\xi) - \xi = b(\epsilon_1 - 1)t^3\tau - c(\epsilon_1 + 1)t\tau^3 + \text{sum of terms} \quad \varphi'_{4i}(t,\tau), \quad i \geqslant 2.$$

Since the quotient $[\sigma(\eta) - \eta]/[\sigma(\xi) - \xi]$ must be integral over $k[[x_1,x_2]]$, this quotient must be a formal power series in t, τ. This implies, at the least, that the leading form $b_1(\epsilon_1 - 1)t^3\tau - c_1(\epsilon_1 + 1)t\tau^3 - 2e_1t^2\tau^2$ of the power series $\sigma(\eta) - \eta$ must be divisible by the leading form $b(\epsilon_1 - 1)t^3\tau - c(\epsilon_1 + 1)t\tau^3$ of $\sigma(\xi) - \xi$. This implies that $e_1 = 0$.

Proof of b. The tensor product $\bar{\mathfrak{o}} \otimes_k \bar{\mathfrak{o}}$ is the set of all power series in 4 indeterminates t, τ, t', τ', of the form $\Sigma\psi_{4i,4j}(t,\tau;\, t',\tau')$, whence $\psi_{4i,4j}$ is a doubly homogeneous form in the two sets of variables $\{t,\tau\}$ and $\{t',\tau'\}$, of degrees $4i$ and $4j$ respectively ($i \geqslant 0, j \geqslant 0$). The kernel of the canonical map $\bar{\mathfrak{o}} \otimes_k \bar{\mathfrak{o}} \to \bar{\mathfrak{o}} \otimes_{\mathfrak{o}} \bar{\mathfrak{o}}$ is the ideal I generated in $\bar{\mathfrak{o}} \otimes_k \bar{\mathfrak{o}}$ by the following 4 elements:

$$\alpha_1 = t^4 - t'^4, \quad \alpha_2 = t^3\tau - t'^3\tau', \quad \alpha_3 = t\tau^3 - t'\tau'^3, \quad \alpha_4 = \tau^4 - \tau'^4.$$

Let $\beta = t^2\tau^2 - t'^2\tau'^2$. To prove assertion (b) we have only to show that β belongs to the integral closure of I in $\bar{\mathfrak{o}} \otimes_k \bar{\mathfrak{o}}$. We use the valuation-theoretic definition of the integral closure of an ideal (see C.A. 2, p. 347, Definition 1, and p. 350, Theorem 1). Let v be any valuation of $k((t,\tau,t',\tau'))$ such that $v(t)$, $v(\tau)$, $v(t')$, $v(\tau')$ are positive. We have to show that $v(\beta) \geqslant v(I)$. Without loss of generality we may assume that $v(\tau) = \min \cdot \{v(t), v(\tau), v(t'), v(\tau')\}$. Let

$$\xi = t/\tau, \quad \eta = t'/\tau, \quad \zeta = \tau'/\tau,$$

whence $v(\xi)$, $v(\eta)$ and $v(\zeta)$ are non-negative. We find easily that

$$\beta = (\xi + \eta\zeta^3)\alpha_3 - \eta^2\zeta^2\alpha_4.$$

Hence $v(\beta) \geqslant \min \cdot \{v(\alpha_3), v(\alpha_4)\} \geqslant v(I)$, as asserted.

1. Saturation and Localization. All rings considered in this paper are commutative and have an identity. Let A be a ring and let F be the total ring of quotients of A. We assume that the following conditions are satisfied:

1. A is free from nilpotent elements (different from zero).
2. F contains as subring a field K of characteristic zero, K contains the identity of F and F is a finite vector space over K.
3. If R denotes the intersection $A \cap K$, then A is integral over R.

From the fact that the identity 1 of A belongs to K follows that all elements of K, different from zero, are regular elements of F (i.e., are not zero divisors in F). Since F is a finite vector space over K, it follows from condition (1) that F is a finite direct sum of fields, say of h fields $F\epsilon_i$, where $1 = \epsilon_1 + \epsilon_2 + \cdots + \epsilon_h$, $\epsilon_i^2 = \epsilon_i$ and $\epsilon_i\epsilon_j = 0$ if $i \neq j$. Furthermore, each field $F\epsilon_i$ is a finite (separable) extension of the field $K\epsilon_i$. Thus, the 5 condition a.-e. of our paper SES III (p. 962) are satisfied, and the saturation $\tilde{A}_K$ of A with respect to K is defined. In the sequel, as long as K is fixed, we shall write $\tilde{A}$ instead of $\tilde{A}_K$.

We fix an algebraic closure Ω of K and we denote by H the set of K-homomorphisms of F into Ω. Then H is a finite set consisting, say, of n elements $\psi_1, \psi_2, \ldots, \psi_n$. We recall from SES III that given two elements ξ, η of F we say that η *dominates* ξ (with respect to the pair $\{A, K\}$) if for any two indices i, j in the set $\{1, 2, \ldots, n\}$ such that $\psi_i(\xi) \neq \psi_j(\xi)$, the quotients $[\psi_i(\eta) - \psi_j(\eta)]/[\psi_i(\xi) - \psi_j(\xi)]$ are integral over the intersection ring R $(= A \cap K)$, while $\psi_i(\eta) = \psi_j(\eta)$, if $\psi_i(\xi) = \psi_j(\xi)$. We note that for any ψ in H the ring $\psi(A)$ is integral over R $(= \psi(R))$.

We denote by $\bar{A}$ and $\bar{R}$ respectively the integral closure of A in F and the integral closure of R in K. We have obviously $\bar{R} = \bar{A} \cap K$. It is easy to see that

$$\tilde{A} \cap K = \bar{R}. \tag{1}$$

In fact, since $\tilde{A} \subset \bar{A}$ (by the definition of saturation), we have $\tilde{A} \cap K \subset \bar{R}$. On the other hand, since for any ψ in H the restriction $\psi/\bar{R}$ is the identity, the elements of $\bar{R}$ dominate the elements of A, and hence $\bar{R} \subset \tilde{A}$ since $\bar{R} \subset \bar{A}$. This proves (1).

Lemma 1.1. *Let A' be a ring between A and F, and assume that A' is integral over the intersection ring $R' = A' \cap K$ (whence the saturation $\widetilde{A'_K}$ is defined). Then $\tilde{A} \subset \widetilde{A'}$.*

Proof. Let B be the intersection of the two subrings $\tilde{A}$ and $\widetilde{A'}$ of F. We have $A \subset B \subset \tilde{A}$, and thus B is integral over R. On the other hand, we have $B \cap K = (\tilde{A} \cap K) \cap (\widetilde{A'} \cap K) = \bar{R} \cap \bar{R}'$ (by (1)) $= \bar{R}$. Hence the saturation of B with respect to K is defined. *We assert that B is saturated with respect to K.* For, let ξ be any element of B and let η be any element of the integral closure $\bar{B}$ of B in F which dominates ξ with respect to $\{B, K\}$. We have $\xi \epsilon \tilde{A}$ (since $B \subset \tilde{A}$) and η also dominates ξ with respect to $\{\tilde{A}, K\}$ (since $B \cap K = \tilde{A} \cap K = \bar{R}$). Hence $\eta \in \tilde{A}$. On the other hand, we have that *a fortiori* η dominates ξ with respect to $\{A', K\}$ (since $B \subset A'$). Since $\xi \in \widetilde{A'}$ it follows that also η belong to the ring $\widetilde{A'}$. Thus $\eta \in \tilde{A} \cap \widetilde{A'} = B$, which proves our assertion that B is saturated with respect to K. Since $A \subset B \subset \bar{A}$, it follows from the minimality property of saturation (SES III, Proposition-Definition II, p. 963) that $\tilde{A} \subset B$, which in conjunction with the inclusions $B \subset \tilde{A}$, $B \subset \widetilde{A'}$ implies that $\tilde{A} = B \subset \widetilde{A'}$.

Let $\mathfrak{p}$ be a prime ideal in R and let M be the multiplicative set $R - \mathfrak{p}$. Let $A' = A[M^{-1}]$ be the localization of A with respect to M. Since every element of M is regular in F, A' is a subring of F, and $A \subset A'$. If we set $R' = A' \cap K$, it is immediate that $R' = R[M^{-1}]$ $(= R_{\mathfrak{p}})$. It is also obvious that A' is integral over R' (since A is integral over R). Thus the saturation $\widetilde{A'}$ of A' with respect to K is defined.

Proposition 1.2 (*Permutability of saturation and localization*). *We have*

$$\widetilde{A[M^{-1}]} = \tilde{A}[M^{-1}]. \tag{2}$$

Proof. Since $A[M^{-1}]$ is a ring between A and F which is integral over the intersection ring $A[M^{-1}] \cap K (= R[M^{-1}])$, it follows from Lemma 1.1 that $\tilde{A} \subset \widetilde{A[M^{-1}]}$ and that consequently

$$\tilde{A}[M^{-1}] \subset \widetilde{A[M^{-1}]}. \tag{3}$$

Now, both rings in (3) have the same integral closure in F, namely $\bar{A}[M^{-1}]$. Hence, by the minimality property of the saturation of $A[M^{-1}]$, the proposition will be established as a consequence of (3) *if we show that* $\tilde{A}[M^{-1}]$ *is saturated with respect to* K. Let then ξ' be any element of $\tilde{A}[M^{-1}]$ and let η' be any element of the integral closure $\bar{A}[M^{-1}]$ of the ring $\tilde{A}[M^{-1}]$, such that η' dominates ξ' with respect to $\{\tilde{A}[M^{-1}], K\}$. We can write ξ' and η' in the form $\xi'=\xi/a$, $\eta'=\eta/b$, where $\xi\in\tilde{A}$, $\eta\in\bar{A}$ and where a,b are elements of M. Upon replacing ξ and η by $b\xi$ and $a\eta$ respectively, we have a similar representation of ξ',η', with a common denominator ab. We may therefore assume that $a=b$. From our assumption that η' dominates ξ' it follows at once that if ψ_i,ψ_j are any two homomorphisms in the set H and if $\psi_i(\xi)\neq\psi_j(\xi)$ then the quotient

$$[\psi_i(\eta)-\psi_j(\eta)]/[\psi_i(\xi)-\psi_j(\xi)]$$

is of the form ξ_{ij}/α_{ij}, where ξ_{ij} is integral over R and α_{ij} is an element of M. Using a common denominator in M for all these quotients ξ_{ij}/α_{ij} we may assume that $\alpha_{ij}=\alpha\in M$ for all $i,j\in\{1,2,\ldots,n\}$ such that $\psi_i(\xi)\neq\psi_j(\xi)$. On the other hand, if $\psi_i(\xi)=\psi_j(\xi)$ then $\psi_i(\xi')=\psi_j(\xi')$ (since $\psi_i(a)=\psi_j(a)=a$), and therefore $\psi_i(\eta')=\psi_j(\eta')$ (by our assumption of domination $\eta'>\xi'$), i.e., $\psi_i(\eta)=\psi_j(\eta)$. From all this we can conclude that the element $\alpha\eta$ of $\bar{A}$ dominates the element $\alpha\xi$ of $\tilde{A}$ (with respect to $\{\tilde{A},K\}$). Therefore $\alpha\eta\in\tilde{A}$, and since we have $\eta'=\eta/a=\alpha\eta/\alpha a$, it follows that $\eta'\in\tilde{A}[M^{-1}]$. This proves that $\tilde{A}[M^{-1}]$ is saturated with respect to K and completes the proof of the proposition.

LEMMA 1.3. *Let* S *denote the set of all prime ideals* $\mathfrak{p}$ *in* A *such that* $\mathfrak{p}\cap R$ *is a given prime ideal* $\mathfrak{P}$ *of* R, *and let* $M=R-\mathfrak{P}$, $M'=A-\cup_{\mathfrak{p}\in S}\mathfrak{p}$. *Then*

$$A[M^{-1}]=A[M'^{-1}]. \tag{4}$$

Proof. Let A' denote, as above, the ring $A[M^{-1}]$. If $\mathfrak{p}'$ is any prime ideal in A' and if we set $\mathfrak{p}'\cap A=\mathfrak{p}$, then $\mathfrak{p}'=A'\mathfrak{p}$, the prime ideal $\mathfrak{p}$ is disjoint from M, and, conversely, if $\mathfrak{p}$ is a prime ideal in A which is disjoint from M then there exists in A' one and only one prime ideal $\mathfrak{p}'$ (namely the ideal $A'\mathfrak{p}$) such that $\mathfrak{p}'\cap A=\mathfrak{p}$ (see C.A. 1, Corollary 1 of Theorem 16, p. 224). The *maximal* prime ideals $\mathfrak{p}'$ of A' are therefore the ideals of the form $A'\mathfrak{p}$, where $\mathfrak{p}$ ranges over the set of maximal elements of the set of prime ideals in A which are disjoint from M. Now, a prime ideal $\mathfrak{p}$ in A is disjoint from M if and only if $\mathfrak{p}\cap R\subset\mathfrak{P}$. Since A is integral over R, it follows from the "going-up" theorem (see C.A. 1, Theorem 3, p. 257 and the Corollary on p. 259) that the maximal elements of the set of prime ideals $\mathfrak{p}$ in A which are disjoint from M are precisely the prime ideals $\mathfrak{p}$ in A which contract to $\mathfrak{p}$ in R. We have therefore shown that the set of prime ideals $A'\mathfrak{p}$, $\mathfrak{p}\in S$, is the set of *all* maximal prime ideals of A'. If then x is an element of M', we have $x \notin \mathfrak{p} = A'\mathfrak{p} \cap A$, for all $\mathfrak{p}$ in S, i.e., x belongs to no maximal ideal of A'. Therefore x is a unit in A'. This shows that $A' = A'[M'^{-1}]$. On the other hand we have $A'[M'^{-1}] \supset A[M'^{-1}]$ (since $A' \supset A$) and $A[M'^{-1}] \supset A[M^{-1}]$ (since $M' \supset M$). We have therefore $A' = A'[M'^{-1}] \supset A[M'^{-1}] \supset A[M^{-1}] = A'$, and this completes the proof of the lemma.

Note. The only assumptions used in this proof are the following: (1) R is an integral domain and is a subring of A; (2) A is integral over R; (3) every element of R, different from zero, is regular in A.

COROLLARY 1.4. *If $\mathfrak{P}$ is the zero ideal (whence M is the set of all non-zero elements of R) then*

$$A[M^{-1}] = F. \tag{5}$$

For, F being a direct sum of h fields, the zero-ideal of F is a finite intersection of h prime ideals $\mathfrak{p}_1^*, \mathfrak{p}_2^*, \ldots, \mathfrak{p}_h^*$ (which are the only prime ideals of F, all minimal in F), and if we set $\mathfrak{p}_i = \mathfrak{p}_i^* \cap A$, then $\mathfrak{p}_1 \cap \mathfrak{p}_2 \cap \cdots \cap \mathfrak{p}_h$ is an irredundant representation of the zero ideal in A as intersection of (distinct) prime ideals. (This follows from the fact that F is the localization of A with respect to the multiplicative system of all regular elements of A). Since all the nonzero elements of K are regular in F, all nonzero elements of R are regular in A. This implies that $\mathfrak{p}_i \cap R = (0)$, $i = 1, 2, \ldots, h$. Since A is integral over R, all elements of the set S are in the present case minimal prime ideals in A. Hence S is actually the set of all minimal prime ideals of A, and the set M' is now the set of all regular elements of A.

[*Note.* The following is a direct proof of (5) which does not make use of the general theory of localization and of the "going-up" theorem for pairs of integrally dependent rings. Let M' now denote the set of all regular elements of A, whence $F = A[M'^{-1}]$. Given any element ξ of M', let $\xi^n + a_1\xi^{n-1} + \cdots + a_n = 0$, $a_i \in R$, be an equation of integral dependence for ξ over R, *of least degree* n. Since ξ is not a zero-divisor of A, we have necessarily $a_n \neq 0$ (in view of our assumption that n is minimum), i.e., $a_n \in M$. We have therefore a relation of the form $\alpha\xi \in M$, with $\alpha = \xi^{n-1} + a_1\xi^{n-2} + \cdots + a_{n-1} \in A$, namely: $\alpha\xi = -a_n$. Hence $1/\xi \in A[M^{-1}]$, and this establishes (5)].

Let K' be the field of fractions of R. Since $K' = R[M^{-1}]$, where $M = R - \{0\}$, it follows from Corollary 1.4 that

$$AK' = F. \tag{6}$$

Since A is integral over R, hence also over K', it follows from (6) that F is integral over K'. *In particular, the field K is an algebraic extension of K'.* In the special case in which A is saturated with respect to K we have $K' = K$, since R is then integrally closed in K, by (1).

Let S be an arbitrary set (finite or infinite) of prime ideals $\mathfrak{p}_\alpha$ in A. We set

$$M'_\alpha = A - \mathfrak{p}_\alpha, \qquad A'_\alpha = A[M'^{-1}_\alpha];$$

$$M' = A - \bigcup_{\mathfrak{p}_\alpha \in S} \mathfrak{p}_\alpha = \cap M'_\alpha, \qquad A' = A[M'^{-1}].$$

Let

$$h_\alpha : A \to A'_\alpha \tag{7}$$

be the canonical homomorphism of A into A'_α. It is easily seen that if x is a regular element of A then $h_\alpha(x)$ is a regular element of A'_α. From this it follows that if F'_α denotes the total ring of quotients of A'_α then h_α can be extended to a homomorphism

$$\bar{h}_\alpha : F \to F'_\alpha \tag{8}$$

and that the extension $\bar{h}_\alpha$ (which is not necessarily surjective) is uniquely determined. We now define the intersection of the rings A'_α to be the following subring of F:

$$\bigcap_{\mathfrak{p}_\alpha \in S} A'_\alpha = \left\{ \xi \in F \mid \bar{h}_\alpha(\xi) \in A'_\alpha, \text{ for all } \mathfrak{p}_\alpha \text{ in } S \right\}. \tag{9}$$

LEMMA 1.5. *Assume that S is the set of all prime ideals $\mathfrak{p}_\alpha$ in A which contract in R to a given prime ideal $\mathfrak{P}$ of R. Then*

$$\bigcap_{\mathfrak{p}_\alpha \in S} A'_\alpha = A[M^{-1}] = A[M'^{-1}], \tag{10}$$

where $M = R - \mathfrak{p}$ (compare with Lemma 1.3).

Proof. We denote by A' the ring $A[M'^{-1}]$. Let ξ be an element of $F: \xi = y/z$, where $y, z \in A$ and z is a regular element of A. Assume that $\xi \in A'$. Then we must have $y/z = x/m'$, where $x, m' \in A$ and $m' \in M'$. Since we have for each α: $M' \subset M'_\alpha$, it follows that $m' \in M'_\alpha$ and $\bar{h}_\alpha(\xi) = h_\alpha(y)/h_\alpha(m') \in A'_\alpha$. This shows that $A' \subset \cap_{\mathfrak{p}_\alpha \in S} A'_\alpha$.

We next prove the opposite inclusion. Assume that $\xi \in \cap_{\mathfrak{p}_\alpha \in S} A'_\alpha$. Then for each $\mathfrak{p}_\alpha$ in S there exist elements x_α, m'_α in A such that $m'_\alpha \in M'_\alpha$ and $\bar{h}_\alpha(y/z) = \bar{h}_\alpha(x_\alpha/m'_\alpha)$. This implies that $ym'_\alpha - zx_\alpha$ belongs to the kernel of h_α, i.e., there exists an element m''_α in M'_α such that $ym'_\alpha m''_\alpha = zx_\alpha m''_\alpha$. If we now set $\bar{m}_\alpha = m'_\alpha m''_\alpha$, then $\bar{m}_\alpha$ is still an element of M'_α, i.e., $\bar{m}_\alpha \not\in \mathfrak{p}_\alpha$, and the product $y\bar{m}_\alpha$ belongs to the principal ideal Az. Let $\mathfrak{p}'_\alpha$ be the extension of $\mathfrak{p}_\alpha$ to the ring A'. Then $\mathfrak{p}'_\alpha$ is a prime ideal in A', and $\mathfrak{p}'_\alpha \cap A = \mathfrak{p}_\alpha$. Hence $\bar{m}_\alpha \not\in \mathfrak{p}'_\alpha$, and since $y\bar{m}_\alpha \in Az \subset A'z$, *it follows that the ideal* $A'z$: $A'y$ *is not contained in* $\mathfrak{p}'_\alpha$. Since the set of prime ideals $\mathfrak{p}'_\alpha$ coincides with the set of all maximal ideals of A' (see proof of Lemma 1.3), we conclude that $A'z$: $A'y$ must be the unit ideal in A', i.e., we must have $y \in A'z$, and hence $\xi \in A'$. This completes the proof of the lemma.

COROLLARY 1.6. *Let* Σ_0 *be an arbitrary set of prime ideals of* R, *and let* $\tilde{\Sigma}$ *be the set of prime ideals* $\mathfrak{P}$ *in* $\tilde{A}$ *such that* $\mathfrak{P} \cap R \in \Sigma_0$. *Then*

$$\bigcap_{\mathfrak{P} \in \tilde{\Sigma}} \tilde{A}_{\mathfrak{P}} = \bigcap_{\mathfrak{P} \in \Sigma_0} \widetilde{A[M(\mathfrak{P})^{-1}]}, \tag{11}$$

where $M(\mathfrak{P}) = R - \mathfrak{P}$.

For, in the first place, we have by Proposition 1.2:

$$\widetilde{A[M(\mathfrak{P})^{-1}]} = \tilde{A}[M(\mathfrak{P})^{-1}]. \tag{12}$$

In the second place, by Lemmas 1.3 and 1.5, applied to $\tilde{A}$ instead of to A (this application is permissible; see Note following immediately the proof of Lemma 1.3), we have:

$$\tilde{A}[M(\mathfrak{P})^{-1}] = \bigcap_{\mathfrak{P} | \mathfrak{P} \cap R = \mathfrak{P}} \tilde{A}_{\mathfrak{P}}. \tag{13}$$

The Corollary now follows from (12) and (13).

The above corollary will be applied later on in the special case in which R is integrally closed and Σ_0 is the set of all prime ideals of R which have height 1. In that case, both the "going-up" *and* the "going-down" theorem are applicable to the pair of rings $\{A, R\}$ (see C.A. 1, Theorem 6, p. 262), and therefore $\tilde{\Sigma}$ is then the set of all prime ideals in $\tilde{A}$ which have height 1.

Quite generally, if A is a ring and Σ is the set of all prime ideals of A which have height 1, *we shall say that A satisfies condition* (S) *if* $\bigcap_{\mathfrak{p} \in \Sigma} A_{\mathfrak{p}} = A$ [here the intersection $\bigcap A_{\mathfrak{p}}$ has been defined earlier; see (9)]. We note that if, in particular, A is a Krull domain then A satisfies condition (S) and is integrally closed (see C.A. 2, Theorem 26, p. 83).

LEMMA 1.7. *Let R be an integral domain, let Σ_0 be the set of prime ideals of R which have height 1 and let L be a field containing R as a subring. If R satisfies condition* (S) *and is integrally closed, and if for any prime ideal $\mathfrak{P}$ in Σ_0 we denote by $\bar{R}_{\mathfrak{P}}$ the integral closure in L of the localization $R_{\mathfrak{P}}$ of R at $\mathfrak{P}$, then $\cap_{\mathfrak{P}\in\Sigma_0}\bar{R}_{\mathfrak{P}}$ = integral closure of R in L.*

Proof. Let K be the field of fractions of R in L, and let x be any element of L which belongs to all the rings $\bar{R}_{\mathfrak{P}}$, $\mathfrak{P}\in\Sigma_0$. Since $R_{\mathfrak{P}}$ is integrally closed, the monic irreducible polynomial of x over K has its coefficients in $\cap_{\mathfrak{P}\in\Sigma_0} R_{\mathfrak{p}}$ (see C.A. 1, Theorem 4, p. 260), hence in R, and this completes the proof.

We now deal with the principal result of this section, a result which will be used extensively in the next two sections.

THEOREM 1.8. *Let A, F and R satisfy conditions* (1), (2) *and* (3) *stated in the beginning of this section. Assume furthermore that* (a) *R is integrally closed and that* (b) *A is a simple ring extension of R, say $A=R[y]$. Then K is necessarily the field of fractions of R. If, in addition, R satisfies condition* (S), *then both A and $\tilde{A}$ satisfy conditions* (S), *and we have*

$$\tilde{A}=\bigcap_{\mathfrak{P}\in\Sigma_0}\widetilde{A[(R-\mathfrak{P})^{-1}]},\tag{14}$$

where Σ_0 is the set of prime ideals of R, of height 1.

Proof. We have seen earlier [see (6)] that if K' denotes the field of fractions of R then K is an algebraic extension of K'. Moreover, from (6) and from assumption (b) of the theorem it follows that $F=K'[y]$. Since y is integral over R [condition (3)] and R is integrally closed [assumption (a) of the theorem], the minimal (monic) polynomial of y over K' has its coefficients in R. If n is the degree of that polynomial then F is a vector space over K', of dimension n. Suppose that the first assertion of the theorem is false, i.e., suppose that K' is a proper subfield of K. Let z be an element of K, not in K' and let $\varphi(Y)$ be the uniquely determined polynomial in $K'[Y]$, of degree $<n$, such that $z=\varphi(y)$. Upon multiplying z by a suitable element of R we may assume that the coefficients of φ are in R, without invalidating the condition that z is not in K'. Then $z=\varphi(y)\in R[y]=A$, whence $z\in A\cap K=R\subset K'$, a contradiction.

We now add the additional assumption that R satisfies condition (S). Let Σ (respectively, $\tilde{\Sigma}$) denote the set of prime ideals in A (respectively, in $\tilde{A}$) which are of height 1. Since both A and $\tilde{A}$ are integral over R and since R is integrally closed, Σ (respectively, $\tilde{\Sigma}$) consists of those and only those prime ideals in A

(respectively, $\tilde{A}$) which contract in R to prime ideals of height 1. We have therefore, by Lemma 1.5 and in view of the relation $A = R[y]$:

$$\bigcap_{\mathfrak{p}\in\Sigma} A_{\mathfrak{p}} = \bigcap_{\mathfrak{P}\in\Sigma_0} A\left[(R-\mathfrak{P})^{-1}\right] = \bigcap_{\mathfrak{P}\in\Sigma_0} R_{\mathfrak{P}}[y]. \tag{15}$$

We also have, by Corollary 1.6 and Proposition 1.2:

$$\bigcap_{\tilde{\mathfrak{p}}\in\tilde{\Sigma}} \tilde{A}_{\tilde{\mathfrak{p}}} = \bigcap_{\mathfrak{P}\in\Sigma_0} \widetilde{A\left[(R-\mathfrak{P})^{-1}\right]} = \bigcap_{\mathfrak{P}\in\Sigma_0} \tilde{A}\left[(R-\mathfrak{P})^{-1}\right]. \tag{16}$$

Now, in (15), each ring $R_{\mathfrak{P}}[y]$ is a free $R_{\mathfrak{p}}$-module, with $\{1, y, y^2, \ldots, y^{n-1}\}$ as free basis, and, similarly, the ring $F = K[y]$ is a free K-module with the same free basis $\{1, y, y^2, \ldots, y^{n-1}\}$. If then z is an element of F and if we write $z = a_0 + a_1 y + \cdots + a_{n-1} y^{n-1}$, $a_i \in K$, then $z \in R_{\mathfrak{P}}[y]$ if and only if all the a_i are in $R_{\mathfrak{P}}$. Therefore z belongs to the intersection of the rings $R_{\mathfrak{P}}[y]$ ($\mathfrak{P}$ ranges over Σ_0) if and only if all the elements $a_0, a_1, \ldots, a_{n-1}$ are in R (since R satisfies condition (S)). This proves, in view of (15), that A satisfies condition (S).

In view of the second equality in (16) (which is merely a restatement of Proposition 1.2) we have

$$\tilde{A} \subset \bigcap_{\mathfrak{P}\in\Sigma_0} \widetilde{A\left[(R-\mathfrak{P})^{-1}\right]}.$$

Therefore, in view of (16), both remaining assertions of the theorem (i.e., the equality (14) and the assertion that also $\tilde{A}$ satisfies condition (S)) will be proved if we establish the validity of the opposite inclusion:

$$\bigcap_{\mathfrak{P}\in\Sigma_0} \widetilde{A\left[(R-\mathfrak{P})^{-1}\right]} \subset \tilde{A}. \tag{17}$$

Let then ξ be any element belonging to the intersection ring on the left hand side of (17). We have $A[(R-\mathfrak{P})^{-1}] = R_{\mathfrak{P}}[y]$. Since ξ belongs to the saturation of $R_{\mathfrak{P}}[y]$ with respect to K, it follows from well known facts on saturation (see SES III, Lemma 1.4, p. 970) that given any two distinct K-homomorphisms ψ_1, ψ_2 of F into Ω, the quotient

$$x = [\psi_1(\xi) - \psi_2(\xi)]/[\psi_1(y) - \psi_2(y)]$$

is integral over $R_{\mathfrak{P}}$. (Note that since $F = K[y]$ and $\psi_1 \neq \psi_2$, we must have

$\psi_1(y) \neq \psi_2(y)$.) Since this is true for any prime ideal $\mathfrak{P}$ of height 1 (in R) and since R satisfies condition (S) and is integrally closed, it follows by Lemma 1.7 that the above quotient x is *integral over* R. Therefore ξ dominates y with respect to the ring $\{A,K\}$, and thus $\xi \in \tilde{A}$. This proves the inclusion (17) and completes the proof of the theorem.

We shall conclude this section by proving a result which will be needed in Section 3.

Let A, F and R satisfy conditions 1), 2) and 3) stated in the beginning of this section. *Assume that A is a (finite) direct sum of q subrings $A_1, A_2, \ldots, A_q$.* Let ϵ'_α be the identity of A_α, let F_α be the total ring of quotients of A_α (in F), let $K_\alpha = K\epsilon'_\alpha$ and let $\bar{A}_\alpha$ be the integral closure of A_α in F_α. We have then $F = \oplus_{\alpha=1}^q F_\alpha$, $\bar{A} = \oplus_{\alpha=1}^q \bar{A}_\alpha$. Since all nonzero elements of K are regular elements of F, the mapping $\theta_\alpha : K \to K_\alpha$ defined by $\theta_\alpha(a) = a\epsilon'_\alpha$, $a \in K$, is an isomorphism, and hence K_α is a field contained in F_α. Since F is a finite-dimensional vector space over K, F_α is finite-dimensional over K_α. Since A is integral over R, A_α is integral over its subring R_α, where $R_\alpha = R\epsilon'_\alpha \subset K_\alpha$. By a stronger reason, A_α is integral over the intersection ring $A_\alpha \cap K_\alpha$. Thus A_α, F_α and K_α satisfy the conditions (1), (2) and (3) stated in the beginning of this section, and the saturation $\tilde{A}_\alpha$ of A_α with respect to K_α is well defined as a subring of $\bar{A}_\alpha$ (and hence also of $\bar{A}$). Note that since A_α is integral over R_α, the intersection $A_\alpha \cap K_\alpha$ is also integral over R_α.

Let H_α be the (finite) set of K-homomorphisms ψ of F into Ω such that $\psi(\epsilon'_\alpha) = 1$. Thus, the set H of all K-homomorphisms of F into Ω is the disjoint union of the q subset $H_1, H_2, \ldots, H_q$ of H. If Ω_α denotes an algebraic closure of K_α, then there exists an isomorphism $\theta^*_\alpha : \Omega \xrightarrow{\sim} \Omega_\alpha$ the restriction of which to K is θ_α. As ψ ranges over H_α, the product $\theta^*_\alpha \cdot \psi$ (restricted to F_α) ranges over the entire set of K_α-homomorphisms of F_α into Ω_α. We can say therefore that given any pair of elements ξ_α, η_α of F_α, η_α dominates ξ_α, with respect to the pair $\{A_\alpha, K_\alpha\}$ if for any two elements ψ, ψ' of H_α we have that the quotient

$$[\psi'(\eta) - \psi(\eta)]/[\psi'(\xi) - \psi(\xi)] \tag{18}$$

is integral over $(\theta^*_\alpha)^{-1}(A_\alpha \cap K_\alpha)$, if $\psi'(\xi) \neq \psi(\xi)$, while $\psi'(\eta) = \psi(\eta)$ whenever $\psi'(\xi) = \psi(\xi)$. If we now recall that $A_\alpha \cap K_\alpha$ is integral over R_α and that $\theta_\alpha^{-1}(R_\alpha) = R$, we see that to say that the quotient (18) is integral over $(\theta^*_\alpha)^{-1}(A_\alpha \cap K_\alpha)$ *is the same as saying that it is integral over* R. Now, using this fact and observing that if ξ is any element of F and if ψ if any homomorphism belonging to H_α then $\psi(\xi) = \psi(\xi_\alpha)$, where $\xi_\alpha = \xi\epsilon'_\alpha$, we conclude that *if ξ, η are any two elements of F such that η dominates ξ with respect to the pair $\{A, K\}$, then for each $\alpha = 1, 2, \ldots, q$ the element η_α dominates ξ_α with respect to the pair*

$\{A_\alpha, K_\alpha\}$. From this it follows that *the subring* $\tilde{A}_1 \oplus \tilde{A}_2 \oplus \cdots \oplus \tilde{A}_q$ *of* $\bar{A}$ *is saturated with respect to the pair* $\{A, K\}$, for if $\xi = \xi\epsilon'_1 + \xi\epsilon'_2 + \cdots + \xi\epsilon'_q \in \oplus_{\alpha=1}^{q} \tilde{A}_\alpha$ and η is an element of $\bar{A}$ which dominates ξ (with respect to the pair $\{A, K\}$), then $\eta\epsilon'_\alpha$ dominates $\xi\epsilon'_\alpha$ with respect to the pair $\{A_\alpha, K_\alpha\}$, and hence $\eta\epsilon'_\alpha$ belongs to $\tilde{A}_\alpha$, since $\xi\epsilon'_\alpha \in \tilde{A}_\alpha$ and $\eta\epsilon'_\alpha \in \bar{A}_\alpha$. Since A is a subring of $\oplus\tilde{A}_\alpha$, it follows from the minimality property of the saturated ring $\tilde{A}$ *that* $\tilde{A}$ *is a subring of* $\oplus\tilde{A}_\alpha$:

$$\tilde{A} \subset \bigoplus_{\alpha=1}^{q} \tilde{A}_\alpha. \tag{19}$$

PROPOSITION 1.9. *The rings A, F, R being as above, assume that* (a) *A is a (finite) direct sum of q local (not necessarily noetherian) rings* A_α *and* (b) *that R is integrally closed in K. Then in* (19) *the equality sign holds*:

$$\tilde{A} = \bigoplus_{\alpha=1}^{q} \tilde{A}_\alpha.$$

Proof. We shall use the representation of $\tilde{A}$ as a union $\cup_{\nu=0}^{\infty} A^{(\nu)}$ of an ascending sequence of subrings $A^{(\nu)}$, where (see SES III, Introduction)

$$A^{(0)} = A, \qquad A^{(\nu)} = A^{(\nu-1)}[L^{(\nu)}], \qquad \nu = 1, 2, \ldots,$$

and where the $L^{(\nu)}$ are subsets of $\bar{A}$ defined inductively as follows: $L^{(\nu)}$ is the set of elements η of $\bar{A}$ such that η dominates some element ξ (which may depend on η) of $A^{(\nu-1)}$ with respect to the pair $\{A, K\}$ (and hence also with respect to the pair $\{A^{(\nu-1)}, K\}$, since $A^{(\nu-1)} \cap K$ is a subring of $\bar{A} \cap K$ and is therefore integral over R). Similarly, we have $\tilde{A}_\alpha = \cup_{\nu=0}^{\infty} A_\alpha^{(\nu)}$, where

$$A_\alpha^{(0)} = A_\alpha, \qquad A_\alpha^{(\nu)} = A_\alpha^{(\nu-1)}[L_\alpha^{(\nu)}],$$

and where $L_\alpha^{(\nu)}$ is the set of all elements of $\bar{A}_\alpha$ which dominate some element of $A_\alpha^{(\nu-1)}$, with respect to the pair $\{A_\alpha, K_\alpha\}$. The proposition will be established if we prove that we have for each $\nu = 1, 2, \ldots$:

$$A^{(\nu)} = A_1^{(\nu)} \oplus A_2^{(\nu)} \oplus \cdots \oplus A_q^{(\nu)}. \tag{20_ν}$$

We show first that the (20_ν) will follow, by induction on ν, for all ν, if $(20)_\nu$ is proved for $\nu = 1$. To see this we first observe that the assumption (b) of the proposition implies that $R\epsilon'_\alpha$ is integrally closed in the field $K\epsilon'_\alpha$ (for $\alpha = 1, 2, \ldots, q$) since $\theta_\alpha : K \to K\epsilon'_\alpha$ is an isomorphism which transforms R onto $R\epsilon'_\alpha$. By

Theorem 4.1 of SES III it follows therefore that $\tilde{A}_\alpha$ is a *local* (not necessarily noetherian) ring. Since $A_\alpha^{(1)}$ is a ring between the two local rings A_α and $\tilde{A}_\alpha$ and since $\tilde{A}_\alpha$ is integral over A_α, it follows that $A_\alpha^{(1)}$ has only one maximal ideal, hence is also a local (not necessarily noetherian) ring. Thus, if (20_1) is true then $A^{(1)}$ is a direct sum of the q local rings $A_\alpha^{(1)}$. Furthermore, since $A^{(1)}$ is integral over R, $A^{(1)} \cap K$ is integral over R, whence $A^{(1)} \cap K = R$, since R is integrally closed in K. Thus, the two assumptions (a) and (b) of the proposition are satisfied if the rings A, R and A_α are replaced respectively by the rings $A^{(1)}$, R and $A_\alpha^{(1)}$. This shows that if (20_1) is proved then (20_ν) and the proposition will follow by induction on ν.

To prove (20_1) it will be sufficient to prove that

$$L^{(1)} = L_1^{(1)} + L_2^{(1)} + \cdots + L_q^{(1)}, \tag{21}$$

where we now agree to include in each set $L_\alpha^{(1)}$ the element zero. It is obvious that $L^{(1)} \subset L_1^{(1)} + L_2^{(1)} + \cdots + L_q^{(1)}$, since—as has already been pointed out earlier—the domination $\eta \succ \xi$ (with respect to the pair $\{A, K\}$) is a consequence of the q relations $\eta\epsilon'_\alpha \succ \xi\epsilon'_\alpha$ (dominations with respect to the pair $\{A_\alpha, K_\alpha\}$). Now, to prove the opposite inclusion it will be sufficient to show that we have $L_\alpha^{(1)} \subset L^{(1)}$, for $\alpha = 1, 2, \ldots, q$. Let us show, for instance, that

$$L_1^{(1)} \subset L^{(1)}. \tag{22}$$

Let F^* be the compositum of the fields $\psi(F)$, $\psi \epsilon H$, and let A^* be the integral closure of R in F^*. For any $\psi \in H$ we have $\psi(A_\alpha) = 0$ (if $\psi \not\in H_\alpha$) or $\psi(A_\alpha)$ is a homomorphic image of A_α (if $\psi \in H_\alpha$) and is therefore a local ring (not necessarily noetherian). Since, in the latter case, the ring $\psi(A_\alpha)$ contains R as a subring, we see that *A^* is integral over $\psi(A_\alpha)$, whenever $\psi \in H_\alpha$*. Therefore, if we denote by $\mathfrak{m}_\alpha$ the maximal ideral of A_α, we find that *every maximal ideal of A^* lies over the maximal ideal $\psi(\mathfrak{m}_\alpha)$ of $\psi(A_\alpha)$*, if $\alpha \in H_\alpha$ (see C.A. 1, p. 259, Complement (2) to Theorem 3).

With these observations in mind, we proceed to the proof of (22). Let η_1 be an element of $L_1^{(1)}$. There exist then an element of A_1 which is dominated by η_1, with respect to the pair $\{A_1, K_1\}$. We fix one such element, say ξ_1. We have to show that there exists an element ξ in A such that η_1 dominates ξ with respect to the pair $\{A, K\}$. We shall take for ξ an element of A of the form

$$\xi = \xi_1 + \xi_2 + \cdots + \xi_q, \tag{23}$$

where ξ_1 is the above element of A_1 which is dominated by η_1, while the remaining $q-1$ components $\xi_2, \ldots, \xi_q$ of ξ are elements of $A_2, \ldots, A_q$ defined as follows:

$$\xi_\alpha = \epsilon'_\alpha \text{ if } \xi_1 \in \mathfrak{m}_1; \tag{24}$$

$$\xi_\alpha \quad \text{an arbitrary element of } \mathfrak{m}_\alpha \text{ if } \xi_1 \not\subset \mathfrak{m}_1 \qquad \alpha = 2, 3, \ldots, q \tag{25}$$

We now prove that with this choice of $\xi_2, \ldots, \xi_q$ we have

$$\eta_1 \succ \xi \quad (\text{with respect to } \{A, K\}). \tag{26}$$

Let ψ and ψ' be any two homomorphisms in the set H. We shall distinguish 3 cases: (1) $\psi, \psi' \in H_1$; (2)$\psi \not\in H_1$, $\psi' \not\in H_1$; (3)$\psi \in H_1$, $\psi' \not\in H_1$.

Case 1. In this case we have $\psi(\xi) = \psi(\xi_1)$, $\psi'(\xi) = \psi'(\xi_1)$, and the condition which (26) imposes on the pair $\{\psi, \psi'\}$ is then a consequence of the domination relation $\eta_1 \succ \xi_1$ (with respect to $\{A_1, K_1\}$.)

Case 2. In this case we have $\psi(\eta_1) = \psi'(\eta_1) = 0$ and the condition which (26) imposes on the pair $\{\psi, \psi'\}$ is now trivially satisfied.

Case 3. Let, say, $\psi' \in H_\alpha$, $\alpha \neq 1$. We consider the two cases: (3a) $\xi_1 \in \mathfrak{m}_1$ and (3b) $\xi_1 \not\in \mathfrak{m}_1$.

Case 3a. $\xi_1 \in \mathfrak{m}_1$. In this case we have, by (24), $\psi'(\xi) = \psi'(\epsilon'_\alpha) = 1$. Since $\psi(\xi)$ $(= \psi(\xi_1))$ belongs in this case to the maximal ideal $\psi(\mathfrak{m}_1)$ of the ring $\psi(A_1)$, it follows that the difference $\psi(\xi) - \psi'(\xi)(= \psi(\xi_1) - 1)$ is a unit in $\psi(A_1)$; it is therefore different from zero, and the quotient $[\psi(\eta_1) - \psi'(\eta_1)]/[\psi(\xi) - \psi'(\xi)]$, i.e., the quotient $\psi(\eta_1)/(\psi(\xi_1) - 1)$, is an element of $\psi(A_1)$ and is therefore integral over R.

Case 3b. $\xi_1 \not\in \mathfrak{m}_1$. In this case ξ_1 is a unit in A_1, and therefore $\psi(\xi)$ $(= \psi(\xi_1))$ is a unit in $\psi(A_1)$. Therefore $\psi(\xi)$ *belongs to none of the maximal ideals of* A^*. On the other hand, we have $\psi'(\xi) = \psi'(\xi_\alpha)$, and by (25), $\psi'(\xi_\alpha)$ belongs to the maximal ideal of $\psi'(A_\alpha)$ and therefore also *to each of the maximal ideals of* A^*. Thus, $\psi(\xi) - \psi'(\xi)$ *belongs to none of the maximal ideals of* A^*. The difference $\psi(\xi) - \psi'(\xi')$ is therefore a unit in A^*, and therefore the quotient

$$[\psi(\eta_1) - \psi'(\eta_1)]/[\psi(\xi) - \psi'(\xi_1)] \quad (= \psi(\eta_1)/[\psi(\xi) - \psi'(\xi_1)])$$

belong to A^* and is therefore integral over R. This completes the proof of (26) and hence also of (22), and the proof of the proposition is now complete.

2. Saturation and Completion. In this section we shall deal exclusively with *semilocal* (noetherian) rings $\mathfrak{o}$ which satisfy the following three conditions:

(A) *The semi-local ring* $\mathfrak{o}$ *has no nilpotent elements different from zero.*

(B) $\mathfrak{o}$ *contains a subring* R *which is a regular, equicharacteristic zero, local domain, and* $\mathfrak{o}$ *is a finite* R*-module.*

(c) *R is a pseudo-geometric ring* (see Nagata, LR, Section 36, p. 131).

The following are important examples of pairs of rings $(\mathfrak{o}, R)$ satisfying conditions (A), (B) and (C):

1. $\mathfrak{o}$ is the (complete) local ring of the closed point of an unmixed r-dimensional algebroid variety V, defined over a ground field k of characteristic zero; $\{x_1, x_2, \ldots, x_r\}$ is a set of parameters of $\mathfrak{o}$, and R is the formal power series ring $k[[x_1, x_2, \ldots, x_r]]$ (canonically embedded in $\mathfrak{o}$; see C.A. 2, Theorem 21, Corollary 2, p. 293, and the "Remark" following that corollary).

2. V is an affine algebraic variety, of pure dimension r, defined over a ground field k of characteristic zero, A is the affine ring of V; $\{x_1, x_2, \ldots, x_r\}$ are elements of A such that A is integral over the polynomial ring $A_0 = k[x_1, x_2, \ldots, x_r]$ [see the C.A. 1, Theorem 25 ("Normalization theorem"), p. 200], R is the localization of A_0 at a prime ideal $\mathfrak{p}_0$ of A_0, and $\mathfrak{o}$ is the localization $A[M^{-1}]$, where $M = A_0 - \mathfrak{p}_0$. In this case, $\mathfrak{o}$ is the semi-local ring whose maximal ideals are associated with the points of V which project into the point of the affine r-space, associated with $\mathfrak{P}_0$.

3. $\mathfrak{o}$ is a semi-local ring satisfying conditions (A) and (B), and of Krull dimension 1. In this case, the regular local domain R has only two prime ideals: the zero ideal and the maximal ideal $\mathfrak{M}$. The ring $R/\mathfrak{M}$ is a field and thus automatically satisfies the finiteness condition for integral extensions, while the ring R itself also satisfies that condition since R is noetherian, integrally closed of characteristic zero (see C.A. 1, Theorem 7, Corollary 1, p. 265). Thus, in this case R is necessarily a pseudo-geometric ring, and condition (C) is satisfied.

Let $\mathfrak{M}$ be the maximal ideal in R. Since $\mathfrak{o}$ is integral over R, each of the (finitely many) maximal ideals $\mathfrak{m}_\nu$ of $\mathfrak{o}$ contracts to $\mathfrak{M}$ in R, and any strictly descending chain of prime ideals in $\mathfrak{o}$, beginning with a given $\mathfrak{m}_\nu$, contracts in R to a strictly descending chain of prime ideals beginning with $\mathfrak{M}$. Hence if r denotes the Krull dimension of R ($r = h(\mathfrak{M})$ = height of $\mathfrak{M}$) then we must have $h(\mathfrak{m}_\nu) \leqslant r$, for any of the $\mathfrak{m}_\nu$. However, since R is integrally closed, every strictly descending chain of prime ideals in R, beginning with $\mathfrak{M}$, can be lifted to a strictly descending chain of prime ideals in $\mathfrak{o}$ whose first term is any of the $\mathfrak{m}_\nu$ (see C.A. 1, Theorem 6, p. 262). Therefore, all the maximal ideals $\mathfrak{m}_\nu$ of $\mathfrak{o}$ have the same height $r(= h(M))$, or, equivalently, all the local rings $\mathfrak{o}_{\mathfrak{m}_\nu}$ has the same Krull dimension r.

In SES III (p. 964) we called a *local* ring $\mathfrak{o}$ *equidimensional* if all the minimal prime ideals of $\mathfrak{o}$ have the same dimension (that dimension is then necessarily equal to the Krull dimension of $\mathfrak{o}$). We now can easily see that *all the above local rings* $\mathfrak{o}_{\mathfrak{m}_\nu}$ *are equidimensional.* For, by property (A), the zero ideal in $\mathfrak{o}$ is a finite intersection of prime ideals:

$$(0) = \mathfrak{p}_1 \cap \mathfrak{p}_2 \cap \cdots \cap \mathfrak{p}_h, \tag{1}$$

where we assume that no $\mathfrak{p}_i$ is redundant. The h prime ideals $\mathfrak{p}_i$ are the minimal prime ideals of $\mathfrak{o}$ and hence we have (always using the fact that R is integrally closed and that $\mathfrak{o}$ is integral over R):

$$\mathfrak{p}_i \cap R = (0), \qquad i = 1, 2, \ldots, h. \tag{2}$$

Each of the semi-local domain $\mathfrak{o}/\mathfrak{p}_i$ contains therefore R as a subring and is integral over R. Therefore, also $\mathfrak{o}/\mathfrak{p}_i$ has Krull dimension r, and, in fact, from the preceding considerations it follows that each of the maximal ideals $\mathfrak{m}_\nu/\mathfrak{p}_i$ of $\mathfrak{o}/\mathfrak{p}_i$ ($\mathfrak{m}_\nu$, any of the maximal ideals of $\mathfrak{o}$ containing $\mathfrak{p}_i$) has dimension r. We conclude therefore that *each of the h minimal prime ideals $\mathfrak{p}_i$ of $\mathfrak{o}$ has dimension r and all the local rings $\mathfrak{o}_{\mathfrak{m}_\nu}$ are equidimensional (of Krull dimension r).*

From (2) it follows that every nonzero element of R is regular in $\mathfrak{o}$. Hence, if F denotes the total ring of quotients of $\mathfrak{o}$ then F contains (canonically) the field of fractions K of R, and every nonzero element of K is regular in F. In particular, the identity of K is also the identity of F. Since $\mathfrak{o}$ is a finite R-module, F is a finite-dimensional vector space over K. Finally, the ring $\mathfrak{o} \cap K$ is integral over R and therefore $\mathfrak{o} \cap K = R$ since R is integrally closed in K. Thus the three conditions (1), (2) and (3) stated in the beginning of section 1 are satisfied if A is replaced by $\mathfrak{o}$. Therefore the saturation $\widetilde{\mathfrak{o}_K}$ of $\mathfrak{o}$ with respect to K is defined. We observe that so far we have made no use of condition (C). We omit the subscript K and we write $\tilde{\mathfrak{o}}$ instead of $\widetilde{\mathfrak{o}_K}$.

We know that F is a direct sum of h fields $F\epsilon_i$. If $\bar{\mathfrak{o}}$ denotes the integral closure of $\mathfrak{o}$ in F, then $\bar{\mathfrak{o}}\epsilon_i$ is the integral closure of $\mathfrak{o}\epsilon_i$ in $F\epsilon_i$, and $\bar{\mathfrak{o}} = \sum_{i=1}^{h} \bar{\mathfrak{o}}\epsilon_i$ (see GTS II, Section 1, relation (5)). The canonical homomorphisms $R \rightarrow R\epsilon_i$, $K \rightarrow K\epsilon_i$ are isomorphisms, and $K\epsilon_i$ is the field of fractions of $R\epsilon_i$. Since $\mathfrak{o}\epsilon_i$ is integral over $R\epsilon_i$, $\bar{\mathfrak{o}}\epsilon_i$ is also the integral closure of $R\epsilon_i$ in $F\epsilon_i$. Since $R\epsilon_i$ is integrally closed in its field of fractions $K\epsilon_i$ and since the field $F\epsilon_i$ is a finite (separable) extension of $K\epsilon_i$, it follows that $\bar{\mathfrak{o}}\epsilon_i$ is a finite $R\epsilon_i$-module, and hence also a finite R-module. Consequently, also $\bar{\mathfrak{o}}$ is a finite R-module and is a (noetherian) semi-local ring. Since $\tilde{\mathfrak{o}}$ is a ring between $\mathfrak{o}$ and $\bar{\mathfrak{o}}$, it follows that also $\tilde{\mathfrak{o}}$ is a finite R-module and a (noetherian) semi-local ring. From Theorem 4.1 of SES III (p. 997) we deduce that $\mathfrak{o}$ and $\tilde{\mathfrak{o}}$ have the same number of maximal ideals, that corresponding maximal ideals $\mathfrak{m}_\nu$ and $\tilde{\mathfrak{m}}_\nu$ of $\mathfrak{o}$ and $\tilde{\mathfrak{o}}$ (i.e., such that $\mathfrak{m}_\nu = \tilde{\mathfrak{m}}_\nu \cap \mathfrak{O}$) have the same residue field and that all the local rings $\tilde{\mathfrak{o}}_{\tilde{\mathfrak{m}}_\nu}$ are equidimensional, of dimension r.

By (5) and (6) of Section 1 we have $F = \mathfrak{o}[M^{-1}] = \mathfrak{o}K$, where M is the set of nonzero elements of R. (Note that the field K' which occurs in (6), i.e., the field of fractions of R, now is K itself). We can also use the multiplicative system M

to localize $\mathfrak{o}$ *as an R-module*. This gives rise to the ring $\mathfrak{o}\otimes_R K$ (see Bourbaki [1], p. 81, Definition 3). The two rings $\mathfrak{o}[M^{-1}]$ and $\mathfrak{o}\otimes_R K$ are, however, canonically isomorphic both as K-modules and as $\mathfrak{o}$-modules (Bourbaki [1], Ch. II, p. 84, Proposition 6). Since both rings are generated by their subrings K and $\mathfrak{o}$, it follows that the canonical isomorphism between F and $\mathfrak{o}\otimes_R K$ is the identity on both K and $\mathfrak{o}$. We shall therefore identify these two rings and write

$$F=\mathfrak{o}\underset{R}{\otimes}K. \tag{3}$$

We note also that since $\mathfrak{o}$, $\tilde{\mathfrak{o}}$ and $\bar{\mathfrak{o}}$ are finite R-modules it follows from condition (C) that each of these three rings is pseudo-geometric.

LEMMA 2.1. *Any pseudo-geometric semi-local ring $\mathfrak{o}$, free from nilpotent elements (different from zero), is analytically unramified* (i.e., *the completion $\hat{\mathfrak{o}}$ of $\mathfrak{o}$ is also free from nilpotent elements*).

Proof. The lemma is well known if $\mathfrak{o}$ is an integral domain (see Zariski [11]; see also Nagata, L. R., p. 132). Assume that $\mathfrak{o}$ has zero divisors. The zero ideal of $\mathfrak{o}$ is an intersection of minimal prime ideals $\mathfrak{p}_1, \mathfrak{p}_2, \ldots, \mathfrak{p}_h$ of $\mathfrak{o}$. Each of the semi-local *domains* $\mathfrak{o}/\mathfrak{p}_i$ is pseudo-geometric and is therefore analytically unramified. In other words (since $\widehat{\mathfrak{o}/\mathfrak{p}_i}\cong\hat{\mathfrak{o}}/\hat{\mathfrak{o}}\mathfrak{p}_i$), each of the ideals $\hat{\mathfrak{o}}\mathfrak{p}_i$ is an intersection of prime ideals. Now, by a well-known theorem concerning (the so-called) Zariski rings (see C.A. 2, p. 266, Corollary 2, where we identify A and E with $\mathfrak{o}$ and take for F and G any two ideals in $\mathfrak{o}$), we have:

$$\hat{\mathfrak{o}}\cdot(\mathfrak{p}_1\cap\mathfrak{p}_2\cap\cdots\cap\mathfrak{p}_h)=\hat{\mathfrak{o}}\mathfrak{p}_1\cap\hat{\mathfrak{o}}\mathfrak{p}_2\cap\cdots\cap\hat{\mathfrak{o}}\mathfrak{p}_h.$$

This shows that the zero-ideal in $\hat{\mathfrak{o}}$ is a finite intersection of prime ideals. Q.E.D.

From the above lemma follows that the rings $\mathfrak{o}$, $\tilde{\mathfrak{o}}$ and $\bar{\mathfrak{o}}$ are analytically unramified.

The completion $\hat{R}$ of R is a formal power series ring over a field (of characteristic zero) and is therefore pseudo-geometric. By a well-known theorem (see Bourbaki [2], p. 68, Theorem 3, part (ii); see also C.A. 2, p. 265, Corollary 1 of Theorem 11 and the observation at the beginning of p. 266) we have that $\mathfrak{o}$ and $\hat{R}$ are subrings of $\hat{\mathfrak{o}}$, that

$$\hat{\mathfrak{o}}=\hat{R}\mathfrak{o}=\hat{R}\underset{R}{\otimes}\mathfrak{o}, \tag{4}$$

and that every nonzero element of $\hat{R}$ is regular in $\hat{\mathfrak{o}}$ (see C.A. 2, Theorem 16, part (b), p. 277). Since $\mathfrak{o}$ is a finite R-module it follows from (4) that $\hat{\mathfrak{o}}$ is a finite

$\hat{R}$-module. Since also the semi-local ring $\bar{\mathfrak{o}}$ is a finite R-module and is analytically unramified, conclusions similar to (4) hold for the ring $\hat{\bar{\mathfrak{o}}}$ and its subring $\hat{R}$:

$$\hat{\bar{\mathfrak{o}}} = \hat{R}\bar{\mathfrak{o}} = \hat{R} \otimes_{R} \bar{\mathfrak{o}}, \tag{5}$$

every nonzero element of $\hat{R}$ is regular in $\hat{\bar{\mathfrak{o}}}$, and $\hat{\bar{\mathfrak{o}}}$ is a finite $\hat{R}$-module. Thus, both pairs of rings $(\hat{\mathfrak{o}}, \hat{R})$ and $(\hat{\bar{\mathfrak{o}}}, \hat{R})$ satisfy conditions (A), (B) and (C) stated in the beginning of this section [with $(\mathfrak{o}, R)$ replaced by $(\hat{\mathfrak{o}}, \hat{R})$ and $(\hat{\bar{\mathfrak{o}}}, \hat{R})$ respectively]. Consequently, if $\hat{K}$ denotes the field of fractions of $\hat{R}$, both pairs of rings $(\hat{\mathfrak{o}}, \hat{K})$ and $(\hat{\bar{\mathfrak{o}}}, \hat{K})$ satisfy conditions (1), (2) and (3) stated in the beginning of Section 1 [with (A, K) replaced by $(\hat{\mathfrak{o}}, \hat{K})$ and $(\hat{\bar{\mathfrak{o}}}, \hat{K})$ respectively].

From the above considerations we proceed to derive a number of consequences.

In the first place it follows that if we denote by $\hat{F}$ and $\hat{\bar{F}}$ the total ring of quotients of $\hat{\mathfrak{o}}$ and $\hat{\bar{\mathfrak{o}}}$ respectively and by $\hat{M}$ the multiplicative set of the nonzero elements of $\hat{R}$, then [see (5), Section 1 and (3) of this Section]:

$$\hat{F} = \hat{\mathfrak{o}}[\hat{M}^{-1}] = \hat{\mathfrak{o}} \otimes_{\hat{R}} \hat{K},$$

$$\bar{F} = \hat{\bar{\mathfrak{o}}}[\hat{M}^{-1}] = \hat{\bar{\mathfrak{o}}} \otimes_{\hat{R}} \hat{K}. \tag{6}$$

Our second consequence concerns *the equality of the two rings* $\hat{\bar{\mathfrak{o}}}$ *and* $\bar{\hat{\mathfrak{o}}}$ (see Corollary 2.3 below), where $\bar{\hat{\mathfrak{o}}}$ is the integral closure of $\hat{\mathfrak{o}}$ (in $\hat{F}$).

Let $\mathfrak{m}$ and $\bar{\mathfrak{m}}$ be the Jacobson radicals of $\mathfrak{o}$ and $\bar{\mathfrak{o}}$ respectively. Since $\bar{\mathfrak{o}}$ is a finite $\mathfrak{o}$-module, $\bar{\mathfrak{m}}$ is the radical of $\bar{\mathfrak{o}}\mathfrak{m}$. Hence the ($\bar{\mathfrak{m}}$-adic) completion $\hat{\bar{\mathfrak{o}}}$ of the semi-local ring $\bar{\mathfrak{o}}$ coincides with $\mathfrak{m}$-adic completion of the $\mathfrak{o}$-*module* $\bar{\mathfrak{o}}$. Therefore (see C.A. 2, Theorem 16, part (a), p. 277, where we identify A and B with $\mathfrak{o}$ and $\bar{\mathfrak{o}}$ respectively), $\hat{\mathfrak{o}}$ is a subring $\hat{\bar{\mathfrak{o}}}$:

$$\hat{\mathfrak{o}} \subset \hat{\bar{\mathfrak{o}}}. \tag{8}$$

(We observe that in view of (4) and (5), the inclusion (8) is equivalent to the statement that $\hat{R}$ is a flat R-module). It now follows from (6), (7) and (8) that $\hat{F}$ *is a subring of* $\hat{\bar{F}}$. By (7), $\hat{\bar{F}}$ is generated by its two subrings $\hat{\bar{\mathfrak{o}}}$ and $\hat{K}$, and hence, by (5), we have $\bar{F} = \hat{\bar{\mathfrak{o}}}\hat{K}$. Since every regular element of $\bar{\mathfrak{o}}$ is also regular in $\hat{\bar{\mathfrak{o}}}$, it follows that F is canonically a subring of $\hat{\bar{F}}$ and that $\hat{\bar{F}} = F\hat{K}$. We now refer to C.A. 2, Corollary 1 of Theorem 11, p. 265. That corollary states a strong form of "linear disjointness" for any Zariski ring A and for any finite A-module

E (strong in the sense that the statement of that corollary remains valid and is nontrivial even if E has no elements which are free with respect to A). In our present particular case that corollary implies that $\hat{R}$ and $\bar{\mathfrak{o}}$ are linearly disjoint over R, in $\hat{\bar{\mathfrak{o}}}$. Passing to the total quotient rings $\hat{K}$ and F of $\hat{R}$ and $\bar{\mathfrak{o}}$, and taking into account that $F = \mathfrak{o} \cdot K$, we find that $\hat{K}$ and F are linearly disjoint over K, in $\hat{\bar{F}}$. Since we have seen above that $\hat{\bar{F}} = F\hat{K}$, it follows that $\hat{\bar{F}} = F \otimes_K \hat{K}$. In a similar way, upon replacing $\hat{\bar{\mathfrak{o}}}$ by $\hat{\mathfrak{o}}$, we find also that $\hat{F} = F \otimes_K \hat{K}$. Thus

$$\hat{\bar{F}} = \hat{F} = F \underset{K}{\otimes} \hat{K}, \tag{9}$$

a relation which, in view of (8), is equivalent to the inclusion

$$\hat{\bar{\mathfrak{o}}} \subset \hat{F}. \tag{10}$$

LEMMA 2.2. *If a pair of rings* $\{\mathfrak{o}, R\}$, *consisting of a semi-local ring* $\mathfrak{o}$ *and a subring* R *of* $\mathfrak{o}$, *satisfies conditions* (A), (B), (C) *stated in the beginning of this section, and if* $\mathfrak{o}$ *is integrally closed in its total ring of quotients* F, *then the completion* $\hat{\mathfrak{o}}$ *of* $\mathfrak{o}$ *is also integrally closed in its total ring of quotients. Furthermore, if* $\mathfrak{m}_i$ *is any of the maximal ideals of* $\mathfrak{o}$ *then both the local ring* $\mathfrak{o}_{\mathfrak{m}_i}$ *and its completion* $\widehat{\mathfrak{o}_{\mathfrak{m}_i}}$ *are local domains, and* $\hat{\mathfrak{o}}$ *is the direct sum of the local domains* $\widehat{\mathfrak{o}_{\mathfrak{m}_i}}$.

Proof. That $\hat{\mathfrak{o}}$ is the direct sum of the complete local rings $\widehat{\mathfrak{o}_{\mathfrak{m}_i}}$ is well known (see C.A. 2, p. 283, Remark following Corollary 2; see also Nagata L.R., p. 56, (17.7)). The ring F is a direct sum of fields $F\epsilon_i$, and the integral closure of $\mathfrak{o}$ in F, i.e., $\mathfrak{o}$ itself, is the *direct sum* of the integrally closed semi-local *domains* $\mathfrak{o}\epsilon_i$. Hence the localization $\mathfrak{o}_{\mathfrak{m}_i}$ is isomorphic to the localization of one of the semi-local *integrally closed domains* $\mathfrak{o}\epsilon_i$. This shows, in the first place that each $\mathfrak{o}_{\mathfrak{m}_i}$ is itself an integrally closed local domain. To complete the proof of the lemma we now have only to show that each completion $\widehat{\mathfrak{o}_{\mathfrak{m}_i}}$ is an integrally closed integral domain. This was essentially proved in Zariski [12], and is stated and proved with all necessary generality in Nagata, L.R., pp. 136–137 (37.3). In referring to Nagata's cited theorem (37.3) we have only to make the following observations: (1) the normal local ring R of Nagata's theorem is in our case one of the isomorphic regular local domains $R\epsilon_i$ and is therefore analytically normal. (2) The field L of Nagata's theorem is now the field of fractions $F\epsilon_i$ of $\mathfrak{o}\epsilon_i$, and this field is a finite (separable) extension of the field of fractions $K\epsilon_i$ of $R\epsilon_i$. (3) The ring $\mathfrak{o}\epsilon_i$, as a finite $R\epsilon_i$-module, is pseudo-geometric (since $R\epsilon_i$ is pseudo-geometric, by condition (C)), and therefore, by Lemma 2.1, *all* prime ideals $\mathfrak{p}$ of $\mathfrak{o}\epsilon_i$ (and not only the prime ideals of $\mathfrak{o}\epsilon_i$ which have height one, as required by Nagata's theorem) are analytically unramified. The conclusion of Nagata's theorem is that under all these assumptions it is true that for every maximal ideal $\mathfrak{m}_j\epsilon_i$ of $\mathfrak{o}\epsilon_i$ the completion $\widehat{(\mathfrak{o}\epsilon_i)}_{\mathfrak{m}_j\epsilon_i}$ (which is isomorphic to $\hat{\mathfrak{o}}_{\mathfrak{m}_j}$) is an integrally closed local domain. This completes the proof of the lemma.

If in the statement of Lemma 2.2 we drop the assumption that $\mathfrak{o}$ is integrally closed and if $\bar{\mathfrak{o}}$ denotes the integral closure of $\mathfrak{o}$ in F, then we know that the pair of rings $\{\bar{\mathfrak{o}}, R\}$ satisfies conditions (A), (B) and (C). Hence the conclusions of the lemma are applicable in any case to $\bar{\mathfrak{o}}$ and $\hat{\bar{\mathfrak{o}}}$. But we can draw another consequence concerning $\hat{\bar{\mathfrak{o}}}$:

Corollary 2.3. *If the pair of rings* $\{\mathfrak{o}, R\}$ *satisfies conditions* (A), (B) *and* (C), *then*

$$\hat{\bar{\mathfrak{o}}} = \bar{\hat{\mathfrak{o}}}, \tag{11}$$

where $\bar{\hat{\mathfrak{o}}}$ *denotes the integral closure of* $\hat{\mathfrak{o}}$ *in its total ring of quotients* $\hat{F}$.

For, by (8) and (9), $\hat{\mathfrak{o}}$ is a subring of $\hat{\bar{\mathfrak{o}}}$, and both rings have the same total ring of quotients F. By Lemma 2.2, $\hat{\bar{\mathfrak{o}}}$ is integrally closed in $\hat{F}$. Hence $\bar{\hat{\mathfrak{o}}} \subset \hat{\bar{\mathfrak{o}}}$. On the other hand, the relation $\hat{\bar{\mathfrak{o}}} = \bar{\mathfrak{o}} \otimes_R \hat{R}$ [see (5)] implies that $\bar{\mathfrak{o}}$ is generated by its subrings $\bar{\mathfrak{o}}$ and $\hat{\mathfrak{o}}$, and since $\bar{\mathfrak{o}} \subset \bar{\hat{\mathfrak{o}}}$ we have the opposite inclusion $\hat{\bar{\mathfrak{o}}} \subset \bar{\hat{\mathfrak{o}}}$, and (11) follows.

With $\mathfrak{o}, R, \hat{\mathfrak{o}}$ being as above and with K, $\hat{K}$ being the fields of fractions of R and $\hat{R}$ respectively, the basic question which we wish to analyze in this section is that of the relationship between the two rings $(\widehat{\bar{\mathfrak{o}}})_K$ and $(\widetilde{\hat{\mathfrak{o}}})_K$. We shall require some preliminary facts.

We fix an algebraic closure $\hat{\Omega}$ of $\hat{K}$ and we denote by Ω the algebraic closure of K in $\hat{\Omega}$. If ψ is any K-homomorphism of F into $\hat{\Omega}$ then $\psi(F) \subset \Omega$ (since $\psi(F)$ is algebraic over K), and thus ψ is a K-homomorphism of F into Ω. From known properties of tensor products it follows from (9) that ψ has a unique extension to a $\hat{K}$-homomorphism $\hat{\psi}$ of $\hat{F}$ into $\hat{\Omega}$. Conversely, if $\hat{\psi}$ is any $\hat{K}$-homomorphism of $\hat{F}$ into $\hat{\Omega}$, then the restriction ψ of $\hat{\psi}$ to F is a K-homomorphism of F into Ω. Hence if $\{\hat{\psi}_1, \hat{\psi}_2, \ldots, \hat{\psi}_n\}$ is the full (necessarily finite) set $\hat{H}$ of $\hat{K}$-homomorphisms of $\hat{F}$ into $\hat{\Omega}$ and ψ_i denotes the restriction of $\hat{\psi}_i$ to F, then $\{\psi_1, \psi_2, \ldots, \psi_n\}$ is the full set H of K-homomorphisms of F into Ω, and we have $\psi_i \neq \psi_j$ if $i \neq j$.

We set $F_i = \psi_i(F)$, $\hat{F}_i = \hat{\psi}_i(\hat{F})$. The n (not necessarily distinct) fields F_i are finite (separable) extensions of K, and they distribute into a certain number of complete sets of conjugate fields over K. A similar statement holds true for the fields $\hat{F}_1, \hat{F}_2, \ldots, \hat{F}_n$ and $\hat{K}$. Let F^* denote the compositum of the fields F_i and let $\hat{F}^*$ denote the compositum of the fields $\hat{F}_i$. Then F^* and $\hat{F}^*$ are finite Galois extensions of K and $\hat{K}$ respectively. Upon applying ψ_i to both sides of (9) we find that $\hat{F}_i = F_i\hat{K}$, and hence

$$\hat{F}^* = F^*\hat{K}. \tag{12}$$

Let $\mathfrak{o}^*$ and $\hat{\mathfrak{o}}^*$ denote the integral closures of R and $\hat{R}$ in F^* and $\hat{F}^*$ respectively. Since we are in characteristic zero, $\mathfrak{o}^*$ is a finite R-module, $\hat{\mathfrak{o}}^*$ is a finite $\hat{R}$-module, and both $\mathfrak{o}^*$ and $\hat{\mathfrak{o}}^*$ are semi-local domains satisfying conditions (A), (B) and (C) (with $\{\mathfrak{o}, R\}$ replaced respectively by $\{\mathfrak{o}^*, R\}$ and $\{\hat{\mathfrak{o}}^*, \hat{R}\}$). We can easily see that $\hat{\mathfrak{o}}^*$ is a *complete local* domain. For, if $\hat{\mathfrak{M}}$ is the maximal ideal of $\hat{R}$ then the intersection of the maximal ideals of the semi-local ring $\hat{\mathfrak{o}}^*$ is the radical of the ideal $\hat{\mathfrak{o}}^*\hat{\mathfrak{M}}$. Therefore the completion of the semi-local domain $\hat{\mathfrak{o}}^*$ coincides with $\hat{\mathfrak{M}}$-adic completion of the $\hat{R}$-module $\hat{\mathfrak{o}}^*$. Hence the completion of $\hat{\mathfrak{o}}^*$ is $\hat{R}\hat{\mathfrak{o}}^*$ (see C.A. 2, Theorem 5, p. 256), i.e., $\hat{\mathfrak{o}}^*$ itself. Thus $\hat{\mathfrak{o}}^*$ is a *complete* semi-local ring, and since it is an integral domain, it must be a *local* ring (see C.A. 2, Corollary 2 of Theorem 18, p. 283; see also Nagata, L.R., p. 56, (17.7)).

Let $\hat{\mathfrak{m}}^*$ be the maximal ideal of $\hat{\mathfrak{o}}^*$. Since $\hat{\mathfrak{m}}^* \cap \hat{R} = \hat{\mathfrak{M}}$, it follows that $\hat{\mathfrak{m}}^* \cap R = \mathfrak{M}$ (= maximal ideal of R). Hence the ideal $\hat{\mathfrak{m}}^* \cap \mathfrak{o}^*$ lies over $\mathfrak{M}$ and is therefore equal to one of the maximal ideals of the semi-local domain $\mathfrak{o}^*$. Let, say, $\hat{\mathfrak{m}}^* \cap \mathfrak{o}^* = \mathfrak{m}_1^*$. Clearly, the local domain $\mathfrak{o}^*_{\mathfrak{m}_1^*}$ is a subring of $\hat{\mathfrak{o}}^*$.

LEMMA 2.4. *The local domain $\hat{\mathfrak{o}}^*$ is the completion of $\mathfrak{o}^*_{\mathfrak{m}_1^*}$.*

Proof. Let $\mathfrak{m}_2^*, \mathfrak{m}_3^*, \ldots, \mathfrak{m}_q^*$ be the maximal ideals of $\mathfrak{o}^*$ other than $\mathfrak{m}_1^*$. By Lemma 2.2, the completion $\widehat{\mathfrak{o}^*}$ of $\mathfrak{o}^*$ is the direct sum of the q complete *integrally closed* local *domains* $\widehat{\mathfrak{o}^*_{\mathfrak{m}_i^*}}$. Let $\widehat{F^*}$ denote the total ring of quotients of $\widehat{\mathfrak{o}^*}$. Then we have $\widehat{F^*} = \widehat{F_1^*} \oplus \widehat{F_2^*} \oplus \cdots \oplus \widehat{F_q^*}$, where $\widehat{F_i^*}$ is the field of fractions of $\widehat{\mathfrak{o}^*_{\mathfrak{m}_i^*}}$. We also have, by applying (9) to the total rings of quotients of $\mathfrak{o}^*$ and $\widehat{\mathfrak{o}^*}$ (and therefore replacing F and $\hat{F}$ by F^* and $\widehat{F^*}$ respectively):

$$\widehat{F^*} = F^* \underset{K}{\otimes} \hat{K}. \tag{13}$$

Let φ_1 be the projection of $\widehat{F^*}$ onto its direct component $\widehat{F_1^*}$. We have

$\varphi_1(\widehat{\mathfrak{o}^*}) = \mathfrak{o}^*_{\mathfrak{m}_1^*}$, and the restrictions of φ_1 to $\hat{R}$ and $\mathfrak{o}^*$ are isomorphisms. We can therefore identify $\varphi_1(\hat{R})$ and $\varphi_1(\mathfrak{o}^*)$ with $\hat{R}$ and $\mathfrak{o}^*$ respectively. With these identifications, the fields $\hat{K}$ and F^* become subfields of the field $\widehat{F_1^*}$. Since φ_1 is now the identity on $\hat{K}$ and F^*, it follows from (13) that $\widehat{F_1^*}$ $(=\varphi_1(\widehat{F^*}))$ is generated by its two subfields F^* and $\hat{K}$. Furthermore, since $\widehat{F_1^*}$ is algebraic over $\hat{K}$, we may assume that $\widehat{F_1^*}$ is contained in $\hat{\Omega}$. But then it follows from (12) that $\widehat{F_1^*} = \hat{F}^*$. Thus, by our identifications and by the definition of $\hat{\mathfrak{o}}^*$, both rings $\hat{\mathfrak{o}}^*$ and $\widehat{\mathfrak{o}^*_{\mathfrak{m}_1^*}}$ are the integral closures of $\hat{R}$ in F^*. Hence $\hat{\mathfrak{o}}^* = \widehat{\mathfrak{o}^*_{\mathfrak{m}_1^*}}$. Q.E.D

Coming now to the basic question of the relationship between the two rings $\widehat{(\tilde{\mathfrak{o}})_K}$, $\widetilde{(\hat{\mathfrak{o}})_{\hat{K}}}$, the question that we wish to investigate is the following: *are these two rings equal*? (assuming always that the pair $\{\mathfrak{o}, R\}$ consisting of the semi-local ring $\mathfrak{o}$ and of the regular local domain R satisfies conditions (A), (B) and (C)). In other words: *do the two operations of saturation and completion permute*? In this section, we shàll answer this question in the affirmative *only in the case in which* R *(and hence also* $\mathfrak{o}$*) has dimension* $r=1$. This case is all we shall need for the applications to saturation of algebroid hypersurfaces which we shall give in the next section.

Before taking up the case $r=1$ we shall show that the conjectured equality

$$\widehat{(\tilde{\mathfrak{o}})_K} = \widetilde{(\hat{\mathfrak{o}})_{\hat{K}}}, \tag{14}$$

in the general case of r-arbitrary, has two implications. Then we shall prove, independently of (14), the first of these two implications in the general case (see Proposition 2.5 below). Next we shall show that the second implication is, in fact, equivalent to (14) (see Proposition 2.6 below). We shall then conclude this section by proving that the second implication (i.e., the hypothesis made in Proposition 2.6) is in fact valid in the case $r=1$ (see Theorem 2.7 below), and this will establish the validity of (14) in the case $r=1$.

First implication of (14): *if* $\hat{\mathfrak{o}}$ *is saturated with respect to* $\hat{K}$ *then* $\mathfrak{o}$ *is saturated with respect to* K. For, under the assumption that $\hat{\mathfrak{o}}$ is saturated with respect to $\hat{K}$, it follows from (14) that $\widehat{(\tilde{\mathfrak{o}})_K} = \hat{\mathfrak{o}}$. Thus $\tilde{\mathfrak{o}}_K \subset \hat{\mathfrak{o}} \cap F$. Now, it is a well-known fact that $\hat{\mathfrak{o}} \cap F = \mathfrak{o}$.(*) Hence $\tilde{\mathfrak{o}}_K = \mathfrak{o}$, i.e., $\mathfrak{o}$ is saturated with respect to K.

Second implication of (14): *if* $\mathfrak{o}$ *is saturated with respect to* K, *then* $\hat{\mathfrak{o}}$ *is saturated with respect to* $\hat{K}$. This is obvious, for if $\mathfrak{o} = \tilde{\mathfrak{o}}_K$ then (14) yields the relation $\hat{\mathfrak{o}} = \widetilde{(\hat{\mathfrak{o}})}_{\hat{K}}$.

PROPOSITION 2.5. *If* $\hat{\mathfrak{o}}$ *is saturated with respect to* $\hat{K}$ *then* $\mathfrak{o}$ *is saturated with respect to* K.

Proof. We have to show that if ζ is an element of $\mathfrak{o}$ and η is any element of $\bar{\mathfrak{o}}$ which dominates ζ with respect to the pair $\{\mathfrak{o}, K\}$, then $\eta \in \mathfrak{o}$. We are therefore assuming that if ψ_i, ψ_j are any K-homomorphism of F into Ω and if $\psi_i(\zeta) \neq \psi_j(\zeta)$, then we have

$$[\psi_i(\eta) - \psi_j(\eta)]/[\psi_i(\zeta) - \psi_j(\zeta)] \in \mathfrak{o}^*,$$

where, we recall, $\mathfrak{o}^*$ is the integral closure of R in F^*, while if $\psi_i(\zeta) = \psi_j(\zeta)$ then also $\psi_i(\eta) = \psi_j(\eta)$. With our previous notations we have $\hat{\psi}_i | F = \psi_i$, $i = 1, 2, \ldots, n$. Hence η also dominates ζ with respect to $\{\hat{\mathfrak{o}}, \hat{K}\}$ (note also that $\bar{\mathfrak{o}} \subset \bar{\hat{\mathfrak{o}}}$ since $\mathfrak{o} \subset \hat{\mathfrak{o}}$), and therefore $\eta \in \hat{\mathfrak{o}}$ (since $\hat{\mathfrak{o}}$ is saturated with respect to $\hat{K}$). Thus $\eta \in \hat{\mathfrak{o}} \cap F = \mathfrak{o}$ (see above footnote (*)).

PROPOSITION 2.6. *Assume that the following is true*: *whenever* $\mathfrak{o}$ *is saturated with respect to* K, *also* $\hat{\mathfrak{o}}$ *is saturated with respect to* $\hat{K}$. *Then equality* (14) *holds true for any pair of rings* $\{\mathfrak{o}, R\}$ *satisfying conditions* (*A*), (*B*) *and* (*C*).

Proof. Since $\tilde{\mathfrak{o}}_K$ is a finite module over $\mathfrak{o}$, having the same total ring of quotients F as $\mathfrak{o}$, it follows by well-known facts about completion, that $\widehat{(\tilde{\mathfrak{o}})_K} = \hat{\mathfrak{o}} \cdot \tilde{\mathfrak{o}}_K$, and from this it follows that $\hat{F}$ (the total ring of quotients of $\hat{\mathfrak{o}}$) is also the total ring of quotients of $\widehat{(\tilde{\mathfrak{o}})_K}$. Now, since $\bar{\mathfrak{o}}$ is a finite $\tilde{\mathfrak{o}}_K$-module, the completion $\hat{\bar{\mathfrak{o}}}$ of the semilocal ring $\bar{\mathfrak{o}}$ is a finite module over $\widehat{(\tilde{\mathfrak{o}})_K}$ and is therefore integral over $\widehat{(\tilde{\mathfrak{o}})_K}$. By Corollary 2.3 we have $\hat{\bar{\mathfrak{o}}} = \bar{\hat{\mathfrak{o}}}$. It follows

*If η is any element of $\hat{\mathfrak{o}} \cap F$ then there exists a regular element a of $\mathfrak{o}$ such that $a\eta \in \mathfrak{o}$, and since $\eta \in \hat{\mathfrak{o}}$ we have $a\eta \in \mathfrak{o} \cap \hat{\mathfrak{o}}a$. Now, it is well known that $\hat{\mathfrak{o}}a \cap \mathfrak{o} = \mathfrak{o}a$ (see C.A. 2, p. 257, Theorem 5, Corollary 2, as applied to the case $E = A = \mathfrak{o}$, $F = \mathfrak{o}a$). Hence $a\eta \in \mathfrak{o}a$, and since a is a regular element of $\mathfrak{o}$, it follows that $\eta \in \mathfrak{o}$.

therefore that $\hat{\mathfrak{o}}$ and $\widehat{(\tilde{\mathfrak{o}})}_K$ not only have the same total ring of quotients $\hat{F}$ but have also the same integral closure in $\hat{F}$. Now, since $(\tilde{\mathfrak{o}})_K$ is saturated with respect to K, we have, by the asumption made in the theorem, that also $\widehat{(\tilde{\mathfrak{o}})}_K$ is saturated with respect to $\hat{K}$. Hence, by the minimality property of the saturation of $\hat{\mathfrak{o}}$ (and observing that $\hat{\mathfrak{o}} \subset \widehat{(\tilde{\mathfrak{o}})}_K$, since $\mathfrak{o} \subset \tilde{\mathfrak{o}}_K$), it follows that

$$\widetilde{(\hat{\mathfrak{o}})}_{\hat{K}} \subset \widehat{(\tilde{\mathfrak{o}})}_K . \tag{15}$$

We shall now show that the following inclusion holds quite generally, i.e., independently of the assumption made in the theorem:

$$\tilde{\mathfrak{o}}_K \subset \widetilde{(\hat{\mathfrak{o}})}_{\hat{K}} , \tag{16}$$

and this will establish the inclusion opposite to (15) (and will complete the proof of the proposition), since the ring $\tilde{\mathfrak{o}}_K$ is a subspace of the semi-local ring $\bar{\mathfrak{o}}$, which in its turn is a subspace of $\bar{\mathfrak{o}}$, and since, on the other hand, $\widetilde{(\hat{\mathfrak{o}})}_{\hat{K}}$ is a *closed* subspace of $\hat{\bar{\mathfrak{o}}} (= \bar{\hat{\mathfrak{o}}})$.

There exists a finite sequence of rings between $\mathfrak{o}$ and $\tilde{\mathfrak{o}}_K$, say

$$\mathfrak{o} = \mathfrak{o}_0 \subset \mathfrak{o}_1 \subset \mathfrak{o}_2 \subset \cdots \subset \mathfrak{o}_N = \tilde{\mathfrak{o}}_K,$$

such that

$$\mathfrak{o}_\alpha = \mathfrak{o}_{\alpha-1}[L_{\alpha-1}], \qquad \alpha = 1, 2, \ldots, N,$$

where $L_{\alpha-1}$ is the set of all elements η of $\bar{\mathfrak{o}}$ with the property that for some element ζ of $\mathfrak{o}_{\alpha-1}$ (depending on η) it is true that

$$[\psi_i(\eta) - \psi_j(\eta)]/[\psi_i(\zeta) - \psi_j(\zeta)] \in \mathfrak{o}^*,$$

for all i, j such that $\psi_i(\zeta) \neq \psi_j(\zeta)$, while $\psi_i(\eta) = \psi_j(\eta)$ if $\psi_i(\zeta) = \psi_j(\zeta)$ $(i, j = 1, 2, \ldots, n)$. Assume that it has already been shown that $\mathfrak{o}_{\alpha-1} \subset \widetilde{(\hat{\mathfrak{o}})}_{\hat{K}}$ (certainly, we have $\mathfrak{o}_0 = \mathfrak{o} \subset \widetilde{(\hat{\mathfrak{o}})}_{\hat{K}}$). Then $\zeta \in \widetilde{(\hat{\mathfrak{o}})}_{\hat{K}}$, and hence by the above property of the pair of elements $\{\zeta, \eta\}$ (and recalling that the extensions $\hat{\psi}_1, \ldots, \hat{\psi}_n$ of $\psi_1, \ldots, \psi_n$ to $\hat{F}$ give all the $\hat{K}$-homomorphisms of $\hat{F}$ into $\hat{\Omega}$), it follows that η belongs to the saturated ring $\widetilde{(\hat{\mathfrak{o}})}_{\hat{K}}$. This shows that $\mathfrak{o}_\alpha \subset \widetilde{(\hat{\mathfrak{o}})}_K$ and complete the proof of (16) and of the proposition.

We now prove the final result of this section:

THEOREM 2.7. *If R has Krull dimension $r=1$ and if $\mathfrak{o}$ is saturated with respect to K, then also $\hat{\mathfrak{o}}$ is saturated with respect to $\hat{K}$ (and hence, by Proposition 2.6, the equality (14), i.e., the permutability of saturation and completion, holds in the case $r=1$).*

Proof. We have to show the following: if $\hat{\zeta}$, $\hat{\eta}$ are elements of $\hat{\mathfrak{o}}$ and $\bar{\hat{\mathfrak{o}}}$ respectively and if for any two $\hat{K}$-homomorphisms $\hat{\psi}_i$, $\hat{\psi}_j$ of $\hat{F}$ into $\hat{\Omega}$ we have

$$[\hat{\psi}_i(\hat{\eta})-\hat{\psi}_j(\hat{\eta})]/[\hat{\psi}_i(\hat{\zeta})-\hat{\psi}_j(\hat{\zeta})]\in\hat{\mathfrak{o}}^*, \quad \text{if } \hat{\psi}_i(\hat{\zeta})\neq\hat{\psi}_j(\hat{\zeta}), \tag{17}$$

while

$$\hat{\psi}_i(\hat{\eta})=\hat{\psi}_j(\hat{\eta}) \quad \text{whenever } \hat{\psi}_i(\hat{\zeta})=\hat{\psi}_j(\hat{\zeta}), \tag{18}$$

then

$$\hat{\eta}\in\hat{\mathfrak{o}}. \tag{19}$$

The local domains $\hat{\mathfrak{o}}^*$ and $\mathfrak{o}^*_{\mathfrak{m}_1^*}$ are integrally closed, of Krull dimension 1. Hence they are valuation rings of discrete, rank 1 valuations $\hat{v}$ and v of their fields of fractions $\hat{F}^*$ and F^* respectively. Furthermore, since the maximal ideal $\hat{\mathfrak{m}}^*$ of $\hat{\mathfrak{o}}^*$ contracts to the maximal ideal of $\mathfrak{o}^*_{\mathfrak{m}_1^*}$ (see Lemma 2.4), v is the restriction of $\hat{v}$ to F^*.

Let $\mathfrak{m}'$, $\hat{\mathfrak{m}}'$ and $\bar{\hat{\mathfrak{m}}}'$ denote the Jacobson radicals of $\mathfrak{o}$, $\hat{\mathfrak{o}}$ and $\bar{\hat{\mathfrak{o}}}$ respectively. We can then write, for any positive integer N:

$$\hat{\zeta}=\zeta_N+\hat{\zeta}_N, \quad \text{with } \zeta_N \text{ in } \mathfrak{o} \text{ and } \hat{\zeta}_N\in(\hat{\mathfrak{m}}')^N. \tag{20}$$

By Corollary 2.3 we have $\eta\in\bar{\mathfrak{o}}$ and we can also write, for any positive integer N:

$$\hat{\eta}=\eta_N+\hat{\eta}_N, \quad \text{with } \eta_N\in\bar{\mathfrak{o}} \text{ and } \hat{\eta}_N\in(\bar{\hat{\mathfrak{m}}}')^N. \tag{21}$$

For any index $i=1,2,\ldots,n$ we have $\psi_i(\bar{\mathfrak{o}})\subset\mathfrak{o}^*$, $\hat{\psi}_i(\bar{\hat{\mathfrak{o}}})\subset\hat{F}_i\subset\hat{F}^*$, and $\hat{\psi}_i(\bar{\hat{\mathfrak{o}}})(=\hat{\psi}_i(\bar{\hat{\mathfrak{o}}}))$ is integral over $\hat{R}$. Hence

$$\hat{\psi}_i(\hat{\mathfrak{m}}'),\ \hat{\psi}_i(\bar{\hat{\mathfrak{m}}}')\subset\hat{\mathfrak{m}}^*, \qquad i=1,2,\ldots,n. \tag{22}$$

Since $\hat{\psi}_i|F=\psi_i$ and since $\psi_i(\bar{\mathfrak{o}})\subset\mathfrak{o}^*$, it follows from (22) and from the relations $\hat{\mathfrak{m}}^*\cap\mathfrak{o}^*=\mathfrak{m}_1^*$, $\bar{\hat{\mathfrak{m}}}'\cap\mathfrak{o}=\mathfrak{m}'$, that

$$\psi_i(\mathfrak{m}')\subset\mathfrak{m}_1^*, \qquad i=1,2,\ldots,n. \tag{23}$$

After these preliminaries we proceed to prove (19) first *under the following additional assumption*:

$$\hat{\psi}_i(\hat{\zeta}) \neq \hat{\psi}_j(\hat{\zeta}) \quad \text{for all } i \neq j \quad (i, j = 1, 2, \ldots, n). \tag{24}$$

Under this assumption we will have $\hat{v}[\hat{\psi}_i(\hat{\zeta}) - \hat{\psi}_j(\hat{\zeta})] < \hat{v}(\hat{\mathfrak{m}}^{*N})$ for all large N and for all pairs of distinct indices i, j $(i, j = 1, 2, \ldots, n)$. Since $\mathfrak{m}_1^* \subset \hat{\mathfrak{m}}^*$ it follows from (20), (22) and (23) that

$$v[\psi_i(\zeta_N) - \psi_j(\zeta_N)] = \hat{v}\left[\hat{\psi}_i(\hat{\zeta}) - \hat{\psi}_j(\hat{\zeta})\right],$$

$$\text{for all } N \text{ large and all } i \neq j. \tag{25}$$

In a similar way it follows from (21) and (22) that

$$v[\psi_i(\eta_N) - \psi_j(\eta_N)] = \hat{v}\left[\hat{\psi}_i(\hat{\eta}_N) - \hat{\psi}_j(\hat{\eta}_N)\right],$$

$$\text{for all } N \text{ large,} \tag{26}$$

provided $\hat{\psi}_i(\hat{\eta}) \neq \hat{\psi}_j(\hat{\eta})$. From (17), (25) and (26) follows that

$$v[\psi_i(\eta_N) - \psi_j(\eta_N)] \geqslant v[\psi_i(\zeta_N) - \psi_j(\zeta_N)], \quad \text{for all large } N, \tag{27}$$

provided $\hat{\psi}_i(\hat{\eta}) \neq \hat{\psi}_j(\hat{\eta})$. Now, for those pairs of distinct indices i, j for which $\hat{\psi}_i(\hat{\eta}) = \hat{\psi}_j(\hat{\eta})$, we have by (21) and (22): $\psi_i(\eta_N) - \psi_j(\eta_N) = -[\hat{\psi}_i(\hat{\eta}_N) - \psi_j(\hat{\eta}_N)] \in \hat{\mathfrak{m}}^{*N}$, and hence by (25) and by our assumption (24), the inequalities (27) hold also if $\hat{\psi}_i(\hat{\eta}) = \hat{\psi}_j(\hat{\eta})$, provided N is large. Thus the (27) hold true for all large N and for all pairs of distinct indices i, j in the set $\{1, 2, \ldots, n\}$. We have therefore

$$[\psi_i(\eta_N) - \psi_j(\eta_N)] / [\psi_i(\zeta_N) - \psi_j(\zeta_N)] \in \mathfrak{o}^*_{\mathfrak{m}_1^*}, \tag{28}$$

for all large N and all $i \neq j$ $(i, j = 1, 2, \ldots, n)$.

Now, let $\mathfrak{m}^*$ be *any* maximal ideal of $\mathfrak{o}^*$. Since F^* is a Galois extension of K and since $\mathfrak{o}^*$ is the integral closure of R in F^*, any two maximal ideals of $\mathfrak{o}^*$ are conjugate over K (see C.A. 2, p. 28, Corollary 3 of Theorem 12). Let then τ be a K-automorphism of F^* such that $\tau(\mathfrak{m}_1^*) = \mathfrak{m}^*$. Applying τ to both sides of (28) we obtain a relation similar to (28), in which ψ_i and ψ_j are replaced by $\tau\psi_i$ and $\tau\psi_j$, while $\mathfrak{o}^*_{\mathfrak{m}_1^*}$ is replaced by $\mathfrak{o}^*_{\mathfrak{m}^*}$. But as ψ_i ranges over the set $H = \{\psi_1, \psi_2, \ldots, \psi_n\}$ of all K-homomorphisms of F into Ω, the product $\tau\psi_i$ ranges over this same set H. It follows that the quotients on the left-hand side of (28) belong to $\mathfrak{o}^*_{\mathfrak{m}^*}$ for *every maximal ideal* $\mathfrak{m}^*$ *of* $\mathfrak{o}^*$. Therefore all these quotients belong to $\mathfrak{o}^*$

(for large N and all $i \neq j$). In other words, η_N *dominates* ζ_N (for N large) *with respect to the pair* $\{\mathfrak{o}, K\}$. Since $\zeta_N \in \mathfrak{o}$ and since $\mathfrak{o}$ is saturated with respect to K, *it follows that* $\eta_N \in \mathfrak{o}$, *for N large*. Since we have by (21): $\eta = \lim \hat{\eta}_N$ in $\hat{\bar{\mathfrak{o}}}$, we conclude that $\hat{\eta}$ belongs to the closure of $\mathfrak{o}$ in $\hat{\bar{\mathfrak{o}}}$. Since $\bar{\mathfrak{o}}$ is a finite $\mathfrak{o}$-module, the closure of $\mathfrak{o}$ in $\hat{\bar{\mathfrak{o}}}$ is $\hat{\mathfrak{o}}$ (see C.A. 2, Theorem 15 on p. 276 and Theorem 16, part (a), on p. 277). This establishes (19) *under the assumption* (24).

We now drop the assumption (24). We begin by proving first the following: *there exist elements* $\hat{\xi}$ *in* $\hat{\mathfrak{o}}$ *such that* $\hat{\psi}_i(\hat{\xi}) \neq \hat{\psi}_j(\hat{\xi})$ *for all* $i \neq j$ $(i,j = 1, 2, \ldots, n)$. To see this, we consider the $\hat{K}$-vector space $\hat{K}\hat{\mathfrak{o}}$ spanned by $\hat{\mathfrak{o}}$ over $\hat{K}$. This is a finite-dimensional vector space (since $\hat{\mathfrak{o}}$ is a finite $\hat{R}$-module). The $\hat{K}$-homomorphisms $\hat{\psi}_1, \hat{\psi}_2, \ldots, \hat{\psi}_n$ induce *distinct* $\hat{K}$-linear maps of $\hat{K}\hat{\mathfrak{o}}$ into $\hat{\Omega}$. For any given pair of distinct indices i, j we denote by V_{ij} the subspace of $\hat{K}\hat{\mathfrak{o}}$ consisting of those elements $\hat{u}$ for which $\hat{\psi}_i(\hat{u}) = \hat{\psi}_j(\bar{u})$. Then V_{ij} is a *proper* subspace of $\hat{K}\hat{\mathfrak{o}}$, and since $\hat{K}$ is an infinite field, the union of the subspaces V_{ij} is a proper subset of $\hat{K}\hat{\mathfrak{o}}$. Let $\hat{y}$ be an element of the complement of this proper subset. We can write $\hat{y}$ in the form $\hat{\xi}/\hat{a}$, where $\hat{a} \in \hat{R}$ and $\hat{\xi} \in \hat{\mathfrak{o}}$ [see (6)], and then $\hat{\xi}$ is an element of $\hat{\mathfrak{o}}$ having the desired property.

After multiplying $\hat{\xi}$ by an element of $\hat{R}$ which belongs to a very high power of the maximal ideal of $\hat{R}$ we may assume that $\hat{\xi}$ belongs to a high power of the Jacobson radical $\hat{\mathfrak{m}}'$ of $\hat{\mathfrak{o}}$. We can therefore assume that $\hat{\xi}$ satisfies also the following condition:

$$\hat{v}\left(\hat{\psi}_\nu(\hat{\xi})\right) > \hat{v}\left[\hat{\psi}_i(\hat{\zeta}) - \hat{\psi}_j(\hat{\zeta})\right], \tag{29}$$

for $\nu = 1, 2, \ldots, n$ and for all $i, j \in \{1, 2, \ldots, n\}$ such that $\hat{\psi}_i(\hat{\zeta}) \neq \hat{\psi}_j(\hat{\zeta})$. We now set

$$\hat{z} = \hat{\zeta} + \hat{\xi}. \tag{30}$$

If for a given pair of *distinct* indices i, j we have $\hat{\psi}_i(\hat{\zeta}) \neq \hat{\psi}_j(\hat{\zeta})$ then it follows from (29) that $\hat{v}[\hat{\psi}_i(\hat{z}) - \hat{\psi}_j(\hat{z})] = \hat{v}[\hat{\psi}_i(\hat{\zeta}) - \hat{\psi}_j(\hat{\zeta})]$, and hence we have

$$\hat{\psi}_i(\hat{z}) \neq \hat{\psi}_j(\hat{z}). \tag{31}$$

If, however, $\hat{\psi}_i(\hat{\zeta}) = \hat{\psi}_j(\hat{\zeta})$, then $\psi_i(\hat{z}) - \hat{\psi}_j(\hat{z}) = \hat{\psi}_i(\hat{\xi}) - \hat{\psi}_j(\hat{\xi})$, and hence, by our choice of $\hat{\xi}$, the inequality (31) continues to hold also in this case. *Thus, the (31) hold for all pairs of distinct indices* i, j; in other words, the element $\hat{z}$ satisfies the condition analogous to (24). We have just pointed out that we have $\hat{v}[\hat{\psi}_i(\hat{z}) - \hat{\psi}_j(\hat{z})] = \hat{v}[\hat{\psi}_i(\hat{\zeta}) - \hat{\psi}_j(\hat{\zeta})]$ whenever $\hat{\psi}_i(\hat{\zeta}) \neq \hat{\psi}_j(\hat{\zeta})$. Therefore it follows from (17) that

$$\hat{v}\left[\hat{\psi}_i(\hat{\eta})-\hat{\psi}_j(\hat{\eta})\right] \geqslant \hat{v}\left[\hat{\psi}_i(\hat{z})-\hat{\psi}_j(\hat{z})\right], \tag{32}$$

for all i, j such that $\hat{\psi}_i(\hat{\zeta}) \neq \hat{\psi}_j(\hat{\zeta})$. If, however, we have $\hat{\psi}_i(\hat{\zeta}) = \hat{\psi}_j(\hat{\zeta})$, then, by the assumption made in the theorem, we have also $\hat{\psi}_i(\hat{\eta}) = \hat{\psi}_j(\hat{\eta})$. Thus, the inequalities (32) hold true for all pairs of indices i, j. In other words, we have now

$$\left[\hat{\psi}_i(\hat{\eta})-\hat{\psi}_j(\hat{\eta})\right]/\left[\hat{\psi}_i(\hat{z})-\hat{\psi}_j(\hat{z})\right] \in \hat{\mathfrak{o}}^*,$$

for all pairs of distinct indices i, j. Since the element $\hat{z}$ of $\hat{\mathfrak{o}}$ satisfies (31) for all $i \neq j$, we conclude, by the first part of the proof, that $\hat{\eta} \in \hat{\mathfrak{o}}$. This completes the proof of the theorem.

Note. The question of permutability of saturation and completion has a bearing upon another question which we shall now discuss.

In SES III we gave a general definition of saturation of a ring A with respect to a subfield K of the total ring of quotients F of A, under the assumption that certain 5 conditions are satisfied (see SE III, Introduction, p. 962, conditions a.-e.). An important case in which these conditions are satisfied is the one in which A is a complete, equidimensional, equicharacteristic zero local ring $\mathfrak{o}$, of Krull dimension r, free from nilpotent elements different from 0, and $K = k((x_1, x_2, \ldots, x_r))$ is the field of fractions of the power series $k[[x_1, x_2, \ldots, x_r]$, where $x_1, x_2, \ldots, x_r$ are parameters of $\mathfrak{o}$, and k is a coefficient field of $\mathfrak{o}$. Suppose now that we are dealing with a local ring $\mathfrak{o}$ which is *not complete* but which otherwise has all the properties just stated above, i.e., $\mathfrak{o}$ is equidimensional, of Krull dimension r, is equicharacteristic zero and has no nilpotent elements different from zero. We shall not assume that $\mathfrak{o}$ has a coefficient field, but we *do* assume that $\mathfrak{o}$ contains a subfield k_0 such that *the residue field* $\mathfrak{o}/\mathfrak{m}$ *of* $\mathfrak{o}$ *is a finite algebraic extension of* k_0. We fix a set of parameters $x_1, x_2, \ldots, x_r$ of $\mathfrak{o}$. To define, also in this case, the *k_0-saturation of* $\mathfrak{o}$ *with respect to the parameters* $x_1, x_2, \ldots, x_r$, to be denoted by $\mathfrak{o}^{k_0}_{\{x_1, x_2, \ldots, x_r\}}$, we could proceed as follows:

The completion $\hat{\mathfrak{o}}$ of $\mathfrak{o}$ is a complete, equidimensional, equicharacteristic zero, local ring of Krull dimension r, and $x_1, x_2, \ldots, x_r$ are parameters of $\hat{\mathfrak{o}}$. Since we are in characteristic zero, the field k_0 is contained in a (unique) coefficient field k of $\mathfrak{o}$ (k = algebraic closure of k_0 in $\mathfrak{o}$). The ring $\hat{\mathfrak{o}}$ is a finite module over its subring $k[[x_1, x_2, \ldots, x_r]]$, and hence is also a finite module over the subring $\hat{R}_0 = k_0[[x_1, x_2, \ldots, x_r]]$ (since k is finite algebraic over k_0). In view of the equidimensionality of $\mathfrak{o}$, also $\hat{\mathfrak{o}}$ is equidimensional, and therefore all nonzero elements of R are regular in $\hat{\mathfrak{o}}$. Hence the field of fractions $\hat{K}_0$ of $\hat{R}_0$ is canonically a subfield of the total ring of quotients $\hat{F}$ of $\hat{\mathfrak{o}}$. The pair $(\hat{\mathfrak{o}}, K_0)$ satisfies conditions a.-e. of SES, Introduction (p. 962), and thus the saturation $\widetilde{(\hat{\mathfrak{o}})}_{\hat{K}_0}$ is defined. Since every regular element of $\mathfrak{o}$ is also regular in $\hat{\mathfrak{o}}$, F is a subring of $\hat{F}$. We define the k_0-saturation of $\mathfrak{o}$ with respect to the parameters x_1, $x_2, \ldots, x_r$ as follows:

$$\tilde{\mathfrak{o}}^{(k_0)}_{\{x_1, x_2, \ldots, x_r\}} = \widetilde{(\hat{\mathfrak{o}})}_{\hat{K}_0} \cap F. \tag{33}$$

This definition can be used, for instance, for the local ring of any point P of an (unmixed) algebraic variety V, taking for k_0 any field containing a field of definition of V and such that P is algebraic over k_0.

Suppose, however, that the above (non-complete) local ring $\mathfrak{o}$ happens also to contain as subring a local domain R_0 such that: (1) $\mathfrak{o}$ is a finite module over R_0; (2) k_0 is a coefficient field of R_0; and (3) the parameters x_1, $x_2, \ldots, x_r$ belong to R_0 and generate there the maximal ideal of R_0. [Such a situation arises, for instance, if we have a k_0-rational point P of an (unmixed) affine r-dimensional algebraic variety V, defined over k_0, immersed in an n-dimensional affine space, with $k_0[x_1, x_2, \ldots, x_N]$ as affine coordinate ring of V. If P is the origin and if we assume that the linear space line $X_1 = X_2 = \cdots = X_r = 0$ meets V only at P, then the local ring $\mathfrak{o}$ of P on V will contain a subring R_0 satisfying the above conditions (1), (2) and (3), provided we take for R_0 the localization of the (polynomial) ring $k_0[x_1, x_2, \ldots, x_r]$ at the origin]. Under these conditions, the saturation $\tilde{\mathfrak{o}}_{K_0}$ is defined (in the sense of the general definition of saturation given in our paper SES III), where K_0 is the field of fractions of R_0. One would wish then to make sure that the above definition of the saturation of $\mathfrak{o}$ (via the completion $\hat{\mathfrak{o}}$) and the present direct definition of the saturation of $\mathfrak{o}$ yield the same ring, i.e., that we have, in the present case:

$$\widetilde{(\hat{\mathfrak{o}})}_{\hat{K}_0} \cap F = \tilde{\mathfrak{o}}_{K_0}. \tag{34}$$

Now, we certainly have $\widehat{\tilde{\mathfrak{o}}_{K_0}} \cap F = \tilde{\mathfrak{o}}_{K_0}$. Therefore, (34) would follow if we knew that the rings $\widetilde{(\hat{\mathfrak{o}})}_{\hat{K}_0}$ and $\widehat{\tilde{\mathfrak{o}}_{K_0}}$ coincide, i.e., if we knew that saturation and completion commute. We do know that, by Theorem 2.7, (34) holds true in the case $r = 1$.

3. Saturation of Hypersurfaces. In this section we maintain the assumption that the pair $\{\mathfrak{o}, R\}$, consisting of a semi-local ring $\mathfrak{o}$ and of a subring R of $\mathfrak{o}$, satisfies the conditions (A), (B) and (C) stated in the beginning of Section 2. If $\mathfrak{p}$ is any prime ideal in $\mathfrak{o}$ we denote by $h_\mathfrak{p}$ the canonical homomorphism of $\mathfrak{o}$ into $\mathfrak{o}_\mathfrak{p}$ and by $\bar{h}_\mathfrak{p}$ the unique extension of $h_\mathfrak{p}$ to a homomorphism of F into the total ring of quotients $F_\mathfrak{p}$ of $\mathfrak{o}_\mathfrak{p}$:

$$h_\mathfrak{p} : \mathfrak{o} \rightarrow \mathfrak{o}_\mathfrak{p}, \tag{1}$$

$$\bar{h}_\mathfrak{p} : F \rightarrow F_\mathfrak{p}. \tag{2}$$

If we have a set $\{\mathfrak{p}_\alpha\}$ of prime ideals in $\mathfrak{o}$, we shall write h_α and $\bar{h}_\alpha$ instead of $h_{\mathfrak{p}_\alpha}$ and $\bar{h}_{\mathfrak{p}_\alpha}$; this is the same notation that we have used in (7) and (8) of Section 1, the ring A which occurs in (7), Section 1, now being replaced by the semi-local ring $\mathfrak{o}$.

The local ring $\mathfrak{o}_\mathfrak{p}$ is canonically a subring of its completion $\hat{\mathfrak{o}}_\mathfrak{p}$, and since every regular element of $\mathfrak{o}_\mathfrak{p}$ is also regular in $\hat{\mathfrak{o}}_\mathfrak{p}$, $F_\mathfrak{p}$ is canonically a subring of the total ring of quotients $\hat{F}_\mathfrak{p}$ of $\hat{\mathfrak{o}}_\mathfrak{p}$. Thus $h_\mathfrak{p}$ and $\bar{h}_\mathfrak{p}$ can be regarded as homomorphisms of $\mathfrak{o}$ and F into $\hat{\mathfrak{o}}_\mathfrak{p}$ and $\hat{F}_\mathfrak{p}$ respectively:

$$h_\mathfrak{p} : \mathfrak{o} \rightarrow \hat{\mathfrak{o}}_\mathfrak{p}, \tag{3}$$

$$\bar{h}_\mathfrak{p} : F \rightarrow \hat{F}_\mathfrak{p}. \tag{4}$$

Let $\mathfrak{P}$ be a given prime ideal in R and let $S(\mathfrak{P})$ be the (finite) set of prime ideals in $\mathfrak{o}$ which contract to $\mathfrak{P}$ in R. We set

$$\mathfrak{o}(\mathfrak{P}) = \mathfrak{o}\left[M(\mathfrak{P})^{-1}\right],$$

where

$$M(\mathfrak{P}) = R - \mathfrak{P}.$$

Since every element of $M(\mathfrak{P})$ is regular in $\mathfrak{o}$, $\mathfrak{o}$ is a subring of $\mathfrak{o}(\mathfrak{P})$, while $\mathfrak{o}(\mathfrak{P})$ is a semi-local ring contained in F. The maximal ideals $\mathfrak{m}_\alpha$ of $\mathfrak{o}(\mathfrak{P})$ are in (1, 1) correspondence with the prime ideal $\mathfrak{p}_\alpha$ belonging to the set $S(\mathfrak{P})$, the correspondence being the one of contraction and extension: $\mathfrak{m}_\alpha \cap \mathfrak{o} = \mathfrak{p}_\alpha$, $\mathfrak{o}(\mathfrak{P}) \cdot \mathfrak{p}_\alpha = \mathfrak{m}_\alpha$. Furthermore, the localizations $\mathfrak{o}(\mathfrak{P})_{\mathfrak{m}_\alpha}$ and $\mathfrak{o}_{\mathfrak{p}_\alpha}$ are canonically isomorphic, in the sense that if h_α is the canonican homomorphism $\mathfrak{o} \rightarrow \mathfrak{o}_{\mathfrak{p}_\alpha}$ and h'_α is the canonical homomorphism $\mathfrak{o}(\mathfrak{P}) \rightarrow \mathfrak{o}(\mathfrak{P})_{\mathfrak{m}_\alpha}$, there exists a unique isomorphism $\theta_\alpha : \mathfrak{o}(\mathfrak{P})_{\mathfrak{m}_\alpha} \xrightarrow{\sim} \mathfrak{o}_{\mathfrak{p}_\alpha}$ such that h_α is the restriction of $\theta_\alpha \cdot h'_\alpha$ to $\mathfrak{o}$. We shall therefore identify $\mathfrak{o}(\mathfrak{P})_{\mathfrak{m}_\alpha}$ with $\mathfrak{o}_{\mathfrak{p}_\alpha}$ for $\mathfrak{p}_\alpha \in S(\mathfrak{P})$ and $\mathfrak{m}_\alpha = \mathfrak{o}(\mathfrak{P}) \cdot \mathfrak{p}_\alpha$.

The completion of the semi-local ring $\mathfrak{o}(\mathfrak{P})$ can then be identified canonically with the direct sum of the completions of the local rings $\mathfrak{o}_{\mathfrak{p}_\alpha}$:

$$\widehat{\mathfrak{o}(\mathfrak{P})} = \bigoplus_{\mathfrak{p} \in S(\mathfrak{P})} \hat{\mathfrak{o}}_{\mathfrak{p}}, \tag{5}$$

and we have $1 = \Sigma_{\mathfrak{p} \in S(\mathfrak{P})} \epsilon_{\mathfrak{p}}$, where $\epsilon_{\mathfrak{p}}$ is the identity of $\mathfrak{o}_{\mathfrak{p}}$.

If $\hat{F}_{\mathfrak{P}}$ denotes the total ring of quotients of $\widehat{\mathfrak{o}(\mathfrak{P})}$, and $\hat{F}_{\mathfrak{p}}$ is the total ring of quotients of $\hat{\mathfrak{o}}_{\mathfrak{p}}$, then we have by (5):

$$\hat{F}_{\mathfrak{P}} = \bigoplus_{\mathfrak{p} \in S(\mathfrak{P})} \hat{F}_{\mathfrak{p}}. \tag{5'}$$

Let $\hat{K}$ be the field of fractions of the completion $\hat{R}$ of R and let $\hat{K}_{\mathfrak{P}}$ be the field of fractions of the completion $\hat{R}_{\mathfrak{P}}$ of the (regular) local domain $R_{\mathfrak{P}}$. By section 1, the saturation of $\widehat{\mathfrak{o}(\mathfrak{P})}$ with respect to $\hat{K}_{\mathfrak{P}}$ is well defined. Similarly, the saturation of $\hat{\mathfrak{o}}_{\mathfrak{p}}$ with respect to the field $\hat{K}_{\mathfrak{P}} \cdot \epsilon_{\mathfrak{p}}$ is well defined. If $\hat{g}_{\mathfrak{p}}$ denotes the natural projection of $\hat{F}_{\mathfrak{P}}$ onto $F_{\mathfrak{p}}$ ($\mathfrak{p} \in S(\mathfrak{P})$; see (5')), then it follows from Proposition 1.9 that there is an isomorphism

$$\widetilde{\left(\widehat{\mathfrak{o}(\mathfrak{P})}\right)}_{\hat{K}_{\mathfrak{P}}} \underset{\oplus \hat{g}_{\mathfrak{p}}}{\overset{\approx}{\longrightarrow}} \bigoplus_{\mathfrak{p} \in S(\mathfrak{P})} \widetilde{(\hat{\mathfrak{o}}_{\mathfrak{p}})}_{\hat{K}_{\mathfrak{P}} \cdot \epsilon_{\mathfrak{p}}}. \tag{6}$$

Note that F is canonically identifiable with a subring of $\hat{F}_{\mathfrak{P}}$ (since every regular element of $\mathfrak{o}$ is also regular in $\widehat{\mathfrak{o}(\mathfrak{P})}$). Similarly, $F_{\mathfrak{p}}$ is canonically identifiable with a subring of $\hat{F}_{\mathfrak{p}}$. It is then easily seen that the restriction of $\hat{g}_{\mathfrak{p}}$ to F coincides with $\bar{h}_{\mathfrak{p}}$ (see (2)). We have the inclusion $\tilde{\mathfrak{o}}_K \subset \widetilde{\mathfrak{o}(\mathfrak{P})}_K \subset \widetilde{(\widehat{\mathfrak{o}(\mathfrak{P})})}_{\hat{K}_{\mathfrak{P}}}$ (for the second of these inclusions see (16), Section 2, where $\mathfrak{o}$ should be replaced by $\mathfrak{o}(\mathfrak{P})$). *If we now assume that* $\mathfrak{P}$ *has height* 1 (and that consequently $\mathfrak{o}(\mathfrak{P})$ has Krull dimension 1) we can use the permutability of saturation and completion (see Theorem 2.7) and, consequently, we can rewrite (6) in that case as follows:

$$\widehat{\widetilde{\mathfrak{o}(\mathfrak{P})}_K} \underset{\oplus \hat{g}_{\mathfrak{p}}}{\overset{\approx}{\longrightarrow}} \bigoplus_{\mathfrak{p} \in S(\mathfrak{P})} \widetilde{(\hat{\mathfrak{o}}_{\mathfrak{p}})}_{\hat{K}_{\mathfrak{P}} \cdot \epsilon_{\mathfrak{p}}}, \quad \text{if } h(\mathfrak{P}) = 1. \tag{7}$$

We have therefore, in particular, by restricting $\hat{g}_{\mathfrak{p}}$ to $\tilde{\mathfrak{o}}_K$, the following homomorphisms:

$$\hat{g}_{\mathfrak{p}} | \tilde{\mathfrak{o}}_K = \bar{h}_{\mathfrak{p}} | \tilde{\mathfrak{o}}_K : \tilde{\mathfrak{o}}_K \rightarrow \widetilde{(\hat{\mathfrak{o}}_{\mathfrak{p}})}_{\hat{K}_{\mathfrak{p}} \cdot \epsilon_{\mathfrak{p}}}, \qquad \mathfrak{p} \in S(\mathfrak{P}), h(\mathfrak{P}) = 1. \tag{8}$$

We shall write (8) symbolically as follows (compare with (9), Section 1):

$$\tilde{\mathfrak{o}}_K \subset \bigcap_{\mathfrak{p}|h(\mathfrak{p})=1} \widetilde{\left(F \cap (\hat{\mathfrak{o}}_{\mathfrak{p}})_{\hat{K}_{\mathfrak{P}}\cdot\epsilon_{\mathfrak{p}}}\right)} \qquad (\mathfrak{P} = \mathfrak{p} \cap R). \tag{9}$$

As is clearly indicated in (9), the intersection symbol $\cap$ is taken over the set of prime ideal $\mathfrak{p}$ in $\mathfrak{o}$ which have height 1, and for every such $\mathfrak{p}$ the corresponding prime ideal $\mathfrak{p}$ in R is the contraction $\mathfrak{p} \cap R$.

We shall now prove the following theorem:

THEOREM 3.1. *If $\mathfrak{o}$ is a simple ring extension of R, then in* (9) *the equality sign holds true*:

$$\tilde{\mathfrak{o}}_K = \bigcap_{\mathfrak{p}|h(\mathfrak{p})=1} \left(F \cap \widetilde{(\hat{\mathfrak{o}}_{\mathfrak{p}})_{\hat{K}_{\mathfrak{P}}\cdot\epsilon_{\mathfrak{p}}}}\right) \qquad (\mathfrak{P} = \mathfrak{p} \cap R). \tag{10}$$

In other words, an element ξ of F belongs to $\tilde{\mathfrak{o}}_K$ if and only if

$$\bar{h}_{\mathfrak{p}}(\xi)(=\hat{g}_{\mathfrak{p}}(\xi)) \in \widetilde{(\hat{\mathfrak{o}}_{\mathfrak{p}})_{\hat{K}_{\mathfrak{P}}\cdot\epsilon_{\mathfrak{p}}}}$$

for every prime ideal $\mathfrak{p}$ of $\mathfrak{o}$ which has height 1. (*Here, for any given $\mathfrak{p}$, the prime ideal $\mathfrak{P}$ of R is the contraction $\mathfrak{p} \cap R$*).

Proof. Since R is regular local domain, it satisfies condition S (see Section 1). Therefore, by Theorem 1.8, as applied to the ring $A = \mathfrak{o}$, we have

$$\tilde{\mathfrak{o}}_K = \bigcap_{\mathfrak{P} \in \Sigma_0} \widetilde{\mathfrak{o}(\mathfrak{P})_K}, \tag{11}$$

where Σ_0 is the set of all prime ideals $\mathfrak{P}$ of R which have height 1. The total ring of quotients of $\mathfrak{o}(\mathfrak{P})$ (and hence also of $\widetilde{\mathfrak{o}(\mathfrak{P})_K}$) is the same as the total ring of quotients F of $\mathfrak{o}$ (since every non-zero element of R is regular in $\mathfrak{o}$). By a well-known result on completions, already used by us in Section 2 [see Section 2, first implication of (14) and the related footnote *], we have

$$\widehat{\widetilde{\mathfrak{o}(\mathfrak{P})_K}} \cap F = \widetilde{\mathfrak{o}(\mathfrak{P})_K}.$$

We can therefore rewrite (11) as follows:

$$\tilde{\mathfrak{o}}_K = \bigcap_{\mathfrak{P} \in \Sigma_0} \widetilde{\left(\mathfrak{o}(\mathfrak{P})_K \cap F\right)}.$$

Since $\mathfrak{o}(\mathfrak{P})$ has Krull dimension 1, Theorem 2.7 is applicable and we can therefore write

$$\tilde{\mathfrak{o}}_K = \bigcap_{\mathfrak{p}\in\Sigma_0} \left(\widetilde{\widehat{\mathfrak{o}(\mathfrak{P})}_{\hat{K}_\mathfrak{p}} \cap F}\right).$$

In other words, $\tilde{\mathfrak{o}}_K$ is the set of all elements ξ of F such that ξ belongs to each of the rings $\widetilde{(\widehat{\mathfrak{o}(\mathfrak{P})})}_{\hat{K}_\mathfrak{P}}$, $\mathfrak{P}\in\Sigma_0$. If we now use the representation (5) of $\widehat{\mathfrak{o}(\mathfrak{P})}$ as a direct sum, we conclude from (6) and from the fact that $\hat{g}_\mathfrak{p}|\tilde{\mathfrak{o}}_K = \bar{h}_\mathfrak{p}|\tilde{\mathfrak{o}}_K$ [see (8)], that $\tilde{\mathfrak{o}}_K$ is the set of all elements ξ of F such that

$$\bar{h}_\mathfrak{p}(\xi) \in \widetilde{(\hat{\mathfrak{o}}_\mathfrak{p})}_{\hat{K}_\mathfrak{P}\cdot\epsilon_\mathfrak{p}} \qquad (\mathfrak{P}=\mathfrak{p}\cap R)$$

for all prime ideals $\mathfrak{p}$ in $\mathfrak{o}$ which have height 1. This completes the proof of the theorem.

We shall now apply Theorem 3.1 to the case in which $\mathfrak{o}$ is the local ring of the closed point Q of an algebroid hypersurface H of dimension r, defined over a ground field k of characteristic zero. In the case, k is, by definition, a coefficient field of $\mathfrak{o}$, and any minimal basis of the maximal ideal $\mathfrak{m}$ of $\mathfrak{o}$ consists of $r+1$ elements (we exclude the case in which Q is a simple point of H, i.e., the case in which $\mathfrak{o}$ is a regular ring). Any minimal basis $\{x_1, x_2,\ldots,x_r, x_{r+1}\}$ of $\mathfrak{m}$ determines an embedding of H in the affine $(r+1)$-space (over k), in which H will then be defined by a single equation

$$H: f(X_1, X_2,\ldots,X_r,X_{r+1})=0, \tag{12}$$

where f is a formal power series with coefficients in k, free from multiple factors and zero at the origin Q. Without affecting the embedding (12) we may operate on the basis elements $x_1, x_2,\ldots,x_r, x_{r+1}$ by an arbitrary non-singular linear homogeneous transformation, with coefficients in k. Since k is an infinite field, we may therefore assume that $f(0, 0,\ldots,0, 1)\neq 0$. That is equivalent to assuming that the first r elements $x_1, x_2,\ldots,x_r$ of the minimal basis are parameters of $\mathfrak{o}$, and when that is so then, by the Weierstrass preparation theorem, we may also assume that f is a monic *polynomial* in X_{r+1}, say of degree n, and that $f(0, 0,\ldots,0, X_{r+1}) = X_{r+1}^n$.

The r parameters $x_1, x_2,\ldots,x_r$ are analytically independent over k, and the formal power series ring $R=k[[x_1, x_2,\ldots,x_r]]$ is a subring of $\mathfrak{o}$. The pair $\{\mathfrak{o}, R\}$ satisfies conditions (A), (B) and (C) stated in the beginning of Section 2, and furthermore, $\mathfrak{o}$ is a simple ring extension of R, namely, $\mathfrak{o}=R[x_{r+1}]$. Thus

Theorem 3.1 is applicable to the hypersurface H and to the parameters $x_1, x_2, \ldots, x_r$.

We now go back, for a moment, to the more abstract situation with which we were dealing in Theorem 3.1. If $\mathfrak{p}$ is any prime ideal in $\mathfrak{o}$, of height 1, then the contraction $\mathfrak{P} = \mathfrak{p} \cap R$ is a principal ideal in R. If ξ is any generator of the principal ideal $\mathfrak{P}$ then ξ is clearly a parameter of the one-dimensional local ring $\mathfrak{o}_\mathfrak{p}$. [Note that the restriction to R of the canonical homomorphism $h_\mathfrak{p}: \mathfrak{o} \rightarrow \mathfrak{o}_\mathfrak{p}$ is an injection and that therefore we can identify (canonically) R with a subring of $\mathfrak{o}_\mathfrak{p}$.]

Definition 3.2. *The ring R is said to be transversal with respect to* $\mathfrak{p}$ *if* ξ *is a transversal parameter of* $\mathfrak{o}_\mathfrak{p}$. *In the special case in which* $\mathfrak{o}$ *is the* (*complete*) *local ring of an algebroid variety V and k is a coefficient field of* $\mathfrak{o}$, *we say that a set of parameters* $x_1, x_2, \ldots, x_r$ *is k-transversal with respect to* $\mathfrak{p}$ *if the formal power series ring* $k[[x_1, x_2, \ldots, x_r]]$ *is transversal with respect to* $\mathfrak{p}$.

[In Appendix B we shall extend the definition of k-transversality of a set of parameters $x_1, x_2, \ldots, x_r$, with respect to $\mathfrak{p}$, to prime ideals $\mathfrak{p}$ in $\mathfrak{o}$ of arbitrary height. We shall show in Appendix B (see Theorem B11) that if k is an infinite field, then the condition of k-transversality of a set of parameters $x_1, x_2, \ldots, x_r$ with respect to a given prime ideal $\mathfrak{p}$ in $\mathfrak{o}$ is—what we call—an "open condition" (see Appendix B, Definition B7), and from this it will follow, in particular, that if $\mathfrak{p}$ is given and if $x_1, x_2, \ldots, x_r$ are "generic" parameters, then $x_1, x_2, \ldots, x_r$ are k-transversal with respect to $\mathfrak{p}$].

Since $\mathfrak{o}$ is a pseudo-geometric ring (in view of conditions (A), (B) and (C) of Section 2), the "singular locus" of **Spec** ($\mathfrak{o}$) is a closed subset of **Spec** ($\mathfrak{o}$), just as it is so in the algebro-geometric case and also in the formal analytic case of an algebroid variety (see Nagata [5]). Hence there is at most a finite number of prime ideals $\mathfrak{p}$ in $\mathfrak{o}$, *of height* 1, such that $\mathfrak{o}_\mathfrak{p}$ is not a regular ring. Let $\mathfrak{p}_1, \mathfrak{p}_2, \ldots, \mathfrak{p}_q$ be these (singular) prime ideals of $\mathfrak{o}$, of height 1. Since $\mathfrak{o}$ satisfies condition S (see Theorem 1.8) we have $q \geqslant 1$, unless $\mathfrak{o}$ is integrally closed, in which case $\tilde{\mathfrak{o}}_K = \mathfrak{o}$ (K, field of fractions of R). We shall exclude the trivial case in which $\mathfrak{o}$ is integrally closed. We therefore asume that $q \geqslant 1$.

Definition 3.3. *The subring R of* $\mathfrak{o}$ *is said to be transversal in codimension* 1 *if R is transversal with respect to each singular prime ideal of* $\mathfrak{o}$, *of height* 1. *In the special case in which* $\mathfrak{o}$ *is the local ring of an algebroid variety V and k is a coefficient field of* $\mathfrak{o}$, *we say that a set of parameters* $x_1, x_2, \ldots, x_r$ *of* $\mathfrak{o}$ *is k-transversal in codimension* 1 *if these parameters are k-transversal with respect to each singular prime ideal of* $\mathfrak{o}$, *of height* 1.

We now prove the following theorem:

THEOREM 3.4. *Let R, R' be two subrings of a semi-local ring $\mathfrak{o}$ and assume that the following conditions are satisfied*: (1) *both pairs of rings* $\{\mathfrak{o}, R\}$ *and* $\{\mathfrak{o}, R'\}$ *satisfy conditions* (*A*), (*B*) *and* (*C*) *stated in the beginning of Section* 2; (2) $\mathfrak{o}$ *is a simple ring extension of both* R *and* R'; (3) *both rings* R, R' *are transversal in codimension* 1. *Then, if* K, K' *denote the field of fractions of* R *and* R' *respectively, we have*

$$\tilde{\mathfrak{o}}_K = \tilde{\mathfrak{o}}_{K'}. \tag{13}$$

The following is a reformulation of the above theorem in the special case of algebroid hypersurfaces:

COROLLARY 3.5. *Let* $\{x_1, x_2, \ldots, x_r\}$ *and* $\{x'_1, x'_2, \ldots, x'_r\}$ *be two sets of parameters of the local ring* $\mathfrak{o}$ *of an algebroid hypersurface, both contained in minimal bases of the maximal ideal* $\mathfrak{m}$ *of* $\mathfrak{o}$, *and let* k, k' *be two coefficient field of* $\mathfrak{o}$. *If the parameters* $x_1, x_2, \ldots, x_r$ *are* k*-transversal in codimension* 1 *and the parameters* $x'_1, x'_2, \ldots, x'_r$ *are* k'*-transversal in codimension* 1, *then the* k*-saturation of* $\mathfrak{o}$ *with respect to the parameters* $x_1, x_2, \ldots, x_r$ *coincides with the* k'*-saturation of* $\mathfrak{o}$ *with respect to the parameters* $x'_1, x'_2, \ldots, x'_r$:

$$\tilde{\mathfrak{o}}^{(k)}_{\{x_1, x_2, \ldots, x_r\}} = \tilde{\mathfrak{o}}^{(k')}_{\{x'_1, x'_2, \ldots, x'_r\}}. \tag{14}$$

Proof. Conside one of the two rings R, R', say R. If $\mathfrak{p}$ is a non-singular prime ideal of $\mathfrak{o}$, of height 1, then $\mathfrak{o}_\mathfrak{p}$ is a regular local domain (of Krull dimension 1), hence is integrally closed. In this case the saturation $\widetilde{(\hat{\mathfrak{o}}_\mathfrak{p})}_{\hat{K}_\mathfrak{P}\cdot\epsilon_\mathfrak{p}}$ (where $\mathfrak{P} = \mathfrak{p} \cap R$) coincides with $\hat{\mathfrak{o}}_\mathfrak{p}$ and is therefore independent of the choice of the choice of R. Assume now that $\mathfrak{p}$ is one of the singular prime ideals $\mathfrak{p}_1, \mathfrak{p}_2, \ldots, \mathfrak{p}_q$ (of height 1). For simplicity of notation we identify the field $\hat{K}_\mathfrak{P}\cdot\epsilon_\mathfrak{p}$ with the isomorphic field $\hat{K}_\mathfrak{p}$. If ξ is a generator of the principal ideal $\mathfrak{P}$ in R, then, by assumption, ξ is a transversal parameter of $\mathfrak{o}_\mathfrak{p}$, and hence also of $\hat{\mathfrak{o}}_\mathfrak{p}$. We fix a coefficient field Δ_0 of $\hat{R}_\mathfrak{P}$. Then $\hat{R}_\mathfrak{P} = \Delta_0[[\xi]]$ and $\hat{K}_\mathfrak{P} = \Delta_0((\xi))$. Let Δ be the algebraic closure of Δ_0 in $\hat{\mathfrak{o}}_\mathfrak{p}$. Then Δ is a coefficient field of $\hat{\mathfrak{o}}_\mathfrak{p}$ and is a finite algebraic extension of Δ_0. By Lemma A8, part 2, of Appendix A, the saturation of $\hat{\mathfrak{o}}_\mathfrak{p}$ with respect to $\hat{K}_\mathfrak{P}$ $[=\Delta_0((\xi))]$ coincides with the saturation of $\hat{\mathfrak{o}}_\mathfrak{p}$ with respect to the field $\Delta((\xi))$. Since ξ is a transversal parameter of $\hat{\mathfrak{o}}_\mathfrak{p}$, the saturation of $\hat{\mathfrak{o}}_\mathfrak{p}$ with respect to $\Delta((\xi))$ is the absolute saturation of $\hat{\mathfrak{o}}_\mathfrak{p}$ and depends only on $\mathfrak{o}$ and $\mathfrak{p}$ (see Corollary A7 in Appendix A). Now, if we pass from the ring R to the ring R', both the parameter ξ and the field $\hat{K}_\mathfrak{P}$ change: the new field $K'_{\mathfrak{P}'}$ is the field of fraction of the local domain $\hat{R}'_{\mathfrak{P}'}$, where $\mathfrak{P}' = \mathfrak{p} \cap R'$, and the parameter ξ' is a generator of the principal ideal $\mathfrak{P}'$ in R'. We will be dealing with the Δ'-saturation of $\hat{\mathfrak{o}}_{\mathfrak{p}}$ with respect to ξ', where the new coefficient field Δ' of $\hat{\mathfrak{o}}_{\mathfrak{p}}$ is the algebraic closure, in $\hat{\mathfrak{o}}_{\mathfrak{p}}$, of some (arbitrarily chosen) coefficient field Δ_0' of $\widehat{R}_{\mathfrak{p}'}'$. At any rate, what we have said above shows that

$$\widetilde{(\hat{\mathfrak{o}}_{\mathfrak{p}})_{\hat{K}_{\mathfrak{P}}\cdot\epsilon_{\mathfrak{p}}}} = \widetilde{(\hat{\mathfrak{o}}_{\mathfrak{p}})_{\hat{K}_{\mathfrak{P}'}\cdot\epsilon_{\mathfrak{p}}}}, \tag{15}$$

for every singular prime ideal $\mathfrak{p}$ of $\mathfrak{o}$, of height 1. Since the equality (15) is trivial for any non-singular prime ideal $\mathfrak{p}$ (in which case, as was pointed out above, both sides of (15) coincide with $\hat{\mathfrak{o}}_{\mathfrak{p}}$), our theorem follows from Theorem 3.1.

Corollary 3.5 follows directly from Theorem 3.4 upon setting $R = k[[x_1, x_2, \ldots, x_r]]$, $R' = k'[[x_1', x_2', \ldots, x_r']]$.

In the case of algebroid hypersurfaces we shall denote by $\tilde{\mathfrak{o}}$ the saturated ring $\tilde{\mathfrak{o}}^{(k)}_{\{x_1, x_2, \ldots, x_r\}}$, where k is any coefficient field of $\mathfrak{o}$ and where $\{x_1, x_2, \ldots, x_r\}$ is any set of parameters of $\mathfrak{o}$ which belongs to a minimal basis of $\mathfrak{m}$ and which are k-transversal in codimension 1, and we call $\tilde{\mathfrak{o}}$ the *absolute saturation of* $\mathfrak{o}$.

PROPOSITION 3.6. *If* $\mathfrak{o}$ *is the local ring of an algebroid hypersurface, if* $\{x_1, x_2, \ldots, x_r\}$ *is any set of parameters of* $\mathfrak{o}$ *which belongs to a minimal basis of* $\mathfrak{m}$ *and if* k *is any coefficient field of* $\mathfrak{o}$, *then* $\tilde{\mathfrak{o}}$ *is a subring of* $\tilde{\mathfrak{o}}^{(k)}_{\{x_1, x_2, \ldots, x_r\}}$.

Proof. We only have to go back to the proof of Theorem 3.4 and observe that, in the notations of that proof, the saturation $\hat{\mathfrak{o}}_{\mathfrak{p}}$ with respect to the field $\hat{K}_{\mathfrak{P}}$ $(=\Delta_0((\xi)))$ contains the absolute saturation of $\widehat{\mathfrak{o}}_{\mathfrak{p}}$ (see Lemma A9 of Appendix A).

We note that while the k-saturation of $\mathfrak{o}$, with respect to a set of parameters $x_1, x_2, \ldots, x_r$ which are k-transveral in codimension 1, has an invariant character (independent of k and of the parameters), the notion of k-transversality of a set of parameters $\{x_1, x_2, \ldots, x_r\}$ with respect to a given prime ideal $\mathfrak{p}$ of $\mathfrak{o}$ is *not* invariant, i.e., is not independent of k. We shall show namely by examples in Appendix B that if $\mathfrak{p} \neq \mathfrak{m}$ then a set of parameters $\{x_1, \ldots, x_r\}$ can be k-transversal with respect to $\mathfrak{p}$ for some coefficient field k of $\mathfrak{o}$, without being k'-transversal with respect to $\mathfrak{p}$ for some other coefficient field k' of $\mathfrak{o}$. The case $\mathfrak{p} = \mathfrak{m}$ is an exception, for, in that case, *k-transversality is a property which is identical with the property of* $\{x_1, x_2, \ldots, x_r\}$ *being a set of transversal parameters of* $\mathfrak{o}$, i.e., with the property that the multiplicity of the ideal $\sum_{i=1}^{r} \mathfrak{o}x_i$ is the same as the multiplicity of the local ring $\mathfrak{o}$.

4. Invariance of the Multiplicity of a Local Ring Under Saturation. We shall prove in this section the following theorem:

THEOREM 4.1. *Let $\mathfrak{o}$ be a complete, equidimensional, equicharacteristic zero, local ring, free from nonzero nilpotent elements, let $\{x_1, x_2, \ldots, x_r\}$ be a set of* ***transversal*** *parameters of $\mathfrak{o}$ and let k be a coefficient field of $\mathfrak{o}$. Then the multiplicity of the k-saturation $\tilde{\mathfrak{o}}^{(k)}_{\{x_1, x_2, \ldots, x_r\}}$ of $\mathfrak{o}$ with respect to the parameters $x_1, x_2, \ldots, x_r$, is the same as the multiplicity of the local ring $\mathfrak{o}$.*

Proof. We shall first give the proof *under the assumption that $\mathfrak{o}$ is a local domain*. After that is done, the extension to the general case will be relatively a straightforward matter.

Let $R = k[[x_1, x_2, \ldots, x_r]]$ and let $\mathfrak{M}$ be the maximal ideal $\sum_{i=1}^r Rx_i$ of the formal power series ring R. By the definition of transversal parameters (see Appendix B, Definition B3), the multiplicity of the extended ideal $\mathfrak{o}\mathfrak{M}$ is the same as the multiplicity of the maximal ideal $\mathfrak{m}$ of $\mathfrak{o}$:

$$e(\mathfrak{o}\mathfrak{M}) = e(\mathfrak{m}). \tag{1}$$

In the proof of our theorem we shall use two theorems, due respectively to Northcott-Rees [7] and to Rees [8]. We shall state now these two theorems, but only in the special case for which we need them in our proof.

THEOREM OF NORTHCOTT-REES (THEOREM 5). *Let $\mathfrak{o}$ be an equidimensional, equicharacteristic local ring, with maximal ideal $\mathfrak{m}$, such that the residue field $\mathfrak{o}/\mathfrak{m}$ is infinite. Let $x_1, x_2, \ldots, x_r$ be a set of parameters of $\mathfrak{o}$. If the ideal $\mathfrak{m}$ is integrally dependent on the ideal $\sum_{i=1}^r \mathfrak{o}x_i$ (see* C.A. 2, p. 349, *Definition* 2), *then $x_1, x_2, \ldots, x_r$ are transversal parameters, i.e., the two ideals $\sum_{i=1}^r \mathfrak{o}x_i$ and $\mathfrak{m}$ have the same multiplicity (see Appendix B, Definition B3).*

THEOREM OF REES [8]. This is the precise converse of the preceding Northcott-Rees theorem. It says therefore that, *under the same assumptions made above, concerning the local ring $\mathfrak{o}$, and assuming now that $x_1, x_2, \ldots, x_r$ are transversal parameters of $\mathfrak{o}$, the ideal $\mathfrak{m}$ is integrally dependent on the ideal $\sum_{i=1}^r \mathfrak{o}x_i$.*

Together, these two theorems give therefore a *characterization* of transversal parameters of equidimensional, equicharacteristic local rings $\mathfrak{o}$, having an infinite residue field $\mathfrak{o}/\mathfrak{m}$.

It will be essential for our purposes to make use of the following *equivalent* formulation of the property of integral dependence of an ideal $\mathfrak{B}$ in $\mathfrak{o}$ on an ideal $\mathfrak{A}$ contained in $\mathfrak{B}$. (This formulation is in terms of completions of ideals; see C.A. 2, p. 347, Definition 1, and p. 350, Theorem 1). Let, namely, F be the field of fractions of $\mathfrak{o}$ and let S be the set of all valuations of F which are non-negative on $\mathfrak{o}$ (and hence also non-negative on the integral closure $\bar{\mathfrak{o}}$ of $\mathfrak{o}$ in F). Then $\mathfrak{B}$ *is integrally dependent on* $\mathfrak{A}$ *if and only if*

$$v(\mathfrak{A}) = v(\mathfrak{B}), \quad \text{for all } v \text{ in S}. \tag{2}$$

Here $v(\mathfrak{A})$ denotes the minimum of the set $\{v(\xi), \xi \in \mathfrak{A}\}$; and similarly for $v(\mathfrak{B})$.

We shall denote the k-saturation of $\mathfrak{o}$ with respect to the given transversal parameters $x_1, x_2, \ldots, x_r$ by $\tilde{\mathfrak{o}}$. We denote by K the field of fractions of the power series ring $R = k[[x_1, x_2, \ldots, x_r]]$. We recall from an earlier paper of ours (see SES III, Introduction, proof of Proposition-Definition II) that there exists a finite chain of rings $\mathfrak{o}_\alpha$ between $\mathfrak{o}$ and $\tilde{\mathfrak{o}}$:

$$\mathfrak{o} = \mathfrak{o}_0 \subset \mathfrak{o}_1 \subset \mathfrak{o}_2 \subset \cdots \subset \mathfrak{o}_q = \tilde{\mathfrak{o}},$$

where the $\mathfrak{o}_\alpha$ are defined inductively as follows:

Let L_α denote the set of elements ξ of $\bar{\mathfrak{o}}$ such that ξ dominates, with respect to $\{\mathfrak{o}, K\}$, some element ζ of $\mathfrak{o}_\alpha$. (ζ may depend on ξ.) Then $\mathfrak{o}_{\alpha+1} = \mathfrak{o}_\alpha[L_\alpha]$.

Let $\mathfrak{m}_\alpha$ denote the maximal ideal of $\mathfrak{o}_\alpha$. Each $\mathfrak{o}_\alpha$ is a complete, equidimensional, equicharacteristic zero, local domain, with coefficient field k (see SES, Theorem 4.1), and $x_1, x_2, \ldots, x_r$ are parameters of $\mathfrak{o}_\alpha$. Since $\mathfrak{o}$ and $\mathfrak{o}_\alpha$ have the same integral closure $\bar{\mathfrak{o}}$, domination with respect to $\{\mathfrak{o}, K\}$ is the same thing as domination with respect to $\{\mathfrak{o}_\alpha, K\}$. Thus, the definition of $\mathfrak{o}_{\alpha+1}$ in terms of $\mathfrak{o}_\alpha$ is the same for all $\alpha = 0, 1, \ldots, q-1$. Using this definition we shall now prove, by induction on α, that $x_1, x_2, \ldots, x_r$ *are transversal parameters of each* $\mathfrak{o}_\alpha$ and that

$$e(\mathfrak{m}_\alpha) = e(\mathfrak{m}_{\alpha+1}), \qquad \alpha = 0, 1, 2, \ldots, q-1. \tag{3}$$

This will complete the proof of the theorem in the case of a local *domain* $\mathfrak{o}$.

It will be sufficient to prove, for instance, that $x_1, x_2, \ldots, x_r$ are transversal parameters of $\mathfrak{o}_1$ and that

$$e(\mathfrak{m}) = e(\mathfrak{m}_1). \tag{4}$$

If $\mathfrak{M}$ is the maximal ideal $(x_1, x_2, \ldots, x_r)$ in $R\,(=k[[x_1, x_2, \ldots, x_r]])$, we have, of course, $v(\mathfrak{M}) = v(\mathfrak{o}\mathfrak{M})$, for every valuation v in S. Hence, the assumption of transversality of the parameters $x_1, x_2, \ldots, x_r$ implies, by the above theorem of Rees, that

$$v(\mathfrak{M}) = v(\mathfrak{m}), \quad \text{for all } v \text{ in S}. \tag{5}$$

The crucial part of our proof of (4) and of the transversality of the parameters $x_1, x_2, \ldots, x_r$ of $\mathfrak{o}_1$ will consist in establishing the following equalities:

$$v(m_1) = v(\mathfrak{m}), \quad \text{for all } v \text{ in S}, \tag{6}$$

or—equivalently (since S is also the set of all valuations v of F which are non-negative on $\mathfrak{o}_1$)—that the ideal $\mathfrak{m}_1$ is integrally dependent on the ideal $\mathfrak{o}_1\mathfrak{m}$. Assume for a moment that this has been established. Then we have by (5): $v(\mathfrak{m}_1) = v(\mathfrak{o}_1\mathfrak{M})$ for all v in S, and hence, by the above theorem of Rees it will follow that $x_1, x_2, \ldots, x_r$ are transversal parameters of $\mathfrak{o}_1$, i.e., that

$$e(\mathfrak{m}_1) = e(\mathfrak{o}_1\mathfrak{M}). \tag{7}$$

Since $\mathfrak{o}_1\mathfrak{M} \subset \mathfrak{o}_1\mathfrak{m} \subset \mathfrak{m}_1$, we have $e(\mathfrak{o}_1\mathfrak{M}) \geqslant e(\mathfrak{o}_1\mathfrak{m}) \geqslant e(\mathfrak{m}_1)$, and hence we have by (7)

$$e(\mathfrak{m}_1) = e(\mathfrak{o}_1\mathfrak{m}). \tag{8}$$

We now apply the projection formula to the two rings $\mathfrak{o}$ and $\mathfrak{o}_1$ and to their maximal ideals $\mathfrak{m}$ and $\mathfrak{m}_1$ (see C.A. 2, p. 299, Corollary 1 of Theorem 24). Since $\mathfrak{o}$ and $\mathfrak{o}_1$ have the same field of fractions F, we have (in the notations of the cited Corollary 1, C.A. 2, p. 299) that $[\mathfrak{o}_1 : \mathfrak{o}] = 1$. In view of the radicial character of the morphism **Spec** ($\mathfrak{o}_1$)→**Spec** ($\mathfrak{o}$) (see SES III, Theorem 4.1), and since we are in characteristic zero, k is also a coefficient field of $\mathfrak{o}_1$, i.e., we have $[\mathfrak{o}_1/\mathfrak{m}_1 : \mathfrak{o}/\mathfrak{m}] = 1$. Hence the cited projection formula leads in the present case to the equality

$$e(\mathfrak{m}) = e(\mathfrak{o}_1\mathfrak{m}),$$

and our theorem now follows from (8) (always under the assumption that $\mathfrak{o}$ is an inegral domain).

So we now proceed to prove (6), i.e., in view of (5), the equivalent equality

$$v(\mathfrak{m}_1) = v(\mathfrak{M}), \quad \text{for all } v \text{ in S}. \tag{9}$$

Upon replacing each element ξ of L_1 by $\xi - c$, where c is the $\mathfrak{m}_1$-residue of ξ in k, we may assume that $L_1 \subset \mathfrak{m}_1$. Since $\mathfrak{o}_1 = \mathfrak{o}[L_1]$ and $L_1 \subset \mathfrak{m}_1$, to prove (9) it will be sufficient to prove that

$$v(\xi) \geqslant v(\mathfrak{M}), \quad \text{for all } v \text{ in S and all } \xi \text{ in } L_1. \tag{9'}$$

At this stage it will be convenient to introduce the least Galois extension F^* of $K(=k((x_1, x_2, \ldots, x_r)))$ which contains F and the set S^* of all valuations of F^* which are non-negative on the power series ring $R(=k[[x_1, x_2, \ldots, x_r]])$. Since $\mathfrak{o}$ is integral over R, S^* consists of the extensions to F^* of the valuation v in S, and thus, proving (9′) is equivalent to proving that

$$v^*(\xi) \geqslant v^*(\mathfrak{M}), \quad \text{for all } v^* \text{ in } S^* \text{ and all } \xi \text{ in } L_1. \tag{9*}$$

Let S_0 be the set of all valuations of K which are non-negative on R. If $v^* \in S^*$ then $v^*|K$ belongs to S_0, and if $v_0 \in S_0$ then v_0 has a finite number of extensions to F^*; the extensions of a given v_0 in S_0 all belong to S^* and form a complete set of conjugates under the Galois group $G = \mathcal{G}(F^*/K)$ (see C.A. 2, p. 28, Corollary 3 of Theorem 12).

Let now ξ be any element of L_1, and let v_0 be any *given* valuation in S_0. We first prove the following:

(A) *There exists at least one extension v^* of v_0 in S^* such that* $v^*(\xi) \geqq v^*(\mathfrak{M})$. [Note that $v^*(\mathfrak{M}) = \min\cdot\{v^*(x_1), v^*(x_2), \ldots, v^*(x_2)\}$].

We shall give an indirect proof. Assume the contrary:

$$v^*(\xi) < v^*(\mathfrak{M}), \quad \text{for } all \text{ extensions } v^* \text{ of } v_0. \tag{10}$$

Let $n = [F:K]$, $g = [F^*:K]$, and let

$$\xi_0 = \frac{1}{n}\mathrm{Tr}_{F/K}\xi = \frac{1}{g}\mathrm{Tr}_{F^*/K}\xi = \frac{1}{g}\sum_{\tau \in G} \xi^\tau.$$

Since $\xi \in \mathfrak{m}_1$, we must have $\xi_0 \in \mathfrak{M}$, and therefore $v^*(\xi_0) \geqq v^*(\mathfrak{M})$, for all extensions v^* of v_0. By our assumption (10) it follows therefore that $v^*(\xi) < v^*(\xi_0)$, for every extension v^* of v_0. Hence $v^*(\xi - \xi_0) = v^*(\xi)$ for all extensions v^* of v_0, or equivalently: $v^*[\Sigma_{\tau \in G}(\xi^\tau - \xi)] = v^*(\xi)$, for all extensions v^* of v_0. From this we can therefore draw the following conclusion:

(B) *For every extension v^* of v_0 there exists an element τ in the Galois group G (τ may depend on v^*) such that $v^*(\xi^\tau - \xi) \leqq v^*(\xi)$.*

Now we shall make use of the fact that $\xi \in L_1$. There exists then an element ζ of $\mathfrak{o}$ such that $v^*(\xi^\tau - \xi) \geqq v^*(\zeta^\tau - \zeta)$ for *all* v^* in S^* and *all* τ in G (this follows from the fact that ξ dominates ζ with respect to the pair $\{K, \mathfrak{o}\}$). Upon replacing ζ by ζ-c, where c is the $\mathfrak{m}$-residue of ζ in k, we may assume that $\zeta \in \mathfrak{m}$. It follows therefore from the above conclusion (B) that

$$v^*(\xi) \geqq v^*(\zeta^\tau - \zeta), \quad \text{for } every \text{ extension } v^* \text{ of } v_0, \\ \text{and for a } suitable\ \tau \text{ in } G \text{ (depending on } v^*). \tag{11}$$

We fix an extension v_0^* of v_0 and we let τ_0 be an element of G such that $v_0^*(\zeta^{\tau_0}) = \min_{\tau \in G}\cdot\{v_0^*(\zeta^\tau)\}$. We then set $v_1^* = v_0^*\cdot\tau_0$. Then we have $v_1^*(\zeta) = \min_{\tau \in G}\cdot\{v_1^*(\zeta^{\tau_0^{-1}\cdot\tau})\}$, i.e., $v_1^*(\zeta) = \min_{\tau \in G}\cdot\{v_1^*(\zeta^\tau)\}$. Therefore, we find that $v_1^*(\zeta^\tau - \zeta) \geqq v_1^*(\zeta)$, for all τ in G. But, by (11), we have, for *some* τ in G: $v_1^*(\xi) \geqq v_1^*(\zeta^\tau - \zeta)$. Therefore, we have $v_1^*(\xi) \geqq v_1^*(\zeta)$, and since $\zeta \in \mathfrak{M}$, we also have $v_1^*(\zeta) \geqslant v_1^*(\mathfrak{M})$. Hence $v_1^*(\xi) \geqslant v_1^*(\mathfrak{M})$, in contradiction with our assumption (10) (since v_1^* is an extension of v_0). This proves the above assertion (A). Observe that so far we have made no use of the fact that $x_1, x_2, \ldots, x_r$ are *transversal* parameters of $\mathfrak{o}$.

We now consider *arbitrary* extensions v^* of v_0 and use our assumption that $x_1, x_2, \ldots, x_r$ are transversal parameters of $\mathfrak{o}$. We have then (5), and since $\zeta \in \mathfrak{m}$, it follows that $v^*(\zeta) \geqslant v^*(\mathfrak{M})$. Since any τ in G leaves invariant $\mathfrak{M}$, it follows that $v^*(\zeta^\tau) \geqslant v^*(\mathfrak{M})$ for all v^* in S^* and all τ in G. Hence $v^*(\zeta^\tau - \zeta) \geqslant v^*(\mathfrak{M})$, for all v^* in S^* and all τ in G. Since ξ dominates ζ, it follows that

$$v^*(\xi^\tau - \xi) \geqslant v^*(\mathfrak{M}), \qquad \text{for all } v^* \in S^* \text{ and all } \tau \in G. \tag{12}$$

Now, by the assertion (A) just proved above, we have $v_1^*(\xi) \geqslant v_1^*(\mathfrak{M})$, for some extension v_1^* of v_0. If we combine this inequality with the inequality (12) for $v^* = v_1^*$, we find

$$v_1^*(\xi^\tau) \geqslant v_1^*(\mathfrak{M}), \quad \text{for all } \tau \text{ in } G,$$

or, equivalently,

$$(v_1^* \cdot \tau)(\xi) \geqslant v_1^*(\mathfrak{M}), \quad \text{for all } \tau \text{ in } G.$$

Since $v_1^*(\mathfrak{M}) = (v_1^* \cdot \tau)(\mathfrak{M})$ (in view of the equality $\tau(\mathfrak{M}) = \mathfrak{M}$) and since the valuation $v_1^* \cdot \tau$ ranges over the set of all extensions of v_0 as τ ranges over G, (9*) is proved. This completes the proof of the theorem in the case of a local *domain* $\mathfrak{o}$.

We now assume that $\mathfrak{o}$ has zero divisors. If $(0) = \mathfrak{p}_1 \cap \mathfrak{p}_2 \cap \cdots \cap \mathfrak{p}_h$ is the irredundant representation of the zero ideal of $\mathfrak{o}$ as intersection of prime ideals, then the total ring of quotients F of $\mathfrak{o}$ is a direct sum of h fields: $F = F\epsilon_1 \oplus F\epsilon_2 \oplus \cdots \oplus F\epsilon_h$, where ϵ_α is the identity of the field $F\epsilon_\alpha$. *The elements* $x_1\epsilon_\alpha, x_2\epsilon_\alpha, \ldots, x_r\epsilon_\alpha$ *are transversal parameters of the local domain* $\mathfrak{o}\epsilon_\alpha$. For, we have $\mathfrak{o}\epsilon_\alpha \cong \mathfrak{o}/\mathfrak{p}_\alpha$, and therefore (see Northcott-Rees [5], Lemma 1, p. 358, as applied to $\mathfrak{q} = \mathfrak{m}$) $e(\mathfrak{m}) = \Sigma_{\alpha=1}^h e(\mathfrak{m}/\mathfrak{p}_\alpha) = \Sigma_{\alpha=1}^h e(\mathfrak{m}\epsilon_\alpha)$. Here $\mathfrak{m}\epsilon_\alpha$ is the maximal ideal of $\mathfrak{o}\epsilon_\alpha$. Similarly, if we set $\mathfrak{M}_\alpha = \mathfrak{M}\epsilon_\alpha$ and apply the cited Lemma 1 of Northcott-Rees ([5], p. 358) to the ideal $\mathfrak{q} = \mathfrak{o}\mathfrak{M}$, we find that $e(\mathfrak{o}\mathfrak{M}) = \Sigma_{\alpha=1}^h e(\mathfrak{o}\epsilon_\alpha \cdot \mathfrak{M}_\alpha)$. Since $e(\mathfrak{m}) = e(\mathfrak{o}\mathfrak{M})$, it follows that

$$\sum_{\alpha=1}^h e(\mathfrak{m}\epsilon_\alpha) = \sum_{\alpha=1}^h e(\mathfrak{o}\epsilon_\alpha \cdot \mathfrak{M}_\alpha) = \sum_{\alpha=1}^h e\left(\sum_{i=1}^r \mathfrak{o}\epsilon_\alpha \cdot x_i\epsilon_\alpha\right). \tag{13}$$

Since $\mathfrak{m}\epsilon_\alpha \supset \mathfrak{o}\epsilon_\alpha \cdot \mathfrak{M}_\alpha$, we have $e(\mathfrak{m}\epsilon_\alpha) \leqslant e(\mathfrak{o}\epsilon_\alpha \cdot \mathfrak{M}_\alpha)$, for $\alpha = 1, 2, \ldots, h$. Hence it follows from (13) that $e(\mathfrak{m}\epsilon_\alpha) = e(\mathfrak{o}\epsilon_\alpha \cdot \mathfrak{M}_\alpha)$, $\alpha = 1, 2, \ldots, h$, which proves that $x_1\epsilon_\alpha, x_2\epsilon_\alpha, \ldots, x_r\epsilon_\alpha$ are transversal parameters of the local domain $\mathfrak{o}\epsilon_\alpha$.

We now use Corollary 3.6 of our paper GTS II (p. 899). By that corollary, the ring $\tilde{\mathfrak{o}}\epsilon_\alpha$ is the $k\epsilon_\alpha$-saturation of the local domain $\mathfrak{o}\epsilon_\alpha$, with respect to the parameters $x_1\epsilon_\alpha, x_2\epsilon_\alpha, \ldots, x_r\epsilon_\alpha$. Since we have just shown that these parameters are transversal, it follows from our theorem (which we have already proved for local domains) that

$$e(\tilde{\mathfrak{m}}\epsilon_\alpha) = e(\mathfrak{m}\epsilon_\alpha), \qquad \alpha = 1, 2, \ldots, h. \tag{14}$$

We have just seen above (as a consequence of the cited Lemma 1 of Northcott-Rees [7], p. 358) that $e(\mathfrak{m}) = \sum_{\alpha=1}^{h} e(\mathfrak{m}\epsilon_\alpha)$. In a similar way it follows from the same lemma that $e(\tilde{\mathfrak{m}}) = \sum_{\alpha=1}^{h} e(\tilde{\mathfrak{m}}\epsilon_\alpha)$. Hence, by (14), we conclude that $e(\tilde{\mathfrak{m}}) = e(\mathfrak{m})$, and this completes the proof of our theorem.

We note the following immediate consequence of our theorem:

Corollary 4.2. *The assumptions being the same as in Theorem* 4.1, *any set of transversal parameters* $z_1, z_2, \ldots, z_r$ *of* $\mathfrak{o}$ *is also a set of transversal parameters of the k-saturation* $\tilde{\mathfrak{o}}$ $(= \tilde{\mathfrak{o}}^{(k)}_{\{x_1, x_2, \ldots, x_r\}})$ *of* $\mathfrak{o}$ *with respect to the transversal parameters* $x_1, x_2, \ldots, x_r$.

For, by assumption, we have $e(\mathfrak{m}) = e(\sum_{i=1}^{r} \mathfrak{o}z_i)$. Hence, by our theorem, we have $e(\tilde{\mathfrak{m}}) = e(\sum_{i=1}^{r} \mathfrak{o}z_i)$. Now, by the projection formula (C.A. 2, p. 299, Corollary 1), as applied to the local rings $A = \mathfrak{o}$, $B = \tilde{\mathfrak{o}}$ and to the ideal $\mathfrak{q} = \sum_{i=1}^{r} \mathfrak{o}z_i$, it follows at once that $e(\sum_{i=1}^{r} \mathfrak{o}z_i) = e(\sum_{i=1}^{r} \tilde{\mathfrak{o}}z_i)$ (since we have in our present case: $[B:A] = 1$, $B\mathfrak{q} = \sum_{i=1}^{r} \tilde{\mathfrak{o}}z_i$ is primary for $\tilde{\mathfrak{m}}$, and $[\tilde{\mathfrak{o}}/\tilde{\mathfrak{m}} : \mathfrak{o}/\mathfrak{m}] = 1$). Thus, $e(\tilde{\mathfrak{m}}) = e(\sum_{i=1}^{r} \tilde{\mathfrak{o}}z_i)$, which proves the corollary.

The question remains open whether Theorem 4.1 is true also if the parameters $x_1, x_2, \ldots, x_r$ are not transversal.

Note. Lipman called our attention to the fact that in our proof of Theorem 4.1 the separate consideration of the case in which $\mathfrak{o}$ has zero divisors can be avoided by using the fact that our valuation-theoretic criterion for the integral closure of an ideal in an integral domain (C.A. 2, p. 350, Theorem 1) has been extended to noetherian rings with zero divisors by Nagata in his paper "Note on a paper of Samuel concerning asymptotic properties of ideals," *Memoirs of the College of Science, University of Kyoto, Series* A, **XXX**, Mathematics No. 2 (1957), pp. 165–175. Since in our separate treatment of rings $\mathfrak{o}$ with zero divisor we use a far from trivial result from our paper GTS II (namely Corollary 3.6, GTS II), the elimination of this separate treatment is a definite simplification of the proof of Theorem 4.1.

Appendix A. Complement to My Paper GTS II.

In this appendix we shall extend to one-dimensional local rings with zero-divisors some results which we have proved in GTS I for one dimensional local *domains* and which we neglected to prove in GTS II. These results are needed and used in Section 3 of the present paper. The bulk of this appendix is taken up by the extension of Theorem 9.1 of GTS I, while only the last few pages deal with the extension of two other results, namely of Lemma 2.2 and Corollary 7.4 of GTS I.

We recall for the benefit of the reader the first of these results.

(*Theorem* 9.1 *of* GTS I). Let k, k' be two coefficient fields of a complete, equicharacteristic zero, equidimensional local *domain* $\mathfrak{o}$, *of Krull dimension 1, let* θ *be the canonical isomorphism* $k \xrightarrow{\sim} k'$ (i.e., the isomorphism defined by $\theta(c) - c \in \mathfrak{m} =$ maximal ideal of $\mathfrak{o}$, for all $c \in k$), and let x be a transversal parameter of $\mathfrak{o}$. Assume that $\mathfrak{o}$ is k-saturated with respect to x. Then there exists an automorphism φ of $\mathfrak{o}$ such that $\varphi | k = \theta$ and $\varphi(x) = x$ (and hence $\mathfrak{o}$ is also k'-saturated with respect to x).

We have also shown in GTS I the following:

(*Theorem* 7.2 *of* GTS I). If $\mathfrak{o}$ is a local domain as above and if $\mathfrak{o}$ is k-saturated, i.e., if $\mathfrak{o}$ is k-saturated with respect to some parameter of $\mathfrak{o}$, then $\mathfrak{o}$ is also k-saturated with respect to any *transversal* parameter.

From these two theorems followed easily (see GTS I, Corollary 9.5) that if $\mathfrak{o}$ is a local domain as above (but not necessarily saturated) and if k, x are respectively a coefficient field and a transversal parameter of $\mathfrak{o}$, then the k-saturation of $\mathfrak{o}$ with respect to x is a ring $\tilde{\mathfrak{o}}$ which is independent of the choice of k and x (this ring $\tilde{\mathfrak{o}}$ is, by definition of saturation, a certain subring of the integral closure $\bar{\mathfrak{o}}$ of $\mathfrak{o}$ in the field of fractions F of $\mathfrak{o}$). We have also shown in GTS I that the automorphism φ of $\mathfrak{o}$, the existence of which is asserted in the above quoted Theorem 9.1 of GTS I, can be chosen so as to satisfy the following condition [see GTS I, p. 640, relation (3)]: if η is any element of F, different from zero, and v is the canonical valuation of the (local) field F, then

$$v(\varphi(\eta)/\eta - 1) > 0. \tag{1}$$

Here φ stands, of course, for the unique extension of the automorphism φ of $\mathfrak{o}$ to an automorphism of F.

If GTS II we were dealing with one-dimensional local rings satisfying the same conditions as above (i.e., complete, equicharacteristic zero and equidimensional) but which may have zero-divisors (they are required, though, to be free from nonzero nilpotent elements). In this more general case, we have shown that the exact analog of the second of the above quoted theorem (Theorem 7.2 of GTS I) still holds true (see GTS II, Theorem 4.5, p. 905). As to the generalization of Theorem 9.1 of GTS I, we carried it out in GTS II *only under the additional assumption* that the residue field of $\mathfrak{o}$ is algebraically closed*. In this appendix we shall prove this generalization *without* the above additional assumption. We shall now state explicitly this generalized theorem, first because condition (1) above has a slightly different formulation in the generalization to local rings with zero divisors, and second because the inductive nature of our present proof requires that we show the existence of an automorphism φ which satisfies an extra condition (a condition which is automatically satisfied if $\mathfrak{o}$ is a local *domain*).

Let F be the total ring of quotients of $\mathfrak{o}$ and let $\bar{\mathfrak{o}}$ be the integral closure of $\mathfrak{o}$ in F. We know that F is a finite direct sum of h fields $(h > 1)$:

$$F = F\epsilon_1 \oplus F\epsilon_2 \oplus \cdots \oplus F\epsilon_h,$$

where ϵ_i is the identity of the field $F\epsilon_i$ and $1 = \epsilon_1 + \epsilon_2 + \cdots + \epsilon_h$. If k is a coefficient field of $\mathfrak{o}$ then $k\epsilon_i$ is a coefficient field of the (complete) local domain $\mathfrak{o}\epsilon_i$, and if $\bar{k}$ is the algebraic closure of k in F, then the field $\bar{k}\epsilon_i$ is the algebraic closure of $k\epsilon_i$ in $F\epsilon_i$, and $\bar{k}$ is the direct sum of the h fields $\bar{k}\epsilon_i$. The integral closure of the local domain $\mathfrak{o}\epsilon_i$ in its field of fractions $F\epsilon_i$ is a complete local domain, equal to $\bar{\mathfrak{o}}\epsilon_i$, and $\bar{k}\epsilon_i$ is a coefficient field of $\bar{\mathfrak{o}}\epsilon_i$. If k' is another coefficient field of $\mathfrak{o}$ and $\bar{k}'$ is the algebraic closure of k' in F, then similarly, $k'\epsilon_i$ is a coefficient field of $\mathfrak{o}\epsilon_i$ and $\bar{k}'\epsilon_i$ is a coefficient field of $\bar{\mathfrak{o}}\epsilon_i$. If θ denotes the canonical isomorphism $k \xrightarrow{\sim} k'$ then the isomorphism $\theta_i : k\epsilon_i \xrightarrow{\sim} k'\epsilon_i$ defined by $\theta_i(c\epsilon_i) = \theta_i(c)\epsilon_i$ $(c \in k)$ is the canonical isomorphism between the two coefficient fields $k\epsilon_i$ and $k'\epsilon_i$ of $\mathfrak{o}\epsilon_i$. If $\bar{\theta}_i$ denotes the canonical isomorphism $\bar{k}\epsilon_i \xrightarrow{\sim} \bar{k}'\epsilon_i$ between the two coefficient fields $\bar{k}\epsilon_i$ and $\bar{k}'\epsilon_i$ of the local domain $\bar{\mathfrak{o}}\epsilon_i$, then $\bar{\theta}_i$ is an extension of θ_i. If we set $\bar{\theta} = \bar{\theta}_1 + \bar{\theta}_2 + \cdots + \bar{\theta}_h$ then $\bar{\theta}$ is the canonical isomorphism $\bar{k} \xrightarrow{\sim} \bar{k}'$ of the two coefficient rings $\bar{k}$, $\bar{k}'$ of the (complete) semi-local ring $\bar{\mathfrak{o}}$, and $\bar{\theta}$ is an extension of θ. We note that by the definition of θ_i we have $\theta_i(\epsilon_i) = \epsilon_i$ and hence also $\bar{\theta}(\epsilon_i) = \epsilon_i$, $i = 1, 2, \ldots, h$. We note also that any automorphism φ of F such that $\varphi | \bar{k} = \bar{\theta}$ satisfies the conditions $\varphi(\epsilon_i) = \epsilon_i$, $\varphi(F\epsilon_i) = F\epsilon_i$, and hence the restriction of φ to $F\epsilon_i$ is an automorphism of $F\epsilon_i$.

The theorem which we wish to prove in this appendix, and which

*With this additional assumption we were actually able to prove a stronger result, namely that the automorphism φ in question exists for *any* isomorphism of k onto k'. In GTS I we have shown by examples that this stronger result may be definitely false (even in the case of local *domains*, of dimension 1) if the residue field of $\mathfrak{o}$ is not algebraically closed.

generalized Theorem 9.1 of GTS I stated above, is the following:

THEOREM A1. *Let k, k' be two coefficient field of a complete, equicharacteristic zero, equidimensional local ring $\mathfrak{o}$, of Krull dimension 1, free from nilpotent elements different from zero, and let x be a transversal parameter of $\mathfrak{o}$. If $\mathfrak{o}$ is k-saturated (and hence is also k-saturated with respect to x; see GTS II, Theorem 4.5, p. 905), then there exists an automorphism φ of F such that $\varphi(\mathfrak{o})=\mathfrak{o}$, $\varphi(x)=x$ and such that φ satisfies the following two additional conditions*:

1. $\varphi|\bar{k}=\bar{\theta}=$ *canonical isomorphism* $\bar{k}\xrightarrow{\sim}\bar{k}'$.
2. *For any* $i=1,2,\ldots,h$ *and for any element* η_i *of* $F\epsilon_i$, *different from zero, we have*

$$\varphi(\eta_i)\cdot\eta_i^{-1}-\epsilon_i\in\bar{\mathfrak{m}}_i,$$

where $\bar{\mathfrak{m}}_i$ *is the maximal ideal of the local domain* $\bar{\mathfrak{o}}\epsilon_i$. *(Here* η_i^{-1} *denotes, of course, the inverse of* η_i *in the field* $F\epsilon_i$, *i.e., we have* $\eta_i\cdot\eta_i^{-1}=\epsilon_i$*).*

The proof will be induction on the number h of direct field components $F\epsilon_i$ of F. At certain stages the proof will be similar to the one we gave in GTS I in the case $h=1$. However, an essential new ingredient of our present proof will be represented by the use of the general structure theorem for saturated local rings (of dimension 1) which we have derived in GTS II. We refer primarily to what we have called there the second *formulation* of that theorem (GTS II, Theorem 3.4). We begin by recalling some definitions and results from GTS II, in order to be in position to give here, for the benefit of the reader, the full statement of the structure theorem and its background. We shall divide the exposition of these preliminaries in seven parts (a)–(g) below.

(a) We set $K=k((x))$ and we consider, as in section 2, the set H of all K-homomorphisms of F into a given algebraic closure Ω of K (the role of the regular local domain R is now played by the ring of formal power series $k[[x]]$). For any ψ in H we have $\psi(\epsilon_i)=1$ for some $i=1,2,\ldots,h$, and $\psi(\epsilon_j)=0$ for $j\neq i$; and then the restriction of ψ to $F\epsilon_i$ is an injection into Ω. Whether $\mathfrak{o}$ is or is not saturated with respect to K, there alwyas exist elements y in $\mathfrak{o}$ with the property that the saturation $\tilde{\mathfrak{o}}_K$ coincided with saturation, with respect to K, of the (local) subring $k[[x]][y]$ of $\mathfrak{o}$. We have called such elements *x-saturators* of $\mathfrak{o}$, relative to k (see GTS II, Definition 1.1 and Proposition 1.2). If y is an x-saturator of $\mathfrak{o}$, relative to k, then $F=K[y]$, and $\tilde{\mathfrak{o}}_K$ is the set of all elements η of $\bar{\mathfrak{o}}$ which *dominate* y, i.e., such that for any pair of elements ψ, ψ' of H we have (see GTS II, Section 1, inequality (17)):

$$v(\psi(\eta)-\psi'(\eta))\geqslant v(\psi(y)-\psi'(y)).$$

Here v stands for the natural valuation of the (local) subfield F^* of Ω which is the compositum of all the fields $\psi(F)$, $\psi \in H$. [The fields F^* and $\hat{F}^*$ introduced in Section 2 (see relation (12) of that section) are now the same, since $\mathfrak{o}$ is complete; similarly, the valuations v, $\hat{v}$ introduced in the course of the proof of Theorem 2.7 (Section 2) are now equal].

(b) From now on we assume that $\mathfrak{o}$ is saturated with respect to the field K [$=k((x))$]. We have defined in Section 2 (formula (3)) of GTS II a positive integer γ by the following property: if y is an x-saturator of $\mathfrak{o}$ relative to k and if H_i is the subset of H consisting of those ψ in H for which $\psi(\epsilon_i)=1$, then for any pair of distinct indices i, j in the set $\{1, 2, \ldots, h\}$ it is true that

$$\min_{\substack{i \neq j \\ 1 \leqslant i, j \leqslant h}} \ \max_{\substack{\psi \in H_i \\ \psi' \in H_j}} \cdot \{v(\psi(y) - \psi'(y))\} \geqslant \gamma,$$

with equality for at least one pair of distinct indices i, j. The integer γ depends only on $\mathfrak{o}$, x and k, being independent of the choice of the x-saturator y.

(c) We have defined in GTS II *normal subsets* $\{\psi_1, \psi_2, \ldots, \psi_h\}$ of H by the following two properties (see the definition on p. 886 of GTS II):

(i) $\psi_i \in H_i$ $(i=1, 2, \ldots, h)$; (ii) $v(\psi_i(y) - \psi_j(y)) \geqslant \gamma$ for $i \neq j$, and we have shown the existence of normal subsets of H.

(d) If $\{\psi_1, \psi_2, \ldots, \psi_h\}$ is any subset of H such that $\psi_i \in H_i$ for $i=1, 2, \ldots, h$ (but not necessarily a normal subset) and if we set $y_i = \psi_i(y)$, then by GTS I, Section 3, p. 589, the saturated local domain

$$\mathfrak{o}_i = \widetilde{k[[x]][y_i]_K}$$

determines a strictly ascending chain of *intermediate saturation fields* K_{i,α_i} of $\mathfrak{o}_i$ relative to K:

$$K = K_{i,0} < K_{i,1} < \cdots < K_{i,g_i} = \psi_i(F), \tag{2}$$

and the maximal ideal $\mathfrak{m}_i$ of $\mathfrak{o}_i$ has an associated standard decomposition (GTS I, Theorem 4.1)

$$\mathfrak{m}_i = \mathfrak{M}_{i,0} + M_{i,1}^{m_{i,1}} + \cdots + \mathfrak{M}_{i,g_i}^{m_{i,g_i}}, \qquad i = 1, 2, \ldots, h.$$

Here $\mathfrak{M}_{i,\alpha_i}$ is the prime (maximal) ideal of the valuation ring of the natural valuation $v|K_{i,\alpha_i}$ oif the local field K_{i,α_i}; in particular, $\mathfrak{M}_{i,0} = \mathfrak{M} = xk[[x]]$. The integers $g_i, m_{i,1}, m_{i,2}, \ldots, m_{i,g_i}$ depend only on $\mathfrak{o}$, x and i $(1 \leqslant i \leqslant h)$, while if we replace ψ_i by some other element of H_i then the new chain of fields analogous to (2) will be obtained by applying to the chain (2) a suitable K-isomorphism of $\psi_i(F)$ into F^*. We set

$$\beta_{i,\alpha_i} = v(\mathfrak{M}_{i,\alpha_i}^{m_{i,\alpha_i}}) = e_{i,\alpha_i} \cdot m_{i,\alpha_i}, \qquad \begin{array}{l} i = 1, 2, \ldots, h; \\ \alpha_i = 1, 2, \ldots, g_i, \end{array} \tag{3}$$

where e_{i,α_i} is the ramification exponent of $F^*/K_{i,\alpha_i}$. It is convenient to define the β_{i,α_i} also for $\alpha_i = 0$ and for $\alpha_i > g_i$; we set, namely, $\beta_{i,0} = 1$, $\beta_{i,\alpha_i} = +\infty$ if $\alpha_i > g_i$. The integers $\beta_{i,0}, \beta_{i,1}, \ldots, \beta_{i,g_i}$ form a strictly ascending sequence:

$$0 = \beta_{i,0} < \beta_{i,1} < \cdots < \beta_{i,g_i}. \tag{4}$$

We have also defined in GTS II, p. 886 (relation (4)) another positive integer q, as follows: q is the greatest integer with the property that

$$\gamma > \min \cdot \{ \beta_{1,q-1}, \beta_{2,q-1}, \ldots, \beta_{h,q-1} \}. \tag{5}$$

(e) We have proved in GTS II (Theorem 2.1) that

$$\beta_{1,\alpha} = \beta_{2,\alpha} = \cdots = \beta_{h,\alpha} \quad (= \beta_\alpha, \text{ say}), \quad \text{for } \alpha = 0, 1, \ldots, q-1, \tag{6}$$

and that, furthermore, *if the set of K-homomorphisms* $\{\psi_1, \psi_2, \ldots, \psi_h\}$ *is normal*, then

$$K_{1,\alpha} = K_{2,\alpha} = \cdots = K_{h,\alpha} \quad (= K_\alpha, \text{ say}), \quad \text{for } \alpha = 0, 1, \ldots, q-1, \tag{7}$$

and hence $e_{1,\alpha} = e_{2,\alpha} = \cdots = e_{h,\alpha}$ for $\alpha = 0, 1, \ldots, q-1$, and

$$m_{1,\alpha} = m_{2,\alpha} = \cdots = m_{h,\alpha} \quad (= m_\alpha, \text{ say}), \quad \text{for } \alpha = 0, 1, \ldots, q-1. \tag{8}$$

We write $\mathfrak{M}_\alpha$ instead of $\mathfrak{M}_{i,\alpha}$ if $0 \leqslant \alpha \leqslant q-1$:

$$\mathfrak{M}_{1,\alpha} = \mathfrak{M}_{2,\alpha} = \cdots = \mathfrak{M}_{h,\alpha} = \mathfrak{M}_\alpha, \qquad \alpha = 0, 1, \ldots, q-1. \tag{9}$$

(f) If $\{\psi_1, \psi_2, \ldots, \psi_h\}$ is a *normal* set of K-homomorphisms of F into Ω, then the restriction of ψ_i^{-1} to K_α $(0 \leqslant \alpha \leqslant q-1)$ is an injection ω_α of K_α into F which is independent of i (see GTS II, p. 895). Clearly, ω_α is the restriction of ω_{q-1} to K_α. We have also shown (see GTS II, Proposition 3.3) that the ascending chain of fields $L_\alpha = \omega_\alpha(K_\alpha)$ $(0 \leqslant \alpha \leqslant q-1;\ L_0 = K)$ is independent of the choice of the normal set $\{\psi_1, \psi_2, \ldots, \psi_h\}$. In this appendix we shall find it convenient to identify, via ω_α, each L_α with its isomorphic image K_α in Ω. Thus, Ω is now an algebraic closure of the *subfield* K_{q-1} *of* F, and the chain $K_0 < K_1 < \cdots < K_{q-1}$ [see (2) and (7)] consists now of subfields of F (here

$K_0 = K$). *The normal set* $\{\psi_1, \psi_2, \ldots, \psi_h\}$ *being fixed once and for always*, the h homomorphisms ψ_i are now K_{q-1}-homomorphisms of F into Ω.

(g) We are now in a position to state the structure theorem (see GTS II, Theorem 3.4). In that theorem there appears a certain partition $\{I_1, I_2, \ldots, I_r\}$ $(r > 1)$ of the set of indices $\{1, 2, \ldots, h\}$, a partition that we have called the *saturation partition with respect to the pair* $\{\mathfrak{o}, x\}$ (the coefficient field k being fixed in advance). That partition was defined as follows:

Two indices i, j in the set $\{1, 2, \ldots, h\}$ belong to one and the same set I_ρ if there exist K-homomorphisms ψ, ψ' of F into Ω such that $\psi \in H_i$, $\psi' \in H_j$ (i.e., $\psi(\epsilon_i) = \psi'(\epsilon_j) = 1$) and such that $v(\psi(y) - \psi'(y)) > \gamma$; here y is an x-saturator of $\mathfrak{o}$ relative to K, and γ is the integer defined in step (b) above. By the definition of γ we have necessarily $r > 1$ if $h > 1$. Let

$$\epsilon'_\rho = \sum_{i \in I_\rho} \epsilon_i, \qquad \rho = 1, 2, \ldots, r,$$

and let $\mathfrak{o}'_\rho$ denote the saturation of the local ring $\mathfrak{o}\epsilon'_\rho$ with respect to the field $K\cdot\epsilon'_\rho$. Then $\mathfrak{o}'_\rho$ is a subring of the total ring of quotients $F\cdot\epsilon'_\rho$ of $\mathfrak{o}\epsilon'_\rho$, hence is at any rate a subring of F:

$$\mathfrak{o}'_\rho = \widetilde{(\mathfrak{o}\epsilon'_\rho)}_{K\epsilon'_\rho} \subset F\epsilon'_\rho \subset F, \qquad \rho = 1, 2, \ldots, r. \tag{10}$$

Let $\mathfrak{A}'_\rho$ be the following ideal in the ring $\mathfrak{o}'_\rho$:

$$\mathfrak{A}'_\rho = \{\xi \in \mathfrak{o}'_\rho \mid v(\psi_i(\xi)) \geqslant \gamma, \text{ for all } i \text{ in } I_\rho\}. \tag{11}$$

The structure theorem 3.4 of STS II consists in the assertion that the maximal ideal $\mathfrak{m}$ of our saturated local ring $\mathfrak{o}$ is given by the following expression:

$$\mathfrak{m} = \mathfrak{M}_0 + \mathfrak{M}_1^{m_1} + \cdots + \mathfrak{M}_{q-1}^{m_{q-1}} + \sum_{\rho=1}^{r} \mathfrak{A}'_\rho, \tag{12}$$

where q is the integer defined in step (d) (see (5)) and where the $\mathfrak{m}_\alpha$ and $\mathfrak{M}_\alpha$ have been defined in (8) and (9). We have shown (GTS II, Proposition 3.5) that

$$\mathfrak{o}'_\rho = \mathfrak{o}\epsilon'_\rho, \qquad \rho = 1, 2, \ldots, r; \tag{13}$$

$$\widetilde{(\mathfrak{o}\epsilon_i)}_{K\epsilon_i} = \mathfrak{o}\epsilon_i, \qquad i = 1, 2, \ldots, h. \tag{14}$$

These two equalities (13) and (14) will be of crucial importance in our present proof; they have been derived in GTS II as consequences of the structure theorem.

After these preliminaries we proceed to the proof of our theorem. By our induction hypothesis, the theorem is true for each of the r saturated local rings $\mathfrak{o}'_\rho$, since r is *greater* than 1 and since therefore the total ring of quotients $F\epsilon'_\rho$ of each of the r rings $\mathfrak{o}'_\rho$ is a direct sum of less than h fields, namely, $F\epsilon'_\rho = \oplus_{i\in I_\rho} F\epsilon_i$. Now, if k and k' are two coefficient fields of $\mathfrak{o}$, then $k\epsilon'_\rho$ and $k'\epsilon'_\rho$ are coefficient fields of $\mathfrak{o}'_\rho$ (in view of (13)), and the isomorphism $\theta'_\rho : k\epsilon'_\rho \xrightarrow{\sim} k'\epsilon'_\rho$ defined by $\theta'_\rho(c\epsilon'_\rho) = \theta(c)\epsilon'_\rho$ $(c \in k)$ is obviously the canonical isomorphism of $k\epsilon'_\rho$ onto $k'\epsilon'_\rho$. Furthermore, since x is a transversal parameter of $\mathfrak{o}$, $x\epsilon'_\rho$ is a transversal parameter of $\mathfrak{o}'_\rho$ (see GTS II, p. 900 and take into account the above equality (13)), and by Definition (10) of $\mathfrak{o}'_\rho$ the ring $\mathfrak{o}'_\rho$ is $k\epsilon'_\rho$-saturated with respect to $x\epsilon'_\rho$. There exists therefore, by our induction hypothesis, an automorphism φ'_ρ of $F\epsilon'_\rho$ which induces an automorphism of $\mathfrak{o}'_\rho$ and which, moreover, satisfies the conditions

$$\varphi'_\rho | k\epsilon'_\rho = \theta'_\rho, \qquad \varphi'_\rho(x\epsilon'_\rho) = x\epsilon'_\rho;$$

$$\varphi'_\rho(\epsilon_i) = \epsilon_i \quad \text{for all } i \text{ in } I_\rho,$$

and

$$\varphi'_\rho(\eta_i)\cdot\eta_i^{-1} = \epsilon_i \in \bar{\mathfrak{m}}_i, \quad \text{for all } i \text{ in } I_\rho.$$

Here η_i is any nonzero element of the field $F\epsilon_i$, η_i^{-1} is its inverse in that field, and $\bar{\mathfrak{m}}_i$ is the maximal ideal in the local domain $\bar{\mathfrak{o}}\epsilon_i$. If we now set

$$\varphi = \varphi'_1 + \varphi'_2 + \cdots + \varphi'_r,$$

then φ will be an automorphism of F satisfying conditions 1 and 2 of the theorem, and, furthermore, we will have $\varphi(x) = x$. We have therefore only to show (and this is the really nontrivial part of the theorem) that $\varphi(\mathfrak{o}) = \mathfrak{o}$. It will be sufficient to prove that

$$\varphi(\mathfrak{o}) \subset \mathfrak{o}, \tag{15}$$

since the integral closure $\bar{\mathfrak{o}}$ of $\mathfrak{o}$ in F is also the integral closure of both subrings $k[[x]]$, $k'[[x]]$ of $\mathfrak{o}$ and since the $\mathfrak{o}$-module $\bar{\mathfrak{o}}/\mathfrak{o}$ and the $\varphi(\mathfrak{o})$-module $\bar{\mathfrak{o}}/\varphi(\mathfrak{o})$ have the same length (in view of the fact that $\varphi(k[[x]]) = k'[[x]]$ and that consequently $\varphi(\bar{\mathfrak{o}}) = \bar{\mathfrak{o}}$). To prove (15) it will be sufficient to prove that $\varphi(\mathfrak{m}) \subset \mathfrak{o}$, since $\mathfrak{o} = k + \mathfrak{m} = k' + \mathfrak{m}$ and $\varphi(k) = k'$.

Let then η be any element of $\mathfrak{m}$. We use the structure theorem, as expressed by (12), *and we first consider the case in which* $\eta \in \sum_{\rho=1}^{r} \mathfrak{A}'_\rho$. We have then $\eta = \eta'_1 + \eta'_2 + \cdots + \eta'_r$, $\eta'_\rho \in \mathfrak{A}'_\rho$, and $\varphi(\eta) = \varphi'_1(\eta'_1) + \varphi'_2(\eta'_2) + \cdots +$

$\varphi_r'(\eta_r')$. We shall show that $\varphi_\rho'(\eta_\rho')\in\mathfrak{A}_\rho'$ (for $\rho=1,2,\ldots,r$), and this will establish the inclusion $\varphi(\eta)\in\mathfrak{m}$.

Since $\eta_\rho'\in\mathfrak{A}_\rho'\subset\mathfrak{o}_\rho'$ and $\varphi_\rho'(\mathfrak{o}_\rho')=\mathfrak{o}_\rho'$, it follows that $\varphi_\rho'(\eta_\rho')\in\mathfrak{o}_\rho'$. Hence we have only to show (in view of the definition (11) of $\mathfrak{A}_\rho'$) that

$$v\big(\psi_i\big(\varphi_\rho'(\eta_\rho')\big)\big)\geqslant\gamma,\quad\text{for all } i \text{ in } I_\rho. \tag{16}$$

We have $F\epsilon_i=F\epsilon_\rho'\cdot\epsilon_i$ and $\varphi_\rho'(\epsilon_i)=\epsilon_i$, if $i\in I_\rho$. Hence φ_ρ' induces an automorphism of the field $F\epsilon_i$, for all i in I_ρ. The restriction of ψ_i to $F\epsilon_i$ is an isomorphism $F\epsilon_i\xrightarrow{\sim}F_i(=\psi_i(F))$. Also the product $\psi_i\cdot\varphi_\rho'$ induces an isomorphism $F\epsilon_i\xrightarrow{\sim}F_i$. It follows that for any index i in I_ρ there exists an automorphism λ_i of F_i such that

$$\psi_i\cdot\varphi_\rho'=\lambda_i\cdot\psi_i. \tag{17}$$

Since $\varphi_\rho'(x\epsilon_i)=x\epsilon_i$ for all i in I_ρ and since $\psi_i(x\epsilon_i)=x$, it follows from (17) that

$$\lambda_i(x)=x,\quad\text{for all } i \text{ in } I_\rho. \tag{18}$$

If c is any element of k and if $c'=\theta(c)$, then $\varphi_\rho'(c\epsilon_i)=c'\epsilon_i$ if $i\in I_\rho$. Now, c' belongs to $\mathfrak{o}$ and hence is integrally dependent on $k[[x]]$. Therefore $c'\epsilon_i$ is integrally dependent on $k\epsilon_i[[x\epsilon_i]]$. From this it follows that $\psi_i(c'\epsilon_i)$ is integrally dependent on $\psi_i(k\epsilon_i[[x\epsilon_i]])$, i.e., on $k[[x]]$. Since $\psi_i(c'\epsilon_i)=(\psi_i\cdot\varphi_\rho')(c\epsilon_i)=(\lambda_i\cdot\psi_i)(c\epsilon_i)$ [by (17)] $=\lambda_i(c)$, we see that every element of the field $\lambda_i(k)$ is integral over $k[[x]]$. This, together with (18), shows that the automorphism λ_i of F_i induces an automorphism of the integral closure of $k[[x]]$ in F_i. Now, F_i is a local field, and the valuation ring of its natural valuation is precisely the integral closure of $k[[x]]$ in F_i. The natural valuation of F_i is the restriction to F_i of the natural valuation v of F^*. We have therefore proved that if ξ_i is any element of F_i then $v(\lambda_i(\xi_i))=v(\xi_i)$. By (17) this is equivalent to saying that we have

$$v\big(\psi_i\big(\varphi_\rho'(\xi_\rho')\big)\big)=v\big(\psi_i(\xi_\rho')\big),\quad\begin{array}{l}\text{for all } \xi_\rho' \text{ in } F\epsilon_\rho'\\ \text{and all } i \text{ in } I_\rho.\end{array} \tag{19}$$

Now, let us apply (19) to our element η_ρ' of $\mathfrak{A}_\rho'$. Since we have $v(\psi_i(\eta_\rho'))\geqslant\gamma$ for all i in I_ρ (by the definition of $\mathfrak{A}_\rho'$), (16) follows directly from (19) if we set $\xi_\rho'=\eta_\rho'$. This completes the proof of the inclusion $\varphi(\eta)\in\mathfrak{o}$ in the case under consideration (i.e., in the case in which $\eta\in\sum_{\rho=1}^{r}\mathfrak{A}_\rho'$).

To complete the proof of our theorem we have now only to show that

$$\varphi(\mathfrak{M}_\alpha^{m_\alpha})\subset\mathfrak{o},\quad\text{for } \alpha=0,1,\ldots,q-1. \tag{20}$$

The inclusion (20) is obvious if $\alpha=0$, since $\mathfrak{M}_0=xk[[x]]$ and $\varphi(\mathfrak{M}_0)=xk'[[x]]\subset\mathfrak{o}$. So we shall assume that $\alpha\geqslant1$. To prove (20) we have to show the following: if y is an x-saturator of $\mathfrak{o}$ relative to k, and if ψ, ψ' are any two K-homomorphisms of F into Ω then

$$v[\psi(\varphi(\eta))-\psi'(\varphi(\eta))] \geqslant v[\psi(y)-\psi'(y)],$$
$$\text{for all } \eta \text{ in } \mathfrak{M}_{\alpha}^{m_{\alpha}},\ \alpha=1, 2,\ldots,q-1. \tag{21}$$

It is sufficient to prove (21) for one fixed element ψ in H, while ψ' is allowed to range over H. For simplicity of notations, we shall take for ψ the first element ψ_1 of our normal set $\{\psi_1, \psi_2,\ldots,\psi_h\}$. Let $\psi' \in H_i$ $(1 \leqslant i \leqslant h)$. Then we can write ψ' in the form: $\psi' = \tau\cdot\psi_i$, where τ is a suitable K-isomorphism of the field F_i $(=\psi_i(F))$ into F^*. Let k_α denote the algebraic closure of k in K_α. By our previous identification of K_α with the subfield L_α of F [see step (f)], φ acts on k_α. We set

$$k'_\alpha = \varphi(k_\alpha), \qquad \alpha = 0, 1,\ldots,q-1. \tag{22}$$

We denote by G_α the Galois group of F^*/K_α. In the sequel α stands for any of the integers $0, 1,\ldots,q-1$ (note that $k_0 = k$ and $k'_0 = k'$). We shall prove several lemmas.

Lemma A2. *If $c' \in k'_\alpha$ and $\tau \in G_\alpha$, then*

$$v[\psi(c')-\psi'(c')] \geqslant v[\psi(y)-\psi'(y)].$$

(Here ψ is ψ_1, and ψ' is of the form $\tau\cdot\psi_i$, indicated above).

Proof. By (22) and by the definition of k_α, the field k'_α is algebraic over the field k' $(=\theta(k)=\varphi(k))$. Let

$$f(c') = \sum_\nu a'_\nu c'^\nu = 0, \qquad a'_\nu \in k', \tag{23}$$

be the monic irreducible equation for c' over k', and let $c' = \varphi(c)$, $c \in k_\alpha$. We know that $\varphi(\bar{\mathfrak{o}}) = \bar{\mathfrak{o}}$, and since $\varphi(\epsilon_i) = \epsilon_i$ for all $i = 1, 2,\ldots,h$, we have also $\varphi(\bar{\mathfrak{o}}\epsilon_i) = \bar{\mathfrak{o}}\epsilon_i$ and $\varphi(\bar{\mathfrak{m}}_i) = \bar{\mathfrak{m}}_i$, where $\bar{\mathfrak{m}}_i$ is the maximal ideal of the local domain $\bar{\mathfrak{o}}\epsilon_i$. Since $\varphi|\bar{k}\epsilon_i = \bar{\theta}_i$, it follows that

$$c' - c \in \bar{\mathfrak{m}}_1 + \bar{\mathfrak{m}}_2 + \cdots + \bar{\mathfrak{m}}_h. \tag{24}$$

We have $\psi(\bar{\mathfrak{m}}_1 + \bar{\mathfrak{m}}_2 + \cdots + \bar{\mathfrak{m}}_k) = \psi(\bar{\mathfrak{m}}_1) \subset \mathfrak{m}^*$, where $\mathfrak{m}^*$ is the prime ideal of the valuation v. Similarly, we have $\psi'(\bar{\mathfrak{m}}_1 + \bar{\mathfrak{m}}_2 + \cdots + \bar{\mathfrak{m}}_h) = \psi'(\bar{\mathfrak{m}}_i) \subset \mathfrak{m}^*$. Therefore, by (24), we have $v[\psi(c')-\psi(c)] > 0$ and $v[\psi'(c')-\psi'(c)] > 0$. Now, $\psi(c) = \psi_1(c) = c$, since $\psi_1|K_\alpha$ is the identity. Similarly, we have $\psi_i(c) = c$, and since $\psi' = \tau\cdot\psi_i$ and τ belongs to G_α (by the assumption made in the lemma), it follows that also $\psi'(c) = c$. Therefore the above inequalities become now: $v[\psi(c') - c] > 0$, $v[\psi'(c') - c] > 0$, and hence we have

$$v[\psi(c') - \psi'(c')] > 0. \tag{25}$$

We set $F(u) = f^{\psi'}(u) = \Sigma\psi'(a'_\nu)u^\nu$, where u is an indeterminate. By (23) we have $F(\psi'(c')) = 0$. Thus, we can write:

$$\begin{aligned} F(\psi(c')) &= F(\psi(c')) - F(\psi'(c')) \\ &= [\psi(c') - \psi'(c')]\{F'(\psi'(c')) + \\ &\qquad \tfrac{1}{2}[\psi(c') - \psi'(c')]F''(\psi'(c')) + \cdots\}. \end{aligned} \tag{26}$$

The successive derivatives $F'(\psi'(c'))$, $F''(\psi'(c'))$, $\cdots$ are elements of the subfield $\psi'(k')[\psi'(c')]$ of F^*, and all the elements of this subfield are integral over the power series ring $k[[x]]$, since k' is integral over k, ψ' is the identity on $k[[x]]$, and $\psi'(c')$ is integral over $\psi'(k')$. Therefore all the nonzero elements of the field $\psi'(k')[\psi'(c')]$ have value zero in the valuation v. Since $F(u)$ is the minimal polynomial of $\psi'(c')$ over $\psi'(k')$ and since we are in characteristic zero, it follows that $v(F'(\psi'(c')) = 0$, and hence, in view of (25), we can conclude that the expression in the curly brackets on the right-hand side of (26) has value zero in v. Therefore it follows from (26) that

$$v(F(\psi(c'))) = v[\psi(c') - \psi'(c')]. \tag{27}$$

By (23) we have $\Sigma_\nu\psi(a'_\nu)[\psi(c')]^\nu = 0$, and hence we can write

$$\sum_\nu [\psi(a'_\nu) - \psi'(a'_\nu)][\psi(c')]^\nu + F(\psi(c')) = 0. \tag{28}$$

Since the a'_ν belong to k', and $k' \subset \mathfrak{o}$, we have that $v[\psi(a'_\nu) - \psi'(a'_\nu)] \geqslant v[\psi(y) - \psi'(y)]$. Furthermore, we also have $v(\psi(c')) \geqslant 0$, since $\psi(c')$ is algebraic over the field $\psi(k')$ and since the elements of this field are integral over $k[[x]]$. Therefore it follows from (28) that $v(F(\psi(c'))) \geqslant v[\psi(y) - \psi'(y')]$, and from this our lemma follows in view of (27).

LEMMA A3. *Under the assumptions made in Lemma* A2, *we have* $v[\psi(c') - \psi'(c')] \geqslant \beta_{\alpha+1}$ *if* $\alpha < q-1$, *and* $v[\psi(c') - \psi'(c')] \geqslant \gamma$ *if* $\alpha = q-1$. (*Here* γ *is the integer introduced in step* (b) *of the preliminaries.*)

Proof. We have assumed that $\psi = \psi_1$ and that $\psi' = \tau\cdot\psi_i$. If we set $y_1 = \psi_1(y)$ and $y_i = \psi_i(y)$, then by Theorem 2.1 and Corollary 2.2 of GTS II we have the following standard decompositions of y_1 and y_i:

$$y_1 = y^{(0)} + y^{(1)} + \cdots + y^{(q-1)} + y_{1,q} + \cdots + y_{1,g_1},$$

$$y_i = y^{(0)} + y^{(1)} + \cdots + y^{(q-1)} + \bar{y}_{i,q-1} + y_{i,q} + \cdots + y_{i,g_i}.$$

Here $y_{i,\alpha_i} \in \mathfrak{M}_{i,\alpha_i}^{m_{i,\alpha_i}}$, $v(y_{i,\alpha_i}) = \beta_{i,\alpha_i}$, $v(\bar{y}_{i,q-1}) \geqslant \gamma$, and, in particular, if $\alpha < q-1$ then $y_{1,\alpha} = y_{i,\alpha} = y^{(\alpha)} \in M_\alpha^{m_\alpha}$ and $v(y^{(\alpha)}) = \beta_\alpha$ [see (3), (6) and (8)]. For $\alpha = q-1$ we have $y_{1,q-1} = y^{(q-1)}$, $y_{i,q-1} = y^{(q-1)} + \bar{y}_{i,q-1}$, and $v(\bar{y}_{i,q-1}) \geqslant \gamma > \beta_{q-1}$. Since $\tau \in G_\alpha$, we have

$$\psi(y) - \psi'(y) = \psi_1(y) - (\tau \cdot \psi_i)(y) = [y^{(\alpha+1)} - \tau(y^{(\alpha+1)})] + \cdots$$
$$+ [y^{(q-1)} - \tau(y^{(q-1)})] - \tau(\bar{y}_{i,q-1}) + [y_{1,q} - \tau(y_{i,q})] + \cdots, \quad \text{if } \alpha < q-1, \tag{29}$$

and

$$\psi(y) - \psi'(y) = -\bar{y}_{i,q-1} + [y_{1,q} - \tau(y_{i,q})] + \cdots, \quad \text{if } \alpha = q-1. \tag{30}$$

From (29) and (30) follow respectively the following inequalities:

$$v[\psi(y) - \psi'(y)] \geqslant \beta_{\alpha+1}, \quad \text{if } \alpha < q-1, \tag{29'}$$

$$v[\psi(y) - \psi'(y)] \geqslant \gamma > \beta_{q-1}, \quad \text{if } \alpha = q-1, \qquad \begin{matrix} \psi = \psi_1 \\ \psi' = \tau \cdot \psi_i, \ \tau \in G_\alpha. \end{matrix} \tag{30'}$$

In regard to (29′) we have only to recall the inequalities (4) and to note that $v(\tau(\bar{y}_{i,q-1})) = v(\bar{y}_{i,q-1}) \geqslant \gamma > \beta_{q-1}$. As to (30′) we have only to recall that by the definition of the integer q we have [see (5)] $\gamma \leqslant \min \cdot \{\beta_{1,\alpha}, \beta_{2,\alpha}, \ldots, \beta_{h,\alpha}\}$ for all $\alpha \geqslant q$. The lemma now follows from Lemma A2 and the inequalities (29′), (30′).

We observe that in (29′) the *equality sign holds* if $\tau \in G_\alpha$ and $\tau \notin G_{\alpha+1}$:

$$v[\psi(y) - \psi'(y)] = \beta_{\alpha+1}, \quad \text{if } \alpha < q-1 \ \text{ and } \ \tau \in G_\alpha, \quad \tau \notin G_{\alpha+1}. \tag{29''}$$

For if $\tau \notin G_{\alpha+1}$ then $v[y^{(\alpha+1)} - \tau(y^{(\alpha+1)})] = \beta_{\alpha+1}$ (see equalities (35) in GTS II, p. 882).

Lemma A3 is essentially a generalization of Lemma 9.2 of our paper GTS I, with g and β_g of that lemma being replaced here by q and γ respectively.

LEMMA A4. *If η is any regular element of F (hence a unit in F) then*

$$[\varphi(\eta)-\eta]/\eta \in \bar{\mathfrak{m}} = \bar{\mathfrak{m}}_1 + \bar{\mathfrak{m}}_2 + \cdots + \bar{\mathfrak{m}}_h,$$

where $\bar{\mathfrak{m}}_i = (=\bar{\mathfrak{m}}\epsilon_i)$ is the maximal ideal of the local domain $\bar{\mathfrak{o}}\epsilon_i$ (and hence $\bar{\mathfrak{m}}$ is the intersection of the maximal ideals of the semi-local ring $\bar{\mathfrak{o}}$).

Proof. We have $\varphi(\eta)=\Sigma_{i=1}^h \varphi(\eta\epsilon_i)$ and therefore $\varphi(\eta)-\eta=\Sigma_{i=1}^h[\varphi(\eta\epsilon_i)-\eta\epsilon_i]$. Since φ satisfies condition 2 of the theorem, it follows that $\varphi(\eta)-\eta=\Sigma_{i=1}^h \eta\epsilon_i\cdot\bar{\zeta}_i$, where $\bar{\zeta}_i\in\bar{\mathfrak{m}}_i$, i.e., $[\varphi(\eta)-\eta]/\eta=\Sigma_{i=1}^h\bar{\zeta}_i\in\bar{\mathfrak{m}}$. Q.E.D.

The following lemma is a generalization of Lemma 9.3 of GTS I:

LEMMA A5. *If $\eta\in\mathfrak{o}\cap K_{q-1}$ $(\eta\neq 0)$ and $v(\eta)\geqslant\beta_\alpha$ for some $\alpha=0, 1,\ldots,q-1$, then the element $\eta'=\varphi(\eta)$ satisfies the following inequality: $v[\psi(\eta')-\psi'(\eta')]>\beta_\alpha$, where it is assumed (as in Lemma A2) that if $\psi'=\tau\cdot\psi_i$ then $\tau\in G_\alpha$.*

Proof. We have $\psi(\eta')-\psi'(\eta')=[\psi(\eta')-\psi(\eta)]+[\psi(\eta)-\psi'(\eta)]+[\psi'(\eta)-\psi'(\eta')]$. Since $0\neq\eta$ and η belongs to the *field* K_{q-1}, η is a unit in F. Therefore, by Lemma A4, and taking into account that $\psi(\eta)=\eta$ and $\psi'(\eta)=\tau(\eta)$ (since $\eta\in K_{q-1}$, and ψ,ψ_i are K_{q-1}-homomorphisms), we find that $v[\psi(\eta')-\psi(\eta)]>v(\psi(\eta))=v(\eta)\geqslant\beta_\alpha$, and that similarly $v[\psi'(\eta)-\psi'(\eta')]>v(\psi'(\eta))=v(\tau(\eta))=v(\eta)\geqslant\beta_\alpha$. Since $\eta\in\mathfrak{o}$ we have $v[\psi(\eta)-\psi'(\eta)]\geqslant v[\psi(y)-\psi'(y)]$, and since $\tau\in G_\alpha$ it follows from the inequalities (29′) and (30′) that $v[\psi(\eta)-\psi'(\eta)]>\beta_\alpha$. Thus we have shown that $\psi(\eta')-\psi'(\eta')$ is a sum of three elements each of which has value greater than β_α in the valuation v. This completes the proof of the lemma.

Finally, we prove the following lemma which is a generalization of Lemma 9.4 of GTS I:

LEMMA A6. *Let $x_0(=x), x_1,\ldots,x_{q-1}$ be uniformizing parameters of $K, K_1,\ldots,K_{q-1}$ respectively, such that for each $\alpha=1, 2,\ldots,q-1$ we have* (compare with GTS I, (10), p. 642)

$$x_{\alpha-1}=a_\alpha x_\alpha^{n_\alpha}, \qquad a_\alpha\in k_\alpha(=\textit{algebraic closure of } k \textit{ in } K_\alpha). \tag{31}$$

Let $x'_\alpha=\varphi(x_\alpha)$ and let α, α' be any two integers such that $0\leqslant\alpha\leqslant\alpha'\leqslant q-1$. If ψ, ψ' are as above (i.e., $\psi=\psi_1, \psi'=\tau\cdot\psi_i$) and if $\tau\in G_{\alpha'}$, then

$$v\left[\frac{\psi(x'_\alpha)}{\psi'(x'_\alpha)}-1\right]\geqslant v[\psi(y)-\psi'(y)].$$

Proof. The lemma is trivial if $\alpha=0$, for in that case $x_\alpha=x'_\alpha=x$ and $\psi(x'_\alpha)=\psi'(x'_\alpha)=x$. We shall therefore use induction on α and assume that $\alpha>0$. If we set $a'_\alpha=\varphi(a_\alpha)$ then we have by (31): $x'_{\alpha-1}=a'_\alpha x'^{n_\alpha}_\alpha$, and thus

$$\frac{\psi(x'_{\alpha-1})}{\psi'(x'_{\alpha-1})}=\frac{\psi(a'_\alpha)}{\psi'(a'_\alpha)}\cdot\left[\frac{\psi(x'_\alpha)}{\psi'(x'_\alpha)}\right]^{n_\alpha}. \tag{32}$$

By our induction hypothesis we have:

$$v\left[\frac{\psi(x'_{\alpha-1})}{\psi'(x'_{\alpha-1})}-1\right]\geqslant v[\psi(y)-\psi'(y)]. \tag{33}$$

Since $a'_\alpha\in k'_\alpha\subset k'_{\alpha'}$ and since $\tau\in G_{\alpha'}$, it follows from Lemma A2 that $v[\psi(a'_\alpha)-\psi'(a'_\alpha)]\geqslant v[\psi(y)-\psi'(y)]$, and therefore

$$v\left[\frac{\psi(a'_\alpha)}{\psi'(a'_\alpha)}-1\right]\geqslant v[\psi(y)-\psi'(y)], \tag{34}$$

since, as we have already had occasion to point out in the course of the proof of Lemma A2, the element $\psi'(a'_\alpha)$ is a unit in the valuation ring of v [since $0\neq a_\alpha\in k_\alpha$, we have $0\neq a'_\alpha\in k'_\alpha$, and we have shown in the course of the proof of Lemma A2 that if c' is any nonzero element of k'_α then $v(\psi'(c'))=0$]. From (32), (33) and (34) follows that

$$v\left(\left[\frac{\psi(x'_\alpha)}{\psi'(x'_\alpha)}\right]^{n_\alpha}-1\right)\geqslant v[\psi(y)-\psi'(y)]>0. \tag{35}$$

We now set $\eta=x_\alpha^{m_\alpha}$. Then η is an element of $\mathfrak{o}$, by (12), and belongs to the field K_α, hence also to the field K_{q-1}. Since $\alpha\leqslant\alpha'$, we have $G_{\alpha'}\subset G_\alpha$, and hence $\tau\in G_\alpha$. Therefore, by Lemma A5, we have

$$v[\psi(\eta')-\psi'(\eta')]>\beta_\alpha. \tag{36}$$

We now make the following observation:

If ξ is any element of F and ψ is any K-homomorphism of F into Ω, then

$$v((\psi\cdot\varphi)(\xi))=v(\psi(\xi)). \tag{37}$$

For, if, say, $\psi \in H_j$, then the kernel of ψ is $\oplus_{\nu \neq j} F\epsilon_\nu$, and the kernel of $\psi \cdot \varphi$ is $\varphi^{-1}(\mathrm{Ker}\,\psi)$. Since $\varphi(F\epsilon_\nu) = F\epsilon_\nu$ for all $\nu = 1, 2, \ldots, h$, it follows that $\mathrm{Ker}(\psi \cdot \varphi) = \mathrm{Ker}\,\psi$. Consequently there exists an isomorphism λ of $\psi(F)$ into Ω such that $\psi \cdot \varphi = \lambda \cdot \psi$. Now, we have $k' = \varphi(k) \subset \mathfrak{o}$, k' is integral over $k[[x]]$, and therefore $\psi(k')[[x]]$ is integral over $k[[x]]$ (since ψ is the identity on $k[[x]]$). From the equality $\psi \cdot \varphi = \lambda \cdot \psi$ follows that the λ-image of the subring $k[[x]]$ of $\psi(F)$ is the subring $\psi(k'[[x]])$ of $\lambda(\psi(F))$, and this subring-as we have just pointed out-is integral over $k[[x]]$. Now, the integral closure of $k[[x]]$ in $\psi(F)$ is the valuation ring V of the valuation of $\psi(F)$ induced by v. Since we have just seen that the ring $\lambda(k[[x]])$ is integral over $k[[x]]$ (and is therefore contained in the valuation ring of v), it follows that the field isomorphism λ: $\psi(F) \xrightarrow{\sim} (\lambda \cdot \psi)(F)$ sends the restriction of v to $\psi(F)$ into the restriction of the *same* valuation v to $(\lambda \cdot \psi)(F)$. From this (37) follows.

Applying (37) to the element $\xi = \eta = x_\alpha^{m_\alpha}$ and to the K-homomorphism ψ', we find that $v(\psi'(x_\alpha'^{m_\alpha})) = v(\psi'(x_\alpha^{m_\alpha})) = v(\tau(x_\alpha^{m_\alpha})) = \beta_\alpha$. Therefore, by (36), we find that

$$v\left(\left[\frac{\psi(x'_\alpha)}{\psi'(x'_\alpha)}\right]^{m_\alpha} - 1\right) > 0. \tag{38}$$

Since m_α and n_α are relatively prime [see (42) on p. 980 of SES III), we conclude from (35) and (38) that $v[\psi(x'_\alpha)/\psi'(x'_\alpha) - 1] > 0$. Let us denote the quotient $\psi(x'_\alpha)/\psi'(x'_\alpha)$ by ζ. We have then $u(\zeta - 1) > 0$, i.e., the v-residue of ζ is 1. We also have by (35): $v(\zeta^{n_\alpha} - 1) \geqslant v[\psi(y) - \psi'(y)]$. Since $\zeta^{n_\alpha} - 1 = (\zeta - 1)(\zeta^{n_\alpha - 1} + \zeta^{n_\alpha - 2} + \cdots + 1)$ and since the v-residue of ζ is 1, it follows that the v-residue of $\zeta^{n_\alpha - 1} + \zeta^{n_\alpha - 2} + \cdots + 1$ is $n_\alpha \neq 0$. Therefore, $v(\zeta - 1) = v(\zeta^{n_\alpha - 1})$, i.e., $v(\zeta - 1) \geqslant v[\psi(y) - \psi'(y)]$. This completes the proof of the lemma.

We can now complete the proof of the inclusion (20). We have to show that if c is any element of k_α $(0 \leqslant \alpha \leqslant q-1)$ and μ is any integer $\geqslant m_\alpha$, and if we set $c' = \varphi(c)$, $\varphi(x_\alpha) = x'_\alpha$, then

$$v[\psi(c'x_\alpha'^{\mu}) - \psi'(c'x_\alpha'^{\mu})] \geqslant v[\psi(y) - \psi'(y)], \tag{39}$$

where ψ is our fixed K-homomorphism ψ_1 of the preassigned normal set $\{\psi_1, \psi_2, \ldots, \psi_h\}$ and ψ' is an arbitrary element $\tau \cdot \psi_i$ of H. Let α' be the greatest integer such that $0 \leqslant \alpha' \leqslant q-1$ and $\tau \in G_{\alpha'}$. [Note that τ always belongs at least to $G_0(= \mathcal{G}(F^*/K))$]. We consider two cases: (1) $\alpha' < \alpha$ and (2) $\alpha \leqslant \alpha'$.

Case 1. $\alpha' < \alpha$. In that case $\alpha' < q-1$ and hence we have by (29''): $v[\psi(y) - \psi'(y)] = \beta_{\alpha'+1}$. On the other hand, since $\mu \geqslant m_\alpha$ it follows that $v(cx_\alpha^{\mu})$

$\geqslant \beta_\alpha \geqslant \beta_{\alpha'+1}$, and since by (37) we have $v(\psi(c'x_\alpha'^\mu)) = v(\psi(cx_\alpha^\mu)$, it follows that $v(\psi(c'x_\alpha'^\mu)) \geqslant \beta_\alpha$ (since $\psi|K_\alpha$ is the identity, whence $\psi(cx_\alpha^\mu) = cx_\alpha^\mu$). In a similar way it follows from (37) that $v(\psi'(c'x_\alpha'^\mu) \geqslant \beta_\alpha$. The inequality (39) follows now from this last two inequalities and from $v[\psi(y) - \psi'(y)] = \beta_{\alpha'+1} \leqslant \beta_\alpha$.

Case 2. $\alpha \leqslant \alpha'$. In this case we write the difference $\psi(c'x_\alpha'^\mu) - \psi'(c'x_\alpha'^\mu)$ in the form $[\psi(c') - \psi'(c')]\psi(x_\alpha'^\mu) + \psi'(c')[\psi(x_\alpha'^\mu) - \psi'(x_\alpha'^\mu)]$, and we note that since $\alpha \leqslant \alpha'$ and $c' \in k_\alpha'$ we have a fortiori $c' \in k_{\alpha'}'$. Therefore, by Lemma A2 (where α is to be replaced by α') we have $v[\psi(c') - \psi'(c')] \geqslant v[\psi(y) - \psi'(y)]$. Since $v(\psi(x_\alpha'^\mu)) \geqslant 0$ and $v(\psi'(c')) \geqslant 0$, we see from the above expression of $\psi(c'x_\alpha'^\mu) - \psi'(c'x_\alpha'^\mu)$ that in order to prove (39) it is sufficient to prove (39) in the special case $c' = 1$. In this special case, the inequality (39) is a direct consequence of Lemma A6. This completes the proof of the theorem.

COROLLARY A7. *If k, k' are two coefficient fields of a complete, equicharacteristic zero, equidimensional local ring* $\mathfrak{o}$, *of Krull dimension* 1, *then the k-saturation of $\mathfrak{o}$ (i.e., the k-saturation of $\mathfrak{o}$ with respect to any transversal parameter; see GTS II, Proposition* 4.6) *coincides with the k'-saturation of* $\mathfrak{o}$, *and is therefore a ring between $\mathfrak{o}$ and $\bar{\mathfrak{o}}$ which is independent of the choice of the coefficient field k (this ring is denoted by $\tilde{\mathfrak{o}}$ and is called the (absolute) saturation of* $\mathfrak{o}$).

This is an immediate consequence of the theorem and is an exact generalization of Corollary 9.5 of GTS I (in this latter corollary, $\mathfrak{o}$ is a local *domain*).

The rest of the appendix will deal with the generalization of two results proved in GST I in the case in which $\mathfrak{o}$ is a local *domain*, namely of Lemma 2.2, GST I (p. 583) and Corollary 7.4, GST I (p. 629).

LEMMA A8 (*generalization of Lemma* 2.2 *of GST I*). *Let x be a parameter of our complete local ring $\mathfrak{o}$ (of Krull dimension 1) and let k be a coefficient field of $\mathfrak{o}$. Let $K = k((x))$ and let K_0 be a subfield of K such that the relative degree $[K:K_0]$ is finite. Then*:

1. *We have $\tilde{\mathfrak{o}}_{K_0} \subset \tilde{\mathfrak{o}}_K$; and, in particular, if $\mathfrak{o}$ is saturated with respect to K, $\mathfrak{o}$ is also saturated with respect to K_0.*

2. *In the special case in which K_0 is a field of the form $k_0((x))$, where k_0 is a subfield of k such that $[k:k_0]$ is finite, we have $\tilde{\mathfrak{o}}_{K_0} = \tilde{\mathfrak{o}}_K$, and, in particular, $\mathfrak{o}$ is saturated with respect to K if and only $\mathfrak{o}$ is saturated with respect to K_0.*

Proof. (1) We first show that the saturation of $\mathfrak{o}$ with respect to K_0 is defined. For this we have only to show that $\mathfrak{o}$ is integral over the ring $\mathfrak{o} \cap K_0$.

Now, $\mathfrak{o} \cap K_0 = \mathfrak{o} \cap K \cap K_0 = k[[x]] \cap K_0$, and $\mathfrak{o}$ is integral over $k[[x]]$. *So we have only to show that* $k[[x]]$ *is integral over* $k[[x]] \cap K_0$.

Let v be the natural valuation of the local field K $(= k((x)))$. Then $k[[x]]$ is the valuation ring of v, and $k[[x]] \cap K_0$ is the valuation ring of the restriction $v|K_0$. Since we are in characteristic zero, we can apply Proposition 1.4 of GTS I (p. 578), in which the pair of fields (K, F) is now to be replaced by the pair (K_0, K). By that proposition, v is the only extension of $v|K_0$ to K. Hence the valuation ring $k[[x]]$ of v is integral over the valuation ring $k[[x]] \cap K_0$ of $v|K_0$, as was asserted.

The second part of the statement (1) of the lemma (beginning with the words "if $\mathfrak{o}$ is saturated $\cdots$") is a consequence of the first part, and, conversely, the inclusion $\tilde{\mathfrak{o}}_{K_0} \subset \tilde{\mathfrak{o}}_K$ is a consequence of the second part of statement (1). The proof is word by word a repetition of the similar proof given on p. 584 of GTS I, and is—at any rate—straightforward. The only change to be made in the proof given in GTS I is that in asserting that k is also a coefficient field of $\tilde{\mathfrak{o}}_K$ we cannot make use of Proposition 2.1 of GTS I (which concerns only the case of a local *domain*), but must fall back on the more general result proved in an earlier paper of ours, namely on Theorem 4.1 of SES III.

We now proceed to prove the second part of statement (1) of the lemma. We denote by H and H_0, respectively, the set of all K-homomorphisms and the set of all K_0-homomorphisms of F into Ω (where Ω is an algebraic closure of K). We have $H \subset H_0$, and also H_0 is a finite set (since $[K : K_0]$ is finite). We denote by F^* the compositum of all the subfields $\psi(F)$ of Ω, $\psi \in H$, and by F_0^* the compositum of all the subfields $\psi_0(F)$ of Ω, $\psi_0 \in H_0$. We have then $K_0 \subset K \subset F^* \subset F_0^*$, F^* and F_0^* are local fields which are Galois extensions of K and K_0 respectively. We denote by v the natural valuation of F^*. By part (3) of Lemma 1.2 of GTS I, v has a unique extension to F_0^*. We shall denote that extension by v_0.

Assume that $\mathfrak{o}$ is saturated with respect to K. Let ζ be any element of $\mathfrak{o}$ and let η be an element of $\tilde{\mathfrak{o}}$ which dominates ζ with respect to a field K_0, i.e., such that for any two elements ψ_0, ψ_0' of H_0 we have that the $v_0[\psi_0(\eta) - \psi_0'(\eta)] \geqslant v_0[\psi_0(\zeta) - \psi_0'(\zeta)]$. Since $H \subset H_0$, it follows that η also dominates ζ with respect to K, i.e., that $v[\psi(\eta) - \psi'(\eta)] \geqslant v[\psi(\zeta) - \psi'(\zeta)]$ for any pair of elements ψ, ψ' of H. Since $\mathfrak{o}$ is saturated with respect to K, it follows that $\eta \in \mathfrak{o}$, showing that $\mathfrak{o}$ is also saturated with respect to K_0.

(2) To prove the second part of the lemma it is now only necessary to show that if $\mathfrak{o}$ is saturated with respect to K_0 it is also saturated with respect to K. Let k^* be the least Galois extension of k_0 containing k. Then F_0^*, as a Galois extension of $k_0((x))(= K_0)$, is also a Galois extension of $k^*((x))$. We next observe

that if ψ_0 is any element of H_0 and if, say, $\psi_0(\epsilon_i)=1$, where $i\in\{1, 2,\ldots,h\}$, we can fix an element ψ of H such that $\psi(\epsilon_i)=1$, and then the restrictions of ψ_0 and ψ to $F\epsilon_i$ are isomorphisms $F\epsilon_i \xrightarrow{\sim} F_{i,0}$, $F\epsilon_i \xrightarrow{\sim} F_i$, where we have set $F_{i,0}=\psi_0(F)$, $F_i=\psi(F)$. We note that both isomorphisms have the same restriction to $k_0((x))\epsilon_i$, namely they send $\alpha_0\epsilon_i$, $\alpha_0\in k_0((x))$, into α_0. There exists therefore a $k_0((x))$ – isomorphism τ: $F_i \xrightarrow{\sim} F_{i,0}$ such that $\psi_0=\tau\cdot\psi$. Clearly, τ can be extended to a $k_0((x))$ automorphism of F_0^*. We continue to denote by τ some specific extension of τ to F_0^*.

Let now ζ be an element of $\mathfrak{o}$, let η be an element of $\bar{\mathfrak{o}}$ and assume that η dominates ζ with respect to the field K. We have to show that η belongs to $\mathfrak{o}$. If c is the m-residue of ζ in k, we may replace ζ by $\zeta-c$, without affecting the assumption that η dominates ζ with respect to K. We may therefore assume that $\zeta\in\mathfrak{m}$ and that therefore $v_0(\psi_0(\zeta))>0$, for all ψ_0 in H_0. We fix a primitive element α of k over k_0 and we set $\xi=\zeta+\alpha$. Then $\zeta\in\mathfrak{o}$ and η still dominates ξ with respect to K. We shall now prove that η *also dominates* ξ *with respect to* K_0, and this will prove that η belongs to $\mathfrak{o}$ and will complete the proof of the lemma.

Let ψ_0, ψ_0' be any two elements of H_0. We can write (as was pointed out above): $\psi_0=\tau\cdot\psi$, $\psi_0'=\tau'\cdot\psi'$, where ψ and ψ' are suitable elements of H, and τ, τ' are $k_0((x))$ automorphisms of F_0^*. If we set $\sigma=\tau'^{-1}\cdot\tau$, then $\psi_0-\psi_0'=\tau'(\sigma\cdot\psi-\psi')$; here σ is again a $k_0((x))$-automorphism. The natural valuation v_0 of F_0^* is invariant under both τ' and σ. We have therefore: $v_0[\psi_0(\xi)-\psi_0'(\xi)]=v_0[\sigma\cdot\psi)(\xi)-\psi'(\xi)]=v_0[(\sigma\cdot(\psi-\psi'))(\xi)+(\sigma\cdot\psi'-\psi')(\xi)]$. Now, we have $(\sigma\cdot\psi'-\psi')(\xi)=(\sigma\cdot\psi')(\zeta)-\psi'(\zeta)+(\sigma\cdot\psi')(\alpha)-\psi'(\alpha)$. Since $\psi'\in H$ and $\alpha\in k\subset K$, we have $\psi'(\alpha)=\alpha$, and hence $(\sigma\cdot\psi')(\alpha)=\sigma(\alpha)$. If $\sigma(\alpha)=\alpha$, then $\sigma|k$ is the identity and $\sigma\cdot\psi\in H$. In that case, we have $v_0[\psi_0(\eta)-\psi_0'(\eta)]=v[(\sigma\cdot\psi)(\eta)-\psi'(\eta)]\geqslant v[(\sigma\cdot\psi)(\xi)-\psi'(\xi)]$ (since η dominates ξ with respect to K) $=v_0[\psi_0(\xi)-\psi_0'(\xi)]$, i.e., in this case we have

$$v_0[\psi_0(\eta)-\psi_0'(\eta)]\geqslant v_0[\psi_0(\xi)-\psi_0'(\xi)]. \tag{40}$$

Assume now that $\sigma(\alpha)\neq\alpha$. In that case $(\sigma\cdot\psi')(\alpha)-\psi'(\alpha)$ is a nonzero element $\sigma(\alpha)-\alpha$ of k^* and its v_0-value is zero. Since, on the other hand, $\zeta\in\mathfrak{m}$, we have $v((\sigma\cdot\psi')\zeta)>0$ and $v(\psi'(\zeta))>0$. Hence it follows from the above expression of $(\sigma\cdot\psi'-\psi')(\xi)$ that $v_0[(\sigma\cdot\psi'-\psi')(\xi)]=0$. On the other hand, we have $(\sigma\cdot(\psi-\psi'))(\xi)=(\sigma\cdot(\psi-\psi'))(\zeta)$, and hence $v_0[(\sigma\cdot(\psi-\psi'))(\xi)]>0$, since $\zeta\in\mathfrak{m}$. In conclusion, we see that from the above expression of $v_0[\psi_0(\xi)-\psi_0'(\xi)]$ it follows that in the present case (i.e., in the case in which $\sigma(\alpha)\neq\alpha$) we have $v_0[\psi_0(\xi)-\psi_0'(\xi)]=0$. In this case, the inequality (40) follows from the fact that

$\eta \in \bar{\mathfrak{o}}$. Thus (40) holds for any pair of elements ψ_0, ψ_0' of H_0, i.e., η dominates ξ also with respect to K_0. Hence $\eta \in \mathfrak{o}$, and this completes the proof of the lemma.

We now conclude this appendix by extending Corollary 7.4 of GST I to the case in which $\mathfrak{o}$ has zero-divisors (all the previous assumptions remaining in force: $\mathfrak{o}$ is complete, equicharacteristic zero, equidimensional, of Krull dimension 1 and free from nonzero nilpotent elements). By the saturation $\tilde{\mathfrak{o}}$ of $\mathfrak{o}$ we mean the k-saturation of $\mathfrak{o}$ with respect to a transversal parameter x, k being a coefficient field of $\mathfrak{o}$ (see Corollary A7).

LEMMA A9. *The ring $\tilde{\mathfrak{o}}$ is the smallest saturated local ring between $\mathfrak{o}$ and $\bar{\mathfrak{o}}$.*

Proof. Let $F = \oplus_{i=1}^{h} F\epsilon_i$, where F is the total ring of quotients of $\mathfrak{o}$. If z is any parameter of $\mathfrak{o}$, then $z\epsilon_i$ is a parameter of $\mathfrak{o}\epsilon_i$, and if x is our transversal parameter of $\mathfrak{o}$ then, by our definition of transversal parameters (see GTS II, p. 902), $x\epsilon_i$ is a transversal parameter of the local domain $\mathfrak{o}\epsilon_i$. Therefore $x\epsilon_i$ is also a transversal parameter of the saturation $\widetilde{(\mathfrak{o}\epsilon_i)}_{k((z))\cdot\epsilon_i}$ of $\mathfrak{o}\epsilon_i$ with respect to the field $k((z))\cdot\epsilon_i$ (see GTS I, Proposition 2.5). By formula (26) of GTS II (p. 899) we have $\widetilde{(\mathfrak{o}\epsilon_i)}_{k((z))\cdot\epsilon_i} = \widetilde{\mathfrak{o}_{k((z))}}\cdot\epsilon_i$. Thus, $x\epsilon_i$ is a transversal parameter of $\widetilde{\mathfrak{o}_{k((z))}}\cdot\epsilon_i$, for $i = 1, 2, \ldots, h$. Hence, again by our definition of transversal parameters (GTS II, p. 902), x is a transversal parameter of $\widetilde{\mathfrak{o}_{k((z))}}$. But then the saturated local ring $\widetilde{\mathfrak{o}_{k((z))}}$ is also k-saturated with respect to x (GTS II, Theorem 4.5). By the minimality property of saturation it follows then that $\tilde{\mathfrak{o}}(=\widetilde{\mathfrak{o}_{k((x))}}) \subset \widetilde{\mathfrak{o}_{k((z))}}$. We have therefore shown that $\tilde{\mathfrak{o}}$ *is the smallest saturation of* $\mathfrak{o}$.

Now, let $\mathfrak{o}'$ be any saturated *local* ring between $\mathfrak{o}$ and $\bar{\mathfrak{o}}$ (not necessarily the saturation of $\mathfrak{o}$ with respect to some parameter; note that any local ring between $\mathfrak{o}$ and $\bar{\mathfrak{o}}$ is necessarily complete, of dimension 1, and is equidimensional).

Let x' be a transversal parameter of $\mathfrak{o}'$ and let k' be the algebraic closure of k in $\mathfrak{o}'$. Then k' is a finite algebraic extension of k and is a coefficient field of $\mathfrak{o}'$. Since $\mathfrak{o}'$ is saturated with respect to the field $k'((x'))$ and since for any positive integer the field $k'((x'))$ is a finite algebraic extension of $k((x'^q))$, it follows from part (1) of Lemma A7 that $\mathfrak{o}'$ is also saturated with respect to $k((x'^q))$. For q large the element x'^q belongs to $\mathfrak{o}$ (since $\mathfrak{o}'$ is a finite $\mathfrak{o}$-module and since therefore high powers of the maximal ideal of $\mathfrak{o}'$ are contained in $\mathfrak{o}$). Since $\mathfrak{o} \subset \mathfrak{o}'$, we have therefore $\widetilde{\mathfrak{o}_{k((x'^q))}} \subset \mathfrak{o}'$; and, on the other hand, we have already proved that $\tilde{\mathfrak{o}} \subset \widetilde{\mathfrak{o}_{k((x'^q))}}$. Hence $\tilde{\mathfrak{o}} \subset \mathfrak{o}'$, and this completes the proof.

Appendix B. Open Conditions on Parameters in Local Rings and, in Particular, Transversality Conditions with Respect to Prime Ideals

Throughout this appendix (unless otherwise specified), $\mathfrak{o}$ stands for an arbitrary local ring, of dimension $r \geqslant 1$, with maximal ideal $\mathfrak{m}$. We shall also assume throughout this appendix that the residue field $k_0 = \mathfrak{o}/\mathfrak{m}$ of $\mathfrak{o}$ is infinite (a condition which is always satisfied if $\mathfrak{o}$ is equicharacteristic zero). For any integer $s \geqslant 0$ we set $\bar{V}_s = \mathfrak{m}^s/\mathfrak{m}^{s+1}$ and $G_{\mathfrak{m}}(\mathfrak{o}) = \oplus_{s=0}^{\infty} \bar{V}_s$. Then $\bar{V}_s$ is a vector space over k_0, and one defines a suitable multiplication in the additive group $G_{\mathfrak{m}}(\mathfrak{o})$ in such a way that $\bar{V}_s \bar{V}_t \subset \bar{V}_{s+t}$. The graded ring $G_{\mathfrak{m}}(\mathfrak{o})$ thus obtained is the associated graded ring of $\mathfrak{o}$ (see C.A. 2, p. 248). The nonzero elements of $\bar{V}_s$ are the homogeneous elements of $G_{\mathfrak{m}}(\mathfrak{o})$, of degree s.

The dimension N of $\bar{V}_1$ is the so-called *embedding dimension* of $\mathfrak{o}$ and is an integer characterized by the property that any minimal basis of $\mathfrak{m}$ consists of exactly N elements. The subset $\bar{V}_1^s$ of V_s spans the entire space $\bar{V}_s$, and from this it follows that any basis $\{x_1, x_2, \ldots, x_M\}$ of $\mathfrak{m}$ (and, in particular, any minimal basis $\{x_1, x_2, \ldots, x_N\}$) leads canonically to a representation of $G_{\mathfrak{m}}(\mathfrak{o})$ as a residue class ring $k_0[x_1, x_2, \ldots, X_M]$ of polynomials in M indeterminates, with coefficients in k_0, modulo a homogeneous ideal $\mathfrak{A}$. Here the $\mathfrak{A}$-residues $\bar{x}_1, \bar{x}_2, \ldots, \bar{x}_M$ of the X_i are precisely the $\mathfrak{m}^2$-residues of $x_1, x_2, \ldots, x_M$, the $\bar{x}_i$ span $\bar{V}_1$ over k_0, and we have $G_{\mathfrak{m}}(\mathfrak{o}) = k_0[\bar{x}_1, \bar{x}_2, \ldots, \bar{x}_M]$.

If z is any element of $\mathfrak{m}$, different from zero, there exists an integer $s \geqslant 0$ such that $z \in \mathfrak{m}^s$, $z \notin \mathfrak{m}^{s+1}$ (here $\mathfrak{m}^s = \mathfrak{o}$ if $s = 0$). We then denote by $\bar{z}$ the $\mathfrak{m}^{s+1}$-residue of z. Thus $\bar{z}$ is a nonzero element of the vector space $\bar{V}_s$. We shall denote by $\bar{\mathfrak{X}}$ the (irrelevant) prime ideal of $G_{\mathfrak{m}}(\mathfrak{o})$ generated by all (homogeneous) elements of $G_{\mathfrak{m}}(\mathfrak{o})$ of positive degree.

We recall the concept of a superficial element z of $\mathfrak{o}$ (with respect to $\mathfrak{m}$), due to Samuel (see C.A. 2, p. 285). If $z \in \mathfrak{m}^s$, $z \notin \mathfrak{m}^{s+1}$ and if $s \geqslant 1$, then z is said to be superficial (of order s) for $\mathfrak{m}$ if there exists an integer $c \geqslant 0$ such that

$$(\mathfrak{m}^n : \mathfrak{o}z) \cap \mathfrak{m}^c = \mathfrak{m}^{n-s} \tag{1}$$

for all large n.

Lemma $B1$. *If $z \in \mathfrak{m}^s$, $z \notin \mathfrak{m}^{s+1}$ $(s \geqslant 1)$, then z is superficial for $\mathfrak{m}$ if and only if there exists an integer $c \geqslant 0$ such that*

$$((0):(\bar{z})) \cap \bar{\mathfrak{X}}^c = (0), \tag{2}$$

where (0) *denotes the zero ideal in* $G_{\mathfrak{m}}(\mathfrak{o})$. *Equivalently, z is not superficial for*

$\mathfrak{m}$ *if and only if*

$$((0):(\bar{z}))\cap\bar{\mathfrak{X}}^c\neq(0) \tag{3}$$

for all $c\geqslant 0$.

Proof. The sufficiency of condition (2) has been established implicitly in C.A. 2 in the course of the proof of Lemma 5, p. 286. For the convenience of the reader we give this proof here. We shall show, more precisely, that if (2) holds for a given integer $c\geqslant 0$ then (1) holds for that particular integer c and for all integers $n\geqslant s+c$. To see this, we first observe that since $z\in\mathfrak{m}^s$, we have $\mathfrak{m}^{n-s}\cdot\mathfrak{o}z\subset\mathfrak{m}^n$, i.e., $\mathfrak{m}^{n-s}\subset\mathfrak{m}^n:\mathfrak{o}z$, and, furthermore, if $n\geqslant s+c$ then we also have $\mathfrak{m}^{n-s}\subset\mathfrak{m}^c$. Thus, $\mathfrak{m}^{n-s}\subset(\mathfrak{m}^n:\mathfrak{o}z)\cap\mathfrak{m}^c$ for all $n\geqslant s+c$. We now show that if $n\geqslant s$ and if (2) holds then $(\mathfrak{m}^n:\mathfrak{o}z)\cap\mathfrak{m}^c\subset\mathfrak{m}^{n-s}$, and this will establish the sufficiency of condition (2). Let ξ be a nonzero element of $(\mathfrak{m}^n:\mathfrak{o}z)\cap\mathfrak{m}^c$ and let, say, $\xi\in\mathfrak{m}^q$, $\xi\not\in\mathfrak{m}^{q+1}$. We have to prove that $\xi\in\mathfrak{m}^{n-s}$, i.e., that $q\geqslant n-s$. We have $q\geqslant c$ (since $\xi\in\mathfrak{m}^c$), and hence $\bar{\xi}\in\bar{\mathfrak{X}}^c$. Since $\bar{\xi}\neq 0$ it follows from (2) that $\bar{\xi}\bar{z}\neq 0$. This inequality signifies that the homogeneous element $\overline{\xi z}$ of $G_{\mathfrak{m}}(\mathfrak{o})$, associated with the product ξz, must be precisely of degree $q+s$, i.e., we must have $\xi z\in\mathfrak{m}^{q+s}$, $\xi z\not\in\mathfrak{m}^{q+s+1}$. Since $\xi z\in\mathfrak{m}^n$ (in view of our choice of ξ, we have $\xi\in\mathfrak{m}^n:\mathfrak{o}z$), we must have therefore $n\leqslant q+s$, i.e., $q\geqslant n-s$, as asserted.

We now prove the necessity of condition (2) by showing the following: *if* (3) *holds for all* $c\geqslant 0$ *then given any integer* $c\geqslant 0$ *and any integer* $n_0\geqslant c+s-1$, *there exists an integer* $n>n_0$ *such that* (1) *is not satisfied*. To see this, we pick a *nonzero homogeneous* element $\bar{\xi}$ of the ideal $((0):(\bar{z}))\cap\bar{\mathfrak{X}}^{n_0}$ (note that this ideal is homogeneous since both $(\bar{z})$ and $\bar{\mathfrak{X}}^{n_0}$ are homogeneous ideals in $G_{\mathfrak{m}}(\mathfrak{o})$). Let q be the degree of $\bar{\xi}$ and let ξ be a representative of $\bar{\xi}$ in $\mathfrak{o}$ (whence $\xi\in\mathfrak{m}^q$, $\xi\not\in\mathfrak{m}^{q+1}$). Since $\bar{\xi}\in\bar{\mathfrak{X}}^{n_0}$, we have $q\geqslant n_0$. Since $\bar{\xi}\bar{z}=0$ it follows that the product ξz must belong to $\mathfrak{m}^{q+s+1}$. Thus $\xi\in(\mathfrak{m}^{q+s+1}:\mathfrak{o}z)\cap\mathfrak{m}^c$ (since $q\geqslant n_0\geqslant c+s-1\geqslant c$). Therefore, if we set $n=q+s+1$ we have $n>n_0$ (since $q\geqslant n_0$) and (1) is not satisfied (since $\mathfrak{m}^{n-s}=\mathfrak{m}^{q+1}$ and $\xi\not\in\mathfrak{m}^{q+1}$). This completes the proof.

COROLLARY $B2$. *Let*

$$(0)=\bar{\mathfrak{Q}}_1\cap\bar{\mathfrak{Q}}_2\cap\cdots\cap\bar{\mathfrak{Q}}_h, \tag{4}$$

or

$$(0)=\bar{\mathfrak{Q}}_1\cap\bar{\mathfrak{Q}}_2\cap\cdots\cap\bar{\mathfrak{Q}}_h\cap\bar{\mathfrak{Q}} \tag{4'}$$

be an irredundant primary decomposition of the zero ideal in $G_{\mathfrak{m}}(\mathfrak{o})$, *where* (4) *refers to the case in which the irrelevant prime ideal* $\bar{\mathfrak{X}}$ *of* $G_{\mathfrak{m}}(\mathfrak{o})$ *is not a prime ideal of the zero ideal and where, in the contrary case, (4′) is to be used, and in that case* $\bar{\mathfrak{Q}}$ *is a primary ideal, with* $\bar{\mathfrak{X}}$ *as associated prime. Let* $\bar{\mathfrak{P}}_j$ *be the prime ideal of* $\bar{\mathfrak{Q}}_j$ $(j=1, 2,\ldots,h)$. *A necessary and sufficient condition that an element* z *of* $\mathfrak{m}$ *be superficial for* $\mathfrak{m}$ *is that the associated homogeneous element* $\bar{z}$ *of* $G_{\mathfrak{m}}(\mathfrak{o})$ *should not belong to any of the prime ideals* $\bar{\mathfrak{P}}_1, \bar{\mathfrak{P}}_2,\ldots,\bar{\mathfrak{P}}_h$.

This is an immediate consequence of Lemma B1, for (2) holds for some integer $c \geqslant 0$ if and only if $\bar{z} \not\subseteq \cup_{j=1}^{h}\bar{\mathfrak{P}}_j$.

We recall that given any ideal $\mathfrak{v}$ in $\mathfrak{o}$ which is primary for the maximal ideal $\mathfrak{m}$ of $\mathfrak{o}$, the length $\lambda(\mathfrak{v}^n)$, for n large, is a polynomial of degree r in n, denoted by $\bar{P}_{\mathfrak{v}}(n)$, whose leading coefficient is of the form $e(\mathfrak{v})/r!$, $e(\mathfrak{v})$—a positive integer (see C.A. 2, p. 294). This integer $e(\mathfrak{v})$ is called the *multiplicity of the ideal* $\mathfrak{v}$. The multiplicity $e(\mathfrak{m})$ of the maximal ideal $\mathfrak{m}$ is called the multiplicity of the local ring $\mathfrak{o}$. Since $\mathfrak{v}^n \subset \mathfrak{m}^n$, we have $\lambda(\mathfrak{v}^n) \geqslant \lambda(\mathfrak{m}^n)$ for all n, and hence $e(\mathfrak{v}) \geqslant e(\mathfrak{m})$ for any ideal $\mathfrak{v}$, primary for $\mathfrak{m}$.

If $z_1, z_2,\ldots,z_r$ are parameters of $\mathfrak{o}$ then the ideal $\Sigma_{i=1}^{r}\, \mathfrak{o}z_i$ is primary for $\mathfrak{m}$, and the multiplicity $e(\Sigma_{i=1}^{r}\, \mathfrak{o}z_i)$ is defined.

*Definition B*3. *The parameters* $z_1, z_2,\ldots,z_r$ *are transversal if* $e(\Sigma_{i=1}^{r}\mathfrak{o}z_i) = e(\mathfrak{m})$.

Proposition *B*4. *If* $\mathfrak{o}$ *is complete, equicharacteristic and equidimensional (see Section 2; see also SES III, p. 964) and if* k *is a coefficient field of* $\mathfrak{o}$, *then for any set of parameters* $z_1, z_2,\ldots,z_r$ *of* $\mathfrak{o}$, *the multiplicity of the ideal* $\Sigma_{i=1}^{r}\mathfrak{o}z_i$ *is equal to the maximal number of elements of* $\mathfrak{o}$ *which are linearly independent over the subring* $k[[z_1, z_2,\ldots,z_r]]$ *of* $\mathfrak{o}$:

$$e\left(\sum_{i=1}^{r} \mathfrak{o}z_i\right) = [\mathfrak{o} : k[[z_1, z_2,\ldots,z_r]]]. \tag{5}$$

In particular, the parameters $z_1, z_2,\ldots,z_r$ *are transversal if and only if*

$$[\mathfrak{o} : k[[z_1, z_2,\ldots,z_r]]] = e(\mathfrak{m}). \tag{6}$$

Proof. The equidimensionality of $\mathfrak{o}$ implies that every nonzero element of $k[[z_1, z_2,\ldots,z_r]]$ is regular in $\mathfrak{o}$. Hence Theorem 24 of C.A. 2, p. 297, is applicable to the case in which the rings A and B of that theorem are respectively the formal power series ring $k[[z_1, z_2,\ldots,z_r]]$ and the local ring $\mathfrak{o}$, and in which the ideal q of A in that theorem is the maximal ideal $\mathfrak{M}$ of the local ring $k[[z_1, z_2,\ldots,z_r]]$. We find then that the two polynomials

$$[\mathfrak{o} : k[[z_1, z_2,\ldots,z_r]]]\bar{P}_{\mathfrak{M}}(n) \quad \text{and} \quad \bar{P}_{\mathfrak{o}\mathfrak{M}}(n)$$

have the same degree and the same leading term. Now, we have clearly: $\bar{P}_{\mathfrak{M}}(n)=\binom{n+r-1}{r}=\frac{n^r}{r!}+\cdots$, while $\mathfrak{o}\mathfrak{M}$ is the ideal $\Sigma_{i=1}^r \mathfrak{o}z_i$ and the leading term of $\bar{P}_{\mathfrak{o}\mathfrak{M}}(n)$ is $\frac{e(\mathfrak{o}M)}{r!}n^r$. This yields (5) and completes the proof of the proposition.

We note that if F is the total ring of quotients of $\mathfrak{o}$ and K is the field of fractions $k((z_1, z_2,\ldots,z_r))$ of the complete regular local domain $k[[z_1, z_2,\ldots,z_r]]$, the assumption of equidimensionality of $\mathfrak{o}$ implies that K is a subfield of F. The integer $[\mathfrak{o}:k[[z_1, z_2,\ldots,z_r]]]$ is simply the dimension of F, regarded as a vector space over K.[1]

Proposition $B5$. *If $z_1, z_2,\ldots,z_r$ are transversal parameters of our local ring $\mathfrak{o}$, then $z_i \notin \mathfrak{m}^2$ $(i=1, 2,\ldots,r)$, and the $\mathfrak{m}^2$-residues $\bar{z}_1, \bar{z}_2,\ldots,\bar{z}_r$ of the z_i are linearly independent over the residue field $k_0(=\mathfrak{o}/\mathfrak{m})$ of $\mathfrak{o}$.*

Proof. Let $\mathfrak{v}$ be the ideal $\Sigma_{i=1}^r\mathfrak{o}z_i$. We consider the associated graded algebra $G_{\mathfrak{v}}(\mathfrak{o})=\oplus_{n=0}^{\infty}\mathfrak{v}^n/\mathfrak{v}^{n+1}$ (see C.A. 2, p. 248; $G_{\mathfrak{v}}(\mathfrak{o})$ is a finitely generated graded algebra over the ring $\mathfrak{o}/\mathfrak{v}$). Since k_0 is an infinite field, there exist elements $z_u=\Sigma_{i=1}^r u_i z_i$ $(u_i \in \mathfrak{o})$ in $\mathfrak{v}$ which are superficial of order 1 for $\mathfrak{v}$ (see C.A. 2, p. 287, Remark (2) following Lemma 5). The existence proof given in the cited reference to C.A. 2 shows that if we set $E=\mathfrak{v}/\mathfrak{v}^2$ and $E'=\mathfrak{m}\mathfrak{v}/\mathfrak{v}^2$ (so that $E'\subset E$ and E/E' is a vector space over k_0), there exists a finite number of proper subspaces L_ν of E/E' such that if $\bar{\xi}$ is any element of E/E' which does not belong to any of the L_ν and if $\bar{z}$ is any representative of $\bar{\xi}$ in E, then any

[1]In GTS I we gave a definition of a transversal parameter x of a complete, equicharacteristic zero and equidimensional local domain $\mathfrak{o}$, *which has Krull dimension* 1 (see GTS I, Definition 3.10, p. 593), and in GTS II, p. 902, we extended this definition to the case in which $\mathfrak{o}$ has zero-divisors (but is free from nonzero nilpotent elements). It is easy to see that in the case $r=1$, our present definition is equivalent with the one given in GTS II. For let $\bar{\mathfrak{o}}$ be the integral closure of $\mathfrak{o}$ in its total ring of quotients F. Then $\bar{\mathfrak{o}}$ is a direct sum of h discrete valuation rings $\bar{\mathfrak{o}}_i$, where $\bar{\mathfrak{o}}_i=\bar{\mathfrak{o}}\epsilon_i$ and $F=\oplus_{i=1}^h F\epsilon_i$. Let v_i be the valuation of the field $F\epsilon_i$ defined by $\bar{\mathfrak{o}}_i$, and let $\bar{k}_i$ be the algebraic closure of $k\epsilon_i$ in $\bar{\mathfrak{o}}_i$. If we apply formula (8) of C.A. 2, p. 299, to the case $A=\mathfrak{o}$, $B=\bar{\mathfrak{o}}$, $\mathfrak{q}=\mathfrak{m}$, then we notice that $[B:A]=1$, $e(\mathfrak{q})=e(\mathfrak{m})$, $B/\mathfrak{p}_i=\bar{k}_i$, $A/\mathfrak{m}=k$, and $e(\mathfrak{q}_i)=e(\bar{\mathfrak{o}}\mathfrak{m}\cdot\epsilon_i)=s_i$, where s_i is the reduced multiplicity of the local domain $\mathfrak{o}\epsilon_i$ (see GTS I, p. 586, proof of the Proposition 2.5 and footnote 2). Thus $e(\mathfrak{m})=\Sigma_{i=1}^h[\bar{k}_i:k\epsilon_i]s_i$. If we now apply the same formula (8) of C.A. 2, p. 299 to the case $A=\mathfrak{o}$, $B=\bar{\mathfrak{o}}$ and $q=\mathfrak{o}x$, then we find that

$$e(\mathfrak{o}x)=\sum_{i=1}^{h}[\bar{k}_i:k\epsilon_i]v_i(x\epsilon_i).$$

Hence $e(\mathfrak{o}x)=e(\mathfrak{m})$ if and only if $v_i(x\epsilon_i)=s_i$ for $i=1, 2,\ldots,h$, i.e., if and only if $x\epsilon_i$ is a transversal parameter of the local domain $\mathfrak{o}\epsilon_i$, for $i=1, 2,\ldots,h$, in the sense of our Definition 3.10 given in GTS I.

representative z of $\bar{z}$ in $\mathfrak{v}$ is superficial of order one for $\mathfrak{v}$. This shows that there exists a finite number of *proper* subspaces $\bar{L}_\nu$ of the vector space k_0^r of all r-tuples $\{\bar{u}_1, \bar{u}_2, \ldots, \bar{u}_r\}$ ($\bar{u}_i \in k_0$), with the property that if $\{u_1, u_2, \ldots, u_r\}$ is any r-tuple of elements of $\mathfrak{o}$ such that the r-tuple $\{\bar{u}_1, \bar{u}_2, \ldots, \bar{u}_r\}$ of their $\mathfrak{m}$-residues does not belong to any of the subspaces $\bar{L}_\nu$, then it is true that the element $z_u = \sum_{i=1}^{r} u_i z_i$ is superficial of order 1 for $\mathfrak{v}$. This result implies that there exists a basis $z_1', \ldots, z_r'$ of $\mathfrak{v}$ ($z_i' = \sum_{j=1}^{r} u_{ij} z_j$, $i = 1, 2, \ldots, r$; $u_{ij} \in \mathfrak{o}$, $\det \| u_{ij} \|$—a unit in $\mathfrak{o}$) such that each z_i' is superficial of order 1 for $\mathfrak{v}$. We may therefore assume that $z_1, z_2, \ldots, z_r$ are superificial of order 1 for $\mathfrak{v}$. Actually, we shall only need the assumption that *one* of the z_i, say z_1, is superficial of order 1 for $\mathfrak{v}$.

We set $\mathfrak{o}^* = \mathfrak{o}/\mathfrak{o}z_1$, $\mathfrak{m}^* = \mathfrak{m}/\mathfrak{o}z_1$, $z_i^* = \mathfrak{o}z_1$-residue of z_i ($i = 2, 3, \ldots, r$). It is clear that $z_2^*, z_3^*, \ldots, z_r^*$ are parameters of $\mathfrak{o}^*$ since $\mathfrak{o}^*$ is of Krull dimension $r-1$ (see C.A. 2, Corollary 1 on p. 291). It is known (see C.A. 2, Lemma 3 on p. 285) that

$$\bar{P}_{\mathfrak{m}^*}(n) = \bar{P}_{\mathfrak{m}}(n) - \lambda(\mathfrak{m}^n : 0z_1).$$

Let $z_1 \in \mathfrak{m}^s$, $z_1 \notin \mathfrak{m}^{s+1}$. Then $\mathfrak{m}^n : \mathfrak{o}z_1 \supset \mathfrak{m}^{n-s}$, $\lambda(\mathfrak{m}^n : \mathfrak{o}z_1) \leqslant \lambda(\mathfrak{m}^{n-s})$, and hence $\bar{P}_{\mathfrak{m}*}(n) \geqslant \bar{P}_{\mathfrak{m}}(n) - \bar{P}_{\mathfrak{m}}(n-s) = \dfrac{es}{(r-1)!} n^{r-1} +$ terms of degree $< r-1$ in n. It follows that

$$e(\mathfrak{m}^*) \geqslant e(\mathfrak{m}) \cdot s. \tag{7}$$

If we set $\mathfrak{v}^* = \sum_{i=2}^{r} \mathfrak{o}^* z_i^*$ $(= \mathfrak{v}/\mathfrak{o}z_1)$ then it follows from the fact that z_1 is superficial of order 1 for $\mathfrak{v}$, that $e(\mathfrak{v}^*) = e(\mathfrak{v})$ (see C.A. 2, second part of the proof of Theorem 22 on p. 295, where the induction from $r-1$ to r, $r \geqslant 2$, is carried out). Since $e(\mathfrak{v}^*) \geqslant e(\mathfrak{m}^*)$, we have therefore, by (7), the following inequalities (using the assumption of transversality of the parameters $z_1, z_2, \ldots, z_r$):

$$e(\mathfrak{m}) = e(\mathfrak{v}) = e(\mathfrak{v}^*) \geqslant e(\mathfrak{m}^*) \geqslant e(\mathfrak{m}) \cdot s. \tag{8}$$

We conclude therefore that $s = 1$, i.e., that $z_1 \notin \mathfrak{m}^2$. Now, from (8) it also follows that $e(\mathfrak{v}^*) = e(\mathfrak{m}^*)$, i.e., that $z_2^*, \ldots, z_r^*$ are transversal parameters of $\mathfrak{o}^*$. We now complete the proof of our proposition by induction on r, since in the case $r = 1$ the proof has already been completed when we have shown that $z_1 \not\subset \mathfrak{m}^2$. We may therefore assume that the $\mathfrak{m}^{*2}$-residues of $z_2^*, \ldots, z_r^*$ are linearly independent over k_0 $(= \mathfrak{o}/\mathfrak{m} = \mathfrak{o}^*/\mathfrak{m}^*)$. Now, suppose we have a relation of the form $\sum_{i=1}^{r} a_i z_i \in \mathfrak{m}^2$, $a_i \in \mathfrak{o}$. We have to show that $a_i \in \mathfrak{m}$ for $i = 1, 2, \ldots, r$. The above relation implies the relation $\sum_{i=2}^{r} a_i^* z_i^* \in \mathfrak{m}^{*2}$, where a_i^* is the $\mathfrak{o}z_1$-residue of a_i.

Hence, by our induction hypothesis, we have $a_i^* \in \mathfrak{m}^*$, $i=2,\ldots,r$, i.e., $a_i \in \mathfrak{m}$, $i=2, 3,\ldots,r$. Therefore the relation $\Sigma_{i=1}^r a_i z_i \in \mathfrak{m}^2$ implies that $a_1 z_1 \in \mathfrak{m}^2$, and since $z_1 \notin \mathfrak{m}^2$, we conclude that also $a_1 \in \mathfrak{m}$. This completes the proof.

PROPOSITION $B6$. *Given r elements $z_1, z_2,\ldots,z_r$ in $\mathfrak{m}$ and denoting by $\bar{z}_i$ the $\mathfrak{m}^2$-residue of z_i, a sufficient condition that the elements z_i be transversal parameters of $\mathfrak{o}$ is that for each of the isolated prime ideals $\overline{\mathfrak{P}}_i$ of the zero ideal in $G_{\mathfrak{m}}(\mathfrak{o})$ (and therefore, a fortiori, also for all the h prime ideals $\overline{\mathfrak{P}}_i$ of the zero ideal; see (4) or (4′)) the ideal $\overline{\mathfrak{P}}_i + \Sigma_{i=1}^r G_{\mathfrak{m}}(\mathfrak{o})\bar{z}_i$ be primary for the irrelevant prime $\overline{\mathfrak{X}}$.*

Proof. As in the beginning of this appendix, we fix a minimal basis $x_1, x_2,\ldots,x_N$ of $\mathfrak{m}$ and we use this basis to represent the associated graded algebra $G_{\mathfrak{m}}(\mathfrak{o})$ as a homomorphic image of the polynomial ring $k_0[X_1, X_2,\ldots,X_N]$:

$$G_{\mathfrak{m}}(\mathfrak{o}) = k_0[X_1, X_2,\ldots,X_N]/\mathfrak{A}, \tag{9}$$

where $\mathfrak{A}$ is a homogeneous polynomial ideal. If $\mathfrak{P}_i$, $\mathfrak{Q}_i$ are the full inverse images in $k[X_1, X_2,\ldots,X_N]$ of $\overline{\mathfrak{P}}_i$ and $\overline{\mathfrak{Q}}_i$, respectively, while $\mathfrak{X}$ and $\mathfrak{Q}$ are the full inverse images of the irrelevant prime ideal $\overline{\mathfrak{X}}$ and of the primary ideal $\overline{\mathfrak{Q}}$ for $\overline{\mathfrak{X}}$ [see (4) and (4′)), then

$$\mathfrak{A} = \mathfrak{Q}_1 \cap \mathfrak{Q}_2 \cap \cdots \cap \mathfrak{Q}_h, \tag{10}$$

or

$$\mathfrak{A} = \mathfrak{Q}_1 \cap \mathfrak{Q}_2 \cap \cdots \cap \mathfrak{Q}_h \cap \mathfrak{Q} \tag{10'}$$

will be in irredundant primary decomposition of the ideal $\mathfrak{A}$ (according as we have (4) or (4′) in Corollary B2). The projective dimension of $\mathfrak{A}$ is $r-1$ (see C.A. 2, p. 291, Remark (2)). Since we are assuming throughout this appendix that $r \geqslant 1$, the irrelevant prime $\mathfrak{X}$ is not an isolated prime of $\mathfrak{A}$, or—equivalently—$\overline{\mathfrak{X}}$ is not an isolated prime of the zero ideal in $G_{\mathfrak{m}}(\mathfrak{o})$. Thus, the assumption made in the theorem implies at any rate that *not all $\bar{z}_i$ are zero*, i.e., *that not all z_i are in* $\mathfrak{m}^2$. (Actually, the fact that at least one isolated prime ideal $\mathfrak{P}_i$ of $\mathfrak{A}$ has projective dimension $r-1$, implies already, in conjunction with the assumption that $\overline{\mathfrak{P}}_i + \Sigma_{i=1}^r G_{\mathfrak{m}}(\mathfrak{o})\bar{z}_i$ is primary for $\overline{\mathfrak{X}}$, that the r linear forms $\bar{z}_1, \bar{z}_2,\ldots,\bar{z}_r$ are linearly independent, and hence no z_i belongs to $\mathfrak{m}^2$. We shall, however, make no use of this observation.)

If $r=1$, we are dealing with a single element $z(=z_1)$ of $\mathfrak{m}$, $z \notin \mathfrak{m}^2$, such that $\bar{z}$ does not belong to any of the h prime ideals $\overline{\mathfrak{P}}_i$. Hence, by Corollary B2, z is superficial for $\mathfrak{m}$, of order 1. In this case, the assertion that z is a transversal parameter of $\mathfrak{o}$ has been established in the course of the proof of Theorem 22 in C.A. 2, pp. 294–295.

We now proceed by induction from $r-1$ to r. Our assumption implies that for any given $j=1, 2,\ldots,h$, the elements $\bar{z}_1, \bar{z}_2,\ldots,\bar{z}_r$ do not all belong to $\bar{\mathfrak{P}}_j$. Hence, for given j, the r-tuples $\{\bar{u}_1, \bar{u}_2,\ldots,\bar{u}_r\}$ of elements of k_0, such that $\Sigma_{i=1}^r \bar{u}_i\bar{z}_i \in \bar{\mathfrak{P}}_j$, form a proper subspace $\bar{L}_j$ of k_0^r. Since k_0 is an infinite field it follows that there exists an element $z_u = u_1z_1 + \cdots + u_rz_r$ $(u_i \in \mathfrak{o})$ such that $z_u \not\in \mathfrak{m}^2$ and such that $\bar{z}_u$ $(=\bar{u}_1\bar{z}_1 + \cdots + \bar{u}_r\bar{z}_r$, where $\bar{u}_i$ is the $\mathfrak{m}$-residue of u_i in $k_0)$ does not belong to any of the prime ideal $\bar{\mathfrak{P}}_j$. Such an element z_u is then superficial of order 1 for $\mathfrak{m}$ (Corollary B2). Since $z_u \not\in \mathfrak{m}^2$, not all the u_i are in $\mathfrak{m}$. Let, say, $u_1 \not\in \mathfrak{m}$. Then also the elements $z_u, z_2,\ldots,z_r$ generate the ideal $\Sigma_{i=1}^r \mathfrak{o}z_i$, and also the elements $\bar{z}_u, \bar{z}_2,\ldots,\bar{z}_r$ generate the ideal $\Sigma_{i=1}^r G_m(\mathfrak{o})\bar{z}_i$. We may therefore assume that z_1 *is superficial of order* 1 *for* $\mathfrak{m}$.

As in the proof of Proposition B5, we set $\mathfrak{o}^* = \mathfrak{o}/\mathfrak{o}z_1$. Let x_ν^* be the $\mathfrak{o}z_1$-residue of x_ν $(\nu=1, 2,\ldots,N)$. Then $\{x_1^*, x_2^*,\ldots,x_N^*\}$ is a basis of the maximal ideal $\mathfrak{m}^*$ of $\mathfrak{o}^*$, and this basis yields a representation of the associated graded algebra $G_{m*}(\mathfrak{o}^*)$ as a homomorphic image of the polynomial ring $k_0[X_1, X_2,\ldots,X_N]$:

$$G_{\mathfrak{m}*}(\mathfrak{o}^*) = k_0[X_1,X_2,\ldots,X_N]/\mathfrak{A}^*,$$

where $\mathfrak{A}^*$ is a homogeneous polynomial ideal, containing the ideal $\mathfrak{A}$ which appears in (9). If $\bar{z}_i = c_{i,1}\bar{x}_1 + c_{i,2}\bar{x}_2 + \cdots + c_{i,N}\bar{x}_N$, $c_{i,\nu} \in k_0$ $(i=1, 2,\ldots,r)$, then the $c_{1,\nu}$ are not all zero, and if we set $l_i(x) = c_{i1}X_1 + c_{i2}X_2 + \cdots + c_{iN}X_N$, then it is clear that the linear form $l_1(X) = c_{11}X_1 + c_{12}X_2 + \cdots + c_{1N}X_N$ belongs to $\mathfrak{A}^*$. Hence

$$\mathfrak{A}^* \supset \mathfrak{A} + k_0[X_1,X_2,\ldots,X_N]\cdot l_1(X). \tag{11}$$

Since z_1 is superficial of order one for $\mathfrak{m}$, it follows, as in the proof of Proposition B5 (with $\mathfrak{m}$ and $\mathfrak{m}^*$ now playing the roles of the ideals $\mathfrak{v}$ and $\mathfrak{v}^*$ of that proof), that $e(\mathfrak{m}^*) = e(\mathfrak{m})$. The assumption of the theorem is equivalent to the assumption that the ideal $(\bar{\mathfrak{Q}}_1 \cap \bar{\mathfrak{Q}}_2 \cap \cdots \cap \bar{\mathfrak{Q}}_h) + \Sigma_{i=1}^r G_{\mathfrak{m}}(\mathfrak{o})\bar{z}_i$ is primary for $\bar{\mathfrak{X}}$, and this is equivalent to assuming that the ideal $(\mathfrak{Q}_1 \cap \mathfrak{Q}_2 \cap \cdots \cap \mathfrak{Q}_h) + \Sigma_{i=1}^r k[X_1, X_2,\ldots,X_N]\cdot l_i(X)$ is primary for the irrelevant prime ideal $\mathfrak{X} = (X_1,X_2,\ldots,X_N)$ in $k[X_1,X_2,\ldots,X_N]$. Thus the assumption of the theorem is equivalent to assuming that there exists an integer $\rho \geqslant 0$ such that

$$\mathfrak{X}^\rho \subset \mathfrak{A} + \sum_{i=1}^r k[X_1,X_2,\ldots,X_N]l_i(X). \tag{12}$$

By (11) it then follows that

$$\mathfrak{X}^\rho \subset \mathfrak{A}^* + \sum_{i=2}^r k[X_1,X_2,\ldots,X_N]l_i(X),$$

i.e., the assumption of the theorem is satisfied if we replace the ring $\mathfrak{o}$ and the elements $z_1, z_2, \ldots, z_r$ by the ring $\mathfrak{o}^*$ and by the $\mathfrak{o}z_1$-residues $z_2^*, z_3^*, \ldots, z_r^*$ of $z_2, z_3, \ldots, z_r$. Hence, by our induction hypothesis, the r-1 elements $z_2^*, z_3^*, \ldots, z_r^*$ are transversal parameters of $\mathfrak{o}^*$, i.e., we have $e(\sum_{i=2}^r \mathfrak{o}^* z_i^*) = e(\mathfrak{m}^*)$, or $e(\sum_{i=2}^r \mathfrak{o}^* z_i^*) = e(\mathfrak{m})$, since we have just observed above that $e(\mathfrak{m}^*) = e(\mathfrak{m})$. Now, it was already established in the course of the proof of Proposition B5 (as a consequence of the fact that z_1 is superficial of order 1 for $\mathfrak{m}$) that $e(\mathfrak{v}^*) = e(\mathfrak{v})$, where $\mathfrak{v}^* = \sum_{i=2}^r \mathfrak{o}^* z_i^*$ and $\mathfrak{v} = \sum_{i=1}^r \mathfrak{o} z_i$. Hence $e(\sum_{i=1}^r \mathfrak{o} z_i) = e(\mathfrak{m})$, and this completes the proof of the proposition.

We now wish to introduce and study suitably defined, generalized transversality conditions which one may impose on sets of parameters of $\mathfrak{o}$. In this connection we wish first to define what we mean by an *open condition of a given order s*, which may be imposed on sets of a given number ρ of elements of $\mathfrak{m}$.

If $N_s = \dim_{k_0} \bar{V}_s$, where $\bar{V}_s = \mathfrak{m}^s/\mathfrak{m}^{s+1}$, we denote by $\mathbf{P}^{(s)}$ the (universal) projective space of dimension $N_s - 1$. If $\bar{z}$ is any vector in $\bar{V}_s$, different from zero, we denote by $\pi_s(\bar{z})$ the corresponding (k_0-rational) point of $\mathbf{P}^{(s)}$. Let Gr $(\mathbf{P}^{(s)}, \rho - 1)$ be the Grassmann variety of the $(\rho-1)$-dimensional linear subspaces of $\mathbf{P}^{(s)}$.

Definition B7. Given a condition C on sets of ρ elements of $\mathfrak{m}^s$, we say that C is an open condition (of order s), if there exists a proper subvariety W of Gr $(\mathbf{P}^{(s)}, \rho-1)$, *defined over k_0, having the following property: if $\bar{z}_1, \bar{z}_2, \ldots, \bar{z}_\rho$ are independent vectors in $\bar{V}^{(s)}$ such that the $(\rho-1)$-space $L(\bar{z}_1, \bar{z}_2, \ldots, \bar{z}_\rho)$ spanned by the corresponding points $\pi_s(\bar{z}_1), \pi_s(\bar{z}_2), \ldots, \pi_s(\bar{z}_\rho)$ of $\mathbf{P}^{(s)}$ does not belong to W, then any set of ρ elements $z_1, z_2, \ldots, z_\rho$ of $\mathfrak{m}^s$ such that $\bar{z}_i = \mathfrak{m}^{s+1}$-residue of z_i $(i = 1, 2, \ldots, \rho)$ satisfies condition C.*

We may add the stipulation that the condition C be called *algebraic* (of order s) if there is a W such as above having the stronger property that a set $\{z_1, z_2, \ldots, z_\rho\}$ of elements of $\mathfrak{m}^s$ such that the $\mathfrak{m}^{s+1}$-residues $\bar{z}_1, \bar{z}_2, \ldots, \bar{z}_\rho$ are independent vectors of $\bar{V}_s$, satisfies condition C if *and only* if the representative point of the $(\rho-1)$-space $L(\bar{z}_1, \bar{z}_2, \ldots, \bar{z}_\rho)$ does not belong to W. For instance, in the case $\rho = 1$, the condition that an element z of $\mathfrak{m}^s$, not in $\mathfrak{m}^{s+1}$, be superficial for $\mathfrak{m}$, is an algebraic condition (of order s). For, if we set $W = \cup_{j=1}^h \pi_s(\bar{V}_s \cap \bar{\mathfrak{P}}_j)$ (both $\bar{V}_s$ and $\bar{\mathfrak{P}}_j$ being subsets of $G_{\mathfrak{m}}(\mathfrak{o})$), then W is the union of linear subspaces of $\mathbf{P}^{(s)}$, *defined over* k_0, and z is superficial for $\mathfrak{m}$ if and only if the point $\pi_s(\bar{z})$ does not belong to W (see Corollary B2).

Proposition $B8$. *The condition that a set of parameters $z_1, z_2, \ldots, z_r$ of $\mathfrak{o}$ be transversal is an open condition of order* 1.

Proof. We take for $\{x_1, x_2, \ldots, x_N\}$ a minimal basis of $\mathfrak{m}$ (whence $N_1 = N = \dim \bar{V}_1$, where $\bar{V}_1 = \mathfrak{m}/\mathfrak{m}^2$) and use the representation (9) of $G_{\mathfrak{m}}(\mathfrak{o})$. We call $\mathbf{P}^{(1)*}$ the projective space in which $X_1, X_2, \ldots, X_N$ are homogeneous coordinates, and we identify $\mathbf{P}^{(1)*}$ with the dual of the projective space $\mathbf{P}^{(1)}$ introduced above (we are now in the special case $s = 1$). A point $\pi_1(\bar{z})$ in $\mathbf{P}^{(1)}$ $(0 \neq \bar{z} \in \bar{V}_1)$ determines (up to a factor in k_0) a linear form $l_{\bar{z}}(X)$. Namely if $\bar{z} = c_1\bar{x}_1 + \cdots + c_N\bar{x}_N$, $c_\nu \subset k_0$, then the c_ν are uniquely determined by $\pi_1(\bar{z})$, up to a factor (since $x_1, x_2, \ldots, x_N$ form a *minimal* basis of $\mathfrak{m}$), and we then set $l_{\bar{z}}(X) = c_1X_1 + \cdots + c_NX_N$. We denote by $L_{\bar{z}}$ the hyperplane $l_{\bar{z}}(X) = 0$; this is then a point of the dual space $\mathbf{P}^{(1)*}$. We shall denote this point by $\pi_1^*(\bar{z})$.

Given an $(r-1)$-dimensional subspace $L(\bar{z}_1, \bar{z}_2, \ldots, \bar{z}_r)$ of $\mathbf{P}^{(1)}$, there corresponds to it, by duality, a linear subspace $L^*(\bar{z}_1, \bar{z}_2, \ldots, \bar{z}_r)$ of dimension $N-1-r$ in $\mathbf{P}^{(1)*}$, namely the subspace defined by the r equations $l_{\bar{z}_1}(X) = l_{\bar{z}_2}(X) = \cdots = l_{\bar{z}_r}(X) = 0$. We now use Proposition B6, which tells us that a sufficient condition that r elements $z_1, z_2, \ldots, z_r$ of $\mathfrak{m}$ be transversal parameters is that (12) be satisfied for some integer $\rho \geq 0$, where $l_i(X)$ now stands for $l_{\bar{z}_i}(X)$. Geometrically speaking, (12) signifies that *the* $(n-1-r)$*-space which we have denoted above by* $L^*(\bar{z}_1, \bar{z}_2, \ldots, \bar{z}_r)$ *does not meet the variety of the ideal* $\mathfrak{A}$. Now, since the projective dimension of $\mathfrak{A}$ is $r-1$, the $(N-1-r)$-spaces in $\mathbf{P}^{(1)*}$ which do meet the variety of $\mathfrak{A}$ form a *proper* algebraic subvariety W^* (defined over k_0) of Gr $(\mathbf{P}^{(1)*}, N-1-r)$. By duality, W^* is represented by a proper algebraic subvariety (defined over k_0) of Gr $(\mathbf{P}^{(1)}, r-1)$, and we have therefore that if the representative point (in Gr $(\mathbf{P}^{(1)}, r-1)$ of the $(r-1)$-space $L(\bar{z}_1, \bar{z}_2, \ldots, \bar{z}_r)$ does not belong to W, then $z_1, z_2, \ldots, z_r$ are transversal parameters of $\mathfrak{o}$. Q.E.D.

Let C be again a condition on sets of ρ elements of $\mathfrak{m}$. Let k be a field contained in $\mathfrak{o}$, and assume that k is an infinite field. We shall say that the condition C is *projectively k-stable*, if whenever a set of ρ elements $z_1, z_2, \ldots, z_\rho$ of $\mathfrak{m}$ satisfies condition C then also any set of ρ elements $z_1', z_2', \ldots, z_\rho'$ obtained from the set $\{z_1, z_2, \ldots, z_\rho\}$ be a non-singular linear homogeneous transformation, with coefficients in k, also satisfies condition C. For instance, it is obvious that the condition that r elements $z_1, z_2, \ldots, z_r$ of $\mathfrak{m}$ be transversal parameters of $\mathfrak{o}$ is projectively k-stable.

Our next definition refers to the case in which $\mathfrak{o}$ is a *complete* local ring and k *is a coefficient field of* $\mathfrak{o}$.

Let $z_1, z_2, \ldots, z_r$ be parameters of $\mathfrak{o}$. Then $\mathfrak{o}$ contains the power series ring $R = k[[z_1, z_2, \ldots, z_r]]$, and $\mathfrak{o}$ is a finite R-module. Let $\mathfrak{p}$ be a prime ideal in $\mathfrak{o}$, of dimension ρ (and hence of height $r-\rho$), and let $\mathfrak{P} = \mathfrak{p} \cap R$. Then also $\mathfrak{P}$ is of height $r-\rho$. Let $(0) = \mathfrak{q}_1 \cap \mathfrak{q}_2 \cap \cdots \cap \mathfrak{q}_h$ be an irredundant primary decomposition of the zero ideal in $\mathfrak{o}$, let $\mathfrak{p}_\nu =$ Radical of $\mathfrak{q}_\nu$, and let, say, $\mathfrak{p}_\nu \subset \mathfrak{p}$ for $\nu = 1$,

$2,\ldots,g$, while $\mathfrak{p}_\nu \not\subset \mathfrak{p}$ for $\nu = g+1, g+2,\ldots,h$. Then, if $h_\mathfrak{p}$ is the canonical homomorphism of $\mathfrak{o}$ into $\mathfrak{o}_\mathfrak{p}$, we have Ker $h_\mathfrak{p} = \mathfrak{q}_1 \cap \mathfrak{q}_2 \cap \cdots \cap \mathfrak{q}_g$ (C.A. 1, Theorem 19, p. 228). Now, $\mathfrak{p}$ contains at least one isolated prime ideal of (0). Let, say, $\mathfrak{p}_1$ be an isolated prime ideal of (0) such that $\mathfrak{p}_1 \subset \mathfrak{p}$. Then $h(\mathfrak{p}_1) = 0$, and therefore $\mathfrak{p}_1 \cap R = (0)$ (since R is an *integrally closed* domain and since therefore also $\mathfrak{p}_1 \cap R$ has height zero). This shows that $(\mathrm{Ker}\, h_\mathfrak{p}) \cap R = (0)$, i.e., the restriction of $h_\mathfrak{p}$ to R is an injection. We therefore regard R as a subring of the ring $\mathfrak{o}' = h_\mathfrak{p}(\mathfrak{o})$. Let $h_\mathfrak{p}(\mathfrak{p}) = \mathfrak{p}'$. Then $\mathfrak{p}'$ is a prime ideal in $\mathfrak{o}'$, we have $\mathfrak{p}' \cap R = \mathfrak{p} \cap R = \mathfrak{P}$, every element of $\mathfrak{o}' - \mathfrak{p}'$ is regular in $\mathfrak{o}'$, and we have $\mathfrak{o}_\mathfrak{p} = \mathfrak{o}'_{\mathfrak{p}'}$. In particular, every element of $R - \mathfrak{p}$ is regular in $\mathfrak{o}'$, and the localization $R_\mathfrak{P}$ is canonically a subring of $\mathfrak{o}_\mathfrak{p}$. Since $\mathfrak{p}' \cap R = \mathfrak{P}$, the maximal ideal $\mathfrak{m}_\mathfrak{p}$ of $\mathfrak{o}_\mathfrak{p}$ contracts in $R_\mathfrak{P}$ to the maximal ideal $\mathfrak{P}R_\mathfrak{P}$ of $R_\mathfrak{P}$. *Therefore the ideal $\mathfrak{o}_\mathfrak{p}\mathfrak{P}$ is primary for $\mathfrak{m}_\mathfrak{p}$.*

Definition B9. The parameters $z_1, z_2,\ldots,z_r$ are said to be k-transversal with respect to the prime ideal $\mathfrak{p}$ if the multiplicity of the ideal $\mathfrak{o}_\mathfrak{p}\mathfrak{p}$ is equal to the multiplicity $e(\mathfrak{m}_\mathfrak{p})$ of the local ring $\mathfrak{o}_\mathfrak{p}$.

It is clear that the condition of k-transversality with respect to a given prime ideal $\mathfrak{p}$ in $\mathfrak{o}$ is projectively k-stable.

Before we state and prove the final (and main) result of this section, we need a lemma.

LEMMA B10. *Let $\mathfrak{o}$ be a local ring, containing an infinite coefficient field k (which we shall identify canonically with the residue field $k_0 = \mathfrak{o}/\mathfrak{m}$ of $\mathfrak{o}$). Let $N = \dim_k \mathfrak{m}/\mathfrak{m}^2$, let $\{x_1, x_2,\ldots,x_{\bar N}\}$ be a basis of $\mathfrak{m}$ (not necessarily minimal) and let ρ be a positive integer $\leqslant N$. We consider a projective space $\bar{\mathbf{P}}$, of dimension $\bar N\rho - 1$, in which the homogeneous coördinates are the entries $u_{i\nu}$ of a matrix $\|u_{i\nu}\|$ with ρ rows and $\bar N$ columns (the $u_{i\nu}$ being indeterminates; $i = 1, 2,\ldots,\rho$; $\nu = 1, 2,\ldots,\bar N$). Let C be a given condition on sets of ρ elements of $\mathfrak{m}$. Consider the following two statements:*

(1) *C is an open condition of order* 1.

(2) *There exists a proper projective subvariety $\bar\Gamma$ of $\bar{\mathbf{P}}$, defined over k, such that if $\{\alpha_{i\nu}\}$ is any k-rational point of $\bar{\mathbf{P}}$ not belonging to $\bar\Gamma$ and if $\{z_1, z_2,\ldots,z_\rho\}$ is any set of ρ elements of $\mathfrak{m}$ such that*

$$\bar z_i = \sum_{\nu=1}^{\bar N} \alpha_{i\nu}\bar x_\nu, \qquad i = 1,2,\ldots,\rho, \tag{13}$$

where $\bar z_i$ and $\bar x_\nu$ denote the $\mathfrak{m}^2$-residues of z_i and x_ν respectively, then the set $\{z_1, z_2,\ldots,z_\rho\}$ satisfies condition C.

Then (1) *implies* (2). *Furthermore, if C is projectively k-stable and if* $\{x_1, x_2, \ldots, x_{\bar{N}}\}$ *is a minimal basis of* $\mathfrak{m}$, *then, conversely,* (2) *implies* (1).

Proof. Assume (1). To prove statement (2) it is permissible to operate on the given basis $\{x_1, x_2, \ldots, x_{\bar{N}}\}$ of $\mathfrak{m}$ by any transformation of the form $x'_\nu = \sum_{\mu=1}^{N} \xi_{\nu\mu} x_\mu$ ($\nu = 1, 2, \ldots, \bar{N}$) such that the $\xi_{\nu\mu}$ are elements of $\mathfrak{o}$ and the determinant $|\xi_{\nu\mu}|$ is a unit in $\mathfrak{o}$. In fact, any such transformation determines a corresponding *non-singular* linear homogeneous transformation on the $\mathfrak{m}^2$-residues $\bar{x}_1, \bar{x}_2, \ldots, \bar{x}_{\bar{N}}$ of the x_ν, namely: $\overline{x'_\nu} = \sum_{\mu=1}^{N} \bar{\xi}_{\nu\mu} \bar{x}_\mu$, where the $\bar{\xi}_{\nu\mu}$ are the $\mathfrak{m}$-residues of the $\xi_{\nu\mu}$, in k. This transformation, in its turn, determines a projective self-transformation σ of $\bar{\mathbf{P}}$, defined as follows: $u'_{i\nu} = \sum_{\mu=1}^{\bar{N}} \overline{\xi'_{\mu\nu}} u_{i\mu}$ ($i = 1, 2, \ldots, \rho$; $\nu = 1, 2, \ldots, \bar{N}$), where the matrix $\| \overline{\xi'_{\mu\nu}} \|$ is the inverse of the matrix $\|\xi_{\nu\mu}\|$. If we set $\alpha = \{\alpha_{i\nu}\}$, $\alpha' = \sigma(\alpha) = \{\alpha'_{i\nu}\}$, where $\alpha'_{i\nu} = \sum_{\mu=1}^{\bar{N}} \overline{\xi'_{\mu\nu}} \alpha_{i\mu}$, then we find from (13) that $\bar{z}_i = \sum_{\nu=1}^{\bar{N}} \alpha'_{i\nu} \overline{x'_\nu}$, and if there exists a proper subvariety $\overline{\Gamma'}$ in $\bar{\mathbf{P}}$, defined over k, such that statement (2) is true with respect to the new basis $\{x'_1, x'_2, \ldots, x'_{\bar{N}}\}$ of $\mathfrak{m}$ (i.e., if the set of elements $z_1, z_2, \ldots, z_\rho$ satisfies condition C whenever the point α' does not belong to $\bar{\Gamma}'$), then the variety $\bar{\Gamma} = \sigma^{-1}(\bar{\Gamma}')$ will have all the properties required in statement (2).

Therefore, if the basis $\{x_1, x_2, \ldots, x_{\bar{N}}\}$ is not minimal (i.e., if $\bar{N} > N$) then we may assume, for the purposes of our proof, that the last $\bar{N} - N$ elements $x_{N+1}, x_{N+2}, \ldots, x_{\bar{N}}$ of that basis belong to $\mathfrak{m}^2$, and that therefore the first N elements $x_1, x_2, \ldots, x_N$ form a *minimal* basis of $\mathfrak{m}$. This being assumed, let $\mathbf{P}$ be the projective subspace of $\bar{\mathbf{P}}$ defined by the equations $u_{i,N+a} = 0$ ($i = 1, 2, \ldots, \rho$; $a = 1, 2, \ldots, \bar{N} - N$). We apply the definition B7 to our given condition C, where we now have $s = 1$. We consider the rational map

$$T: \mathbf{P} \rightarrow \mathrm{Gr}(\mathbf{P}^{(1)}, \rho - 1) \qquad (\dim \mathbf{P}^{(1)} = N - 1) \tag{14}$$

defined generically by the condition that the T-image of the generic point $\{u_{iq}\}$ of $\mathbf{P}$ ($i = 1, 2, \ldots, \rho$; $q = 1, 2, \ldots, N$) is the $(\rho - 1)$-subspace of $\mathbf{P}^{(1)}$ whose Grassmann coordinates are the ρ-rowed minors of the matrix $\|u_{iq}\|$ (this matrix is formed by the first N columns of the matrix $\|u_{i\nu}\|$). Let Γ_1 be the full inverse image $T^{-1}\{W\}$ of the variety W (defined over k) which appears in Definition B7 in regard to the given *open* condition C, of order 1. Then Γ_1 is a proper subvariety of $\mathbf{P}$ (since T is surjective) and is, of course, defined over k. Let Γ_2 be the variety in $\mathbf{P}$ defined by the condition that the matrix $\|u_{iq}\|$ has rank less than ρ. We set $\Gamma = \Gamma_1 \cup \Gamma_2$. Then Γ is defined over k, and is a proper subvariety of $\mathbf{P}$. It is clear that statement (2) is true if we take for $\bar{\Gamma}$ the hypercone in $\bar{\mathbf{P}}$

which projects Γ from the subspace of $\bar{\mathbf{P}}$ defined by the equations $u_{iq}=0$ ($i=1, 2,\ldots,\rho$; $q=1, 2,\ldots,N$).

Now, let us assume (2) and that the basis $\{x_1, x_2,\ldots,x_{\bar{N}}\}$ is minimal (whence $\bar{N}=N$). We also assume that the condition C is projectively k-stable. In this case we shall write $\mathbf{P}$ instead of $\bar{\mathbf{P}}$ and Γ instead of $\bar{\Gamma}$. If $X_1, X_2,\ldots,X_N$ are indeterminates and if we set $Z_i=\sum_{\nu=1}^{N}u_{i\nu}X_\nu$ ($i=1, 2,\ldots,\rho$) and regard $Z_1, Z_2,\ldots,Z_\rho$ as homogeneous coördinates in a projective $(\rho-1)$-space $\mathbf{P}'$, then any non-singular projective self-transformation $\varphi': Z_i'=\sum_{j=1}^{\rho} a_{ij}Z_j$ of $\mathbf{P}'$, *defined over* k (i.e., such that the a_{ij} belong to k), induces a non-singular projective self-transformation $\varphi: u_{i\nu}'=\sum_{j=1}^{\rho} a_{ij}u_{j\nu}$ ($i=1, 2,\ldots,\rho$; $\nu=1,2,\ldots,N$) of $\mathbf{P}$, also defined over k, and these induced transformations form a group $H(k)$. Since our condition C is projectively k-stable, statement (2) remains true if we replace Γ by the (possibly smaller) variety $\bigcap_{\varphi\in H(k)}\varphi(\Gamma)$, which is also defined over k. *Hence we may assume that Γ is invariant under all the transformations φ in $H(k)$.*

We introduce the larger group $H(\Omega)$ consisting of all projective self-transformations φ of $\mathbf{P}$ which are induced by the non-singular projective transformations $\varphi': Z_i'=\sum_{j=1}^{\rho} a_{ij}Z_j$ ($i=1, 2,\ldots,\rho$) of $\mathbf{P}'$ which are defined over some universal domain Ω containing k (i.e., the a_{ij} are now allowed to range over Ω). *We assert that Γ is also invariant under all the transformations φ in $H(\Omega)$.* To see this, we set $u_i=\{u_{i1},u_{i2},\ldots,u_{i,N}\}$ and we consider any homogeneous polynomial $f(u)=f(u_1,u_2,\ldots,u_\rho)$ in the $u_{i\nu}$, with coefficients in k, which vanishes on Γ. Let $\bar{u}=\{\bar{u}_1,\bar{u}_2,\ldots,\bar{u}_\rho\}=\{\bar{u}_{i\nu}\}$ be any point of Γ ($\bar{u}_{i\nu}\in\Omega$). Let $\|\lambda_{ij}\|$ be a ρ-rowed square matrix, with indeterminate entries λ_{ij}, and let $g(\{\lambda_{ij}\})$ be the polynomial in the λ_{ij}, with coefficients in $k(\{\bar{u}_{i\nu}\})$, defined by $g(\{\lambda_{ij}\})=f(\sum_{j=1}^{\rho}\lambda_{1j}\bar{u}_j, \sum_{j=1}^{\rho}\lambda_{2j}\bar{u}_j,\ldots,\sum_{j=1}^{\rho}\lambda_{\rho j}\bar{u}_j)$. We have then $g(\{\bar{\lambda}_{ij}\})=0$ for all $\bar{\lambda}_{ij}$ in k such that $\det(\bar{\lambda}_{ij})\neq 0$, since the point $\{\bar{u}_{i\nu}\}$ belongs to Γ and since Γ is invariant under all transformations φ in $H(k)$. Since we have assumed that k is an infinite field, it follows that the polynomial $g(\{\lambda_{ij}\})$ must be identically zero, i.e., also the point $\varphi_{\bar{\lambda}}(\bar{u})=\{\sum_{j=1}^{\rho}\bar{\lambda}_{ij}\bar{u}_{j\nu}\}$ ($i=1, 2,\ldots,\rho$; $\nu=1, 2,\ldots,N$) lies on Γ, for all $\bar{\lambda}_{ij}$ in Ω such that $\det(\bar{\lambda}_{ij})\neq 0$. Since φ_λ ranges over $H(\Omega)$ as the $\bar{\lambda}_{ij}$'s range over Ω (subject to the condition $\det(\bar{\lambda}_{ij})\neq 0$), our assertion follows.

We now set $W=T\{\Gamma\}$ (= *total* transform of Γ), where T is the rational transformation (14). Then W is a subvariety of $\mathrm{Gr}(\mathbf{P}^{(1)}, \rho-1)$, defined over k and W is a *proper* subvariety of Gr ($\mathbf{P}^{(1)}$, $\rho-1$), for in the contrary case Γ would contain a point $\{\bar{u}_{i\nu}\}$ such that $T(\{\bar{u}_{i\nu}\})=T(\{u_{i\nu}\})$, where the $u_{i\nu}$ are indeterminates, and this would imply that Γ contains the general point $\{u_{i\nu}\}$ of $\mathbf{P}$

(since the relation $T(\{\bar{u}_{i\nu}\}) = T(\{u_{i\nu}\})$ is equivalent with the assertion that the two points $\{\bar{u}_{i\nu}\}$ and $\{u_{i\nu}\}$ are conjugate under the group $H(\Omega)$), contrary to the fact that Γ is a proper subvariety of **P**. Since it is clear that the condition of the Definition B7 is satisfied, for the given condition C, by the above variety W, the lemma is proved.

We now state and prove the final result of this appendix.

THEOREM $B11$. *Let $\mathfrak{p}$ be a prime ideal in a complete local ring and let k be an infinite coefficient field of $\mathfrak{o}$. Then the condition that a set of parameters $z_1, z_2, \ldots, z_r$ of $\mathfrak{o}$ be k-transversal with respect to $\mathfrak{p}$ (in the sense of Defintion* B9) *is an open condition of order* 1.

Proof. In the special case $\mathfrak{p} = \mathfrak{m}$, k-transversality of $z_1, z_2, \ldots, z_r$ with respect to $\mathfrak{p}$ means simply that $z_1, z_2, \ldots, z_r$ are transversal parameters of $\mathfrak{o}$ (see Definition B3 and observe that in Definition B9, in the special case $\mathfrak{p} = \mathfrak{m}$, the ideal $\mathfrak{P}$ is the ideal generated in $k[[z_1, z_2, \ldots, z_r]]$ be the elements $z_1, z_2, \ldots, z_r$). In this special case the theorem has already been proved (Proposition B8). We note that in this special case ($\mathfrak{p} = \mathfrak{m}$) the property of k-transversality is independent of the choice of the coefficient field k. We shall therefore assume that $p < m$.

The theorem is essentially trivial in the other extreme case, i.e., in the case in which $\mathfrak{p}$ is a minimal prime ideal in $\mathfrak{o}$ (equivalently: $\mathfrak{p}$ has height zero). For, in that case the local ring $\mathfrak{o}_{\mathfrak{p}}$ has Krull dimension zero, and the multiplicity e' of the maximal ideal of $\mathfrak{o}_{\mathfrak{p}}$ is equal to the length of $\mathfrak{o}_{\mathfrak{p}}$ regarded as an $\mathfrak{o}_{\mathfrak{p}}$-module (see C.A. 2, p. 294, proof of Theorem 22, case $d = 0$). Thus, in this case e' is also the length of the zero ideal of $\mathfrak{o}_{\mathfrak{p}}$. On the other hand, we know that if $\{z_1, z_2, \ldots, z_r\}$ is any set of parameters of the complete local ring $\mathfrak{o}$ and k is any coefficient field of $\mathfrak{o}$, then $\mathfrak{p} \cap k[[z_1, z_2, \ldots, z_r]]$ is also a minimal prime ideal in $k[[z_1, z_2, \ldots, z_r]]$, i.e., the zero ideal. Thus we see that *any set of parameters $z_1, z_2, \ldots, z_r$ of $\mathfrak{o}$ is k-transversal with respect to any minimal prime ideal $\mathfrak{p}$ of* 0. The condition of k-transversality with respect to a minimal prime ideals is therefore vacuous, and vacuous conditions are, of course, open conditions.

So from now on we assume that if ρ denotes the height of $\mathfrak{p}$ then $0 < \rho < r$.

We consider the local domain $\mathfrak{o}^* = \mathfrak{o}/\mathfrak{p}$. This is a complete local domain, of Krull dimension ρ, with maximal ideal $\mathfrak{m}^* = \mathfrak{m}/\mathfrak{p}$ and k as coefficient field. We fix a *minimal* basis $x_1, x_2, \ldots, x_N$ of $\mathfrak{m}$ and we denote by x_ν^* the $\mathfrak{p}$-residue of x_ν. Thus, both $\mathfrak{o} = k[[x_1, x_2, \ldots, x_N]$ and $\mathfrak{o}^* = k[[x_1^*, x_2^*, \ldots, x_N^*]]$ are homomorphic images of the formal power series ring $k[[X_1, X_2, \ldots, X_N]]$ in N variables $X_1, X_2, \ldots, X_N$. Let **P** be the projective space introduced in Lemma B10, with ρ

replaced by r. Hence the homogeneous coordinates in **P** are the entries $u_{i\nu}$ of a matrix with r rows and N columns (the $u_{i\nu}$ being indeterminates). We denote by $\mathbf{P}^*$ and $\mathbf{P}'$ the two (disjoint) subspaces of **P** defined respectively by the equations

$$\mathbf{P}^*: u_{\rho+n,\nu}=0, \qquad n=1,2,\ldots,r-\rho; \quad \nu=1,2,\ldots,N; \tag{15*}$$

$$\mathbf{P}': u_{m,\nu}=0, \qquad m=1,2,\ldots,\rho; \quad \nu=1,2,\ldots,N. \tag{15'}$$

Since transversality of parameters (with respect to $\mathfrak{m}$) is an open condition, we can use Lemma B10 and fix a proper algebraic subvariety Γ of **P**, defined over k, such that if a k-rational point $\{\alpha_{i\nu}\}$ of **P** does not belong to Γ and if z_1, $z_2,\ldots,z_r$ are elements of $\mathfrak{m}$ such that

$$\bar{z}_i=\sum_{\nu=1}^{N}\alpha_{i,\nu}\bar{x}_\nu, \qquad i=1,2,\ldots,r, \tag{16}$$

then z_1, $z_2,\ldots,z_r$ are transversal parameters of $\mathfrak{o}$. Here $\bar{z}_i$ and $\bar{x}_\nu$ are the $\mathfrak{m}^2$-residues of z_i and x_ν respectively. Similarly, applying Lemma B10 to the complete local domain $\mathfrak{o}^*$, we fix a proper algebraic subvariety Γ^* of $\mathbf{P}^*$, defined over k, such that if a k-rational point $\{\alpha_{m\nu}\}$ of $\mathbf{P}^*$ ($m=1,2,\ldots,\rho$; $\nu=1,2,\ldots,N$) does not belong to Γ^* and if z_1^*, $z_2^*,\ldots,z_\rho^*$ are elements of $\mathfrak{m}^*$ such that

$$\bar{z}_m^*=\sum_{\nu=1}^{N}\alpha_{m\nu}\bar{x}_\nu^*, \qquad m=1,2,\ldots,\rho, \tag{16*}$$

then z_1^*, $z_2^*,\ldots,z_\rho^*$ are transversal parameters of $\mathfrak{o}^*$. Here $\bar{z}_m^*$ and $\bar{x}_\nu^*$ are the $\mathfrak{m}^{*2}$-residues of z_m^* and x_ν^* respectively. Actually, we shall make no use of the condition that z_1^*, $z_2^*,\ldots,z_\rho^*$ are *transversal* parameters; all we shall use is that they are *parameters* of $\mathfrak{o}^*$.

Let $h_\mathfrak{p}$ be the canonical homomorphism of $\mathfrak{o}$ into $\mathfrak{o}_\mathfrak{p}$, let $h_\mathfrak{p}(\mathfrak{o})=\mathfrak{o}'$, $h_\mathfrak{p}(\mathfrak{p})=\mathfrak{p}'$, $h_\mathfrak{p}(x_\nu)=x_\nu'$ ($\nu=1,2,\ldots,N$). The (complete) local ring $\mathfrak{o}'$ has the same Krull dimension r as $\mathfrak{o}$ (since $\operatorname{Ker} h_\mathfrak{p}$ is contained in at least one minimal prime of $\mathfrak{o}$), the maximal ideal $\mathfrak{m}'$ of $\mathfrak{o}'$ is generated by x_1', $x_2',\ldots,x_N'$, k can be canonically identified with a coefficient field of $\mathfrak{o}'$, both $\mathfrak{p}$ and $\mathfrak{p}'$ have height ρ, and $\mathfrak{o}_\mathfrak{p}=\mathfrak{o}'_{\mathfrak{p}'}$. The local ring $\mathfrak{o}_\mathfrak{p}$ has Krull dimension $r-\rho$, and its residue field $\Delta_\mathfrak{p}^*$ is the field of fractions of the local domain $\mathfrak{o}'/\mathfrak{p}'$, the latter being canonically isomorphic to $\mathfrak{o}/\mathfrak{p}$ (since $\operatorname{Ker} h_\mathfrak{p}\subset\mathfrak{p}$), i.e., to $\mathfrak{o}^*=k[[x_1^*,x_2^*,\ldots,x_N^*]]$. Thus, we may write $\Delta_\mathfrak{p}^*=k((x_1^*,x_2^*,\ldots,x_N^*))=$ field of formal meromorphic functions of $x_1^*,\ldots,x_N^*$, over k. We pass to the completion $\hat{\mathfrak{o}}_\mathfrak{p}$ of $\mathfrak{o}_\mathfrak{p}$. In $\hat{\mathfrak{o}}_\mathfrak{p}$ we may identify $\Delta_\mathfrak{p}^*$ with a coefficient field of $\hat{\mathfrak{o}}_\mathfrak{p}$ containing k. [Note that we have $k\subset\mathfrak{o}'\subset\mathfrak{o}'_{\mathfrak{p}'}\subset\hat{\mathfrak{o}}'_{\mathfrak{p}'}$]. The maximal ideal $\hat{\mathfrak{m}}'$ of $\hat{\mathfrak{o}}_{\mathfrak{p}}$ is $\mathfrak{p}'\hat{\mathfrak{o}}_{\mathfrak{p}}$. Now, we have $\mathfrak{o}' = k[[x'_1, x'_2, \ldots, x'_N]]$, and thus $\mathfrak{p}'$ is the set of elements $f(x'_1, x'_2, \ldots, x'_N)$ in $\mathfrak{o}'$ such that the element $f(x^*_1, x^*_2, \ldots, x^*_N)$ (of $\mathfrak{o}^*$) is zero. The maximal ideal $\hat{\mathfrak{m}}'$ is therefore generated in $\hat{\mathfrak{o}}_{\mathfrak{p}}$ by all the formal power series $f(x'_1, x'_2, \ldots, x'_N)$ (with coefficients in k) such that $f(x^*_1, x^*_2, \ldots, x^*_N) = 0$. Since $\Delta^*_{\mathfrak{p}} = k((x^*_1, x^*_2, \ldots, x^*_N)) \subset \hat{\mathfrak{o}}_{\mathfrak{p}}$, the differences $x'_1 - x^*_1$, $x'_2 - x^*_2, \ldots, x'_N - x^*_N$ are elements of $\hat{\mathfrak{o}}_{\mathfrak{p}}$ which belong to $\hat{\mathfrak{m}}'$, since x^*_ν is the $\mathfrak{p}'$-residue of x'_ν in $\mathfrak{o}'$ and is therefore also the $\hat{\mathfrak{m}}'$-residue of x'_ν. It follows that the N elements

$$x'_1 - x^*_1,\ x'_2 - x^*_2, \ldots, x'_N - x^*_N$$

form a basis of $\hat{\mathfrak{m}}'$ (not necessarily minimal). Applying again Lemma B10—this time to the complete local ring $\hat{\mathfrak{o}}_{\mathfrak{p}}$ of Krull dimension $r-\rho$ and to the coefficient field $\Delta^*_{\mathfrak{p}}$ of $\hat{\mathfrak{o}}_{\mathfrak{p}}$—we fix a proper subvariety Γ'^* of $\mathbf{P}'$ [see (15′)], defined over $\Delta^*_{\mathfrak{p}}$, such that if $\{\beta_{\rho+n,\nu}\}$ $(n = 1, 2, \ldots, r-\rho;\ \nu = 1, 2, \ldots, N)$ is any $\Delta^*_{\mathfrak{p}}$-rational point of $\mathbf{P}'$ which does not belong to Γ'^* and if $\{z'_{\rho+1}, z'_{\rho+2}, \ldots, z'_r\}$ is any set of $r-\rho$ elements of $\hat{\mathfrak{m}}'$ such that

$$\overline{z'}_{\rho+n} = \sum_{\nu=1}^{N} \beta_{\rho+n,\nu} \overline{(x'_\nu - x^*_\nu)} \qquad (n = 1, 2, \ldots, r-\rho) \tag{16′}$$

then $z'_{\rho+1}, z'_{\rho+2}, \ldots, z'_r$ are transversal parameters of $\hat{\mathfrak{o}}_{\mathfrak{p}}$. Here $\overline{z'}_{\rho+n}$ and $\overline{x'_\nu - x^*_\nu}$ denote the $\hat{\mathfrak{m}}^{*2}$-residues of $z'_{\rho+n}$ and $x'_\nu - x^*_\nu$ (the space $\hat{\mathfrak{m}}'/\hat{\mathfrak{m}}'^2$ being regarded as a vector space over $\Delta^*_{\mathfrak{p}}$).

We shall now associate with Γ'^* a certain *proper* subvariety Γ' of $\mathbf{P}'$, *defined over* k, which we shall call *the reduction of* Γ'^* **mod** $\mathfrak{m}^*$; in symbols: $\Gamma' = \Gamma'^*/\mathfrak{m}^*$.

Quite generally, suppose we have a local *domain* $\mathfrak{o}$ (not necessarily complete), with maximal ideal $\mathfrak{m}$ and residue field $k = \mathfrak{o}/\mathfrak{m}$. Let K be the field of fractions of $\mathfrak{o}$ and let $\mathbf{P}_n$ be an n-dimensional projective space. Let V be a *proper* subvariety of $\mathbf{P}_n$, defined over K. We shall associate with V a well-defined *proper* subvariety $\bar{V}$ of $\mathbf{P}_n$, *defined over* k, called the *reduction of* V **mod** $\mathfrak{m}$; in symbols: $\bar{V} = V/\mathfrak{m}$. We shall then apply this construction to the case in which $\mathfrak{o}$ is our complete local ring $\mathfrak{o}^*$, k is the given coefficient field of $\mathfrak{o}^*$, $\mathbf{P}_n$ is the above projective space $\mathbf{P}'$ (whence $n = (r-\rho)N - 1$), and V is Γ'^*.

The homogeneous coordinates $(a_0, a_1, \ldots, a_n)$ of any point P of $\mathbf{P}_n$ can always be assumed to be elements of $\mathfrak{o}$ (since K is the field of fractions of $\mathfrak{o}$). *If there exists a set of homogeneous coordinates* $a_0, a_1, \ldots, a_n$ *of* P *in* $\mathfrak{o}$, *such that* *not all the coordinates a_i belong to* $\mathfrak{m}$, then we define *the reduction of P* **mod** $\mathfrak{m}$, and we denote it by $P/\mathfrak{m}$, the k-rational point $\bar{P}(\bar{a}_0, \bar{a}_1, \ldots, \bar{a}_n)$, where the $\bar{a}_i$ are the $\mathfrak{m}$-residues of the a_i. It is immediately seen that if P admits a reduction mod $\mathfrak{m}$ (i.e., if there exists a set of homogeneous coordinates $a_0, a_1, \ldots, a_n$ of P ($a_i \in \mathfrak{o}$) such that not all the a_i are in $\mathfrak{m}$), the reduced point $\bar{P} = P/\mathfrak{m}$ is uniquely determined (i.e., it is independent of the choice of the homogeneous coordinates $a_0, a_1, \ldots, a_n$, always subject to the condition that not all the a_i are in $\mathfrak{m}$).

Now, given the subvariety V of $\mathbf{P}$, we denote by $\bar{V}_0$ the set of all points $\bar{P} = P/\mathfrak{m}$ obtained by letting P range over the set of all points P of V which admit a reduction modulo $\mathfrak{m}$. The variety $\bar{V} = V/\mathfrak{m}$ in $\mathbf{P}_n$, defined over k, which we associate with V is the smallest variety in $\mathbf{P}_n$, *defined over* k, which contains the set $\bar{V}_0$. (Observe that all points of $\bar{V}_0$ are, by definition, k-rational points). We shall now prove that if V is a *proper* subvariety of $\mathbf{P}_n$, also $\bar{V}$ is a *proper* subvariety of $\mathbf{P}_n$.

We first observe that if we have a representation of $\bar{V}$ as intersection $V_1 \cap V_2 \cap \cdots \cap V_q$ of varieties V_i, all defined over K, then $\bar{V}_0 \subset \cap_{i=1}^q \bar{V}_{i,0}$ and hence $V/\mathfrak{m} \subset \cap_{i=1}^h V_i/\mathfrak{m}$. Therefore it is sufficient to consider the case in which V *is a hypersurface*.

Let $f(X_0, X_1, \ldots, X_n) = 0$ be an equation of the hypersurface V, where we may assume that the coefficient of f are in the local ring $\mathfrak{o}$. Let s be the greatest integer such that all the coefficients of f are in $\mathfrak{m}^s$ (thus not all the coefficients of f are in $\mathfrak{m}^{s+1}$). We fix in the set of coefficients of f a maximal subset $\{b_1, b_2, \ldots, b_N\}$ of coefficients whose $\mathfrak{m}^{s+1}$-residue $\bar{b}_1, \bar{b}_2, \ldots, \bar{b}_N$ (in the graded algebra $G_{\mathfrak{m}}(\mathfrak{o})$ of $\mathfrak{o}$) are linearly independent (over k). Then it follows that we can write f in the following form:

$$f = b_1 f_1 + b_2 f_2 + \cdots + b_N f_N + g(X_0, X_1, \ldots, X_N),$$

where $f_1, f_2, \ldots, f_N$ are (homogeneous) polynomials with coefficients in $\mathfrak{o}$ such that in each of these polynomials at least one of the coefficients is equal to 1 (namely, if ω_i is the nomonial in $X_0, X_1, \ldots, X_N$ whose coefficient in f is b_i then ω_i will occur in f_i with coefficients 1), *while g is a form with coefficients* $\mathfrak{m}^{s+1}$. It follows, in the first place, that the polynomials $\bar{f}_i$ obtained from f_i by reducing all the coefficients of f_i mod $\mathfrak{m}$ is a form $\bar{f}_i$ *with coefficients in* k, which is not identically zero. In the second place, if a point $P(a_0, a_1, \ldots, a_n)$ on V admits a reduction $\bar{P} = P/\mathfrak{m} = (\bar{a}_0, \bar{a}_1, \ldots, \bar{a}_n)$, mod $\mathfrak{m}$, then the fact that $f(a_0, a_1, \ldots, a_n) = 0$ implies that $b_1 f_1(a_0, a_1, \ldots, a_N) + b_2 f_2(a_0, a_1, \ldots, a_N) + \cdots + b_N f_N(a_0, a_1, \ldots, a_N) \in \mathfrak{m}^{s+1}$, or, equivalently, $\bar{b}_1 \bar{f}_1(\bar{a}_0, \bar{a}_1, \ldots, \bar{a}_N) + \bar{b}_2 \bar{f}_2(\bar{a}_0, \bar{a}_1, \ldots, \bar{a}_N) + \cdots + \bar{b}_N \bar{f}_N(\bar{a}_0, \bar{a}_1, \ldots, \bar{a}_N) = 0$. This implies that $\bar{f}_i(\bar{a}_0,$

$\bar{a}_1,\ldots,\bar{a}_N)=0$, for $i=1, 2,\ldots,N$, showing that $V/\mathfrak{m}$ is contained in the variety defined by the equations $\bar{f}_i(X_0,X_1,\ldots,X_n)=0$, $i=1, 2,\ldots,N$. This proves our assertion that $V/\mathfrak{m}$ is a proper subvariety of $\mathbf{P}_n$.

At the risk of repeating the obvious, we wish nevertheless to emphasize again that, by its very definition, the variety $V/\mathfrak{m}$ has the following property: *if a k-rational point P_0 of $\mathbf{P}_n$ does not lie on $V/\mathfrak{m}$, then none of the K-rational points of $\mathbf{P}_n$ whose reduction mod $\mathfrak{m}$ is P_0 lies on V.*

We have now introduced altogether *three* varieties, *defined over k:* (1) the variety Γ in the $(rN$-1)-demensional projective space $\mathbf{P}$, introduced in connection with the *transversality* requirement on the r elements (parameters) z_1, $z_2,\ldots,z_r$ of $\mathfrak{m}$ [see (16)]; (2) the variety Γ^* in the $(\rho N$-1)-dimensional linear subspace $\mathbf{P}^*$ of $\mathbf{P}$ [see (15*)], introduced in connection with the requirement that the elements $z_1^*, z_2^*,\ldots,z_\rho^*$ be *parameters* of $\mathfrak{o}^*$ [see (16*), where we, for convenience, have used a sufficient condition for the elements $z_1^*, z_2^*,\ldots,z_\rho^*$ to be *transversal* parameters); and (3) the variety $\Gamma'=\Gamma'^*/\mathfrak{m}^*$ in the $((r-\rho)N-1)$-dimensional subspace $\mathbf{P}'$ of $\mathbf{P}$ [see (15′)], which is the $\mathfrak{m}^*$-reduction of the variety Γ'^* (defined over $\Delta_\mathfrak{p}^*=k((x_1^*,\ldots,x_N^*)))$ introduced in connection with the *transversality* requirement on the $r-\rho$ parameter $z'_{\rho+1}, z'_{\rho+2},\ldots,z'_r$ of $\hat{\mathfrak{o}}_\mathfrak{p}$ [see (16′)]. The spaces $\mathbf{P}^*$ and $\mathbf{P}'$ are disjoint subspaces of $\mathbf{P}$, of complementary dimensions $\rho N-1$ and $(r-\rho)N-1$ respectively, and each of the three varieties Γ, Γ^*, Γ' is a proper subvariety of its ambient projective space.

We now denote by Γ_1 the hypercone which projects Γ^* from $\mathbf{P}'$ and by Γ_2 the hypercone which projects Γ' from $\mathbf{P}^*$, and we set $\Gamma_0=\Gamma\cup\Gamma_1\cup\Gamma_2$. Then Γ_0 is a proper subvariety of $\mathbf{P}$, defined over k. We now proceed to show that *if in Lemma* B10 *we take for the local ring $\mathfrak{o}$ our present complete local ring $\mathfrak{o}$ and for $\{x_1, x_2,\ldots,x_{\bar{N}}\}$ our present **minimal** basis $\{x_1,x_2,\ldots,x_N\}$ of $\mathfrak{m}$, while we take as condition C the condition of k-transversality of r parameters $z_1, z_2,\ldots,z_r$ of $\mathfrak{o}$ with respect to the given prime ideal $\mathfrak{p}$ of $\mathfrak{o}$, then statement* (2) *given in that lemma is valid when $\bar{\mathbf{P}}$ is our present projective space $\mathbf{P}$ (of dimension $rN-1$) and $\bar{\Gamma}$ is the variety Γ_0 just introduced above.* Since k-transversality with respect to $\mathfrak{p}$ is a projectively k-stable condition and since we are dealing with a *minimal* basis $\{x_1, x_2,\ldots,x_N\}$ of $\mathfrak{m}$, it will follow from Lemma B10 that statement (1) of that lemma holds true if condition C is the condition of k-transversality with respect to $\mathfrak{p}$, and this will complete the proof of the theorem.

Let then $\{\alpha_{i\nu}\}$ be a k-rational point of $\mathbf{P}$ which does not belong to the variety Γ_0 and let $z_1, z_2,\ldots,z_r$ be any set of r elements of $\mathfrak{m}$ satisfying the relations (16). We have to show that these elements are k-transversal with respect to the given prime ideal $\mathfrak{p}$ of $\mathfrak{o}$.

Since the point $\{\alpha_{i\nu}\}$ is not on Γ_0, it is a fortiori not on the subvariety Γ of Γ_0. Therefore, the relations (16) imply that the elements $z_1, z_2, \ldots, z_r$ are *transversal* parameters of $\mathfrak{o}$. Since $\mathfrak{o}$ is complete we can identify (canonically) the power series ring $k[[z_1,z_2,\ldots,z_r]]$ with a subring R of $\mathfrak{o}$:

$$R = k[[z_1,z_2,\ldots,z_r]] \subset \mathfrak{o}. \tag{17}$$

We are also given that the point $\{\alpha_{i\nu}\}$ does not lie on the hypercone Γ_1 projecting Γ^* from the subspace $\mathbf{P}'$ of $\mathbf{P}$. In other words, the point $\{\alpha_{m\nu}\}$ of $\mathbf{P}^*$ $(m=1, 2,\ldots,\rho;\ \nu=1, 2,\ldots,N)$ does not lie on Γ^*. Therefore it follows from (16*) that the elements $z_1^*, z_2^*,\ldots,z_\rho^*$ are (transversal) parameters of $\mathfrak{o}^*$. They are therefore analytically independent over k, and the restriction to $k[[z_1,z_2,\ldots,z_\rho]]$ of the canonical homomorphism $\mathfrak{o}\to\mathfrak{o}^*=\mathfrak{o}/\mathfrak{p}$ is an injection. We can (and shall) therefore identify the ρ elements $z_1, z_2,\ldots,z_\rho$ with their $\mathfrak{p}$-residues $z_1^*, z_2^*,\ldots,z_\rho^*$ and the subring $k[[z_1^*, z_2^*,\ldots,z_\rho^*]]$ of $\mathfrak{o}^*$ with the subring $k[[z_1, z_2,\ldots,z_\rho]]$ of R [see (17)].

We know that the restriction to R of the canonical homomorphism $h_{\mathfrak{p}}:\mathfrak{o}\to\mathfrak{o}_{\mathfrak{p}}$ is an injection. Therefore we can identify the r elements $z_1, z_2,\ldots,z_r$ with their $h_{\mathfrak{p}}$-images $z_1', z_2',\ldots,z_r'$ in $\hat{\mathfrak{o}}_{\mathfrak{p}}$.

Let

$$z_i = g_i(x_1,x_2,\ldots,x_N), \qquad i=1,2,\ldots,r \tag{18}$$

where g_i is a formal power series with coefficients in k. By our identification $z_i=z_i'$ for $i=1, 2,\ldots,r$, we can therefore write:

$$z_i' = z_i = g_i(x_1',x_2',\ldots,x_N'), \qquad i=1, 2,\ldots,r. \tag{18'}$$

The $\hat{\mathfrak{m}}_{\mathfrak{p}}$-residue z_i^* of z_i' is equal to the element $g_i(x_1^*, x_2^*,\ldots,x_N^*)$ of $\mathfrak{o}^*$. We note that, by (16) and (18), the sum of the first degree terms in $g_i(x_1',x_2',\ldots,x_N')$ is the linear form $\sum_{\nu=1}^N \alpha_{i\nu}x_\nu'$. Hence, using the Taylor expansion of the power series $g_i(x_1',x_2',\ldots,x_N')$ as a power series in the differences $x_1'-x_1^*$, $x_2'-x_2^*,\ldots,x_N'-x_N^*$, we will have:

$$(z_i - z_i^*) - \left[\sum_{\nu=1}^{N}(\alpha_{i\nu}+\pi_{i\nu}^*)(x_\nu'-x_\nu^*)\right] \in \hat{\mathfrak{m}}_{\mathfrak{p}}^2 \qquad (i=1, 2,\ldots,r) \tag{19}$$

where the $\pi_{i\nu}^*$ are elements of the maximal ideal $\mathfrak{m}^*$ of $\mathfrak{o}^*$. By assumption, the point $\{\alpha_{i\nu}\}$ does not belong to the hypercone Γ_2 which projects Γ' from the subspace $\mathbf{P}^*$ of $\mathbf{P}$ [see (15*)]. In other words, the k-rational point $\{\alpha_{\rho+n,\nu}\}$ of $\mathbf{P}'$ $(n=1, 2,\ldots,r-\rho;\ \nu=1, 2,\ldots,N)$ does not belong to Γ' $(=\Gamma'^*/\mathfrak{m}^*)$. Since this point is the $\mathfrak{m}^*$-reduction of the $\Delta_{\mathfrak{p}}^*$-rational point $\{\alpha_{\rho+n,\nu}+\pi_{\rho+n,n}^*\}$, *it follows* *that the point* $\{\alpha_{\rho+n,\nu},\pi^*_{\rho+n,\nu}\}$ *of* $\mathbf{P}'$ $(n=1,2,\ldots,r-\rho;\nu=1,2,\ldots,N)$ *does not belong to* Γ'^*. Hence, by the definition of Γ'^* [see (16*)] and using the last $r-\rho$ of the relations (19) (i.e., for the $i=\rho+1,\ \rho+2,\ldots,r$), *we conclude that the* $(r-\rho)$ *elements* $z_{\rho+n}-z^*_{\rho+n}$ $(=z'_{\rho+n}-z^*_{\rho+n};\ n=1,2,\ldots,r-\rho)$ *are transversal parameters of* $\hat{\mathfrak{o}}_{\mathfrak{p}}$, i.e., we have

$$e\left(\sum_{n=1}^{r-\rho}\left(\hat{\mathfrak{o}}_{\mathfrak{p}}\cdot(z_{\rho+n}-z^*_{\rho+n})\right)\right)=e(\hat{\mathfrak{m}}_{\mathfrak{p}}). \tag{20}$$

Since

$$\hat{\mathfrak{m}}_{\mathfrak{p}}\subset\sum_{i=1}^{r}\hat{\mathfrak{o}}_{\mathfrak{p}}(z_i-z_i^*)\subset\sum_{n=1}^{r-\rho}\hat{\mathfrak{o}}_{\mathfrak{p}}(z_{\rho+n}-z^*_{\rho+n}),$$

it follows that

$$e(\hat{\mathfrak{m}}_{\mathfrak{p}})\leqslant e\left(\sum_{i=1}^{r}\hat{\mathfrak{o}}_{\mathfrak{p}}(z_i-z_i^*)\right)\leqslant\sum e\left(\sum_{n=1}^{r-\rho}\hat{\mathfrak{o}}_{\mathfrak{p}}(z_{\rho+n}-z^*_{\rho+n})\right),$$

and therefore we conclude from (20) that

$$e(\hat{\mathfrak{m}}_{\mathfrak{p}})=e\left(\sum_{i=1}^{r}\hat{\mathfrak{o}}_{\mathfrak{p}}(z_i-z_i^*)\right). \tag{21}$$

Now, we turn our attention to the formal power series ring $R=k[[z_1,z_2,\ldots,z_r]]$. This is a subring of $\mathfrak{o}$, and $\mathfrak{o}$ is a finite R-module. Let $\mathfrak{p}\cap R=\mathfrak{P}$. Since we have already identified $z_1,z_2,\ldots,z_r$ with their $h_{\mathfrak{p}}$-images $z_1',z_2',\ldots,z_r'$ in the subring $\mathfrak{o}'=h_{\mathfrak{p}}(\mathfrak{o})$ of $\mathfrak{o}_{\mathfrak{p}}$, the localization $R_{\mathfrak{p}}$ is a subring of $\mathfrak{o}_{\mathfrak{p}}$, and $\mathfrak{o}_{\mathfrak{p}}$ is a finite $R_{\mathfrak{p}}$-module. Thus, the completion $\hat{R}_{\mathfrak{p}}$ of the local domain $R_{\mathfrak{p}}$ is a subring and a subspace of $\hat{\mathfrak{o}}_{\mathfrak{p}}$. The subfield $k((z_1^*,z_2^*,\ldots,z_r^*))$ is a coefficient field of the (regular) complete local domain $R_{\mathfrak{p}}$, since the $\mathfrak{p}$-residues $z_1^*,z_2^*,\ldots,z_r^*$ of $z_1,z_2,\ldots,z_r$ are also the $\mathfrak{p}$-residues of $z_1,z_2,\ldots,z_r$, and since therefore $R/\mathfrak{P}\cong k[[z_1^*,z_2^*,\ldots,z_r^*]]$. It follows[2] that

$$\hat{R}_{\mathfrak{P}}=k((z_1^*,z_2^*,\ldots,z_r^*))[[z_1-z_1^*,z_2-z_2^*,\ldots,z_r-z_r^*]] \tag{22}$$

[2]If we denote by A the (complete) power series ring which appears on the right-hand side of (22), then, using the Taylor expansion of any element $f(z_1,z_2,\ldots,z_r)$ of R as a power series in $z_1-z_1^*,z_2-z_2^*,\ldots,z_r-z_r^*$, we see that $R\subset A$. We have $f(z_1,z_2,\ldots,z_r)\in\mathfrak{P}$ if and only if $f(z_1^*,z_2^*,\ldots,z_r^*)=0$, i.e., if and only if f belongs to the maximal ideal of A (generated by $z_1-{}^*_1,z_2-z_2^*,\ldots,z_r-z_r^*$). Hence $A\subset R_{\mathfrak{p}}$, and since $\mathfrak{P}$ is the contraction of the maximal of A, it follows that $A\subset\hat{R}_{\mathfrak{P}}$. On the other hand, it is obvious that the complete local ring $\hat{R}_{\mathfrak{P}}$ contains A. This proves (22).

and that the maximal ideal $\mathfrak{P}\hat{R}_{\mathfrak{P}}$ of $\hat{R}_{\mathfrak{P}}$ is generated by the r elements $z_1 - z_1^*, z_2 - z_2^*, \ldots, z_r - z_r^*$. It follows therefore, by (21), that

$$e(\hat{\mathfrak{m}}_{\mathfrak{p}}) = e\left(\hat{\mathfrak{o}}_{\mathfrak{p}}\mathfrak{P}\right). \tag{23}$$

Now, since $\mathfrak{o}_{\mathfrak{p}}\mathfrak{P}$ is primary for $\mathfrak{m}_{\mathfrak{p}}$, the completion $\hat{\mathfrak{o}}_{\mathfrak{p}}$ of the local ring $\mathfrak{o}_{\mathfrak{p}}$ coincides with $\mathfrak{o}_{\mathfrak{p}}\mathfrak{P}$-adic completion of $\mathfrak{o}_{\mathfrak{p}}$ (see C.A.2, p. 256). Applying Corollary 1 of Theorem 6 of C.A.2, p. 258, to the case $A = \mathfrak{o}_{\mathfrak{p}}$ and $\mathfrak{m} = \mathfrak{o}_{\mathfrak{p}}\mathfrak{P}$, we find that the graded algebras $G_{\mathfrak{o}_{\mathfrak{p}}\mathfrak{P}}(\mathfrak{o}_{\mathfrak{p}})$ and $G_{\mathfrak{o}_{\mathfrak{p}}\mathfrak{P}}(\hat{\mathfrak{o}}_{\mathfrak{p}})$ coincide. We have therefore $e(\mathfrak{o}_{\mathfrak{p}}\mathfrak{P}) = e(\hat{\mathfrak{o}}_{\mathfrak{p}}\mathfrak{P})$. Similarly, applying the above-cited Corollary to the case $A = \mathfrak{o}_{\mathfrak{p}}$, $\mathfrak{m} = \mathfrak{m}_{\mathfrak{p}}$, we find that $e(\mathfrak{m}_{\mathfrak{p}}) = e(\hat{\mathfrak{m}}_{\mathfrak{p}})$. Therefore (23) implies that

$$e(\mathfrak{m}_{\mathfrak{p}}) = e(\mathfrak{o}_{\mathfrak{p}}\mathfrak{P}) \qquad (\mathfrak{P} = \mathfrak{p} \cap k[[z_1, z_2, \ldots, z_r]])$$

showing that the parameters $z_1, z_2, \ldots, z_r$ are k-transversal with respect to the given prime ideal $\mathfrak{p}$. This completes the proof of our theorem.

We shall conclude this section by giving an example which will illustrate our contention (see Section 3) that the property of k-transversality of a set of parameters $z_1, z_2, \ldots, z_r$ of $\mathfrak{o}$ with respect to a given prime ideal $\mathfrak{p}$ (different from $\mathfrak{m}$ and of positive height) is *not* necessarily invariant under change of coefficient field k. We shall actually illustrate this fact by two examples. In the first example, the parameters $z_1, z_2, \ldots, z_r$ will be *transversal* parameters of $\mathfrak{o}$. In the second example, $\mathfrak{o}$ will be a complete *regular* ring, but in that case the parameters $z_1, z_2, \ldots, z_r$ will (necessarily) not be transversal. We say "necessarily," for if $\mathfrak{o}$ is regular, then $e(\mathfrak{m}) = 1$, and hence, if $z_1, z_2, \ldots, z_r$ are transversal parameters of $\mathfrak{o}$ we have, by (6): $[\mathfrak{o} : k[[z_1, z_2, \ldots, z_r]]] = 1$. Since $\mathfrak{o}$ is an integral domain which is integral over $k[[z_1, z_2, \ldots, z_r]]$, and since $k[[z_1, z_2, \ldots, z_r]]$ is integrally closed in its field of fractions, we see that (6) implies in this case that $\mathfrak{o} = k[[z_1, z_2, \ldots, z_r]]$. This shows that $z_1, z_2, \ldots, z_r$ are necessarily regular parameters of $\mathfrak{o}$ and are k-transversal with respect to *any* prime ideal $\mathfrak{p}$ in $\mathfrak{o}$, for *any* coefficient field k.

Example 1. Let k be a pure transcendental extension $k_0(u, v)$ of a field k_0, of transcendence degree 2 over k_0 (whence u and v are algebraically independent transcendentals over k_0). Consider the surface F in affine 3-space, defined, over k, by the following equation:

$$F: X_1^{\rho-1}[(2u - v + X_1)X_2 + (u - v)X_1X_3] + X_3^{\rho} = 0,$$

where ρ is an integer > 1. Let $\mathfrak{o}$ be the (complete) local ring of F at the origin,

F now being regarded as an *algebroid* surface. We denote by x_1, x_2, x_3 the F-residues of X_1, X_2, X_3, whence $\mathfrak{o} = k[[x_1, x_2, x_3]]$. If we set $u' = u - x_1, v' = v - x_1$, then u', v' are also algebraically independent transcendentals over k_0, and if we set $k' = k_0(u', v')$, then also k' is a coefficient field of $\mathfrak{o}$.

We take as prime ideal $\mathfrak{p}$ in $\mathfrak{o}$ the ideal generated by x_2 and x_3 (this ideal defines the line $X_2 = X_3 = 0$, which lies entirely on our surface F). We set

$$z_1 = ux_1 + x_2 + x_1x_3, \qquad z_2 = vx_1 + 2x_2 + x_1x_3. \tag{24}$$

Since the tangent cone of our surface F is the cone $(2u - v)X_1^{p-1}X_2 + X_3^p = 0$ while the tangent line $X_1 = X_2 = 0$ of the curve defined on F by the ideal $\mathfrak{o}z_1 + \mathfrak{o}z_2$ (i.e., the curve defined by the equations $uX_1 + X_2 + X_1X_3 = 0$, $vX_1 + 2X_2 + X_1X_3 = 0$) does not lie on that tangent cone, it follows that z_1, z_2 are *transversal parameters of* $\mathfrak{o}$ (this is a consequence, expressed in geometric language, of Proposition B6). We shall show that *the parameters* z_1, z_2 *are* k-*transversal with respect to* $\mathfrak{p}$ *but are not* k'-*transversal with respect to* $\mathfrak{p}$.

If $\mathfrak{P} = \mathfrak{p} \cap k[[z_1, z_2]]$, then $\mathfrak{P}$ is the principal ideal generated in $R = k[[z_1, z_2]]$ by the element $vz_1 - uz_2$:

$$\mathfrak{P} = R\cdot(vz_1 - uz_2) = R\cdot[(v - 2u)x_2 + (v - u)x_1x_3].$$

Now, the local ring $\mathfrak{o}_\mathfrak{p}$ is the local ring of the origin on the plane curve Γ, defined, over $k((x_1))$, in the (X_2, X_3)-plane, by the equation

$$\Gamma : x_1^{p-1}[(2u - v + x_1)X_2 + (u - v)x_1X_3] + X_3^p = 0. \tag{25}$$

The tangent line of Γ, at the origin, is $(2u - v + x_1)X_2 + (u - v)x_1X_3 = 0$, while the curve defined, over $k((x_1))$, by the principal ideal $(vz_1 - vz_2)$, is the line $(v - 2u)X_2 + (v - u)x_1X_3 = 0$, which is distinct from the tangent line of Γ. Hence $vz_1 - uz_2$ is a transversal parameter of $\mathfrak{o}_\mathfrak{p}$, showing that z_1, z_2 are k-transversal with respect to $\mathfrak{p}$.

The equation (25) of the curve Γ shows that the origin $X_2 = X_3 = 0$ is a simple point of Γ, and hence $\mathfrak{o}_\mathfrak{p}$ is a regular ring, of Krull dimension 1, and each of the elements x_2, x_3 is a regular parameter of $\mathfrak{o}_\mathfrak{p}$, while both fields $k((x_1))$ and $k'((x_1))$ are coefficient fields of $\mathfrak{o}_\mathfrak{p}$. Thus $\mathfrak{o}_\mathfrak{p} = k'((x_1))[[x_2]]$. Now, from (24) and from the definition $u' = u - x_1$, $v' = v - x_1$ of u', v', we find that

$$z_1 = (u' + x_1)x_1 + x_2 + x_1x_3, \qquad z_2 = (v' + x_1)x_1 + 2x_2 + x_1x_3. \tag{26}$$

Let $\mathfrak{P}' = \mathfrak{p} \cap k'[[z_1, z_2]]$. To find a generator of the principal ideal $\mathfrak{P}'$ in $k'[[z_1, z_2]]$ we have to construct an irreducible power series $\xi' = \varphi(z_1, z_2)$, with coefficients in k', such that if ξ' is viewed as a power series in x_1, x_2, x_3, with coefficients in

k', ξ' belong to the ideal $\mathfrak{o}x_2+\mathfrak{o}x_3$, or—equivalently—that $\varphi((u'+x_1)x_1,(v'+x_1)x_1)$ be identically zero. To find φ, we can proceed as follows:

We set $x_1'=(u'+x_1)x_1$. Then x_1' is a regular parameter of the power series ring $k'[[x_1]]$, and $(v'+x_1)x_1$ is an element of this ring. Therefore $(v'+x_1)x_1$ is a power series in x_1', with coefficients in k'. Let

$$(v'+x_1)x_1=\sum_{i=1}^{\infty} c_1' x_1'^i, \qquad c_1'\in k'. \tag{27}$$

Then it is clear we can take for ξ' the power series

$$\xi'=z_2-\sum_{i=1}^{\infty} c_i' z_1^i. \tag{28}$$

From (25) we can find the expression of x_3 as an element of $k'((x_1))[[x_2]]$:

$$x_3=\frac{2u'-v'+2x_1}{(v'-u')x_1}x_2+\text{terms of degree}\geqslant 2 \text{ in } x_2.$$

Thus, by (26), we can write:

$$z_1=x_1'+\frac{u'+2x_1}{v'-u'}\cdot x_2+\text{terms of degree}\geqslant 2\text{ in } x_2$$

$$z_2=(v'+x_1)x_1+\frac{v'+2x_1}{v'-u'}x_2+\text{terms of degree}\geqslant 2\text{ in } x_2.$$

Substituting in (28) and taking into account (27), we can find the expression of ξ' as a power series in x_2, with coefficients in $k'((x_1))$. One finds then that the coefficient of x_2 in that power series is equal to

$$\frac{v'+2x_1}{v'-u'}-\sum_{i=1}^{\infty} ic_i'x_1'^{i-1}\cdot\frac{u'+2x_1}{v'-u'}, \tag{29}$$

where x_1' has been defined above: $x_1'=(u'+x_1)x_1$. By (27), we have that

$$\sum_i ic_i'x_1'^{i-1}=\frac{d((v'+x_1)x_1)}{dx_1'}=\frac{d(v'x_1+x_1^2)}{dx_1}\bigg/\frac{d(u'x_1+x_1^2)}{dx_1}$$

$$=(v'+2x_1)/(u'+2x_1).$$

Substituting this expression of $\sum ic_i'x_1'^{i-1}$ in (29), we find that the coefficient (29) of x_2 in the expression of ξ' is equal to zero. Thus, ξ', as an element of $\mathfrak{o}_\mathfrak{p}$ $(=k'((x_1))[[x_2]])$, is divisible by x_2^2, and therefore ξ' is not a transversal parameter of $\mathfrak{o}_{\mathfrak{p}}$. This shows that z_1, z_2 are not k'-transversal with respect to our prime ideal $\mathfrak{p} = \mathfrak{o}x_2 + \mathfrak{o}x_3$.

Example 2. With the same field $k = k_0(u, v)$ as in example 1, we now take for $\mathfrak{o}$ *the power series ring* $k[[x_1, x_2]]$ *of two independent variables* x_1, x_2 *over* k. Thus, we are now dealing with the (completed) local ring of the origin of the affine (x_1, x_2)-plane, over k. As in Example 1, we set $u' = u - x_1$, $v' = v - x_1$ and take for k' the field $k_0(u', v')$, which is then another coefficient field of $\mathfrak{o}$. We take as prime ideal $\mathfrak{p}$ the principal ideal $\mathfrak{o}x_2$. Then $\mathfrak{o}_{\mathfrak{p}} = k((x_1))[[x_2]] = k'((x_1))$ $[[x_2]]$.

We now set

$$z_1 = ux_1 + (u + x_1 + x_2)x_2 = (u' + x_1)x_1 + (u' + 2x_1 + x_2)x_2,$$

$$z_2 = vx_1 + (v + x_1 + 2x_2)x_2 = (v' + x_1)x_1 + (v' + 2x_1 + 2x_2)x_2. \tag{30}$$

We have $z_1 = (x_1 + x_2)(x_1 + x_2 + u')$, $z_2 = (x_1 + x_2)(x_1 + x_2 + v') + x_2^2$. Hence $\mathfrak{o}z_1 = \mathfrak{o} \cdot (x_1 + x_2)$, and therefore $x_2^2 \in \mathfrak{o}z_1 + \mathfrak{o}z_2$. Thus, both $x_1 + x_2$ and x_2^2 belong to the ideal $\mathfrak{o}z_1 + \mathfrak{o}z_2$. This implies that the elements x_1^2, x_1x_2, x_2^2 belong to $\mathfrak{o}z_1 + \mathfrak{o}z_2$, *and that therefore* z_1, z_2 *are parameters of* $\mathfrak{o}$.

From the expressions (30) of z_1 and z_2 it follows that the principal ideal $\mathfrak{o}x_2 \cap k[[z_1, z_2]]$ is generated by the element

$$\xi = vz_1 - uz_2 = [(v - u)x_1 + (v - 2u)x_2]x_2,$$

and this element is therefore a regular parameter of the ring $\mathfrak{o}_{\mathfrak{p}}$ $(= k((x_1))[[x_2]])$. Thus z_1, z_2 are k-transversal with respect to the prime ideal $\mathfrak{p} = \mathfrak{o}x_2$.

To find the generator ξ' of the principal ideal $\mathfrak{o}x_2 \cap k'[[z_1, z_2]]$ we proceed as in Example 1. We set namely $x_1' = (u' + x_1)x_1$ and we consider the expansion (27) of $(v' + x_1)x_1$ as a power series in x_1', with coefficients in k'. Then taking into account the expressions (30) of z_1, z_2 as elements of $k'[[x_1, x_2]]$, we conclude again that ξ' is given by (28), and we can find the expression of ξ' as a power series in x_2, with coefficients in $k'((x_2))$. One finds then that the coefficient of x_2 in that power series is equal to

$$(v' + 2x_1) - \sum_{i=1}^{\infty} ic_i' x_1'^{i-1}(u' + 2x_1). \tag{31}$$

As in Example 1, so also here, we have that $\sum ic_i' x_1'^{i-1} = (v' + 2x_1)/(u' + 2x_1)$. Hence the coefficient (31) is zero, ξ' is divisible by x_2^2, showing that z_1, z_2 are not k'-transversal with respect to the prime ideal $\mathfrak{o}x_2$.

HARVARD UNIVERSITY

REFERENCES.

[1] N. Bourbaki, *Algèbre Commutative*, Ch. II. Localization.

[2] N. Bourbaki, *Algèbre Commutative*, Ch. III. Graduations, filtrations et topologies.

[3] J. Lipman, "Relative Lipschitz-saturation," *American Journal of Mathematics*, to appear in *American Journal of Mathematics*, Vol. **XCVII** (1975) No. 3.

[4] J. Lipman, "Absolute saturation of one-dimensional local rings," *American Journal of Mathematics*, to appear in *American Journal of Mathematics*, Vol. **XCVII** (1975) No. 3.

[5] M. Nagata, "On the closedness of singular loci," *Publs. Math. Inst. Haut. Etud. Sci.*, **2** (1959), pp. 29–36.

[6] M. Nagata, *Local Rings*, Interscience, New York, London, 1962. This book will be referred to as Nagata, L. R.

[7] D. G. Northcott and D. Rees, "A note on reduction of ideals with an application to the generalized Hilbert function," *Proceedings of the Cambridge Philosophical Society*, **50** (1954), pp.353–359.

[8] D. Rees, "A-transforms of local rings and a theorem on multiplicities of ideals," *Proceedings of the Cambridge Philosophical Society*, **57** (1961), pp. 8–17.

[9] F. Pham and B. Teissier, *Fractions lipschitziennes d'une algèbre analytique complexe et saturation de Zariski*, Centre de Mathématique de l'Ecole Polytechnique, Paris, Juin 1969, no. MI7.0669.

[10] F. Pham, *Fractions lipschitziennes et saturation de Zariski des algèbres analytiques*, Actes du Congrès International des Mathematiciens (Nice, 1970), Tome 2, 649–654, Gauthier-Villars, Paris 1971.

[11] O. Zariski, "Analytical irreducibility of normal varieties," *Annals of Mathematics*, **49** (1948), pp. 352–361.

[12] O. Zariski, "Sur la normalité analytique des variétés normales," *Annals Institut Fourier*, **2** (1950), pp. 161-164.

[13] O. Zariski and P. Samuel, *Commutative Algebra*, volumes 1 and 2 (referred to as C. A. 1 and C. A. 2, respectively).

[14] O. Zariski, "Studies in equisingularity III. Saturation of local rings and equisingularity," *American Journal of Mathematics*, **90** (1968), pp. 961–1023 (referred to as SES. III).

[15] O. Zariski, "General theory of saturation and of saturated local rings I. Saturation of complete local domains of dimension one having arbitrary coefficient fields (of characteristic zero)," *American Journal of Mathematics*, **93** (1971), pp. 573–648 (referred to as GTS I).

[16] O. Zariski, "General theory of saturation and of saturated local rings II. Saturated local rings of dimension 1," *American Journal of Mathematics*, **93** (1971), pp. 872–964 (referred to as GTS II).

On Equimultiple Subvarieties of Algebroid Hypersurfaces

(critical variety of a projection/equimultiplicity)

OSCAR ZARISKI

Department of Mathematics, Harvard University, Science Center, 1 Oxford Street, Cambridge, Massachusetts 02138

Contributed by Oscar Zariski, January 20, 1975

ABSTRACT We deal with a finite projection π_V of a hypersurface V in $(r+1)$-space onto an affine r-space A_r and with the critical variety Δ of π_V in A_r. We show that if Δ is equimultiple along a smooth subvariety W_0, then W_0 admits a unique lifting to a subvariety W of V (necessarily isomorphic to W_0) and that also V is then equimultiple along W.

Let P be the closed point of an algebroid variety V (or, in particular, let P be any closed point of an algebraic, or complex-analytic variety V), of pure dimension r, defined over an algebraically closed ground field k of characteristic zero. We denote by $\mathfrak{o}$ the (formally complete) local ring of V (or, in particular, if V is algebraic, the formal completion of the local ring of V at P), and by $\mathfrak{m}$ the maximal ideal of $\mathfrak{o}$. We assume that the embedding dimension of V (locally at P) is $r+1$, i.e., that $\mathfrak{m}$ has a minimal basis of $r+1$ elements. (We thus exclude the trivial case in which P is a simple point of V.) Let $(x_1,x_2,\ldots x_r,z)$ be a minimal basis of $\mathfrak{m}$, such that $(x_1,x_2,\ldots x_r)$ is a set of parameters of $\mathfrak{o}$. Then V is defined (locally at P) by an equation of the form:

$$V: f(X_1,X_2,\ldots,X_r;Z) = 0,$$

where f is a monic polynomial in Z, of a certain degree $n > 1$, with coefficients in the ring $k[[X_1,X_2,\ldots,X_r]]$ of formal power series in the r variables $X_1,X_2,\ldots X_r$, with coefficients in k. Furthermore, f is free from multiple factors, and $f(0,0,\ldots,0,Z) = Z^n$.

We denote by π the projection of the ambient affine $(r+1)$-space $A_{r+1}(X;Z)$ of V onto the affine r-space $A_r(X)$ of the r variables $X_1,X_2,\ldots,X_r$. Thus, π is defined generically by $\pi(X_1,X_2,\ldots,X_r,Z) = (X_1,X_2,\ldots,X_r)$. The restriction π_V of π to V is finite, of degree n, the full inverse image of the general point $(x_1,x_2,\ldots x_r)$ of $A_r(X)$, under π_V^{-1} being the set of n distinct points $(x_1,x_2,\ldots, x_r,z^{(\alpha)})$, $\alpha = 1,2,\ldots,n$, the $z^{(\alpha)}$'s being the roots of $f(x_1,x_2,\ldots,x_r,Z)$. If P_0 denotes the origin of $A_r(X)$, then $\pi_V^{-1}\{P_0\}$ consists of the point P only (counted to the multiplicity n).

We shall denote by Δ the *reduced* critical variety of π_V in $A_r(X)$, i.e., the reduced variety defined by the equation $D(X_1,X_2,\ldots,X_r) = 0$, where $D(X)$ is the Z-discriminant of f. Thus, Δ is an algebroid variety (locally, at P_0), of pure dimension $r-1$ and of embedding dimension $\leq r$, with equality if and only if Δ has a simple point at the origin P_0 of $A_r(X)$.

If W is an irreducible algebroid subvariety of V/k (passing through P), we say that V is *equimultiple at P, along W*, if V has at P the same multiplicity as it has at the general point of W.

Let W_0 be an irreducible algebroid subvariety of the critical variety Δ (passing through P_0), of codimension $s-1$ on Δ ($s \geq 1$; hence of dimension $r-s$). *We assume that P_0 is a simple point of W_0.* The theorem that we wish to prove is the following:

THEOREM. *If Δ is equimultiple at P_0, along W_0, then the full inverse image of W_0 on V (locally at P), under π_V^{-1}, is an irreducible subvariety W of V, having at P a simple point. Furthermore. the restriction of π_V to W is an isomorphism between W and W_0, and V is equimultiple at P, along W.*

Proof: Without loss of generality we may assume that W_0 is the linear sub-space $X_1 = X_2 = \ldots = X_s = 0$ of $A_r(X)$. We denote the affine spaces $A_{r+1}(X,Z)$ and $A_r(X)$ by S and S_0, respectively. We apply to S_0 the monoidal transformation $T_0: S_0' \to S_0$ with center W_0. In terms of the three regular (rational) maps

$$\pi_V: V \to S_0;\ \pi: S \to S_0;\ T_0: S_0' \to S_0,$$

we can define the two formal (completed) fiber products V' and S', where

$$V' = V \underset{S_0}{\times} S_0' \subset S \underset{S_0}{\times} S_0' = S'.$$

We denote by $T, T_{V'}$, π' and $\pi'_{V'}$ the associated regular maps $T: S' \to S$, $T/V' = T_{V'}: V' \to V$, $\pi': S' \to S_0'$, $\pi'_{V'} = \pi'/V': V' \to S_0'$. It is obvious that π' is a *finite* map, generically of degree n. In the above, S_0' is an (open) nonsingular, irreducible algebraic variety, of dimension r. It contains the $(r-1)$-dimensional, nonsingular, irreducible variety $\mathcal{E}_0' = T_0^{-1}\{W_0\}$, which is the exceptional divisor into which W_0 is blown up by T_0. This exceptional divisor $\mathcal{E}_0'$ is fibered by $(s-1)$-dimensional projective spaces $\mathbf{P}_{s-1}$, the base space of this fiber bundle being W_0. We shall be particularly interested in the fiber $\Gamma_0' = T_0^{-1}\{P_0\}$ and with the full inverse image $\Gamma' = P \times \Gamma_0'$ of Γ_0' on V', under $\pi'_{V'}$. The variety Γ' is isomorphic to Γ_0', i.e., Γ' is a projective $(s-1)$-space, the closed points of V' are the closed points of Γ', and V' is a formal (complete) reduced scheme, defined along Γ'.

Any closed point P' of Γ' is of the form $P' = P \times P_0'$, where P_0' is a closed point of Γ_0'. Given such a point P', we may assume, without loss of generality, that the following r quantities $x_1', x_2', \ldots, x_r'$ are regular parameters of the local ring of the (simple) point P_0' of S_0':

$$x_1' = x_1, x_2' = x_2/x_1, \ldots, x_s' = \frac{x_s}{x_1}; \; x_i' = x_i, \quad i = s+1, s+2, \ldots, r.$$

This being so, the algebroid formal scheme V' is defined, locally at $P'(= P \times P_0')$, by the equation

$$V': f'(X_1', X_2', \ldots, X_r'; Z) = 0,$$

where

$$f'(X_1', X_2', \ldots, X_r'; Z) = f(X_1', X_1'X_2', \ldots, X_1'X_s', X'_{s+1}, \ldots, X_r'; Z).$$

The exceptional divisor $\mathcal{E}_0'$ and the fiber Γ_0' will be defined, locally at P_0', by the equations:

$$\mathcal{E}_0': X_1' = 0; \; \Gamma_0': X_1' = X'_{s+1} = \ldots = X_r' = 0,$$

while Γ $(= P \times \Gamma_0')$ is defined (locally at P') by the equations:

$$\Gamma: X_1' = X'_{s+1} = \ldots = X_r' = Z = 0.$$

If Δ' denotes the critical variety (in S_0') of the (finite) map $\pi'_V{}': V' \to S_0'$, then $\Delta' = \mathcal{E}_0' \cup \Delta_1'$, where Δ_1' is the proper transform $T_0^{-1}[\Delta]$ of Δ.

Since Δ is equimultiple at P_0 along W_0, it is well known that $\Gamma_0' \not\subset \Delta_1'$. So we may assume that the above closed point P_0' or Γ_0' does not belong to Δ_1'. *Therefore, P_0' is a simple point of Δ'*, since locally, at P_0', Δ' is identical with $\mathcal{E}_0'$. Therefore, by our theory of equisingularity in codimension 1 (see ref. 1, Theorem 4.4), the full inverse $\mathcal{E}'$ of $\mathcal{E}_0'$ on V', under $\pi'_{V'}$, has a simple point at P', and $\pi'_{V'}$, induces an isomorphism between $\mathcal{E}'$ and $\mathcal{E}_0$. Now, the equations of $\mathcal{E}'$ are clearly the following:

$$\mathcal{E}': X_1' = 0, \; f'(0,0,\ldots,0,X'_{s+1},X'_{s+2},\ldots,X_r';Z) = 0,$$

where we note that

$$f'(0,0,\ldots,0,X_{s+1},X_{s+2},\ldots X_r;Z) = f(0,0,\ldots,0;X_{s+1},X_{s+2},\ldots,X_r;Z).$$

Therefore, the monic polynomial $f'(0,0,\ldots,0,X'_{s+1}, X'_{s+2},\ldots,X_r';Z)$ must be the n-th power of $Z - \psi(X'_{s+1},X'_{s+2},\ldots,X_r')$, where ψ is a formal power series, with coefficients in k, zero at the origin. Since W, by its very definition, is defined by the equations

$$X_1 = X_2 = \ldots X_s = f(0,0,\ldots,0, X_{s+1},X_{s+2},\ldots,X_r;Z) = 0,$$

it follows that W is also defined by the equations

$$W: X_1 = X_2 = \ldots = X_s = Z - \psi(X_{s+1},X_{s+2},\ldots,X_r) = 0.$$

This shows that W is irreducible, has a simple point at the origin P, and the projection π_V induces an isomorphism between W and W_0. Upon replacing z by $z - \psi(x_{s+1},x_{s+2},\ldots,x_r)$ one may assume that W is defined by the equations $X_1 = X_2 = \ldots = X_s = Z = 0$.

By the theory of equisingularity in codimension 1, as applied to V' and the critical variety Δ' in S_0' (we recall that $\Delta' = \mathcal{E}_0'$, locally at P_0'), we have

$$e_{P'}(V') = e_{\mathcal{E}'}(V'), \qquad [1]$$

$e_{P'}(V')$ and $e_{\mathcal{E}}'(V')$ denote the multiplicities, on V', of the point P' and of the general point of $\mathcal{E}'$, respectively. Now, if $M = X_1^{\alpha_1}X_2^{\alpha_2}\ldots X_r^{\alpha_r}Z^{\beta}$ is any monomial that occurs in $f(X_1,X_2,\ldots,X_r;Z)$, then the corresponding monomial M' in f' will be given by $M' = X_1'^{\alpha_1+\alpha_2+\ldots+\alpha_s} X_2'^{\alpha_2}\ldots Z^{\beta}$, and hence degree of $M' \geq$ degree of M. This shows that

$$e_{P'}(V') \geq e_P(V). \qquad [2]$$

On the other hand, the degree of M' in X_1' and Z is $\alpha_1 + \alpha_2 + \ldots + \alpha_s + \beta$. Since $\mathcal{E}'$ is defined by the equations $X_1' = Z = 0$, we conclude by [1] and [2] that we must have, for all monomials M that actually occur in f: $\alpha_1 + \alpha_2 + \ldots + \alpha_s + \beta \geqq e_P(V)$. This shows that $e_W(V) \geq e_P(V)$ (since W is defined by $X_1 = X_2 = \ldots = X_s = Z = 0$). Therefore, $e_W(V) = e_P(V)$, which completes the proof of the theorem.

We point out the following obvious corollary of our theorem:

COROLLARY. *If W is an irreducible subvariety of V, having at P a simple point, and if the projective $\pi_V: V \to S_0$ is transversal to W at P [whence $W_0 = \pi_V(W)$ is isomorphic to W], then the equimultiplicity of Δ along W_0, at P_0, implies that W is the full inverse image of W_0, under π_V, and that V is equimultiple at P, along W.*

Note: Teissier and Lê Dũng Tráng have generalized our theorem to the case in which π is a (linear) projection of our hypersurface $X \subset \mathbb{C}^{N+1}$ onto a $\mathbb{C}^k$, $k < N$. For a precise formulation of this generalization see Teissier's lectures "Introduction to Equisingularity Problems" in the Proceedings of the Summer Institute in Algebraic Geometry held in Arcata (California) in 1974.

This research was supported in part by NSF Contract no. GP-30854X.

1. Zariski, O. (1965) "Studies in equisingularity II," *Amer. J. Math.* 87, no. 4, pp. 972–1006.

THE ELUSIVE CONCEPT OF EQUISINGULARITY AND RELATED QUESTIONS.

By Oscar Zariski.

Introduction. In this lecture I will present a *tentative* theory of equisingularity for *algebraic hypersurfaces* V. I say *tentative*, because our theory, while it leads to an *equisingular stratification* of V having many of the desirable properties which one expects equisingularity to possess, it also has some shortcomings. For instance, it is not an *a priori biregularily invariant* theory, nor is it based on purely *local* considerations, whereas equisingularity is in principle a local phenomenon. On the other hand, while our definition of equisingularity implies the Whitney conditions A and B, our equisingular stratification of V is definitely finer than the Whitney stratification. This represents a definite gain, for we shall show by an example (given by Briancon and Speder) that the Whitney conditions A and B are definitely not acceptable as a criterion of equisingularity.[1]

1. I have studied the problem of equisingularity in a number of papers during the past decade, in particular in 3 papers published in the American Journal of Mathematics in 1965 and 1968, under the common title "Studies of Equisingularity" (see [2], [3] and [4]; these papers will be referred to as SES I, II, III respectively). Other papers of mine on this subject are [1] and [6], published respectively in 1964 and 1971, and the exposition [5] presented in September, 1969, at the summer meeting of the "Centro Internazionale Matematico Estivo" in Varenna, Italy. The question, informally speaking, is the following.

We are given a (reduced) r-dimensional hypersurface V immersed in an $(r+1)$-dimensional affine or projective space, defined over an algebraically closed field k of characteristic zero. Let Ω be a universal domain

[1] *Note.* I wish to point out that two statements on the *stable* behavior of our definition of equisingularity under monoidal (blowing up) transformations, which were formulated in the lecture as *theorems*, are formulated in this paper only as *conjectures*, because I have found belatedly that my proofs of those two statements are incomplete. I will show also that Whitney's conditions are definitely not "stable" under blowing up transformations. On the other hand I am now in possession of a slightly modified definition and theory of equisingularity (for hypersurfaces) which is biregularily invariant (and which I intend to present in a separate paper).

containing k (Ω = an algebraically closed extension field of k, of infinite transcendence degree over k). We shall denote by $V(k)$ the set of k-rational points of V, while V will stand for the set of all geometric points of V, i.e., all points of V with coordinates in Ω. Let W be an *irreducible* subvariety of V/k, contained in the singular locus V_{sing} of V (since k is of characteristic zero, the singular locus of V is independent of the choice of the field of definition of V). Let Q be a k-rational point of W. One wishes to give a precise meaning to an intuitive statement such as the following: the singularity which V has at Q is not any worse than—or is "equivalent to"—the singularity which V has at the general point of W/k. If k is the field $\mathbf{C}$ of complex numbers then the reference to the general point of $W/\mathbf{C}$ can be replaced by a reference to the set of points of W which are sufficiently near Q. We express this question formally in terms of equisingularity as follows:[2] *what does it mean when we say that V is equisingular at Q along W?*

We should add that it is desirable to have a definition of equisingularity which is independent of the choice of the (common) algebraically closed field of definition k of V and W. Whatever definition is agreed upon, one could use the symbol $\mathrm{Eqs}(W/V)$ to denote the set of all (geometric) points Q of W such that V is equisingular at Q along W, while the symbol $\mathrm{Eqs}_k(W/V)$ may denote the set of k-rational points of $\mathrm{Eqs}(W/V)$, where k is a common (algebraically closed) field of definition of V and W (we recall that we have assumed W to be irreducible over some algebraically closed field of definition of W, and hence W will be irreducible over any algebraically closed field of definition of W).

Despite the global setting of the above formulation of our question, equisingularity is *in principle* a local property, and should be really formulated in terms of the completion $\hat{\mathfrak{o}}$ of the local ring $\mathfrak{o} = \mathfrak{o}_Q(V)$ of V at Q and of the prime ideal $\hat{\mathfrak{p}}$ of W in $\hat{\mathfrak{o}}$ (assuming, as we may, that W is *analytically irreducible at* Q; in fact, we shall be primarily interested in the case in which Q is a *simple* point of W). More generally, instead of starting with a (global) hypersurface (in affine or projective space), we could start with an arbitrary complete local ring $\mathfrak{o}$, of dimension r, satisfying the following conditions: $\mathfrak{o}$ is equicharacteristic, of characteristic zero; $\mathfrak{o}$ has no nilpotent elements (different from zero) and $\mathfrak{o}$ is equidimensional (i.e., all the prime ideals of the zero ideal of $\mathfrak{o}$ have the same

[2]We have assumed that W is contained in the singular locus of V. If the contrary case, the general point of W is simple for V, and in that case equisingularity of V at Q along W can only mean that also Q is a simple point of V.

dimension r); and finally we assume that the embedding dimension of $\mathfrak{o}$ is $r+1$, i.e., the maximal ideal $\mathfrak{m}$ of $\mathfrak{o}$ has a minimal basis of $r+1$ elements (we exclude therefore the trivial case of a regular ring $\mathfrak{o}$). In this local formulation the role of V is played by the formal algebroid variety $V=\text{Spec}(\mathfrak{o})$, the role of Q is played by the maximal ideal $\mathfrak{m}$ (considered as a "point" of Spec($\mathfrak{o}$)), while W would the variety $\mathcal{V}(\mathfrak{p})(=\text{Spec}(\mathfrak{o}/\mathfrak{p})$ of some prime ideal $\mathfrak{p}$ in $\mathfrak{o}$. We have a partial treatment of this more general, but purely local, case, but it is not yet complete and we shall not present it in this paper. At present we wish only to stress two questions which are left unsolved in our treatment:

(1) Our definition of equisingularity in the local set up *a priori* depends on the choice of a coefficient field of $\mathfrak{o}$ (we assume that the residue field $\mathfrak{o}/\mathfrak{m}$ is algebraically closed). We do not know whether a change of the coefficient field of $\mathfrak{o}$ affects our definition of equisingularity.

(2) In the global case of an algebraic hypersurface it is an open question whether our definition of equisingularity in that case is equivalent to the one we propose in the abstract local case, for in the global case the treatment is based on a *particular* choice of parameters in the local ring of V at Q.

2. Since we have assumed that $W\subset V_{\text{sing}}$ it follows that $\dim W\leqslant r-1$. In our paper SES II we gave a complete treatment of the case $\dim W=r-1$. In that case there are no open questions left outstanding, and our treatment, even in the global case, is purely local. The basic reason why the case $\dim W=r-1$ lends itself to a complete treatment is the fact that the concept of *equivalent singularities* of plane algebroid curves is well established and is classical (while no such definition exists for singularities of hypersurfaces of dimension $\geqslant 2$). That concept is intrinsic (i.e., it is independent of the choice of the coefficient field of the complete 1-dimensional local ring, of embedding dimension 2), for it is based solely on the resolution of the singularity of the curve by successive blowing-ups of the origin of the plane curve and of its successive transforms. This being so, given a hypersurface V (globally or locally), and given the simple point Q of an $(r-1)$–dimensional irreducible subvariety W of V, we can consider, in the ambient space $\mathbf{A}_{r+1}$ of V, algebroid *surfaces* Φ_2 (defined over k) having at Q a simple point and such that the tangent spaces L_{r-1} and L_2 of W and Φ_2 at Q are independent (and hence meet only at Q). Then the intersection $\Gamma=\Phi_2\cap V$ is an algebroid curve, of embedding dimension 2 (since $\Gamma\subset\Phi_2$ and Q is a simple point of Φ_2).

Any such Γ is called a *W-transversal section of* V, at Q. If P is a general (geometric) point of W, and if we consider the field $k(P)$, then the equation of V, *considered over* $k(P)$, yields a plane algebroid curve Γ_P, defined over $k(P)$, and this is the only W-transversal section of V at P [(see SES II, p. 980)]. Our definition of equisingularity of V at Q along W was then the following: *V is equisingular at Q along W, if there exists a W-transversal section Γ of V at Q such that the curves Γ and Γ_P have equivalent singularities* (*at Q and P respectively*).

If, however, $\dim W = r - s$, $s > 1$, W-transversal sections of V at Q would be embedded (locally at Q) hypersurfaces Γ of dimension s, [see SES II] and similarily for Γ_P. Since the general concept of equivalent singularities of hypersurfaces of dimension $s > 1$ is not available, the above definition of equisingularity cannot be extended to the case in which $\dim W < r-1$. However, in the case $\dim W = r-1$, we have proved in our paper SES II two criteria of equisingularity, and each of these two criteria can be *formulated* in the general case of a W of any dimension and thus can be used as a basis of a possible generalized definition of equisingularity.

The first of these two criteria is differential-geometric in nature and is as follows:

Criterion 1. *If $\dim W = r-1$, then V is equisingular at Q along W if and only if the triplet $(V-W, W, Q)$ satisfies the Whitney conditions A and B* (see SES II, Theorem 8.1, p. 1004).

The second criterion deals with locally finite (at Q) projections of V onto an affine r-space. Let $f(X_1, X_2, \ldots, X_r, X_{r+1}) = 0$ be a local equation of V at Q (we assume that Q is the origin), and let $x_1, x_2, \ldots, x_{r+1}$ be the f-residues of the X_i. We assume that $x_1, x_2, \ldots, x_r$ are parameters of the local ring $\mathfrak{o}$ of V at Q (geometrically speaking this means that the line $X_1 = X_2 = \cdots = X_r = 0$ does not belong to V). By the Weierstrass preparation theorem we may assume that f is a monic polynomial in X_{r+1} (of degree $n \geqslant$ multiplicity of V at $Q \geqslant 2$), with coefficients which belong to the power series ring $k[[X_1, X_2, \ldots, X_r]]$ and which vanish at the origin. The mapping $(X_1, X_2, \ldots, X_r, X_{r+1}) \to (X_1, X_2, \ldots, X_r)$ induces a finite projection $\pi (= \pi_{(x)})$ of V (locally at Q) onto the affine space $\mathbf{A}_r$ of the indeterminates $X_1, X_2, \ldots, X_r$ (considered locally at the origin $Q_0 : X_1 = X_2 = \cdots = X_r = 0$). The critical locus Δ_π of π is a hypersurface $D(X_1, X_2, \ldots, X_r) = 0$, in $\mathbf{A}_r / k$, where D is the reduced discriminant of f with respect to X_{r+1}, and this hypersurface Δ_π contains the origin $Q_0 = \pi(Q)$. We shall now also assume that the direction $X_1 = X_2 = \cdots = X_r = 0$

of the projection π is not tangent to W at Q. We shall express this assumption by saying that π is a W-*permissible* projection. Under this assumption the projection $W_0 = \pi(W)$ of W will be isomorphic to W, and thus it will be of the same dimension as W and will have at Q_0 a simple point. Moreover, W_0 will be contained in Δ_π, since W was assumed to be contained in V_{sing}. But since $\dim W_0 = \dim W = r-1 = \dim \Delta_\pi$, it follows that W_0 is an irreducible component of Δ_π.

CRITERION 2. *If* $\dim W = r-1$, *then* V *is equisingular at* Q *along* W *if and only if there exists* (*locally at* Q) *a* W-*permissible projection* π *of* V *onto* $\mathbf{A}_r$ *such that* Δ_π *has a simple point at* $Q_0(=\pi(Q))$; *equivalently, if and only if*, Δ_π *coincides with* $W_0(=\pi(W))$, *locally at* Q_0.

Note that since in our case ($\dim W = r-1$) W_0 is an irreducible component of Δ_π and has a simple point at Q, to say that Δ_π has a simple point at Q is the same as saying that Δ_π *is equisingular at* Q *along* W_0. We say that the projection π_X is V-*transversal* at Q if the direction $X_1 = X_2 = \cdots = X_r = 0$ is not tangent to V at Q. It is clear that any projection π_X which is V-transversal at Q is necessarily also W-permissible. We say that a W-permissible projection π *is equisingular at* Q if the associated critical locus Δ_π is equisingular at $Q_0(=\pi(Q))$ along W_0 $(=\pi(W))$. Thus, our Criterion 2 states that *if* $\dim W = r-1$ *then* V *is equisingular at* Q *along* W *if and only if there exists a* W-*permissible projection* π *of* V *at* Q *which is equisingular at* Q. In our cited paper SES II we have proved that *if* V *is equisingular at* Q *along* W *then every projection* π *which is* V-*transversal at* Q *is equisingular* (always under the assumption that $\dim W = r-1$).

The differential-geometric Criterion 1 might easily tempt a differential geometer to give the following general definition of equisingularity: *If* W *is any irreducible subvariety of* V, *having at* Q *a simple point, then* V *is equisingular at* Q *along* W *if the Whitney conditions A and B are satisfied by the triplet* $(V-W, W, Q)$.

The second criterion suggests the following general definition of equisingularity, which is *inductive* with respect to the integer $s = \mathrm{cod}_V(W) = r - \dim W$ (since $\mathrm{cod}_{\Delta_\pi}(W_0) = r-1-\dim W_0 = r-1-\dim W = s-1$): V *is equisingular at* Q *along* W *if there exists a finite* W-*permissible* (*local*) *projection* π *of* V *at* Q *such that the critical locus* Δ_π *of* π *is equisingular at* $Q_0(=\pi(Q))$ *along* $W_0(=\pi(W))$. This was the definition that we gave in 1971 in our paper [6].

Now, one test that any definition of equisingularity must meet is the test of its *stable behavior along* W *under blowing-up of* W. The following example (Briançon-Speder) will elucidate the meaning of "stability along W" and will show that neither one of the above two definitions meets the test of stability and therefore neither one is a "good" definition of equisingularity.

Consider the following hypersurface V in affine 4-space:

$$V: z^6 + y^3 + tx^4y + x^6 = 0.$$

The singular locus of V is the line $x = y = z = 0$. It can be easily shown (see Briançon-Speder's preprint [8]) that $(V - W, W, Q)$ satisfies Whitney's conditions, *for any point* Q *of* W. Thus the Whitney stratification of V consists of the two strata $V - W$ and W. Now let V^* be the variety obtained from V by blowing up W and let $T^*: V^* \to V$ be the associated birational regular map (or morphism). The variety V^* is covered by two affine hypersurfaces V' and V'', where the induced morphism $T' = T^*|V'$ and $T'' = T^*|V''$ are defined by the equations:

$$T': x = x', y = x'y', z = x'z'$$

and

$$T'': x = x''z'', y = y''z'', z = z''$$

respectively. The equations of V' and V'' are respectively

$$V': x'^3z'^6 + y'^3 + tx'^2y' + x'^3 = 0;$$
$$V'': z''^3 + y''^3 + tx''^4y''z''^2 + x''^6z''^3 = 0.$$

The affine hypersurface V' contains the line $l': x' = y' = z' = 0$, the affine hypersurface V'' contains the line $l'': x'' = y'' = z'' = 0$, and we have

$$V' \cap V'' = V' - l' = V'' - l'';$$
$$V^* = V' \cup l'' = V'' \cup l'.$$

For our purpose it will be sufficient to examine one of the hypersurfaces V', V'', say V'. The transform $T'^{-1}\{W\}$ of W on V' is the plane $\mathfrak{S}_2': x' = y' = 0$, which is also the singular locus of V' (a three-fold plane). We are now in the case in which the singular locus is an irreducible non-singular variety, of codimension 1 on V'. The projection $\pi': (x', y', z', t) \to (x', z', t)$ is not only $\mathfrak{S}_2'$-permissible, but is even V'-transversal at every point of $\mathfrak{S}_2'$. We find that the critical locus of π' is the surface

$$\Delta_{\pi'}: x'\left[27(1 + z'^6)^2 + 4t^3\right] = 0.$$

The projection $\pi'(\mathfrak{S}_2')$ of $\mathfrak{S}_2'$ is the plane $x'=0$, which is one of the two irreducible components of $\Delta_{\pi'}$, the second component being the surface $27(1+z'^6)^2+4t^3=0$. Since π' is V'-transversal at every point of $\mathfrak{S}_2'$, it follows from the above cited theorem that the set $\mathrm{Eqs}(\mathfrak{S}_2'|V')$ consists of all points of the plane $\mathfrak{S}_2'$ which do not lie on the curve

$$\mathcal{C}' : x' = y' = 27(1+z'^6)^2 + 4t^3 = 0.$$

Therefore, by Criterion 1, the sets $V' - \mathfrak{S}_2'$ and $\mathfrak{S}_2' - \mathcal{C}'$ constitute two of the strata of the Whitney stratification of V'. Now, the curve $\mathcal{C}'$ has 6 cusps Q_i at the points where $t=0$, $z'^6 = -1$. So each of these 6 cusps must necessarily constitute by itself one stratum or the Whitney stratification of V'. All the remaining points of $\mathcal{C}'$ are simple for $\mathcal{C}'$. For the purposes of the conclusion we wish to arrive at, it is not necessary for us to check whether or not $\mathcal{C}' - \cup_{i=1}^6 Q_i'$ constitutes a single stratum in the Whitney stratification of V'. The important fact to be stressed is that while $\mathcal{C}'$ is a 12-fold covering of W, there are Whitney strata on V', namely the 6 strata Q_i' (and also the strata the union of which is $\mathcal{C}' - \cup_{i=1}^6 Q_i'$), which are not mapped *onto* W by T'; and in fact, there are strata which are mapped by T' onto the single point Q of W, where Q is the origin $x=y=z=t=0$. This represents what we call a lack of stability of the Whitney conditions *along* W under the blowing-up transformation T, namely the stability property fails at least at the point Q of W. We could hardly accept in this case the definition of equisingularity based on the Whitney condition. In fact, we would be inclined to say, without further *ado*, that V is definitely *not* equisingular at Q along W, in view of the special role that Q plays in the blowing-up transformation T. Namely, while the fibres F_R' in $\mathcal{C}'$ which correspond to points R of W, $R \neq Q$, consist of 12 distinct points, the fibre F_Q' consists of 6 points (the 6 cusps Q_i'; the 12-fold covering $\mathcal{C}'$ of W is ramified at Q).

This example shows that also the second tentative definition of equisingularity (based on Criterion 2) is not acceptable, for in our example the projection $\pi : (x,y,z,t) \rightarrow (x,y,t)$ is W-permissible and is equisingular at Q, the associated critical locus Δ_π being the surface $y^3 + tx^4y + x^6 = 0$, and it is easily seen that this surface is equisingular at the origin $x=y=t=0 (=\pi(Q))$ along the line $W_0 = \pi(W) : x = y = 0$ (this assertion follows from the fact that the critical locus of the W_0-permissible projection $(x,y,t) \rightarrow (x,t)$ of Δ_π is the curve $x(4t^3+27)=0$, and this curve has a simple point at the origin $x=t=0$).

Note that the projection π, while W-permissible at Q, is not V-transversal at Q. One would be tempted to give a definition of equisingularity which consists in requiring that there exist a *V-transversal equisingular* projection at Q, along W. However, we doubt that the existence of such a projection necessarily implies the Whitney condition at $(V-W, W, Q)$, and any "good" definition of equisingularity should, we believe, *imply* the Whitney conditions. In his paper [9], Speder proves that if the *generic* (linear) projection π of V is equisingular at Q along W (i.e., if V is equisingular at Q, along W, in the sense of the definition that we gave in 1963 in [1]), then the Whitney conditions were satisfied. We shall now outline a treatment of the problem of equisingularity which is based on the old definition we gave in 1963.

3. Let V be defined by an equation (in affine $\mathbf{A}_{r+1}$)

$$V: f(X_1, X_2, \dots, X_r, X_{r+1}) = 0,$$

where f is a polynomial, whose coefficients in the universal domain Ω. This polynomial f is uniquely determined, up to an arbitrary non-zero factor in Ω, and the subfields k of Ω over which V is defined are those which contain the subfield k_0 of Ω which is generated by the ratios of the coefficients of f. We may therefore assume that the coefficients of f belong to k_0, which is the least subfield of Ω over which V is defined.

Let $x_1, x_2, \dots, x_{r+1}$ be the f-residues of the X_i. We set

$$X_i^* = \sum_{j=1}^{r+1} u_{ij} X_j, \qquad x_i^* = \sum_{j=1}^{r+1} u_{ij} x_j; \qquad i = 1, 2, \dots, r, \tag{1}$$

where the u_{ij} are indeterminates over Ω. The elements $X_1^*, \dots, X_r^*, X_{r+1}$ form a new system of coordinates, and in these coordinates the equation of V will take the form

$$f^*(X_1^*, \dots, X_r^*, X_{r+1}) = 0,$$

where the coefficients of f^* belong to the field $k_0^* = k_0(\{u_{ij}\})$. We take as new universal domain the algebraic closure Ω^* of $\Omega(\{u_{ij}\})$. We consider the projection $\pi^*: (X_1, X_2, \dots, X_{r+1}) \to (X_1^*, \dots, X_r^*)$ of V onto the affine r-space. The critical locus Δ^* of this projection will be defined by an equation of the form

$$D^*(X_1^*, \dots, X_r^* : \{u_{ij}\}) = 0,$$

where D^*—the reduced discriminant of f^* with respect to X_{r+1}—is a polynomial in the X_i^*, with coefficients in k_0^*. Let $W_0^*=\pi^*(W)$, $Q_0^*=\pi^*(Q)=$ the origin $X_1^*=\cdots=X_r^*=0$. The hypersurface Δ^* is defined over k_0^*, and when dealing with Δ^* we take Ω^* as our universal domain. We now give the following definition of equisingularity, which is similar to the tentative definition we gave in **2** (using Criterion II) but in which the W-transversal projection π is replaced by the above *generic linear* projection π^*:

Definition 1. *V is equisingular at Q along W if Δ^* is equisingular at Q_0^* along W_0^*.*

As was observed earlier, this definition uses induction with respect to the integer $s=\text{cod}_V(W)$. We also note that our definition is independent of the choice of the field of definition k of $V(k\subset\Omega)$.

Our first theorem is the following:

THEOREM 1. *If k is any field of definition of W and V, then the set* $W-\text{Eqs}(W/V)$ *is a proper subvariety of W, defined over k.*

The proof is by induction with respect to s. By the induction hypothesis, the set $W_0^*-\text{Eqs}(W^*/\Delta^*)$ is a proper subvariety of W^*/k^*, where $k^*=k(\{u_{ij}\})$. This subvariety is therefore defined by a set of equations

$$\varphi_\alpha(X_1^*,\ldots,X_r^*;\{u_{ij}\})=0,\qquad \alpha=1,2,\ldots,N, \tag{2}$$

where the coefficients of the φ_α may be assumed to belong to the polynomial ring $k[\{u_{ij}\}]$. A geometric point $\{a_1,a_2,\ldots,a_{r+1}\}$ of $W(a_i\in\Omega)$ belongs to $W-\text{Eqs}(W/V)$ if and only if the π^*-projection or that point belongs to the variety defined by (2). Hence the point (a) must satisfy the equations.

$$\varphi_\alpha\left(\sum_{j=1}^{r+1}u_{1j}a_j,\sum_{j=1}^{r+1}u_{2j}a_j,\ldots,\sum_{j=1}^{r+1}u_{rj}a_j;\{u_{ij}\}\right)=0. \tag{3}$$

Since the u_{ij} are algebraically independent over Ω, hence also over $k(a_1,a_2,\ldots,a_{r+1})$, the left-hand-side of (3) must consist of polynomials in the u_{ij}'s which are identically zero. This leads to a set of algebraic relations between $a_1,a_2,\ldots,a_{r+1}$, with coefficients in k. It remains to show that these relations define a *proper* subvariety of W. To see these, we have only to observe that the π^*-projection of a general point $(\xi_1,\xi_2,\ldots,\xi_{r+1})$ of $W/k(\xi_i\in\Omega)$ is a general point $(\xi_1^*,\xi_2^*,\ldots,\xi_r^*)$ of W_0^*/k^*, and since (ξ^*), by our induction hypothesis, does not belong to $W^* - \text{Eqs}(W_0^*/\Delta^*)$, it follows that the point (ξ) does not belong to $W - \text{Eqs}(W/V)$.

The next theorem is basic and establishes the existence (and unicity) of an *equisingular stratification of* V.

THEOREM 2. *There exists a stratification of* V *onto strata* S_i *having the usual properties of a stratification* and having furthermore the following properties:*

(1) *Each stratum* S_i *is a (non-empty) Zariski open subset of an absolutely irreducible subvariety* $\bar{S}_i$ *of* V/k *(where* k *is any algebraically closed field of definition of* $k \subset \Omega$*).*
(2) *Each point of* S_i *is a simple point of* $\bar{S}_i$.
(3) *The boundary* $\bar{S}_i - S_i$ *of* S_i*, if not empty, is a union of strata.*
(4) *If* W *is any absolutely irreducible subvariety of* V *then the set* Eqs(W/V) *is the intersection of* W *with some stratum* S_i *(which is uniquely determined, since the set* Eqs(W/V) *is non-empty, by Theorem* 1).

The following are straightforward consequences of the theorem:

(a) *For any stratum we have* $S_i = \text{Eqs}(\bar{S}_i/V)$. This follows by applying property (4) to the case $W = \bar{S}_i$.
(b) *The strata* S_i *are uniquely determined.* Obvious and follows from (4).
(c) *The singular locus* V_{sing} *of* V *is a union of strata, while* $V - V_{\text{sing}}$ *is a union of* q *strata, where* q *is the number of absolutely irreducible components of* V.

For if $V_1, V_2, \ldots, V_q$ are the absolutely irreducible components of V, then we have by (4): $\text{Eqs}(V_\alpha/V) = V_\alpha \cap S_{i(\alpha)}$, where $i(\alpha)$ depends on α. Since $\text{Eqs}(V_\alpha/V)$ is a non-empty Zariski open subset of V_α it follows that the absolutely irreducible subvariety $\bar{S}_{i(\alpha)}$ of V contains V_α, and therefore coincides with V_α, since V_α is an absolutely irreducible component of V. We have therefore $\text{Eqs}(V_\alpha/V) = S_{i(\alpha)}$, and since $\text{Eqs}(V_\alpha/V)$ consists of the points of V_α which are simple for V, it follows that $V - V_{\text{sing}} = \cup_{\alpha=i}^{q} S_{i(\alpha)}$, and this proves (c).

*The properties alluded to are the following: any two strata are disjoint, V_r is the union of all the strata, and the "boundary" of any stratum is a union of strata.

We shall not give here the proof of Theorem 2, but we shall indicate the crucial point of the proof. The proof is by induction on r, since the theorem is trivial if $r=0,1$. By the induction hypothesis, we have an equisingular stratification $\{\Sigma_j^*\}$ of Δ^*, satisfying the conditions of Theorem 2. We recall that the universal domain for Δ^* is given by $\Omega^*=$ algebraic closure of $\Omega(\{u_{ij}\})$. We temporarily extend the universal domain Ω for V to Ω^*, and we denote by V^* the set of points, *rational over* Ω^*, which satisfy the equation $f(X_1,X_2,\ldots,X_{r+1})=0$ of V. If $V_1,V_2,\ldots,V_q$ are the absolutely irreducible components of V, then we introduce the q disjoint strata $S_\alpha=V_\alpha-(V_\alpha\cap V_{\text{sing}})$ whose union is $V-V_{\text{sing}}$. It is clear that these q strata satisfy condition (2) of the theorem and also condition (4) whenever $W\not\subset V_{\text{sing}}$. So we must now stratify only V_{sing}. For each stratum Σ_j^* of Δ^* we have the least, absolutely irreducible subvariety $\bar{\Sigma}_j^*$ of Δ^*, containing Σ_j^*. The full inverse images $\pi^{*-1}(\bar{\Sigma}_j^*)$ and $\pi^{*-1}(\Sigma_j^*)$ on V^* are such that the first is a variety defined over Ω^*, while the second is a non-empty Zariski open subset of the first. By induction on r we may assume that on Δ^* equisingularity implies equimultiplicity. Thus Δ^* is equimultiple along $\bar{\Sigma}_j^*$ at each point of Σ_j^*. By our PNAS Note [7] it follows that V^* is equimultiple along $\pi^{*-1}(\bar{\Sigma}_j^*)$ at each point of $\pi^{*-1}(\Sigma_j^*)$, and that at any point Q^* of $\pi^{*-1}(\Sigma_j^*)$ π^* induces an isomorphism (over Ω^*) between $\pi^{*-1}(\bar{\Sigma}_j^*)$ and $\bar{\Sigma}_j^*$, locally at Q^* and $\pi^*(Q^*)$. From this it follows that each point of $\pi^{*-1}(\Sigma_j^*)$ is simple for $\pi^{*-1}(\bar{\Sigma}_j^*)$. It follows also that $\pi^{*-1}(\Sigma_j^*)$ is a disjoint union of sets $S_{j,\nu}^*$ having the following properties:

(a) each $S_{j,\nu}^*$ is a non-empty Zariski open subset of the least (absolutely) irreducible subvariety $\bar{S}_{j,\nu}^*$ of V^* containing $S_{j,\nu}^*$;
(b) $\dim\bar{S}_{j,\nu}^*=\dim\bar{\Sigma}_j^*$, and the $\bar{S}_{j,\nu}^*$ are the (absolutely) irreducible components of $\pi^{*-1}(\bar{\Sigma}_j^*)$;
(c) each point of $S_{j,\nu}^*$ is simple for $\bar{S}_{j,\nu}^*$.

From what was said above follows also that if $S_{j,\nu}^*$ contains a singular point of V^*, then all points of $S_{j,\nu}^*$ are singular points (of the same multiplicity) of V^*. *From now on we shall consider only those sets* $S_{j,\nu}^*$ *which contain singular points of* V^*, and we shall denote these sets by S_i^* $(i=q+1,q+2,\ldots,N)$, where q is the number of absolutely irreducible components of V. We have therefore $\cup S_i^*=V_{\text{sing}}^*$, and any two S_i^* are disjoint. Furthermore, S_i^* is a non-empty Zariski open subset of the least (absolutely) irreducible subvariety $\bar{S}_i^*$ of V_{sing}^* containing S_i^*, and each point of S_i^* is simple for $\bar{S}_i^*$. From the fact that condition (3) of Theorem 3 is satisfied for the stratification $\{\Sigma_i^*\}$ of Δ^*, follows easily that also the similar condition is satisfied by the S_i^*, namely $\bar{S}_i^* - S_i^*$ is a union of sets S_i^*.

The crucial point of our proof is the following: for each $i = q+1, q+2, \ldots, N$, the sets $\bar{S}_i^*(\Omega)$ are varieties *defined and irreducible over any algebraically closed field k of definition of* V. Once this is proved, then it is easy to show that the strata $S_1, S_2, \ldots, S_q$ of simple points of V (introduced earlier) and the sets $S_i = \bar{S}_i^*(\Omega) \cap S_i^*$ $(i = q+1, q+2, \ldots, N)$ give the desired stratification of V, satisfying the conditions stated in Theorem 3.

We shall just give an idea of the proof of the above statement. Let $r' = \dim V_{\text{sing}}$. Then we have also $r' = V_{\text{sing}}^*$, and since $\cup S_i^* = \cup \bar{S}_i^* = V_{\text{sing}}^*$, we must have $\dim \bar{S}_i^* \leqslant r'$ for all $i = q+1, q+2, \ldots, N$, with equality for some values of i. Consider an $\bar{S}_i^*$ whose dimension is exactly r', say, let $\dim \bar{S}_{q+1}^* = r'$. Then $\bar{S}_{q+1}^*$ is an irreducible component of $V_{\text{sing}}^*/\Omega^*$. It is then clear that $\bar{S}_{q+1}^*(\Omega)$ is an absolutely irreducible compound of V_{sing}/Ω, and is therefore defined over any algebraically closed field k of definition of V.

If $\bar{S}_{q+1}^*, \bar{S}_{q+2}^*, \ldots, \bar{S}_{q+q_1}^*$ are the $\bar{S}_i^*$ which have dimension r', then $V_{\text{sing}} - \cup_{\alpha=1}^{q_1} \bar{S}_{q+\alpha}^*(\Omega) = \cup_{\beta > q_1} \bar{S}_{q+\beta}^*(\Omega)$ is a subvariety of V_{sing}, of $\dim r'' < r'$. Thus $\dim \bar{S}_{q+\beta}^* \leqslant r''$, for all β, with equality for some values of β. It follows in a similar way as above, that all the varieties $\bar{S}_{q+\beta}^*(\Omega)$ whose dimension is equal to r'' must be defined over k. Continuing in this fashion to strata of decreasing dimension, the key point of the proof is established.

We conclude this paper by stating two conjectures which bear upon the question of stability of our definition of equisingularity.

Given a subvariety Γ of V (not necessarily irreducible) and given the equisingular stratification $\{S_i\}$ of V, those sets $S_i \cap \Gamma$ which are not empty give a decomposition of Γ into disjoint subsets $S_{i,\Gamma}$ which are open subvarieties of Γ and have the property that each is the largest open subvariety of Γ such that V is equisingular along it at each of its points (the $S_{i,\Gamma}$ are not necessarily irreducible nor are they necessarily non-singular). We call the $S_{i,\Gamma}$ the *V-equisingularity strata of* Γ.

Now let W be an absolutely irreducible subvariety of V, and let $T^*: V^* \to V$ be the transformation of V obtained by blowing-up W. Let $\mathfrak{S}_{r-1}^*$ be the open subvariety of V^* which is given by $T^{*-1}(\text{Eqs}(W/V))$. Let us denote the open subset Eqs. (W/V) of W by W_1. Since V is equimultiple along W_1, $\mathfrak{S}^*_{r-1}$ is pure $(r-1)$–dimensional, and each irreducible component of $\mathfrak{S}^*_{r-1}$ is mapped by T^* onto W_1. Let $\{S^*_{i,\mathfrak{S}_{r-1}}\}$ be the V^*-equisingularity strata of $\mathfrak{S}^*_{r-1}$.

CONJECTURE 1. *T^* maps each stratum $S^*_{i,\mathfrak{S}^*_{r-1}}$ surjectively onto W_1. Furthermore, all points of $S^*_{i,\mathfrak{S}_{r-1}}$ are simple points of the stratum, and each stratum has locally, at each of its points Q^*, one and the same dimension (which we shall denote by ρ^*_i).*

For any point Q of W_1, let $L^*(Q)$ be the corresponding fiber $T^{*-1}(Q)$ on $\mathfrak{S}^*_{r-1}$. Let Q^* be any point of $L^*(Q)$, and let $S^*_{i,\mathfrak{S}^*_{r-1}}$ be the stratum (in $\mathfrak{S}^*_{r-1}$) containing Q^*. In the complex domain, we propose the following:

CONJECTURE 2. *There exist:* (1) *a neighborhood $N_{L^*_Q}(Q^*)$ of Q^* in $L^*(Q)\cap S^*_{i,\mathfrak{S}^*_{r-1}}$,* (2) *a neighborhood $N_{W_1}(Q)$ of Q on W_1,* (3) *a neighborhood $N_{S^*_{i,\mathfrak{S}^*_{r-1}}}(Q^*)$ of Q^* in $S^*_{i,\mathfrak{S}^*_{r-1}}$ and* (4) *an isomorphic analytical map*

$$\varphi : N_{W_1}(Q)\times N_{L^*(Q)}(Q^*) \overset{\varphi}{\underset{\sim}{\rightarrow}} N_{S^*_{i,\mathfrak{S}^*_{r-1}}}(Q^*),$$

such that the following diagram is commutative:

$$\begin{array}{ccc} N_{W_1}(Q)\times N_{L^*(Q)}(Q^*) & \overset{\varphi}{\rightarrow} & N_{S^*_{i,\mathfrak{S}^*_{r-1}}}(Q^*) \\ & \psi_1 \searrow & \downarrow T^* \\ & & N_{W_1(Q)} \end{array}$$

Here ψ_1 is the projection of the direct product onto the first factor.

REFERENCES.

[1] O. Zariski, "Equisingular points on algebraic varieties," *Seminari dell'Istituto Nazionale di Alta Matematica, 1962–63*, Edizioni Cremonese, Roma, 1964, pp. 164–177.

[2] ———, "Studies in equisingularity I. Equivalent singularities of plane algebroid curves," *Amer. J. Math.*, 87 (1965), pp. 507–536.

[3] ———, "Studies in equisingularity II. Equisingularity in co-dimension 1 (and characteristic zero)," *Amer. J. Math.*, 87 (1965), pp. 972–1006.

[4] ———, "Studies in equisingularity III. Saturation of local rings and equisingularity," *Amer. J. Math.*, 90 (1968), pp. 961–1023.

[5] ———, "Contributions to the problem of equisingularity," *Centro Internazionale Matematico Estivo* (*C.M.I.E.*), *Questions on Algebraic Varieties. III ciclo*, Varenna, 7–17 Settembre 1969, Edizioni Cremonese, Roma, 1970, pp. 261–343.

[6] ———, "Some open questions in the theory of singularities," *Bull. Amer. Math. Soc.*, **77** (1971), pp. 481–491.

[7] ———, "On Equimultiple Subvarieties of Algebroid Hypersurfaces," *Proc. Nat. Acad. Sci. USA*, **72**, No. 4 (1975), pp. 1425–1426.

[8] J. Briançon et J. P. Speder, *Les conditions de Whitney n'impliquent pas l'equisingularité au sens de Zariski*, Univ. de Nice, Dept. de Mathematiques, Janvier 1975.

[9] J. P. Speder, "Equisingularité et conditions de Whitney," *Amer. J. Math.*, **97** (1975), pp. 571–588.

A NEW PROOF OF THE TOTAL EMBEDDED RESOLUTION THEOREM FOR ALGEBRAIC SURFACES (BASED ON THE THEORY OF QUASI-ORDINARY SINGULARITIES).*

By Oscar Zariski

Introduction. In this paper we shall deal only with (complete) algebraic surfaces F (not necessarily irreducible) defined over an algebraically closed ground field k of characteristic zero and *embedded in an irreducible,* NON-SINGULAR, *three-dimensional (complete) variety* M (also defined over k and embedded in some projective space). We may refer to such surfaces as *globally embeddable* surfaces. Our treatment could be extended, with some minor modifications, to surfaces F which are only *locally embeddable*, at each closed (i.e., k-rational) point P of the surface, in an irreducible, non-singular, three-dimensional variety M_P (depending on P). That means that if $\mathfrak{o}$ is the local ring of F at P and if P is a *singular* point of F, then the maximal ideal $\mathfrak{m}$ of $\mathfrak{o}$ has a basis of three elements (such a basis is necessarily minimal, since P is a singular point of F). However, for the sake of simplicity of the exposition we shall restrict ourselves to globally embeddable surfaces.

Starting with our three-dimensional, non-singular, irreducible variety M, we shall deal with sequences $T, T_1, T_2, \ldots, T_{N-1}$ of successive monoidal transformations

$$T_i : M_i \to M_{i+1}, \qquad i = 0, 1, 2, \ldots, N-1;$$
$$T_0 = T, \quad M_0 = M, \tag{1_i}$$

where the center Γ_i of T_i in M_i is either a point or a *non-singular irreducible* curve (whence the three-dimensional irreducible varieties $M_1, M_2, \ldots, M_N$ are all non-singular). We agree to denote the center Γ_0 of T by Γ.

Given our surface F in M, we denote by F_1 and F_1^* respectively the proper T-transform $T[F]$ and the total T-transform $T\{F\}$ of F in M_1 (for the definition of $T[F]$ and $T\{F\}$ see Section 1). More generally, we define, by induction on i,

Manuscript received June 28, 1977.
*This work was supported by a National Science Foundation Grant No. MC576-80402.

the surfaces F_i and F_i^* in M_i:

$$F_i = T_{i-1}[F_{i-1}], \qquad F_i^* = T_{i-1}\{F_{i-1}\},$$
$$i = 1, 2, \ldots, N; \qquad T_0 = T, \quad F_0 = F = F_0^*. \tag{2_i}$$

All the varieties F_i, F_i^* are of pure dimension 2. Moreover, we have $F_i^* = F_i$ if and only if $\Gamma_{i-1} \not\subset F_{i-1}$.

The *total embedding resolution theorem* (TERT) for F (proved by Hironaka in [1] not only for surfaces but also for varieties of any dimension) asserts the following:

(TERT). *There exists a sequence* (1_i) *of monoidal transformations such that the last proper and total transforms* F_N *and* F_N^* *of* F *have the following three properties:*

(a) F_N *is non-singular.*
(b) *The only singularities of* F_N^* *are normal crossings.*
(c) *Each irreducible component of* F_N^* *is non-singular.*

In our Lincei note [5] we have given a very simple proof of the existence of a sequence (1_i) such that condition (a) of (TERT) is satisfied. Our proof gave an explicit construction of such a revolving system (1_i) of monoidal transformations T_i. Our chief purpose in the present paper is to develop a proof of (TERT) which uses the results of our cited Lincei note and which is simpler than the original proof given by Hironaka in [1]. In this paper we shall use freely the various results which we have obtained in that Lincei note.

Our proof is based on two results which are stated below as Main Theorem I and Main Theorem II respectively.

Hironaka, to whom we have communicated these two theorems, was able to give a complete and fairly simple proof (unpublished) of Main Theorem I using the results of our Lincei note [5]. Thus, our chief purpose in the present paper will be to give a full proof of Main Theorem II, a proof which is of interest in itself. Actually, we shall prove a *stronger theorem*, stated below as Main Theorem II′. With the aid of this stronger theorem II′ it is possible to prove (TERT) by proving a theorem (stated below as Main Theorem I′) which is weaker than Main Theorem I and which therefore conceivably admits a proof which may be even simpler than the (unpublished) proof of Theorem I which was outlined to us by Hironaka. At any rate, we propose, as a problem to the reader, the finding of as simple a proof as possible of Main Theorem I′.

To state the main theorems I, II and I′, II′, we need two definitions.

Definition 1. The sequence (1_i) of monoidal transformations T_i is said to be *F-permissible* if for $i=0,1,\ldots,N-1$ the following two conditions are satisfied:

(1) The centers Γ_i of T_i belong to F_i.
(2) If Γ_i is a curve, then F_i is equimultiple along Γ_i.

To state our second definition we need some preliminaries.

We denote by E_i^i the irreducible non-singular exceptional surface $T_{i-1}\{\Gamma_{i-1}\}$ in M_i into which the center Γ_{i-1} of T_{i-1} is blown up by T_{i-1}:

$$E_i^i = T_{i-1}\{\Gamma_{i-1}\}, \qquad i=1,2,\ldots,N; \quad \Gamma_0=\Gamma. \tag{3_i}$$

We define, by induction on i, the irreducible surfaces $E_i^1, E_i^2, \ldots, E_i^{i-1}$ in M_i, $i>1$, as follows:

$$E_i^j = T_{i-1}[E_{i-1}^j], \qquad j=1,2,\ldots,i-1; \quad i=2,3,\ldots,N. \tag{4_i}$$

We set

$$\mathcal{E}_i = \bigcup_{j=1}^{i-1} E_i^j, \qquad i=2,3,\ldots,N;$$

$$\mathcal{E}_i^* = \bigcup_{j=1}^{i} E_i^j, \qquad i=1,2,\ldots,N. \tag{5_i}$$

We agree to denote by $\mathcal{E}_1$ also the surface $\mathcal{E}_1^*$: it is the irreducible surface E_1^1. We have therefore:

$$\mathcal{E}_i = T_{i-1}[\mathcal{E}_{i-1}^*], \qquad i=2,3,\ldots,N; \tag{6_i}$$

$$\mathcal{E}_i^* = T_{i-1}\{\mathcal{E}_{i-1}^*\}, \quad i=1,2,\ldots,N; \qquad \mathcal{E}_0^*=\Gamma, \tag{6_i^*}$$

and

$$F_i^* = F_i \cup \mathcal{E}_i^*, \qquad i=1,2,\ldots,N. \tag{7_i}$$

Since in the course of our proof of (TERT) we will have to deal with the decomposition (7_i) of the intermediate total transforms F_i^*, the inductive character of our proof makes it necessary that we modify slightly the initial setup in our three-dimensional variety M. We shall namely assume that our starting point is a surface F^* in M which is a union of two (reduced) surfaces F and $\mathscr{E}$:

$$F^* = F \cup \mathscr{E}. \tag{8}$$

It is assumed that the pair of surfaces $F, \mathscr{E}$ satisfies the following three conditions:

(a_0) *F and $\mathscr{E}$ have no common irreducible components.*

(b_0) *The only singularities of $\mathscr{E}$ are normal crossings.* $\qquad$ (9)

(c_0) *Each irreducible component of $\mathscr{E}$ is non-singular.*

(It is not excluded that $\mathscr{E}$ is the empty set and that consequently $F^* = F$ as in our original setup.) Our inductive definition (2_i) of the F_i and the F_i^* remains unaltered, except that this time, for $i = 0$, we set $F_0 = F$, $F_0^* = F^*(= F \cup \mathscr{E})$. Our definition 3.1 of F-permissible (or of F^*-permissible) sequences of monoidal transformations remains unchanged. A slight modification of the definition (5_i) of the surfaces $\mathscr{E}_i$ and $\mathscr{E}_i^*$ will be necessary if $\mathscr{E}$ is not the empty set. We use (6_i) and (6_i^*) to define inductively the surfaces $\mathscr{E}_i, \mathscr{E}_i^*$, where this time i ranges from 1 to N in *both* (6_i) and (6_i^*), *with the convention that if* $\mathscr{E} \neq \varnothing$, then $\mathscr{E}_0^* = \mathscr{E} \cup \Gamma$ (instead of $\mathscr{E}_0^* = \Gamma$, as previously; naturally, if $\Gamma \subset \mathscr{E}$, then $T\{\mathscr{E} \cup \Gamma\} = T\{\mathscr{E}\}$, whence, in this case, we have $\mathscr{E}_1^* = T\{\mathscr{E}\}$).

Let $E^1, E^2, \ldots, E^q$ be the irreducible components of $\mathscr{E}$ (if $\mathscr{E} \neq \varnothing$). Instead of ($3_i$) and ($4_i$) we shall now have, for each $i = 1, 2, \ldots, N$, a set $q + i$ distinct irreducible surfaces $E_i^1, E_i^2, \ldots, E_i^{q+i-1}, E_i^{q+i}$, where

$$E_i^{q+i} = T_{i-1}\{\Gamma_{i-1}\}, \tag{10_i}$$

while the $q + i - 1$ surfaces $E_i^1, E_i^2, \ldots, E_i^{q+i-1}$ are defined inductively as follows:

$$E_0^\nu = E^\nu, \qquad \nu = 1, 2, \ldots, q; \tag{11_0}$$

$$E_i^j = T_{i-1}[E_{i-1}^j], \qquad j = 1, 2, \ldots, i-1+q;$$
$$i = 1, 2, \ldots, N. \tag{11_i}$$

We have therefore the relations

$$\mathscr{E}_i = \bigcup_{j=1}^{q+i-1} E_i^j,$$

$$\mathscr{E}_i^* = \bigcup_{j=1}^{q+i} E_i^j = \mathscr{E}_i \cup E_i^{q+i} = \mathscr{E}_i + T_{i-1}\{\Gamma_{i-1}\},$$
$$i = 1, 2, \ldots, N, \tag{12_i}$$

which now replace (5_i). The relations (7_i) remain unchanged.

Definition 2. Given the surface $F^* = F \cup \mathfrak{S}$ [see (8)], where $\mathfrak{S}$, if not empty, satisfies conditions (a_0), (b_0) and (c_0) in (9), the sequence (1_i) of monoidal transformations is said to be *$(F, \mathfrak{S})$-permissible* if it is F-permissible and if, furthermore, for each $i = 0, 1, \ldots, N-1$ such that Γ_i is a curve, the following condition is satisfied: if P_i is any point of Γ_i there exists at most one irreducible component E_i^j of $\mathfrak{S}_i^*$ which contains the point P_i and does not contain the curve Γ_i, and if such a component E_i^j does exist, then P_i is a simple intersection of E_i^j and Γ_i. In the special case in which $\mathfrak{S}$ is the empty set we shall use the term *strictly F-permissible* to designate an $(F, \varnothing)$-permissible sequence of monoidal transformations.

PROPOSITION 3. *If the sequence (1_i) of monoidal transformations is $(F, \mathfrak{S})$-permissible, then each surface $\mathfrak{S}_i^*$ $(i = 1, 2, \ldots, N)$ satisfies conditions similar to (a_0), (b_0) and (c_0). Namely:*

(a_i) *F_i and $\mathfrak{S}_i^*$ have no irreducible components in common.*
(b_i) *The only singularities of $\mathfrak{S}_i^*$ are normal crossings.*
(c_i) *Each irreducible component of $\mathfrak{S}_i^*$ is non-singular.*

This proposition is a direct consequence of Lemmas 1.4 and 1.5, proved in Section 1.

MAIN THEOREM I. *Given the surface F^* $(= F \cup \mathfrak{S}$ [see (8)], where $\mathfrak{S}$, if not empty, satisfies conditions (a_0), (b_0) and (c_0), there exists an $(F, \mathfrak{S})$-permissible sequence of monoidal transformations $T, T_1, T_2, \ldots, T_{N-1}$ such that the proper transform F_N of F is a non-singular surface (while, by Proposition 3, the surface $\mathfrak{S}_N^*$ satisfies necessarily conditions (a_N), (b_N) and (c_N) of that Proposition).*

MAIN THEOREM II. *Let F, $\mathfrak{S}$ and $\mathfrak{S}^*$ be three (reduced) surfaces in M, any two of which are free from common irreducible components. We set*

$$F^* = F \cup \mathfrak{S}, \tag{13}$$

and we assume that the three surfaces $F, \mathfrak{S}, \mathfrak{S}^*$ satisfy the following conditions:

(1) *F is non-singular, and all irreducible components of $\mathfrak{S}$ and $\mathfrak{S}^*$ are non-singular.*
(2) *All the singularities of the surfaces $F \cup \mathfrak{S}^*$ and $\mathfrak{S} \cup \mathfrak{S}^*$ are normal crossings.*

Then there exists a sequence $T, T_1, T_2, \ldots, T_{N-1}$ of monoidal transformations which is both $(F^*, \mathfrak{S}^*)$-permissible (see Definition 2) and F-permissible *(see Definition* 1) *such that the last proper transform F_N of F is still non-singular, while all the singularities of the last total transform $F_N \cup \mathfrak{S}_N \cup \mathfrak{S}_N^*$ of $F^* \cup \mathfrak{S}^*$ are normal crossings (here $\mathfrak{S}_N$ is the last proper transform of $\mathfrak{S}$, whence $F_N^* = F_N \cup \mathfrak{S}_N$ is the last proper transform of F^*).*

We now show that (TERT), *in a somewhat stronger formulation*, follows from the above two main theorems; namely, we may add in (TERT) the condition that *the sequence of transformations* (1_i) *be F-permissible*. To see this, we apply Main Theorem I to the case in which $\mathfrak{S}$ is the empty set and we denote the surfaces F_N^* [see (7_i)], F_N and $\mathfrak{S}_N^*$ of that theorem by $\bar{F}^*$, $\bar{F}$ and $\bar{\mathfrak{S}}$ respectively. Thus $\bar{F}^*$ $(=\bar{F} \cup \bar{\mathfrak{S}})$ is the last total transform of F, and $\bar{F}$ is the last proper transform of F and is non-singular, while all the singularities of $\bar{\mathfrak{S}}$ are normal crossings, and each irreducible component of $\bar{\mathfrak{S}}$ is non-singular. The conditions (1), (2) of Main Theorem II are satisfied if the three surfaces F, $\mathfrak{S}$ and $\mathfrak{S}^*$ of that theorem are respectively the surfaces $\bar{F}$, $\bar{\mathfrak{S}}$ and the empty set. We have therefore a second sequence of monoidal transformations $\bar{T}, \bar{T}_1, \ldots, \bar{T}_{\bar{N}-1}$:

$$\bar{T}: \bar{M} \to \bar{M}_1, \qquad \bar{M} = M_N;$$
$$\bar{T}_j : \bar{M}_j \to \bar{M}_{j+1}, \qquad j = 1, 2, \ldots, \bar{N}-1, \tag{14}$$

which is both $\bar{F}$-permissible and strictly $\bar{F}^*$-permissible (see last sentence of Definition 2) and is such that the last proper transform $\bar{F}_{\bar{N}}$ of $\bar{F}$ is non-singular, while all the singularities of the last total transform of $\bar{F}^*$ are normal crossings. Since the sequence $\bar{T}, \bar{T}_1, \ldots, \bar{T}_{\bar{N}-1}$ is $\bar{F}$-permissible, the combined sequence of the $N+\bar{N}$ monoidal transformations $T_i, \bar{T}_j$ is F-permissible, and the surface $\bar{F}_N$ is the last proper transform of F under the combined sequence of transformations, while the total transform of $\bar{F}^*$ is also the total transform of F. This proves (TERT) in the slightly stronger formulation indicated above.

We now state the two Main Theorems I′ and II′, which—as we pointed out earlier—can replace the main theorems I and II for the purpose of proving the total embedding resolution theorem.

MAIN THEOREM I′. *Given the (reduced) surface F, there exists an F-permissible sequence $T, T_1, T_2, \ldots, T_{N-1}$ of monoidal transformations such that if F_N and $F_N \cup \mathfrak{S}_N$ are respectively the proper and the total transforms of F under the product of the N transformations $T, T_1, T_2, \ldots, T_{N-1}$, then all the irreducible components of $F_N \cup \mathfrak{S}_N$ are non-singular surfaces.*

MAIN THEOREM II′. *Let F and $\mathfrak{S}$ be two (reduced) surfaces in M, free from common irreducible components and satisfying the following two condi-*

tions:

Condition 1. *All irreducible components of* F *and* $\mathfrak{S}$ *are non-singular.*
Condition 2. *If* F^{μ} *is any irreducible component of* F, *then the only singularities of the surface* $F^{\mu} \cup \mathfrak{S}$ *are normal crossings.*

Then there exists a sequence (1_i) *of monoidal transformations* $T, T_1, T_2, \ldots, T_{N-1}$ *which is both* $(F, \mathfrak{S})$*-permissible and* F*-permissible such that if* F_N *and* $F_N \cup \mathfrak{S}_N$ *are respectively the proper transform of* F *and the total transform of* $F \cup \mathfrak{S}$ *under the product of the* N *transformations* $T, T_1, T_2, \ldots, T_{N-1}$, *then* F_N *is non-singular, while the only singularities of* $F_N \cup \mathfrak{S}_N$ *are normal crossings.*

The total embedding resolution theorem follows from these two theorems by applying first Main Theorem I′ to our given surface F. Let us denote by $\bar{F}$ the surface $F_N \cup \mathfrak{S}_N$ of Main Theorem I′. All the irreducible components of $\bar{F}$ are then non-singular. The two conditions 1 and 2 of Main Theorem II′ are satisfied if the surfaces $F, \mathfrak{S}$ of that theorem are replaced respectively by the surface $\bar{F}$ and the empty set. Thus, Main Theorem II′ implies that there exists a second sequence (14) of monoidal transformations $\bar{T}, \bar{T}_1, \ldots, \bar{T}_{\bar{N}-1}$ which is $\bar{F}$-permissible and is such that the last proper transform $\bar{F}_{\bar{N}}$ of $\bar{F}$ is non-singular, while all the singularities of the last total transform $\bar{F}_{\bar{N}} \cup \bar{\mathfrak{S}}_{\bar{N}}$ of $\bar{F}_{\bar{N}}$ are normal crossings. The surface $\bar{F}_{\bar{N}} \cup \bar{\mathfrak{S}}_{\bar{N}}$ is also the total transform of F under the product of the $N + \bar{N}$ transformations of the combined sequence $T, T_1, \ldots, T_{N-1}, \bar{T}, \bar{T}_1, \ldots, \bar{T}_{\bar{N}-1}$, while the proper transform of F under the product of these transformations is a union of some of the components of $\bar{F}_{\bar{N}}$ and hence is also non-singular. Note that we cannot assert now that the sequence $T, T_1, \ldots, T_{N-1}, \bar{T}, \bar{T}_1, \ldots, \bar{T}_{\bar{N}-1}$ is F-permissible, because Main Theorem II′ says only that the sequence $\bar{T}, \bar{T}_1, \ldots, \bar{T}_{\bar{N}-1}$ is $\bar{F}$-permissible, i.e., $F_N \cup \mathfrak{S}_N$-permissible, and not necessarily F_N-permissible.

It is clear that Main Theorem I′ is a consequence of Main Theorem I (replace in Main Theorem I the surface $\mathfrak{S}$ by the empty set; the fact that the surface F_N of Main Theorem I is non-singular implies of course that all the irreducible components of F_N are non-singular). On the other hand, if we replace in Main Theorem II′ the surfaces F and $\mathfrak{S}$ by the surfaces $F^* = F \cup \mathfrak{S}$ and $\mathfrak{S}^*$ of Main Theorem II, then condition 1 of II′ is satisfied in view of condition (1) of II, while condition 2 of II′ follows from condition (2) of II. Therefore, using the conclusion of Main Theorem II′ we find that there exists a sequence $T, T_1, \ldots, T_{N-1}$ of monoidal transformations which is both $(F^*, \mathfrak{S}^*)$-permissible and F-permissible such that the last proper transform $F_N \cup \mathfrak{S}_N$ of F^* is non-singular, while all the singularities of the last total transform $F_N \cup \mathfrak{S}_N \cup$

$\mathcal{E}_N^*$ of $F^* \cup \mathcal{E}^*$ are normal crossings. This is the conclusion of Main Theorem II, except that we cannot assert now that the sequence $T, T_1, \ldots, T_{N-1}$ is F-permissible (Theorem II′ tells us only that this sequence is F^*-permissible, i.e., $F \cup \mathcal{E}$-permissible). Thus, apart from this indicated weaker aspect of Main Theorem II′, Main Theorem II is—for the rest of it—a consequence of Main Theorem II′.

1. Some Properties of Monoidal Transformations.[1] Let T be a monoidal transformation of the ambient variety M whose center in M is either a point P or an irreducible, non-singular curve Γ. We denote by M' the T-transform of M. Thus M' is an irreducible, non-singular, three-dimensional variety, and T^{-1} is a regular map (a morphism) of M' onto M. The center Γ of M is blown up by T into a non-singular, irreducible surface E', *the exceptional surface* of T^{-1} in M' (while Γ is *the fundamental locus* of T in M).

We consider first the case in which the center of T is a point P of M. The set of tangential directions of M at P can be identified with the set of lines through P in the linear (three-dimensional) tangent space of M at P. Hence the above set of tangential directions has a canonical structure of a projective plane over k. The points P' of the fundamental surface E' are in $(1,1)$ correspondence with the tangential directions d of M at P. The correspondence is such that if we approach the point P along *any* analytic branch C which is tangent to d at P, then the T-transform C' of C will be an analytic branch which passes through the point P' which corresponds to the tangential direction d. Thus, the exceptional surface E' has in this case a canonical structure of a projective plane. The *lines* in E' correspond to the planes through P which are contained in the tangent three-space of M at P.

If F is any irreducible (algebraic or algebroid) surface contained in M, we shall denote by $T\{F\}$ and $T[F]$ respectively the set of all points P' of M' such that $T^{-1}\{P'\} \in F$ and the irreducible (algebraic or algebroid) surface in M' whose general point is the T-image of a general point of F. These two surfaces will be called respectively the *total* and the *proper* transform of F. They are identical if and only if $P \notin F$. If, however, $P \in F$, then $T\{F\} = T[F] \cup E'$. In this case, the intersection of $T[F]$ with E' is an algebraic curve whose points correspond to the directions of the tangent cone of F at P. In particular, if P is a simple point of F, then $T[F] \cap E'$ is the *line* in E' which corresponds to the directions, at P, of the tangent plane of F at P (this plane being contained, of course, in the tangent 3-space of M at P).

If F is a reducible surface in M, the proper transform $T[F]$ of F will be the

[1]For further information concerning monoidal transformations see our papers [2] and [3].

union of the proper transforms of the irreducible components of F, while the *total transform* $T\{F\}$ will denote again the set of points P' of M' such that $T^{-1}(P')\in F$. Hence $T[F]=T\{F\}$ if and only if $P\notin F$, while $T\{F\}=T[F]\cup E'$ if $P\in F$.

In a similar way we define the symbols $T[C]$ and $T\{C\}$ if C is an algebraic (or algebroid) curve on M. We have, as in the case of surfaces F, that $T[C]$ is always a curve in M', whose irreducible components are the proper T-transforms $T[C_i]$ of the irreducible components of C_i of C. If $P\notin C$ then $T\{C\}=T[C]$, while $T\{C\}=T[C]\cup E'$ if $P\in C$.

We next consider the case in which the center of T is an irreducible non-singular curve Γ. In this case, the exceptional surface E' is a *ruled surface*, birationally equivalent to the direct product of Γ and a projective line. To each point P of Γ there corresponds a fiber γ_P' on E' which is a generator of the ruling on F. Each fiber γ_P' is canonically a projective line, the points of γ_P' being in (1, 1) correspondence with the planes of the pencil of planes contained in the tangent 3-space of M at P and containing the tangent line of Γ at P. If we approach a point P of Γ along *any* analytical branch C such that the tangent line d of C at P is *different* from the tangent line d_0 of Γ at P, then the T-transform C' of C will be an analytic branch which intersects the fiber γ_P' in the point P' which corresponds to the plane determined by the two (*distinct*) lines d, d_0. If, however, we approach P along branches C which are tangent to Γ at P, then the corresponding branches C' in M' will intersect the fiber γ_P' at *variable* points P' (depending on the branch C; see the algebraic proof given later on in this section).

The definition of the proper and total T-transforms $T[\;]$ and $T\{\;\}$ of an algebraic or algebroid surface F or curve C in M is similar to the definition given above, in the case in which the center of T was a point. However, in the present case we have $T[F]=T\{F\}$ if and only if $\Gamma\not\subset F$, while $T\{F\}=T[F]\cup E'$ if $\Gamma\subset F$. Furthermore $T[C]$ is defined only if Γ is not an irreducible component of C. As to Γ itself, only the *total* transform $T\{\Gamma\}$ is defined: it is the exceptional surface E'.

The following special cases will be of particular interest in the sequel, and we state the corresponding results as our first two lemmas.

LEMMA 1.1. *Let T be a monoidal transformation centered at a point P of M, and let F and C be respectively an algebroid surface and an algebroid curve having at P a simple point. Let $F'=T[F]$ and $C'=T[C]$. Then:*

(a) *The intersection of F' with the exceptional surface E' is a line in E', and all the points of that line are normal crossings of the surface $T\{F\}$*

($=F'\cup E'$). Similarly, the intersection C' with E' is a single point P', and P' is a simple intersection of E' and C'.

(b) *If C is tangent to F at P and if P' is the point of E' which corresponds to the direction of the tangent line of C at P (whence P' is the only point of E' which belongs to both F' and C'), then we have $i(F',C';P')=i(F,C;P)-1$, where $i(F',C';P')$ denotes the intersection multiplicity of F' and C' at P' [and similarly for $i(F,C;P)$].*

LEMMA 1.2. *Let T be a monoidal transformation centered at an irreducible non-singular curve Γ, let F and C be respectively an algebroid surface and an algebroid curve in M, and let P be a point of Γ. Let $F'=T[F]$ and $C'=T[C]$.*

(a) *If F does not contain Γ and if P is a simple intersection of F and Γ, then F' contains the fiber γ'_P, and all the points of that fiber are normal crossings of the surface $F'\cup E'$.*

(b) *If F contains Γ and if P is a simple point of F, then F' intersects the fiber γ'_P in a single point P', and this point is a normal crossing of the surface $T\{F\}$. If all the points P of Γ are simple for F, then $F'\cap E'$ is an irreducible non-singular curve Γ' which is a transversal unisecant of the family of fibers γ'_P, and the above mapping $P'\to P$ is a biregular map of Γ' onto Γ.*

(c) *If P is a simple point of C and if C is not tangent to Γ at P, then the point P' of the fiber γ'_P which belongs to C' is a simple intersection of C' and E'.*

The proof will be given later on in this section.

We now present briefly the local algebra which lies behind the above stated properties of monoidal transformations.

Let first T be a monoidal transformation centered at a point P. We fix a set of local regular parameters X,Y,Z of M at P. For simplicity of notation we shall denote by X,Y,Z also the $\mathfrak{M}$-residues of these parameters, where $\mathfrak{M}$ is the maximal ideal of the (complete) local ring $\mathfrak{O}$ of M at P. Hence a change of the local regular parameters amounts to a non-singular linear homogeneous transformation of X,Y,Z (more precisely: of the $\mathfrak{M}$-residues of X,Y,Z), with coefficients in the field k ($=\mathfrak{O}/\mathfrak{M}$). On the exceptional surface E', regarded (canonically) as a projective plane, the elements X,Y,Z are homogeneous coordinates. If a,b,c are the homogeneous coordinates of a point P' of E', and if, say, $a\neq 0$, then $Y/X-b/a, Z/X-c/a$ are regular parameters of M' and P' (where now X, Y and Z stand for regular parameters of M at P and not merely the $\mathfrak{M}$-residues of such parameters). The local equation of E' at P' is then $X=0$ (and this proves incidentally that E' is a non-singular surface). Let F be a surface (algebraic or algebroid) containing P, and let $f(X,Y,Z)=0$ be its local equation at P; here

$$f(X,Y,Z)=f_n(X,Y,Z)+f_{n+1}(X,Y,Z)+\cdots, \tag{1}$$

where f_i is a form of degree i in X,Y,Z, f_n is not identically zero, and n is the multiplicity of the point P of F ($n=1$ if and only if P is a simple point of F). Assuming for simplicity that $a=1$, $b=c=0$, and setting $Y'=Y/X, Z'=Z/X$ (whence X,Y',Z' are regular parameters of M' at P'), the local equation of $T[F]$ at P' (assuming that $P'\in F'$) will be

$$f_n(1,Y',Z')+Xf_{n+1}(1,Y',Z')+X^2f_{n+2}(1,Y',Z')+\cdots=0. \tag{2}$$

Note that the intersection of F' with E' is the *algebraic* curve $f_n(1,Y',Z')=0$, and hence the assumption that $P'\in T[F]$ signifies that the tangent cone $f_n(X,Y,Z)=0$ of F at P contains the line $Y=Z=0$ (since we have assumed that $b=c=0$). Thus we must have $f(1,0,0)=0$.

Assume, in particular, that P is a simple point of F. Let $0=aX+bY+cZ+$ terms of degree $\geqslant 2$ be the local equation of F at P. The intersection of $T[F]$ with E' is then the line $aX+bY+cZ=0$ (X, Y and Z being regarded as homogeneous coordinates in E'). If P' is any point on that line, we may assume that P' is the point $(1,0,0)$, in which case we have $a=0$. The local equation of E' at P' is then $X=0$, while the local equation of $T[F]$ at P' is $0=bY'+cZ'+\alpha X+$(terms of degree $\geqslant 2$) in X,Y',Z', where α is an element of k. Since b, c are not both zero, it follows that P' is a simple point of $T[F]$ and that the tangent plane of $T[F]$ at P' is different from $X=0$.

Let P be a simple point of C. We may assume that the tangent line of C at P is $Y=Z=0$. Then the parametric equations of C at P are of the form $X=t+\cdots$, $Y=\beta_2t^2+\cdots$, $Z=\gamma_2t^2+\cdots$. At the point P' of E', which corresponds to the tangent direction $Y=Z=0$, the elements X, $Y'=Y/X$ and $Z'=Z/X$ are regular local parameters of M'. The parametric equations of C' are then $X=t+\cdots$, $Y'=\beta_2t+\cdots$, $Z'=\gamma_2t+\cdots$, and it is clear that the point $P':X=Y'=Z'=0$ is a simple intersection of C' with the exceptional surface $E':X=0$.

This completes the proof of part (a) of Lemma 1.1.

We now prove part (b) of Lemma 1.1. The notation being as in the above proof of part (a), let $X=X(t)=\alpha_1t+\alpha_2t^2+\cdots$, $Y=Y(t)=\beta_1t+\beta_2t^2+\cdots$, $Z=Z(t)=\gamma_1t+\gamma_2t^2+\cdots$ be parametric equations of C at P. Since we have assumed that the point P' of $C'\cap E'$ is the point $X=1$, $Y'=Z'=0$, we must have $\alpha_1\neq 0$, $\beta_1=\gamma_2=0$, and we may assume that $\alpha_1=1$. The parametric equations of C' at P' are $X = X(t)$, $Y' = Y(t)/X(t) = \beta_2 t + \cdots$, $Z' = Z(t)/X(t) = \gamma_2 t + \cdots$. If n is the intersection multiplicity of C and F at P, then $f(X(t), Y(t), Z(t)) = \delta t^n + \cdots$, $\delta \neq 0$. The local equation of F' at P' is $f(X, XY', XZ')/X = 0$, and since $f(X(t), Y(t), Z(t))/X(t) = \delta t^{n-1} + \cdots$, this proves part (b) of Lemma 1.1.

Now we consider the case of a monoidal transformation T whose center is an irreducible *non-singular* curve Γ. Let P be a point of Γ. We choose regular local parameters X, Y, Z of M at P in such a way that $X = Z = 0$ are local equations of Γ at P. On the fiber γ'_P on E', considered in its canonical structure of a projective line, the parameters X, Z (or—more precisely—the $\mathfrak{M}$-residues of X, Z, where $\mathfrak{M}$ is the maximal ideal of the (complete) local ring $\mathfrak{O}$ of M at P) are homogeneous coordinates. If P' is a point of γ'_P, and if a, c are homogeneous coordinates of P', then, assuming that $a \neq 0$, the elements $X, Y, Z/X - c/a$ are regular local parameters of M' at P'. Given the point P' we may assume (after replacing X and Z by suitable linear homogeneous combinations) that $a = 1, c = 0$. We then set $Z' = Z/X$, so that now X, Y, Z' are regular local parameters of M' at P'. The local equation, at P', of the exceptional surface E' is $X = 0$ (which proves that E' is again a non-singular surface), while the local equations, at P', of the fiber γ'_P are $X = Y = 0$.

If F is a surface in M, containing the point P of Γ *and not containing* Γ, and if $f(X, Y, Z) = 0$ is the local equation of F at P, where f is given by (1), then the local equation of $T[F]$ at P' is as follows:

$$f_n(X, Y, XZ') + f_{n+1}(X, Y, XZ') + \cdots = 0, \tag{3}$$

which shows that the entire fiber $\gamma'_P : X = Y = 0$ is contained in $T[F]$. Note that X is not a factor of the left-hand side of (1) since $F \not\supset \Gamma$ and since therefore the power series $f_n(0, Y, 0) + f_{n+1}(0, Y, 0) + \cdots$ is not identically zero. In particular, assume that P is a simple point of F and that F is not tangent to Γ at P. Then $n = 1$, $f_n(X, Y, Z) = aX + bY + cZ$, and $b \neq 0$. In that case, the local equation (3) of $T[F]$ takes the form

$$0 = aX + bY + cXZ' + (\text{terms of degree} \geqslant 2).$$

The tangent plane of $T[F]$ at P' is the plane $aX + bY = 0$, and since $b \neq 0$ this plane is different from the plane $E' : X = 0$. This shows that P' is a normal crossing of the surface $T[F] \cup E'$. Since P' is an arbitrary point of γ'_P, this proves part (a) of Lemma 1.2.

We next prove part (b) of Lemma 1.2. Since $\Gamma \subset F$ and P is a simple point of F, the local equation of F at P is of the form $A(X, Y, Z)X + B(X, Y, Z)Z = 0$, where A and B are power series in X, Y, Z and where $a = A(0,0,0)$ and $b = B(0,0,0)$ are not both zero. We may assume that $b = 1$, and upon replacing $aX + Z$ by Z, that $a = 0$. We set $Z' = Z/X$. The equation of F' is then $A(X, Y, XZ') + B(X, Y, XZ')Z' = 0$, the equations of the fiber γ'_P are $X = Y = 0$, and therefore the intersection of F' with the fiber γ'_P (leaving aside, for the moment, the point of the fiber γ'_P where $Z' = \infty$) is the point $X = Y = Z' = 0$ (since $a = 0$ and $b = 1$). The tangent plane of F' at this point is $\alpha X + \beta Y + bZ' = 0$, where α and β are certain elements of k (they are the coefficients of X^2 and XY in the local equation of F at P), while the local equation, at P', of the exceptional surface E' is $X = 0$. Since $b \neq 0$, the assertion that P' is a normal crossing of $F' \cup E'$ is proved. As to the point P'_∞ of the fiber γ'_P where $Z' = \infty$, we set $X' = X/Z$, so that X', Y, Z are regular local parameters of M' at that point. The left-hand side of the local equation of F at P, after dividing through by the factor Z, yields the power series $A(X'Z, Y, Z)X' + B(X'Z, Y, Z)$, and the value of this power series at $X' = Y = Z = 0$ is the non-zero constant b. This shows that $P'_\infty \notin F'$.

The local equation of $\Gamma'(= F' \cap E')$ at P' is given by $X = 0$, $A(0, Y, 0) + B(0, Y, 0)Z' = 0$. Since $A(0,0,0) = a = 0$ and $B(0,0,0) = b = 1$, the second of these two equations has the form $0 = \gamma Y + Z' +$ (terms of degree $\geqslant 2$ in Y, Z'). This shows that Γ' has a simple point at P' and is transversal to the fiber $\gamma'_P : X = Y = 0$. That the mapping $P' \to P$ of Γ' onto Γ is biregular is obvious. This completes the proof of part (b) of Lemma 1.2.

We now prove part (c) of Lemma 1.2. Let $X = a_1 t + a_2 t^2 + \cdots$, $Y = b_1 t + b_2 t^2 + \cdots$, $Z = c_1 t + c_2 t^2 + \cdots$ be parametric equations of C, locally at P. Since P is a simple point of C, the coefficients a_1, b_1, c_1 are not all zero. The tangent line of C at P is defined by the equations $X : Y : Z = a_1 : b_1 : c_1$. Since this line is different from the "line" $\Gamma : X = Z = 0$, it follows that a_1 and c_1 are not both zero. We may assume that, say, $a_1 = 1$. Upon replacing $Y - b_1 X$ and $Z - c_1 X$ by Y and Z respectively, we may assume that $b_1 = c_1 = 0$. Then, upon setting $Z' = Z/X$, it is clear that X, Y, Z' are regular local coordinates of M' at the point P' where C' intersects the exceptional surface $E' : X = 0$. The parametric equations of C' at P' are $X = t + a_2 t^2 + \cdots$, $Y = b_2 t^2 + \cdots$, $Z' = c_2 t + \cdots$. Hence the tangent line to C' at P' is the line $Y = Z' - c_2 X = 0$, and this line is clearly transversal at P' to the surface $E' : X = 0$. This completes the proof of Lemma 1.2.

If, on the contrary, we consider branches C through P which are tangent to Γ at P, then the parametric equations of any such branch can be assumed to be of the form $X = a_m t^m + \cdots$, $Y = t^n + \cdots$, $Z = c_m t^m + \cdots$, where $m > n$ and a_m, c_m are not both zero. The point P' of the fiber γ'_P in which C' meets that fiber is the point where $X:Z=a_m:c_m$, and this can be any point of the fiber (as a_m and c_m vary).

The next lemma also deals with monoidal transformations centered at an irreducible non-singular curve Γ and will be useful in the sequel. Incidentally, part (a) of that lemma will show how essential for the conclusion of part (a) of Lemma 1.2 was the assumption that P is a *simple* intersection of F and Γ.

LEMMA 1.3. *Let T be a monoidal transformation of M, centered at an irreducible non-singular curve Γ, let P be a point of Γ, and let F be a surface having at P a simple point. Let $F'=T[F]$ and $E'=T\{\Gamma\}$.*

(a) *If F does not contain Γ but is tangent to Γ at P, then F' contains the fiber γ'_P; all points of that fiber, except one, are simple points of F'; and the surfaces F' and E' have the same tangent plane at each of these points. At exactly one point of the fiber γ'_P, namely at the point which corresponds to the directions in the tangent plane of F at P, F' has a double point.*

(b) *Let C be an algebraic (or algebroid) curve having at P a simple point. Assume that $C\not\subset F$, that P is a simple intersection of C and F, and that the curves C and Γ have distinct tangents at P. Let P' be the point of the fiber γ'_P where the proper T-transform C' of C meets that fiber. Then $P'\notin F'$ if $\Gamma\subset F$, while P' is a simple intersection of C' and F' if $\Gamma\not\subset F$.*

(c) *Let Γ' be an algebraic curve in E' which is a transversal (non-singular) unisecant of the family of fibers γ'_P. Assume that P is a simple intersection of F and Γ. Then the point P' where Γ' meets the fiber γ'_P is a simple intersection of F' and Γ'.*

Proof. We assume throughout the proof that the local equations of Γ at P are $X=Z=0$.

(a) The local equation of F at P is of the form

$$F: aX+cZ+(\text{terms of degree} \geqslant 2 \text{ in } X,Y,Z)=0.$$

If we set $Z'=Z/X$ then the local equation of F' at any point P' of γ'_P where $Z'=\alpha\neq\infty$ is

$$F':(a+c\alpha)X+c(Z'-\alpha)X+\left[\text{a sum of terms } b_{(i)}X^{i_1}Y^{i_2}(Z'-\alpha)^{i_3}\right]=0,$$
$$(i_1+i_2\geqslant 2),$$

and $X,Y,Z'-\alpha$ are regular local parameters of M' at P'. This shows that $\gamma'_P\subset F'$. Now, without loss of generality we may assume that $c\neq 0$. The point P' is then simple for F', unless $\alpha=-a/c$, and the tangent plane of F' at P' is then $X=0$, i.e., E'. If $\alpha=-a/c$, then P' is a double point of F'. This particular point of γ'_P corresponds to directions, about P, in the tangent plane $aX+cZ=0$ of F at P. If $\alpha=\infty$, then regular local parameters of M' at P' are the elements X', Y, Z, where $X'=X/Z$, and the local equation of F' at P' is then

$$aX'Z+cZ+(\text{terms of degree} \geqslant 2 \text{ in } X,Y,Z)=0,$$

and since $c\neq 0$, we see that P' is a simple point of F', and that the tangent plane of F' at P' is the plane $Z=0$, i.e., E'. This completes the proof of part (a).

(b) We may assume that the local equations of C at P are $Y=Z=0$ (since Γ and C have distinct tangent lines at P). Let

$$F: aX+bY+cZ+(\text{terms of degree} \geqslant 2)=0 \tag{4}$$

be the local equation of F at P. Since P is a simple intersection of F and C, we must have $a\neq 0$. At the point P' where C' meets the fiber γ'_P we must have $X=Y=Z'=0$, where $Z'=Z/X$. If $\Gamma \subset F$, then necessarily $b=0$, and the equation of F' is then $0=a+cZ'+(\text{terms of positive degree in } X,Y,Z')$. Since $a\neq 0$, it follows that $P'\notin F'$. Assume now that $\Gamma \not\subset F$. Then the local equation of F' at P' is

$$aX+bY+cXZ'+(\text{a sum of terms } b_{(i)}X^{i_1}Y^{i_2}Z'^{i_3})=0, \qquad (i_1+i_2 \geqslant 2). \tag{4'}$$

Since the local equations of C' at P' are $Y=Z'=0$ and since $a\neq 0$, it follows that P' is a simple intersection of C' and F.

(c) We may assume that X, Y and $Z'(=Z/X)$ are regular local parameters of M' at P' and that the local equations of the fiber γ'_P are $X=Y=0$; here $X=0$ is the local equation of E' at P', while (4) is the local equation of F at P. Since P is a simple intersection of F and Γ, we must have $b\neq 0$ in (4). The equation of the unisecant Γ' may be assumed to be $X=Z'=0$. It follows then from (4′) that P' is a simple intersection of F' and Γ'. This completes the proof of the lemma.

The next two lemmas deal with the effect of monoidal transformations T of M on surfaces F all singularities of which are normal crossings.

LEMMA 1.4. *Let F be a surface in M all singularities of which are normal crossings. If T is any monoidal transformation $M\rightarrow M'$ of M centered at a point P, then the total T-transform $F^*=T\{F\}$ of F also has only normal crossings.*

Proof. The assertion of the lemma is trivially true if $P\notin F$. Assume then that $P\in F$. Let $T\{P\}=E'$ and $T[F]=F'$, whence $F^*=F'\cup E'$. It is sufficient to show that all the points P' of $F'\cap E'$ are normal crossings of F^*, for at any point P' of F^* which does not belong to the exceptional surface E' the assertion that P' is at most a normal crossing of F^* follows from the biregularity of T^{-1} at P', while if $P' \in E'$ and $P' \notin F'$, then P' is a simple point of F^*. We shall consider three cases, according as P is a simple point, a double point or a triple point of F.

Case 1. *P is a simple point of F.* In this case the lemma follows directly from part (a) of Lemma 1.1.

Case 2. *P is a double point of F.* Let F_1 and F_2 be the two irreducible branches of F which pass through P, let $F_1' = T[F_1]$, $F_2' = T[F_2]$, and let $F_i' \cap E_i' = p_i'$, $i = 1,2$. The lines p_1', p_2' are distinct, since the tangent planes of F_1 and F_2 at P are distinct. We have $F' = T[F] = F_1' \cup F_2'$, locally, at any point P' of the intersection $p_1' \cup p_2'$ of F' with E'. If P' is different from the intersection of p_1' and p_2' (say, P' belongs p_1' and not to p_2'), then $F' \cup E'$ coincides locally at P', with $F_1' \cup E'$, and our assertion follows then from part (a) of Lemma 1.1. Assume now that P' is the common point of p_1' and p_2'. In this case, $F' \cup E'$ coincides, locally at P', with the surface $F_1' \cup F_2' \cup E'$, and P' is then a triple point of that surface. By part (a) of Lemma 1.1 the tangent plane of E' is distinct from the tangent plane of F_i' at P' $(i = 1,2)$. The intersection of the two surfaces F_1, F_2 is a curve C having at P a simple point. Therefore the intersection of F_1', F_2'—which, locally at P', is a curve which coincides with the proper T-transform C' of C—has, by part (a) of Lemma 1.1, a simple point at P' and is *not tangent to E' at P'*. This implies that the tangent planes of F_1', F_2' and E' are independent (since the double curve of $F_1' \cup F_2' \cup E'$ consists, locally at P', of the three curves p_1', p_2', C'; here p_1' and p_2' belong to E', while the tangent of C' at P' is not in E'). Thus P' is a normal crossing of F^*.

Case 3. *P is a triple point of F.* Let F_1, F_2, F_3 be the irreducible branches of F at P, and let $F_i' = T[F_i]$, $p_i' = F_i' \cap E'$. The three lines p_i' form a triangle in E'. If P' is any point of this triangle, other than a vertex, and if, say, $P' \in p_1'$, then F^*, locally at P', coincides with $F_1' \cup E'$, and thus P' is a normal crossing of F^*, by case 1. If P' is a vertex of the above triangle, say $P' = p_1' \cap p_2'$, then F^*, locally at P', coincides with $F_1' \cup F_2' \cup E'$, and the point P' is a triple point of F^* and is a normal crossing of F^*, by Case 2.

This completes the proof of the lemma.

LEMMA 1.5. *Let F be an algebraic surface in M such that all singularities of F are normal crossings, and let T be a monoidal transformation $M \to M'$ of M, centered at an irreducible non-singular curve Γ. In order that also the total T-transform F^* of F have only normal crossings it is necessary and sufficient that the following condition be satisfied at any point P of Γ: there exists at most one branch F_i of F at P such that $\Gamma \not\subset F_i$, and if such a branch F_i does* *exist, then P is a simple intersection of F_i and Γ. Furthermore, if the stated condition is satisfied and if each irreducible component of F is non-singular, then also each irreducible component of F^* is non-singular.*

Proof. We first prove the sufficiency of the condition stated in the lemma. Let F' denote the proper transform $T[F]$ of F. We have then $F^* = F'$ if $\Gamma \not\subset F$ and $F^* = F' \cup E'$ if $\Gamma \subset F$; here E' is the exceptional surface $T\{\Gamma\}$ in M'. We have only to show that *every point P' of the intersection $F' \cap E'$ is a normal crossing of F^**, since if P' is a point of F^* which does not belong to E', then F^* coincides, locally at P', with F', and if $P = T^{-1}(P')$, then T is biregular at P and the local ring of F at P is isomorphic to the local ring of F' at P'.

Let then P' be a point of $F' \cap E'$, and let $T^{-1}(P') = P$, where P is then a point of $\Gamma \cap F$. We consider three cases, according as P is a simple, a double or a triple point of F.

Case 1. *P is a simple point of F.* If $\Gamma \subset F$, then the assertion that P' is a normal crossing of $F' \cup E'$ follows directly from Lemma 1.2, part (b). If $\Gamma \not\subset F$, then, by assumption, P is a simple intersection of Γ and F, and the assertion that P' is a normal crossing of $F' \cup E'$ follows from Lemma 1.2, part (a).

Case 2. *P is a double point of F.* In this case, the assumption of our lemma is that at least one of the two branches F_1, F_2 of F at P must contain Γ. We have here two possible cases: either Γ is the double curve of F (whence $\Gamma = F_1 \cap F_2$, locally at P), or Γ belongs to only one of the two branches F_1, F_2, say $\Gamma \subset F_1$, and in that case the condition stated in the lemma says that P is a simple intersection of Γ and F_2. If Γ is a double curve of F, then the tangent planes of F_1 and F_2 are two distinct planes through the tangent line of Γ at P, and therefore the proper transforms $T[F_1]$ and $T[F_2]$ intersect the fiber γ'_P in distinct points P'_1, P'_2. By Lemma 1.2, part (b) (or, also, by the preceding case 1 of the proof), both points P'_1, P'_2 are normal crossings of $F' \cup E'$. Suppose now that $\Gamma \subset F_1$, $\Gamma \not\subset F_2$. In that case $T[F_1]$ meets the fiber γ'_P is a single point P'_1, while, by Lemma 1.2, part (a), $T[F_2]$ contains the entire fiber γ'_P. If P' is any point of γ'_P, different from P'_1, then P' is a double point of $F' \cup E'$, and is a normal crossing of this surface, by Lemma 1.2, part (a). As to P'_1, this point is a triple point of $F' \cup E'$. The surface $F' \cup E'$ coincides, locally at P'_1, with $T[F_1] \cup T[F_2] \cup E'$. The tangent planes of $T[F_1]$ and of $T[F_2]$ at P'_1 are distinct from the tangent plane of E' at P'_1, by Lemma 1.2, parts (a) and (b). If C denotes now the double curve of F, locally at P, then the tangent line of C at P is different from the tangent line of Γ at P (since $C \subset F_2$ and since Γ meets F_2 transversally at P). Hence if C' denotes the proper transform $T[C]$ of C, then C' is the intersection of $T[F_1]$ and $T[F_2]$, locally at P'_1, and by Lemma 1.2, part (c), P_1' is a simple intersection of E' and C'. This implies that the tangent planes of $T[F_1]$, $T[F_2]$ and E' at P_1' are independent, showing that P_1' is a normal crossing of $F' \cup E'$.

Case 3. *P is a triple point of F.* If F_1, F_2, F_3 are the three branches of F through P, then the condition of our lemma implies that Γ must be a double curve of F, say $\Gamma = F_1 \cap F_2$ and $\Gamma \not\subset F_3$. In that case, P is necessarily a simple intersection of Γ and F_3. The total transform $T[F_3]$ contains the entire fiber γ_P', while the total transforms $T[F_1]$ and $T[F_2]$ meet γ_P' in two distinct points P_1', P_2'. By case 1, every point P' of γ_P', different from P_1', P_2', is a normal crossing (a double point) of $F' \cup E'$, while by case 2 each of the points P_1', P_2' is also a normal crossing (a triple point) of $F' \cup E'$.

If each irreducible component of F is non-singular, then it follows, by applying case 1 of the proof of sufficiency of the condition to each irreducible component of F, that also each irreducible component of F^* is non-singular.

This completes the proof of the sufficiency of the condition.

We now prove the necessity of the condition stated in the lemma. Assume then that the condition is not satisfied. Then for some point P of Γ it is true that either Γ is tangent at P to one of the branches of F at P (and is not contained in that branch), or there exist at least two branches of F at P, say F_1 and F_2, such that P is a simple intersection of Γ with F_i $(i=1,2)$. In the first case it follows from Lemma 1.3 that the entire fiber is contained in F^* but that *no point of that fiber* is a normal crossing of F^*. Let us now consider the second case. Let C be the double curve $F_1 \cap F_2$ of F. It is clear that C and Γ have distinct tangents at P. Let P' be *the* point of the fiber γ_P' which corresponds to the tangent line of C at P [see Lemma 1.2, part (c)]. We know that the fiber γ_P' belongs to both surfaces $T[F_1]$ and $T[F_2]$ [Lemma 1.2, part (a)]. We shall show that these surfaces have the same tangent plane at P', and this will complete the proof of the necessity of the condition stated in the lemma.

We may assume, as usual, that $X = Z = 0$ are the local equations of Γ at P. If $0 = a_iX + b_iY + c_iZ +$ (terms of degree $\geqslant 2$) is the local equaton of F_i at P, then we have $b_i \neq 0$ $(i=1,2)$, since P is a simple intersection of F_i and Γ. Upon replacing $a_1X + b_1Y + c_1Z$ by Y, we may assume that $a_1 = c_1 = 0$ and $b_1 = 1$. Since F_1 and F_2 have distinct tangent planes at P, it follows now not only that we have $b_2 \neq 0$ but also that a_2 and c_2 are not both zero. Let, say, $c_2 \neq 0$. Upon replacing $a_2X + c_2Z$ by Z, we may now assume that $a_2 = 0$. Thus, the local equations of F_1 and F_2 at P are now respectively $0 = Y +$ (terms of degree $\geqslant 2$), $0 = b_2Y + Z +$ (terms of degree $\geqslant 2$). The tangent line of the double curve C is now therefore defined by the local equations $Y = Z = 0$. If we set $Z' = Z/X$, then X, Y, Z' will be zero at the point P' and will be regular local parameters of

M' at P'. The local equations of $T[F_1]$ and $T[F_2]$ at P' are respectively $0 = Y +$ (terms of degree $\geqslant 2$) in X, Y, Z', and $0 = b_2 Y + XZ' +$ (terms of degree $\geqslant$ 2) in X, Y, Z'. The plane $Y = 0$ is thus the tangent plane of both $T[F_1]$ and $T[F_2]$ at P. This completes the proof.

We shall conclude this section with one more lemma which we shall have occasion to use in the next section.

LEMMA 1.6. *Let T be a monoidal transformation centered at a point P, and let F be a surface having at P an n-fold point. Let $F' = T[F]$, and let E' be the exceptional surface $T\{P\}$ in M'. Then the following is true:*

(a) *If $E' \cap F'$ contains two points which are n-fold for F', then $E' \cap F'$ is the line joining these points in (the projective plane) E', and this line is an n-fold intersection of E' and F' (equivalently: the tangent cone of F at P is a plane counted n times).*

(b) *If an irreducible component Γ_0' of the intersection curve $E' \cap F'$ is n-fold for F', then Γ_0' is a line in E', and we have $E' \cap F' = \Gamma_0'$.*

(c) *If the n-fold point P of F is a quasi-ordinary isolated n-fold point of F, then $E' \cap F'$ contains at most two points which are n-fold for F', and any such point of F' is necessarily quasi-ordinary and isolated.*

Proof. (a) Let P' be a point of $E' \cap F'$ which is an n-fold for F'. We may choose the local regular parameters X, Y, Z of M at P in such a way that the corresponding homogeneous coordinates of P' in E' will be $1, 0, 0$. Hence, if we set $Y' = Y/X$, $Z' = Z/X$, then X, Y', Z' are local regular coordinates of M' at P'. The local equation of F' at P' is given by (2), and the fact that the origin $X = Y' = Z' = 0$ is n-fold for F' implies that $f_n(1, Y', Z')$ must be homogeneous, of degree n, in Y' and Z'. Since the algebraic curve $F' \cap E'$ is defined by the equations $X = 0$, $f_n(1, Y', Z') = 0$, it follows that this curve is a set of n lines through P. If, then, there is also a second n-fold point Q' of F' in E', then $F' \cap E'$ must be the line $P'Q'$, counted n times. This proves part (a) of the lemma.

(b) This is a direct consequence of (a).

(c) This assertion is identical with Proposition 2.6 of our Lincei note [5].

2. Proof of Main Theorem II′. We begin by restating the theorem.

MAIN THEOREM II′. *Let F and $\mathfrak{G}$ be two (reduced) surfaces in M, free from common irreducible components and satisfying the following two conditions:*

Condition 1. *All irreducible components of F and $\mathfrak{G}$ are non-singular.*

Condition 2. *If F^{μ} is any irreducible component of F, then the only singularities of the surface $F^{\mu} \cup \mathfrak{S}$ are normal crossings.*

Then there exists a sequence of monoidal transformations $T, T_1, \ldots, T_{N-1}$ which is $(F, \mathfrak{S})$-permissible, such that the proper transform F_N of F under the product of the transformations $T, T_1, \ldots, T_{N-1}$ is non-singular, while the total transform $F_N \cup \mathfrak{S}_N$ of $F \cup \mathfrak{S}$ under that product of these transformations has the property that its only singularities are normal crossings.

We define the integer $n = e(F)$:

$$n = e(F) = \max_{P \in F} \{e_P(F)\}, \tag{1}$$

where $e_P(F)$ denotes the multiplicity of the point P of F.

We observe at once that if $n = 1$, then already the given surface $F \cup \mathfrak{S}$ has only normal crossings (while F, of course, is non-singular). For if P is any point of $F \cup \mathfrak{S}$, we have two cases to consider: either $P \in F$ or $P \not\in F$. If $P \in F$, then P belongs to a uniquely determined irreducible component F^{μ} of F (since $n = 1$), and in that case the surface $F \cup \mathfrak{S}$ coincides, locally at P, with the surface $F^{\mu} \cup \mathfrak{S}$, and has therefore at P nothing worse than a normal crossing. If, however, $P \not\in F$, then $F \cup \mathfrak{S}$ coincides, locally at P, with the surface $\mathfrak{S}$, and the conclusion is the same. (Note that condition 2 implies that all the singularities of $\mathfrak{S}$ are also normal crossings.) We shall therefore assume from now on that $n > 1$.

We next prove that *the above conditions 1 and 2 are invariant under monoidal transformations T centered at points P of F*. More precisely, if we set

$$T[F] = F_1, T\{P\} = E_1, \mathfrak{S}_1 = E_1 \cup T[\mathfrak{S}], \tag{2}$$

then conditions 1 and 2 are satisfied, with F and $\mathfrak{S}$ replaced by F_1 and $\mathfrak{S}_1$ respectively. This is obvious as far as condition 1 is concerned. We now observe that if F_1^{μ} is any irreducible component of F_1, then F_1^{μ} is the proper T-transform of an irreducible component F^{μ} of F. If $P \in F^{\mu} \cup \mathfrak{S}$, then $F_1^{\mu} \cup \mathfrak{S}_1$ is the total T-transform of $F^{\mu} \cup \mathfrak{S}$. If $P \not\in F^{\mu} \cup \mathfrak{S}$, then $F_1^{\mu} \cup \mathfrak{S}_1 = T\{F^{\mu} \cup \mathfrak{S}\} \cup E_1$, and $T\{F^{\mu} \cup \mathfrak{S}\} \cap E_1 = \emptyset$. In either case our assertion that the only singularities of $F_1^{\mu} \cup \mathfrak{S}_1$ are normal crossings follows from Lemma 1.4.

Let $T, T_1, \ldots, T_{N-1}$ be a sequence of monoidal transformations:

$$T : M \to M_1, \tag{3_0}$$

$$T_i : M_i \to M_{i+1}, \qquad i = 1, 2, \ldots, N-1, \tag{3_i}$$

where the center Γ_i of T_i is either a point of M_i or an irreducible *non-singular* curve Γ_i in M_i (here $M_0 = M$, $T_0 = T$, and Γ_0 will be denoted by Γ). Let $E^1, E^2, \ldots, E^q$ be the irreducible components of $\mathcal{E}$:

$$\mathcal{E} = \bigcup_{\nu=1}^{q} E^{\nu}. \tag{4}$$

We define in M_i, by induction on i, the surface F_i, the $q+i$ irreducible surfaces E_i^j ($j = 1,2,\ldots,q+i$) and the surface $\mathcal{E}_i$ as follows:

$$F_i = T_{i-1}[F_{i-1}] \qquad (F_0 = F), \tag{5_i}$$

$$E_i^j = T_{i-1}[E_{i-1}^j], \quad j = 1,2,\ldots,q+i-1 \qquad (E_0^j = E^j), \tag{6_i}$$

$$E_i^{q+i} = T_{i-1}\{\Gamma_{i-1}\}, \tag{7_i}$$

$$\mathcal{E}_i = \bigcup_{j=1}^{q+i} E_i^j, \tag{8_i}$$

where $i = 1,2,\ldots,N$. It is clear that we have

$$\mathcal{E}_i = T_{i-1}[\mathcal{E}_{i-1}] \cup E_i^{q+i} \qquad (\mathcal{E}_0 = \mathcal{E}). \tag{9_i}$$

The irreducible component E_i^{q+i} of $\mathcal{E}_i$ is the exceptional surface in M_i into which the center Γ_{i-1} of T_{i-1} is blown up by T_{i-1}.

From Proposition 1.2 of our Lincei note [5] it follows that there exists an F-permissible sequence $T, T_1, \ldots, T_{N-1}$ of monoidal transformations such that each T_i is centered at an n-fold point of F_i and such that either $e(F_N) < n$ or $e(F_N) = n$ and all the exceptional n-fold points of F_N are quasi-ordinary. If $e(F_N) < n$, we have achieved a reduction of the numerical character $e(F)$ of F, while at the same time the new pair of surfaces $F_N, \mathcal{E}_N$ still satisfies conditions 1 and 2 (since all the T_i are centered at points). Therefore we may assume that $e(F_N) = n$, and thus—upon replacing F and $\mathcal{E}$ by F_N and $\mathcal{E}_N$—*we may assume from now on that our original surface F satisfies the condition that all its exceptional n-fold points are quasi-ordinary*. The validity of this condition is tacitly assumed in the remainder of this section. We remind the reader that also this condition is invariant under monoidal transformations centered at n-fold points of F and of the successive proper transforms F_i of F [as long as $e(F_i)$ remains equal to n].

PROPOSITION 2.1. *The set of n-fold points of F is of pure dimension* 1.

Proof. Let P be an n-fold point of F. If P is not an exceptional n-fold point of F, it belongs, by definition of exceptional singular points, to the n-fold curve of F, and is in fact a simple point of that curve (while F is equisingular at P along the n-fold curve of F). Assume now that P is exceptional. Since P is then a quasi-ordinary n-fold point of F, there exist regular local parameters X, Y, Z of M at P, such that the local equation of F at P has the following form:

$$F: \prod_{\alpha=1}^{n} \left[Z - X^{\lambda_1} Y^{\lambda_2} G_\alpha(X, Y) \right] = 0. \tag{10}$$

Furthermore, the F-residues x, y of X, Y are *transversal* parameters of F at P. Here, in general, λ_1, λ_2 are non-negative rational numbers, $\lambda_1 + \lambda_2 \geqslant 1$, while the G_α are fractional power series in X, Y. (However, in our present case, since all the irreducible components of F are non-singular, λ_1 and λ_2 are non-negative *integers*, while the G_α are integral power series in X and Y.) Furthermore, for each index α there exists an index β such that $G_\alpha(0,0) \neq G_\beta(0,0)$ (see our Lincei note [5], relations (7) and (8)). As in that note, so also here we denote by $\lambda(P)$ the rational number $\lambda_1 + \lambda_2$. The numerical character $\lambda(P)$ of P is a rational number (an integer, in our present case) which, when written in minimal terms, has a denominator less than or equal to n. By Corollary 2.5 of our note, P is an isolated n-fold point if and only if $\lambda_1 < 1$ and $\lambda_2 < 1$, and this is impossible in the present case, since λ_1 and λ_2 are non-negative *integers* and $\lambda_1 + \lambda_2 \geqslant 1$. This completes the proof of the proposition.

We shall denote by $S(F; n)$ the n-fold curve of F. By Proposition 2.2 of our Lincei note [5], the only possible singularities of the curve $S(F; n)$ are ordinary double points. Each such double point P is necessarily an exceptional n-fold point of F, and hence a quasi-ordinary singular point of F. If (10) is the local equation of F at P, then we must have necessarily $\lambda_1 \geqslant 1$ and $\lambda_2 \geqslant 1$, and the two branches of $S(F; n)$ at P are defined, locally at P, by the equations $X = Z = 0$ and $Y = Z = 0$. We show next that it is permissible to assume that every double point of $S(F; n)$ is an intersection of two irreducible components of the algebraic curve $S(F; n)$ *and that consequently each irreducible component of $S(F; n)$ is a curve free from singularities*. For, assume that P is a double point of an irreducible component Γ of $S(F; n)$. We apply to F the monoidal transformation T centered at P. The proper T-transform Γ' of Γ is then an n-fold curve of the proper T-transform F' of the surface F, and the ordinary double point P of Γ is then resolved into two simple points P_1', P_2' of Γ. However, the transformation T creates a new n-fold curve on F'; namely, if E' is the exceptional surface $T\{P\}$, then $F' \cap E'$ is the line $P_1' P_2'$ in E', and this line is n-fold for F'. To see this, we have only to show [see Lemma 1.6, part (b)] that $F' \cap E'$ contains an irreducible component Γ_0' which is n-fold for F'. If we set $Y' = Y/X$, $Z' = Z/X$, then the origin $P_1': X = Y' = Z' = 0$ is a point of F', and the local equation of F' at P' is given by

$$F': \prod_{\alpha=1}^{n} \left[Z' - X^{\lambda_1 + \lambda_2 - 1} Y'^{\lambda_2} G_\alpha (X, XY') \right] = 0. \tag{11}$$

The local equation of E' at P_1' is $X = 0$, and the line $X = Z' = 0$ is n-fold for F' (since $\lambda_1 + \lambda_2 - 1 \geqslant 1$); this proves our assertion.

It follows that by applying monoidal transformations centered at double points of irreducible components of the curve $S(F; n)$ we can reach a situation in which all the irreducible n-fold curves of the surface F are free from singularities. We assume from now on that this setup prevails already on our original surface F.

In the sequel we use the notation introduced in the Introduction [see especially (2_i), (6_i), (6_i^*), (7_i) and (12_i) of the Introduction].

LEMMA 2.2. *There exists a sequence of monoidal transformations* $T, T_1, \ldots, T_{N-1}$, *centered at n-fold points of F and of the successive proper transforms* $F_1, F_2, \ldots, F_{N-1}$ *of F, such that the last total transform* $F_N^* (= F_N \cup \mathfrak{S}_N^*)$ *of* $F^* (= F \cup \mathfrak{S})$ *satisfies the following condition*:

Condition 3. *If* E_N *is any irreducible component of* $\mathfrak{S}_N^*$ *and* Γ_N *is any irreducible n-fold curve of* F_N, *then either* $\Gamma_N \subset E_N$ *or all the intersections of* Γ_N *and* E_N *are simple.*

Proof. Assuming that the given surface $F \cup \mathfrak{S}$ does not satisfy the condition of the lemma, we denote by Σ the set of triads $\{E, \Gamma, P\}$, where E is an irreducible component of $\mathfrak{S}$, Γ is an irreducible n-fold curve of F such that $\Gamma \not\subset E$, and P is a common point of E and Γ. We denote by $m(F, \mathfrak{S})$ the maximum of the intersection multiplicities $i(E, \Gamma; P)$ as the triad $\{E, \Gamma, P\}$ ranges over the above set Σ, and by $l(F, \mathfrak{S})$ the number of triads $\{E, \Gamma, P\}$ in Σ such that $i(E, \Gamma;\ P) = m(F, \mathfrak{S})$. Since we have assumed that the surface $F \cup \mathfrak{S}$ does not satisfy the condition of the lemma, it follows that the set Σ is non-empty and that $m(F, \mathfrak{S}) \geqslant 2$. We fix a triad $\{\bar{E}, \bar{\Gamma}, P\}$ in Σ such that $i(\bar{E}, \bar{\Gamma};\ P) = m(F, \mathfrak{S})$, and we apply to $F \cup \mathfrak{S}$ the monoidal transformation T centered at P. Let $T[F] = F_1$ and $T\{F \cup \mathfrak{S}\} = F_1^* = F_1 \cup \mathfrak{S}_1^*$. Our lemma will be proved if we show that

$$m(F_1, \mathfrak{S}_1^*) \leqslant m(F, \mathfrak{S}), \tag{12}$$

and that

$$m(F_1, \mathfrak{S}_1^*) = m(F, \mathfrak{S}) \quad \text{implies} \quad l(F_1, \mathfrak{S}_1^*) \leqslant l(F, \mathfrak{S}) - 1. \tag{13}$$

Since we have assumed that F carries n-fold curves, it follows that $e(F_1) = n$. In the notation of (12_i) of the Introduction, we have $\mathfrak{S}_1^* = \cup_{j=1}^{q+1} E_1^j$, where E_1^{q+1} is the exceptional surface $T\{P\}$ in M_1. Let Σ_1 be the set of triads $\{E_1, \Gamma_1, P_1\}$ on $F_1 \cup \mathfrak{S}_1$ similar to the set Σ on $F \cup \mathfrak{S}$. The n-fold curve Γ_1 of F_1 is either the proper T-transform of an n-fold curve Γ of F or—possibly—a new n-fold curve of F_1, namely the intersection $F_1 \cap E_1^{q+1}$, and in the later case this intersection is a line in E_1^{q+1} [see Lemma 1.6, part (b)]. In the first case, we have to consider two subcases: (a) $E_1 \neq E_1^{q+1}$ and (b) $E_1 = E_1^{q+1}$. We set $T^{-1}\{P_1\} = Q$. In case (a) we have $i(E_1, \Gamma_1;\, P_1) = i(E, \Gamma;\, Q)$ if $P_1 \not\in E_1^{q+1}$ (equivalently: if $Q \neq P$), and $i(E_1, \Gamma_1;\, P_1) = i(E, \Gamma;\, Q) - 1$ if $P_1 \in E_1^{q+1}$ [see Lemma 1.1, part (b)]. In case (b) we have $i(E_1, \Gamma_{1,};\, P_1) = 1$, by Lemma 2.1, part (a)(since Γ is non-singular). In the case in which Γ_1 is the line $F_1 \cup E_1^{q+1}$, we have necessarily $E_1 \neq E_1^{q+1}$, $Q = P$. In that case we shall show now that $i(E_1, \Gamma_1;\, P_1) = 1$. The intersection of E_1 with E_1^{q+1} is a line p_1, and all points of that line are normal crossings of E_1 and E_1^{q+1} [Lemma 1.1, part (a)]. This line p_1 is distinct from the line Γ_1 (since $\Gamma_1 \not\subset E_1$). Therefore the point P_1 of p_1 must be a simple intersection of Γ_1 and E_1, as was asserted.

The above considerations prove the inequality (12). They also show that the triads $\{E_1, \Gamma_1, P_1\}$ in Σ_1 such that $i\{E_1, \Gamma_1;\, P_1\} = m(F, \mathfrak{S})$ correspond, in $(1,1)$ fashion, to the triads $\{E, \Gamma;\, Q\}$ in Σ such that $Q = T^{-1}(P_1) \neq P$ and $i(E, \Gamma;\, Q) = m(F, \mathfrak{S})$. This proves also the implication (13), and completes the proof of the lemma.

In view of Proposition 2.2 it will be sufficient to prove Main Theorem II′ under the additional assumption that the two surfaces F and $\mathfrak{S}$ satisfy also condition 3 of that proposition. *We assume from now on that condition 3 is satisfied.* It is a trivial matter to show that condition 3 is also invariant under monoidal transformations centered at n-fold points of F.

PROPOSITION 2.3. *There exists a sequence of monoidal transformations $T, T_1, \ldots, T_{N-1}$, centered at n-fold points of F and of the successive proper transforms $F_1, F_2, \ldots, F_{N-1}$ of F, such that the last total transform $F_N \cup \mathfrak{S}_N$ of $F \cup \mathfrak{S}$ satisfies the following condition*:

Condition 4. *If Γ' is any irreducible component of $S(F'; n)$ and P' is any point of Γ', then there exists at most one irreducible component E' of $\mathfrak{S}_N$ which contains P' and does not contain Γ'.*

Proof. Given a pair $\{\Gamma, P\}$, where Γ is an irreducible component of $S(F; n)$ and P is a point of Γ, we shall say that *the pair* $\{\Gamma, P\}$ *is* $\mathfrak{S}$*-permissible* if condition 4 is satisfied by that pair. We shall say that Γ is $\mathfrak{S}$-permissible if $\{\Gamma, P\}$ is $\mathfrak{S}$-permissible for any point P of Γ. Consider the set Σ of all pairs $\{\Gamma, P\}$ which are *not* $\mathfrak{S}$-permissible, and let $\rho = \rho(F, \mathfrak{S})$ be the number of such pairs. Assuming that the surface $F \cup \mathfrak{S}$ does not satisfy condition 4 of the proposition, we have $\rho \geqslant 1$. We fix a pair $\{\Gamma, P\}$ in the set Σ. Note that there cannot exist more than two distinct pairs $\{\Gamma^i, P^i\}$ in Σ such that $P^i = P$, since the curve $S(F; n)$ has only ordinary double points. We apply the monoidal transformation T centered at P and use the notation of (4) and (5_1)–(8_1). We also set $T[\Gamma^i] = \Gamma_1^i$. We know that the surface $F_1 \cup \mathfrak{S}_1$ still satisfies conditions 1, 2 and 3 (with F and $\mathfrak{S}$ replaced by F_1 and $\mathfrak{S}_1$). We shall show now that $\rho(F_1, \mathfrak{S}_1) \leqslant \rho(F, \mathfrak{S}) - 1$, and this will prove our proposition.

Consider any of the pairs $\{\Gamma^i, P^i\}$ in the set Σ such that $P^i = P$. The point P must then belong to at least two irreducible components of $\mathfrak{S}$, say E^1 and E^2, such that $\Gamma^i \not\subset E^\nu$, $\nu = 1, 2$. Since $P \in F$, and since $F \cup \mathfrak{S}$ has only normal crossings, P cannot be a triple point of $\mathfrak{S}$. Therefore E^1 and E^2 are the only irreducible components of $\mathfrak{S}$ which contain P. Since P is a simple intersection of Γ^i with both E^1 and E^2 (condition 3), none of the intersections of Γ_1^i with E_1^1 or E_1^2 belongs to E_1^{q+1}. This shows (taking into account the biregularity of T at all points of M different from P) that the pairs $\{\Gamma_1^i, Q_1^i\}$ which are not $\mathfrak{S}_1$-permissible and such that Γ_1^i is the proper T-transform of an irreducible component Γ^i of $S(F; n)$ are in $(1,1)$ correspondence with the pairs $\{\Gamma^i, Q^i\}$, $Q^i \neq P$, which are not $\mathfrak{S}$-permissible, the number of these latter pairs being at most $\rho - 1$. To complete the proof we have only to consider the case in which T introduces a new n-fold curve Γ_1^0 of $F_1 \cup \mathfrak{S}_1$ and prove that in that case the curve Γ_1^0 is necessarily $\mathfrak{S}_1$-permissible. Now, the curve Γ_1^0 is contained in the exceptional surface E_1^{q+1} and also in F_1, and hence also in at least one irreducible component F_1^μ of F_1. Since $F_1^\mu \cup \mathfrak{S}_1$ has only normal crossings, no point of Γ_1^0 can belong to more than two irreducible components of $\mathfrak{S}_1$. Since one of these components, namely E_1^{q+1}, contains Γ_1^0, it follows that Γ_1^0 is $\mathfrak{S}_1$-permissible, as asserted.

From now on we shall assume that the surface $F \cup \mathfrak{S}$ satisfies the four conditions 1–4 given above.

We shall now prove that also condition 4 is invariant under monoidal transformations T centered at points P of the curve $S(F; n)$. Let Γ_1 be any irreducible component of $S(F_1; n)$, let P_1 be any point of Γ_1, and let $Q = T^{-1}(P_1)$. We consider two cases:

Case 1. $\Gamma_1 = T[\Gamma]$, where Γ is an irreducible component of $S(F; n)$.

Case 2. Γ_1 is a new n-fold curve of F_1 created by T [whence Γ_1 is a line in the projective plane E_1^{q+1}; see Lemma 1.6, part (b)].

In both cases we have to prove that *the pair* $\{\Gamma_1, P_1\}$ *is E_1-permissible.*

Case 1. If $P_1 \notin E_1^{q+1}$, or—equivalently—if $Q \neq P$, then the assertion is trivial, in view of the biregularity of T at P. Assume then that $P_1 \in E_1^{q+1}$, whence $Q = P$. Clearly Γ_1 is not contained in E_1^{q+1}. Therefore we must show that if $P_1 \in E_1^{\nu}$, for some $\nu = 1, 2, \ldots, q$, then necessarily $\Gamma_1 \subset E_1^{\nu}$. Assume the contrary: $P_1 \in E_1^{\nu}$ and $\Gamma_1 \not\subset E_1^{\nu}$. Then $P \in E^{\nu}$ and $\Gamma \not\subset E^{\nu}$. Hence, by condition 3, P is a simple intersection of Γ and E^{ν}. Therefore the intersection P_1 of Γ_1 with E_1^{q+1} does not lie on the line $E_1^{\nu} \cap E_1^{q+1}$, i.e., $P_1 \notin E_1^{\nu}$—a contradiction.

Case 2. In this case we have $\Gamma_1 \subset E_1^{q+1}$. Therefore we have only to show that the point P_1 belongs to at most one of the remaining q irreducible components E_1^{ν} of $\mathcal{E}_1$. By Lemma 1.6, parts (a) and (b), the tangent cone of F at P is a plane counted n times. Since all irreducible components of F are non-singular, exactly n irreducible components F^{μ} of F contain the point P, and they must have the same tangent plane π at P. Therefore, by condition 2, any irreducible component E^{ν} of $\mathcal{E}$ which passes through P must have at P a tangent plane which is different from π. Since $F \cup \mathcal{E}$ satisfies condition 4, there is *at most* one irreducible component E^{ν} of $\mathcal{E}$ which contains the point P and does not contain Γ. If $P \in E^{\nu}$ and $E^{\nu} \supset \Gamma$, then the tangent plane of E^{ν} at P is a plane which contains the tangent line of Γ at P but is different from the plane π. Therefore the point in which E_1^{ν} meets the fiber γ_p' (in E_1^{q+1}) is different from the point P_1 (since $P_1 \in \Gamma_1$ and since therefore P_1 is the point of γ_p' which corresponds to the plane π). This proves our assertion that P_1 belongs to at most one of the irreducible components $E_1^1, E_1^2, \ldots, E_1^q$ of $\mathcal{E}_1$ other than E_1^{q+1} (here we recall that $E_1^{\nu} = T[E^{\nu}]$, $\nu = 1, 2, \ldots, q$.)

There is one more condition which we wish to impose on the surface $F \cup \mathcal{E}$, and that is the following:

Condition 5.   *If P is a double point of the n-fold curve $S(F; n)$ of F and if Γ^1, Γ^2 are the two irreducible components of $S(F; n)$ which contain P, then any irreducible component of $\mathcal{E}$ which contains P contains at least one of the two curves Γ^1, Γ^2.*

The next lemma will show that the monoidal transformations centered at the double points of the n-fold curve $S(F; n)$ of F lead to a total transform $F_N \cup \mathcal{E}_N$ of $F \cup \mathcal{E}$ (where F_N is the proper transform of F) which satisfies condition 5 (with $F, \mathcal{E}$ replaced by $F_N, \mathcal{E}_N$).

LEMMA 2.4. *Let P be any point of $S(F;\, n)$, and let T be the monoidal transformation centered at P. If P_1 is a point of the exceptional surface E_1^{q+1} such that P_1 is a double point of the curve $S(F_1;\, n)$, then condition 5 is satisfied when F, $\mathfrak{S}$ and P are replaced by F_1, $\mathfrak{S}_1$ and P_1.*

Proof. Let Γ_1^1, Γ_1^2 be two irreducible components of $S(F_1;n)$ which contain the point P_1. We can conclude at once the case in which both Γ_1^1 and Γ_1^2 are the proper transforms of two irreducible components Γ^1 and Γ^2 of $S(F;\, n)$. For in that case, both Γ^1 and Γ^2 contain the point P, and since they have distinct tangents at P, their proper T-transforms Γ_1^1, Γ_1^2 cannot intersect E_1^{q+1} in one and the same point P_1. Thus, we must assume that one of the two curves Γ_1^1, Γ_1^2, say Γ_1^2, is a new n-fold curve of F_1 created by T, and is therefore a line in E_1^{q+1}. Consider any of the remaining q irreducible components E_1^ν of $\mathfrak{S}_1$. Suppose that for a given $\nu = 1,2,\ldots,q$ we have $P_1 \in E_1^\nu$. Then $P \in E^\nu$. If $E_1^\nu \not\supset \Gamma_1^1$, then $E^\nu \not\supset \Gamma^1$, where $\Gamma_1^1 = T[\Gamma^1]$. By condition 3, P is then a simple intersection of E^ν and Γ^1. This implies that the point $\Gamma_1^1 \cap E_1^{q+1}$ does not belong to the line $E_1^\nu \cap E_1^{q+1}$—a contradiction, since $P_1 \in \Gamma_1^1 \cap E_1^\nu \cap E_1^{q+1}$. This completes the proof of Lemma 2.4.

We observe that Lemma 2.4 implies that also condition 5 is invariant under monoidal transformations centered at points of $S(F;\, n)$.

For the convenience of the reader we now summarize the results obtained so far in this section.

In proving Main Theorem II′ we may assume that the surface $F \cup \mathfrak{S}$ satisfies the following 5 conditions (the first two being conditions 1 and 2 assumed in Main Theorem II′):

Condition 1. *All the irreducible components of F and $\mathfrak{S}$ are non-singular.*

Condition 2. *If F^μ is any irreducible component of F, then all the singularities of the surface $F^\mu \cup \mathfrak{S}$ are normal crossings.*

Condition 3. *If Γ is any irreducible component of the n-fold curve $S(F;\, n)$ of F, and E is any irreducible component of $\mathfrak{S}$ which does not contain Γ, then all the intersections of Γ and E are simple.*

Condition 4. *If Γ is as in Condition* 3, *then Γ is $\mathfrak{S}$-permissible.*

Condition 5. *If P is a common point of two irreducible components Γ^1, Γ^2 of the curve $S(F;\, n)$, then any irreducible component of $\mathfrak{S}$ which contains the point P contains at least one of the two curves Γ^1, Γ^2.*

Furthermore, the following condition is satisfied:

Condition 1′. *All the points of the curve* $S(F;\, n)$ *which are exceptional n-fold points of F are quasi-ordinary. Furthermore, the set of n-fold points of F is of pure dimension* 1 (*it is the curve* $S(F;\, n)$), *all its irreducible components are non-singular, and the only singularities of that curve are ordinary double points.*

Finally, we recall that we have proved that conditions 1–5 and 1′ are invariant under monoidal transformations centered at points of $S(F;\, n)$.

For every irreducible n-fold curve Γ of F, we denote by $\lambda(\Gamma)$ the number of successive monoidal transformations, centered at Γ and at the successive proper transforms of Γ on the proper transforms of F, which are needed in order to reduce the multiplicity n of curve Γ. If the local equation of F at a point P of Γ is given by (10), and Γ is the curve $X=Z=0$ (whence necessarily $\lambda_1 \geqslant 1$), then $\lambda(\Gamma)$ is the integral part $[\lambda_1]$ of λ_1. Since in our present case all the irreducible components of F are non-singular, λ_1 is an integer, and $\lambda(\Gamma)=\lambda_1$. We set

$$\lambda(F) = \max_{\Gamma \subset S(F;\, n)} (\lambda(\Gamma)), \tag{14}$$

and we denote by $l(F)$ the number of irreducible n-fold curves Γ of F such that $\lambda(\Gamma)=\lambda(F)$.

We now come to the last step of the proof of Main Theorem II′. This step is represented by the next proposition, of which Main Theorem II′ will be an immediate consequence.

Proposition 2.5. *Let* Γ *be an irreducible component of* $S(F;\, n)$ *such that* $\lambda(\Gamma)=\lambda(F)$, *and let* T *be the monoidal transformation centered at* Γ. *The notation being the same as in* (5_1)–(8_1) (*except that this time the exceptional surface* E_1^{q+1} *in* M_1 *is the total T-transform* $T\{\Gamma\}$ *of the curve* Γ), *we have* $e(F_1) \leqslant n$, *and the surface* $F_1 \cup \mathfrak{S}_1$ *satisfies conditions* 1 *and* 2 (*with* $F, \mathfrak{S}$ *being replaced respectively by* $F_1, \mathfrak{S}_1$). *Furthermore, if* $e(F_1)=n$, *then the surface* $F_1 \cup \mathfrak{S}_1$ *satisfies also conditions* 3, 4, 5 *and condition* 1′, *and we have either* $\lambda(F_1)<\lambda(F)$ *or* $\lambda(F_1)=\lambda(F)$ *and* $l(F_1)=l(F)-1$.

Proof. The last part of the proposition is obvious. Also the inequality $e(F_1) \leqslant n$ is self-evident. Since every irreducible component of F either contains Γ or has no intersections with Γ, while, by condition 3, every irreducible component of $\mathfrak{S}$ either contains Γ or has only simple intersections with Γ, it follows from Lemma 1.2, parts (a) and (b), that condition 1 is satisfied, with $F, \mathfrak{E}$ replaced respectively by $F_1, \mathfrak{E}_1$. That $F_1, \mathfrak{E}_1$ also satisfy the condition similar to condition 2 follows from Lemma 1.5 and from the fact that $F, \mathfrak{E}$ satisfy conditions 2, 3 and 4.

Assume now that $e(F_1, \mathfrak{E}_1) = n$ $[= e(F, \mathfrak{E})]$. It follows from Proposition 2.2 of our Lincei note [5] that the pair of surfaces $F_1, \mathfrak{E}_1$ satisfies condition 1′, with $F, \mathfrak{E}$ replaced by $F_1, \mathfrak{E}_1$. It is only necessary to take into account the fact that if Γ^1 is any irreducible component of $S(F; n)$ which is different from Γ, then every common point of Γ and Γ^1 is an ordinary double point of the curve $\Gamma \cup \Gamma^1$ and that therefore $T[\Gamma^1]$ is still non-singular. Another fact that has to be taken into account is that if $E_1^{q+1} \cap F_1$ contains an n-fold curve Γ_1 of F_1, then $E_1^{q+1} \cap F_1 = \Gamma_1$ and the two curves Γ and Γ_1 are biregularly equivalent. This follows from Lemma 1.2, part (b), by observing that exactly n irreducible components of F contain Γ and that the assumption that $F_1 \cap E_1^{q+1}$ contains an n-fold curve of F_1 signifies that these n irreducible components of F have the same tangent plane at each point of Γ and that therefore their proper T-transforms intersect E_1^{q+1} along the curve Γ_1, this curve being then a transversal unisecant of the family of fibers γ'_P on E_1^{q+1} (P varies on Γ).

We now prove that the surface $F_1 \cup \mathfrak{E}_1$ satisfies conditions 3, 4, 5.

We begin with the proof of condition 3. Let Γ_1 be an irreducible component of $S(F_1; n)$, and let E_1 be any irreducible component of $\mathfrak{E}_1$ which does not contain Γ_1. We have to show that all the intersections of Γ_1 and E_1 are simple.

We first consider the case in which Γ_1 is the proper transform Γ_1^1 of an irreducible component Γ^1 of $S(F; n)$, where $\Gamma^1 \neq \Gamma$. In that case the exceptional surface E_1^{q+1} certainly does not contain Γ_1^1, and if we apply part (c) of Lemma 1.2, with Γ^1 playing the role of the curve C of that lemma, we conclude that all the intersections of E_1^{q+1} and Γ_1^1 are simple (since at each of their common points the curves Γ and Γ^1 have distinct tangents). Consider now any of remaining q irreducible components E_1^ν of $\mathfrak{E}_1$ [see (8_1)] such that $E_1^\nu \not\supset \Gamma_1^1 (= \Gamma_1)$. Then $E^\nu \not\supset \Gamma^1$, and, by condition 3, all intersections of Γ^1 and E^ν are simple. It follows then from Lemma 1.3, part (b), that also all the intersections of E_1^ν and Γ_1^1 $(= \Gamma_1)$ are simple.

We next consider the case in which $T^{-1}\{\Gamma_1\} = \Gamma$. In that case Γ_1 is a non-singular transversal unisecant of the family of fibers γ'_P in E_1^{q+1}. We have $\Gamma_1 \subset E_1^{q+1}$, and hence we have only to prove that if $\Gamma_1 \not\subset E_1^\nu$ where $1 \leqslant \nu \leqslant q$, then all the intersections of Γ_1 and E_1^ν are simple. We consider separately the two cases: (1) $E^\nu \supset \Gamma$ and (2) $E^\nu \not\supset \Gamma$. If $E^\nu \supset \Gamma$, then at each point P of Γ the tangent plane of E^ν at P and the (unique, n-fold) tangent plane of F at P both contain the tangent line of Γ at P, and are *distinct* planes (in view of condition 2 of Main Theorem II'). Hence E_1^ν and Γ^1 meet the fibre γ_P^1 in *distinct* points. Since this is so for *any* point P of Γ, it follows that in this case E_1^ν *and* Γ^1 *have no common points*. If, however, $E^\nu \not\supset \Gamma$, then our assertion that all the intersections of E_1^ν and Γ^1 are simple follows from part (c) of Lemma 1.3, with the surface E^ν playing the role of the surface F of that lemma, since all the intersections of Γ and E^ν are simple (condition 3).

This completes the proof as far as condition 3 is concerned.

We now give the proof of the validity of condition 4 for the surface $F_1 \cup \mathfrak{E}_1$. Again we first consider the case $\Gamma_1 = \Gamma_1^1 = T[\Gamma^1]$, where Γ^1 is an irreducible component of $S(F; n)$, different from Γ. Let P_1 be a point of Γ_1^1. We have to show that there is at most only one irreducible component of $\mathfrak{E}_1$ which contains P_1 and does not contain Γ_1^1. The case in which $P_1 \notin E_1^{q+1}$ being trivial (in view of the biregularity of T^{-1} at P_1 and in view of the validity of condition 4 for the surface $F \cup \mathfrak{E}$), we assume that $P_1 \in E_1^{q+1}$, whence $P = T^{-1}(P_1) \in \Gamma \cap \Gamma^1$. Since $P_1 \in E_1^{q+1} \not\supset \Gamma_1^1$, we must show that if any of the remaining q irreducible components E_1^ν of $\mathfrak{E}_1$ contains the point P_1 then it must contain Γ_1^1. *We shall show that if* $E_1^\nu \not\supset \Gamma_1^1$, *then* $P_1 \notin E_1^\nu$. Since $E_1^\nu \not\supset \Gamma_1^1$, it follows that $E^\nu \not\supset \Gamma^1$. Since P is a common point of Γ and Γ^1, it follows from condition 5 that $E^\nu \supset \Gamma$. Thus the tangent plane of E^ν at P contains the tangent line of Γ at P. Since $\Gamma^1 \not\subset E^\nu$, P is a simple intersection of Γ^1 and E^ν. Therefore, the plane determined by the (distinct) tangent lines of Γ and Γ^1 at P is different from the tangent plane of E^ν at P. This shows that E_1^ν and Γ_1^1 intersects the fiber γ_P' in distinct points, and that therefore $P_1 \notin E_1^\nu$, as was asserted.

We now consider the case in which Γ_1 corresponds to Γ itself, i.e., $\Gamma = T^{-1}(\Gamma_1)$[this case can only arise if $\lambda(\Gamma) \geqslant 2$]. In this case we have $\Gamma_1 = F_1 \cap E_1^{q+1}$. By Lemma 1.2, part (b), Γ_1 is a transversal unisecant of the family of fibers γ_P' in E_1^{q+1} (here P ranges over Γ). Let P_1 be any point of Γ_1, and let $P = T^{-1}(P_1)$ (whence $P \in \Gamma$). Since in the present case we have $\Gamma_1 \subset E_1^{q+1}$, we have only to prove that there is at most one irreducible component E_1^ν of $\mathfrak{E}_1$ $(\nu = 1, 2, \ldots, q)$ such that $P_1 \in E_1^\nu$ and $\Gamma \not\subset E_1^\nu$. We shall show even more: namely, that at most one of the q surfaces E_1^ν can contain P_1. Now, if $\Gamma \subset E^\nu$, then, by condition 2, the (n-fold) tangent planes of F at P and the tangent plane of E^ν at P are *distinct* planes, both containing the tangent line of Γ at P, and therefore F_1 and E_1^ν meet the fiber γ_P' in two distinct points. It follows that $P_1 \notin E_1^\nu$ for all $\nu = 1, 2, \ldots, q$ such that $\Gamma \subset E^\nu$. Since by condition 4 there is at most only one irreducible component E^ν of $\mathfrak{E}$ which contains P and does not contain Γ, our assertion is proved (what we have shown in fact is that if $\Gamma \subset E^\nu$, then E_1^ν does not meet Γ_1).

This completes the proof of the validity of condition 4 for the surface $F_1 \cup \mathcal{E}_1$.

We shall now prove that the surface $F_1 \cup \mathcal{E}_1$ satisfies also the analog of condition 5, and this will complete the proof of the proposition.

Let P_1 be a common point of two irreducible components Γ_1^1, Γ_1^2 of $S(F_1; n)$, and let E_1^ν be an irreducible component of $\mathcal{E}_1$ which contains P_1 ($\nu = 1, \ldots, q, q+1$; $E_1^{q+1} = T\{\Gamma\}$). We have to show that

$$P_1 \in E_1^\nu \quad \Rightarrow \quad E_1^\nu \supset \Gamma_1^1 \text{ or } E_1^\nu \supset \Gamma_1^2. \tag{15}$$

The case in which $\Gamma_1^i = T[\Gamma^i]$, $\Gamma^i \neq \Gamma$ ($i \neq 1, 2$) is trivial, for in that case the point $P = T^{-1}(P_1)$ is a common point of Γ^1 and Γ^2 and therefore does not belong to Γ [since $S(F; n)$ has only double points]. Thus T is then biregular at P, and (15) follows in this case from the validity of condition 5 for the surface $F \cup \mathcal{E}$. So we have only to consider the case in which one of the two curves Γ_1^1, Γ_1^2 corresponds to the curve Γ itself. Let, say, $\Gamma_1^2 = \Gamma_1$, with $T^{-1}(\Gamma_1) = \Gamma$. We now have $E_1^{q+1} \supset \Gamma_1$. Therefore we have only to prove (15) in the case in which $1 \leqslant \nu \leqslant q$. Now, if $E^\nu \supset \Gamma^1$, then $E_1^\nu \supset \Gamma_1^1$. Suppose therefore that $P \in E^\nu$ but $\Gamma^1 \not\subset E^\nu$. We shall show that this assumption is in contradiction with the assumption $P_1 \in \Gamma_1^2$. Since P is a common point of Γ and Γ^1, it follows from condition 5 that necessarily $\Gamma \subset E^\nu$. Thus the point P_1 is the point of the fiber γ_P' which corresponds to the tangent plane of E^ν at P (since $\Gamma \subset E^\nu$, this plane contains the tangent line of Γ at P). Since Γ^1 is transversal to E^ν at P (by condition 3, since we have assumed that $\Gamma^1 \not\subset E^\nu$), it follows that P_1 cannot belong to Γ_1^2, and this is the desired contradiction. This completes the proof of the proposition.

If $e(F_1, \mathcal{E}_1) = n$, we apply to $F_1 \cup \mathcal{E}_1$ a monoidal transformation T_1 centered at an irreducible component Γ_1 of $S(F_1; n)$, and so we continue. The last part of Proposition 2.5 guarantees that after a finite number of such successive transformations T_i we will get a surface $F_N \cup \mathcal{E}_N$ such that $e(F_N) < n$. Since, by Proposition 2.5, all the intermediate surfaces $F_i \cup \mathcal{E}_i$ ($0 \leqslant i \leqslant N-1$) satisfy conditions 3 and 4, and since all the transformations used in order to secure the validity of conditions 3, 4, and 5 were monoidal transformations centered at points of our original surface F and of the successive proper transforms of F, it follows that the combined sequence of monoidal transformations is F-permissible. It is also $(F, \mathcal{E})$-permissible, since all the intermediate surfaces $F_i \cup \mathcal{E}_i$ ($0 \leqslant i \leqslant N-1$) satisfy conditions 3 and 4. This completes the proof of Main Theorem II'.

HARVARD UNIVERSITY

REFERENCES.

[1] H. Hironaka, "Resolution of singularities of an algebraic variety over a field of characteristic zero: I-II," *Ann. of Math.* **79**, No. 2 (1964), pp. 109–326.

[2] O. Zariski, "Foundations of a general theory of birational correspondences," *Trans. Amer. Math. Soc.* **53** (1943), pp. 490–542.

[3] ———, Introduction to the problem of minimal models in the theory of algebraic surfaces, *Publ. Math. Soc. Japan*, No. 4 (1958), pp. 1–89.

[4] ———, "Studies in equisingularity II. Equisingularity in co-dimension 1 (and characteristic zero)," *Amer. Journal Math.* **87** (1965), pp. 972–1006.

[5] ———, "Exceptional singularities of an algebraic surface and their reduction," *Atti Accad. Naz. Lincei Rend. Cl. Sci. Fis. Mat. Natur.*, serie VIII, **43**, fasc. 3–4 (Settembre–Ottobre 1967), pp. 135–196.

FOUNDATIONS OF A GENERAL THEORY OF EQUISINGULARITY ON r-DIMENSIONAL ALGEBROID AND ALGEBRAIC VARIETIES, OF EMBEDDING DIMENSION r + 1.*

By Oscar Zariski

Introduction. We develop in this paper a general (local) theory of equisingularity on algebroid hypersurfaces V, and a general (global) theory of equisingularity on r-dimensional algebraic varieties V, defined over an algebraically closed ground field k (of characteristic zero) and having local embedding dimension $r + 1$ at each of their singular k-rational points. In preceding papers, especially in [1], we dealt only with equisingularity of an algebraic variety V, of dimension r, along singular irreducible subvarieties W of V, of codimension 1. In the present paper we deal with singular subvarieties of any codimension s $(1 \leq s \leq r)$.

In the absence of any general theory of equivalence of singularities of varieties of dimension >1, any definition of equisingularity of V along singular subvarieties W of codimension >1 is bound to be to some extent a matter of individual choice and perception. We cannot, nor do we intend to, claim that our definition is a definitive one. We can say, nevertheless, that our definition is not trivially too strong, as for instance one based on analytical equivalence would be, nor is it trivially too weak as a definition based on equimultiplicity would be. A definitely positive aspect of our theory is that it leads to a substantial collection of results which are both non-trivial and geometrically sound, including a complete theory of stratification of our varieties V into equisingularity strata.

Since one of the principal conditions which we wanted our concept of equisingularity to satisfy is that it be analytically intrinsic (and not only biregularily invariant), we decided to develop first a *local* theory of equisingularity of *algebroid* hypersurfaces. The greatest part of our paper (five out of a total of six sections) is dedicated to this local theory

---

*This work was supported by a National Science Foundation Grant.

[Reprinted from *American Journal of Mathematics*, Vol. 101, No. 2, pp. 453-514.]

After this local theory is developed, it is applied in section 6 to the "global" theory of equisingularity of algebraic varieties, including the stratification of these varieties into equisingularity strata. Such a stratification is also developed for algebroid varieties (in section 5), but in that case the strata are easier to introduce and have a more formal character. Nevertheless, some of the properties of the formal strata in the algebroid case are essential for the theory of stratification of algebraic varieties [see, for instance, the use we make, in section 6, of part (d) of Theorem 5.1].

Our theory of equisingularity is based on the concept of the *dimensionality type* s of a singular point of V; here $1 \leq s \leq r$ (but *simple* points has dimensionality type zero). Our singularity strata are maximal open irreducible subvarieties S of V all points of which have the same dimensionality type s, and the closure of S (i.e., the smallest algebraic subvariety $\bar{S}$ of V containing S) is an irreducible variety of dimension $r - s$, while S is a Zariski dense open subset of $\bar{S}$ (i.e., $\bar{S} - S$ is a proper subvariety of $\bar{S}$). An informal preparatory discussion of the concept of dimensionality type and of the nature of the strata S is to be found in section 3. We advise the reader to read section 3 before reading the more technical sections 1 and 2.

We call attention of the reader to the fact that one crucial result in section 6, namely the proof that each equisingularity stratum S is a Zariski dense open subset of its algebraic closure $\bar{S}$ has been proved by H. Hironaka in his paper [5] published in the present issue of the American Journal of Mathematics.

1. Algebroid hypersurfaces, their projections and critical loci. Let V be a formal (reduced) algebroid variety, of pure dimension r, defined over a field k of characteristic zero. Let P be the center (i.e., the closed point) of V and let $\mathfrak{o}$ be the (complete) local ring of V at P. Thus $\mathfrak{o}$ is a noetherian local ring, of dimension r, free from nilpotent elements different from zero, and having k as coefficient field. We assume that *the embedding dimension* of V is $r + 1$, i.e., that the maximal ideal $\mathfrak{m}$ of $\mathfrak{o}$ has a minimal basis of $r + 1$ elements (equivalently: the k-vector space $\mathfrak{m}/\mathfrak{m}^2$ has dimension $r + 1$). Thus V can be represented (up to an arbitrary analytical isomorphism over k) as a hypersurface in an $(r + 1)$-dimensional affine $(r + 1)$-space $\mathbf{A}_{r+1}/k$, and P is necessarily a *singular* point of V. We have thus excluded the case in which $\mathfrak{o}$ is a regular local ring, in which case every minimal basis of $\mathfrak{m}$ would consist of r elements and P would be a simple point of V.

The variety V is to be thought of as being the set *Spec* ($\mathfrak{o}$), i.e., the set of all prime ideals $\mathfrak{p}$ of $\mathfrak{o}$. We shall refer to any such ideal $\mathfrak{p}$ as a *point of* V. In particular, the closed P of V is the maximal ideal $\mathfrak{m}$ of $\mathfrak{o}$.

Given two points R_1 and R_2 of V, corresponding to two prime ideals $\mathfrak{p}_1$ and $\mathfrak{p}_2$ of $\mathfrak{o}$, we say that R_2 is an *analytic specialization of* R_1 if $\mathfrak{p}_2 \supset \mathfrak{p}_1$. We shall usually omit the word "analytic" and will say simply that R_2 is a specialization R_1 if $\mathfrak{p}_2 \supset \mathfrak{p}_1$. In particular, the closed point P of V is a specialization of any point R of V.

The set of all points which are specializations of a given point R of V shall be referred to as the *irreducible* (algebroid) *subvariety W of V having R as general point.* This algebroid variety W can be canonically identified with the algebroid (irreducible) variety Spec ($\mathfrak{o}/\mathfrak{p}$), where $\mathfrak{p}$ is the prime ideal associated with the point R, the local ring $\mathfrak{o}/\mathfrak{p}$ being a complete local *domain*, having k as coefficient field. The customary notation for this variety W is $V(\mathfrak{p})$ (*the variety of* $\mathfrak{p}$). More generally, if $\mathfrak{A}$ is any ideal in $\mathfrak{o}$ then the set of prime ideals $\mathfrak{p}$ in $\mathfrak{o}$ which contain $\mathfrak{A}$ is the variety $V(\mathfrak{A})$ of the ideal $\mathfrak{A}$. We have $V(\mathfrak{A}) = V(\mathit{Rad}\ \mathfrak{A})$, and $V(\mathit{Rad}\ \mathfrak{A})$ is canonically identifiable with the algebroid variety Spec ($\mathfrak{o}/\mathit{Rad}\ \mathfrak{A}$), which is a reduced variety since the local ring $\mathfrak{o}/\mathit{Rad}\ \mathfrak{A}$ is free from non-zero nilpotent elements. The ideal $\mathit{Rad}\ \mathfrak{A}$ is an irredundant (finite) intersection of prime ideals in $\mathfrak{o}$, and the associated varieties of these prime ideals are the irreducible components of $V(\mathit{Rad}\ \mathfrak{A})$. In particular, the zero ideal (0) in $\mathfrak{o}$ coincides with its own radical (since V is reduced), and if

$$(0) = \mathfrak{p}_1 \cap \mathfrak{p}_2 \cap \cdots \cap \mathfrak{p}_q \quad (\mathfrak{p}_i \not\supset \mathfrak{p}_j \quad \text{if} \quad i \neq j) \tag{1}$$

is the irredundant representation of (0) as an intersection of prime ideals, then the q irreducible varieties $V_i = V(\mathfrak{p}_i)$ are the irreducible components of V, and we have

$$V = V_1 \cup V_2 \cup \cdots \cup V_q. \tag{2}$$

Our assumption that V is of pure dimension r signifies that each V_i has dimension r, or—equivalently—that each of the q prime ideals $\mathfrak{p}_i$ in (1) has dimension r (or—what is the same—depth r). These prime ideals are, of course, all of height zero, i.e., they are the minimal prime ideals of $\mathfrak{o}$.

The (well known) terms and concepts introduced above are independent of the choice of the coefficient field k of $\mathfrak{o}$. From now on, however, the chosen coefficient field k will begin to play an essential role. In fact, the entire theory of equisingularity which we develop in this paper depends *a priori* on the choice of the coefficient field k of $\mathfrak{o}$. It is a basic open question whether our equisingularity concept is or is not actually independent of the choice of the coefficient field k of our complete local ring $\mathfrak{o}$.

Given a minimal basis $\{x_1, x_2, \ldots, x_{r+1}\}$ of the maximal ideal $\mathfrak{m}$ of $\mathfrak{o}$, it is well known that $\mathfrak{o}$ is a homomorphic image of the power series ring $\mathfrak{O} = k[[X_1, X_2, \ldots, X_{r+1}]]$ in $r + 1$ independent variables over k and that the kernel of this homomorphism is a principal ideal generated by a power series $f(X)(=f(X_1, X_2, \ldots, X_{r+1}))$ which is free from multiple factors (since V is reduced) and which is uniquely determined by the basis $\{x_1, x_2, \ldots, x_{r+1}\}$ of $\mathfrak{m}$ and by the coefficient field k up to an arbitrary unit factor in $\mathfrak{O}$. The elements $x_1, x_2, \ldots, x_{r+1}$ are the $f(X)$-residues of $X_1, X_2, \ldots, X_{r+1}$. If $f^{(1)}(X), f^{(2)}(X), \ldots, f^{(q)}(X)$ are the irreducible factors of $f(X)$ then the q prime ideals $\mathfrak{p}_i$ in (1) are the principal ideals $\mathfrak{o}\cdot f^{(i)}(x)$. We refer to the equation $f(X_1, X_2, \ldots, X_{r+1}) = 0$ as the equation of V (relative to the given minimal basis $\{x_1, x_2, \ldots, x_{r+1}\}$ of $\mathfrak{m}$ and the given coefficient field k):

$$V: f(X_1, X_2, \ldots, X_{r+1}) = 0. \tag{3}$$

The equation of the irreducible component V_i of V is then $f^{(i)}(X_1, X_2, \ldots, X_{r+1}) = 0$, relative to the minimal basis $[x_1^{(i)}, x_2^{(i)}, \ldots, x_{r+1}^{(i)}\}$ of the maximal ideal $\mathfrak{m}/\mathfrak{p}_i$ of the local domain $\mathfrak{o}/\mathfrak{p}_i$; here $x_j^{(i)}$ is the $\mathfrak{p}_i$-residue of x_j, or—what is the same thing—the $f_i(X)$-residue of X_j.

A set of r elements $x_1, x_2, \ldots, x_r$ of $\mathfrak{m}$ constitute a set of *parameters* of $\mathfrak{o}$ if these elements generate in $\mathfrak{o}$ an ideal which is primary for $\mathfrak{m}$. *We shall deal from now on only with such sets of elements* $x_1, x_2, \ldots, x_r$ of $\mathfrak{m}$ *that are parameters and that can be extended to a minimal basis* $\{x_1, x_2, \ldots, x_r, x_{r+1}\}$ *of* $\mathfrak{m}$. If $\{x_1, x_2, \ldots, x_r, x_{r+1}\}$ is a minimal basis of $\mathfrak{m}$ and if (3) is the equation of V relative to that basis, then $x_1, x_2, \ldots, x_r$ are parameters of $\mathfrak{o}$ if and only if $f(0, 0, \ldots, 0, X_{r+1})$ is not identically zero (equivalently: if the line $X_1 = X_2 = \cdots = X_r = 0$ does not belong to V).

If $f_n(X_1, X_2, \ldots, X_{r+1})$ is the leading form of the power series $f(X)$ in (3), then $n > 1$ (since P is a singular point of V), and this integer n is independent of the choice of the coefficient field k and of the basis $\{x_1, x_2, \ldots, x_{r+1}\}$ of $\mathfrak{m}$. The integer n is the *multiplicity* of the local ring $\mathfrak{o}$ and is also the multiplicity of the singular point P of V. The r elements $x_1, x_2, \ldots, x_r$ of the minimal basis $\{x_1, x_2, \ldots, x_r, x_{r+1}\}$ are said to be *transversal* parameters if the leading term of the non-zero power series $f(0, 0, \ldots, 0, X_{r+1})$ is of degree n, or—equivalently—if the monomial $X_{r+1}{}^n$ actually occurs in the leading form $f_n(X_1, X_2, \ldots, X_r, X_{r+1})$ of $f(X)$, i.e., if $f_n(0, 0, \ldots, 0, X_{r+1}) \neq 0$. It is clear that if $r + 1$ elements $x_1', x_2', \ldots, x_r', x_{r+1}'$ of $\mathfrak{o}$ are defined implicitly by equations of the form

$$x_i = \sum_{j=1}^{r+1} c_{ij}x_j' + \text{terms of degree} > 1 \text{ in the } x\text{'s},$$

$$(i = 1, 2, \ldots, r + 1;\ c_{ij} \in k)$$

where we assume that the determinant $|c_{ij}|$ is different from zero, then $\{x_1', x_2', \ldots, x_r', x_{r+1}'\}$ is a basis of $\mathfrak{m}$ (necessarily minimal), and $x_1', x_2', \ldots, x_r'$ are transversal parameters if and only if $f_n(c_{1,r+1}, c_{2,r+1}, \ldots, c_{r+1,r+1}) \neq 0$.

We assume now that $x_1, x_2, \ldots, x_r$ are parameters of $\mathfrak{o}$ (not necessarily transversal) and that $\{x_1, x_2, \ldots, x_r, x_{r+1}\}$ is a basis of $\mathfrak{m}$. Let $X_{r+1}{}^m$ be the least power of X_{r+1} which actually occurs in $f(0, 0, \ldots, 0, X_{r+1})$. We have them $m \geq n$, with equality if and only if $x_1, x_2, \ldots, x_r$ are *transversal* parameters. Since the power series $f(X)$ is only defined to within an arbitrary unit factor in the power series ring $k[[X_1, X_2, \ldots, X_r, X_{r+1}]]$, we may assume, by the Weierstrass preparation theorem, that $f(X)$ is a monic polynomial in X_{r+1}, of degree m, with coefficients in $k[[X_1, X_2, \ldots, X_r]]$:

$$f(X_1, X_2, \ldots, X_r, X_{r+1}) = X_{r+1}{}^m + a_1(X_1, X_2, \ldots, X_r)X_{r+1}{}^{m-1} + \cdots + \alpha_m(X_1, X_2, \ldots, X_r), \quad (4)$$

where the leading form of each power series $a_i(X_1, X_2, \ldots, X_r)$ is of positive degree, and, furthermore, for $i = m - n + 1, m - n + 2,$

$\ldots, m$, the leading form of $a_i(X)$ is of degree $\geq n - m + i$, with equality for at least one of these n values of i.

The parameters $x_1, x_2, \ldots, x_r$ are analytically independent over k, and thus the subring

$$A = k[[x_1, x_2, \ldots, x_r]] \tag{5}$$

of $\mathfrak{o}$ is canonically isomorphic to the power series ring $k[[X_1, X_2, \ldots, X_r]]$. We denote by $\hat{\mathbf{A}}_r/k$ the algebroid (irreducible) r-dimensional variety Spec (A); this variety is merely the r-dimensional affine space $\hat{\mathbf{A}}_r$ over k, considered locally at the origin P_0, as a formal algebroid variety (having at P_0 a simple point). We have a natural projection

$$\pi_A: V \rightarrow \hat{\mathbf{A}}_r/k \tag{6}$$

of V onto $\hat{\mathbf{A}}_r/k$, defined by

$$\pi_A(\mathfrak{p}) = \mathfrak{p} \cap A = \mathfrak{p}_0; (\mathfrak{p} \in \text{Spec}(\mathfrak{o}), \mathfrak{p}_0 \in \text{spec}(A)). \tag{7}$$

Using the geometric language of points, we shall write (7) in the form

$$\pi_A(R) = R_0, \tag{7$'$}$$

where R and R_0 are the points of V and of $\hat{\mathbf{A}}_r/k$ represented respectively by the prime ideals $\mathfrak{p}$ and $\mathfrak{p}_0$ of $\mathfrak{o}$ and A. That the mapping π_A is surjective follows from the fact that, in view of (4), $\mathfrak{o}$ is integral over A, and that consequently over every prime ideal $\mathfrak{p}_0$ of A there lies at least one prime ideal $\mathfrak{p}$ of $\mathfrak{o}$. Furthermore, the fact that $\mathfrak{o}$ is a finite A-module implies that the number of such prime ideals $\mathfrak{p}$ is *finite* (in fact, not greater than m). For this reason we shall refer to the projection π_A as a *finite* projection.

We observe that the projection π_A depends only on the (regular) local ring A and not on the choice of the parameters $x_1, x_2, \ldots, x_r$ of $\mathfrak{o}$ which form a basis of the maximal ideal of A. On the other hand, if the parameters $x_1, x_2, \ldots, x_r$ are given then the projection π_A depends on the choice of the coefficient field k, since $A = k[[x_1, x_2, \ldots, x_r]]$. We say that the projection π_A is *transversal at* P if the integer m in (4) is equal to n, i.e., if the parameters $x_1, x_2, \ldots, x_r$ are transversal. The coördinates of P being all zero, to say that the projection π_A is transversal at P is the same as saying that $X_{r+1} = 0$ is an n-field root of $f(0, 0, \ldots, 0, X_{r+1})$; or — equivalently — that the line $X_1 = X_2 = \cdots = X_r = 0$ has exactly an n-fold intersection with V at P (and therefore does not belong to the tangent hypercone of V at P). More generally, if R is any point of V and if $\xi_1, \xi_2, \ldots, \xi_r, \xi_{r+1}$ are the coördinates of R (i.e., the $\mathfrak{p}$-residues of R in the local domain $\mathfrak{o}/\mathfrak{p}$, where $\mathfrak{p}$ is the prime ideal of $\mathfrak{o}$ associated with the point R), then we say that π_A is transversal at R if the multiplicity of the root $X_{r+1} = \xi_{r+1}$ of the polynomial $f(\xi_1, \xi_2, \ldots, \xi_r, X_{r+1})$ is equal to the multiplicity of the point R of V [The multiplicity of R at V is the multiplicity of the local ring of quotients $\mathfrak{o}_\mathfrak{p}$; it can be proved (and is well known) that ν is the multiplicity of V at R if and only if all the partial derivatives of $f(X_1, X_2, \ldots, X_{r+1})$, of order $\leq \nu - 1$, vanish at R, but not all partial derivatives of f, of order ν, vanish at R]. The following observation is obvious: we have always $\nu \leq n$ for any point R of V, and if $\nu = n$ then every projection π_A which is transversal at P is also transversal at R (the converse is not necessarily true).

Let R and R_0 be as in (7′) and let $\xi_1, \xi_2, \ldots, \xi_r$ be the coördinates of R_0, i.e., the $\mathfrak{p}_0$-residues of $x_1, x_2, \ldots, x_r$ in the local (regular) *domain* $A/\mathfrak{p}_0$. It follows from (7) that $\xi_1, \xi_2, \ldots, \xi_r$ can be identified also with the $\mathfrak{p}$-residues of $x_1, x_2, \ldots, x_r$ in the local domain $\mathfrak{o}/\mathfrak{p}$. If then ξ_{r+1} is the $\mathfrak{p}$-residue of x_{r+1} then $\xi_1, \xi_2, \ldots, \xi_r, \xi_{r+1}$ are the coördinates of R, and we must have

$$f(\xi_1, \xi_2, \ldots, \xi_r, \xi_{r+1}) = 0.$$

Conversely, if the point $R_0 = (\xi_1, \xi_2, \ldots, \xi_r)$ is given and if ξ_{r+1} is any root of the polynomial $f(\xi_1, \xi_2, \ldots, \xi_r, X_{r+1})$ in the algebraic closure of the field $k((\xi_1, \xi_2, \ldots, \xi_r))$ of meromorphic functions of $\xi_1, \xi_2, \ldots, \xi_r$ (over k), then the point R whose coördinates are $\xi_1, \xi_2, \ldots, \xi_r, \xi_{r+1}$ belongs to V, and R_0 is the π_A-projections of R. Two distinct roots ξ_{r+1}, ξ_{r+1}' of $f(\xi_1, \xi_2, \ldots, \xi_r, X_{r+1})$ yield the same point R of V if and only if these roots are conjugate over the field $k((\xi_1, \xi_2, \ldots, \xi_r))$, for it is in this and only in this case that the ideal $\mathfrak{p}$ of those elements $g(x_1, x_2, \ldots, x_r, x_{r+1})$ of $\mathfrak{o}$ which yield true relations $g(\xi_1, \xi_2, \ldots, \xi_r, \xi_{r+1}) = 0$ coincides with the ideal $\mathfrak{p}'$ of those elements $g(x_1, x_2, \ldots, x_{r+1})$ of $\mathfrak{o}$ which yield true relations $g(\xi_1, \xi_2, \ldots, \xi_r, \xi_{r+1}') = 0$. If μ is the number of distinct roots of $f(\xi_1, \xi_2, \ldots, \xi_r, X_{r+1})$ which are conjugate to ξ_{r+1}, we call μ *the relative degree of the point R over its* π_A*-projection* R_0. This integer μ is also the relative degree $[k((R)) : k((R_0))]$, where $k((R))$ and $k((R_0))$ are respectively the fields of fractions $k((\xi_1, \xi_2, \ldots, \xi_r, \xi_{r+1}))$, $k((\xi_1, \xi_2, \ldots, \xi_r))$ of the complete local domains $\mathfrak{o}/\mathfrak{p}$ and $A/\mathfrak{p}_0$. Let then $R_1, R_2, \ldots, R_g$ be the *distinct* points of V which have the same projection R_0, let μ_j be the relative degree of R_j over R_0, let $\xi_{r+1}{}^{(j)}$ be a root of $f(\xi_1, \xi_2, \ldots, \xi_r, X_{r+1})$ which yields the point R_j, and let e_j be the multiplicity of that root. Then we have

$$e_1\mu_1 + e_2\mu_2 + \cdots + e_g\mu_g = m, \tag{8}$$

where—we recall—m is the degree of $f(X_1, X_2, \ldots, X_r, X_{r+1})$ in X_{r+1} [see (4)]. It is clear, from the very definition of the integer g and of the g integers μ_j, that these integers are independent of the choice of the element x_{r+1} which together with the elements $x_1, x_2, \ldots, x_r$ gives a basis of $\mathfrak{m}$. It is also easy to see that the integers e_j are also independent of the choice of the element x_{r+1}; the proof is left to the reader.

We call the integer e_j the *ramification index* of R_j relative to the projection π_A, and we write $e_j = e(R_j; \pi_A)$. We say that a point R of V is a *ramified* point of the projection π_A if $e(R; \pi_A) > 1$. We call a point R_0 of $\hat{\mathbf{A}}_r/k$ a *critical* point of π_A if at least one of the points of V which project into R_0 under π_A is a ramified point of π_A. The set of critical points of π_A is an algebroid hypersurface in $\hat{\mathbf{A}}_r/k$, for this set is defined by the equation

$$D(X_1, X_2, \ldots, X_r) = 0,$$

where D is the X_{r+1}-discriminant of $f(X_1, X_2, \ldots, X_r, X_{r+1})$. We call this hypersurface the *critical variety* of the projection π_A of V and we denote it by $\Delta(\pi_A)$. Since we wish to regard $\Delta(\pi_A)$ as a *reduced* variety, we replace the power series $D(X_1, X_2, \ldots, X_r)$ by the product of its distinct (i.e., non-associated) factors in $k[[X_1, X_2, \ldots, X_r]]$. We denote this product by $\Delta_A(X_1, X_2, \ldots, X_r)$. Thus the reduced critical variety $\Delta(\pi_A)$ is defined by the equation

$$\Delta_A(X_1, X_2, \ldots, X_r) = 0. \tag{9}$$

The critical variety $\Delta(\pi_A)$ is uniquely determined by the two local rings

$\mathfrak{o}$ and A (and hence depends also on the choice of the coefficient field k of $\mathfrak{o}$).

The set of ramified points R of the projection π_A is an algebroid subvariety of V, of pure dimension $r - 1$, since this set is the intersection of V with the hypersurface $F_{X_{r+1}}'(X_1, X_2, \ldots, X_r, X_{r+1}) = 0$, this latter hypersurface having no irreducible components in common with V, since $F(X)$ is free from multiple factors. In particular, all *singular* points of V are ramified points of *any* projection π_A.

A *simple* point R of V, having coördinates $\xi_1, \xi_2, \ldots, \xi_{r+1}$, is a ramified point of π_A if and only if the "line" $X_i = \xi_i$ $(i = 1, 2, \ldots, r)$ is tangent to V at R. Actually, the term "line" is somewhat misleading in the present context, unless R is the closed point P of V (in which case all the ξ_i are zero). If $R \neq P$ and if ρ is the dimension of the irreducible subvariety W_ρ of V having R a general point, then the irreducible subvariety $W_\rho^{(0)}$ of $\Delta(\pi_A)$, having as general point the projection R_0 of R, is also of dimension ρ, and in the ambient space $\hat{\mathbf{A}}_{r+1}/k$ of V the equations $X_i = \xi_i$ $(i = 1, 2, \ldots, r)$ define an algebroid *cylinder* $L_{\rho+1}$ of dimension $\rho + 1$, having $(\xi_1, \xi_2, \ldots, \xi_r, t)$ as general point over k, where t is an analytical transcendental over the field $k((\xi_1, \xi_2, \ldots, \xi_r))$ of meromorphic functions of $\xi_1, \xi_2, \ldots, \xi_r$. The variety $W_\rho^{(0)}$ is the base of this cylinder $L_{\rho+1}$, while the X_{r+1}-axis is its axis. If $R_1(= R)$, $R_2, \ldots, R_g$ are the points of V which project into the point R_0, then the intersection of V with $L_{\rho+1}$ is the union of g irreducible subvarieties $W_{\rho,1}(= W_\rho), W_{\rho,2}, \ldots, W_{\rho,g}$, of dimension ρ, R_j being the general point of $W_{\rho,j}$. Since we have assumed that R is a simple point of V and a ramified point of π_A, the variety W_ρ is a *simple* subvariety of V, and the cylinder $L_{\rho+1}$ is tangent to V at the point R and, consequently, also at each point of W_ρ which is simple for V (note that since R is a simple point of V, *almost all points* of W_ρ are simple points of V; namely, if V_{sing} denote the set of singular points of V then the intersection $W_\rho \cap V_{sing}$ is a proper subvariety of the irreducible variety W_ρ).

For any prime ideal $\mathfrak{p}_0$ of A, or for the corresponding point of R_0 of Spec (A), the primary irredundant decomposition of the extended ideal $\mathfrak{o}\mathfrak{p}_0$ is essentially well known and is completely determined by the factorization of the polynomial $f(\xi_1, \xi_2, \ldots, \xi_r, X_{r+1})$ into irreducible factors over the field $k((\xi_1, \xi_2, \ldots, \xi_r))$ of meromorphic functions of $\xi_1, \xi_2, \ldots, \xi_r$; here $\xi_1, \xi_2, \ldots, \xi_r$ are the $\mathfrak{p}_0$-residues of $x_1, x_2, \ldots, x_r$. Namely, if $R_1, R_2, \ldots, R_g$ are the points of V whose π_A-projection is the point $R_0(=\mathfrak{p}_0)$, and if μ_j and e_j are respectively the relative degree and the ramification index of R_j over R_0 ($j = 1, 2, \ldots, g$), then the monic polynomial $f(\xi_1, \xi_2, \ldots, \xi_r, X_{r+1})$ in X_{r+1} factors into irreducible factors as follows:

$$f(\xi_1, \xi_2, \ldots, \xi_r, X_{r+1}) = \prod_{j=1}^{g} [F_j(\xi_1, \xi_2, \ldots, \xi_r, X_{r+1})]^{e_j}.$$

Here $F_j(\xi_1, \xi_2, \ldots, \xi_r, X_{r+1})$ is a monic polynomial in X_{r+1}, of degree μ_j, with coefficient in $k[[\xi_1, \xi_2, \ldots, \xi_r]]$. The coördinates of the point R_j are $\xi_1, \xi_2, \ldots, \xi_r, \xi_{r+1}^{(j)}$, where $\xi_{r+1}^{(j)}$ can be any of the μ_j roots of the polynomial $F_j(\xi_1, \xi_2, \ldots, \xi_r, X_{r+1})$. The prime ideal $\mathfrak{p}_j$ in $\mathfrak{o}$ which is associated with the point R_j is the ideal

$$\mathfrak{p}_j = \mathfrak{o}\cdot\mathfrak{p}_0 + \mathfrak{o}\cdot f_j(x_1, x_2, \ldots, x_r, x_{r+1}). \tag{10}$$

The primary irredundant decomposition of the ideal $\mathfrak{o}\cdot\mathfrak{p}_0$ is then the following:

$$\mathfrak{o}\cdot\mathfrak{p}_0 = \bigcap_{j=1}^{g} \mathfrak{q}_j, \tag{11}$$

where

$$\mathfrak{q}_j = \mathfrak{o}\cdot\mathfrak{p}_0 + \mathfrak{o}\cdot(f_j(x_1, x_2, \ldots, x_r, x_{r+1})^{e_j}). \tag{12}$$

2. The generic projection of the variety V. If $\{x_1, x_2, \ldots, x_{r+1}\}$ is a (necessarily) minimal basis of the maximal ideal $\mathfrak{m}$ of our local ring $\mathfrak{o}$, every element z of $\mathfrak{m}$ is a power series in $x_1, x_2, \ldots, x_{r+1}$ with coefficients in k and constant term zero:

$$z = \sum_{j=1}^{r+1} c_j x_j + \sum_{1\le j_1\le j_2\le r+1} c_{j_1j_2}x_{j_1}x_{j_2}$$
$$+ \cdots + \sum_{1\le j_1\le j_2\le\cdots\le j_N\le r+1} c_{j_1j_2\cdots j_N}x_{j_1}x_{j_2}\cdots x_{j_N} + \cdots,$$

where the $c_j, c_{j_1j_2}, \ldots, c_{j_1j_2\cdots j_N}, \cdots$ are elements of k. The basic idea of this section is to consider power series x^* in $x_1, x_2, \ldots, x_{r+1}$ with *indeterminate* coefficients u_j, $u_{j_1j_2}$, ... Actually we shall introduce r such power series x_1^*, x_2^*, ..., x_r^*, with indeterminate coefficients $u_j^{(i)}$, $u_{j_1j_2}^{(i)}$, ..., $u_{j_1j_2\cdots j_N}^{(i)}$, ... ($i = 1, 2, \ldots, r$; $1 \leq j_1 \leq j_2 \leq \cdots \leq j_N \leq r + 1$):

$$\begin{aligned} x_i^* &= \varphi_i(x_1, x_2, \ldots, x_{r+1}; \{u\}) \\ &= \sum_{j=1}^{r+1} u_j^{(i)} x_j + \sum_{1 \leq j_1 \leq j_2 \leq r+1} u_{j_1j_2}^{(i)} x_{j_1} x_{j_2} + \cdots \\ &\quad + \sum_{1 \leq j_1 \leq j_2 \leq \cdots \leq j_N \leq r+1} u_{j_1j_2\cdots j_N}^{(i)} x_{j_1} x_{j_2} \cdots x_{j_N} + \cdots. \end{aligned} \tag{1}$$

We denote by k^* the field generated over k by the infinitely many indeterminates $u_{\{j\}}^{(i)}$ and introduce the local ring $\mathfrak{o}^* = k^*[[x_1, x_2, \ldots, x_{r+1}]]$ obtained from $\mathfrak{o}$ by the coefficient field extension $k \to k^*$. When we say that the u's are indeterminates we use this term not merely in an algebraic sense (i.e., that the u's are algebraically independent over $\mathfrak{o}$) but in a stronger analytic sense. Namely, we mean to stipulate that if R is any point of V, the associated prime ideal of $\mathfrak{o}$ being $\mathfrak{p}$, and if $\xi_1, \xi_2, \ldots, \xi_{r+1}$ are the coördinates of R (i.e., the $\mathfrak{p}$-residues of $x_1, x_2, \ldots, x_{r+1}$), then any analytic relation $G^*(\xi_1, \xi_2, \ldots, \xi_{r+1}) = 0$, with coefficients in k^*, is a consequence of analytic relations $g(\xi_1, \xi_2, \ldots, \xi_{r+1}) = 0$ between the ξ's *over* k (where therefore $g(x_1, x_2, \ldots, x_{r+1}) \in \mathfrak{p}$). That means that we must have

$$G^*(x_1, x_2, \ldots, x_{r+1}) \in \mathfrak{o}^*\mathfrak{p} \quad \text{if} \quad G^*(\xi_1, \xi_2, \ldots, \xi_{r+1}) = 0, \tag{2}$$

where $\mathfrak{o}^* = k^*[[x_1, x_2, \ldots, x_{r+1}]]$.

A formal and intrinsic definition of the ring $\mathfrak{o}^*$ is the following:

We consider the tensor product

$$\mathfrak{o}_0^* = \mathfrak{o} \otimes_k k^*.$$

The ring $\mathfrak{o}_0^*$ is generated by its two subrings $\mathfrak{o}$ and k^*, and these two subrings are linearly disjoint over k (see [3], CA1, Ch. III, Section 14). In the ring $\mathfrak{o}_0^*$ the extended ideal $\mathfrak{o}_0^*\mathfrak{m}$ of the maximal ideal $\mathfrak{m}$ of $\mathfrak{o}$ is a maximal ideal, and $\mathfrak{o}^*$ is defined as the $\mathfrak{o}_0^*\mathfrak{m}$-adic completion of $\mathfrak{o}_0^*$:

$$\mathfrak{o}^* = \mathfrak{o}_0^*\mathfrak{m}\text{-}\textit{adic completion of } \mathfrak{o}_0^*. \tag{3}$$

It is then clear that if $\{x_1, x_2, \ldots, x_{r+1}\}$ is *any* basis of $\mathfrak{m}$ in $\mathfrak{o}$, then $\mathfrak{o}^*$ can be canonically identified with $k^*[[x_1, x_2, \ldots, x_{r+1}]]$:

$$\mathfrak{o}^* = k^*[[x_1, x_2, \ldots, x_{r+1}]]$$
$$\cong k^*[[X_1, X_2, \ldots, X_{r+1}]]/(f(X_1, X_2, \ldots, X_{r+1})). \quad (3')$$

The ring $\mathfrak{o}^*$ is a complete local ring, of dimension r, free from non-zero nilpotent elements, and $\mathfrak{o}^*\mathfrak{m}$ is the maximal ideal $\mathfrak{m}^*$ of $\mathfrak{o}^*$. Furthermore, every minimal basis of $\mathfrak{m}$ is also a minimal basis of $\mathfrak{o}^*$. All these statements follow directly from well known properties of tensor products and of completions.

The algebroid variety $V^* = \operatorname{Spec}(\mathfrak{o}^*)$ is defined by the same equation (3) (Section 1) by which V is defined, and since k^* is a pure transcendental extension of k, the irreducible factors of $f(X_1, X_2, \ldots, X_{r+1})$ remain irreducible over k^*. Hence V^* is still of pure dimension r and has the same number q of irreducible components as V. While V^* can be regarded as being obtained from V by the ground field extension $k \to k^*$, it is important to point out that *V is canonically a subset of V^**. Namely, if $\mathfrak{p}$ is any prime ideal of $\mathfrak{o}$, then it follows from results of Serre and Grothendieck on "excellent schemes" (see [4], Ch. IV, second part, Section 7.9 and especially Corollary 7.9.9, p. 222) that the extended ideal $\mathfrak{o}^*\mathfrak{p}$ is a prime ideal in $\mathfrak{o}^*$ and that $\mathfrak{o}^*\mathfrak{p} \cap \mathfrak{o} = \mathfrak{p}$. It is clear that the multiplicity n of the closed point P of V is also the multiplicity of the closed point P of V^*. From now on we shall identify any point R of V, represented by a prime ideal $\mathfrak{p}$ of $\mathfrak{o}$, with the point of V^* represented by the prime ideal $\mathfrak{o}^*\mathfrak{p}$ of $\mathfrak{o}^*$.

We now consider the elements $x_1^*, x_2^*, \ldots, x_r^*$ of $\mathfrak{o}^*$ introduced in (1). Since all the r-rowed determinants of the r by $r+1$ matrix $|u_j^{(i)}|$ of the coefficients of the linear terms in (1) are different from zero and since each coefficient $u_j^{(i)}$ is different from zero, it follows, for any $j = 1, 2, \ldots, r+1$, that the elements of $x_1^*, x_2^*, \ldots, x_r^*, x_j$ form a basis of $\mathfrak{m}^*$. Furthermore, it is easily seen that $x_1^*, x_2^*, \ldots, x_r^*$ are *transversal parameters* of $\mathfrak{o}^*$. To see this we consider, say, the basis $x_1^*, x_2^*, \ldots, x_r^*, x_{r+1}$ of $\mathfrak{m}^*$. Upon solving the r equations (1) with respect to $x_1, x_2, \ldots, x_r$ we get

$$x_j = \sum_{i=1}^{r+1} U_j^{(i)} x_i^* + \text{terms of degree} > 1 \text{ in the } x_i^*,$$
$$(j = 1, 2, \ldots, r;\ x_{r+1}^* = x_{r+1}) \quad (4)$$

where the $U_j^{(i)}$ are the cofactors of the $(r+1)$-rowed square matrix obtained from the matrix $\|u_j^{(i)}\|$ by adding to this r-rowed matrix an extra $(r+1)$-th row consisting of the elements $0, 0, \ldots, 0, 1$ (note that we have then $U_{r+1}^{(i)} = 0$ for $i = 1, 2, \ldots, r$, while $U_{r+1}^{(r+1)} = 1$).

It is obvious that the $r(r+1)$ elements $U_j^{(i)}$ in (4) are also generators of the field $k(\{u_j^{(i)}\})$ and therefore are algebraically independent over k. Now if $f^*(X_1^*, X_2^*, \ldots, X_r^*, X_{r+1}) = 0$ is the equation of V^* relative to the basis $x_1^*, x_2^*, \ldots, x_r^*, x_{r+1}$ of $\mathfrak{m}^*$, then it follows from (4) that the coefficient of $X_{r+1}{}^n$ in the leading form $f_n^*(X_1^*, X_2^*, \ldots, X_r^*, X_{r+1})$ of f^* is $f_n(U_1^{(r+1)}, U_2^{(r+1)}, \ldots, U_r^{(r+1)}, 1)$ and is therefore different from zero. This proves our assertion that $x_1^*, x_2^*, \ldots, x_r^*$ are transversal parameters of $\mathfrak{o}^*$. Furthermore, we recall that these parameters belong to a minimal basis of $\mathfrak{m}^*$, namely that for any $j = 1, 2, \ldots, r+1$ the set $\{x_1^*, x_2^*, \ldots, x_r^*, x_j\}$ is a basis of $\mathfrak{m}^*$. Thus the parameters $x_1^*, x_2^*, \ldots, x_r^*$ belong to the category of parameters of $\mathfrak{o}^*$ similar to the one we agreed to deal with exclusively in the preceding section (in the case of the local ring $\mathfrak{o}$).

The r transversal parameters $x_1^*, x_2^*, \ldots, x_r^*$ of $\mathfrak{o}^*$ will play an important role in the sequel. For this reason we shall show at once that these parameters are *essentially* independent of the choice of the basis $\{x_1, x_2, \ldots, x_{r+1}\}$ of $\mathfrak{m}$ which served to define $x_1^*, x_2^*, \ldots, x_r^*$ in (1). To see this, let $\{y_1, y_2, \ldots, y_{r+1}\}$ be any other basis of $\mathfrak{m}$. We will have then

$$x_j = \psi_j(y_1, y_2, \ldots, y_{r+1}) = \sum_{\nu=1}^{r+1} c_\nu^{(j)} y_\nu$$
$$+ \text{ terms of degree } > 1 \text{ in the } y_\nu\text{'s} \quad (j = 1, 2, \ldots, r+1),$$

where the ψ_j are power series with coefficients in k and where the $(r+1)$-rowed determinant $|c_\nu^j|$ is different from zero. Substituting these expressions of the x_j into (1) we find that the expressions of the r elements x_i^* as power series in $y_1, y_2, \ldots, y_{r+1}$ can be obtained by replacing in (1) the indeterminates $\{u_j^{(i)}\}$, $\{u_{j_1j_2}^{(i)}\}$ $\cdots$ (i-fixed) by quantities $\{v_j^{(i)}\}$, $\{v_{j_1j_2}^{(i)}\}$, $\cdots$ which are polynomials in the $\{u_j^{(i)}\}$, $\{u_{j_1j_2}^{(i)}\}$, $\cdots$ with coefficients in k, and that for each N *the polynomial ring* $k[\{v_j^{(i)}\}, \{v_{j_1j_2}^{(i)}\}, \ldots, \{v_{j_1j_2\cdots j_N}^{(i)}\}]$ *coincides with the polynomial ring* $k[\{u_j^{(i)}\}, \{u_{j_1j_2}^{(i)}\}, \ldots, \{u_{j_1j_2\cdots j_N}^{(i)}\}]$. Hence also the v's are indeterminates, and generate over k the same field k^* which is generated by the u's.

We now set

$$A^* = k^*[[x_1^*, x_2^*, \ldots, x_r^*]] \tag{5}$$

and we apply to the algebroid hypersurface V^*/k^* the considerations developed in Section 1, replacing the rings $\mathfrak{o}$ and A of Section 1 by the rings $\mathfrak{o}^*$ and A^*, with k^* playing the role of the coefficient field k. We therefore have now a projection π_{A^*}, which we shall denote by π_u^*:

$$\pi_u^*: V^* \to \hat{\mathbf{A}}_r/k^*, \tag{6}$$

defined by

$$\pi_u^*(\mathfrak{p}^*) = \mathfrak{p}^* \cap A^* = \mathfrak{p}_0^*, \qquad (p^* \in Spec\,(\mathfrak{o}^*), \mathfrak{p}_0^* \in Spec\,(A^*)). \tag{7}$$

Since V is canonically a subset of V^*, the restriction of π_u^* to V maps V *into* $\hat{\mathbf{A}}_r/k^*$. We call this restriction of π_u^* *the generic projection of* V and we denote it by π_u. This generic projection π_u of V is defined explicitly as follows:

$$\pi_u(\mathfrak{p}) = \mathfrak{o}^*\mathfrak{p} \cap A^* = \mathfrak{p}_0^*, \qquad (\mathfrak{p} \in \mathrm{Spec}\,(\mathfrak{o}), \mathfrak{p}_0^* \in \mathrm{Spec}\,(A^*)), \tag{8}$$

or—using the geometric language of points:

$$\pi_u(R) = R_0^*, \qquad (R \in V, R_0^* \in \hat{\mathbf{A}}_r/k^*), \tag{8$'$}$$

where R and R_0^* are the points represented by the prime ideals $\mathfrak{p}$ and $\mathfrak{p}_0^*$ respectively.

Since the generic projection π_u depends only on the two local rings $\mathfrak{o}^*$ and A^*, we can replace the regular parameters $x_1^*, x_2^*, \ldots, x_r^*$ of A^* [see (5)] by any other set of regular parameters of A^*, without affecting the generic projection π_u. We proceed to replace the x_i^* by other regular parameters of A^* which have a somewhat simple expression as power series in $x_1, x_2, \ldots, x_{r+1}$ (with coefficients in k^*). This will somewhat simplify the algebra of some proofs given in the sequel.

From the definition of the cofactors $U_j^{(i)}$ which appear in (4) it follows that the cofactors $U_j^{(i)}$ in which $1 \le i \le r$ are also the cofactors of the *square* matrix $\|u_j^{(i)}\|$ in which both j *and* i take the values 1, 2, $\ldots, r$. It follows that if we set

$$\tilde{x}_i^* = U_i^{(1)}x_1^* + U_i^{(2)}x_2^* + \cdots + U_i^{(r)}x_r^*, \qquad (i = 1, 2, \ldots, r) \tag{9}$$

then also the r elements $\tilde{x}_1^*, \tilde{x}_2^*, \ldots, \tilde{x}_r^*$ are regular parameters of A^* (since the determinant of the coefficients of the elements $x_1^*, x_2^*,$

$\ldots, x_r^*$ in (9) is different from zero), and the expression of each element $\tilde{x}_i^*$ as a power series $\tilde{\varphi}_i$ in $x_1, x_2, \ldots, x_r, x_{r+1}$ has $x_i + v^{(i)}x_{r+1}$ as leading form of degree 1, where $v^{(i)} = -\bar{U}_r^{(r+1)}$. It is easy to see that the r quantities $v^{(i)}$ and all the coefficients v of the terms of degree ≥ 2 in $\bar{\varphi}_i$, for $i = 1, 2, \ldots, r$, are algebraically independent over $\mathfrak{o}$. We now replace the elements x_i^* by the elements $\tilde{x}_i^*$, and we use the old letters x_i^* to denote the new generators $\tilde{x}_i^*$ of A^*. We shall use the letters u to denote the coefficients v in the terms of degree ≥ 2 in $\bar{\varphi}_i$, and we shall use the letter $u^{(i)}$ to denote the coefficient $v^{(i)}$ of x_{r+1}. Thus, the definition (1) of the x_i^* is now as follows:

$$x_i^* = x_i + u^{(i)}x_{r+1} + \sum_{1 \leq j_1 \leq j_2 \leq r+1} u_{j_1 j_2}{}^{(i)} x_{j_1} x_{j_2} + \cdots + \sum_{1 \leq j_1 \leq j_2 \leq \cdots \leq j_N} u_{j_1 j_2 \cdots j_N}{}^{(i)} x_{j_1} x_{j_2} \cdots x_{j_N} + \cdots \qquad (i = 1, 2, \ldots, r). \tag{10}$$

From now on we shall denote by k^* the field generated over k by the indeterminates u which appear in (10).

From (10) it follows that $x_1, x_2, \ldots, x_r$ are power series in $x_1^*, x_2^*, \ldots, x_r^*, x_{r+1}$, with coefficients which are elements of the *polynomial ring* $k[\{u\}]$. Therefore the equation of V^* is now of the form

$$V^*: f^*(X_1^*, X_2^*, \ldots, X_r^*, X_{r+1}) = 0, \tag{11}$$

where f^* is a power series with coefficients in the polynomial ring $k[\{u\}]$. The leading form f_n^* of f^* is still of degree n, and the coefficient of $X_{r+1}{}^n$ in f_n^* is equal to $f_n(-u^{(1)}, -u^{(2)}, \ldots, -u^{(r)}, 1)$ and is therefore different from zero. This coefficient is a polynomial in the $u^{(i)}$.

From the fact that the coefficient $a^*(u)$ of $X_{r+1}{}^n$ in f^* is different from zero and that all the coefficients of the power series f^* belong to the polynomial ring $k[\{u\}]$ follows, by the Weierstrass preparation theorem, that f^*, after multiplication by a suitable unit factor in $k^*[[X_1^*, X_2^*, \ldots, X_r^*, X_{r+1}]]$, *having coefficients in* $k[\{u\}]_{(a^*)}$, where $k[\{u\}]_{(a^*)}$ denotes the ring of quotients of $k[\{u\}]$ with respect to the multiplicative system consisting of all the positive powers of $a^*(u)$, may be assumed to be a monic polynomial in X_{r+1} of degree n, with coefficient in $k[\{u\}]_{(a^*)}[[X_1^*, X_2^*, \ldots, X_r^*]]$:

$$f^*(X_1^*, X_2^*, \ldots, X_r^*, X_{r+1}) = X_{r+1}{}^n + a_1^*(X_1^*, X_2^*, \ldots, X_r^*)X_{r+1}{}^{n-1} + \cdots + a_n^*(X_1^*, X_2^*, \ldots, X_r^*), \tag{12}$$

where $a_\nu^*(X_1^*, \ldots, X_r^*)$ is a power series in $k[\{u\}]_{(a^*)}[[X_1^*, X_2^*, \ldots, X_r^*]$ having a leading form of degree $\geq \nu$.

Let $D^*(X_1^*, X_2^*, \ldots, X_r^*)$ be the X_{r+1}-discriminant of f^*. The coefficients of the power series D^* belong in $k[\{u\}]_{(a^*)}$, and we have $D^*(0, 0, \ldots, 0) = 0$ since the origin P is a singular point of V^*. Apart from the fact that D^* may have multiple factors, the equation $D^*(X_1^*, X_2^*, \ldots, X_r^*) = 0$ yields the critical hypersurface of the projection π_u^* of V^* onto $\hat{\mathbf{A}}_r/k^*$ [see (6)]. We shall denote this critical hypersurface by Δ_u^*.

We consider this hypersurface as defined over k^*. Let m be the degree of the leading form of D^*. Upon replacing, if necessary, the variables $X_1^*, X_2^*, \ldots, X_r^*$ by suitable linear form in these variables, with coefficients in k, we may assume that the monomial X_r^{*m} actually occurs in D^*. If b^* is the coefficient of X_r^{*m} in D^* (where $b^* \in k[\{u\}]_{(a^*)}$), then by the Weierstrass preparation theorem it follows that after multiplying D^* by a suitable unit factor in $k[\{u\}]_{(\nu^*)}[[X_1^*, \ldots, X_r^*]]$, where $\nu^* = a^*b^*$, we may assume that D^* is a monic polynomial in X_r^*, of degree m, with coefficients in $k[\{u\}]_{(\nu^*)}[[X_1^*, X_2^*, \ldots, X_{r-1}]]$.

The roots of the polynomial D^* in X_r^* are therefore integral over the integral domain $k[\{u\}]_{(\nu^*)}[[X_1^*, X_2^*, \ldots, X_{r-1}^*]]$. Now, the polynomial ring $k[\{u\}]$ is an integrally closed domain, and therefore also the ring of quotients $k[\{u\}]_{(\nu^*)}$ is integrally closed. *It is known that this implies that also the ring $k[\{u\}]_{(\nu^*)}[[X_1^*, X_2^*, \ldots, X_{r-1}^*]]$ is integrally closed in its quotient field.* For this ring is obviously a Krull ring, and it is known that any power series ring $A[[X_1, X_2, \ldots, X_n]]$ over a Krull ring A is itself a Krull ring and is therefore integrally closed (See Bourbaki, *Algèbre Commutative,* Chapter 7, p. 83, exercise 9(b)).

Now, let $D_1^*(X_1^*, X_2^*, \ldots, X_r^*)$ be any *monic* polynomial in X_r^* which is an *irreducible* factor of D^* (over k^*) and let g_1 be the degree of D_1^* on X_r^*. The roots $x_r^{(1)}, x_r^{(2)}, \ldots, x_r^{(g_1)}$ of D_1^* are then integral over $k[\{u\}]_{(\nu^*)}[[X_1^*, \ldots, X_{r-1}^*]]$, and therefore the elementary symmetric functions of these roots must be elements of the ring $k[\{u\}]_{(\nu^*)}[[X_1^*, X_2^*, \ldots, X_{r-1}^*]]$ (since they belong to the field of fractions of that *integrally closed* ring). We have therefore shown that *D_1^* is a monic polynomial in X_r^* with coefficients in $k[\{u\}]_{(\nu^*)}$.* Since this is true for any irreducible factor of D^*, it follows that the product $\Delta_u^*(X_1^*, X_2^*, \ldots, X_r^*)$ of the *distinct* irreducible factors of D^* is a monic polynomial in X_r^* with coefficients in the ring $k[\{u\}]_{(\nu^*)}[[X_1^*, X_2^*, \ldots, X_{r-1}^*]]$. Thus, the (reduced) equation of the (reduced) critical hypersurface $\Delta_u{}^*$ of the generic projection $\pi_u{}^*$ is given by

$$\Delta_u{}^*: \Delta_u{}^*(X_1{}^*, X_2{}^*, \ldots, X_r{}^*) = 0, \tag{13}$$

where

$$\Delta_u{}^*(X_1{}^*, X_2{}^*, \ldots, X_{r-1}{}^*, X_r{}^*) = X_r{}^{*g} + \sum_{\nu=1}^{g} b_\nu(X_1{}^*, X_2{}^*, \ldots, X_{r-1}{}^*)X_r{}^{*g-\nu}, \tag{14}$$

and where

$$b_\nu(X_1{}^*, X_2{}^*, \ldots, X_{r-1}{}^*) \in k[\{u\}]_{(\nu^*)}[[X_1{}^*, X_2{}^*, \ldots, X_{r-1}{}^*]]. \tag{15}$$

We note also that since D^* is a monic polynomial in $X_r{}^*$ of degree m, where m is the degree of the leading form of D^*, it follows that also the degree g of $\Delta_u{}^*(X_1{}^*, X_2{}^*, \ldots, X_{r-1}{}^*, X_r{}^*)$ in $X_r{}^*$ is equal to the degree of the leading form of $\Delta_u{}^*(X_1{}^*, X_2{}^*, \ldots, X_r{}^*)$. In other words, the elements $x_1{}^*, x_2{}^*, \ldots, x_{r-1}{}^*$ are transversal parameters of $\Delta_u{}^*$, and the origin P_0 is a g-fold point of the hypersurface $\Delta_u{}^*$.

3. Introductory informal remarks on equisingularity and on the dimensionality type of the points of V. We shall introduce informally in this section a basic concept of our theory of equisingularity: *the dimensionality type* of any point Q of our algebroid r-dimensional variety V. This will be a numerical character of the point Q which we shall denote by $d.t.(V, Q)$. It will be an *integer* σ satisfying the inequalities $0 \leq \sigma \leq r$. Since this integer will depend *a priori* not only on V and Q but also on the choice of the coefficient field k of the complete local ring $\mathfrak{o}$ of V at its origin P, we should actually use the notation $d.t.(V, Q)/k$, but since the coefficient field k for V is assumed in this work to have been fixed once and for always, we shall omit the reference to k and use the notation $d.t.(V, Q)$. However, in the course of the exposition other algebroid varieties (related to V) will come to the fore, with coefficient fields which are field extensions of k, and in those cases the notation will have to include the reference to the associated coefficient fields, as indicated in the above notation $d.t.(V, Q)/k$.

We give in the next section the formal definition of the integer $d.t.(V, Q)$. In this section we shall develop the intuitive idea which lies behind this concept. We begin with simple points of V.

If Q is a simple point of V, then Q is represented by a prime ideal $\mathfrak{p}$ of $\mathfrak{o}$ such that the localization $\mathfrak{o}_\mathfrak{p}$ of $\mathfrak{o}$ at $\mathfrak{p}$ is a *regular* local ring. It is well known that this is equivalent to assuming that not all the $r + 1$ partial derivatives of $f(X_1, X_2, \ldots, X_{r+1})$ vanish at the point Q. For that it is necessary and sufficient that if $f_1, f_2, \ldots, f_q$ are the irreducible factors of f then only one of the q polynomials f_i vanish at Q and that if, say, f_i vanishes at Q then not all the $r + 1$ partial derivatives of f_i vanish at Q. Geometrically speaking, this signifies that *Q belongs to only one of the q irreducible components $V_1, V_2, \ldots, V_q$ of V* [see (2), Section 1] *and is a simple point of the component.** In particular the general point of any component V_i of V is a simple point of V. We have thus, in the special case of a *simple* point Q of V, that the following is true: there exists one and only one *maximal irreducible* algebroid subvariety $W(Q)$ of V containing Q and *such that the type of "singularity" which V has at Q is the same as the type of "singularity" which V has at the general point of* $W(Q)$. The variety $W(Q)$ is that irreducible component of V which contains the point Q. (We have put the term "singularity" in quotation marks, because in the present case both Q and the general point of $W(Q)$ are not singular points at all).

Since the codimension of each V_i is zero we prefer to attach to any *simple* point Q of V the *dimensionality type zero* rather than r (which is the dimension of V_i), the idea being that the smaller the dimensionality type of a point the "simpler" is the singularity type of that point (simple points being non-singular all).

Our geometric intuition, or—at least—our geometrically motivated expectation, suggests that the following be true *for any point Q of V* (this time we assume that Q is a singular point): there exists a *uniquely determined maximal irreducible* algebroid subvariety $W(Q)$ of V containing Q such that the singularity which V has at Q is of the same type

---

*From an ideal-theoretic point of view this conclusion follows from well-known properties of quotient rings $\mathfrak{o}_\mathfrak{p}$. Namely, assuming, if possible, that $\mathfrak{p}$ contains more than one of the q prime ideals $\mathfrak{p}_i$ of the zero ideal of $\mathfrak{o}$, say $\mathfrak{p} \supset \mathfrak{p}_i$, $i = 1, 2, \ldots, \rho$, $\rho > 1$, $\mathfrak{p} \not\supset \mathfrak{p}_i$ if $i > \rho$, then it is well known [see C.A. vol. I, Ch. IV, Theorems 18 and 19, pp. 225–228) that the ρ extended ideals $\mathfrak{o}_\mathfrak{p} \cdot h(\mathfrak{p}_i)$ (where h is the canonical homomorphism of $\mathfrak{o}$ into $\mathfrak{o}_\mathfrak{p}$), $i = 1, 2, \ldots, \rho$, are distinct and are the prime ideals of the zero ideal in $\mathfrak{o}_\mathfrak{p}$. Since $\rho > 1$, this would imply that $\mathfrak{o}_\mathfrak{p}$ is not an integral domain, which is in contradiction with the regularity of $\mathfrak{o}_\mathfrak{p}$.

Furthermore, if, say, $\mathfrak{p} \supset \mathfrak{p}_1$ (whence $\mathfrak{p} \not\supset \mathfrak{p}_i$, $i > 1$), then the two local rings $\mathfrak{o}_\mathfrak{p}$ and $(\mathfrak{o}/\mathfrak{p}_1)_{\mathfrak{p}/\mathfrak{p}_1}$ are canonically isomorphic. Since $\mathfrak{o}/\mathfrak{p}_1$ is the local ring of V_1 and the prime ideal $\mathfrak{p}/\mathfrak{p}_1$ in $\mathfrak{o}/\mathfrak{p}_1$ corresponds to the point Q, considered as a point of the algebroid variety V_1, it follows that Q is also a simple point of V_1.

as, or equivalent (in some sense) to, the singularity which V has at the general point of $W(Q)$ (see Proposition 5.4). The difficulty of giving this intuitive statement a strictly logical and formal foundation lies in the absence, in the theory of singularities, of any general criterion of equivalence of two singular points of a given variety of dimension greater than 1 (or of two varieties of that dimension). We would also expect Q to be a *simple* point of the variety $W(Q)$, because at any singular point Q' of $W(Q)$ one would *expect* V to have (*because of the maximality property of* $W(Q)$) a more complicated singularity than it has at the general point of $W(Q)$. Again here we are dealing with a statement which is not obvious but which will be proved in this paper (see Section 5, Theorem 5.4). For the moment we were able to illustrate this statement only in the trivial case of a simple point Q of V. In that case $W(Q)$ is the irreducible component V_i of V which contains Q, and Q is then certainly a simple point of V_i. If, however, $r > 1$, then there is nothing to prevent a simple point Q from being a singular point of some *proper* subvariety of V_i, say, of a curve contained in V_i and passing through Q. This shows the essential role which the maximality property of $W(Q)$ is bound to play in any proof of the claim that Q must be a simple point of $W(Q)$.

Once a strictly logical and formal foundation is provided for this intuitive outline of our theory, it will be possible to define the dimensionality type of any singular point Q of V and to define also the concept of equisingularity on V. Informally speaking, this could be done as follows:

If the dimension of the irreducible variety $W(Q)$ is $r - \sigma$, then we can say that the dimensionality type of the point Q is σ. Here, if Q is a singular point, then necessarily σ is *positive,* because the irreducible variety $W(Q)$ must be contained in the singular locus V_{sing} of V (since the general point of $W(Q)$ must also be singular for V), whence $r - \sigma < r$. (This shows incidentally that a point Q is simple for V if and only if its dimensionality type is zero). On the other hand, we must have also $r - \sigma \geq 0$, i.e., $\sigma \leq r$, and the maximum $\sigma = r$ is reached if and only if $W(Q)$ consists of the point Q only. In this case we are dealing with an *isolated* singularity Q of V. Here the term "isolated" is not used only in the set-theoretic sense, i.e., we do not mean to say that Q is necessarily a *component* of singular locus of V. We only mean to say that Q is isolated as far as the complexity of the singularity which V possesses at Q. On the other hand, it is clear that for any point Q of V which *is* a component of the singular locus of V we will have $W(Q) = Q$ and hence $d.t.(V, Q) = r$.

In the actual treatment given in this paper the procedure will be just *the opposite* of the one just outlined above. We will first *define* the integer $\sigma = d.t.(V, Q)$ and then we shall prove the existence and uniqueness of the irreducible variety $W(Q)$. We shall then also prove that $W(Q)$ has dimension $r - \sigma$ and that Q is a simple point of $W(Q)$.

Once the integer $\sigma = d.t.(W, Q)$ is defined and the existence of the variety $W(Q)$ is proved, one can consider the subset of $W(Q)$ consisting of points of dimensionality type σ. We denote this subset by $S(Q)$. We shall prove in the case of algebroid varieties that the set $S(Q)$ is a Zariski dense open subset of $W(Q)$, i.e., that the complement of $S(Q)$ in $W(Q)$ is a *proper subvariety of* $W(Q)$. In the case of algebraic varieties this has been proved by Hironaka; see Introduction and Section 6. Hence $W(Q)$ is the closure of the set $S(Q)$. If Q' is any point of $S(Q)$ then from the unicity of the variety $W(Q')$ it follows at once that the two (open) varieties $S(Q)$ and $S(Q')$ coincide (and we have $W(Q) = W(Q'))$, and that every point on the boundary $W(Q) - S(Q)$ of the open variety $S(Q)$ has dimensionality type *greater* than σ.

We shall call the open variety $S(Q)$ an *equisingularity stratum* S of V, and we shall say that V *is equisingular along* $W(Q)$ *at each point* Q' *of* $S(Q)$. We shall also prove (both in the algebroid and in the algebraic case) that the boundary of a stratum is a union of strata. This will give a *covering* of V by a *finite* number of equisingularity strata.

More generally, given any irreducible subvariety Z of V, we can consider the unique equisingularity stratum S of V which contains the general point of Z. Then $S \cap Z$ is a Zariski dense open subset of Z, and we shall say that V *is equisingular along* Z *at a point* Q *of* Z *if and only if* $Q \in S \cap Z$.

As we have mentioned earlier, the difficulty of giving out intuitive program of work a strictly logical and formal foundation lies in the absence of a general definition of equivalent singularities. This difficulty is not present only in the case of plane curves. For this reason, the notion of equisingularity can be introduced quite rigorously in an intuitive manner in the *case* $s = 1$. In this special case, using the above informal introduction of the variety $W(Q)$, we require that there exist a unique $(r - 1)$*-dimensional* irreducible subvariety $W(Q)$ of V such that the singularities which V has at Q and at the general point R of $W(Q)$ be equivalent in some sense. In this special case we can consider the section of V with a generic plane L_2 through the point Q, or—to be even more specific—a section with a plane L_2 which is *transversal* to the tangent linear $(r - 1)$-space T_{r-1} of $W(Q)$ at Q (we assume here that Q is a simple point of $W(Q)$). The section of L_2 with V is a plane algebroid curve $C_1(L_2)$ in L_2, having at Q a singular point. In a similar manner we obtain a plane algebroid curve $C_1(L_2^*)$, having a singular point at the general point R of $W(Q)$. Since we know what it means to say that the two plane algebroid curves $C_1(L_2)$ and $C_1(L_2^*)$ have equivalent singularities at Q and R respectively, the irreducible variety $W(Q)$ is well defined *if the dimensionality type of Q is 1.* Namely, $W(Q)$ must be an $(r - 1)$-dimensional irreducible component of V_{sing} which has at Q a simple point and is such that the two generic plane sections $C_1(L_2)$ and $C_1(L_2^*)$ have equivalent singularities. (Actually, it must be required that Q be a simple point of V_{sing}, so that Q must lie on *only one* $(r - 1)$-dimensional irreducible components of V_{sing}). We have developed a complete theory of the singularities of V which have dimensionality type 1 in our three papers SES I, II, III, especially in [1]. In these papers, more precisely in SES II, we prove that $C_1(L_2)$ and $C_1(L_2^*)$ have equivalent singularities (and that consequently $d.t.(V, Q) = 1$) if and only if there exists a projection π of V onto affine $\mathbf{A}_r$ which is, locally at Q, transversal to V (i.e., the direction of the projection is not tangent to the tangent hypercone of V at Q) and which is such that the point $\pi(Q)$ in $\mathbf{A}_r$ is a *simple* point of the critical hypersurface Δ_π of π in $\mathbf{A}_r$. In other words, the point $\pi(Q)$ of Δ_π must have dimensionality type 0, hence *one* less than the dimensionality type of Q on V. We have also proved in SES II that if $\pi(Q)$ is a simple point of Δ_π for *one* locally transversal projection π, at Q, then the same will also be true for *any* projection π', locally transversal to V at Q.

Our idea is to generalize this method of projections, involving the singularities of the critical hypersurface Δ of the projection π. Since Δ has dimension $r - 1$, an induction procedure becomes available in the general case. But, as can be shown by examples, it is not sufficient, in the case of points of dimensionality of type ≥ 2, to use arbitrary transversal projections. We find that in the *first* step, meaning in the *definition* of dimensionality type, we must use the *generic* projections introduced in Section 2. This we proceed to do in the next section.

4. Formal definition and some preliminary properties of the concept of dimensionality type of points of V. Let π_u be the generic projection of our r-dimensional algebroid variety V onto the affine space $\mathbf{A}_r/k^*$ [see (8) and (8′), Section 2] and let Q be a singular point of V. We may assume, by induction on r, that the dimensionality type of points of algebroid varieties of dimension ρ less than r (and of embedding dimension $\rho + 1$) has already been defined, since for $r = 1$ everything is straightforward. Namely, if the origin P of the *plane* algebroid curve V is a simple point of V, then V is irreducible and consists of two points only: the point P and the general point of V. Both these points are *simple* points of V and therefore have dimensionality type zero (as was stipulated in Section 3). If, however, P is a singular point of V, it is an isolated singular point of V and therefore has dimensionality type 1 $(=r)$, while the remaining q points of V (where q is the number of the irreducible branches V_i of V) are simple points of V (they are the general points of $V_1, V_2, \ldots, V_q$) and have dimensionality type zero.

Let $\Delta_u{}^*$ be the (reduced) critical hypersurface of the generic projection π_u of V onto $\mathbf{A}_r/k^*$ and let $Q_0{}^* = \pi_u(Q)$, $P_0{}^* = \pi_u(P)$. Since Q was assumed to be a singular point of V, the point $Q_0{}^*$ belongs to $\Delta_u{}^*$, while $P_0{}^*$ is the origin of $\Delta_u{}^*$. We take the field $k^*(=k(\{u\}))$ as coefficient field of the local ring of $\Delta_u{}^*$ at $P_0{}^*$ (see Section 2). By our induction hypothesis, the integer $d.t.(\Delta_u{}^*, Q_0{}^*)/k^*$ is well defined.

Definition 4.1. The dimensionality type σ of the point Q of V is equal to $1 + d.t.(\Delta_u{}^*, Q_0{}^*)/k^*$:

$$\sigma = d.t.(V, Q)/k = 1 + d.t.(\Delta_u{}^*, Q_0{}^*)/k^*. \tag{1}$$

Since, by this definition, we have that $d.t.(\Delta_u{}^*, Q_0{}^*)k^* = \sigma - 1$, and since, by induction on r, we have $0 \leq d.t.(\Delta_u{}^*, Q_0{}^*)/k^* \leq r - 1$ for any point $Q_0{}^*$ of $\Delta_u{}^*$ (these inequalities being trivially true if $\Delta_u{}^*$ is a curve, i.e., if $r = 2$), it follows from (1) that

$$1 \leq d.t.(V, Q) \leq r \tag{2}$$

for any *singular* point of V. We recall, from Section 3, that simple points of V have dimensionality type 0. It is easy to see that our convention of assigning to simple points of V of dimensionality type 0 is in agreement with (1) of Definition 4.1, if we agree to assign to any point of $\mathbf{A}_r$, which *does not belong to the critical hypersurface $\Delta_u{}^*$, the dimensionality type* -1. This follows from the (intuitively obvious) fact that if Q is a simple point of V then the π_u projection $Q_0{}^*$ of Q does not belong to the critical hypersurface Δ_u^*. The proof is straightforward, and is as follows:

Let $\xi_1, \xi_2, \ldots, \xi_{r+1}$ be the coördinates of the point Q. The indeterminates u are to be regarded as indeterminates over the field $k((\xi_1, \xi_2, \ldots, \xi_{r+1}))$ of meromorphic functions of the ξ's (except for the single relation $u^{(1)}u^{(2)} \cdots u^{(r)} = 1$ [see (11), Section 2]. The X^*-coördinates of Q_0^* are $\xi_1^*, \xi_2^*, \ldots, \xi_r^*$, where the ξ^*'s are obtained by replacing in (10) of Section 2 the x's by the ξ's. We have to show two things: (1) the projecting "line" QQ_0^* is not tangent to V at any simple point of V and (2) that this line does not pass through any singular point of V. The prove (1) it will be sufficient to prove that the "line" QQ_0^* is not tangent to V at Q. Now, the equations of the "line" QQ_0^* in the coördinate system of $X_1, X_2, \ldots, X_{r+1}$ are the following [see (10), Section 2]:

$$X_{r+1} - \xi_{r+1} + u^{(i)}(X_i - \xi_i) + (\text{terms of degree} \geq 2 \text{ in } X_1 - \xi_1, X_2 - \xi_2, \ldots, X_{r+1} - \xi_{r+1}) = 0 \qquad (i = 1, 2, \ldots, r). \quad (3)$$

The curve defined by (3) has at Q a simple point and its tangent line at Q is the line $X_{r+1} - \xi_{r+1} + u^{(i)}(X_i - \xi_i) = 0$, $i = 1, 2, \ldots, r$. Thus, along this tangent line the quantities $X_1 - \xi_1, X_2 - \xi_2, \ldots, X_{r+1} - \xi_{r+1}$ are proportional to $1/u^{(1)}, 1/u^{(2)}, \ldots, 1/u^{(r)}, -1$. The tangent hyperplane of V at Q is given by the equation

$$\sum_{i=1}^{r+1} \frac{\partial f}{\partial \xi_i}(X_i - \xi_i) = 0.$$

Since $u^{(1)}, u^{(2)}, \ldots, u^{(r)}$ satisfy only the relation $u^{(1)}u^{(2)} \cdots u^{(r)} = 1$, it follows that

$$\sum_{i=1}^{r} \frac{1}{u^{(i)}} \frac{\partial f}{\partial \xi_i} - \frac{\partial f}{\partial \xi_{r+1}}$$

is different from zero (since the $r + 1$ quantities $\partial f/\partial \xi_i$ are not all zero). This proves that the line QQ_0^* is not tangent to V at Q.

To prove part (2) of our assertion it is sufficient to observe that the dimension of V_{sing} is $\leq r - 1$, and that V_{sing} (and also V_{sing}^*) is defined over k, while the curve defined by the equations (3) is essentially a generic "line" through R in the affine space $\mathbf{A}_{r+1}$ and therefore does not meet V_{sing}. As an immediate corollary of Definition 4.1, we have the following

Proposition 4.2. *If Q, Q' are two points of V and Q' is a specialization of Q over k, then $d.t.(V, Q') \geq d.t.(V, Q)$.*

Proof. Let $\sigma = d.t.(V, Q)$. The assertion is trivial if $\sigma = 0$ (i.e., if Q is a simple point of V). It is also trivial if $\sigma = 1$, for in that case Q is a singular point of V, whence also Q' is a singular point of V, and therefore $d.t.(V, Q') \geq 1$. We shall therefore use induction on σ, assuming that the proposition is true *for any dimension* r whenever $d.t.(V, Q)$ is less than σ.

Let $\mathfrak{p}$ and $\mathfrak{p}'$ be the prime ideals of $\mathfrak{o}$ which are associated with the points Q and Q' respectively. Let $\pi_u(Q) = Q_0{}^*$, $\pi_u(Q') = Q_0'{}^*$, and let $\mathfrak{p}_0{}^*$ and $\mathfrak{p}_0'{}^*$ be the prime ideals of A^* associated with $Q_0{}^*$ and $Q_0'{}^*$ respectively. We have then [see (8) and (8′), Section 2]: $\mathfrak{p}_0{}^* = \mathfrak{o}^*\mathfrak{p} \cap A^*$, $\mathfrak{p}_0'{}^* = \mathfrak{o}^*\mathfrak{p}' \cap A^*$. Since Q' is a specialization of Q over k, we have $\mathfrak{p} \subset \mathfrak{p}'$. Therefore $\mathfrak{p}_0{}^* \subset \mathfrak{p}_0'{}^*$, i.e., $Q_0'{}^*$ is a specialization of $Q_0{}^*$ over k^*. Since $d.t.(\Delta_u{}^*, Q_0{}^*)/k^* = \sigma - 1$, it follows by our induction hypothesis, that $d.t.(\Delta_u{}^*, Q_0'{}^*)/k^* \geq d.t.(\Delta_u{}^*, Q_0{}^*)/k^*$. Since $d.t.(V, Q') = 1 + d.t.(\Delta_u{}^*, Q_0'{}^*)$ and $d.t.(V, Q) = 1 + d.t.(\Delta_u{}^*, Q_0{}^*)$, the proposition follows.

Another consequence of our Definition 4.1 is the following

Proposition 4.3. *The dimensionality type of any point of V is invariant under any extension of the coefficient field k.*

Proof. Let k' be any field containing k. By the complete local ring $\mathfrak{o}'$ which is obtained from $\mathfrak{o}$ by the coefficient field extension $k \to k'$ we mean the analog of the local ring $\mathfrak{o}^*$ defined in Section 2 by means of the coefficient field extension $k \to k^*$. Namely, $\mathfrak{o}'$ is the completion of the ring $\mathfrak{o} \otimes_k k'$ with respect to the (maximal) ideal in $\mathfrak{o} \otimes_k k'$ which is the extension of the maximal ideal $\mathfrak{m}$ of $\mathfrak{o}$. We now introduce the set of indeterminates $\{u\}$ and we set (as in Section 2) $k^* = k(\{u\})$. We consider the u's also as indeterminates over k' and we set $k'^* = k'(\{u\})$. It is clear that k'^* is also an extension $k^{*\prime}$ of the field k^*, obtained by extending the subfield k of k^* to k'. We can write therefore:

$$k'^* = k'(\{u\}) = k^{*\prime} \tag{4}$$

We set

$$\mathfrak{o}^{*\prime} = \widehat{\mathfrak{o}^* \bigotimes_{k^*} k^{*\prime}}, \tag{5}$$

$$\mathfrak{o}'^{*} = \widehat{\mathfrak{o}' \bigotimes_{k'} k'^{*}}, \tag{5'}$$

where the symbol $\widehat{\quad}$ indicates completion with respect to the maximal ideals in $\mathfrak{o}^* \otimes_{k^*} k^{*\prime}$ and $\mathfrak{o}' \otimes_{k'} k'^{*}$ which are the extensions of the maximal ideal $\mathfrak{m}^*$ of $\mathfrak{o}^*$ and of the maximal ideal $\mathfrak{m}'$ of $\mathfrak{o}'$ respectively. Now, since both $\mathfrak{m}^*$ and $\mathfrak{m}'$ are the extensions (in $\mathfrak{o}^*$ and $\mathfrak{o}'$ respectively) of the maximal ideal $\mathfrak{m}$ in $\mathfrak{o}$, it follows that two completions in (5) and (5′) are both with respect to the extensions of $\mathfrak{m}$ in the two rings $\mathfrak{o}^* \otimes_{k^*} k^{*\prime}$ and $\mathfrak{o}' \otimes_{k'} k'^{*}$. Now it is easy to see that these two completions coincide and that consequently

$$\mathfrak{o}^{*\prime} = \mathfrak{o}'^{*} \tag{6}$$

To see this, we observe that we have

$$\mathfrak{o}^* \bigotimes_{k^*} k^{*\prime} = (\widehat{\mathfrak{o} \bigotimes_{k} k^*}) \bigotimes_{k^*} k^{*\prime}$$

$$= k^*[[x_1, x_2, \ldots, x_{r+1}]] \bigotimes_{k^*} k^{*\prime},$$

whence

$$\mathfrak{o}^{*\prime} = k^{*\prime}[[x_1, x_2, \ldots, x_{r+1}]], \tag{7}$$

where the defining relation between the $+ 1$ basis elements $x_1, x_2, \ldots, x_{r+1}$ of $\mathfrak{m}$ is the original relation $f(x_1, x_2, \ldots, x_{r+1}) = 0$ which defines our variety V. On the other hand, we have

$$\mathfrak{o}' \bigotimes_{k'} k'^{*} = (\widehat{\mathfrak{o} \bigotimes_{k} k'}) \bigotimes_{k'} k'^{*} = k'[[x_1, x_2, \ldots, x_{r+1}]] \bigotimes_{k'} k'^{*},$$

and hence

$$\mathfrak{o}'^{*} = k'^{*}[[x_1, x_2, \ldots, x_{r+1}], \tag{7'}$$

where the defining relation between the elements $x_1, x_2, \ldots, x_{r+1}$ is still $f(x_1, x_2, \ldots, x_{r+1}) = 0$. Now (6) follows from (7), (7′) and (4). Let now V^* be the algebroid variety Spec ($\mathfrak{o}^*$) and let $V^{*\prime} =$ Spec ($\mathfrak{o}^{*\prime}$). Similarly, let $V' =$ Spec ($\mathfrak{o}'$), $V'^* =$ Spec ($\mathfrak{o}'^*$). We have then by (6): $V^{*\prime} = V'^*$. Since Proposition 4.3 is trivially true in the case $r = 1$, we can use induction on r. We assume therefore that the proposition is true for algebroid varieties of dimension $r - 1$. We observe now that the critical hypersurface $\Delta_u'^*$ of the generic projection π_u' of V' is obviously obtainable from Δ_u^* by the coefficient field extension $k^* \to k^{*\prime}$. Let now Q be any point of V. The proposition is trivially true if Q is a simple point of V, for in that case Q is also a simple point of V', whence $d.t.(V, Q)/k = d.t.(V', Q)/k' = 0$. Assume now that Q is a singular point of V (and therefore also a singular point of V'). Then $\pi_u(Q) = Q_0^*$ is a point of $\Delta_u'^*$; therefore, by our induction hypothesis, we have $d.t.(\Delta_u'^*, Q_0^*)/k^{*\prime} = d.t.(\Delta_u^*, Q_0^*)/k^*$. Since $d.t.(V, Q)/k = 1 + d.t.(\Delta_u^*, Q_0^*)/k^*$ and $d.t.(V', Q)/k' = d.t.(\Delta_u'^*, Q_0^*)/k'^*$, and since $k^{*\prime} = k'^*$[by (4)], the equality $d.t.(V, Q)/k = d.t.(V', Q)/k'$ follows, and the proposition is proved.

5. The basic theorem on the equisingularity stratification of *V*. We shall denote by s the dimensionality type of the closed point P of V. Since P is a specialization of every point of V, it follows from Proposition 4.2 that the dimensionality type σ of any point Q of V satisfies the inequality $0 \leq \sigma \leq s$. We shall denote by $V(\sigma)$ the set of points of V which have dimensionality type σ (for a given σ this set may be empty if σ is positive). We have therefore

$$\bigcup_{\sigma=0}^{s} V(\sigma) = V; \qquad \bigcup_{\sigma=1}^{s} V(\sigma) = V_{sing}. \tag{1}$$

We shall denote by $V\{\sigma\}$ the set of points of V which have dimensionality type $\geq \sigma$. We have therefore

$$V\{\sigma\} = V(\sigma) \cup V\{\sigma + 1\}; \tag{2}$$

$$V\{0\} = V, \qquad V\{1\} = V_{sing}. \tag{2'}$$

The set $V(0)$ is the set of simple points of V. If $V_1, V_2, \ldots, V_q$ are the irreducible components of V, then $V(0)$ is the disjoint union of q sets

$S_{i,0}$, where $S_{i,0} = V_i - (V_i \cap V_{sing})$ $(i = 1, 2, \ldots, q)$. The algebroid closure $\overline{S_{i,0}}$ of $S_{i,0}$ (i.e., the least algebroid subvariety of V which contains $S_{i,0}$) is the irreducible variety V_i, of dimension r, and $S_{i,0}$ is a Zariski open (dense) subset of V_i, i.e., the boundary $\overline{S_{i,0}} - S_{i,0}$ of $S_{i,0}$ is a *proper* algebroid subvariety of V_i, namely $V_i \cap V_{sing}$. Moreover, each point of $S_{i,0}$ is a simple point of $\overline{S_{i,0}}$.

Our study of the case $\sigma = 1$ (see [1], SEQ, II) shows that the set-up in this case is quite similar to the above set-up in the case $\sigma = 0$, except that $V(1)$ *may be empty*. If $V(1)$ is non-empty, the closure $\overline{V(1)}$ of $V(1)$ is a subvariety of V, of pure dimension $r - 1$. The set $V(1)$ is non-empty if and only if V_{sing} has dimension $r - 1$. When that is so, then $V(1)$ is the disjoint union of q_1 subsets $S_{i_1,1}$, where q_1 is the number of irreducible $(r - 1)$-dimensional components of V_{sing}, and these components are precisely the algebroid closures $\overline{S_{i_1,1}}$ of the q_1 sets $S_{i_1,1}$. Each set $S_{i_1,1}$ is a Zariski dense open subset of $\overline{S_{i_1,1}}$, the boundary $\overline{S_{i,1}} - S_{i,1}$ being a proper algebroid subvariety of $\overline{S_{i_1,1}}$ (and therefore of dimension $< r - 1$). Moreover, *each point of $S_{i_1,1}$ is a simple point of* $\overline{S_{i_1,1}}$, and in SES II we have shown that each such point is even a *simple point of V_{sing}* (hence, in particular, $\overline{S_{i_1,1}}$ is the only $(r - 1)$-dimensional component of V_{sing} which contains that point).

We call the $q + q_1$ open (irreducible) smooth varieties $S_{i,0}$, $S_{i_1,1}$ the *equisingularity strata* of V, of dimensionality type 0 and 1 respectively. We repeat that the strata $S_{i_1,1}$ are present if and only if V_{sing} has dimension $r - 1$.

Since the boundary of $S_{i,0}$ is the variety $V_i \cap V_{sing}$ and since the union of these q varieties is V_{sing}, it follows that each of the q_1 $(r - 1)$-dimensional irreducible components $\overline{S_{i_1,1}}$ of V_{sing}, and therefore also each stratum $S_{i_1,1}$, belongs to the boundary $V_i \cap V_{sing}$ of at least one r-dimensional stratum $S_{i,0}$. Moreover, since each point $S_{i_1,1}$ is a simple point of V_{sing}, it follows easily that *if a point Q of a stratum $S_{i_1,1}$ belongs to the boundary of a stratum $S_{i,0}$ then the entire stratum $S_{i_1,1}$ belongs to the boundary of $S_{i,0}$*. For, let say Q be a common point of a stratum $S_{1,1}$ and of the boundary $V_1 - S_{1,0}$ of a stratum $S_{1,0}$. Assuming, if possible, that $S_{1,1} \not\subset V_1 - S_{1,0}$, we must have $S_{1,1} \subset V_i \cap V_{sing}$ for some $i \neq 1$, say $S_{1,1} \subset V_2 \cap V_{sing}$. The intersection $V_1 \cap V_2$ must have at least one irreducible component $\overline{S_{i,1}}$ of dimension $r - 1$, passing through Q, and this component is different from $\overline{S_{1,1}}$, since $\overline{S_{1,1}} \not\subset V_1$, while $S_{i,1} \subset V_1$. Hence Q lies on at least *two* distinct irreducible $(r - 1)$-dimensional components of V_{sing} and is therefore *not* a simple point of V_{sing}, in contradiction with what was said above (namely, that the points of each stratum $S_{i,1}$ are simple points of V_{sing}).

The main object of this work is to prove the following basic theorem, which generalizes to any value of σ the results stated above for the cases $\sigma = 0, 1$:

THEOREM 5.1.

(*a*) *For any integer* σ, such that the set $V(\sigma)$ *is non-empty* ($0 \leq \sigma \leq s$), $V(\sigma)$ *is a disjoint union of a certain number* q_σ *of sets* $S_{i_\sigma,\sigma}$ ($i_\sigma = 1, 2, \ldots, q_\sigma \geq 1$), *called strata in the sequel, where each* $S_{i_\sigma,\sigma}$ *is an irreducible* $(r - \sigma)$*-dimensional Zariski dense open subset of its algebroid closure* $\overline{S}_{i_\sigma,\sigma}$ *(equivalently,* $\overline{S}_{i_\sigma,\sigma}$ *is an irreducible* $(r - \sigma)$*-dimensional algebroid subvariety of V, and* $\overline{S}_{i_\sigma,\sigma} - S_{i_\sigma,\sigma}$ *is a proper subvariety of* $\overline{S}_{i_\sigma,\sigma}$*).*

(*b*) *Every point of each* $S_{i_\sigma,\sigma}$ *is a simple point of* $\overline{S}_{i_\sigma,\sigma}$*, and—furthermore—V is equimultiple along* $S_{i_\sigma,\sigma}$*.*

(*c*) *The stratum W(P) which contains the closed point P of V is a closed subvariety of V (i.e., its boundary is empty) and coincides with the set V(s) of all points of V which have the maximum dimensionality type s.*

(*d*) *The boundary of each stratum is a union of strata.*

We call each stratum $S_{i_\sigma,\sigma}$ (σ being such that $V(\sigma)$ is non-empty) an *equisingularity stratum* of *V*, of *dimensionality type* σ. Theorem 5.1 yields a stratification of *V* by equisingularity strata, satisfying the usual properties of a stratification, in view of the disjointness of any two distinct strata (part (*a*) of the theorem) and in view of parts (*b*) and (*d*).

This section is devoted largely to the proof of Theorem 5.1. The proof will require a preliminary basic lemma which we shall state and prove below and which concerns an algebroid hypersurface V_t^* whose equation has coefficients in a field $k^* = k(\{t\})$ generated over *k* by a (finite or infinite) set $\{t\}$ of parameters. This lemma will then be applied to the critical hypersurface Δ_u^* of the generic projection π_u of our original *r*-dimensional algebroid variety V/k, the parameters *t* being in that case the indeterminates *u*. The reason why we use the letter *t* for the parameters which appear in the equation of the hypersurface V_t^* of the lemma (instead of using the letter *u*) is that, in the course of the proof of the lemma, we shall have to consider the generic projection of V_t^*, and, following the notations which we have used in the preceding sections, we will want to continue using the letter u for the parameters of the generic projection π_u^* of V_t^*, the u's being then indeterminates over the coefficient field $k(\{t\})$ of the local ring of V_t^*. We recall (see Section 2) that the coefficients of the equation of Δ_u^* were not arbitrary elements of the field $k(\{u\})$ but were elements of the ring of quotients $k[\{u\}]_{(v^*)}$ of the *polynomial ring* $k[\{u\}]$ with respect to the multiplicative system (v^*) consisting of all the positive integral powers of some non-zero element v^* of the polynomial ring $k[\{u\}]$ [see (15), Section 2]. We also recall that the equation of Δ_u^* was given by (14), Section 2, the left-hand side of that equation being a monic *polynomial* in one of variables $X_1^*, X_2^*, \ldots, X_r^*$ (namely in X_r^*; see (14) and (15), Section 2). For this reason we impose a similar restriction on the class of r-dimensional hypersurfaces V_t^* which our lemma will deal with. We assume namely that V_t^* is defined by an equation

$$V_t^*: f^*(X_1, X_2, \ldots, X_{r+1}; \{t\}) = 0, \tag{3}$$

where f^* is a monic polynomial in X_{r+1} and a power series in $X_1, X_2, \ldots, X_r$ (zero at the origin $X_1 = X_2 = \cdots = X_{r+1} = 0$), and that the coefficients of f^* belong to the ring of quotients $k[\{t\}]_M$ of the polynomial ring $k[\{t\}]$ with respect to the multiplicative system M consisting of the positive powers of some non-zero element $a(\{t\})$ of $k[\{t\}]$:

$$M = \text{set of positive powers of a non-zero polynomial } a(\{t\}). \tag{4}$$

In the applications of our lemma to the case in which V_t^* is the critical hypersurface Δ_u^* (and therefore r is to be then replaced by $r - 1$) we shall be dealing only with those points of Δ_u^* which are π_u-projections of (necessarily singular) points of V. Now, the coördinates $\xi_1^*, \xi_2^*, \ldots, \xi_r^*$ of the π_u^*-projection Q_0^* of a point $Q(\xi_1, \xi_2, \ldots, \xi_{r+1})$ of V are given by (1), Section 2, where $x_1, x_2, \ldots, x_{r+1}$ are to be replaced by $\xi_1, \xi_2, \ldots, \xi_{r+1}$. The coördinates $\xi_1^*, \xi_2^*, \ldots, \xi_r^*$ of Q_0^* are therefore elements of the tensor product $k[\{u\}] \otimes_k k[[\xi_1, \xi_2, \ldots, \xi_{r+1}]]$, namely power series in the ξ's, with coefficients in $k[\{u\}]$ and constant term zero, the u's being indeterminates over the complete local domain $k[[\xi_1, \xi_2, \ldots, \xi_{r+1}]]$. We shall therefore be interested only in a special class of points of V_t^*, namely in the points Q^* whose coördinates $\xi_1^*, \xi_2^*, \ldots, \xi_{r+1}^*$ belong to the tensor product (over k) of $k[\{t\}]$ with a complete local domain $k[[\xi_1, \xi_2, \ldots, \xi_N]]$; here, $\xi_1, \xi_2, \ldots, \xi_N$ are the coördinates of a variable point (ξ) of a *fixed* algebroid variety Z embedded in some affine N-space and defined over k. We shall denote the associated point of V_t^* by Q_ξ^*. We furthermore assume that the coördinates ξ_i^* of Q_ξ^* are given by the following expressions:

$$\xi_i^* = \varphi_i(\xi_1, \xi_2, \ldots, \xi_N; \{t\}), \qquad (i = 1, 2, \ldots, r+1) \tag{5}$$

where the φ_i are power series in the ξ's, with *fixed* coefficients in $k[[t]]$ (i.e., independent of the variable point (ξ) of Z) and constant term zero. Here the t's are considered as indeterminates over the local domain $k[[\xi_1, \xi_2, \ldots, \xi_N]]$. More precisely (and in complete agreement with the definition of tensor products), let $\mathfrak{p}$ be the prime ideal in $k[[Y_1, Y_2, \ldots, Y_N]]$ which is associated with the point $(\xi_1, \xi_2, \ldots, \xi_N)$, and let $\mathfrak{p}^*$ be the prime ideal in $k^*[[X_1, X_2, \ldots, X_{r+1}]]$ associated with the point Q_ξ^*. Let h be the homomorphism $k^*[[X_1, X_2, \ldots, X_{r+1}]] \to k^*[[Y_1, Y_2, \ldots, Y_N]]$ defined by $h(X_i) = \varphi_i(Y_1, Y_2, \ldots, Y_N; \{t\})$. Then $h(\mathfrak{p}^*)$ is the extension to $k^*[[Y_1, Y_2, \ldots, Y_N]]$ of the ideal $k[\{t\}] \otimes_k \mathfrak{p}$ in the subring $k[\{t\}] \otimes_k k[[Y_1, Y_2, \ldots, Y_N]$ of $k^*[[Y_1, Y_2, \ldots, Y_N]]$.

We shall denote by $S_k(V_t^*)$ the set of points Q_ξ^* described above.

We shall consider specializations $\{t\} \to \{\bar{t}\}$ of the parameters t, where the $\bar{t}$'s are arbitrary elements of $k[\{t\}]$. For the proof of Theorem 5.1 the important case will be the one in which the $\bar{t}$'s belong to k. We shall also consider specializations $\{\xi_1, \xi_2, \ldots, \xi_N\} \xrightarrow{k} \{\zeta_1, \zeta_2, \ldots, \zeta_N\}$, where $(\zeta_1, \zeta_2, \ldots, \zeta_N)$ is therefore an arbitrary point of the (irreducible) algebroid subvariety of Z having $(\xi_1, \xi_2, \ldots, \xi_N)$ as general point. It is clear that the point Q_ζ^* is a specialization of the point Q_ξ^*, over k^*, and belongs to $S_k(V_t^*)$.

We assume, of course, that V_t^* is a *reduced* hypersurface, i.e., that the power series f^* in (3) has no multiple factors. Furthermore, we assume that if n is the degree of the leading form of f^* $(n \geq 1)$ then f^* is a monic polynomial in X_{r+1}, of degree n (this agrees with the condition which is satisfied by Δ_u^*; see the end of Section 2).

We shall say that a given property, which may or may not hold true for some specializations $\{t\} \to \{\bar{t}\}$, *is true for almost all specializations* $\{t\} \to \{\bar{t}\}$, if there exists a non-zero polynomial $g(\{t\})$ in $k[\{t\}]$ such that that property holds true for all specializations $\{t\} \to \{\bar{t}\}$ such that $g(\{\bar{t}\}) \neq 0$.

If Q_ξ^* is any point in $S_k(V_t^*)$ and $\{t\} \to \{\bar{t}\}$ is any specialization of the parameters t, we shall denote by $Q_{\xi,\bar{t}}^*$ the point whose coördinates $\bar{\xi}_1^*, \bar{\xi}_2^*, \ldots, \bar{\xi}_{r+1}$ are obtained from (5) by replacing the parameters t by the $\bar{t}$'s. Thus, we have:

$$\bar{\xi}_i^* = \varphi_i(\xi_1, \xi_2, \ldots, \xi_N; \{\bar{t}\}). \qquad (i = 1, 2, \ldots, r + 1) \qquad (6)$$

Before we state our basic lemma we introduce some notations. Given an algebroid variety $V = \text{Spec}(\mathfrak{o})$ and given a point Q of V, *we denote by* V_Q *the set of points* R *of* V *such that* Q *is a specialization of* R. As was explained earlier in Section 1, if $\mathfrak{p}$ is the prime ideal in $\mathfrak{o}$ which is associated with Q then V_Q is the set of prime ideals of $\mathfrak{o}$ which are contained in $\mathfrak{p}$. Since there is a canonical (1, 1)-correspondence between the prime ideals in $\mathfrak{o}$ which are contained in $\mathfrak{p}$ and the prime ideals of the quotient ring $\mathfrak{o}_\mathfrak{p}$, it follows that V_Q *can be canonically identified with the variety* Spec $(\mathfrak{o}_\mathfrak{p})$. However, this latter variety is not, strictly speaking, an algebroid variety (in the sense in which we use this term), since, in general, the local ring $\mathfrak{o}_\mathfrak{p}$ is not complete (if the dimension r of $\mathfrak{o}$ is ≥ 3 and the dimension ρ of the prime ideal $\mathfrak{p}$ satisfies the inequalities $0 < \rho \leq r - 2$, then $\mathfrak{o}_\mathfrak{p}$ is not complete). *We next denote by the symbol* $\text{EQS}_k(V, Q)$ (where k is the fixed coefficient field of $\mathfrak{o}$) the set of points R of V_Q such that $d.t.(V, R) = d.t.(V, Q)$.

We can now state our basic lemma.

Lemma 5.2. *Let* V_t^* *be the (reduced) algebroid variety defined by* (3) *and let* Q_ζ^* *be a point of* $S_k(V_t^*)$. *Then for almost all specializations* $\{t\} \to \{\bar{t}\}$, *where the* $\bar{t}$'s *are in* $k[\{t\}]$ (*in particular, for almost all specializations* $\{t\} \to \{\bar{t}\}$, *the* $\bar{t}$'s *in* k) *the equation*

$$V_{\bar{t}}^*: f^*(X_1, X_2, \ldots, X_{r+1}; \{\bar{t}\}) = 0 \qquad (7)$$

is meaningful [*i.e., we have* $a(\{\bar{t}\}) \neq 0$, *see* (4)], *defines a reduced algebroid variety* $V_{\bar{t}}^*$ (*over the field* $k(\{\bar{t}\})$ *which contains the specialization* $R_{\xi,\bar{t}}^*$ *of any point* R_ξ^* *of* $S_k(V_t^*)$, *and furthermore the following two conditions are satisfied*:

(*a*) $d.t.(V_{\bar{t}}^*, Q_{\zeta,\bar{t}}^*)/k(\{\bar{t}\}) = d.t.(V_t^*, Q_\zeta^*)/k^*$;

(*b*) *if* R_ξ^* *is any point of* $S_k(V_t^*)$ *such that the point* $(\zeta_1, \zeta_2, \ldots, \zeta_N)$ *is a specialization, over* k, *of the point* $(\xi_1, \xi_2, \ldots, \xi_N)$, *then*

$$R_{\xi,\bar{t}}^* \in \mathrm{EQS}_{k(\{\bar{t}\})}(V_{\bar{t}}^*, Q_{\zeta,\bar{t}}^*) \leftrightarrow$$
$$R_{\xi}^* \in \mathrm{EQS}_{k^*}(V_t^*, Q_{\zeta}^*).$$

Proof. One condition that we must impose on our specializations $\{t\} \rightarrow \{\bar{t}\}$ is the one mentioned in the statement of the lemma, namely the inequality

$$a(\{\bar{t}\}) \neq 0. \tag{8}$$

This insures that the power series $f^*(\{X\}; \{\bar{t}\})$ in $X_1, X_2, \ldots, X_{r+1}$ is well defined and has coefficients in the ring $k[\{\bar{t}\}]_{\bar{M}}$, where $\bar{M}$ is the multiplicative system of the non-negative powers of $a(\{\bar{t}\})$. In particular, if the $\bar{t}$'s belong to k, then the coefficients of $f^*(\{X\}; \{\bar{t}\})$ belong to k. Furthermore, the power series $f^*(\{X\}; \{\bar{t}\})$ is a monic polynomial in X_{r+1}, of degree n, and its leading form is also of degree n.

Let $D^*(X_1, X_2, \ldots, X_r; \{t\})$ be the X_{r+1}-discriminant of $f^*(X_1, X_2, \ldots, X_{r+1}; \{t\})$. This discriminant is not identically zero (since $f^*(X_1, X_2, \ldots, X_{r+1}; \{t\})$ has no multiple factors) and is a power series in $X_1, X_2, \ldots, X_r$ with coefficients in the ring of quotients $k[\{t\})_M$ (see (4)). If (8) is satisfied then $D^*(X_1, X_2, \ldots, X_r; \{\bar{t}\})$ is well defined and is the X_{r+1}-discriminant of $f^*(X_1, X_2, \ldots, X_{r+1}; \{\bar{t}\})$ (since $f^*(X_1, X_2, \ldots, X_{r+1}; \{\bar{t}\})$ is a monic polynomial in X_{r+1}, of the same degree n as $f^*(X_1, X_2, \ldots, X_{r+1}; \{t\})$). Since $f^*(X_1, X_2, \ldots, X_{r+1}; \{t\})$ has no multiple factors, $D^*(X_1, X_2, \ldots, X_r; \{t\})$ is not identically zero. Furthermore, $f^*(X_1, X_2, \ldots, X_{r+1}; \{\bar{t}\}$ has no factors independent of X_{r+1} (since the degree of the leading form of $f^*(X_1, X_2, \ldots, X_{r+1}; \{\bar{t}\}$ is equal to the degree n of $f^*(X_1, X_2, \ldots, X_{r+1}; \{\bar{t}\})$ in X_{r+1}). Therefore, the power series $f^*(X_1, X_2, \ldots, X_{r+1}; \{\bar{t}\})$ will have no multiple factors if (and only if) $D^*(X_1, X_2, \ldots, X_r; \{\bar{t}\})$ is not identically zero. To insure that this is the case it is sufficient to consider any non-zero coefficient $b(\{t\})/a(\{t\})^{\mu}$ of the power series $D^*(X_1, X_2, \ldots, X_r; \{t\})$ and impose on the specialization $\{t\} \rightarrow \{\bar{t}\}$ the inequality

$$b(\{\bar{t}\}) \neq 0. \tag{9}$$

If then both (8) and (9) are satisfied, then the equation (7) will define a *reduced* algebroid hypersurface $V_{\bar{t}}^*$ having at the origin P an n-fold point.

It is easily seen that for any point Q_ξ^* of $S_k(V_t^*)$ and for any specialization $\{t\} \to \{\bar{t}\}$ (satisfying (8)), the point $Q_{\xi,\bar{t}}^*$ belongs to the hypersurface $V_{\bar{t}}^*$. This follows from the fact that the t's are indeterminates over the local domain $k[[\xi_1, \xi_2, \ldots, \xi_N]]$. Namely, from what was pointed out earlier about the relationship between $h(\mathfrak{p}^*)$ and the ideal $k[\{t\}] \otimes_k \mathfrak{p}$, where $\mathfrak{p}^*$ is the prime ideal of Q_ξ^* in $k^*[[X_1, X_2, \ldots, X_{r+1}]]$ and $\mathfrak{p}$ is the prime ideal of the point $(\xi_1, \xi_2, \ldots, \xi_N)$ in $k[[Y_1, Y_2, \ldots, Y_N]]$, it follows that since $f^*(\xi_1^*, \xi_2^*, \ldots, \xi_{r+1}^*; \{t\}) = 0$, we must have an identity of the form

$$f^*(\varphi_1(\{Y\}; \{t\}), \varphi_2(\{Y\}; \{t\}), \ldots, \varphi_{r+1}(\{Y\}; \{t\})) = \sum_{\mu=1}^{m} A_\mu(\{Y\}; \{t\}) g_\mu(\{Y\}), \quad (10)$$

where the A_μ are power series in $Y_1, Y_2, \ldots, Y_N$, with coefficients in $k[\{t\}]_M$, and the $g_\mu(\{Y\}) \in \mathfrak{p}$, i.e., $g_\mu(\xi_1, \xi_2, \ldots, \xi_N) = 0$. From (10) it follows therefore that

$$f^*(\varphi_1(\{\xi\}; \{\bar{t}\}), \varphi_2(\{\xi\}; \{\bar{t}\}), \ldots, \varphi_{r+1}(\{\xi\}; \{\bar{t}\}); \{\bar{t}\}) = 0,$$

i.e., $f^*(\bar{\xi}_1^*, \bar{\xi}_2^*, \ldots, \bar{\xi}_{r+1}^*; \{\bar{t}\}) = 0$ [see (6)], whence $Q_{\xi,\bar{t}}^* \in V_{\bar{t}}^*$, as was asserted.

We first consider the special case in which Q_ζ^* is a *simple* point of V_t^*. To prove part (*a*) of the Lemma in this special case we have to prove that *for almost all specializations* $\{t\} \to \{\bar{t}\}$, *the point* $Q_{\zeta,\bar{t}}$ *is a simple point of* $V_{\bar{t}}^*$. Since Q_ζ^* is a simple point of V_t^*, one of the $r + 1$ partial derivatives $\partial f^*/\partial X_i$ is different from zero at Q_ζ^*. Let, say, $\partial f^*/\partial X_1$ be different from zero at Q_ζ^*. We shall show that for almost all specializations $\{t\} \to \{\bar{t}\}$ the partial derivative $\partial f^*(\{X\}; \{\bar{t}\})/\partial X_1$ is different from zero at $Q_{\xi,\bar{t}}^*$, and this will complete the proof of part (*a*) of the lemma in the special case in which Q_ζ^* is a simple point of V_t^*.

By a suitable linear homogeneous transformation of the coördinates $\zeta_1, \zeta_2, \ldots, \zeta_N$, with coefficients in k, we may assume that if ρ is the dimension of the local domain $\mathfrak{o} = k[[\zeta_1, \zeta_2, \ldots, \zeta_N]]$ then $\zeta_1, \zeta_2, \ldots, \zeta_\rho$ are parameters of that local domain. Since we are dealing with a *complete* (noetherian) local domain it follows that the local domain $k[[\{\zeta\}]]$ is a finite module of the (regular) ρ-dimensional local domain $R = k[[\zeta_1, \zeta_2, \ldots, \zeta_\rho]]$. Let $\omega_1, \omega_2, \ldots, \omega_h$ be an R-module basis of $\mathfrak{o}$, and let d be the relative degree of the quotient field K of $\mathfrak{o}$ over to the quotient field K_0 of R (the former being obviously a finite algebraic extension of the latter). We have then necessarily $h \geq d$, since obviously $\omega_1, \omega_2, \ldots, \omega_h$ is also a K_0-basis of K. We may assume that $\omega_1, \omega_2, \ldots, \omega_d$ are linearly independent over K_0 and therefore form a linearly independent basis of K over K_0. There exists therefore an element $c(\zeta_1, \zeta_2, \ldots, \zeta_\rho)$ in R, different from zero, such that

$$c(\zeta_1, \zeta_2, \ldots, \zeta_\rho) \cdot \omega_{d+j} \in \sum_{\nu=1}^{d} R \cdot \omega_\nu, \qquad j = 1, 2, \ldots, h - d.$$

Therefore

$$c(\zeta_1, \zeta_2, \ldots, \zeta_\rho) \cdot \mathfrak{o} \subset \sum_{\nu=1}^{d} R \cdot \omega_\nu. \tag{11}$$

We denote the partial derivative $\partial f^*(X_1, X_2, \ldots, X_{r+1}; \{t\})/\partial X_1$ by $f_1^*(X_1, X_2, \ldots, X_{r+1}; \{t\})$. If $\zeta_1^*, \zeta_2^*, \ldots, \zeta_{r+1}^*$ are the X-coördinates of Q_ζ^*, then, by assumption, we have

$$f_1^*(\zeta_1^*, \zeta_2^*, \ldots, \zeta_{r+1}^*; \{t\}) \neq 0. \tag{12}$$

Upon replacing the ζ^*'s by their expressions in terms of $\zeta_1, \zeta_2, \ldots, \zeta_N$ and the t's we have that

$$f_1^*(\zeta_1^*, \zeta_2^*, \ldots, \zeta_{r+1}^*; \{t\}) = F(\zeta_1, \zeta_2, \ldots, \zeta_N; \{t\}),$$

where F is a power series in $\zeta_1, \zeta_2, \ldots, \zeta_N$ with coefficients in $k[\{t\}]_M$. From (11) it follows therefore at once that

$$c(\zeta_1, \zeta_2, \ldots, \zeta_\rho) f_1^*(\zeta_1^*, \zeta_2^*, \ldots, \zeta_{r+1}^*; \{t\}) = \sum_{\nu=1}^{d} F_\nu(\zeta_1, \zeta_2, \ldots, \zeta_\rho; \{t\}) \omega_\nu, \tag{13}$$

where the F_ν are power series in $\zeta_1, \zeta_2, \ldots, \zeta_\rho$, with coefficients in $k[\{t\}]_M$. From (12) it follows that we must have $F_\nu(\zeta_1, \zeta_2, \ldots, \zeta_\rho; \{t\}) \neq 0$, for some value of ν. Let, say, $F_1(\zeta_1, \zeta_2, \ldots, \zeta_\rho; \{t\}) \neq 0$, and let $\alpha_{i_1 i_2 \cdots i_n}(\{t\})/a(\{t\})^q$ be a coefficient of the power series F_1 which is different from zero. Here $\alpha_{i_1 i_2 \cdots i_n}(\{t\})$ is a polynomial in the t's, and the above quotient $\alpha_{(i)}/a^q$ is the coefficient of the term $\zeta_{i_1}\zeta_{i_2} \cdots \zeta_{i_n}$, where $1 \leq i_1 \leq i_2 \leq \cdots \leq i_n \leq \rho$, and n is some positive integer. We impose on our specializations $\{t\} \to \{\bar{t}\}$ the following inequality (in addition to the inequalities (8) and (9)):

$$\alpha_{i_1 i_2 \cdots i_n}(\{\bar{t}\}) \neq 0. \tag{14}$$

Since $\zeta_1, \zeta_2, \ldots, \zeta_\rho$ are analytically independent over k they are also analytically independent over $k(\{t\})$ (always by the usual argument based on our assumption that the t's are indeterminates over $k[[\zeta_1, \zeta_2, \ldots, \zeta_N]]$). Since $k(\{\bar{t}\})$ is a subfield of $k(\{t\})$, it follows that $\zeta_1, \zeta_2, \ldots, \zeta_\rho$ are also analytically independent over the field $k(\{\bar{t}\})$. It follows therefore from (14) that $F_1(\zeta_1, \zeta_2, \ldots, \zeta_\rho; \{\bar{t}\}) \neq 0$. Since $\omega_1, \omega_2, \ldots, \omega_d$ are linearly independent over the quotient field K_0 of $k[[\zeta_1, \zeta_2, \ldots, \zeta_\rho]]$, they are also linearly independent over $k(\{t\})[[\zeta_1, \zeta_2, \ldots, \zeta_\rho]]$ and therefore also over $k(\{\bar{t}\})[[\zeta_1, \zeta_2, \ldots, \zeta_\rho]]$. It follows therefore from (13) and from the inequality $F_1(\zeta_1, \zeta_2, \ldots, \zeta_\rho; \{\bar{t}\}) \neq 0$ that $f_1^*(\zeta_1^*, \zeta_2^*, \ldots, \zeta_{r+1}^*; \{\bar{t}\}) \neq 0$. This completes the proof of part (a) of the lemma in the case of a simple point Q_ζ^* of V_t^*. As to part (*b*) of the lemma, it is trivial in the case of a simple point Q_ζ^*. Namely, from the assumption that the point $(\zeta_1, \zeta_2, \ldots, \zeta_N)$ is a specialization, over k, of the point $(\xi_1, \xi_2, \ldots, \xi_N)$, it follows that the point Q_ζ^* is a specialization of R_ξ^*, over k^*. Hence, since Q_ζ^* is a simple point of V_t^* also R_ξ^* is necessarily a simple point of V_t^*. Therefore in this case we always have $R_\xi^* \in \mathrm{EQS}_{k^*}(V_t^*, Q_\zeta^*)$, i.e., the implication $\Rightarrow$ in part (*b*) of the lemma is trivially satisfied. On the other hand, from the fact that Q_ζ^* is a specialization of R_ξ^* over k^* follows trivially that $Q_{\zeta,\bar{t}}^*$ is a specialization of $R_{\xi,\bar{t}}^*$ over $k(\{\bar{t}\})$. Since $Q_{\zeta,\bar{t}}^*$ is a simple point of $V_{\bar{t}}^*$ if (14) holds true, also $R_{\xi,\bar{t}}^*$ is then necessarily a simple point of $V_{\bar{t}}^*$, i.e., $R_{\xi,\bar{t}}^* \in \mathrm{EQS}_{k(\{\bar{t}\})}(V_{\bar{t}}^*, Q_{\zeta,\bar{t}}^*)$. We have therefore proved that, if Q_ζ^* is a simple point of V_t^* then all the conditions stated in the lemma are satisfied for all the specializations $\{t\} \to \{\bar{t}\}$ such that

$$a(\{\bar{t}\}) \cdot b(\{\bar{t}\}) \cdot \alpha_{i_1 i_2 \cdots i_n}(\{\bar{t}\}) \neq 0.$$

From now on we assume that Q_ζ^* is a singular point of V_t^*, whence $\sigma = d.t.(V_t^*, Q_\zeta^*)/k^* \geq 1$. In this case we shall use induction with respect to r. If $r = 1$ $V_t{}^*$ is a plane (reduced) algebroid curve, defined over k^*. The only point of the curve $V_t{}^*$ which can be singular is its closed point P (the origin $X_1 = X_2 = 0$). Therefore $Q_\zeta{}^*$ is necessarily the point P. The remaining points of $V_t{}^*$ are the q general points of the q irreducible branches of $V_t{}^*$, and these are simple points of $V_t{}^*$. Therefore, in the present case the set $(V_t{}^*)_{Q_\zeta{}^*}$ coincides with $(V_t{}^*)_P$, i.e., with $V_t{}^*$, and the set $\mathrm{EQS}_{k^*}(V_t{}^*, P)$ consists of the point P only. If then the inequalities (8) and (9) are satisfied then, as was pointed out earlier (for arbitrary values of r), the reduced curve $V_{\bar{t}}{}^*$ has at P a singular point (in fact, a point of the *same* multiplicity n as the multiplicity of P for $V_t{}^*$). Thus $d.t.(V_{\bar{t}}{}^*, P)/k(\{\bar{t}\}) = d.t.(V_t{}^*, P)/k^*(= 1)$, and condition ($a$) of the lemma is satisfied for all the specializations $\{t\} \to \{\bar{t}\}$ such that $a(\{\bar{t}\}) \cdot b(\{\bar{t}\}) \neq 0$. As to condition ($b$) of the lemma, it is vacuous, since the set $\mathrm{EQS}_{k(\{\bar{t}\})}(V_{\bar{t}}{}^*, P)$ consist of the point P only.

From now on we shall assume that the lemma is true in dimension $r - 1$. We introduce the generic projection $\pi_u{}^*$ of $V_t{}^*/k^*$ onto the affine variety $\mathbf{A}_r/k^{**}$, where $k^{**} = k^*(\{u\})$ [see Section 2, where k is to be replaced by $k^*(= k(\{t\}))$]. Let $\Delta_{t,u}{}^{**}$ be the critical hypersurface of $\pi_u{}^*$, and let $\pi_u{}^*(Q_\zeta{}^*) = Q_{\zeta,u}{}^{**}$. Since $Q_\zeta{}^*$ is a singular point of $V_t{}^*$, the point $Q_{\zeta,u}{}^{**}$ belongs necessarily to $\Delta_{t,u}{}^{**}$, and we have, by the definition of dimensionality type:

$$d.t.(\Delta_{t,u}{}^{**}, Q_{\zeta,u}{}^{**})/k^{**} = d.t.(V_t{}^*, Q_\zeta{}^*)/k^* - 1 = \sigma - 1. \tag{15}$$

We apply our inductive hypothesis to the $(r - 1)$-dimensional variety $\Delta_{t,u}{}^{**}$. In applying our lemma in this case, we replace k by the field $k(\{u\})$, the t's being indeterminates over this field. It follows that for almost all specializations $\{t\} \to \{\bar{t}\} \subset k(\{u\})[\{t\})]$—*and here we shall restrict ourselves to specializations in which the* $\bar{t}$'s *belong to* $k[\{t\}]$—the hypersurface $\Delta_{\bar{t},u}{}^{**}$ is well defined and is reduced, and that, by part (a) of the lemma, and by (15) we have

$$d.t.(\Delta_{\bar{t},u}{}^{**}, Q_{\zeta,\bar{t},u}{}^{**}) = \sigma - 1. \tag{16}$$

Here $Q_{\zeta,\bar{t},u}{}^{**}$ is the specialization of the point $Q_{\zeta,u}{}^{**}$ under $\{t\} \to \{\bar{t}\}$. Now, under this specialization the projection $\pi_u{}^*$ is specialized to the generic projection $\pi_{\bar{t},u}{}^*$ of $V_{\bar{t}}{}^*$, as follows at once from the definition of the generic projection given in Section 2. It is also clear that $\Delta_{\bar{t},u}^{**}$ is the (generic) critical hypersurface of the projection $\pi_{\bar{t},u}^*$ of $V_{\bar{t}}^*$ and that $\pi_{\bar{t},u}^*(Q_{\zeta,\bar{t}}^*) = Q_{\zeta,\bar{t},u}^{**}$. It follows therefore from (16) and from the definition of dimensionality type that

$$d.t.(V_{\bar{t}}^*, Q_{\zeta,\bar{t}}^*)/k(\{\bar{t}\}) = \sigma,$$

and this proves part (*a*) of the lemma.

To prove part (*b*) of the lemma, we set $R_{\xi,u}^{**} = \pi_u^*(R_\xi^*)$. Since $(\zeta_1, \zeta_2, \ldots, \zeta_N)$ is a specialization of $(\xi_1, \xi_2, \ldots, \xi_N)$ over k, it follows that *$Q_{\zeta,u}^{**}$ is a specialization of $R_{\xi,u}^{**}$ over $k^*(\{u\})$*$(= k^{**})$. From the definition of dimensionality type we have

$$d.t.(\Delta_{t,u}^{**}, R_{\xi,u}^{**})/k^{**} = d.t.(V_t^*, R_\xi^*)/k^* - 1,$$

and hence, by (15), it follows that

$$R_{\xi,u}^{**} \in \mathrm{EQS}_{k^{**}}(\Delta_{t,u}^{**}, Q_{\zeta,u}^{**}) \Leftrightarrow R_\xi^* \in \mathrm{EQS}_{k^*}(V_t, Q_\zeta^*). \quad (16')$$

In applying our induction hypothesis to $\Delta_{t,u}^{**}$, we replace V_t^* by $\Delta_{t,u}^{**}$, the fields k and k^* of the lemma by the fields $k(\{u\})$ and $k^{**}[= k(\{u\}, \{t\})]$ respectively, and the point Q_ζ^* by the point $Q_{\zeta,u}^{**}$. Since $(\zeta_1, \zeta_2, \ldots, \zeta_N)$ is a specialization of $(\xi_1, \xi_2, \ldots, \xi_N)$ over k, hence also over $k(\{u\})$, it follows that for almost all specializations $\{t\} \to \{\bar{t}\} \subset k^{**}$, and hence also for almost all specializations $\{t\} \to \{\bar{t}\} \subset k^*(= k(\{t\})$, we have, for all points $R_{\xi,u}^{**}$ such that $(\xi_1, \xi_2, \ldots, \xi_N) \xrightarrow{k} (\zeta_1, \zeta_2, \ldots, \ldots, \zeta_N)$:

$$R_{\zeta,\bar{t},u}^{**} \in \mathrm{EQS}_{k(\{u\},\{\bar{t}\})}(\Delta_{\bar{t},u}^{**}, Q_{\zeta,\bar{t},u}^{**})$$
$$\Leftrightarrow R_{\xi,u}^{**} \in \mathrm{EQS}_{k^{**}}(\Delta_{t,u}^{**}, Q_{\zeta,u}^{**}). \quad (17)$$

Hence, by (16′), we have for almost all specializations $\{t\} \to \{\bar{t}\} \subset k(\{t\})$:

$$R_{\xi,\bar{t},u}^{**} \in \mathrm{EQS}_{k(\{u\},\{\bar{t}\})}(\Delta_{\bar{t},u}^{**}, Q_{\zeta,\bar{t},u}^{**}) \Leftrightarrow R_\xi^* \in \mathrm{EQS}_{k^*}(V_t^*, Q_\zeta^*). \quad (18)$$

Now each of our specializations $\{t\} \to \{\bar{t}\}$ defines a generic projection $\pi_{\bar{t},u}^*$ of $V_{\bar{t}}^{**}$, and the critical hypersurface of $\pi_{\bar{t},u}^*$ is the hypersurface

$\Delta_{\bar{t},u}{}^{**}$. We have also

$$\pi_{\bar{t},u}{}^{*}(Q_{\zeta,\bar{t}}{}^{*}) = Q_{\zeta,\bar{t},u}{}^{**},$$

$$\pi_{\bar{t},u}{}^{*}(R_{\xi\bar{t}}{}^{*}) = R_{\xi,\bar{t},u}.$$

Hence, by (18), and using the very definition of the dimensionality type (based on the use of generic projections), we find that part (b) of the lemma is valid.

This concludes the proof of our basic lemma.

Before we proceed to the proof of the basic Theorem 5.1 we draw some preliminary consequences of Lemma 5.2. We go back to the r-dimensional algebroid (reduced) hypersurface V/k dealt with in the previous sections.

PROPOSITION 5.3. *Given a singular point Q of V, of dimensionality type σ, the following two conditions are satisfied for almost all k-rational projections $\pi_{\bar{u}}$ of V onto $\mathbf{A}_r/k$ and for all points R of V such that Q is a specialization of R over k:*

(a) $d.t.(\Delta_{\bar{u}}, \overline{Q}_0)/k = \sigma - 1$,

(b) $\pi_{\bar{u}}(R) \in \mathrm{EQS}_k(\Delta_{\bar{u}}, \overline{Q}_0) \Leftrightarrow R \in \mathrm{EQS}_k(V, Q)$,

where $\overline{Q}_0 = \pi_{\bar{u}}(Q)$ and $\Delta_{\bar{u}}$ is the critical hypersurface of $\pi_{\bar{u}}$.

Proof. We apply Lemma 5.2 to the case in which the r-dimensional variety V_t is replaced by the $(r - 1)$-dimensional critical hypersurface Δ_u of the generic projection π_u of V, the indeterminates u now playing the role of the parameters t. It follows from Lemma 5.2 that for almost all specializations $\{u\} \to \{\bar{u}\} \subset k$ the specialized hypersurface $\Delta_{\bar{u}}$ is well defined and is reduced. Furthermore, it is clear that for almost all specializations $\{u\} \subset \{\bar{u}\} \subset k$, the specialized quantities $\bar{x}_i{}^*$ (where the r elements $x_i{}^*$ are defined in (1) of Section 2) form a set of parameters of the local ring $\mathfrak{o}$ of V (equivalently: the determinant $|\bar{u}_j{}^{(i)}|$ is different from zero) and hence define a finite projection $\pi_{\bar{u}}$ of V onto $\mathbf{A}_r(k)$.

The hypersurface $\Delta_{\bar{u}}$ is then the critical hypersurface of $\pi_{\bar{u}}$. We take as set $S_k(\Delta_u)$ the set of points of Δ_u which are π_u-projections R_0 of (necessarily singular) points R of V. The $r + 1$ power series $\varphi_i(\xi_1, \xi_2, \ldots, \xi_N; \{t\})$ of Lemma 5.2 [see (5)] and the algebroid variety Z/k are now respectively the r power series given in (1), Section 2 (where $x_1, x_2, \ldots, x_{r+1}$ are to be replaced by $\xi_1, \xi_2, \ldots, \xi_{r+1}$, the integer N now being equal to $r + 1$) and Z being now our original variety V. Our proposition is then a direct consequence of Lemma 5.2, the conditions (*a*) and (*b*) of the proposition being the analogues of the conditions (*a*) and (*b*) of the lemma.

PROPOSITION 5.4 *Let Q be a point of V, of dimensionality type σ, and let $W(Q)$ denote the algebroid closure in V of the set* $\mathrm{EQS}_k(V, Q)$ *(i.e., $W(Q)$ is the smallest algebroid subvariety of V containing the set* $\mathrm{EQS}_k(V, Q)$*). Then $W(Q)$ is an irreducible subvariety of V, of dimension $r - \sigma$, whose general point R belong to the set* $\mathrm{EQS}_k(V, Q)$*, and it is the greatest irreducible subvariety of V which contains Q and which, in addition, has the property that its general point has dimensionality type σ. Furthermore, $W(Q)$ is smooth at Q, and V is equimultiple at Q along $W(Q)$. In particular, we have $W(P) =$* $\mathrm{EQS}_k(V, P)$*, where P is the closed point of V.*

Proof. The proportion is trivial if Q is a simple point of V. In that case there is only one irreducible component V_i of V which contains Q. If, say, V_1 is that component then it is clear that $\mathrm{EQS}_k(V, Q)$ is a subset of $V_1 - (V_1 \cap V_{sing})$ containing the general point R of V_1/k and that therefore $W(Q) = V_1$. Hence the dimension of $W(Q)$ is in this case equal to r, while σ is equal to zero. The general point R of V_1 is of course a simple point of V, and Q is a specialization of R over k. The rest of the proposition is obvious in this case (note that if P is a simple point, then $\mathrm{EQS}_k(V, P) = V$). We assume from now on that Q is a singular point of V. We shall use induction with respect to r, the proposition being trivial in the case $r = 1$, for in that case the point Q, being a singular point of the plane curve V, must be the closed point P of V and hence $W(P) = \mathrm{EQS}_k(V, P) = P$. We assume now that the proposition is true in the case of $(r - 1)$-dimensional algebroid varieties.

We shall say that a *k-rational projection* $\pi_{\bar{u}}$ of V (i.e., a $\pi_{\bar{u}}$ such that the $\bar{u}$'s belong to k) is *Q-permissible* if it satisfies the conditions (*a*) and (*b*) of Proposition 5.3. We fix a Q-permissible k-rational projection $\pi_{\bar{u}}$ of V and use the notations of Proposition 5.3, except that we write π and Δ instead of $\pi_{\bar{u}}$ and $\Delta_{\bar{u}}$. Since Q is a singular point of V, of dimensionality type σ, and since π is a Q-permissible projection of V, it follows that the π-projection $\overline{Q}_0$ of Q belongs to Δ and has dimensionality type $\sigma - 1$ $(\sigma \geq 1)$. Since Δ has dimension $r - 1$ it follows from our induction hypothesis that if we set $Z_0 = \mathrm{EQS}_k(\Delta, \overline{Q}_0)$ then the algebroid closure $\overline{Z}_0$ of Z_0 in Δ is an irreducible subvariety of Δ, of dimension $r - \sigma$ $(= (r - 1) - (\sigma - 1))$, having at $\overline{Q}_0$ a simple point, and that Δ is equimultiple at $\overline{Q}_0$ along $\overline{Z}_0$. It follows therefore from a generalized version of our PNAS Note [2] (see Appendix) that the full inverse image $\pi^{-1}\{\overline{Z}_0\}$ of $\overline{Z}_0$ on V has a simple point at Q and that V is equimultiple at Q along $\pi^{-1}\{\overline{Z}_0\}$. Since π is a finite projection, the set $\pi^{-1}\{\overline{Z}_0\}$ is an algebroid subvariety of V, of pure dimension $r - \sigma$. Since Q is a simple point of that subvariety, there is only one irreducible component of $\pi^{-1}\{\overline{Z}_0\}$ which contain the point Q. We denote by Z that component. Let R be the general point of Z. Then R is a point of V, of dimension $r - \sigma$, and the equimultiplicity of V at Q along $\pi^{-1}\{\overline{Z}_0\}$ signifies that R and Q are points of V *of the same* multiplicity. From the implication $\Leftarrow$ in condition (b) of Proposition 5.3 it follows that the set $\mathrm{EQS}_k(V, Q)$ is contained in $\pi^{-1}\{Z_0\}$. If Q' is any point of that set then Q' must belong to at least one irreducible component of $\pi^{-1}\{Z_0\}$, and any such component must also contain Q, since Q is a specialization of Q', over k. It follows that *the set* $\mathrm{EQS}_k(V, Q)$ *is contained in* Z. Since R is the general point of Z, the point $\overline{R}_0(= \pi(R))$ must be the general point of $\overline{Z}_0$ (since R and R_0 have the same dimension over k, whence the dimension of $\overline{R}_0$ is $r - \sigma$, equal to the dimension of $\overline{Z}_0$). By our induction hypothesis, the general point $\overline{R}_0$ of $\overline{Z}_0$ belongs to the set $\mathrm{EQS}_k(\Delta, \overline{Q}_0)$, i.e., we have $d.t.(\Delta, \overline{R}_0)/k = d.t.(\Delta, \overline{Q}_0)/k = \sigma - 1$. Since Q is a specialization of R over k, it follows from the implication $\Rightarrow$ in Proposition 5.3 that $R \in \mathrm{EQS}_k(V, Q)$, whence $d.t._k(V, R) = \sigma$. From the two facts just established above, namely that 1) $\mathrm{EQS}_k(V, Q) \subset Z$ and that 2) the general point R of Z is contained in $\mathrm{EQS}_k(V, Q)$, follows that Z is the algebroid closure of the set $\mathrm{EQS}_k(V, Q)$, i.e., *we have* $Z = W(Q)$. Thus the variety $W(Q)$ of the proposition has all the properties stated in that proposition, except that we still have to show that $W(Q)$ is the greatest irreducible subvariety of V which contains Q and whose general point has dimensionality type σ. Let then Z' be any irreducible subvariety of V which contains Q and whose general point R' has dimensionality type σ. Since Q is a specialization of R', it follows that $R' \in \mathrm{EQS}_k(V, Q)$. Hence $R' \in W(Q)$, showing that $Z' \subset W(Q)$. This proves the maximality property of the variety $W(Q)$.

Finally, if Q is the point P, then the set $\mathrm{EQS}_k(V, P)$ is the set of *all* points of V which have dimensionality type s, since Q is a specialization of any point of V. Hence $\mathrm{EQS}_k(V, P) = V(s)$. Now, let Q' be any point of $W(P)$. We have then that Q' is a specialization of R and P is a specialization of Q'. Hence, by Proposition 4.2, we have $d.t._k(V, R)$

$\leq d.t._k(V, Q') \leq d.t._k(V, P)$. Since $d.t._k(V, R) = d.t._k(V, P) = s$, it follows that $d.t._k(V, Q') = s$, whence $Q' \in \mathrm{EQS}_k(V, P)$. This shows that $W(P) = \mathrm{EQS}_k(V, P)$.

This completes the proof of the proposition.

We shall explicitly state the following inequality which holds for any point Q of V and which is a direct consequence of Theorem 5.4:

$$\dim Q/k + d.t.(V, Q) \leq r. \tag{19}$$

For, if $d.t.Q = \sigma$ then we have $Q \in W(Q)$ and $\dim (W(Q)) = r - \sigma$, whence $\dim Q/k \leq r - \sigma$, and this is precisely (19).

The inequality (19) will be used very frequently in the next section.

We now proceed to the proof of the basic Theorem 5.1. We begin by observing that *for any integer* σ, $0 \leq \sigma \leq s$, such that $V(\sigma) \neq \emptyset$, *there is only a finite number of points* R *on* V *such that* $d.t._k(V, R) = \sigma$ *and* $\dim R/k = r - \sigma$. This assertion is trivial in the case $r = 1$, for in that case V itself is a finite set of points. It is also trivial for any value of r if $\sigma = 0$. Assuming that this assertion is true for algebroid varieties of dimension $r - 1$, we consider the generic projection π_u of V and the critical hypersurface Δ_u of π_u. By induction, we have that Δ_u has only a finite number of points of a given dimensionality type $\sigma - 1$ $(\sigma \geq 1)$ (over the ground field $k^* = k(\{u\})$) and of dimension $r - \sigma$ $(= (r - 1) - (\sigma - 1))$ over k^*. Now, since π_u is a finite projection, any point of Δ_u is the projection of only a finite number of points of V^*, where V^* is the algebroid variety obtained from V by the ground field extension $k \to k^*$. If R is any point of V, of dimensionality type σ over k and of dimension $r - \sigma$ over k, then R is also a point of V^*, of dimensionality type σ over k^* and of dimension $r - \sigma$ over k^* (see Proposition 4.3), and $\pi_u(R)$ is a point of Δ_u of dimensionality type $\sigma - 1$ and of dimension $r - \sigma$ over k^*. From this our assertion follows.

As a consequence it follows that as the point Q ranges over the entire set $V(\sigma)$ of points of dimensionality type σ (assuming of course that $V(\sigma)$ is non-empty) the variety $W(Q)$ ranges over a *finite set of varieties*. This finite set of varieties consists of all irreducible $(r - \sigma)$-dimensional algebroid subvarieties of V having as general point a point of dimensionality type σ. (Note that if W is any such variety and if R is the general point of W/k then $\mathrm{EQS}(V, R)$ consists of the single point R and $W(R)$ is W.) Let $\{W_1, W_2, \ldots, W_{q_\sigma}\}$, $q_\sigma \geq 1$, be this finite set of varieties and let R_i be the general point of W_i. Since every point

Q of $V(\sigma)$ belongs to one of these q_σ varieties it follows that if we set

$$S_i = W_i \cap V(\sigma), \qquad (i = 1, 2, \ldots, q_\sigma) \tag{20}$$

then

$$V(\sigma) = \bigcup_{i=1}^{q_\sigma} S_i. \tag{21}$$

It is easy to see that $S_i \cap S_j = \emptyset$ if $i \neq j$, and that consequently (21) *is a decomposition of* $V(\sigma)$ *into disjoint sets* S_i. This follows directly from the fact that, by Proposition 5.4, every point Q of $V(\sigma)$ belongs to one and only one irreducible $(r - \sigma)$-dimensional subvariety of V whose general point has dimensionality type σ.

We recall that we agreed to denote by $V\{\sigma\}$ the set of all points of V which have dimensionality type $\geq \sigma$. *We prove now that* $V\{\sigma\}$ *is a subvariety of* V. We use induction from $\sigma + 1$ to σ, since $V\{s\}$ *coincides with* $V(s)$, and $V(s) = \mathrm{EQS}_k(V, P) = W(P)$, by Proposition 5.4. We have $V\{\sigma\} = V(\sigma) \cup V\{\sigma + 1\}$. Since by our induction hypothesis $V\{\sigma + 1\}$ is a variety, we have $\overline{V\{\sigma\}} \subset \overline{V(\sigma)} \cup V\{\sigma + 1\}$. Now, by (21), we have $\overline{V(\sigma)} = \bigcup_{i=1}^{q_\sigma} W_i$ since obviously $\overline{S_i} = W_i$ (the *subset* S_i of the *variety* W_i contains the general point R_i of W_i). Since every point of W_i is a specialization of R_i and since $d.t.(V, R_i) = \sigma$, it follows from Proposition 4.2 that $W_i \subset V\{\sigma\}$. Hence $\overline{V(\sigma)} \subset V\{\sigma\}$, and therefore from the above inclusion $\overline{V\{\sigma\}} \subset \overline{V(\sigma)} \cup V\{\sigma + 1\}$ follows the inclusion $\overline{V\{\sigma\}} \subset V\{\sigma\} \cup V\{\sigma + 1\} = V\{\sigma\}$, and this proves that $\overline{V\{\sigma\}} = V\{\sigma\}$ and that consequently $V\{\sigma\}$ is a variety.

It was pointed out in the above proof that $\overline{S_i} = W_i$. We prove next that $W_i - S_i$ *is a proper subvariety of the (irreducible) variety* W_i *(and that consequently each set* S_i *is a Zariski dense open subset of its algebroid closure* $\overline{S_i}$). We have $\overline{S_i} \not\subset V\{\sigma + 1\}$ (since $R_i \in S_i \subset \overline{S_i}\}$. Since we have just proved that $V\{\sigma + 1\}$ is a variety it follows that $\overline{S_i} \cap V\{\sigma + 1\}$ is a proper subvariety of $\overline{S_i}$. Since $S_i \subset V(\sigma)$ we have $S_i \cap V\{\sigma + 1\} = \emptyset$, and therefore $\overline{S_i} \cap V\{\sigma + 1\} \subset \overline{S_i} - S_i$. On the other hand, if Q' is any point of $W_i - S_i (= \overline{S_i} - S_i)$ then $Q' \in W_i$ and $Q' \neq S_i (= W_i \cap V(\sigma))$, whence $d.t._k(V, Q') > \sigma$ (by Proposition 4.2 we have $d.t._k(V, Q') \geq \sigma$ for any point Q' of W_i, since the general point R_i of W_i has dimensionality type σ). This shows that $Q' \in V\{\sigma + 1\}$, and consequently $\overline{S_i} - S_i \subset V\{\sigma + 1\}$. This, in conjunction with the above inclusion $\overline{S_i} \cap V\{\sigma + 1\} \subset \overline{S_i} - S_i$, implies that $\overline{S_i} - S_i = \overline{S_i} \cap V\{\sigma + 1\}$ = a proper subvariety of $\overline{S_i}$, as was asserted.

This completes the proof of part (*a*) of Theorem 5.1.

Parts (*b*) and (*c*) of Theorem 5.1 are included in the statement of Theorem 5.4 which we have already proved. We now proceed to the proof of part (*d*) of Theorem 5.1.

Let S be a stratum of V, let R be the general point of S (i.e., the general point of the irreducible subvariety $\overline{S}$ of V which is the algebroid closure of S) and let $\sigma = d.t.(V, R)$. Let Q' be any point of the boundary $\overline{S} - S$ of S and let $\sigma' = d.t.(V, Q')$. We have to prove that if S' denotes the stratum containing the point Q' then $S' \subset \overline{S} - S$.

We first observe that, by Proposition 4.2, we have $\sigma' \geq \sigma$ (since Q' is a specialization of R over k). We know that $S = \overline{S} \cap V(\sigma)$ [see (20)]. Since $Q' \in \overline{S} - S$ it follows that $Q' \notin V(\sigma)$. Hence $d.t.(V, Q') \neq \sigma$, and this implies that $\sigma' > \sigma$. Let S' denote the stratum containing the point Q' and let R' be the general point of S'. *We have to prove that $R' \in \overline{S}$,* for once this is proved it will follow that $R' \in \overline{S} - S$ (since $d.t.(V, R') = d.t.(V, Q') = \sigma' > \sigma$) and that therefore the entire stratum S' belongs to the boundary $\overline{S} - S$ of the stratum S.

To prove part (*d*) of Theorem 5.1 we shall use induction on the dimension r of V, since for $r = 1$ part (*d*) of that theorem is trivial (the only point Q' of a plane algebroid curve V which can (and does) belong to the *boundary* of a stratum is *the closed point* P of V, provided P is a singular point of V; in this case the stratum S' containing Q' is the point $Q'(= P)$ itself, and therefore $R' = Q'(= P)$). We therefore assume that part (*d*) of Theorem 5.1 is true in dimension $r - 1$.

We first consider the case $\sigma = 0$, in which case S is a stratum of non-singular points. Therefore, as has been pointed out in the beginning of this section, we have $S = V_1 - (V_1 \cap V_{sing})$, where V_1 is some irreducible component of V. The boundary of S is the variety $V_1 \cap V_{sing}$. We have to show that $R' \in V_1 \cap V_{sing}$. By Theorem 5.4 we know that R' and Q' are points of V of the same multiplicity. Since the multiplicity of R' on V is the sum of the multiplicities $e_{R'}(V_i)$ of R' as a point of V_i ($i = 1, 2, \ldots, q$), and since we have $e_{R'}(V_i) \leq e_{Q'}(V_i)$ for each i (since Q' is a specialization of R'), the equality

$$\sum_{i=1}^{q} e_{R'}(V_i) = \sum_{i=1}^{q} e_P(V_i)$$

implies that $e_{R'}(V_i) = e_{Q'}(V_i)$, for $i = 1, 2, \ldots, q$. Since the point Q' belongs to V_1 it follows that also R' belongs to V_1, and therefore $R' \in V_1 \cap V_{sing}$ (since $\sigma' > 0$), showing that $R' \in \overline{S}(= V_1 - V_{sing})$.

We now deal with the case $\sigma > 0$. In this case the points R, Q' and R' are singular points of V. We consider the generic projection π_u of V onto $\mathbf{A}_r$. Let $\Delta_u{}^*$ be the critical hypersurface of π_u and let

$$\pi_u(R) = R_0{}^*, \qquad \pi_u(Q') = Q_0{}'^*, \qquad \pi_u(R') = R_0{}'^*$$

The points $R_0{}^*$, $Q_0{}'^*$ and $R_0{}'^*$ belong to $\Delta_u{}^*$, since R, Q' and R' are singular points of V. We have

$$\begin{aligned} &d.t.(\Delta_u{}^*, R_0{}^*)/k^* = \sigma - 1, \\ &d.t.(\Delta_u{}^*, Q_0{}'^*) = d.t.(\Delta_u{}^*, R_0{}'^*) = \sigma' - 1, \end{aligned} \tag{22}$$

and since Q' is a specialization of R over k, the point $Q_0{}'^*$ is a specialization of $R_0{}^*$ over k^*; here k^* is the field $k(\{u\})$. We have $\dim R_0{}^*/k^* = \dim R/k = r - \sigma$, $=(r-1) - (\sigma - 1)$, and since $d.t.(\Delta_u{}^*, R_0{}^*)/k^* = \sigma - 1$, it follows that $R_0{}'^*$ is the general point of the stratum $S_0{}^*$ containing $R_0{}^*$. Since $Q_0{}'^*$ is a specialization of $R_0{}^*$ over k^* and since the dimensionality type of $Q_0{}'^*$ on $\Delta_u{}^*$ is $\sigma' - 1$ [see (22)], hence different from $\sigma - 1$, it follows that $Q_0{}'^*$ belongs to the boundary $\overline{S_0}{}^* - S_0{}^*$ of the stratum $S_0{}^*$. By our induction hypothesis it follows then that the entire stratum $S_0{}'^*$ containing the point $Q_0{}'^*$ is contained in $\overline{S_0}{}^* - S_0{}^*$. Since $R_0{}'^*$ belongs to $S_0{}'^*$, and is in fact the general point of $S_0{}'^*$, we have then that $R_0{}'^* \in \overline{S_0}{}^* - S_0{}^*$. We proceed to show that this last inclusion implies the inclusion $R' \in \overline{S} - S$, and this will complete the proof of part (*d*) of Theorem 5.1.

We consider the full inverse image $\pi_u{}^{-1}\{R_0{}'^*\}$ of the point $R_0{}'^*$. This is a finite set of points, and the singular point R' is one point of that set. We shall show now that *all the remaining points of the set* $\pi_u{}^{-1}\{R_0{}'^*\}$ *are simple points of the variety* V/k^*. Once this is proved, the inclusion $R' \in \overline{S} - S$ can be proved as follows:

Since $R_0{}'^* \in \overline{S_0}{}^* - S_0{}^*$, $R_0{}'^*$ is a specialization of $R_0{}^*$ over k^*. Therefore some point $R_i{}'$ of the set $\pi_u{}^{-1}\{R_0{}'^*\}$ must be a specialization of the point R. Since R is a singular point of V, also $R_i{}'$ must be a singular point of V. Therefore $R_i{}'$ must be the point R'. We have therefore that R' is a specialization of R over k^*, and therefore also over k. This proves the desired inclusion $R' \in \overline{S} - S$.

Let $\xi_1, \xi_2, \ldots, \xi_{r+1}$ be the coördinates of R. The r coördinates ξ_i^* of the point R_0^* are the quantities $\varphi_i(\xi_1, \xi_2, \ldots, \xi_{r+1}; \{u\})$ [see (1), Section 2]. We recall that each function $\varphi_i(X_1, X_2, \ldots, X_{r+1}; \{u\})$ depends only on the indeterminates $\{u^{(i)}\}$, each indeterminate $u_{j_1 j_2 \cdots j_N}{}^{(i)}$ being the coefficient of the monomial $X_{j_1} X_{j_2} \cdots X_{j_N}$, N being an arbitrary positive integer, and $1 \leq j_1 \leq j_2 \leq \cdots \leq j_N \leq r + 1$. The points of the set $\pi_u{}^{-1}\{R_0^*\}$ are the intersections of V/k^* with the *algebraic* curve Γ_u^* in $\mathbf{A}_{r+1}/k^*$ defined by the r equations

$$\varphi_i(X_1, X_2, \ldots, X_{r+1}; \{u^{(i)}\}) - \xi_i^* = 0, \qquad i = 1, 2, \ldots, r. \qquad (23_i)$$

Since $\xi_i^* = \varphi_i(\xi_1, \xi_2, \ldots, \xi_{r+1}; \{u^{(i)}\})$, it follows that the lefthand side of (23_i) is a power series in the $X_j - \xi_j$, $j = 1, 2, \ldots, r + 1$, with coefficients in $k[[\{\xi\}]][\{u^{(i)}\}]$ (which are linear in the $u^{(i)}$). Therefore each of the equations (23_i) defined an algebroid hypersurface F_i having R as its closed point, with $k((\{\xi\}))(\{u^{(i)}\})$ as coefficient field. It is clear that each F_i is the general algebroid hypersurface in $\mathbf{A}_{r+1}/k$ having R as its closed point and that, moreover, each F_i is also the general hypersurface in $\mathbf{A}_{r+1}/K_{i-1}^*$ having R as its closed point; here K_{i-1}^* denotes the field $k((\{\xi\}))(\{u^{(1)}\}; \{u^{(2)}\}, \ldots, \{u^{(i-1)}\})$. The points of the set $\pi_u{}^{-1}\{R_0^*\}$ are the intersections of V/K_r^* with the algebroid curve Γ_u^* defined by the r equations (23_i). Since V is defined over k, also V_{sing} is an algebroid variety defined over k. Let ρ be the dimension of V_{sing}. The intersection $V_{sing} \cap F_1$ is clearly an algebroid variety Z_1 of dimension $\rho - 1$, defined over K_1. Since the $u^{(2)}$ are indeterminates over the field K_1, it follows that the intersection $Z_1 \cap F_2$ is an algebroid variety Z_2 of dimension $\rho - 2$, defined over K_2. Proceeding in this fashion we find that the intersection $V_{sing} \cap F_1 \cap F_2 \cap \cdots \cap F_\rho$ is a finite set of points, one of which is the point R. Since $\rho \leq r - 1$, it follows that the intersection $V_{sing} \cap F_1 \cap F_2 \cap \cdots \cap F_\rho \cap F_{\rho+1}$ is the single point R. Therefore $V \cap F_1 \cap F_2 \cap \cdots \cap F_r$ consists of a *finite* set of points all of which, except R, are simple points of V. This completes the proof of part (d) of Theorem 5.1.

6. Foundations of a general (global) theory of equisingularity on r-dimensional algebraic varieties, of embedding dimension $= r + 1$ at each of their singular k-rational points.

Let V be an algebraic variety (in projective or affine space), of pure dimension r, defined over an algebraically closed ground field k of characteristic zero, and of embedding dimension $\leq r + 1$, locally, at each of its k-rational points. If P is any such point of V we consider the completion $\hat{\mathfrak{o}}$ of the local ring $\mathfrak{o}$ of V at P and we denote by $\hat{V}_P$ the algebroid variety Spec ($\hat{\mathfrak{o}}$) having origin at P. The ground field k is a coefficient field of $\hat{\mathfrak{o}}$, and it will be with reference to this ground field k that we shall speak of the dimensionality type of points of $\hat{V}_P$. The point P is singular for V if and only if it is singular for $\hat{V}_P$. If P is singular for V, any minimal basis $\{x_1, x_2, \ldots, x_{r+1}\}$ of the maximal ideal $\mathfrak{m}$ of $\mathfrak{o}$ will be a minimal basis of the maximal ideal $\hat{m}$ of $\hat{\mathfrak{o}}$, and thus also $\hat{V}_P$ is of embedding dimension $r + 1$. We define *the dimensionality type of P on V*, in symbols: $d.t.(V, P)/k$ (or simply $d.t.(V, P)$) as being the dimensionality type of P on $\hat{V}_P$. We have then:

$$0 \leq d.t.(V, P) \leq r,$$

with $d.t.(V, P) = 0$ if and only if P is a simple point of V. This definition is, by its very nature, analytically intrinsic.

If we wish (as we do) to define the dimensionality type of points R of V which are not k-rational (and hence of positive dimension over k), the following procedure suggests itself quite naturally:

We consider the irreducible subvariety W of V having R as general point and we pick a k-rational point P on W at which P is analytically irreducible (in particular, we can take for P any simple point of W). The analytical irreducibility of W at P implies that W determines a unique irreducible *algebroid* subvariety $\hat{W}$ of $\hat{V}_P$, and hence R determines a unique point of $\hat{V}_P$, namely the general point of $\hat{W}$. We continue to denote by R the general point of $\hat{W}$. *Then we can define the dimensionality type* $d.t.(V, R)$ (relative to the given ground field) *as being the dimensionality type of* R *on* $\hat{V}_P$. This immediately raises a question, which will be one of the various questions that we shall study and answer in this section, namely: *is the above definition of* $d.t.(V, R)$ *independent of the choice of the k-rational point P on W? We shall answer this question later on in the affirmative.* The proof is not straightforward and will require a considerable amount of preparatory analysis (unless R is a simple point of V, in which case we obviously have $d.t.(V, R) = 0$).

Another, less important question, is the following: if the irreducible variety W is not analytically irreducible at P and has, say, g irreducible analytical branches at P, do the g distinct points $R_1, R_2, \ldots, R_g$ which correspond on $\hat{V}_P$ to R have the same dimensionality type?

Another paramount question which will be studied and answered (affirmatively) in this section is the following: *Is the irreducible algebroid variety* EQS($\hat{V}_P$, *P*) *algebraic* (this variety has been defined in Section 5, just before Lemma 5.2 was stated; see also the last statement of Theorem 5.4).

We assume from now on that P is a singular point of V (in the case of a simple point P the questions formulated above are trivial). We fix a minimal basis $x_1, x_2, \ldots, x_r, x_{r+1}$ of the maximal ideal $\mathfrak{m}$ of the local ring $\mathfrak{o}$ of V at P. The $r+1$ elements are algebraically dependent over k. We have then

$$f(x_1, x_2, \ldots, x_r, x_{r+1}) = 0, \tag{1}$$

where f is a polynomial with coefficients in k and is uniquely determined (up to an arbitrary non-zero factor in k) by the condition of being of least possible degree. The polynomial f is free from multiple factors, its leading form f_n has degree n, where n is the multiplicity of the singular point P of V. The local ring $\mathfrak{o}$ of V at P is the ring of quotients of the ring $k[X_1, X_2, \ldots, X_{r+1}]/(f(X_1, X_2, \ldots, X_{r+1}))$ with respect to its maximal ideal generated by the f-residues $x_1, x_2, \ldots, x_{r+1}$ of the X_i. As to the completion $\hat{\mathfrak{o}}$ of $\mathfrak{o}$ we have

$$\hat{\mathfrak{o}} = k[[X_1, X_2, \ldots, X_{r+1}]]/(f(X_1, X_2, \ldots, X_{r+1})). \tag{2}$$

Let V_1 be the algebraic hypersurface in affine $\mathbf{A}_{r+1}$ defined by the equation

$$V_1 : f(X_1, X_2, X_{r+1}) = 0. \tag{3}$$

The hypersurface V_1 is birationally equivalent to V, and the birational transformation of V into V_1 is biregular at P, the corresponding point of V_1 being the origin P_1: $X_1 = X_2 = \cdots = X_{r+1} = 0$. We therefore have $\hat{V}_P = \hat{V}_{1,P_1}$.

In the present case of *algebraic* varieties we shall deal primarily with projections $\pi_{\bar{u}}$ of V_1 *which are of finite degree N*. By this we mean a projection $\pi_{\bar{u}}$ of V_1 onto an r-dimensional affine space $\mathbf{A}_r^*$:

$$\pi_{\bar{u}}(X_1, X_2, \ldots, X_{r+1}) = (X_1^*, X_2^*, \ldots, X_r^*),$$

where the X_i^* are polynomials of degree N in $X_1, X_2, \ldots, X_{r+1}$ which are zero at P and have coefficients $(\bar{u})$ in some extension field $k^* = k(\bar{u})$:

$$X_i^* = \varphi_{i,(\bar{u})}(X_1, X_2, \ldots, X_{r+1}) = \sum_{j=1}^{r+1} \bar{u}_j^{(i)}(X_j + \sum_{1 \le j_1 \le j_2 \le r+1} \bar{u}_{j_1 j_2}^{(i)} X_{j_1} X_{j_2}$$
$$+ \cdots + \sum_{1 \le j_1 \le \cdots \le j_N \le r+1} \bar{u}_{j_1 j_2 \cdots j_N}^{(i)} X_{j_1} X_{j_2} \cdots X_{j_N}. \quad (4)$$

The equations (4) define a finite projection $\pi_{\bar{u}}$ of V_1 onto the space $\mathbf{A}_r^*$, in the neighborhood of the origin P_0^* of $\mathbf{A}_r^*$, if and only if the r elements

$$x_{i,(\bar{u})}^* = \varphi_{i,(\bar{u})}(x_1, x_2, \ldots, x_{r+1}), \qquad (i = 1, 2, \ldots, r) \qquad (4')$$

form a set of parameters of the local ring obtained from $\mathfrak{o}$ by the coefficient field extension $k \to k^*$.

We shall say that the projection $\pi_{\bar{u}}$ is a generic projection of V_1, of degree N, if all the coefficients $\bar{u}$ of the r polynomials $\varphi_{i,\bar{u}}$ are indeterminates. In the case of a generic projection we shall use, as in the preceding sections, the letter u instead of $\bar{u}$, and the projection will be therefore denoted by π_u. The r polynomials in (4) will be denoted in that case by $\varphi_{i,(u)}$.

We shall call a projection $\pi_{\bar{u}}$ *k-rational* if all the coefficients $\bar{u}$ belong to the ground field k.

If $\Delta_{\bar{u}}$ is the critical hypersurface of a projection $\pi_{\bar{u}}$ of V_1 and if Q is a singular point of $\hat{V}_P$ we shall say that $\pi_{\bar{u}}$ is *quasi-permissible at Q* if the following equality holds true:

$$d.t.(\hat{V}_P, Q) = d.t.(\Delta_{\bar{u}}, \pi_{\bar{u}}(Q))/k(\{\bar{u}\}) + 1, \qquad (5)$$

or—equivalently, if condition (*a*) of Proposition 5.3 (where V is to be replaced by $\hat{V}_P$) is satisfied by the projection $\pi_{\bar{u}}$. As in section 5, we shall say that $\pi_{\bar{u}}$ is *permissible at Q* if $\pi_{\bar{u}}$ is k-rational and satisfies both condition (*a*) and (*b*) of Proposition 5.3. We know, by Proposition 5.3, that for every sufficiently high integer N almost all projections $\pi_{\bar{u}}$ of degree N are quasi-permissible at Q, and that almost all k-rational projections of degree N are permissible at Q.

For almost all projections $\pi_{\bar{u}}$ of degree N it is true that the inverse image $\pi_{\bar{u}}^{-1}\{P_0^*\}$ of the origin P_0^*: $X_1^* = X_2^* = \cdots = X_r^* = 0$, is a finite set of points of V_1, or—more precisely—a zero-dimensional cycle $Z_{\bar{u}}$ on V_1. The cycle $Z_{\bar{u}}$ is the intersection of V_1 with the curve $\Gamma_{\bar{u}}$ defined by the r equations

$$\Gamma_{\bar{u}}: \varphi_{i,(\bar{u})}(X_1, X_2, \ldots, X_r, X_{r+1}) = 0, \qquad i = 1, 2, \ldots, r. \tag{6}$$

As the $\bar{u}$'s vary, the curve $\Gamma_{\bar{u}}$ varies in an irreducible algebraic system Σ of curves, the general member of this system being the curve Γ_u associated with the generic projection π_u of degree N, and the cycle $Z_{\bar{u}}$ varies in an irreducible algebraic system $\{Z_{\bar{u}}\}$ of zero-dimensional cycles. The system $\{Z_{\bar{u}}\}$ is defined over k, and the cycles $Z_{\bar{u}}$ are of degree mN^r, where m is the degree of the hypersurface V_1 (since N^r is the order of the curve $\Gamma_{\bar{u}}$). The general member of this system is the intersection cycle Z_u of V_1 with the general member Γ_u of Σ.

Proposition 6.1. *The general cycle Z_u of the system $\{Z_{\bar{u}}\}$ consists of the point P_1 counted n times and of $mN^r - n$ distinct (simple) points of V_1 (all counted to multiplicity* 1).

Proof. All the curves $\Gamma_{\bar{u}}$ pass through the origin P_1: $X_1 = X_2 = \cdots = X_r = 0$, and the tangent of the general curve Γ_u at P_1 is the line

$$\sum_{j=1}^{r+1} u_j^{(i)} X_j = 0, \qquad i = 1, 2, \ldots, r.$$

This line obviously does not lie on the tangent hypercone $f_n(X_1, X_2, \ldots, X_{r+1}) = 0$ of V_1 at P_1 (here f_n is the leading form of the polynomial f). Thus P_1 occurs in the general cycle Z_u to multiplicity exactly equal to n.

To prove the proposition it is sufficient to exhibit a particular cycle in the system $\{Z_{\bar{u}}\}$ which consists of the point P_1 counted n times and of $mN^r - n$ distinct (simple) points of V_1, all counted to multiplicity 1. We take for each of the polynomials $\varphi_{i,(\bar{u})}$ the product of N *linear generic* polynomials $\varphi_i^{(\nu)}$ in $X_1, X_2, \ldots, X_{r+1}$ ($\nu = 1, 2, \ldots, N$) subject to the only condition that the hyperplane $H_i^{(1)}$: $\varphi_i^{(1)} = 0$ pass through P_1. With such a choice of the parameter $\bar{u}$ we find that the corresponding curve $\Gamma_{\bar{u}}$ consists of one *generic* line of $\mathbf{A}_{r+1}$ through P_1 and of $N^r - 1$ *generic* lines of $\mathbf{A}_{r+1}$. This proves our proposition.

Corollary 6.2. *For almost all k-rational specializations $\{u\} \to \{\bar{u}\} \subset k$ the cycle $Z_{\bar{u}}$ consists of the point P_1 counted n times and of $mN^r - n$ distinct (simple) points of V_1 (all counted to multiplicity* 1).

Obvious.

We now consider the r elements $x_{i,(u)}^*$ introduced in (4′), which define the *generic* projection π_u of degree N. Each of the $r + 1$ elements $x_1, x_2, \ldots, x_r, x_{r+1}$ is algebraic over the field $K^* = k(u)(x_{1,(u)}^*, \ldots, x_{r,(u)}^*)$, and by Proposition 6.1 the field $K^*(x_1, x_2, \ldots, x_r, x_{r+1})$ is of relative degree mN^r over K^*. It can be easily verified, using the proof of Proposition 6.1, that the $mN^r - n$ points of the general cycle Z_u, different from P_1, have, for each $i = 1, 2, \ldots, r, r + 1$, distinct X_i-coördinates. Let us consider the minimal polynomial $f_{(u)}(x_{1,(u)}^*, x_{2,(u)}^*, \ldots, x_{r,(u)}^*; X_{r+1})$ of X_{r+1} over K^*. Here $f_{(u)}(X_1^*, X_2^*, \ldots, X_r^*; X_{r+1})$ is a polynomial in $X_1^*, X_2^*, \ldots, X_r^*, X_{r+1}$ of degree mN^r in X_{r+1} and with coefficients in the polynomial ring $k[\{u\}]$. (Note that x_{r+1} is integrally dependent on $k^*[x_{1,(u)}^*, x_{2,(u)}^*, \ldots, x_{r,(u)}^*]$). We denote by $a(u)$ the coefficient of $X_{r+1}{}^{mN^r}$ in $f_{(u)}$. Since 0 is an n-fold root of the polynomial $f_{(u)}(0, 0, \ldots, 0; X_{r+1})$, it follows from the Weierstrass preparation theorem that we have an identity of the form

$$\begin{aligned} &f_{(u)}(X_1^*, \ldots, X_r^*; X_{r+1}) \\ &= F_{(u)}(X_1^*, \ldots, X_r^*; X_{r+1})G_{(u)}(X_1^*, \ldots, X_r^*; X_{r+1}), \qquad (7) \end{aligned}$$

where $F_{(u)}$ is a monic polynomial in X_{r+1}, of degree n, with coefficients in the power series ring $k[\{u\}]_{(a(u))}[[X_1^*, \ldots, X_r^*]]$, while $G_{(u)}$ is a power series in $X_1^*, \ldots, X_r^*, X_{r+1}$ with coefficients in $k[\{u\}]_{(a(u))}$, and we have $G_{(u)}(0, \ldots, 0; 0) \neq 0$. Here $k[\{u\}]_{(a(u))}$ denotes the ring of quotients of the polynomial ring $k[\{u\}]$ with respect to the multiplicative system consisting of the non-negative powers of $a(u)$.

The algebroid hypersurface $F_{(u)}(X_1^*, X_2^*, \ldots, X_r^*; X_{r+1}) = 0$ is the original algebroid variety $\hat{V}_P(= V_{1,P_1})$ with $k(\{u\})$ as new (extended) ground field, while the equation $f_{(u)}(X_1^*, \ldots, X_r^*; X_{r+1}) = 0$ defines an algebraic hypersurface $V_{(u)}^*$ in $\mathbf{A}_{r+1}/k(\{u\})$ which is a rational transform of $V_1/k(\{u\})$ (and hence also of $V/k(\{u\})$) and which has the property that if P^* is the origin $X_1^* = X_2^* = \cdots = X_r^* = X_{r+1} = 0$ then $V/k(\{u\})$ and $V_{(u)}^*/k(\{u\})$ are analytically equivalent at P and P^* respectively.

For any specialization $\{u\} \to \{\bar{u}\} \subset k$ such that $a(\bar{u}) \neq 0$ the specialized polynomial $f_{(\bar{u})}$ and the specialized power series $F_{(\bar{u})}$ and $G_{(\bar{u})}$ are well defined, and for almost all such specializations the equation $F_{(\bar{u})} = 0$ defines our original algebroid variety $\hat{V}_P(= \hat{V}_{1,P_1})$, while the equation $f_{(\bar{u})} = 0$ defines a rational transform $V_{(\bar{u})}^*$ of V/k, and the two algebraic varieties $V_{(u)}^*$ and V are analytically equivalent at P.

Let $\Delta_{(\bar{u})}$ and $\hat{\Delta}_{(\bar{u})}$ denote respectively the critical hypersurfaces of the $\pi_{\bar{u}}$-projection of the algebraic hypersurface $V_{(\bar{u})}{}^*$ and of the algebroid hypersurface $\hat{V}_P(=\hat{V}_{1,P_1})$. The two hypersurfaces $\Delta_{(\bar{u})}$ and $\hat{\Delta}_{(\bar{u})}$ are therefore the *reduced* hypersurfaces which are defined by the equations $D_{X_{r+1}}(f_{(\bar{u})}) = 0$ and $D_{X_{r+1}}(F_{(\bar{u})}) = 0$ respectively, where $D_{X_{r+1}}(f_{(\bar{u})})$ and $D_{X_{r+1}}(F_{(\bar{u})})$ are the X_{r+1}-discriminants of $f_{(\bar{u})}$ and $F_{(\bar{u})}$ respectively.

PROPOSITION 6.3. *For almost all specializations $\{u\} \to \{\bar{u}\} \subset k$ the algebraic critical hypersurface $\Delta_{(\bar{u})}$ coincides, locally at the origin $P_0{}^*$ of the image space $\mathbf{A}_r{}^*$ of $\pi_{\bar{u}}$, with the algebroid hypersurface $\hat{\Delta}_{(\bar{u})}$.*

Proof. Let $\bar{x}_{r+1}{}^{(1)}, \bar{x}_{r+1}{}^{(2)}, \ldots, \bar{x}_{r+1}{}^{(mN^r)}$ be the branches of the mN^r-valued algebraic function $\bar{x}_{r+1}$ of $X_1{}^*, X_2{}^*, \ldots, X_r{}^*$ defined by the equation $f_{(\bar{u})}(X_1{}^*, X_2{}^*, \ldots, X_r{}^*; X_{r+1}) = 0$. We consider the values of these branches at the origin $P_0{}^*$. By Corollary 6.2 we know that for almost all specializations $(u) \to (\bar{u}) \subset k$, n and only n of these branches, say $\bar{x}_{r+1}{}^{(1)}, \bar{x}_{r+1}{}^{(2)}, \ldots, \bar{x}_{r+1}{}^{(n)}$, have value zero, while the remaining $mN^r - n$ branches have values which are distinct from each other and different from zero. Since we have

$$D_{X_{r+1}}(f_{(\bar{u})}) = \prod_{1\le i<j\le n} (\bar{x}_{r+1}{}^{(i)} - \bar{x}_{r+1}{}^{(j)})^2$$

$$\cdot\left(\prod_{\nu=1}^{mN^r} \cdot \prod_{\mu(>\nu)=n+1}^{mN^r} (\bar{x}_{r+1}{}^{(\nu)} - \bar{x}_{r+1}{}^{(\mu)})\right)^2,$$

where the second factor (the double product) is different from zero at the origin P^*, while the first factor is equal to $D_{X_{r+1}}(F_{(\bar{u})})$, the proposition is proved.

We are now in a position to answer affirmatively the questions raised in the beginning of this section.

The first question concerns the definition of the dimensionality type of a *singular* point R of V which is not k-rational. We consider the irreducible algebroid subvariety W/k of V having R as general point. We consider on W any two k-rational points P, P', such that W is analytically irreducible at P and also at P' (in particular, we can take for P and P' any two simple points of W). *We have to prove that*

$$d.t.(\hat{V}_P, R)/k = d.t.(\hat{V}_{P'}, R)/k. \tag{8}$$

To prove this we make use of part (a) of Proposition 5.3 and of Proposition 6.3 just proved above. By Proposition 5.3 almost all k-rational projections $\pi_{\bar{u}}$ of $\hat{V}_P$ are quasi-permissible at P *and* at R. Similarly, almost all k-rational projections $\pi_{\bar{u}}'$ of $\bar{V}_{P'}$ are quasi-permissible at P' *and* at R. It follows that for all sufficiently high integers N almost all k-rational projections $\pi_{\bar{u}}$ of $\hat{V}_P$, of degree N, are quasi-permissible at P and R, while also almost all k-rational projections $\pi_{\bar{u}}'$ of $\hat{V}_{P'}$, of the same degree N, are quasi-permissible at P' and R. We fix an affine model of our algebraic variety V such that both P and P' belong to that model. We continue to denote by V that affine model.

Let $k[z_1, z_2, \ldots, z_M]$ be the coördinate ring of V; here M is the dimension of ambient affine space of V. Let $a_1, a_2, \ldots, a_M$ and $a_1', a_2', \ldots, a_M'$ be respectively the coördinates of P and P'. The M elements $z_1 - a_1, z_2 - a_2, \ldots, z_M - a_M$ generate the maximal ideal $\mathfrak{m}$ of the local ring $\mathfrak{o}$ of V at P; and, similarly, $z_1 - a_1', z_2 - a_2', \ldots, z_M - a_M'$ generate the maximal ideal $\mathfrak{m}'$ of the local ring $\mathfrak{o}'$ of V at P'. For almost all sets of $M(r+1)$ elements c_{ij} in k ($i = 1, 2, \ldots, r+1$; $j = 1, 2, \ldots, M$) the $r+1$ elements

$$x_i = \sum_{j=1}^{M} c_{ij}(z_j - a_j), \qquad (i = 1, 2, \ldots, r+1) \tag{9}$$

form a minimal basis of $\mathfrak{m}$, and the $r+1$ elements

$$x_i' = \sum_{j=1}^{M} c_{ij}(z_j - a_j'), \qquad (i = 1, 2, \ldots, r+1) \tag{9'}$$

form a minimal basis of $\mathfrak{m}'$. We have

$$x_i - x_i' = b_i, \qquad (i = 1, 2, \ldots, r+1) \tag{10}$$

where

$$b_i = \sum_{j=1}^{M} c_{ij}(a_j' - a_j). \qquad (i = 1, 2, \ldots, r+1)$$

As in (3) we introduce the (reduced) hypersurface V_1 in $\mathbf{A}_{r+1}$ such that the element $x_1, x_2, \ldots, x_{r+1}$ in (9) are the f-residues of $X_1, X_2, \ldots, X_{r+1}$. The origin P_1 belongs to V_1. Since $f(x_1, x_2, \ldots, x_{r+1}) = 0$, it follows from (10) that $f(x_1' + b_1, x_2' + b_2, \ldots, x_{r+1}' + b_{r+1}) = 0$.

Passing to the $\mathfrak{m}'$-residues we conclude that $f(b_1, b_2, \ldots, b_{r+1}) = 0$, i.e., that the point $P_1'(b_1, b_2, \ldots, b_{r+1})$ also belongs to V_1. It is clear that the birational transformation between V and V_1 is biregular at the corresponding points P, P_1 and is also biregular at the corresponding points P', P_1'. Since P (as well as P') is a specialization of R over k, it follows that the birational transformation of V onto V_1 is also biregular at R. To prove (8) it is therefore permissible to replace in (8) the variety V by the variety V_1, and the points P, P', R be the corresponding points P_1, P_1' and R_1 of V_1. We shall therefore identify from now on our original variety V with the hypersurface V_1, and we shall therefore denote by V the hypersurface defined by (3) and by P, P', R the points P_1, P_1', R_1.

We now fix an integer N sufficiently high so that the following be true: *almost all k-rational projections $\pi_{\bar{u}}$ of V, of degree N, are quasi-permissible at P, P′ and R.* In terms of the coördinates $X_1, X_2, \ldots, X_{r+1}$ which are zero at P the equations of $\pi_{\bar{u}}$ are: $\pi_{\bar{u}}(\{X\}) = \{X^*\}$, where $\{X\} = (X_1, X_2, \ldots, X_{r+1})$, $\{X^*\} = (X_1^*, X_2^*, \ldots, X_r^*)$ and (see (4)):

$$\pi_{\bar{u}}: X_i^* = \varphi_{i,(\bar{u})}(X_1, X_2, \ldots, X_{r+1}). \qquad (i = 1, 2, \ldots, r) \qquad (11)$$

If we use the coördinates $X_1', X_2', \ldots, X_{r+1}'$ which are zero at P' (where $X_i' = X_i - b_i$; see (10)) we must also make a change of coördinates $\{X^*\} \to \{X'^*\}$ of the image space $\mathbf{A}_r$ of $\pi_{\bar{u}}$, namely $X_i'^* = X_i^* - \varphi_{i,(\bar{u})}(b_1, b_2, \ldots, b_{r+1})$, $i = 1, 2, \ldots, r$. Hence in terms of the new coördinate systems in $\mathbf{A}_{r+1}$ and $\mathbf{A}_r^*$, the equations of $\pi_{\bar{u}}$ will be as follows:

$$\pi_{\bar{u}} = \pi_{\bar{v}}': X_i'^* = \varphi_{i,(\bar{v})}(X_1', X_2', \ldots, X_{r+1}'), \qquad (i = 1, 2, \ldots, r) \quad (11')$$

where

$$\varphi_{i,(\bar{v})}(X_1', X_2', \ldots, X_{r+1}') = \varphi_{i,(u)}(X_1' + b_1, X_2' + b_2, \ldots, X_{r+1}' + b_{r+1}) - \varphi_{i,(u)}(b_1, b_2, \ldots, b_{r+1}).$$

It is clear that for almost all k-rational specializations $(u) \to (\bar{u})$ also the specialization $(v) \to (\bar{v})$ will yield a projection $\pi_{\bar{v}}'$ (locally at P') which is quasi-permissible at P'. It is in this sense that we have used the expression "$\pi_{\bar{u}}$ is quasi-permissible at P *and* P'". When we say that $\pi_{\bar{u}}$ is also quasi-permissible at the point R, we actually mean to imply that $\pi_{\bar{u}}$ is quasi-permissible at R when R is considered as a point of $\hat{V}_P$ and that $\pi_{\bar{v}}'$ is quasi-permissible at R when R is considered as a point of $\hat{V}_{P'}$.

We now consider the *algebraic* critical hypersurface $\Delta_{(\bar{u})}(=\Delta_{(\bar{v})})$ of the (global) projection $\pi_{\bar{u}}(=\pi_{\bar{v}})$ of the algebraic hypersurface V. By Proposition 6.3 we can assume that the specialization $(u) \to (\bar{u}) \subset k$ has also the following two properties: $\Delta_{(\bar{u})}$ coincides, locally at the point $P_0 = \pi_u(P)$, with the critical algebroid hypersurface $\hat{\Delta}_{(\bar{u})}$ of the $\pi_{(\bar{u})}$-projection of $\hat{V}_P$, and coincides, locally at the point $P_0' = \pi_{\bar{v}}(P')$, with the critical algebroid hypersurface $\hat{\Delta}_{(\bar{v})}$ of the $\pi_{(\bar{v})}$-projection of $\hat{V}_{P'}$:

$$\widehat{\Delta_{(\bar{u}),P_0}} = \hat{\Delta}_{(\bar{u})}, \qquad (P_0 = \pi_{(\bar{u})}(P)); \tag{12}$$

$$\widehat{\Delta_{(\bar{v}),P_0'}}(=\widehat{\Delta_{(\bar{u}),P_0'}}) = \hat{\Delta}_{(\bar{v})}, \qquad (P_0' = \pi_{\bar{v}}(P')). \tag{12'}$$

We set

$$R_0 = \pi_{(\bar{u})}(R)(=\pi_{\bar{v}}(R)).$$

By definition of the integers $d.t.(\hat{V}_P, R)$, $d.t.(\hat{V}_{P'}, R)$ we have

$$d.t.(\hat{V}_P, R) = d.t.(\widehat{\Delta_{(\bar{u})}, R_0}) + 1, \tag{13}$$

$$d.t.(\hat{V}_{P'}, R) = d.t.(\widehat{\Delta_{(\bar{v})}, R_0}) + 1. \tag{13'}$$

Hence, by (12) and (12′), we have

$$d.t.(\hat{V}_P, R) = d.t.(\widehat{\Delta_{(\bar{u}),P_0}}, R_0) + 1, \tag{14}$$

$$d.t.(\hat{V}_{P'}, R) = d.t.(\widehat{\Delta_{(\bar{u}),P_0'}}, R_0) + 1. \tag{14'}$$

Now, $\Delta_{(\bar{u})}$ is an *algebraic* hypersurface, and the points P_0, P_0' and R_0 belong to $\Delta_{(\bar{u})}$ (R_0 belongs to $\Delta_{(\bar{u})}$ since R was assumed to be a singular point of V), and both P_0 and P_0' are specializations of R_0, over k. We can assume, by induction on r, that (8) is true if V is replaced by the $(r-1)$-dimensional hypersurface $\Delta_{(\bar{u})}$ and P, P', R are replaced by P_0, P_0', R_0. For, if $r = 1$, then (8) is trivial (any point R of an algebraic curve, which is not k-rational, is obviously a simple point of the curve). We have therefore the equality

$$d.t.(\widehat{\Delta_{(\bar{u}),P_0}}, R_0) = d.t.(\widehat{\Delta_{(\bar{u}),P_0'}}, R_0).$$

From this the equality (8) follows in view of (14) and (14′).

For the sake of completeness we shall deal also with the case in which W is analytically reducible at one of the points P, P', say at P, while being analytically irreducible at P'. Let $\hat{W}_1, \hat{W}_2, \dots, \hat{W}_\rho$ be the irreducible analytical components of W at P, and let R_ν be the general point of $\hat{W}_\nu$ $(\nu = 1, 2, \dots, \rho)$. The ρ points R_ν are distinct points of $\hat{V}_P$. We shall continue to denote by R the unique point of $\hat{V}_{P'}$ which is associated with the general point R of W. We fix our k-rational projection $\pi_{\bar{u}}$ of sufficiently high degree N in such a manner that it be quasi-permissible at P and at each point R_ν of $\hat{V}_P$ and that the resulting k-rational projection $\pi_{(\bar{v})}$ of $\hat{V}_{P'}$ be quasi-permissible at P' and R. Furthermore, we assume that the condition of Proposition 6.3 is satisfied both at P and P', or—equivalently—that $\pi_{(\bar{u})}$ and $\pi_{(\bar{v})}$ satisfy (12) and (12′) respectively.

The $\pi_{\bar{u}}$-transform W_0 of W will be a reducible algebroid subvariety of $\widehat{\Delta_{(\bar{u}),P_0}}$ of the same dimension as W, and its ρ irreducible components $W_{0,1}, W_{0,2}, \dots, W_{0,\rho}$ have $R_{0,1}, R_{0,2}, \dots, R_{0,\rho}$ as general points, where $R_{0,\nu} = \pi_{(\bar{u})}(R_\nu)$. Then (13′) is satisfied, and also (13) is satisfied if R is replaced by any of the ρ points R_ν and R_0 is replaced by the corresponding point $R_{0,\nu}$. Note that since R is a singular point, the points $R_{0,\nu}$ belong to $\Delta_{(\bar{u}),P_0}$. By induction on r, we have for each $\nu = 1, 2, \dots, \rho$:

$$d.t.(\widehat{\Delta_{(\bar{u}),P_0}}, R_{0,\nu}) = d.t.(\widehat{\Delta_{(\bar{u}),P_0'}}, R_0),$$

and from these equalities follow the desired equalities which generalize (8), namely

$$d.t.(\hat{V}_P, R_\nu) = d.t.(\hat{V}_{P'}, R), \qquad \nu = 1, 2, \dots, \rho.$$

We shall now deal with the last question raised in the beginning of this section, namely we shall prove the following

THEOREM 6.4. *If V is our original algebraic variety and P is a k-rational point of V, the algebroid variety $W(P) = \mathrm{EQS}_k(\hat{V}_P, P)$ is algebraic, irreducible, of dimension $r - s$, and P is a simple point of $W(P)$. Furthermore, V is equimultiple at P along $W(P)$. Here s is the dimensionality type of P on V.*

Proof. We recall that we have proved in section 5 (see last statement of Theorem 5.4) that the subset $\mathrm{EQS}_k(\hat{V}_P, P)$ of $\hat{V}_P$ is a (closed) algebroid subvariety of $\hat{V}_P$. For the sake of clarity we shall make precise the meaning we attach to a statement that *an algebroid variety W, embedded in an affine space $\mathbf{A}_M$ and defined over k, is algebraic.* Let $X_1, X_2, \ldots, X_M$ be affine coördinates in $\mathbf{A}_M$ and let P: $X_1 = X_2 = \cdots = X_M = 0$ be the origin of W. We say that W is *algebraic* if W is the variety of some ideal $\mathfrak{A}$ in the power series ring $k[[X_1, X_2, \ldots, X_M]]$ such that $\mathfrak{A}$ has a basis consisting of *polynomials* in $X_1, X_2, \ldots, X_M$ (with coefficients in k).

We now consider the hypersurface V_1 in $\mathbf{A}_{r+1}$, defined by the equation (3). We know that the local ring of V_1 at the origin P_1: $X_1 = X_2 = \cdots = X_{r+1} = 0$ coincides with the local ring of V at P. We now consider a k-rational projection $\pi_{(\bar{u})}$ of V_1 onto $\mathbf{A}_r$, of a sufficiently high degree N, such that $\pi_{(\bar{u})}$ is P-permissible (and not only quasi-permissible). Then, by Proposition 5.3, we have that $W(P) = \pi_{(\bar{u})}^{-1}(W_0(P_0))$, where P_0 is the origin $X_1^* = X_2^* = \cdots = X_r^* = 0$ of the image space $\mathbf{A}_r^*$ of $\pi_{(\bar{u})}$ and where $W_0(P_0) = \mathrm{EQS}_k(\hat{\Delta}_{(\bar{u})}, P_0)$. We can now use induction on r, since the theorem is trivial in the case $r = 1$ (in the case $r = 1$, $W(P)$ is either the point P, if P is a singular point of the curve V, and $W(P) = V$ is P is a simple point of V). We can therefore assume that $W_0(P_0)$ is an algebraic variety, since by Proposition 6.3 $\hat{\Delta}_{(\bar{u})}$ is an algebraic hypersurface $\Delta_{\bar{u}}$. Therefore $W_0(P_0)$ can be defined by a set of *polynomial* equations $\psi_\nu(X^*) = 0$ in $X_1^*, X_2^*, \ldots, X_r^*$. Since $W(P) = \pi_{\bar{u}}^{-1}(W_0(P_0))$, it follows that $W(P)$ is the variety defined by the equation $\psi_\nu(X^*) = 0$ and $f(X_1, X_2, \ldots, X_{r+1}) = 0$. Since the X^*'s are polynomials in $X_1, X_2, \ldots, X_{r+1}$, and since the birational transformation of V onto V_1 is given by the equations $X_i = \sum_{j=1}^{M} c_{ij} Z_j$ (where we assume that P is the origin in affine space $\mathbf{A}_M$ in which V is embedded), it follows that $W(P)$ (which we now regard as being embedded in $\mathbf{A}_M$) can be defined by polynomial equations in $Z_1, Z_2, \ldots, Z_M$, and hence is algebraic.

The rest of the theorem is merely a repetition of Theorem 5.4, with the closed point P replacing the point Q.

Our main purpose in the remainder of this section is to introduce the *equisingularity strata* on V and prove that the resulting covering of V by these strata enjoys the usual properties of a stratification. Given any point P of V, of dimensionality type s, we consider the irreducible subvariety $W(P)$ (see Theorem 6.4). The general point R of $W(P)$ also has dimensionality type s (since $W(P) = \mathrm{EQS}(V_P, P)$ and has dimension $r - s$). The first thing that has to be proved is that the set $W(P) \cap V(s)$, i.e., the set of points of $W(P)$ which have the same dimensionality type s as P and R is a Zariski dense open subset of $W(P)$, or—equivalently—that the set of points of $W(P)$ which have dimensionality type $>s$ is a *proper* algebraic subvariety of $W(P)$. This essential and important result is proved by H. Hironaka in the paper [5] cited by us in the Introduction.

We use the result of Hironaka and we denote by $S(P)$ the set of points of $W(P)$ which have dimensionality type s. By Theorem 5.1, the point P is a simple point of $W(P)$. If P' is any other k-rational point contained in $S(P)$, then P' is also a specialization of R, and therefore $W(P)$ coincided with $W(P')$, whence $S(P) = S(P')$. We see therefore that all the points of $S(P)$ are simple points of $W(P)$. We call $S(P)$ the *equisingularity stratum* containing the point P. Thus each equisingularity stratum S, of dimensionality type s, is a smooth Zariski dense open subset of the $(r - s)$-dimensional irreducible subvariety $\overline{S}$ of V ($\overline{S} = W(P)$, for any point P in S). Any two equisingularity strata are obviously disjoint in view of the fact just proved, namely that if $P' \in S(P)$ then $S(P') = S(P)$.

We now prove that *the number of equisingularity strata of given dimensionality type s is finite.* We know that this is trivially true for the set of strata of dimensionality type zero; and also of dimensionality 1 (see our paper [1], SES II). We shall use induction from s to $s + 1$. We use the notations $V(s)$ and $V\{s\}$ introduced in section 5, where now V is our *algebraic* variety (and not an algebroid variety). Let Σ_s denote the (finite) set of equisingularity strata of dimensionality type $\leq s$. We have then

$$V\{s + 1\} = V_{sing} - \bigcup_{S_i \in \Sigma_s} S_i .$$

It is easily seen that $V\{s + 1\}$ is an algebraic subvariety of V. For, if R is any point of $V\{s + 1\}$ then by Proposition 4.2 every specialization of R over k belongs to $V\{s+1\}$. On the other hand, if $R \notin V\{s+1\}$, then $d.t.(V, R) = \sigma \leq s$, and hence, by Hironaka's result, almost all specializations of R over k have dimensionality type σ and therefore do not belong to $V\{s+1\}$. This proves the algebraic character of the set $V\{s+1\}$. No point R of $V\{s+1\}$ can have dimension (over k) $\geq r-s$, for in the contrary case we would have necessarily $d.t.(V, R) \leq s$. Therefore the variety $V\{s+1\}$ has dimension $\leq r-s-1$. We may assume that $V(s+1) \neq \emptyset$ (in the contrary case, we would have $V\{s+1\} = V\{s+2\}$, and we could replace in our induction hypothesis the integer s by the integer $s+1$). From the assumption that $V(s+1) \neq \emptyset$ follows that $V\{s+1\}$ must have dimension exactly equal to $r-s-1$ (and not *less* than $r-s-1$), for $V(s+1) \neq \emptyset$ implies the existence of equisingularity strata S of dimensionality type $s+1$, and any such stratum S has dimension $r-s-1$, and the closure $\overline{S}$ of S must be an irreducible component of $V\{s+1\}$. Conversely, every irreducible component W of $V\{s+1\}$, of dimension $r-s-1$, gives rise to a stratum S of dimensionality type $s+1$; it is the stratum containing the general point R of W, for the inequality $d.t.(V, R) \geq s+1$ together with the equality $dim\ R/k = r-s-1$ imply that $d.t.(V, R) = s+1$. We conclude therefore that the number of strata of dimensionality type $s+1$ is equal to the number of $(r-s-1)$-dimensional components of the variety $V\{s+1\}$, and is therefore finite. This completes our inductive proof of the statement that the set of equisingularity strata of V is finite.

We shall conclude this section (and the paper) by proving that *the boundary of any equisingularity stratum is a union of strata.*

Let S be an equisingularity stratum, of dimensionality type s, let R be the general point of the closure $\overline{S}/k$ of S, and let P' be a k-rational point of the boundary $\overline{S} - S$ of S. We know then that if $d.t.(V, P') = s'$ then $s' > s$. Let $S' = S(P')$ be the equisingularity stratum containing P' and let R' be the general point of $\overline{S'}/k$. What we have to show is that $R' \in \overline{S} - S$, because if that is shown then it will follow that $S' \subset \overline{S} - S$. Let $\hat{V}_{P'}$ be the *algebroid* r-dimensional variety which is defined by V, locally at the k-rational point P', i.e., let $\hat{V}_{P'} = Spec\ (\mathfrak{o}')$, where $\mathfrak{o}'$ is the completion of the local ring of V at P'. We set $W = \overline{S}$ and $W' = \overline{S'}$. The irreducible algebraic variety W', of dimension $r-s'$, is analytically irreducible at P' since P' is a simple point of W'. Therefore the general point R' of W'/k defines a unique point of $\hat{V}_{P'}$. We shall continue to denote that point by R'. Since P' is also a specialization of point R, over k, the point R is associated with a finite number of points $\hat{R}_1, \hat{R}_2, \ldots, \hat{R}_g$ of $\hat{V}_{P'}$, where g is the number of irreducible analytical branches of W at P'. Consider any of these g points $\hat{R}_i$ and denote it by $\hat{R}$. The stratum S having R as general point yields a well-defined stratum $\hat{S}$ of $\hat{V}_{P'}$ having $\hat{R}$ as general point, for we know that $d.t.(\hat{V}_{P'}, \hat{R}) = d.t.(V, R) = s$ (by the definition of the dimensionality type of R given earlier in this section) and we know also that the (analytic) dimension of the point $\hat{R}$ is equal to the (algebraic) dimension of the point R, i.e., is equal to $r - s$. Since $P' \in \overline{S}$, it follows that $P' \in \overline{\hat{S}}$. On the other hand we have $P' \notin \hat{S}$ since $d.t.(V, P') = s' > s = d.t.(V, R) = d.t.(\hat{V}_{P'}, \hat{R})$. Hence P' belongs to the *boundary* of $\hat{S}$. Since the *closed* stratum $\hat{S}'$ of $\hat{V}_{P'}$ (i.e., the stratum of $\hat{V}_{P'}$ containing the closed point P' of $\hat{V}_{P'}$; see Theorem 5.1, part (*c*)) contains the point R', it follows from Theorem 5.1, part (*d*), that $R' \in \overline{\hat{S}}$. We have found therefore that R' is an (analytic) specialization of the general point $\hat{R}$ of $\hat{S}$, and therefore R' is also an (algebraic) specialization of R. This completes the proof of the stratification property stated above.

Appendix

A generalization of a theorem on equimultiple subvarieties of an algebroid hypersurface (proved in [2]).

Our PNAS note [2] deals only with the special case in which the point Q is the closed point of V. This special case is the only case which needs to be considered in our theory of equisingularity of *algebraic* varieties (see section 6). Although our primary purpose in dealing with algebroid varieties is to prepare the ground for an *analytically intrinsic* theory of equisingularity of algebraic varieties, we nevertheless shall give here the generalization of the proof given in the PNAS note to the general case in which Q is an arbitrary point of V. This generalization consists essentially in replacing V by its localization $\tilde{V}_Q$ of V at Q; here $\tilde{V}_Q = \text{Spec}(\hat{\mathfrak{o}}_{\mathfrak{p}})$, where $\hat{\mathfrak{o}}_{\mathfrak{p}}$ is the completion of the (local) ring of quotients $\mathfrak{o}_{\mathfrak{p}}$ of $\mathfrak{o}$ with respect to the prime ideal $\mathfrak{p}$ associated with the point Q. The point Q is now the *closed* point of $\tilde{V}_Q$ and we shall show that the statement used in the text in the case of the pair $\{V, Q\}$ can be derived from our PNAS note, as applied to the closed point Q of $\tilde{V}_Q$.

Our k-rational projection π is defined by a set of parameters $x_1, x_2, \ldots, x_r$ belonging to a minimal basis $\{x_1, x_2, \ldots, x_r, x_{r+1}\}$ of the maximal ideal $\mathfrak{m}$ of $\mathfrak{o}$. Since x_{r+1} is integral over the (regular) local domain $k[[x_1, x_2, \ldots, x_r]]$, it follows that if $\xi_1, \xi_2, \ldots, \xi_r, \xi_{r+1}$ are the coördinates of Q (i.e., the $\mathfrak{p}$-residues of $x_1, x_2, \ldots, x_r, x_{r+1}$), then ξ_{r+1} is integral over the integral domain $k[[\xi_1, \xi_2, \ldots, \xi_r]]$. If ρ is the (analytic) dimension of Q over k, then $\tilde{V}_Q$ is an algebroid variety of dimension ρ defined over the field of fraction $K = k((\xi_1, \xi_2, \ldots, \xi_{r+1}))$ of the integral domain $\mathfrak{o}/\mathfrak{p}$ (K is a coefficient field of $\hat{\mathfrak{o}}_\mathfrak{p}$). Necessarily ρ of the r elements $\xi_1, \xi_2, \ldots, \xi_r$, say $\xi_1, \xi_2, \ldots, \xi_\rho$, are analytically independent over k, while K is an algebraic extension of the field $k((\xi_1, \xi_2, \ldots, \xi_\rho))$ of meromorphic functions of $\xi_1, \xi_2, \ldots, \xi_\rho$ (we could, if necessary, identify $\xi_1, \xi_2, \ldots, \xi_\rho$ with $x_1, x_2, \ldots, x_\rho$).

The equation of $\tilde{V}_Q$ is given by

$$\tilde{V}_Q: \tilde{f}(Y_1, Y_2, \ldots, Y_{r-\rho+1}) = 0, \tag{1}$$

where

$$\tilde{f}(Y_1, Y_2, \ldots, Y_{r-\rho+1}) = f(\xi_1, \xi_2, \ldots, \xi_\rho, Y_1 + \xi_{\rho+1}, \ldots, Y_{r-\rho+1} + \xi_{r+1}),$$

and the maximal ideal of $\hat{\mathfrak{o}}_\mathfrak{p}$ has the minimal basis $y_1, y_2, \ldots, y_{r-\rho+1}$, where $y_i = h(x_{\rho+i}) - \xi_{\rho+i}$, h being the canonical homomorphism of $\mathfrak{o}$ into $\mathfrak{o}_\mathfrak{p}$. Thus, $\tilde{f}(Y)$ is a power series in $Y_1, Y_2, \ldots, Y_{r-\rho+1}$, with coefficients in K, and the Y-coördinates of Q are all zero. The set of points of $\tilde{V}_Q$ consists of those points of V which have Q as a specialization over k. If R is any such point then it is easy to see that the multiplicity $e(V, R)$ of V at R is equal to the multiplicity $e(\tilde{V}_Q, R)$ of $\tilde{V}_Q$ at R:

$$e(\tilde{V}_Q, R) = e(V, R). \tag{2}$$

To see this we first observe that from the very definition of the multiplicity of Q on $\tilde{V}_Q$ (i.e., the multiplicity of the *complete* local ring $\hat{\mathfrak{o}}_\mathfrak{p}$; see C.A.II, p. 214) it follows that the multiplicity $e(\tilde{V}_Q, Q)$ is equal to the multiplicity of Q on V_Q (i.e., to the multiplicity of the local ring $\mathfrak{o}_\mathfrak{p}$). By a similar argument it follows that the multiplicity of R on $\tilde{V}_Q$ is equal to the multiplicity of R on V_Q. Hence in order to prove (2) we have only to show that

$$e(V_Q, R) = e(V, R). \tag{2'}$$

We set $\bar{\mathfrak{o}} = \mathfrak{o}_{\mathfrak{p}}$. Let $\mathfrak{P}$ be the prime ideal in $\mathfrak{o}$ which is associated with the point R of V. Then $e(V, R)$ is the multiplicity of the local ring $\mathfrak{o}_{\mathfrak{p}}$, while $e(V_Q, R)$ is the multiplicity of the local ring $\bar{\mathfrak{o}}_{\mathfrak{p}}$, where $\tilde{\mathfrak{P}} = \bar{\mathfrak{o}}.h(\mathfrak{P})$, h being the canonical homomorphism of $\mathfrak{o}$ into $\bar{\mathfrak{o}}$. From $\tilde{\mathfrak{P}} = \bar{\mathfrak{o}}.h(\mathfrak{P})$ follows at once that the local rings $\bar{\mathfrak{o}}_{\mathfrak{p}}$ and $\mathfrak{o}_{\mathfrak{P}}$ have the same completion and therefore have the same multiplicity. This proves (2′) and hence also (2).

If $D(X_1, X_2, \ldots, X_r)$ denotes the X_{r+1}-discriminant of $f(X_1, X_2, \ldots, X_r, X_{r+1})$ and $\tilde{D}(Y_1, Y_2, \ldots, Y_{r-\rho})$ denotes the $Y_{r-\rho+1}$-discriminant of $\tilde{f}(Y_1, Y_2, \ldots, Y_{r-\rho+1})$, then

$$\tilde{D}(Y_1, Y_2, \ldots, Y_{r-\rho}) = D(\xi_1, \xi_2, \ldots, \xi_\rho, Y_1 + \xi_{\rho+1}, \ldots, Y_{r-\rho} + \xi_{r-\rho}). \quad (3)$$

Our k-rational projection π of V onto $\hat{A}_r/k$ determines a unique projection $\tilde{\pi}$ of $\tilde{V}_Q$ onto $\hat{A}_{r-\rho}/K$ defined by $\tilde{\pi}(\{y_1, y_2, \ldots, y_{r+\rho+1}\}) = \{y_1, y_2, \ldots, y_{r-\rho}\}$. More specifically, if $\tilde{\mathfrak{P}}$ is any prime ideal of $\hat{\mathfrak{o}}_{\mathfrak{p}}(= K[[y_1, y_2, \ldots, y_{r-\rho+1}]]$ and if $\tilde{\mathfrak{P}}_0 = \tilde{\mathfrak{P}} \cap K[[y_1, y_2, \ldots, y_{r-\rho}]]$, then $\tilde{\pi}(R) = R_0$, where R and R_0 are the points of $\tilde{V}_Q$ and of $\hat{A}_{r-\rho}/K$ associated respectively with the prime ideals $\tilde{\mathfrak{P}}$ and $\tilde{\mathfrak{P}}_0$. From this and from (3) it follows therefore that the critical variety $\tilde{\Delta}$ of $\tilde{\pi} \mid \tilde{V}_Q$ is obtained from the critical variety Δ of $\pi \mid V$ by first passing from Δ to the localization $\tilde{\Delta}_{Q_0}$ of Δ at the point Q_0, where $Q_0 = \pi(Q)$, and then by extending the coefficient field $K_0(= k((\xi_1, \xi_2, \ldots, \xi_r))$ of $\tilde{\Delta}_{Q_0}$ to the field K. Since a coefficient field extension does not affect the multiplicity of any point of $\tilde{\Delta}_{Q_0}$, we therefore have, by applying the general result (2) to Δ/k and $\tilde{\Delta}_{Q_0}/K$, the following equality for any point R_0 of $\tilde{\Delta}_{Q_0}/K$:

$$e(\tilde{\Delta}_{Q_0}/K, R_0) = e(\Delta/k, R_0), \qquad (R_0 \in \tilde{\Delta}_{Q_0}). \quad (4)$$

Now, let W_0 be any irreducible subvariety Δ, passing through the point Q_0 (in the text, W_0 is the variety $\overline{Z_0}$). The dimension of W_0 is ρ', where $\rho' \geq \rho$. The case $\rho' = \rho$ being trivial, we may assume that $\rho' > \rho$. The variety W_0 determines an irreducible subvariety $\tilde{W}_0$ of $\tilde{\Delta}_{Q_0}$, of dimension $\rho' - \rho$, and if Q_0 is a simple point of W_0, Q_0 is also a simple point of $\tilde{W}_0$. By assumption (in the text) we have that if R_0 is the general point of W_0/k then $e(\Delta, R_0) = e(\Delta, Q_0)$. Applying (4) to both sides of this equality, where we replace R_0 by Q_0 in applying (4) to the right-hand of the equality, we find that

$$e(\tilde{\Delta}_{Q_0}/K, R_0) = e(\tilde{\Delta}_{Q_0}/K, Q_0). \tag{5}$$

Since Q_0 is the closed point of $\tilde{\Delta}_{Q_0}$ and is a simple point of $\tilde{W}_0$, our theorem of the PNAS note yields the following result:

$\tilde{\pi}^{-1}\{\tilde{W}_0\}$ has a unique irreducible component $\tilde{W}$ at the point Q, and $\tilde{V}_Q$ is equimultiple at Q along $\tilde{W}$, i.e., *we have*

$$e(\tilde{V}_Q, Q) = e(\tilde{V}_Q, R), \tag{6}$$

where R is the general point of $\tilde{W}$, and clearly $\tilde{\pi}(R) = R_0$. Furthermore, Q is a simple point of $\tilde{W}$.

Now $\tilde{W}$ is the Q-localization of a unique irreducible subvariety W of V, and R is also the general point of W. Furthermore, since $\tilde{\pi} = \pi \mid \tilde{V}_Q$, it follows that $\pi(R) = R_0$. Thus W is the irreducible component of $\pi^{-1}\{W_0\}$ which contains the point Q. Since Q is a simple point of $\tilde{W}$, it is also a simple point of W. Applying (2) to (6) we find that $e(V, Q) = e(V, R)$, i.e., V is equimultiple at Q along W. This completes the proof of the generalization of the result established in our PNAS note.

HARVARD UNIVERSITY

REFERENCES

[1] O. Zariski, "Studies in equisingularity II. Equisingularity in codimension 1 (and characteristic zero)," *Amer. J. Math.*, **87** (1965) pp. 972–1006. This paper will be referred to in the text as SEQ II.

[2] ———, "On equimultiple subvarieties of algebroid hypersurfaces," *Proc. Nat. Acad. Sci. U.S.A.*, **72** (1975) pp. 1425–1426.

[3] O. Zariski and P. Samuel, *Commutative Algebra*, v. I(1958) and v. II(1960). These two books will be referred to as C.A.I, and C.A.II respectively.

[4] A. Grothendieck, *Elements de géometrie algébrique*, Chapter IV.

[5] H. Hironaka, *On Zariski dimensionality type*. Amer. J. Math. **101**, No. 2(1979) pp. •••-•••.

Abstract of the paper
"Foundations of a general theory of equisingularity on r-dimensional algebroid and algebraic varieties, of embedding dimension r+1".

By Oscar Zariski

Introduction. In paper [1], having the same title as the present abstract of that paper, we develop for the first time a general (local) theory of equisingularity on algebroid r-dimensional hypersurfaces, having as coefficient field k an algebraically closed field of characteristic 0, and we use this local theory in order to develop (see [1], §6) a general global and analytically intrinsic theory of equisingularity on r-dimensional algebraic varieties V, defined over an algebraically closed ground field k, of characteristic zero, and having, locally, at each of their k-rational singular points Q, embedding dimension r+1. By this we mean that if $\underline{o}$ is the local ring of V at such a point Q, then the maximal ideal $\underline{m}$ of $\underline{o}$ has a basis of $r+1$ elements (any such basis is necessarily minimal, since Q is assumed to be a singular point). This is equivalent to assuming that V is biregularly equivalent, locally at Q, to an algebraic r-dimensional hypersurface (having also a singular point at Q).

Our algebroid varieties V are hypersurfaces defined by a

single equation $f(X_1,X_2,\ldots,X_{r+1}) = 0$, where f is a power series with coefficients in k, free from multiple factors and zero at the origin $P: X_1 = X_2 = \cdots = X_{r+1} = 0$. This point P is the only closed point of V, and we shall sometimes refer to P as <u>the origin of</u> V. The points of V are represented by the prime ideals of the (complete) local ring $\underline{o}$ of V at P; here $\underline{o}$ is the power series ring $k[[x_1,x_2,\cdots,x_{r+1}]]$, where the x_i are the $f(X)$-residues of the X_i. The customary notation for this definition of V as a set of points is $V = \mathrm{Spec}(\underline{o})$. The origin P of V is represented by the maximal ideal $\underline{m} = \sum_{i=1}^{r+1} \underline{o}x_i$ of $\underline{o}$. Since we assumed that P is a singular point, it follows that the leading form of the power series $f(X)$ is of degree $n > 1$; here n denotes the multiplicity of the singular point P of V.

We observe at once that the field k is a coefficient field of the complete local ring $\underline{o}$ of V, i.e., k is a subfield of $\underline{o}$ which is mapped isomorphically onto the residue class field of $\underline{m}$ by the canonical homomorphism of $\underline{o}$ onto $\underline{o}/\underline{m}$. Thus k <u>is not uniquely determined by</u> $\underline{o}$, and this will pose one of the open questions which we shall formulate at the end of this report, namely: <u>is our theory of equisingularity on algebroid varieties</u> V <u>independent of the choice of the coefficient field</u> k <u>of</u> $\underline{o}$? A similar question arises for <u>algebraic</u> varieties V. (See section 3).

§1. The concept of the dimensionality type of points of V.

Our theory of equisingularity is based on the introduction of a fundamental numerical invariant of any point R of our (algebroid or algebraic) variety V/k: the dimensionality type of R on V/k; in symbols: $d.t_k(V,R)$. We first give the definition in the case of algebroid varieties V.

If V is a smooth variety, i.e., if the origin P of V is a simple point, then all points R of V are simple points (i.e., the local ring of V at R is always a regular ring, of dimension $r-\rho$, where ρ is the dimension of the prime ideal $\underline{p}$ in $\underline{o}$ associated with the point R). In this case we define the dimensionality type of V at R as being the integer zero.

If R is a singular point of V, we shall define the integer $d.t_k(V,R)$ by induction on r, except when R is a simple point of V, in which case we set, as before: $d.t._k(V,R) = 0$. Let $(x_1,x_2,\ldots,x_{r+1})$ be a minimal basis of the maximal ideal $\underline{m}$ of the local (complete) local ring $\underline{o}$ of V (at P). For each integer $i = 1,2,\ldots,r$ we introduce a power series x_i^* in $x_1,x_2,\ldots,x_{r+1}$, with indeterminate coefficients $\{u^{(i)}\}$ and zero at the origin. Let $\{u\}$ denote the union of the r sets of indeterminates $\{u^{(i)}\}$, let $k^* = k(\{u\})$ be the field obtained by adjoining to k the infinite set of indeterminates u and let $\underline{o}^*$ denote the complete local ring $k^*[[x_1,x_2,\ldots,x_{r+1}]]$. Then $\underline{o}^*$ is the local ring at P, of the algebroid variety V^*/k^* obtained from B by the coefficient field extension $k \longrightarrow k^*$ (and hence defined, over k^*, by the same equation $f(X_1,X_2,\ldots,X_{r+1}) = 0$ which defined V over k). It is easy to see that $\underline{o}^*$, and hence also V^*, is independent of the choice of the minimal basis $(x_1,x_2,\ldots,x_{r+1})$ of $\underline{m}$, and that this basis is also a minimal basis of the maximal ideal $\underline{m}^*$ of $\underline{o}^*$. It is also easy to see that for any integer $i = 1,2,\ldots,r+1$ we have $\underline{o}^* = k^*[[x_1^*,x_2^*,\ldots,x_r^*,x_i]]$, whence also $(x_1^*,x_2^*,\ldots,x_r^*,x_i)$ is a minimal basis of m^*. Let $f_i^*(X_1^*,X_2^*,\ldots,X_r^*,X_i) = 0$ be the equation of V^*/k^* defined by this basis; here f_i^* is free from multiple factors and is defined uniquely up to an arbitrary unit factor in $k^*[[X_1^*,X_2^*,\ldots,X_r^*,X_i]]$. The leading form of f_i^* has degree n, equal to the multiplicity n of the closed point P of V, and the term X_i^n actually occurs on f_i^*. By the Weierstrass preparation theorem we may assume that f_i^* is a monic polynomial in X_i, of degree n. It can be easily proved that the reduced X_i-discriminant of f_i^* (i.e., the product of the distinct, non-associated, irreducible factors of the X_i-discriminant of f_i^*) is a non-unit element of the power series ring $k^*[[X_1^*,X_2^*,\ldots,X_r^*]]$ which is independent of the value of i $(i = 1,2,\ldots,r+1)$. We denote this reduced discriminant by $\Delta_u^*(X_1^*,X_2^*,\ldots,X_r^*)$.

The local ring $A^* = k^*[[x_1^*, x_2^*, \cdots, x_r^*]]$ is a complete regular ring, of dimension r, and the algebroid variety Spec (A^*) is merely the affine r-dimensional space $\mathbb{A}_r/k^*$, considered as a smooth algebroid variety, locally at the origin P_0^*: $X_1^* = X_2^* = \cdots = X_r^* = 0$. The ring A^* is a subring of o^*, and we have a canonical mapping π_u^* of V^* onto $\mathbb{A}_r^*$ by setting $\pi_u^*(p^*) = p_0^* = p^* \cap A^*$, where p^* is any prime ideal of o^*. In particular, if p is any prime ideal of o, then o^*p is a prime ideal p^* of o^*, and it is easily seen that $o^*p \cap o = p$. Thus π_u^* induces a mapping π_u of V into $\mathbb{A}_r/k^*$. We call π_u the <u>generic projection</u> of V/k <u>into</u> $\mathbb{A}_r/k^*$, and we call the hypersurface Δ_u^* in $\mathbb{A}_r/k^*$ defined by the equation

$$\Delta_u^*(X_1^*, X_2^*, \cdots, X_r^*) = 0,$$

<u>the critical hypersurface</u> of the generic projection π_u of V/k. The projection π_u^* has the obvious property that π_u^{*-1} is finitely valued on $\mathbb{A}_r/k^*$. It can also be proved that if R is any point of V then $R_0^* = \pi_u(R) \in \Delta_u^*$ <u>if and only if</u> R <u>is a singular point of</u> V/k.

We can now give our inductive definition of the integer $d.t._k(V,R)$, for any <u>singular</u> point R of V:

(1) $$d.t._k(V,R) = 1 + d.t._{k^*}(\Delta_u^*, R_0^*) ,$$

where $R_0^* = \pi_u(R)$.

If $r = 1$ then the critical hypersurface Δ_u^* is a single point, namely the origin P_0^* of $\mathbf{A}_r/k^*$ (always assuming that P is a singular point of V). Hence in this case we have $d.t._{k^*}(\Delta_u^*, P_0^*) = 0$ and $d.t._k(V,P) = 1$. All other points of the algebroid curve V are simple points of V: they are the general points of the irreducible analytical branches of V through P. Hence, in the case $r = 1$ we have $0 \leqq d.t._k(V,R) \leqq 1$, for every point R of V. It follows therefore from (1), by induction on r, that we have

$$(2) \qquad 0 \leqq d.t._k(V,R) \leqq r,$$

for any point R of V, with equality $d.t._k(V,R) = 0$ if and only if R is a simple point of V. Actually, the inequalities (2) can be strengthened. Namely, if $s = d.t._k(V,P)$, then we have

$$(3) \qquad 0 \leqq d.t._k(V,R) \leqq s$$

for all points R of V. The inequality $d.t._k(V,R) \leqq s$ is trivial if $r = 1$. Using induction on r, this inequality follows directly from (1), since $d.t._{k^*}(\Delta_u^*, P_0^*) = s-1$, whence, by our induction hypothesis, we have $d.t._k(\Delta_u^*, R_0^*) \leqq s-1$ for every point R_0^* of Δ_u^*, and from this (3) follows, in view of (1).

The inequality $d.t._k(V,R) \leqq d.t._k(V,P)(= s)$ is also a special case of more general result. Namely, if Q and Q' are any two points of V and if $\underline{p}$ and $\underline{p}'$ are the prime ideals of $\underline{o}$ associated with these two points, then we say that Q __is a specialization__ of Q' if $\underline{p} > \underline{p}'$. Then it can be proved, again by induction on r and using the definition (1), that __if__ Q __is a specialization of__ Q' __then__

$$(5) \qquad d.t._k(V,Q') \leqq d.t._k(V,Q) \qquad (Q = \text{a specialization of } Q').$$

Since the origin P of V is a specialization of __any__ point of V, the inequality $d.t._k(V,R) \leqq d.t._k(V,P)$ follows directly from (5).

We now pass to the definition of the dimensionality type of points of __algebraic__ varieties V/k. The case of k-rational points P of V/k is trivial. Namely, if $\underline{o}^*$ is the completion of the local ring $\underline{o}$ of V/k at P, then we consider the __algebroid__ variety $\hat{V}_P = \text{Spec}(\underline{o}^*)$, and we define the integer $d.t._k(V,P)$ as being the integer $d.t._k(\hat{V}_P,P)$:

$$(6) \qquad d.t._k(V,P) = d.t._k(\hat{V}_P,P) .$$

This definition of the dimensionality type of P on V is __analytically intrinsic__, since it depends only on the __completion__ of the local ring of V/k at P (the ground field k being fixed once and for always). Let now R be a point of V which is not k-rational. In this case we consider the irreducible subvariety W of V/k having R as a general point (over k). We next fix any k-rational point P of W, and—for simplicity—we restrict ourselves only to points P of W at which W is analytically irreducbile (in particular, any simple point of W will do). With such a choice of P the point R of V defines a unique point of the algebroid variety $\hat{V}_P$. We shall continue to denote this point of $\hat{V}_P$ by R. This point R of $\hat{V}_P$ is represented by the prime ideal $\underline{o}^*\underline{p}$ of $\underline{o}^*$, where $\underline{p}$ is the prime ideal of W in the local ring $\underline{o}$ of V at P. It can be proved that the integer $d.t._k(\hat{V}_P,R)$ is independent of the choice of the k-rational point P on W. This integer shall be, by definition, the dimensionality type of V at R.

For the sake of completeness we briefly consider the case in which W is analytically reducible at P. In that case the above ideal $\underline{o}^*\underline{p}$ is an intersection of a certain number q $(q > 1)$ of prime ideal $\underline{p}_i^*$ in $\underline{o}^*$, giving rise to q irreducible subvarieties $W_1^*, W_2^*, \ldots, W_q^*$ of $\hat{V}_P$. Let R_i^* be the general point of W_i^*, i.e., R_i^* is the point of $\hat{V}_P$ represented by the prime ideal $\underline{p}_i^*$. It can be proved that the q integers $d.t._k(\hat{V}_P, R_i^*)$ are all equal to each other and are equal to the integer $d.t._k(V,R)$ introduced above.

§2. The equisingularity stratification of V.

Now we shall present the central part of our theory of equisingularity. We first deal with algebroid hypersurfaces V. We call a projection $\pi_{\bar{u}}$ of V onto $\mathbb{A}_r/k$ k-rational if it is obtained from the generic projection π_u by a specialization $\{u\} \longrightarrow \{\bar{u}\}$ of the indeterminates u, in which the $\bar{u}$ belong to k and if - furthermore - the r by r+1 matrix of the coefficients of the linear terms in the (specialized) power series $x_{1,\bar{u}}, x_{2,\bar{u}}, \cdots, x_{r,\bar{u}}$ (obtained from $x_1^*, x_2^*, \cdots, x_r^*$ by the specialization $u \longrightarrow \bar{u}$) has rank r (this second condition is necessary to insure that the r elements $x_{1,\bar{u}}, x_{2,\bar{u}}, \cdots, x_{r,\bar{u}}$ are parameters of the local ring $\underset{\sim}{o}$ of V at the origin P and that therefore $\pi_{\bar{u}}$ be a finite projection of V onto $\mathbb{A}_r/k$). For any point Q of V we denote by $EQS_k(V,Q)$ the set of points R of V such that Q is a specialization of R and $d.t._k(V,R) = d.t._k(V,Q)$. By means of a basic lemma (which we shall not state here) which concerns algebroid hypersurfaces V_t^* depending on parameters t and which is then applied to the case in which V_t is the critical hypersurface Δ_u^* of the generic projection π_u of V onto $\mathbb{A}_r/k^*$ (the indeterminates u playing the role of the parameters t), we can prove the following result:

Proposition 1. Given a singular point Q of V, of dimensionality type σ, the following two conditions are satisfied for almost all k-rational projections $\pi_{\bar{u}}$ of V onto $\mathbb{A}_r/k$ and for all points R of which Q is a specialization:

(a) $\text{d.t.}_k(\Delta_{\bar{u}}, \bar{Q}_0) = \sigma-1, \qquad (\bar{Q}_0 = \pi_{\bar{u}}(Q))$;

(b) $\pi_{\bar{u}}(R) \in EQS_k(\Delta_{\bar{u}},\bar{Q}_0) \Longleftrightarrow R \in EQS_k(V,Q)$.

(We say that a property A holds for almost all specializations $\{u\} \longrightarrow \{\bar{u}\} \subset k^*$, if there exists a non-zero polynomial $g(\{u\})$ in the indeterminates u, with coefficients in k, such that the property A holds for all specializations $\{u\} \longrightarrow \{\bar{u}\} \subset k^*$ such that $g(\{\bar{u}\}) \neq 0$).

From this proposition we can derive the following result:

Proposition 2. Let Q be any point of V, of dimensionality type σ ($\sigma \geqq 0$) and let W(Q) denote the algebroid closure in V of the set $EQS_k(V,Q)$ (i.e., W(Q) is the smallest algebroid subvariety of V which contains the set $EQS_k(V,Q)$). Then W(Q) is an irreducible algebroid variety, of dimension $r-\sigma$. The general point R of W(Q) belongs to the set $EQS_k(V,Q)$, and W(Q) is the greatest irreducible algebroid subvariety of V which has the following two properties: (1) it contains the point Q and (2) its general point R has the same dimensionality type σ as the point Q (in view of (1), condition (2) is equivalent to: $R \in EQS_k(V,Q)$). Furthermore,

$W(Q)$ is smooth at Q, and V is equimultiple at Q along $W(Q)$ (i.e., the points Q and R have the same multiplicity on V). In the special case in which Q is the closed point P of V we have $W(P) = EQS_k(V,P)$ (equivalently: the set $EQS_k(V,P)$ is itself an algebroid subvariety of V). Finally, if $S(Q)$ denotes the set of points of $W(Q)$ which have dimensionality type exactly equal to σ, then $W(Q)-S(Q)$ is a proper algebroid subvariety of $W(Q)$.

Note that in view of (5), §1, all points of $W(Q)-S(Q)$ have dimensionality type greater than σ.

From the properties of the sets $W(Q)$ and $S(Q)$ stated in Proposition 2 follows at once if Q' is any point of $S(Q)$ then $W(Q') = W(Q)$ and $S(Q') = S(Q)$. The subset $S(Q)$ of $W(Q)$ has the following properties: (a) $S(Q)$ is a Zariski dense open (and irreducible) subset of its algebroid closure $W(Q)$; (b) all points of $S(Q)$ are simple points of $W(Q)$; (c) V is equimultiple along $S(Q)$ at each of its points; (d) all points of $S(Q)$ have the same dimensionality type σ; (e) $S(Q)$ is a maximal subset of V having properties (a)-(d). We call the set $S(Q)$ an equisingularity stratum, of dimensionality type σ.

Proposition 2 is trivial if Q is a simple point of V (i.e., if $\sigma = 0$). In that case, Q belongs to only one irreducible component of V, say to V_1, and the set stratum $S(Q)$ is the set $V_1 - (V_1 \cap V_{sing})$, while $W(Q)$ is equal to V_1. Our proof of

Proposition 2 in the case $\sigma > 0$ is based on induction with respect to r, the proposition being trivial in the case $r = 1$. Applying the induction hypothesis to the critical hypersurface $\Delta_{\bar{u}}$ of $\pi_{\bar{u}}|V$, where we assume that $\pi_{\bar{u}}$ satisfies conditions (a) and (b) of Proposition 1, the proof of Proposition 2 is relatively easy, except in one point: to prove that V is equimultiple along $S(Q)$ at each point of $S(Q)$, we have to prove (in an Appendix) a generalized version of a result proved by us in our PNAS Note [2]. We note that *any two distinct equisingularity strata are disjoint*; this follows from the fact that if $Q' \in S(Q)$ then $W(Q) = W(Q')$ and hence $S(Q) = S(Q')$.

For any integer σ, $0 \leqq \sigma \leqq s$ $(= \text{d.t.}_k(V,P))$, we denote by $V(\sigma)$ the set of points of V which have dimensionality σ. By using the above propositions 1 and 2 we can prove the following *basic theorem* on the equisingularity stratification of V:

Theorem on equisingularity stratification of V. 1) *For any integer* σ *such that* $V(\sigma)$ *is non-empty, the set* $V(\sigma)$ *is a disjoint union of a finite number of* $(r-\sigma)$*-dimensional equisingularity strata*. 2) *The boundary* $\bar{S}-S$ *of each equisingularity stratum* S, *of dimensionality type* σ, *is a finite union of equisingularity strata of dimensionality type* $> \sigma$.

Passing now to *algebraic varieties* V, our theory of equisingularity in the algebraic case is based largely on the equisingularity theory of algebroid varieties. Namely, given any k-rational point P of V, of dimensionality type s, the consideration of the algebroid variety $\hat{V}_P$ (see section 1) yields an algebroid subvariety $\widehat{W(P)}$ of $\hat{V}_P$ equal to the set $EQS_k(\hat{V}_P, P)$ which is the (r-s)-dimensional equisingularity stratum of $\hat{V}_P$ containing the point P. If R is the general point of the algebroid variety $\widehat{W(P)}$, then R defines an irreducible (r-s)-dimensional <u>algebraic</u> subvariety W(P) of V having R as general point. While P is the only k-rational point of $\widehat{W(P)}$, the variety W(P) contains infinitely many k-rational points (if $s < r$). One crucial point of the theory in the algebraic case was proved by Hironaka, namely that <u>the set of all points of</u> W(P) <u>which have dimensionality type</u> s <u>is a Zariski dense open subset of</u> W(P) (see Hironaka [4]). Once this is proved, it is possible to generalize to algebraic varieties Proposition 2 and the basic theorem, replacing in the statements of these two results the word "algebroid" by the word "algebraic". The only thing to keep in mind is that in the algebraic case all the equisingularity strata are of the type S(P), where P is a k-rational point of V; and that these strata are not necessarily closed subvarieties of V (while in the algebroid case, the stratum S(P) containing the closed point P of V was the closed algebroid subvariety $EQS_k(V,P)$ of V (see Proposition 2)).

§3. Critical remarks and some open questions.

1) One general question concerning our theory of equisingularity arises in regard to our very definition of equisingularity strata as based on the concept of the dimensionality type of points of V. In our earlier paper [2] we dealt with algebraic varieties and with the special case of (r-1)-dimensional equisingularity strata (hence with points of dimensionality type $\sigma = 1$). We have defined such strata S by the condition that $S \subset V_{sing}$, $\dim \bar{S} = r-1$ ($\bar{S}$ = algebraic closure of S) and that if Q is any k-rational point of S, and R is the general point of the (r-1)-dimensional irreducible component $\bar{S}$ of V_{sing}, then Q is a simple point of $\bar{S}$, and the "plane" algebroid curves Γ_Q and Γ_R (i.e., the curves of embedding dimension 2) which are sections of V with algebroid surfaces F_Q and F_R having simple points at Q and R respectively and transversal to $\bar{S}$ at these points, have equivalent singularities at Q and R respectively. This definition made sense in the case $\sigma = 1$, for the notion of equivalent singularities of plane algebroid curves is well defined and is classical. We have also proved in that paper [2] the following two results: (1) If Γ_Q and Γ_R have equivalent singularities for one choice of the surfaces F_Q and F_R, then the same is true for any choice of F_Q and F_R (always subject to the condition that these surfaces be smooth and transversal to $\bar{S}$ and Q and R respectively); (2) If $\pi_{\bar{u}}$ is any k-rational finite projection

of V onto $\mathbb{A}_r/k$ which is transversal to V, locally at Q (i.e., if the elements $x_{1,\bar{u}}, x_{2,\bar{u}}, \cdots, x_{r,\bar{u}}$ are transversal parameters of the local ring $\underline{o}$ of V at Q), then Γ_Q and Γ_R have equivalent singularities if and only if the point $\pi_{\bar{u}}(Q)$ is a simple point of the critical hypersurface $\Delta_{\bar{u}}$ (whence $\pi_{\bar{u}}(Q)$ must have dimensionality type 0). In particular, it follows that the condition similar to (2) is satisfied by the generic projection π_u.

This last result, in the case $\sigma = 1$, led us to the general definition of the dimensionality type $d.t._k(V,Q)$ given in section 1. In that definition we had to use the generic projection π_u, since it can be shown by examples that in the case $\sigma > 1$ not every k-rational transversal projection $\pi_{\bar{u}}$ satisfies condition (a) of Proposition 1. However, since there does not yet exist a general criterion of equivalent singularities of algebroid hypersurfaces of dimension $r > 1$, there is no guarantee that our definition of equisingularity strata - or, for that matter, any definition of equisingularity strata - of dimension less than r-1, is the "right one". Any such definition is, for the time being, largely a matter of personal perception. One thing we can say in favor of our definition is that it leads to results which are geometrically sound and non-trivial. Also, our definition is not too obviously weak (as a definition based solely on equimultiplicity would have been), nor is it too strong (as a definition based on analytical equivalence of the complete local rings in question would have been).

2) We have already mentioned in our Introduction the question of whether our definition of the dimensionality type of the closed point P of an algebroid variety V is independent of the choice of the coefficient field k of the local ring of V at P. Even if the coefficient field k is fixed in advance, a similar question arises for the points R of V different from P. Namely, does our integer $d.t._k(V,R)$ depend only on k and on the local ring of V at R? This question can be re-phrased as follows: Let o' be the completion of the local ring o_R of R on V, and let k' be any coefficient field of o' which contains k. Is it true that $d.t._k(V,R) = d.t._{k'}(V',R)$, where V' = Spec(o') is the algebroid variety having R as its closed point?

A similar question arises in the case of algebraic varieties V, namely: is our definition of the integer $d.t._k(V,R)$ independent of the choice of the algebraically closed ground field k of definition of V?

References

[1] O. Zariski, "Foundations of a general theory of equisingularity on r-dimensional algebroid and algebraic varieties, of embedding dimension r+1". Amer. J. of Math., v. 101 (April, 1979).

[2] ________, "Studies in equisingularity, II. Equisingularity in codimension 1 (and characteristic zero)". Amer. J. of Math., v. 87 (1965), pp. 972-1006.

[3] ________, "On equimultiple subvarieties of algebroid hypersurfaces", Proc. Nat. Acad. Sci., U.S.A., vol. 72 (1975), pp. 1425-1426.

[4] H. Hironaka, "On Zariski dimensionality type", Amer. J. of Math., vol. 101 (April, 1979).

Printed in the United States
by Baker & Taylor Publisher Services

Printed in the United States
by Baker & Taylor Publisher Services